Inorganic Reactions in Water

First edition

Ronald L. Rich

Inorganic Reactions in Water

First edition

Springer

Dr. Ronald L. Rich
Bluffton University
112 S. Spring St.
Bluffton OH 45817-1112
USA
RichR@bluffton.edu

ISBN 978-3-540-73961-6 e-ISBN 978-3-540-73962-3

DOI 10.1007/978-3-540-73962-3

Library of Congress Control Number: 2007931852

Cover design: KünkelLopka GmbH, Heidelberg

Printed on acid-free paper

9 8 7 6 5 4 3 2 1

springer.com

Dedicated to my mentor Henry Taube
1915–2005
for supreme encouragement and example
in both science and ethics,
for whom Nature often had to make sense.
"You can learn a lot from test-tube chemistry."
Nobel Prize 1983

Preface

Water as a solvent, and the reactions in it, are supremely important in many fields. The excitement of newer fields of chemistry, however, has pushed these reactions down the list of priorities for providing convenient reference works. This excitement has often made "test-tube chemistry" seem to be passé.

The Japanese word *kagaku* for chemistry can be interpreted as change-science. Space does not permit a comprehensive, but only a representative, description of the reactions, changes, of nearly all the elements and their simpler compounds, primarily inorganic ones and primarily in water. Carbonyl complexes, for example, are very far from comprehensively mentioned here.

Sincerest thanks are due: first to my wife, Elaine Sommers Rich, for clarifying the English in the entire manuscript; to Kathrin Engisch, Derek W. Smith and John L. Sommer for reviewing sections of it; to Mark Amstutz, Donald Boyd, Jonathan J. Rich, Jon Stealey and M. D. Wilson, for help with computers; to Daniel J. Berger, Erwin Boschmann, Sebastian Canagaratna, and staff members at universities including Bluffton, Bowling Green State, California (Berkeley), Cincinnati, Ohio Northern, Ohio State, Oregon, Tulane, Virginia Tech, and Wright State, for help with the literature or computers; to Marion Hertel, Jörn Mohr, and Martin Weissgerber for diligent preparations for publishing; and to Gordon Bixel and others for other help. Finally, I thank my family, who generously ask whether this book might have been finished sooner without them!

Table of Contents

Introduction

In general we wish to maximize the number and variety of reactions selected, trimming reluctantly from both the newer and the older literature. To permit many comparisons, we thus often omit the equations and details required to maximize success for the reactions as preparations, as in *Inorganic Syntheses* and the well-known textbooks of inorganic preparations, although a few elaborations are presented for one reason or another. We hope that other variations of style prove that "variety is the spice of life".

Various reagents involved in these reactions, moreover, have only non-aqueous sources, so such sources may be either not given or only hinted at here. With some especially interesting exceptions we also exclude complicated phenomena.

We emphasize similarities in chemical behavior, not electronic structures and theories, although these structures guide us to a basically 18-column form of the Periodic Chart to make each topic easy to locate. The organizing Chart used here, however, has some novelties which may be useful, as discussed below, even after comparison with the numerous previously proposed forms of the chart [1].

We must also note immediately that we have not taken time or space for tens of thousands of relevant primary references, although some secondary sources include many self contradictions, unequal "equations" and subtle errors which we can only hope to reduce. Review journals and serials are few enough to mention [2] for providing relatively concentrated sources for further investigation.

References are of course valuable, but so, for some purposes, are the additional descriptions of actual chemical behavior that compete for the same space. That is, the loss of some data is judged to be outweighed, *for the scope of this work*, by the (still brief) descriptions of a greater variety of reactions.

Some authors [3] understandably include only recent references. Here we include many general references which provide more specifics, along with a few particular ones, consistent with the wish to provide a wide, albeit only representative, sample of the vast current data on inorganic reactions in water. This is to give not only a list but also some perspective, with a few data on especially interesting structures and other aspects.

Many other writers, and our appendices, offer further views on periodicity. Our regrettably small sample includes: a historical and philosophical survey [4]; a hexagonal form with many gaps [5]; data on new elements [6]; periodicity for geologists [7]; a celebration of almost every element [8]; a geological chart [9]; predictions even through element number 1138 [10]; a discussion of semimetallicity [11]; the chemistry of the newest elements [12]; the quantum-mechanical explanation of periodicity [13]; a general survey [14]; and a book, with inexplicably confusing editing, on chemical periodicity [15]. Chemdex and Stanford University list numerous periodic charts [16], although many are not periodic or not chemical or not charts.

Laing has pointed out [17] that we can show more chemical relationships if we assign more than one place to some elements, specifically those in the first two main (eight-column) rows. We have modified his proposal because oxygen and fluorine do not resemble the metals in Groups 6 and 7, even in their oxidation states. Lithium and sodium, likewise, although they share one oxidation state with the copper Group, seem otherwise too different from that Group to honor the slight similarity in this chart. Several further elements receive more than one place here, because of their chemical similarities to others. Still, each element has only one chapter and section number, e.g., 17.2 for Cl (Group 17), in what follows in the Periodic Chart in Table Intro.1, published elsewhere [18] and explained further below. Electronic structure, again, plays second fiddle. The predominant oxidation states are one of the major criteria for our groupings, but the decisions about these relationships still require some judgment calls.

Cronyn reminds us [19] that hydrogen, although often placed with the alkali metals and/or even the halogens, actually resembles carbon in many ways. In modern inorganic and organometallic chemistry the abundance of species containing M-C and M-H bonds (where M is a metal) is quite striking. The nearly equal electronegativities of C and H are important sources of other similarities.

Sanderson also recognized this earlier in his insightful Periodic Chart [20]. Here then, we choose to show the same point by putting hydrogen above carbon, in addition to its other positions. We have not yet, however, honored this principle in our chapter numbering, because most chemists may look for hydrogen with the alkali metals. The similarity to the halogens is real but weaker; the halide ions are stable in water, for example, but the hydride ion is not.

Derek W. Smith has suggested quite appropriately, however [21], "that a modern Periodic Table should emphasise relationships among elements having similar (at least superficially) atomic electron configurations, inviting comparisons among stoichiometries/oxidation states/valences/coordination numbers (see JCE, 2005, 82, 1202). H and C are not comparable in this way."

Jensen notes various resemblances between Be and Mg to Zn, Cd and Hg, more than to Ca through Ra, so that the zinc Group may be treated as non-transitional [22]. Sanderson again offered a creative recognition of this [20]. The Group does resemble the **d**-block, however, in its NH_3 and CN^- complexes, thus supporting the IUPAC numbering and classification. The zinc Group seems to be transitional between the transitional and main Groups! Hereafter we say "**d**-block" or "**f**-block" for "transitional", because the **f**-block has also been called "(inner) transitional".

Table Intro.1. Atomic, Chapter and Section Numbers of the Elements

1													1			1	2
H													H			H	He
1.0																	18.0
3	4	5	6	7						3	4	5	6	7	8	9	10
Li	Be	B	C	N						Li	Be	B	C	N	O	F	Ne
1,1	2,1											13,1	14,1	15,1	16,1	17,1	18,1
11	12	13	14	15	16	17				11	12	13	14	15	16	17	18
Na	Mg	Al	Si	P	S	Cl				Na	Mg	Al	Si	P	S	Cl	Ar
1,2	2,2											13,2	14,2	15,2	16,2	17,2	18,2
19	20	21	22	23	24	25	26	27	28	29	30	31	32	33	34	35	36
K	Ca	Sc	Ti	V	Cr	Mn	Fe	Co	Ni	Cu	Zn	Ga	Ge	As	Se	Br	Kr
1,3	2,3	3.n	4,1	5,1	6,1	7,1	8,1	9,1	10,1	11,1	12,1	13,3	14,3	15,3	16,3	17,3	18,3
37	38	39	40	41	42	43	44	45	46	47	48	49	50	51	52	53	54
Rb	Sr	Y	Zr	Nb	Mo	Tc	Ru	Rh	Pd	Ag	Cd	In	Sn	Sb	Te	I	Xe
1,3	2,3	3.n	4,2	5,2	6,2	7,2	8,2	9,2	10,2	11,2	12,2	13,4	14,4	15,4	16,4	17,4	18,4
55	56	57	58	59	60	61	62	63	64	65	66	67	68	69	70	71	
Cs	Ba	La	Ce	Pr	Nd	Pm	Sm	Eu	Gd	Tb	Dy	Ho	Er	Tm	Yb	Lu	
1,3	2,3	3.n	3.n	3.n	3.n	3.n	3.n	3.n	3.n	3.n	3.n	3.n	3.n	3.n	3.n	3.n	.
	70	71	72	73	74	75	76	77	78	79	80	81	82	83	84	85	86
	Yb	Lu	Hf	Ta	W	Re	Os	Ir	Pt	Au	Hg	Tl	Pb	Bi	Po	At	Rn
			4,2	5,2	6,2	7,2	8,2	9,2	10,2	11,3	12,3	13,5	14,5	15,5	16,4	17,4	18,4
87	88	89	90	91	92	93											
Fr	Ra	Ac	Th	Pa	U	Np											
		89	90	91	92	93	94	95	96	97	98	99	100	101	102	103	
		Ac	Th	Pa	U	Np	Pu	Am	Cm	Bk	Cf	Es	Fm	Md	No	Lr	
		3.n	3.n	3.n	3.n	3.n	3.n	3.n	3.n	3.n	3.n	3.n	3.n	3.n	3.n	3.n	.
	102	103	104	105	106	107	108	109	110	111	112	113	114	115	116	117	118
	No	Lr	Rf	Db	Sg	Bh	Hs	Mt	Ds	Rg	Uub	Uut	Uuq	Uup	Uuh	Uus	Uuo
			4,2	5,2	6,2	7,2	8,2	9,2	10,2	11,3	12,3	13,5	14,5	15,5	16,4	17,4	18,4

n = 1, 2, 3, 4, 5, 6 or 7 for M^0, M^{II}, M^{III}, M^{IV}, M^V, M^{VI} or M^{VII} in turn.

The number of elements in nature is often stated as under 92 because of the short $t_{1/2}$ of Tc and Pm, but some post-92 elements are certainly natural too, in supernova products, and some have long $t_{1/2}$. Also for the important relativistic effects in the high-Z elements see Fig A.1 in Appendix A, and refer to other data [23–26].

The border between physical changes and chemical reactions is open to dispute. Melting sodium seems physical, without change in oxidation number or the nature of the coordination sphere, so that we do not think of it as a change of substance, and cooling easily reverses it. The chemical reaction of sodium with water, however, is more drastic in each respect. But what about the dissolution of sodium in mercury, or of NaCl in water? To make a long story short, it is convenient here not to separate information, often only qualitative, about solubilities from other observations about reactions. A useful rough guide [27a], by the way, could be "soluble" (> 50 g L^{-1}), "moderately soluble" (10–50 g L^{-1}), "slightly soluble" (1–10 g L^{-1}), "moderately insoluble" (0.01–1 g L^{-1}) and "insoluble" (< 0.01 g L^{-1}). To write the last one, for example, as < 1 cg L^{-1}, as preferred elsewhere in this book, saves little space but avoids suggesting the false precision in common reports such as 10 mg L^{-1} (with two significant digits) in these quantities.

Not all change is progress, and some older books [27–37], and numerous other laboratory manuals on qualitative analysis, gave useful descriptions of inorganic reactions in water before new topics crowded them out of our curricula and even encyclopedias [38], largely excepting only the most complete references [39]. Some "qual" books explicitly point out easily missed distinctions among similar elements. We include these, but our attention here is mainly on the reactions themselves, rather than their places in any particular analytical system. One rather compact source [27] is especially valuable in spite of internal contradictions, partly for numerous old references omitted here.

Limitations of space, time and expertise, of course, preclude updating much in these reports here. Our own plan to provide a single modest book, with some emphasis on including older information, makes the choice of recent material even more arbitrary. The completeness of Mellor's (older) reporting, however, does again expose various contradictions, e.g., that Mo dissolves not at all, slowly or quickly in dilute or concentrated H_2SO_4.

A related problem with the older sources then, is that their interpretation of phenomena, even when described well, is at times not just old-fashioned, but inaccurate. One more example is the formulation of aqueous ammonia as NH_4OH instead of NH_3, although Raman spectroscopy disproves the existence or at least the importance of any discrete NH_4OH molecule in water. This and other errors are discussed elsewhere [40], also on boiling point vs mass, with resistance by some writers to correction [41a], on the inductive effect in RCO_2H [41b], on the meaning of acidity [41c], on the Periodic Chart [41d], and on thermodynamics [41e]. A more dangerous example, with details, of a clear obstruction of correction in "the leading US chemical journal", is important enough to justify a brief reminder here [42].

Some rules to help students predict the products of simple inorganic reactions may be helpful [43]. Many references on quantitative analysis [44–49], sometimes especially the older ones that could not depend on "black boxes", provide valuable data on inorganic reactions in water, information summarized here when we mention analytical reactions, but not used in recent treatments dedicated to instrumental analysis and procedures not covered here.

Moreover, for accuracy in equations we want to write the formulas of the actual reacting species. The reaction of aqueous HCl and NaOH, for example, is between hydrated hydrons (a la IUPAC, i.e., H^+, the normal isotopic mixture of $^1H^+$, $^2H^+$ and $^3H^+$, not just the proton), and OH^-, and there are no molecules of HCl, NaOH or NaCl in solution, so that:

$$HCl + NaOH \rightarrow H_2O + NaCl$$

may be better written as the net ionic equation:

$$H^+ + OH^- \rightarrow H_2O$$

Still better might be to recognize that all of the hydrogens in the hydrated H^+ are actually equivalent, so that we write:

$$H_3O^+ + OH^- \rightarrow 2\ H_2O$$

This leaves us with a dilemma, however. Sometimes it is important that the other ions, Cl^- and Na^+ in this example, be those specified even though, or in fact precisely because, they do not participate in this reaction. Suppose we chose to illustrate acid-base neutralization by mixing dilute solutions of $Ba(OH)_2$ and H_2SO_4. Now the equations above would be inadequate because of the simultaneous precipitation of $BaSO_4$. Here we will often resolve such problems by writing the conventional formulas when specific substances are required or mentioned in the text, but usually with the formulas of the predominant active species shown in equations. This last reaction, then, could be written, for the more-dilute acid, as:

$$Ba^{2+} + 2\ OH^- + 2\ H_3O^+ + SO_4^{2-} \rightarrow BaSO_4\downarrow + 4\ H_2O, \text{ or}$$

$$Ba^{2+} + 2\ OH^- + H_3O^+ + HSO_4^- \rightarrow BaSO_4\downarrow + 3\ H_2O$$

in less-dilute acid. We need some flexibility, though, depending on the intended objects of attention. In fact, because the coordinated waters of other ions are normally not shown, we could write simply H^+ instead of H_3O^+. In liquids and solids, however, H^+, unlike other ions, is always covalently bound to something and is not even *relatively* independent, like the ions in, say, solid NaCl, so our preference for accuracy calls for the hydrated formula.

Kauffman [40c] notes that showing the hydration (first sphere) of the other cations would likewise promote accuracy with, e.g., this first equation instead of the less revealing second equation here:

$$[Al(H_2O)_6]^{3+} + NH_3 \rightarrow [AlOH(H_2O)_5]^{2+} + NH_4^+$$

$$Al^{3+} + NH_3 + H_2O \rightarrow AlOH^{2+} + NH_4^+$$

This would be consistent with our preference for writing Hg_2^{2+} instead of Hg^+, and Rh_2^{4+} instead of Rh^{2+} or, at times, even better in the latter case, $[Rh_2(H_2O)_{10}]^{4+}$. The attachment of any metal atom to any other(s) in water, however, usually seems more important than its attachment to the ubiquitous solvent molecules, and our interest in structures is secondary, so we do not always adopt this added complication. This would call for writing sulfur usually as S_8, but its structure as a product can be unclear, so, as others do, we often and reluctantly write only S.

One additional example emphasizes the greater simplicity of net ionic equations, even with the fuller formula for the oxonium ion, in the two equations here for the same reaction:

$$K_2WO_4 + (2 + n)\,HCl \rightarrow WCl_n^{(n\text{-}6)-} + 2\,KCl + (n - 6)\,H^+ + 4\,H_2O$$

$$WO_4^{2-} + 8\,H_3O^+ + n\,Cl^- \rightarrow [WCl_n]^{(n\text{-}6)-} + 12\,H_2O$$

We may simplify further by writing the traditional formulas for the reagents so-called "$(NH_4)_2S$" or "$(NH_4)_2CO_3$", but in quotation marks, to represent the inevitably hydrolyzed and complex mixtures of NH_3, NH_4^+, and HS^- (with only smaller amounts of the nominal S^{2-}) or NH_3, NH_4^+, HCO_3^-, and $CO_2NH_2^-$ (with smaller amounts of the nominal CO_3^{2-} plus CO_2). And when a reaction is described briefly as going with either H_2S or S^{2-}, for example, we should infer that it proceeds also with HS^-.

We often prefer to spell out "water, methanol" etc. when used as solvents, but write "H_2O", e.g., for explicit reactants or products.

Jensen proposes using quotation marks or underlining for non-molecular solids [50]. Thus, "NaCl" or *NaCl* would show that solid sodium chloride does not actually have separate molecules with that formula, unlike, say, P_4. This resembles somewhat the suggestion just above for handling structurally misleading conventional formulae, but we have not yet adopted this promising idea.

For another abbreviation we often write, say, "in NH_3" instead of "in aqueous solutions of NH_3". This book is, after all, about reactions in aqueous solutions. We write "in liquid NH_3" if that is needed.

The earlier collective term "fixed (i.e., non-volatile) alkali" for NaOH or KOH etc. can usually be replaced by "OH^-" when the cation is not crucial, while the simple term "alkali" is still useful, for brevity, to represent any of these or NH_3 (formerly NH_4OH). We often need a similar distinction between the salts of the alkalis, including the more hydrolyzed NH_4^+ species, and those of the fixed alkalis, which can be referred to simply (again when the cation is only a "spectator") as the actually predominant CO_3^{2-}, S^{2-} and so on. "Alkali fluoride" would therefore mean NH_4F, KF or CsF etc., as an example of the former.

The pursuit of accuracy and clarity in chemistry also suggests always calling O_2 or Br_2 dioxygen or dibromine, but usage and convenience have dictated otherwise here and generally.

Surely, however, most numbers require greater consistency, and we want to avoid the very common confusion between a change and its final result, or between addition and multiplication, preferring not to say "The change from four-coordinate Li^+ (0.59 Å) to 12-coordinate Cs^+ (1.88 Å) represents more than a three-fold difference in size; ..." [51] when the change and difference, 1.29 Å, are much less than three-fold, and only the final result is more than three times as great, with all due respect to a most valued compendium.

We normally prefer to go far toward the IUPAC recommendations [52] even with some less familiar names like diazane (N_2H_4) and sulfane (H_2S), often adding parenthesized formulas for clarity. We likewise write of the hydron (see **1.0.1**), not the proton; moreover for the photon we prefer the IUPAC symbol γ instead of $h\nu$ (which actually denotes the energy, not the particle). Also, the small-capital M for the unit, molar, distinguishes it from the M prefix for mega.

Contrarily, the unit equivalent, now often dismissed, is needed at times [53]. Then too the custom and rule, a la IUPAC, of writing, e.g., arsenic(−III) instead of arsenic(III−) would seem better if changed to suggest three physical charge units, not a mathematical subtraction of three, just as we now write S^{2-}, not S^{-2}.

This author notes also that mixed organic and inorganic formulas such as a possible $[Cr(acac)(en)(pn)]^{2+}$ could be written more briefly and still quite clearly as, e.g., $[CrAcacEnPn]^{2+}$. In such cases, the contents of the brackets identify the complex, and its identity does indeed include any *cis, trans, mer* or *fac*, so that $Cs_2[\textit{trans}\text{-}CrCl_2(H_2O)_4]Cl_3$, instead of $Cs_2\text{-}\textit{trans}\text{-}[CrCl_2(H_2O)_4]Cl_3$, is used here.

More seriously, the names and formulas for our relatively simple substances are clear without deciphering the IUPAC nomenclature that is clearly required for complicated cases beyond our scope, and we write $[\textit{trans}\text{-}MClBr(NH_3)_4]^{2+}$ or $[\textit{fac}\text{-}MCl_3Br_3]^{3-}$ as needed.

Following other writers, we write "dismutate", instead of the more customary and somewhat longer "disproportionate". Again following others, we would prefer to use the lower-case (non-IUPAC) (i) and (v) for oxidation numbers to avoid even temporary visual confusion with iodine and vanadium, although we have not done so here. This writer, for example, once briefly interpreted "Cyano-bridged $M(II)_9M(V)_6$ molecular clusters" as involving vanadium, which does occur in clusters. We do write "aq" for an indefinite amount of water in formulas, not counted in writing (balanced) equations; see Appendix D, Abbreviations.

We further use the term "higher-*Z*" non-metals for "heavier" (larger quantum-number *n*) ones below the first main period of the periodic chart, because the mass, as such, is practically irrelevant for their reactivity, in spite of the persistent myths ascribing an important role to it in volatility, for example. The occasionally used terms "higher" or "lower" (halogens) for the *same* meaning can be, in turn, ambiguous. The longer term "more protonic" might be appropriate. The term "heavy metal", although perhaps still useful occasionally, can often be clarified as "**d**- or **p**-block" or something similar.

Here then, we offer much of the descriptive aqueous chemistry in the older sources, with corrections, interpreted with the added insights and improved symbolism of recent decades, plus new information, including reactions of the recently discovered elements, but without many of the older strictly analytical techniques. Still, we mention the Marsh test for arsenic, for example, because not all laboratories around the world have the instruments that give quicker results.

We regret, however, that many appropriate reactions that appear in multiple older sources are neither confirmed nor denied even in Gmelin [39]. We nevertheless include some here if strong reasons for doubt seem absent, and must hope that this may help identify errors.

Publications often give exasperatingly few data, especially but not only in secondary sources, even to identify some important products or conditions for reactions where attention is understandably directed elsewhere, and we in turn certainly cannot present nearly all that is available now. We omit most, but not quite all, of the vast, interesting and useful information on kinetics, mechanisms and equilibria, as well as most molecular structures and hydrothermal syntheses at high *P*.

An example from the chemistry of iron can illustrate the present treatment. Older books included reactions such as that of $BaCO_3$ with aqueous Fe^{III}, used to separate iron from M^{II}. Newer texts [54–57] on the other hand, omit these sometimes-required descriptions in favor of differently useful information on equilibrium constants, some with kinetic data, to interpret the hydrolysis and polymerization [58] that complicate the chemistry of many species, such as Fe^{III}, in water. We offer something of a complement here. Another useful source, [59], should be noted, although it focuses somewhat on qualitative analysis with many specialized organic reagents.

Richens [60] describes hydrated ions, with a modern emphasis on structural and other theoretical aspects rather than many actual reactions. Emsley [61] summarizes conveniently the physical, biological, chemical and geological properties of the elements, in two pages of tables and charts for each one. And Burgess [62] discusses reactions in aqueous systems, particularly the chemistry of metal ions. Marcus [63] had covered some related data. The many sources on reactions in water in the environment are mostly treated without that emphasis here.

"Salt" is a term sometimes avoided nowadays, perhaps because it had been applied to compounds, such as PbS, which do not seem very salt-like even though possibly made from acids and bases. It still has some advantage over "compound", however, to distinguish such substances from even weakly basic or acidic oxides, such as PbO or PbO_2. Some flexibility in this usage seems called for.

We often abbreviate hydrate formulas like $Fe(OH)_3 \cdot xH_2O$ further as $Fe(OH)_3 \cdot aq$, partly to eliminate complicated balancing in equations.

Table Intro.1 shows the Periodic Chart and our over-all organization. The subtopics for each element are arranged by (primary) *reagents* rather than by the *reactant* element's oxidation state (except for Group 3), partly because this evades the ambiguity of oxidation state with ligands such as NO, and partly because this offers some convenient diversity from the usual sequences when a different organization may be more useful.

In some cases convenience reverses the roles of reactant and reagent. Thus in "Dissolved species similar to Cr^{III}, Fe^{III} and Al^{III} are precipitated as hydroxides by $BaCO_3$, while Mn^{2+}, Co^{2+} and Ni^{2+} are not," we can compare the various **3d** cations and Al^{III} with the single reagent $BaCO_3$, reported in **14.1.4 Other reactions** as if for $BaCO_3$ as reactant. Similarly, "The OH^- ion can leach Zr, Hf, Nb, Ta, Mo, W and Al from some ores" under **16.1.4** replaces four statements under ***n*.2.1**.

When several reagents, e.g., Mg, Fe and Zn, cause the same reaction and are listed together, the position within the subsection is normally determined by the first one in the periodic chart, going left to right and then top to bottom. Deviations from this and the other principles of organization may occur, but, we hope, not too often.

We focus first on the interactions of these reagents with the element in question and its hydrides, oxides, hydroxides etc., usually in order of rising average oxidation states. With carbon, however, the catenated HCH_3CO_2 (or CH_3CO_2H) and $H_2C_2O_4$ are taken up after HNCO. Whether to write HCH_3CO_2 or CH_3CO_2H, incidentally, may depend partly on whether the inorganic (as in H_2PHO_3) or organic practice seems to promote clarity in each case, but CH_3COOH might suggest a peroxide!

Also, when a reaction product has been identified, it may be helpful to list the further simple chemical reactivities of that product with important (secondary) reagents, instead of scattering them throughout the list of those reagents. For example, after we see that Ag^+ and SCN^- yield AgSCN, we mention briefly, in the same place, some further chemical properties of that precipitate, with the various relevant important reagents, presented in the usual order for the non-metallic and metallic central or characteristic elements from left to right and top to bottom in the Periodic Chart.

This principle of organization, however, like the more conventional ones, may raise a question of its own. Will a reaction with aqueous HCl be found under H_3O^+ or Cl^-? On the one hand, the dissolution of, say, MnO in HCl has little role for the Cl^- beyond maintaining electrical neutrality, and is therefore treated under H_3O^+, such as from HNO_3 or H_2SO_4. On the other hand, we regard redox processes as primary (absent other special features of interest), and the dissolution of MnO_2 in HCl involves not only the H_3O^+ but also the Cl^-, first as a ligand and then as a reductant, and is therefore treated in the Cl^- subsection.

Of course there is duplication, however; the reaction of MnO_2 with HCl, for example, may still appear under the subchapters for both Mn and Cl if we know products for each reactant, just as in other sources. Alternately, the schönites, "Tutton salts" and alums may be mentioned economically mainly under **16.2.4**

Other reactions of sulfates, not with every metal, and the oxidations of I^- and SCN^- to $[I(SCN)_2]^-$ with various oxidants may be treated efficiently in **17.4.3 Reduced chalcogens.**

Also, making useful comparisons within a narrative may call for some further mention of behavior that would, strictly speaking, be out of place; e.g., see section **14.5.1** on **Oxonium** about Pb species' reactivities with various acids whose anions, not only the H_3O^+, are crucial. Again, no organizational system is always superior.

Sources often classify reactions seemingly inappropriately. As an example, the dissolution of $PbSO_4$ in concentrated solutions of $CH_3CO_2^-$ may be found with the reactions of sulfur, and it may indeed be of interest there at times. The main action, however, is the formation of complexes between Pb^{2+} and $CH_3CO_2^-$, with less change in the SO_4^{2-} going from an ionic salt into solution. With economy of presentation we hope that most such cases are classified here more logically.

Arranging data by reactants makes obvious, for good or ill, the absence of information in most comprehensive compilations about many possible classical reactions, e.g., of aqueous tin species with borate. The extent of this absence and the worse abundance of contradictory reports, already mentioned, continue to be troublesome, and our resolution of some of the latter must be doubtful. We do not assume, however, that older reports are always less reliable.

We include some observations of the visual sensitivities of reactions for detecting species of interest. These are usually based on a few mL of solution in a test tube or small beaker.

The text then describes the various reactions. As the Table of Contents shows, the substances considered as reactants (rather than reagents) are taken up in order from left to right and then top to bottom in the Periodic Chart, with one Section for each element or subgroup of rarer or very similar elements. Because of the importance of redox behavior, in most cases a simple, objective criterion for similarity can be that the electrode potential between the element and its highest oxidation state differs by less than 2 decivolts from that of the immediate neighbor above or below it in the same Group. The lanthanoids and actinoids, however, are arranged by oxidation state.

In each Section we start with reagents derived from H and O, then the other Row-2 (periodic chart) non-metals B through F, separately from Rows 3 through 5 because their reactions are so different, followed by the latter non-metals Si through Xe, and finally the metals Li through U, plus electrons and photons. The (highly radioactive) species with $Z > 92$ are important too, but much more as reactants than as reagents.

For each set of reagents we normally sequence them from the left to the right in the periodic chart, then from top to bottom, and in the order of rising oxidation states (with some exceptions for carbon), likewise for reactants where some are considered together. When groups such as Cl^-, Br^- and I^-, or ClO^-, ClO_2^- and ClO_3^-, or Fe^{2+}, Co^{2+} and Ni^{2+} are discussed at once, the first member is decisive.

An example might be Cl_2, Br_2 and I_2 acting on Pd and Pt, then on Pd^{II} and Pt^{II}, followed by, say, ClO_3^-, BrO_3^- and IO_3^- on Pd and Pt, then on Pd^{II} and Pt^{II}. Sometimes the order is varied, perhaps for comparisons.

A partial rationale for dealing with all of the oxidation states of a particular reactant element for each reagent before going to the next reagents is that a given reagent may yield a similar result for several species of the element. Sulfane, H_2S, can thus produce PbS from PbO, Pb_3O_4, PbO_2, $Pb(CH_3CO_2)_2$ and so on.

The reactions with metal-derived reagents are subdivided differently from the others, as oxidations, reductions and other reactions, because their distinct reactions are fewer in some cases, and their Periodic-Chart Groups are inconveniently numerous. The main order is rising Group number, with period number and oxidation number being secondary.

The order of the reagents for each chapter and section follows, although some unimportant reagents are omitted in particular cases. First we enumerate the element's classical (i.e., mostly excluding organometallic) oxidation states in water (or in contact with it; many are insoluble), often as shown in its hydrides and oxides. Then the subsections of chapter *m*, element section *n*, are, except for omissions:

m.n.1 Reagents Derived from Hydrogen and Oxygen: Dihydrogen; Water (oxidane); Oxonium; Hydroxide; Peroxide; Di- and trioxygen.

m.n.2 Reagents Derived from the Other Row-2 Non-Metals, Boron through Fluorine: Boron species; Carbon oxide species; Cyanide species; Some "simple" organic species; Reduced nitrogen; Elemental and oxidized nitrogen; Fluorine species.

m.n.3 Reagents Derived from Rows 3-to-5 Non-Metals, Silicon through Xenon: Silicon species; Phosphorus species; Arsenic species; Reduced chalcogens; Elemental and oxidized chalcogens; Reduced halogens; Elemental and oxidized halogens; Xenon species.

m.n.4 Reagents Derived from the Metals Lithium through Uranium, plus Electrons and Photons: Oxidation; Reduction; Other reactions. Some borderline chemistry of arsenic may put it here as a metal.

Two subsections are added. One is 6.2.5 on: reactions involving chalcogeno Mo and W clusters, which is subdivided further into polyoxohomopolymetalates; polyoxoheteropolymetalates; chalcogeno (S, Se) cuboidal clusters, general; [S, Se clusters], homometallic; and [S, Se clusters], heterometallic. The second added subsection is 8.1.5 on reactions involving the "Prussian blues".

Some further subdivisions are to avoid confusingly long subsections but without requiring a deeper level of numbering for the entire book; Thus, 8.2.2 separates elemental-nitrogen and nitrogen-fixation-related reactions from others with oxidized nitrogen.

In additon, 6.2.4 reduction is subdivided into metallic species and electrons and photons; 14.1.4 other reactions into carbon monoxide and carbonate species, cyano species, and simple organic species; 15.1.4 reduction into nitrogen(<III), nitrite and nitrate; 15.2.4 other reactions into phosphorus(<V), monophosphates,

poly- and metaphosphates, and using high temperatures; 16.2.4 oxidation into reduced sulfur, thiosulfates and polythionates, and sulfites; 16.2.4 other reactions into sulfides, other reduced sulfur, sulfur and thiosulfates, polythionates, sulfites, and sulfates; 17.2.4 reduction into chlorine and hypochlorite, and chlorine(>I); and 17.2.4 other reactions into chloride, and chlorine and chlorine(>0).

In most cases the "simple" organic reagents above include chains of more than two carbon atoms only when, as with tartaric or citric acids, their traditionally important inorganic reactions are often omitted nowadays. Even small molecules such as $(CH_3)_3P$, which are treated well in many other modern compilations, are de-emphasized here.

The electronegativities of C and of S, Se and Te on the various scales might justify grouping the ligands, reductants or precipitants SCN^--*κS*, $SeCN^-$-*κSe*, CS_2 and so on, with the elemental and oxidized chalcogen species, but their chemistry as soft species often puts them more with the chalcogenides. We treat thiocyanato-*N* etc. with the cyanides, and thiocyanato-*S* (also thiocyanate as reductant, and CS_2 etc.,) with the reduced chalcogens, depending on the more likely site of coordination, but with the cyanides when neither N or S is, or both are, coordinated, or when the site is uncertain. Multiple sites are not similarly separated for cyanato-*N* or cyanato-*O* (both with cyanides) or nitro-*N* or nitrito-*O* (both oxidized nitrogen) or others.

We largely leave out various other topics relevant to aqueous inorganic chemistry, because of the space required. Our own work included a little on clock reactions and gas-releasing oscillators [64a-b], substitution kinetics [64c] (providing, incidentally, early evidence for actual Au^{II} and Pt^{III}, albeit transitory and not isolated), the use of chelating ion-exchange resins as reagents to dissolve difficultly soluble salts [64d], and preserving reactive ions in solid solutions [64e]. See also [65] for far more on chemical oscillations.

Space limitations prompt us to omit other references of relatively narrow scope, and to emphasize those which, regardless of title or age, include considerable information on aqueous inorganic reactions. Many gaps in these descriptions remain, partly due to crowding from an abundance of data, and partly from the opposite problem of missing data. Let us just list now, however, some further more *general* references. Reviews are often preferred here and in the chapters. In too many cases to cite, we imply, "See the references cited therein."

First, from newer to older, there are more encyclopedias [66–77], recent or large general texts but omitting some with much, say, excellent introductory physical, but little descriptive inorganic, chemistry [78–100], and books on various large groups of elements [101–137]. Many other books on broad topics are quite useful and interesting, including [138–270]. References [175a] and [175b] are interesting as more literary than chemical treatments. Reference [210], chapter 6, on reductionism, holism and complementarity is broadly philosophical but may especially interest some.

On the above point on references omitted, we note that much-used compendia [68, 72] can have more than 3300 references in just one chapter, and here we can only marvel at that. For similar reasons, although describing a few syntheses in detail, we only summarize many others and often omit complexes of over two different ligands, absent some special interest.

A few additional relevant articles from serials and journals after 1999, applicable to various chapters, hence not listed with any single ones, offer: developing nuclear chemistry [271]; "exocharmic" reactions [272]; how to predict inorganic-reaction products [273]; Rf, Db and Sg, especially on non-aqueous aspects and theory [274]; a thematic issue on water [275]; a review of main-Group chemistry [276]; and a review of **d**-block chemistry [277].

Further references, from the 1990s, discuss: **d**-block cyanides [278]; relativistic effects [279]; strong closed-shell interactions [280]; metal-ligand multiple bonding [281]; the trans-actinoid elements [282]; relativistic trans-actinoid predictions [283]; **d**-block oxygen kinetics etc. [284]; the thermodynamics of ligands with hydrons and metal cations in water at high temperatures [285]; the structure and dynamics of hydrated ions [286]; large, weakly coordinating anions [287]; inter-metal atom-transfer reactions of O, S, Se and N [288]; and metal-metal dimers and chains [289].

Some references from the 1980s discuss, among other things: ligand design for selective complexation of metal ions in water [290]; triangular, bridged complexes [291]; ionic radii in water [292]; relativistic effects in structures [293]; aqua-ion structures by diffraction [294]; metal-centered oxygen-atom transfers [295]; unusual, but mostly non-aqueous, metal cations [296]; an acidity scale for binary oxides [297]; metallic multiple bonds [298]; empirical thermodynamic rules for the solvation of monoatomic ions [299]; Henry Taube's work on mechanisms [300]; making Hard and Soft Acids and Bases more quantitative [301]; **d**-block metal-metal bonding [302]; heterolytic activation of H2 by **d**-block complexes [303]; and metal-sulfur bond reactivity [304].

Additional references from earlier years discuss: ring, cage and cluster compounds of the main-Group elements, general, emphasizing structure [305]; "equivalent" and "normal" [306]; 7-coordination chemistry [307]; non-adiabatic electron transfer [308]; 1-dimensional inorganic complexes [309]; cyano-complexes of Groups 4–7 [310]; **d**-block photochemistry [311]; hypervalent compounds [312]; platinum-Group thermochemistry and oxidation potentials [313]; the homogeneous catalysis of hydrogenation, oxidation etc. [314]; **d**-block NO complexes [315]; a graphical method for redox free energies [316]; early detection of bridged activated complexes from labile Cr^{2+} and inert Co^{III} [317]; and complex-ion substitution kinetics [318].

For readers interested in certain other inorganic but non-aqueous contributions we list macrocyclic chelates [319], and, beyond reactions, a general model for cubic crystal structures [320], a semi-empirical theory of boiling points [321], and an old note on simplified calculations for the harmonic oscillator and rigid rotator [322].

Faraday Discussion 141 on water, "perhaps the most important chemical substance known", is scheduled for 2008, August 27–29 [323].

Let us now examine the desired properties of each element in turn.

References

1. Mazurs EG (1974) Graphic representations of the periodic system during one hundred years, 2nd edn. University of Alabama, Tuscaloosa
2. Acc Chem Res; Adv Inorg Chem; Adv Trans Metal Coord Chem; Annu Rep Prog Chem Sect A: Inorg Chem; Chem Rev; Chem Soc Rev; Coord Chem Rev; Inorg Synth; MTP Internat Rev Sci; Polyhedron; Prog Inorg Chem; Quart Rev; Spec Per Rep; Top Curr Chem
3. Cotton FA, Wilkinson G, Murillo CA, Bochmann M (1999) Advanced inorganic chemistry, 6th edn. Wiley, New York
4. Scerri ER (2007) The periodic table: its story and its significance. Oxford, Oxford
5. Leach MR, Moran J, Stewart P (2006 Oct 24) New York Times, p D4
6. Winter M (2006) http://www.webelements.com/
7. Goho A (2003) Sci News 164:264
8. Jacobs M (ed) (2003 Sept 8) Chem Eng News 81:27
9. Railsback B (2003) Geology 31:737
10. Karol PJ (2002) J Chem Educ 79:60
11. Hawkes SJ (2001) J Chem Educ 78:1686
12. Hoffman DC, Lee DM (1999) J Chem Educ 76:331
13. Scerri ER (1998) J Chem Educ 75:1384
14. van Spronsen JW (1969) The periodic system of chemical elements. Elsevier, Amsterdam
15. Rich RL (1965) Periodic Correlations. Benjamin Cummings, San Francisco
16. Chemdex: http://www.chemdex.org; Stanford University: http://www-sul.stanford.edu/depts/swain/help/subjectguides/general/desk/toolbox.html#per
17. Laing M (1989) J Chem Educ 66:746
18. Rich RL (2005) J Chem Educ 82:1761
19. Cronyn MW (2003) J Chem Educ 80:947
20. Sanderson RT (1967) Inorganic chemistry. Reinhold, New York
21. Smith D (2006) private communication.
22. Jensen WB (2003) J Chem Educ 80:952
23. Thayer JS (2005) J Chem Educ 82:1721
24. Hess BA (ed) (2003) Relativistic effects in heavy-element chemistry and physics. Wiley, New York
25. Kaldor U, Wilson S (eds) (2003) Theoretical chemistry and physics of heavy and superheavy elements. Springer, Berlin Heidelberg New York
26. Balasubramanian K (1997) Relativistic effects in chemistry. Wiley, New York
27. McAlpine RK, Soule BA (1933) Qualitative chemical analysis. Van Nostrand, New York (a) ibid p 131
28. Emeléus HJ (ed) (1975) Inorganic chemistry, series 2. Butterworth, London
29. Bailar JC Jr, Emeléus HJ, Nyholm R, Trotman-Dickenson AF (eds) (1973) Comprehensive inorganic chemistry. Pergamon, Oxford
30. Sneed MC, Maynard JL, Brasted RC (eds) (1972) Comprehensive inorganic chemistry. RE Krieger Publishing, Huntington NY
31. Durrant PJ, Durrant B (1962) Introduction to advanced inorganic chemistry. Wiley, New York
32. Jacobson CA, Hampel CA (eds) (1946–1959) Encyclopedia of chemical reactions. Reinhold, New York
33. Charlot G (1957) L'analyse qualitative et les réactions en solution. Masson, Paris
34. Remy H (1956) Treatise on inorganic chemistry. Elsevier, Amsterdam

35. Latimer WM, Hildebrand JH (1951) Reference book of inorganic chemistry, 3rd edn. Macmillan, New York
36. Feigl F, Oesper RE (trans) (1940) Specific and special reactions. Elsevier, New York
37. Mellor JW (ed) (1922-) Mellor's comprehensive treatise on inorganic and theoretical chemistry, and supplements. Longman, London
38. King RB (ed) (1994) Encyclopedia of inorganic chemistry. Wiley, New York
39. GMELIN Institute for Inorganic Chemistry of the Max-Planck Society for the Advancement of Science (ed) (1988-) Gmelin Handbook of Inorganic and Organometallic Chemistry, 8th edn. Springer, Berlin Heidelberg New York
40. (a) Hawkes SJ (2004) J Chem Educ 81:1569 (b) Rich RL (1993) J Chem Educ 70:260 (c) Kauffman GB (1991) J Chem Educ 68:534 (d) Yoke JT (1991) J Chem Educ 68:533 (e) Yoke J (1989) J Chem Educ 66:310 (f) Laing M (1988) Spectrum 26(4):11
41. (a) Rich RL (2003) U Chem Educ 7:35 (b) Rich RL (1991) J Chem Educ 68:534 (c) Rich RL (1990) J Chem Educ 67:629 (d) Rich RL (1988 Nov 7) Chem Eng News 66:67 (e) Rich RL (1985 Jun 10) Chem Eng News 63:2
42. Menger FM, Haim A (1992) Nature 359:666
43. DeWit DG (2006) J Chem Educ 83:1625
44. Skoog DA, West DM, Holler FJ (1996) Fundamentals of analytical chemistry, 7th edn. Saunders, Fort Worth
45. Kolthoff IM, Elving PJ (eds) (1961–1980) Treatise on analytical chemistry Part II Vols 1–10 (inorganic), Index Vol 17; (1979) 2nd edn Part I Vol 2. Wiley, New York
46. (a) Inczédy J (1976) Analytical applications of complex equilibria. Ellis Horwood, Chichester (b) Pickering WF (1971) Modern analytical chemistry. Dekker, New York (c) Charlot G (1967) Chimie analytique generale. Masson, Paris (d) Ringbom A (1963) Complexation in analytical chemistry. Interscience, New York
47. Furman NH, Welcher FJ (eds) (1962–1963) Standard methods of chemical analysis, 6th edn, vol 1 and vol 2, part A. D Van Nostrand, Princeton
48. Lundell GEF, Bright HA, Hoffman JI (1953) Applied Inorganic Analysis, 2nd edn. Wiley, New York
49. Treadwell FP, Hall WT (1937–1942) Analytical chemistry, 9th edn. Wiley, New York
50. Jensen WB (2004) J Chem Educ 81:1772
51. Hanusa TP in McCleverty JA, Meyer TJ (eds) (2004) Comprehensive coordination chemistry II, vol 3, p 1. Elsevier, Amsterdam
52. Leigh GJ (ed) (1990) Nomenclature of Inorganic Chemistry. IUPAC, Blackwell, London
53. (a) Rich RL (1986) J Chem Educ 63:785 (b) Westland AD (1980) Chem Brit 16:247 (c) West TS (1980) Chem Brit 16:306
54. Greenwood NN, Earnshaw A (1997) Chemistry of the elements, 2nd edn. Elsevier, Amsterdam
55. Holleman AF, Wiberg E, Wiberg N; Eagleson M, Brewer W (trans); Aylett BJ (rev) (2001) Inorganic chemistry. Academic, San Diego
56. Jones C (2001) D- and f-block chemistry. Royal Society of Chemistry, Cambridge
57. Massey AG (2000) Main group chemistry, 2nd edn. Wiley, New York
58. Baes CF Jr, Mesmer RE (1976) The hydrolysis of cations. Wiley, New York
59. Burns DT, Townshend A, Carter AH (1981) Inorganic reaction chemistry. Ellis Horwood-Wiley, New York
60. Richens DT (1997) The chemistry of aqua ions. Wiley, New York
61. Emsley J (1998) The elements, 3rd edn. Clarendon, Oxford
62. Burgess J. (1999) Ions in solution, 2nd edn. Horwood, Chichester
63 Marcus Y (1985) Ion solvation. Wiley, New York

64. (a) Rich RL, Noyes RM (1990) J Chem Educ 67:606 (b) Kaushik SM, Rich RL, Noyes RM (1985) J Phys Chem 89:5722 (c) Rich RL, Taube H (1954) J Phys Chem 58:1 and 58:6 (d) Rich RL (1963) J Chem Educ 40:414 (e) Rich RL (1975) J Chem Educ 52:805
65. Nicolis G, Portnow J (1973) Chem Rev 73:365

Encyclopedias

66. Considine GD (ed) (2005) Van Nostrand's encyclopedia of chemistry, 5th edn. Wiley, Hoboken, New Jersey
67. Daintith J (ed) (2004) The facts on file dictionary of inorganic chemistry. Market House Books, New York
68. McCleverty JA, Meyer TJ (eds) (2004) Comprehensive coordination chemistry II, 10 vols. Elsevier Pergamon, Amsterdam
69. Patnaik P (2003) Handbook of inorganic chemicals. McGraw-Hill, New York
70. Stwertka A, Stwertka E (1998) A guide to the elements. Oxford University Press, Oxford
71. Macintyre JE (exec ed) (1992) Dictionary of inorganic compounds. Chapman & Hall, London
72. Wilkinson G, Gillard RD, McCleverty JA (eds) (1987) Comprehensive coordination chemistry, vols 1–5. Pergamon, Oxford
73. Grayson M, Eckroth D (eds) (1984) Kirk-Othmer encyclopedia of chemical technology, 3rd edn, 24 vols. Wiley, New York
74. Kragten J (1977) Atlas of metal-ligand equilibria in aqueous solution. Wiley, New York
75. Emeléus HJ (ed) (1972 & 1975) MTP international review of science: inorganic chemistry, ser 1 & 2. Butterworth, London
76. Trotman-Dickenson AF, Bailar JC Jr, Emeléus HJ, Nyholm R (eds) (1973) Comprehensive inorganic chemistry, 5 vols. Pergamon, Oxford
77. Pascal P (ed) (1956–1962) Nouveau traité de chimie minérale, 20 vols; Pacault A, Pannetier G (eds) (1974) Compléments au nouveau traité de chimie minérale. Masson, Paris

Recent or large general texts

78. Kotz JC, Treichel PM, Weaver GC (2006) Chemical reactivity, 6th edn, 3 vols. Brooks/Cole, Belmont, CA
79. Cox PA (2004) Instant notes in inorganic chemistry, 2nd edn. Garland Science, New York
80. Miessler GL, Tarr DA (2004) Inorganic chemistry. Pearson Education, Upper Saddle River, NJ
81. Mackay KM, Mackay RA, Henderson W (2002) Introduction to modern inorganic chemistry, 6th edn. Nelson Thornes, Cheltenham
82. Holleman AF, Wiberg E, Wiberg N (1995) Eagleson M, Brewer W (trans) Aylett BJ (revised) (2001) Inorganic chemistry. Academic, San Diego
83. House JE, House KA (2001) Descriptive inorganic chemistry. Brooks/Cole, Belmont, CA
84. Housecroft CE, Sharpe AG (2001) Inorganic chemistry. Prentice Hall, London

85. Rayner-Canham G (2000) Descriptive inorganic chemistry, 2nd edn. Freeman, New York
86. Schubert U, Hüsing N (2000) Synthesis of inorganic materials. Wiley, New York
87. Winter M, Andrew J (2000) Foundations of inorganic chemistry. Oxford University Press, Oxford
88. Shriver DF, Atkins PW (1999) Inorganic chemistry, 3rd edn. Oxford University Press, Oxford
89. Mingos DMP (1995, 1998) Essentials of inorganic chemistry 1, 2. Oxford University Press, Oxford
90. Rao CNR (1995) Chemical approaches to the synthesis of inorganic materials. Wiley, New York
91. Douglas B, McDaniel D, Alexander J (1994) Concepts and models of inorganic chemistry, 3rd edn. Wiley, New York
92. Bowser JR (1993) Inorganic chemistry. Brooks/Cole, Belmont, CA
93. Huheey JE, Keiter EA, Keiter RL (1993) Inorganic chemistry: principles of structure and reactivity, 4th edn. Benjamin Cummings, San Francisco
94. Porterfield WW (1993) Inorganic chemistry: a unified approach, 2nd edn. Academic, San Diego
95. Sharpe AG (1992) Inorganic chemistry, 3rd edn. Wiley, New York
96. Lee JD (1991) Concise inorganic chemistry, 4th edn. Chapman & Hall, London
97. Moody BJ (1991) Comparative inorganic chemistry. Edward Arnold, London
98. Wulfsberg G (1991) Principles of descriptive inorganic chemistry. University Science Books, New York
99. Durant PJ, Durant B (1962) An introduction to advanced inorganic chemistry. Longman, London
100. Sidgwick NV (1950) Chemical elements and their compounds. Oxford University Press, Oxford

On various large groups of elements

101. Schädel M (ed) (2003) The chemistry of superheavy elements. Springer, Berlin Heidelberg New York
102. Henderson W (2002) Main group chemistry. Wiley, New York
103. Hofman S (2002) On beyond uranium. Taylor and Francis, New York
104. Johnson D (ed) (2002) Metals and chemical change. Royal Society of Chemistry, Cambridge
105. Jones CJ (2001) D- and f-block chemistry. Royal Society of Chemistry, Cambridge
106. Gerloch M, Constable EC (2000) Transition metal chemistry. Wiley, New York
107. Massey AG (2000) Main group chemistry. Wiley, New York
108. Housecroft CE (1999) The heavier d-block metals: aspects of inorganic and coordination chemistry. Oxford University Press, Oxford
109. McCleverty JA (1999) Chemistry of the first-row transition metals. Oxford University Press, Oxford
110. Cotton SA (1997) Chemistry of precious metals. Blackie, London
111. Norman NC (1997) Periodicity and the s- and p-block elements. Oxford University Press, Oxford
112. King RB (1995) Inorganic chemistry of main group elements. VCH, New York
113. Housecroft CE (1994) Cluster molecules of the p-block elements. Oxford University Press, Oxford

114. Gerloch M, Constable EC (1994) Transition metal chemistry: the valence shell in d-block chemistry. VCH, Weinheim
115. Winter MJ (1994) D-block chemistry. Oxford University Press, Oxford
116. Hartley FR (ed) (1991) Chemistry of the platinum group metals, recent developments, in Studies in inorganic chemistry, vol 11. Elsevier, Amsterdam
117. Johnson BFG (1972–1978) Inorganic chemistry of the transition elements, literature review, 6 vols. The Chemical Society, London
118. Parish RV (1977) The metallic elements. Longman, London
119. Steudel R, Nachod FC, Zuckerman JJ (1977) Chemistry of the non-metals. Walter de Gruyter, Berlin
120. Cotton SA, Hart FA (1975) The heavy transition metals. Macmillan, London
121. Livingstone S (1975) The chemistry of ruthenium, rhodium, palladium, osmium, iridium and platinum. Pergamon, New York
122. Sharp DWA (1974) Transition metals – part 1. Buttersworth, London
125. Earnshaw A, Harrington TJ (1973) The chemistry of the transition elements. Clarendon, Oxford
126. Connolly TF (1972) Groups IV, V and VI metals and compounds. Plenum, New York
127. Kepert DL (1972) The early transition metals. Academic, London
128. Songina OA, transl Schmorak J (1970) Rare metals. Israel Program for Scientific Translations, Jerusalem
129. Colton R, Canterford JH (1969) Halides of the first row transition metals. Wiley, New York
130. Canterford JH, Colton R (1968) Halides of the second and third row transition metals. Wiley, New York
131. Griffith WP (1967) The chemistry of the rare platinum metals. Wiley, New York
132. Collman JP (1966) Transition metal chemistry. Dekker, New York
133. Jolly WL (1966) The chemistry of the non-metals. Prentice Hall, Upper Saddle River, NJ
134. Rochow EG (1966) The metalloids. Heath, Boston
135. Sherwin E, Weston GJ (1966) Chemistry of the non-metallic elements. Pergamon, Oxford
136. Larsen EM (1965) Transitional elements. Benjamin Cummings, San Francisco
137. Browning PE (1903) Introduction to the rarer elements. Wiley, New York

On broad topics

138. Scerri ER (2007) The periodic table: its story and its significance. Oxford University Press, Oxford
139. Urben PG (ed) (2007) Bretherick's Handbook of Reactive Chemical Hazards, 7th edn, 2 vols. Academic, Oxford
140. Bagotsky VS, transl Müller K (2006) Fundamentals of electrochemistry, 2nd edn. Wiley, Hoboken NJ
141. Driess M, Noth H (2004) Molecular clusters of the main group elements. Wiley, New York
142. Rayner-Canham GW, Laing M (on periodic patterns), Scerri ER (on the $n + l$ rule), Balasubramanian K (on relativity) in Rouvray DH, King RB (eds) (2004) The periodic table: into the 21st century. Research Studies, Baldock, UK
143. Asperger S (2003) Chemical kinetics and inorganic reaction mechanisms, 2nd edn. Springer, Berlin Heidelberg New York

144. Basolo F, Burmeister JL (2003) On being well coordinated, selected papers of Fred Basolo. World Scientific, Singapore
145. Reinhard PG, Suraud E (2003) An introduction to cluster dynamics. Wiley, New York
146. Woollins JD (2003) Inorganic experiments, 2nd edn. Wiley-VCH, Weinheim.
147. Zanello P (2003) Inorganic electrochemistry: theory, practice and applications. Royal Society of Chemistry, Cambridge
148. Benjamin MM (2002) Water chemistry. McGraw-Hill, Boston
149. Herrmann WA (ed) (1996–2002) Synthetic methods of organometallic and inorganic chemistry, 10 vols. Thieme, New York
150. Bard AJ, Faulkner LR (2001) Electrochemical methods, 2nd edn. Wiley, New York
151. Dell RM, Rand DAJ (2001) Understanding batteries. Royal Society of Chemistry, Cambridge
152. Kubas GJ (2001) Metal dihydrogen and σ-bond complexes. Springer, Berlin Heidelberg New York
153. McCleverty JA, Connelly NG (eds) (2001) Nomenclature of inorganic chemistry II: recommendations 2000. Royal Society of Chemistry, Cambridge
154. Hall N (ed) (2000) The new chemistry. Cambridge University, Cambridge
155. Burgess J (1999) Ions in solution, 2nd edn. Horwood, Chichester
156. Girolami GS, Rauchfuss TB, Angelici RJ (1999) Synthesis and technique in inorganic chemistry: a laboratory manual, 3rd edn. University Science Books, Sausalito, CA
157. Tanaka J, Suib SL (1999) Experimental methods in inorganic chemistry. Prentice Hall, Upper Saddle River, NJ
158. Tobe ML, Burgess J (1999) Inorganic reaction mechanisms. Harlow, Essex
159. Bockris J(O'M), Reddy AKN (1998) Modern electrochemistry, 2nd edn, ionics. Plenum, New York
160. Jordan RB (1998) Reaction mechanisms of inorganic and organometallic systems, 2nd edn. Oxford University Press, Oxford
161. Mingos DMP (1998) Essential trends in inorganic chemistry. Oxford University Press, Oxford
162. vdPut PJ (1998) The inorganic chemistry of materials: how to make things out of elements. Plenum, New York
163. Errington RJ (1997) Advanced practical inorganic and metalorganic chemistry. Nelson Thornes, Cheltenham
164. Norman NC (1997) Periodicity and the s- and p-block elements. Oxford University Press, Oxford
165. Swaddle TW (1997) Inorganic chemistry: an industrial and environmental perspective. Academic, San Diego
166. Bernstein, ER (1996) Chemical reactions in clusters. Oxford University Press, Oxford
167. Compton RG, Sanders GHW (1996) Electrode potentials. Oxford University Press, Oxford
168. Davies JA, Hockensmith CM, Kukushkin VYu, Kukushkin, YuN (1996) Synthetic coordination chemistry. World Scientific, Singapore.
169. Kettle SFA (1996, 1998) Physical inorganic chemistry, a coordination chemistry approach. Oxford University Press, Oxford
170. Martell AE, Hancock RD (1996) Metal complexes in aqueous solutions. Plenum, New York
171. Müller A, Dress A, Vögtle F (eds) (1996) From simplicity to complexity – and beyond. Vieweg, Braunschweig

172. Stumm W, Morgan JJ (1996) Aquatic chemistry, 3rd edn. Wiley, New York
173. Vogel AI (1996) Vogel's qualitative inorganic analysis. Longman, New York
174. Astruc D (1995) Electron transfer and radical processes in transition metal chemistry. VCH, Weinheim
175. (a) Atkins, PW (1995) The periodic kingdom. Basic, New York; (b) Levi P, Rosenthal R (trans) (1984) The periodic table. Schocken Books, New York
176. Weller MT (1995) Inorganic materials chemistry. Oxford University Press, Oxford
177. Blesa MA, Morando PJ, Regazzoni AE (1994) Chemical dissolution of metal oxides. CRC, Boca Raton
178. Housecroft CE (1994) Cluster molecules of the p-block elements. Oxford University Press, Oxford
179. Jolivet JP, Henry M, Livage J (1994) Bescher E (trans) (2000) Metal oxide chemistry and synthesis. Wiley, New York
180. Kauffman GB (ed) (1994) Coordination chemistry: a century of progress. American Chemical Society, Washington
181. Lappin AG (1994) Redox mechanisms in inorganic chemistry. Horwood, New York
182. Pope MT, Miller A (1994) Polyoxometallates: from platonic solids to antiretroviral activity. Springer, Berlin Heidelberg New York
183. Rieger PH (1994) Electrochemistry, 2nd edn. Springer, Berlin Heidelberg New York
184. Cotton FA, Walton RA (1993) Multiple bonds between metal atoms, 2nd edn. Oxford University Press, Oxford
185. Henderson RA (1993) The mechanisms of reactions at transition metal sites. Oxford University Press, Oxford
186. Lappin AG (1993) Redox mechanisms in inorganic chemistry. Ellis Horwood
187. Morel FMM, Hering JG (1993) Principles and application of aquatic chemistry. Wiley, New York
188. Fehlner TP (1992) Inorganometallic chemistry. Plenum, London
189. Martell AE, Motekaitis RJ (1992) Determination and use of stability constants, 2nd edn. Wiley, New York
190. Wilkins RG (1991) Kinetics and mechanism of reactions of transition metal complexes, 2nd edn. VCH, Weinheim
191. Beck MT, Nagypál I (1990) Durham DA (trans) Williams DR (trans ed) Chemistry of complex equilibria. Horwood, Chichester
192. Smith DW (1990) Inorganic substances. Cambridge University, Cambridge
193. Martell AE, Smith RM (1974) Critical stability constants, vol 4 and later supplements. Plenum, New York
194. Sanderson RT (1989) Simple inorganic substances. Robert E. Krieger, Malabar, FL
195. Magini M (ed) (1988) X-ray diffraction of ions in aqueous solution: hydration and complex formation. CRC, Boca Raton
196. Meites L, Zuman P, Rupp EB, Fenner TL, Narayanan A (1988) CRC handbook series in inorganic electrochemistry, 8 vols. CRC, Boca Raton
197. Nugent WA, Mayer JM (1988) Metal-Ligand Multiple Bonds. Wiley, New York
198. Woollins JD (1988) Non metal rings, cages and clusters. Wiley, New York
199. Haiduc I, Sowerby DB (eds) (1987) The chemistry of inorganic homo and heterocycles. Academic, London
200. Katakis D, Gordon G (1987) Mechanisms of Inorganic Reactions. Wiley, New York
201. Bard AJ (ed) (1973–1986) Encyclopedia of electrochemistry of the elements, vols I-IX. Dekker, New York
202. Golub AM, Köhler H, Skopenko VV (eds) (1986) Chemistry of pseudohalides. Elsevier, Amsterdam

203. Bard AJ, Parsons R, Jordan J (eds) (1985) Standard potentials in aqueous solution. Dekker, New York
204. Wells AF (1984) Structural inorganic chemistry, 5th edn. Clarendon, Oxford
205. Braterman PS (ed) (1983) Reactions of coordinated ligands. Plenum, New York
206. Sykes AG (ed) (1982–84) Advances in inorganic and bioinorganic mechanisms. Academic, London
207. Burford N, Chivers T, Rao MNS, Richardson JF (1983) Rings, clusters and polymers of the main group elements, ACS Symposium Series No 232. American Chemical Society, Washington
208. Matthes S (1983) Eine einführung in die spezielle mineralogie, petrologie und lagerstättenkunde. Springer, Berlin Heidelberg New York
209. Pope MT (1983) Heteropoly and isopoly oxometalates. Springer, Berlin Heidelberg New York
210. Primas H (1983) Chemistry, quantum mechanics and reductionism. Springer, Berlin Heidelberg New York
211. Shakhashiri BZ (1983) Chemical demonstrations. University of Wisconsin, Madison
212. Taube H (1983) Electron transfer between metal complexes – retrospective, in Frängsmyr T, Malmström BG (eds) (1992) Nobel lectures, chemistry 1981–1990. World Scientific, Singapore
213. Young CL (ed) (1983) Sulfur dioxide, chlorine, fluorine, and chlorine oxides. Pergamon, Oxford
214. Antelman MS (1982) The encyclopedia of chemical electrode potentials. Plenum, New York
215. Cotton FA, Walton RA (1982) Multiple bonds between metal atoms. Wiley, New York
216. Högfeldt E (1982) Stability constants of metal-ion complexes, part A: inorganic ligands. Pergamon, Oxford
217. Kepert DL (1982) Inorganic stereochemistry. Springer, Berlin Heidelberg New York
218. Spiro TG (ed) (1982) Metal ions in biology. Wiley, New York
219. Stumm W, Morgan JJ (1981) Aquatic chemistry, 2nd edn. Wiley, New York
220. Cannon RD (1980) Electron transfer reactions. Buttersworth, London
221. Choppin GR, Rydberg J (1980) Nuclear chemistry. Pergamon, Oxford
222. Snoeyink VL, Jenkins D (1980) Water chemistry. Wiley, New York
223. Krauskopf KB (1979) Introduction to geochemistry. McGraw-Hill, New York
224. Martell AE (ed) (1978) Coordination chemistry, monograph 174. American Chemical Society, Washington
225. Milazzo G, Caroli S (1978) Tables of standard electrode potentials. Wiley, New York
226. Lang IW (1977) Patterns and periodicity. Wiley, New York
227. Kauffman GB (1968, 1976) Classics in coordination chemistry, part 1, part 2. Dover, New York
228. Sharpe AG (1976) The chemistry of cyano complexes of the transition metals. Academic, London
229. Adamson AW, Fleischauer PD (eds) (1975) Concepts of inorganic photochemistry. Wiley, New York
230. Taylor MJ (1975) Metal-to-metal bonded states of the main group elements. Academic, London
231. Conway BE, Bockris J(O'M) (eds) (1974) Modern aspects of electrochemistry. Plenum, New York
232. Pourbaix M (1974) Franklin JA (trans) Atlas of electrochemical equilibria in aqueous solutions. National Association of Corrosion Engineers, Houston TX

233. Rubin AJ (ed) (1974) Aqueous-environmental chemistry of metals. Ann Arbor Science, Ann Arbor MI
234. Franks F (ed) (1973) Water in crystalline hydrates; aqueous solutions of simple nonelectrolytes. Plenum, New York
235. Pearson RG (ed) (1973) Hard and soft acids and bases. Dowden, Hutchinson & Ross, Stroudsburg PA
236. Feigl F, Anger V, transl Oesper RE (1972) Spot tests in inorganic analysis, 6th edn. Elsevier, Amsterdam
237. Franks F (ed) (1972) Water, a comprehensive treatise, vols 1–3. Plenum, New York
238. Tobe ML (1972) Reaction mechanisms in inorganic chemistry. Buttersworth, London
239. The Chemical Society (1964 and 1971) Stability constants (special publications nos. 17 and 25). The Chemical Society, London
240. Balzani V, Carassiti V (1970) Photochemistry of coordination compounds. Academic, New York
241. Haiduc I (1970) The chemistry of inorganic ring systems, 2 parts. Wiley, New York
242. Martell AE (ed) (1970) Coordination chemistry, monograph 174. American Chemical Society, Washington
243. Taube H (1970) Electron transfer reactions of complex ions in solution. Academic, New York.
244. Angelici RJ (1969) Synthesis and techniques in inorganic chemistry. Saunders, Philadelphia
245. Jörgensen CK (1969) Oxidation numbers and oxidation states. Springer, Berlin Heidelberg New York
246. Marcus Y, Kertes AS (1969) Ion exchange and solvent extraction of metal complexes. Wiley-Interscience, London
247. Samsonov GV (1969) Turton CN, Turton TI (trans) (1973) The oxide handbook. Plenum, New York
248. Brown D (1968) Halides of the transition elements. Wiley, New York
249. Candlin JP, Taylor KA, Thompson DT (1968) Reactions of transition-metal complexes. Elsevier, Amsterdam
250. Basolo F, Pearson RG (1967) Mechanisms of inorganic reactions, 2nd edn. Wiley, New York.
251. Calvert JG, Pitts JN Jr (1966) Photochemistry. Wiley, New York
252. Langford CH, Gray HB (1966) Ligand substitution processes. Benjamin Cummings, San Francisco
253. Lavrukhina AK, Pozdnyakov AA (1966) Kondor R (trans) (1970) Analytical chemistry of technetium, promethium, astatine and francium. Ann Arbor-Humphrey, Ann Arbor
254. Parsons CRH, Dover C (1966) The elements and their order; foundations of inorganic chemistry. Ginn, Boston
255. Reynolds WL, Lumry RW (1966) Mechanisms of electron transfer. Ronald Press, New York
256. Brauer G (ed) (1963–1965) Handbook of preparative inorganic chemistry, 2nd edn, 2 vols. Academic, New York
257. Garrels RM, Christ CL (1965) Solutions, minerals, and equilibria. Harper and Row, New York
258. Lister MW (1965) Oxyacids. Oldbourne, London
259. Wyckoff RWG (1965) Crystal structures, 2nd edn. Interscience, New York
260. Kavanau JL (1964) Water and solute-water interactions. Holden-Day, San Francisco
261. Hunt JP (1963) Metal Ions in Aqueous Solution. Benjamin Cummings, San Francisco
262. Jörgensen CK (1963) Inorganic complexes. Academic, London

263. Souchay P (1963) Polyanions et polycations. Gauthier-Villars, Paris
264. Colburn CB, Gould RF (eds) (1962) Free radicals in inorganic chemistry, monograph 36. American Chemical Society, Washington
264. Jörgensen CK (1962) Absorption spectra and chemical bonding in complexes. Pergamon, London
265. Schlessinger GG (1962) Inorganic laboratory preparations. Chemical Publishing Company, New York
266. Kirschner S (ed) (1961) Advances in the chemistry of coordination compounds. Macmillan, New York
267. Hurd DT (1952) An introduction to the chemistry of the hydrides. Wiley, New York
268. Latimer WM (1952) Oxidation potentials, 2nd edn. Prentice Hall, Upper Saddle River, NJ
269. Noyes AA, Bray WC (1952) A system of qualitative analysis for the rare elements. Macmillan, New York
270. Price TS (1912) Per-acids and their salts. Longmans Green, London

Other articles cited in the text

271. http://livingtextbook.oregonstate.edu/
272. Ramette RW (a) (2007) J Chem Educ 84:16; (b) (1980) J Chem Educ 57:68
273. DeWit DG (2006) J Chem Educ 83:1625
274. Kratz JV (2003) Pure Appl Chem 75:103
275. Pratt LR (2002) Chem Rev 102:2625
276. Greenwood NN (2001) J Chem Soc Dalton Trans 2001:2055
277. Cotton FA (2000) J Chem Soc Dalton Trans 2000:1961
278. Dunbar KR, Heintz RA (1997) in Karlin KD (ed) Prog Inorg Chem 45:283
279. Kaltsoyannis N (1997) J Chem Soc Dalton Trans 1997:1
280. Pyykkö P (1997) Chem Rev 97:597
281. Che CM, Yam VWW (eds) (1996) 1:1 Adv Trans Metal Coord Chem
282. Cotton SA (1996) Chem Soc Rev 25:219
283. Pershina VG (1996) Chem Rev 96:1977
284. Bakac A (1995) in Karlin KD (ed) Prog Inorg Chem 43:267
285. Chen X, Izatt RM, Oscarson JL (1994) Chem Rev 94:467
286. Ohtaki H, Radnai T (1993) Chem Rev 93:1157
287. Strauss SH (1993) Chem Rev 93:927
288. Woo LK (1993) Chem Rev 93:1125
289. Clark RJH (1990) Chem Soc Rev 19:107
290. Hancock RD, Martell AE (1989) Chem Rev 89:1875
291. Cannon RD, White RP (1988) in Lippard SJ (ed) Prog Inorg Chem 36:195
292. Marcus Y (1988) Chem Rev 88:1475
293. Pyykkö P (1988) Chem Rev 88:563
294. Enderby JE et al (1987) J Phys Chem 91: 5851
295. Holm RH (1987) Chem Rev 87:1401
296. O'Donnell TA (1987) Chem Soc Rev 16:1
297. Smith DW (1987) J Chem Educ 64:480
298. Cotton FA, Walton RA (1985) Struct Bonding 62:1
299. Notoya R, Matsuda A (1985) J Phys Chem 89:3922
300. Lippard S (ed) (1983) Prog Inorg Chem, vol 30
301. Pearson RG (1983) J Chem Educ 64:561
302. Chisholm MH, Rothwell IP (1982) in Lippard SJ (ed) Prog Inorg Chem 29:1

303. Brothers PJ (1981) in Lippard SJ (ed) Prog Inorg Chem 28:1
304. Kuehn CG, Isied SS (1980) in Lippard SJ (ed) Prog Inorg Chem 27:153
305. Gillespie RJ (1979) Chem Soc Rev 8:315
306. Irving HMNH (1978) Pure Appl Chem 50:325
307. Drew MGB (1977) in Lippard SJ (ed) Prog Inorg Chem 23:67
308. Taube H (1977) Adv Chem Ser #162:127
309. Miller JS, Epstein AJ (1976) in Lippard SJ (ed) Prog Inorg Chem 20:1
310. Griffith WP (1975) Coord Chem Rev 17:177
311. Waltz WL, Sutherland RG (1972) Chem Soc Rev 1:241
312. Musher JI (1969) Angew Chem Int Edit 8:54
313. Goldberg RN, Hepler LG (1968) Chem Rev 68:229
314. Halpern J (1968) Disc Farad Soc 46:7
315. Johnson BFG, McCleverty JA (1966) in Cotton FA (ed) Prog Inorg Chem 7:277
316. Ebsworth EAV (1964) Educ Chem 1:123
317. Taube H (1956) Rec Chem Prog 17:25
318. Taube H (1952) Chem Rev 50:69
319. Rich RL, Stucky GL (1965) Inorg Nucl Chem Lett 1:61
320. Rich RL (1995) J Chem Educ 72:172
321. (a) Rich RL (2003) Phys Chem Chem Phys 5:2053 (b) Rich RL (1998) J Chem Educ 75:394 (c) Rich RL (1995) J Chem Educ 72:172 (d) Rich RL (1993) Bull Chem Soc Jpn 66:1065
322. Rich RL (1963) J Chem Educ 40:365
323. Heriot-Watt University (2008) www.rsc.org/FD141

1 Hydrogen and the Alkali Metals

1.0 Hydrogen, $_1H$

Oxidation numbers: (-I), (0) and (I) as in SbH_3, H_2 and AsH_3, and H_2O; see Sect. 15.3 for AsH_3. The elementary substances (0) are usually omitted hereafter. We note in passing that the IUPAC name for water, oxidane, is available for future adoption.

1.0.1 Reagents Derived from Hydrogen and Oxygen

Water (oxidane). Water at 25 °C dissolves H_2 up to about 0.8 mM.

Oxonium. No reaction with H_2. Note: the IUPAC term hydron for the normal isotopic mixture of $^1H^+$, $^2H^+$ and $^3H^+$ is more appropriate than the common term proton, and we will call the transfer of H^+ to or from a base hydronation/dehydronation rather than protonation/deprotonation.

Table 1.1 lists the pK_a of common acids in H_2O in the order of falling acidity, defining K_a (and pK_a) as usual in this book, also for H_3O^+, H_2O and OH^- for consistency, as $[H_3O^+][X^-]/[HX]$, with HX as the acid.

In Pauling's rules [1], the pK_a values for polybasic, mononuclear oxoacids rise by about 5 for each successive stage. The first pK_a of $XO_j(OH)_k$ is ≥ 8, ~ 2, ~ -3 or < -8 for $j = 0$, 1, 2 or 3, respectively.

A useful concept, Acidity Grade, AG, log $[H_3O^+]/[OH^-]$, is proposed [2] to replace pH. This gives high values for high acidities, unlike the counterintuitive pH, whose low values stand for high $[H^+]$ or $[H_3O^+]$ despite the "H" in "pH". At 25 °C we then have:

$$AG = 14.00 - 2\ \mathrm{pH}$$

The hydron is dihydrated in $[H(H_2O)_2][Y(C_2O_4)_2]\cdot H_2O$, but oxonium is tetrahydrated in H_2O as $[H_3O(H_2O)_4]^+$.

Because the anions of weak acids hold H^+ more or less firmly, we may rightly expect them to hold other cations more or less firmly also, thus forming insoluble salts, in spite of the near uniqueness of H^+ and variations among the other cations. The resulting solubility rules then make some sense but with complications well elucidated elsewhere [3].

Table 1.1. Inorganic, non-metallic, binary, and mono- and dinuclear oxo acids

	pK_a		pK_a		pK_a
$HClO_4$	–10	H_3PO_4	2.16	$H_4IO_6^-$	8.27
HI	–9.5	H_3AsO_4	2.22	$HSeO_3^-$	8.27
HBr	–9	$H_3[P_2O_7]^-$	2.36	HBrO	8.60
HCl	–7.0	H_2TeO_3	2.46	H_3AsO_3	9.23
$[(-SO_3H)_2]$	–3.4	H_2Te	2.64	H_3BO_3	9.24
H_2SO_4	–3.0	H_2SeO_3	2.64	NH_4^+	9.25
$H_4PO_4^+$	–3	HNO_2	3.14	$H[P_2O_7]^{3-}$	9.25
H_2SeO_4	–3	HF	3.17	H_4SiO_4	9.51
$HClO_3$	–2.7	H_5IO_6	3.29	$H_2XeO_6^{2-}$	10.
H_3O^+	–1.74	H_2CO_3	3.76	HCO_3^-	10.33
HNO_3	–1.37	H_2Se	3.89	HIO	10.64
$H_6IO_6^+$	–0.80	$HO_2^\bullet$	4.45	$XeO_3(aq)$	10.8
$N_2H_6^{2+}$	0.27	HN_3	4.72	HSe^-	11.0
$[(-SO_2H)_2]$	0.35	NH_3OH^+	5.95	HTe^-	11
$H_2S_2O_3$	0.6	$H_3XeO_6^-$	6	$H_5TeO_6^-$	11.00
HIO_3	0.80	$H_2O\cdot CO_2$	6.35	$HAsO_4^{2-}$	11.50
H_2PHO_3	1.20	$H_2[P_2O_7]^{2-}$	6.60	$H_3IO_6^{2-}$	11.60
HPH_2O_2	1.23	$HPHO_3^-$	6.70	H_2O_2	11.65
$H_4[P_2O_7]$	1.52	$H_2AsO_4^-$	6.98	$H_3SiO_4^-$	11.74
H_5IO_6	1.55	H_2S	6.99	HO	11.8
$HSeO_4^-$	1.66	$(=NOH)_2$	7.05	HPO_4^{2-}	12.33
$HS_2O_3^-$	1.74	HSO_3^-	7.1	$H_2BO_3^-$	12.74
$H_2O\cdot SO_2$	1.82	$H_2PO_4^-$	7.21	HS^-	12.89
$HClO_2$	1.94	HClO	7.54	HBO_3^{2-}	13.80
HSO_4^-	1.96	H_6TeO_6	7.61	H_2O	15.74
$H_4[P_2O_6]$	2	$HTeO_3^-$	7.7	NH_3	23
H_4XeO_6	2	$N_2H_5^+$	7.94	OH^-	29

Hydroxide, Peroxide and Dioxygen. Aqueous H_2 does not react with OH^-, H_2O_2, HO_2^- or O_2 (unless catalyzed).

1.0.2 Reagents Derived from the Other 2nd-Period Non-Metals, Boron through Fluorine

Oxidized nitrogen. Free hydrogen does not affect HNO_3 or aqua regia at 25 °C. Hydrogen catalyzed by Pt black, however, reduces dilute HNO_3 to NH_4NO_2, and concentrated HNO_3 to HNO_2, approximately:

$$2\,NO_3^- + 2\,H_3O^+ + 5\,H_2 \rightarrow NO_2^- + NH_4^+ + 6\,H_2O$$

1.0.3 Reagents Derived from the 3rd-to-5th-Period Non-Metals, Silicon through Xenon

Oxidized chalcogens. Free H_2 does not affect H_2SO_4 at ambient *T*.

Elemental and oxidized halogens. Chlorine and bromine combine with free H_2 directly in light, but heat is required to make it react with I_2. Hydrogen with platinum black combines with Cl_2, Br_2 and I_2 in the dark.

Hydrogen with platinum black reduces ClO^- and ClO_3^-, but not ClO_4^-, to Cl^-. Oxo-bromates and iodates are also reduced.

1.0.4 Reagents Derived from the Metals Lithium through Uranium, plus Electrons and Photons

Oxidation. Uncatalyzed H_2 does not reduce $Cr_2O_7^{2-}$, cold $FeCl_3$ or $[Fe(CN)_6]^{3-}$, but acidic solutions of MnO_4^-, $[PdCl_4]^{2-}$, Ag^+, Cu^{2+}, Hg^{2+} and Hg_2^{2+} oxidize H_2 to H_3O^+. Moreover, MnO_4^- in neutral or alkaline solution slowly oxidizes free H_2 (i.e., need not be in a metal; see below).

However, Cu^{2+} catalyzes the reductions of Cr^{VI}, Fe^{III}, Tl^{III}, IO_3^- etc., by (relatively slowly) forming CuH^+, which is then rapidly oxidized:

$$Cu^{2+} + H_2 + H_2O \leftrightarrows CuH^+ + H_3O^+$$

$$\underline{CuH^+ + H_2O \rightarrow Cu^{2+} + 2\,e^- + H_3O^+}$$

$$H_2 + 2\,H_2O \rightarrow 2\,e^- + 2\,H_3O^+$$

Free hydrogen acts very slowly on a neutral solution of $AgNO_3$, precipitating traces of Ag; in a concentrated solution $AgNO_2$ is formed. Solutions of Cu, Au and Pt are also reduced.

Hydrogen in Pd—also see **Other Reactions** below—is oxidized by and completely reduces Cr^{VI} to Cr^{III}, MnO_4^- in acidic solution to Mn^{2+}, and Fe^{III} to Fe^{2+}; and it quantitatively precipitates Pd, Pt, Cu, Ag, Au and Hg, but it does not reduce the alkali or alkaline-earth cations, or the salts of Ce, U, Mo, W, Co, Ni, Zn, Cd, Al, Pb, As, Sb or Bi.

Hydrogen with platinum black reduces $[Fe(CN)_6]^{3-}$ to $[Fe(CN)_6]^{4-}$.

Reduction. Many **d**-block complexes reduce H_2 to H^{-I}. Its oxidation of square-planar $\mathbf{d}^8$ ions to $\mathbf{d}^6$, for example, tends to go more readily for the lower (in the periodic chart) members of each Group, as: $Fe^0 < Ru^0 < Os^0$; $Co^I < Rh^I < Ir^I$; and $Ni^{II} << Pd^{II} << Pt^{II}$.

Many metals generate H_2 from dilute acids (seldom HNO_3). In the laboratory, one may use dilute H_2SO_4 with metallic Zn. Platinized zinc, or an alloy containing, say, 10 % Cu, reduces the superficial overpotential and secures a smooth, even flow of the gas. Adding a small amount of $CuSO_4$ to produce Cu is also satisfactory:

$$Zn + 2\,H_3O^+ \rightarrow Zn^{2+} + H_2\uparrow + 2\,H_2O$$

While being "born" or "nascent", and under proper conditions, such hydrogen combines readily with Si, N_2, P, As, Sb, O_2, S, Se, Te, Cl_2, Br_2, I_2 etc., toward which it is quite inert ordinarily, absent flames and so on.

We note, however, that "nascent H_2" generated by different methods does not reduce the same substances, and that not every case is clear, so that the metal may be responsible, with H_2 merely concomitant. Hydrogen obtained from Al and OH^- does not reduce As^V; that formed by Zn and acids gives AsH_3; Sb^V with sodium amalgam and acids gives Sb; with Zn and acids, SbH_3. Neither electrolytic H_2 nor that from Na_{Hg} (amalgam) and acids reduces chlorates, but Zn and acids rapidly form chlorides. Zinc and acids, but not Na_{Hg}(!), quickly reduce AgX.

In the common electrolytic preparation of NaOH from ordinary salt, or of KOH from KCl, hydrogen is a by-product at the cathode:

$$e^- + H_2O \rightarrow OH^- + {}^1/_2\,H_2\uparrow$$

Beta (e^-) rays, plus alpha and gamma rays, produce $H^\bullet$ radicals in H_2O.

Light (274 nm), H_3O^+ and $[CuCl_n]^{(n-1)-}$ generate H_2.

Other reactions. Metallic Pd dissolves up to 900 volumes of H_2 at 25 °C, or up to 3000 volumes in colloidal Pd. The latter value gives a concentration over 200-M H (> 100-M H_2) in the metal, which has a self-concentration of 113 M (based on its massive density).

Hydrogen also dissolves, albeit less spectacularly, in Fe, Ni, Pt etc., and is thereby activated. In this condition it readily combines with many substances, somewhat as does "nascent hydrogen".

1.1 Lithium, $_3$Li

Oxidation number: (I), as in Li^+.

1.1.1 Reagents Derived from Hydrogen and Oxygen

Water. The hydrated Li^+ ion is $[Li(H_2O)_4]^+$ or, at times, $[Li(H_2O)_6]^+$.

Metallic Li dissolves readily, releasing H_2 and forming LiOH.

The oxide Li_2O dissolves slowly and yields the hydroxide, LiOH. The solubility of LiOH is ~5 M at 10 °C, rising to over 6 M at 100 °C.

Most of the Li salts are soluble in H_2O. Some, including LiCl and $LiClO_3$, are very deliquescent. The hydroxide, carbonate (2 dM at 0 °C), fluoride (1 dM at 18 °C, comparable to NaF) and phosphate, Li_3PO_4 (3 mM), are, like those of the alkaline-earth metals, less soluble than nearly all of the corresponding compounds of the other alkali metals.

Oxonium. Lithium dissolves vigorously in acids and forms salts.

Hydroxide. Aqueous Li_2SO_4 and $Ba(OH)_2$ yield LiOH (and $BaSO_4\downarrow$).

Peroxide. The white peroxide, Li_2O_2, is best obtained by treating aqueous LiOH and H_2O_2 with ethanol, and drying the precipitate.

1.1.2 Reagents Derived from the Other 2nd-Period Non-Metals, Boron through Fluorine

Carbon oxide species. Carbonate ion precipitates Li_2CO_3.

Some "simple" organic reagents. Tartaric acid does not precipitate Li^+ from dilute solution (distinction from K^+, Rb^+ and Cs^+).

Fluorine species. Ammonium fluoride, in excess, precipitates LiF. The separation is more complete from ammoniacal solution.

1.1.3 Reagents Derived from the 3rd-to-5th-Period Non-Metals, Silicon through Xenon

Phosphorus species. Soluble phosphates precipitate lithium phosphate, more soluble in NH_4Cl than in H_2O alone (distinction from Mg^{2+}). In dilute solutions the phosphate is not precipitated until the solution is boiled. The sensitivity of the test is increased by adding NaOH, forming a double phosphate of Na and Li. The phosphate dissolved in HCl is not at once reprecipitated on neutralization with NH_3 (distinction from at least Ca^{2+} through Ra^{2+}). Ethanol promotes precipitation.

1.1.4 Reagents Derived from the Metals Lithium through Uranium, plus Electrons and Photons

Reduction. Charging one kind of "lithium-ion" batteries intercalates the lithium (reversed during discharge) in graphite:

$$Li^+ + 6\,C + e^- \rightarrow LiC_6$$

Other reactions. Aqueous $[PtCl_6]^{2-}$ does not precipitate Li^+ from dilute solution (distinction from K^+, Rb^+ and Cs^+).

1.2 Sodium, $_{11}$Na

Oxidation number: (I), as in Na^+.

1.2.1 Reagents Derived from Hydrogen and Oxygen

Water. Hydrated Na^+ tends to be $[Na(H_2O)_6]^+$.

Sodium decomposes water violently, even at room temperature, releasing H_2, which frequently ignites:

$$Na + H_2O \rightarrow Na^+ + OH^- + {}^1/_2\, H_2\uparrow$$

The monoxide, white, is very hygroscopic, forming NaOH, which is also quite hygroscopic.

Sodium peroxide, Na_2O_2, pale yellow, dissolves with much heating but mainly as HO_2^- and OH^- if cooled well (OH^- catalyzes decomposition to NaOH and O_2) with a slight further hydrolysis, $pK \approx 4$:

$$HO_2^- + H_2O \leftrightarrows H_2O_2 + OH^-$$

Most sodium salts are soluble, except $Na[Sb(OH)_6]$, $Na_2[SiF_6]$ and a number of more complex ones, such as $NaK_2[Co(NO_2)_6]\cdot H_2O$ and $NaMg(UO_2)_3(CH_3CO_2)_9\cdot 6H_2O$, which are only slightly soluble.

The nitrate and chlorate are deliquescent. The hydrated carbonate ($10\,H_2O$), acetate ($3\,H_2O$), phosphate ($12\,H_2O$), sulfite ($8\,H_2O$) and sulfate ($10\,H_2O$) are efflorescent.

Seawater contains $NaHCO_3$ and $NaSO_4^-$ complexes, and Na^+.

Dioxygen. Moist air oxidizes Na rapidly, unless kept under kerosene.

1.2.2 Reagents Derived from the Other 2nd-Period Non-Metals, Boron through Fluorine

Some "simple" organic reagents. Various triple acetates of Na, used in analysis, are relatively insoluble. Zinc uranyl acetate and neutral, not too dilute, Na^+ precipitate $NaZn(UO_2)_3(CH_3CO_2)_9\cdot 6H_2O$, yellow and crystalline. The corresponding Mg and Co (not Ca) salts are similar.

Solutions of $C_2O_4^{2-}$ precipitate, from not too dilute Na^+, crystalline, white sodium oxalate, soluble in inorganic acids.

1.2.3 Reagents Derived from the 3rd-to-5th-Period Non-Metals, Silicon through Xenon

Silicon species. Aqueous H_2SiF_6 precipitates Na_2SiF_6 from not too dilute Na^+. Its solubility is 4 cM at 17.5 °C, or much less in aqueous ethanol.

Reduced halogens. A solution containing Na^+ and Li^+ can be saturated with HCl gas to separate Na^+ as solid NaCl.

1.2.4 Reagents Derived from the Metals Lithium through Uranium, plus Electrons and Photons

Non-redox reactions. Sodium hydroxide, NaOH, can be made by treating a solution of Na_2CO_3 with Ca, Sr or Ba oxide or hydroxide:

$$CO_3^{2-} + Ca(OH)_2 \rightarrow 2\,OH^- + CaCO_3\downarrow$$

Excess $Mg(CH_3CO_2)_2$ (like Zn^{2+} and Co^{2+} salts) plus $UO_2(CH_3CO_2)_2$ precipitate Na^+ for gravimetry as $NaMg(UO_2)_3(CH_3CO_2)_9 \cdot {}^{13}/_2H_2O$.

Aqueous H_2PtCl_6 and Na^+ give reddish crystals of sodium hexachloroplatinate(2−) only from a concentrated solution, readily distinguished from the yellow potassium or ammonium salts.

A solution of $K[Sb(OH)_6]$ produces in neutral or alkaline solutions of Na^+ a slow-forming, white, crystalline precipitate, $Na[Sb(OH)_6]$, slightly soluble in cold H_2O. Precipitation can often be accelerated by rubbing the glass under the surface of the liquid with a stirring rod. Large amounts of K^+ may hinder the reaction; acids and NH_4^+ cause the separation of $H[Sb(OH)_6]$ or less hydrated forms. Most of the other metals interfere. The unstable reagent should be prepared and dissolved only when needed.

Sodium hydroxide is made by the electrolysis of aqueous NaCl in the cathode compartment, but without reduction or oxidation of the Na^+:

$$e^- + H_2O \rightarrow OH^- + {}^1/_2\,H_2\uparrow$$

1.3 Potassium, $_{19}K$; Rubidium, $_{37}Rb$; Cesium, $_{55}Cs$ and Francium, $_{87}Fr$

Oxidation number: (I), as in K^+, Rb^+, Cs^+ and Fr^+.

1.3.1 Reagents Derived from Hydrogen and Oxygen

Water. In this section Alk is K, Rb and/or Cs. The n in $[Alk(H_2O)_n]^+$ is 6 or more. The metals dissolve violently in cold H_2O, yielding H_2 and AlkOH. Both AlkOH and Alk_2O are deliquescent and quite exothermically soluble as AlkOH. Most salts are readily soluble.

Potassium hyperoxide, KO_2 (formed when the metal is heated with an excess of oxygen), is a yellow amorphous powder about the color of $PbCrO_4$, decomposed by H_2O or moist air with evolution of O_2:

$$2\,KO_2 + H_2O \rightarrow HO_2^- + O_2\uparrow + 2\,K^+ + OH^-$$

It is a powerful oxidant, changing Fe, Pt, Cu, Ag, Zn, Sn, As and Sb etc. to the oxides or salts, phosphorus to PO_4^{3-} and sulfur to SO_4^{2-}. The similar RbO_2 and CsO_2 are dark orange and brown, respectively.

Table 1.2 shows the solubilities for some less soluble salts of K, Rb and Cs, a few of which have been used for separations. The data from apparently reliable sources are so mutually discrepant that no more than one significant digit, if that many, is usually justified. We present molarities rather than millimolarities because, say, 400 mM wrongly suggests possibly 100 times as much precision with three significant digits. And scientific symbolism (using E) would be excellent but it weakens the visual impact of differences.

Most of this book uses "Alk" for any of the alkali metals. Salts that are stable with the relatively large NH_4^+ are also often stable with the other large Alk^+, i.e., K^+, Rb^+ or Cs^+, but not Li^+ or Na^+, so we often abbreviate $(K,Rb,Cs,NH_4)X$ as $(Alk,NH_4)X$, but the Alk^+ in AlkX may stand for any of these cations, if appropriate, when we omit details. Just as CN, N_3 and SCN (radicals) have been called pseudo halogens, we could call NH_4 a pseudo alkali metal.

Table 1.2. Solubilities for Certain Difficultly Soluble Salts of K, Rb and Cs

	$c(K^+)$/M		$c(Rb^+)$/M		$c(Cs^+)$/M	
$Alk[BF_4]$	0	.04	0	.04[a]	0	.07
AlkHTart	0	.03	0	.05[b]	0	.25
$AlkClO_4$	0	.12	0	.08	0	.07
$AlkIO_4$	0	.018	0	.02[c]	0	.07[d]
$Alk_4[SiW_{12}O_{40}]$	0	.4	0	.007		
$AlkMnO_4$	0	.4	0	.05	0	.009[e]
$Alk_2[PtCl_6]$	0	.04	0	.001 0	0	.000 3
$AlkAl(SO_4)_2$	0	.22	0	.05	0	.011

All temperatures are 20 °C except for a, 17; b, ?; c, 13; d, 15; and e, 19 °C. Tart is tartrate or $C_4H_4O_6^{2-}$, the silicododecatungstates are $Alk_4[SiW_{12}O_{40}]\cdot 18H_2O$, and the alums, $AlkAl(SO_4)_2$, are $[Alk(H_2O)_6][Al(H_2O)_6](SO_4)_2$.

We see a few substantial differences that can be checked out for use in separations. The data for Alk_2SiF_6 are especially discordant and therefore excluded, but the Cs^+

concentration is said to be nearly 3 M (or 3 N in the convenient, older symbolism for formulas like Alk_2X, where only the Alk^+ is of interest), with 11 mM or much less for the saturated concentrations of the K^+ and Rb^+ salts.

In addition, the complex salts $K_2[TiF_6]\cdot H_2O$, $K_2[ZrF_6]$, $K_3[PMo_{12}O_{40}]$, $K_3[PW_{12}O_{40}]$, $K_3[Co(NO_2)_6]\cdot {}^3/_2H_2O$, $K_2[SiF_6]$ and $K_2[GeF_6]$ are slightly soluble to insoluble in cold water.

Seawater contains KSO_4^- complexes and K^+.

1.3.2 Reagents Derived from the Other 2nd-Period Non-Metals, Boron through Fluorine

Carbon oxide species. The monoxides, Alk_2O, and AlkOH absorb CO_2 from the air, becoming white, soluble Alk_2CO_3.

Some "simple" organic reagents. Tartaric acid, $H_2C_4H_4O_6$, or more readily $NaHC_4H_4O_6$, precipitates, from sufficiently concentrated K^+ solutions, clear, crystalline $KHC_4H_4O_6$. If the solution is initially alkaline (when testing for K^+) it should be acidified with tartaric acid. Cations of only the alkali metals may be present. Precipitation is increased by agitation and by adding ethanol. The precipitate is soluble in inorganic acids, and in alkalis forming the more soluble normal salt, $K_2C_4H_4O_6$, insoluble in 50 % ethanol. Aqueous Rb^+ also precipitates as $RbHC_4H_4O_6$.

1.3.3 Reagents Derived from the 3rd-to-5th-Period Non-Metals, Silicon through Xenon

Silicon species. Hexafluorosilicic acid, $H_2[SiF_6]$, added in excess to a neutral solution containing K^+, gives a gelatinous precipitate of the potassium salt, $K_2[SiF_6]$. Weak bases hydrolyze the reagent, and silicic acid separates, which is easily mistaken for the salt.

Oxidized chalcogens. One can reflux pollucite, approximately $Cs_4H_2Al_4Si_9O_{27}$, 30 hours with 7-M H_2SO_4 to precipitate silica; cooling then yields the alum $[Cs(H_2O)_6][Al(H_2O)_6](SO_4)_2$.

Elemental and oxidized halogens. Refluxing the mineral pollucite, $\sim Cs_4H_2Al_4Si_9O_{27}$, up to 30 hours with concentrated HCl, adding I_2 and HNO_3 to the solute, and evaporating nearly dry, isolates the Cs:

$$3\,Cs^+ + {}^3/_2\,I_2 + 6\,Cl^- + NO_3^- + 4\,H_3O^+ \rightarrow 3\,CsICl_2\downarrow + NO\uparrow + 6\,H_2O$$

A solution of perchloric acid, $HClO_4$, forms with K^+ a crystalline, white precipitate of potassium perchlorate, $KClO_4$. One way to separate Na^+ and Li^+ from K^+, Rb^+ and Cs^+ is to precipitate the latter as perchlorates from an ethanolic solution.

1.3.4 Reagents Derived from the Metals Lithium through Uranium, plus Electrons and Photons

Non-redox reactions. Traces of Cs^+ may be precipitated from a final solute by means of $H_4[SiW_{12}O_{40}]$.

Aqueous $Na_4[Fe(CN)_6]$ and Ca^{2+} precipitate Rb^+ or Cs^+ as ,e.g., $Cs_2Ca[Fe(CN)_6]$. Not precipitated are NH_4^+, Li^+, Na^+ or K^+.

A solution of $Na_3[Co(NO_2)_6]$ gives, with K^+ acidified with acetic acid, a golden yellow precipitate of $K_2Na[Co(NO_2)_6] \cdot H_2O$. In concentrated solution, $K_3[Co(NO_2)_6]$ is formed quickly. Dilute solutions must stand, although warming hastens the separation. Because NH_4^+ gives a similar precipitate, it must first be removed, to detect K^+. Iodides and other reductants must also be absent. Many modifications of this test to increase its sensitivity have been suggested, including the addition of, e.g., Ag^+, Hg_2^{2+} or Pb^{2+}, which enable the detection of less than 3 mM K^+.

The acid $H_2[PtCl_6]$ forms, in concentrated, acidic solutions of K^+, a crystalline, yellow precipitate of $K_2[PtCl_6]$. Although slightly soluble in H_2O, it is practically insoluble in 80 % ethanol. Aqueous NH_4^+ also gives the test. The presence of CN^- or I^- inhibits the reaction. In either case evaporation with concentrated HCl will solve the problem. Large amounts of Na^+ decrease the sensitivity of the test.

The precipitations of K as $K_2Na[Co(NO_2)_6] \cdot H_2O$ and $K_2[PtCl_6]$, as well as $KHC_4H_4O_6$, $K_2[SiF_6]$, $KClO_4$, etc. have been used for the quantitative separation of K under carefully controlled conditions, but perhaps less often for its detection.

Cesium is precipitated by $[SnCl_6]^{2-}$ with concentrated HCl as $Cs_2[SnCl_6]$ (separation from the other Alk^+ and NH_4^+), or by $[SbCl_4]^-$ as $Cs_3[Sb_2Cl_9]$ (separation from all alkali cations but NH_4^+). Thus, refluxing the mineral pollucite, $\sim Cs_4H_2Al_4Si_9O_{27}$, up to 30 h with concentrated HCl dissolves the Cs^+, and adding $SbCl_3$ isolates it:

$$3\,Cs^+ + 2\,SbCl_4^- + Cl^- \rightarrow Cs_3Sb_2Cl_9\downarrow$$

followed by H_2O (and possibly H_2S to remove the Sb more completely):

$$Cs_3Sb_2Cl_9 + 6\,H_2O \rightarrow 3\,Cs^+ + 2\,SbOCl\downarrow + 7\,Cl^- + 4\,H_3O^+$$

Precipitating Rb^+ and Cs^+ (M^+) as the triple nitrites, $M_2Na[Bi(NO_2)_6]$, using $NaNO_2$ and $BiCl_3$, leaves K^+ in the solute to be detected perhaps as the hexanitrocobaltate(3−).

Adding $Na_3[Bi(S_2O_3\text{-}\kappa S,\kappa O)_3]$ to an ethanolic solution of K^+ precipitates yellow $K_3[Bi(S_2O_3\text{-}\kappa S,\kappa O)_3]$. The test is very sensitive, depending, however, on the amount of ethanol present. Apparently NH_4^+, Sr^{2+} and Ba^{2+} interfere. The reagent is unstable.

References

1. Earnshaw A, Greenwood NN (1997) Chemistry of the elements, 2nd edn. Elsevier, Amsterdam, p 54
2. van Lubeck H (1999) J Chem Educ 76:892
3. Wulfsberg G (1987) Principles of descriptive inorganic chemistry. Brooks/Cole, Monterey, CA, chapter 3, p 59

Bibliography

See the general references in the Introduction, and some more-specialized books [4–12]. Some articles in journals discuss: the structures of $H(H_2O)_n^+$ by IR spectroscopy [13a], for $4 \leq n \leq 27$ [13b] and $6 \leq n \leq 27$ [13c]; complexes of **d**-block elements with H_2, which may split homolytically as reductants or heterolytically as acids [14]; and the **d**-block–hydrogen bond [15].

4. Peruzzini M, Poli R (eds) (2001) Recent advances in hydride chemistry. Elsevier, Amsterdam
5. Sapse AM, Schleyer PvR (1995) Lithium chemistry. Wiley, New York
6. Harriman A, West MA (1982) Photogeneration of hydrogen. Academic, London
7. Hajos AH (1979) Complex hydrides. Elsevier, Amsterdam
8. Giguère PA (1975) Compléments au nouveau traité de chimie minéral: peroxyde d'hydrogène et polyoxydes d'hydrogène. Masson, Paris
9. Soustelle M, Adloff JP (1974) Compléments au nouveau traité de chimie minéral: rubidium, césium, francium. Masson, Paris
10. Shaw BL (1967) Inorganic hydrides. Pergamon, Oxford
11. Korenman IM (1964) Kaner N (trans) Slutzkin D (ed) (1965) Analytical chemistry of potassium. Ann Arbor-Humphrey, Ann Arbor
12. Perel'man FM (1960) Towndrow RGP (trans) Clarke RW (ed) (1965) Rubidium and caesium. Macmillan, New York
13. (a) Zwier TS (2004) Science 304:1119 (b) Miyazaki M, Fujii A, Ebata T, Mikami N (2004) ibid:1134 (c) Shin JW et al (2004) ibid:1137
14. Heinekey DM, Oldham WJ Jr (1993) Chem Rev 93:913
15. Pearson RG (1985) Chem Rev 85:41

2 Beryllium and the Alkaline-Earth Metals

2.1 Beryllium, $_4Be$

Oxidation number: (II), as in BeO.

2.1.1 Reagents Derived from Hydrogen and Oxygen

Water. Beryllium is only slightly affected by H_2O; BeO and $Be(OH)_2$ are insoluble in H_2O. The basic carbonate is slightly soluble, the complex fluorides, e.g., Na_2BeF_4, moderately soluble. Salts (all very toxic) such as $[Be(H_2O)_4]SO_4$ exemplify the tetrahedral $[Be(H_2O)_4]^{2+}$.

The halide salts are deliquescent, and (non-aqueous derived) $BeCl_2$ gives $[Be(H_2O)_4]^{2+}$; dehydration forms $Be(OH)_2$ and releases HCl.

Many properties of Be^{II} are like those of Al^{III}, showing the diagonal relationship in some forms of the periodic chart. An important difference is that boiling a solution of beryllate, $[Be(OH)_4]^{2-}$, in water readily precipitates $Be(OH)_2$ (partial similarity with zinc).

Some natural waters may contain $[BeF_4]^{2-}$, and certain hot natural waters may contain Be carbonates.

Oxonium. Beryllium, BeO and $Be(OH)_2$ dissolve readily in H_3O^+, but strongly ignited BeO is insoluble in all common acids except HF.

Hydroxide. Beryllium dissolves easily in OH^-, releasing H_2. Aqueous Be^{2+} and OH^- form, e.g., $[\{Be(H_2O)_3\}_2(\mu\text{-}OH)]^{3+}$, $Be_2O(OH)_2$ and $Be(OH)_2$, amorphous when fresh, soluble as $[Be(OH)_4]^{2-}$ unless aged.

2.1.2 Reagents Derived from the Other 2nd-Period Non-Metals, Boron through Fluorine

Carbon oxide species. From Be^{2+} the alkali carbonates precipitate a basic carbonate, soluble like $Be(OH)_2$, when fresh, in excess of the reagent, saturated $NaHCO_3$ or "$(NH_4)_2CO_3$" solution (containing much NH_3, NH_4^+ and HCO_3^-) (distinctions from Al^{3+}). From these solutions [*tetrahedro*-$Be_4(\mu_4\text{-}O)(\mu\text{-}CO_3)_6]^{6-}$ can be precipitated, e.g.:

$$4\,Be^{2+} + 6\,HCO_3^- + 8\,NH_3 + H_2O \rightarrow [Be_4O(CO_3)_6]^{6-} + 8\,NH_4^+$$

$$4\,Be(OH)_2 + 6\,HCO_3^- \rightarrow [Be_4O(CO_3)_6]^{6-} + 7\,H_2O$$

$$[Be_4O(CO_3)_6]^{6-} + 2\,[Co(NH_3)_6]^{3+} \rightarrow [Co(NH_3)_6]_2[Be_4O(CO_3)_6]\cdot aq\downarrow$$

Some "simple" organic species. Heating BeO, $Be(OH)_2$, $BeCO_3$ or a basic carbonate with various concentrated monobasic organic acids, e.g., acetic, and evaporating, give [*tetrahedro*-$Be_4(\mu_4\text{-}O)(\mu\text{-}RCO_2)_6]$, cf. CO_3^{2-}, remarkably stable, which can be extracted and recrystallized from $CHCl_3$ or even hexane. Dilute acids but not H_2O attack these.

Oxalic acid and $C_2O_4^{2-}$ form no precipitate with Be^{2+}.

Reduced nitrogen. Beryllium hydroxide, $Be(OH)_2\cdot aq$, is precipitated by NH_3 from solutions of Be^{2+}, and is insoluble in excess. The gelatinous product is amphoteric, resembling $Al_2O_3\cdot aq$ in many properties.

Oxidized nitrogen. Beryllium, only slightly affected by cold HNO_3, dissolves readily in hot dilute HNO_3, but concentrated acid passivates it.

Fluorine species. Even strongly ignited BeO dissolves in HF.

2.1.3 Reagents Derived from the 3rd-to-5th-Period Non-Metals, Silicon through Xenon

Phosphorus species. Ammonium phosphate precipitates Be^{2+} from neutral or slightly acidic solutions as $BeNH_4PO_4$.

Reduced chalcogens. Aqueous "$(NH_4)_2S$" precipitates Be^{2+} as $Be(OH)_2$.

Oxidized chalcogens. Beryllium dissolves readily in dilute H_2SO_4, releasing H_2. In hot and concentrated acid, SO_2 is released.

Aqueous SO_4^{2-} precipitates, from sufficiently concentrated Be^{2+}, a crystalline beryllium sulfate.

Reduced halogens. Beryllium dissolves readily in HCl, releasing H_2.

Oxidized halogens. The salt $Be(ClO_4)_2\cdot 2H_2O$ is $[Be(H_2O)_4][Be(ClO_4)_4]$.

2.2 Magnesium, $_{12}$Mg

Oxidation number: (II), as in Mg^{2+}.

Throughout the book we may write "Ae" for any, some or all of the alkaline-earth elements, Mg through Ra.

2.2.1 Reagents Derived from Hydrogen and Oxygen

Water. The hydrated Mg^{2+} is normally $[Mg(H_2O)_6]^{2+}$.

The oxide, MgO, and hydroxide, $Mg(OH)_2$, are insoluble.

In water, the oxide is changed very slowly to the hydroxide.

The acetate, nitrate, chloride, bromide, iodide and chlorate are deliquescent, the sulfate (7 H_2O) slightly efflorescent.

The borate, carbonate, oxalate, fluoride, phosphates ($MgHPO_4 \cdot 3H_2O$, $Mg_3(PO_4)_2 \cdot nH_2O$, $MgNH_4PO_4 \cdot 6H_2O$), arsenite and arsenate are insoluble. The tartrate, phosphite and sulfite are slightly soluble.

Seawater contains $MgCO_3$ (dissolved), $MgHCO_3^+$, $MgSO_4$ and Mg^{2+}.

Oxonium. Magnesium is soluble in acids, and is attacked by various salts that are acidic by hydrolysis; MgO and $Mg(OH)_2$ also dissolve readily, less if aged, in non-precipitating acids:

$$Mg + 2\,H_3O^+ \rightarrow Mg^{2+} + H_2\uparrow + 2\,H_2O$$

Hydroxide. The alkali-metal and other alkaline-earth hydroxides precipitate Mg^{2+} as $Mg(OH)_2$, white, gelatinous. It is insoluble in excess of the reagent. No precipitation occurs in the presence of NH_4^+.

2.2.2 Reagents Derived from the Other 2nd-Period Non-Metals, Boron through Fluorine

Carbon oxide species. Magnesium reacts even with carbonic acid, releasing hydrogen, and is also attacked by HCO_3^-:

$$Mg + CO_2 + H_2O \rightarrow MgCO_3\downarrow + H_2\uparrow$$

$$MgCO_3 + CO_2 + H_2O \rightarrow Mg^{2+} + 2\,HCO_3^-$$

In contact with water MgO very slowly absorbs CO_2 from the air.

Aqueous CO_3^{2-} precipitates, e.g., $Mg_2(CO_3)(OH)_2 \cdot nH_2O$ or, depending on conditions, $Mg_5(CO_3)_4(OH)_2 \cdot nH_2O$. "Ammonium carbonate" with other NH_4^+ salts does not precipitate Mg^{2+}, but a concentrated solution precipitates Mg^{2+} fairly completely in 30–40 % ethanol.

Aqueous HCO_3^- does not precipitate Mg^{2+} in the cold; upon boiling, CO_2 is released and $MgCO_3 \cdot 3H_2O$ appears.

Some “simple” organic species. Magnesium oxalate is insoluble, yet adding $C_2O_4^{2-}$ to Mg^{2+}, either acidified or ammoniacal, gives no precipitate even after some time, although adding an equal volume of ethanol, propanone or concentrated acetic acid quickly precipitates it.

Reduced nitrogen. Ammonia precipitates Mg^{2+} partly as $Mg(OH)_2$:

$$Mg^{2+} + 2\ NH_3 + 2\ H_2O \leftrightarrows Mg(OH)_2\downarrow + 2\ NH_4^+$$

Sufficient initial NH_4^+ prevents precipitation, and such a “magnesia mixture” is used to precipitate and determine phosphate.

The oxide, hydroxide and carbonate dissolve in solutions of NH_4^+, but the phosphates, arsenite and arsenate are insoluble.

2.2.3 Reagents Derived from the 3rd-to-5th-Period Non-Metals, Silicon through Xenon

Phosphorus species. Alkali phosphates, HPO_4^{2-}, precipitate Mg^{2+} as $MgHPO_4$ from neutral solution—if the solution is boiled the precipitate is $Mg_3(PO_4)_2 \cdot 7H_2O$—or $MgNH_4PO_4 \cdot 6H_2O$ by adding NH_3 to a solution containing Mg^{2+} and $H_2PO_4^-$ or H_3PO_4. If the HPO_4^{2-} is added to an ammoniacal solution of Mg^{2+}, $Mg_3(PO_4)_2 \cdot nH_2O$ is precipitated. For $MgNH_4PO_4 \cdot 6H_2O$, which is crystalline, a tendency to supersaturation usually may be overcome by rubbing the test tube or beaker beneath the surface of the liquid with a stirring rod. The presence of NH_4^+ prevents the precipitation of any $Mg(OH)_2$. The precipitate is readily soluble in acetic or oxalic as well as inorganic acids. To detect Mg^{2+} after removing most other metals, one may therefore add $(NH_4)_2HPO_4$ to a cold acidic solution and then make it alkaline with dilute NH_3, while stirring vigorously. The precipitate must be crystalline.

Arsenic species. From As(> 0) in neutral solution Mg precipitates As.

Soluble arsenates precipitate Mg salts.

Reduced chalcogens. Magnesium sulfide is decomposed by H_2O, and Mg^{2+} is not precipitated by H_2S or “$(NH_4)_2S$”. The addition of S^{2-} results in a separation of $Mg(OH)_2$.

Oxidized chalcogens. Soluble sulfates do not precipitate Mg^{2+} (distinction from Ca^{2+}, Sr^{2+}, Ba^{2+} and Ra^{2+}). The anhydrous sulfate, however, is insoluble in H_2O and dilute acids.

From Se(> 0) and Te(> 0) at pH ~7, Mg precipitates Se and Te.

2.2.4 Reagents Derived from the Metals Lithium through Uranium, plus Electrons and Photons

Oxidation. From their various salts in neutral solution Mg, while going to Mg^{II}, precipitates elemental Th, Mn, Fe, Co, Ni, Pd, Pt, Cu, Ag, Au, Zn, Cd, Hg, Tl, Sn, Pb, Sb, Bi etc. Magnesium anodes, e.g., to protect wet Fe cathodically, deliver e^- with much (wasted) H_2.

Other reactions. Adding $K_4[Fe(CN)_6]$ to cold Mg^{2+} precipitates white, crystalline $K_2Mg[Fe(CN)_6]$. Ammonium gives a triple salt. The rate of separation in either case depends largely on the concentration of Mg^{2+}.

Adding either $[Fe(CN)_6]^{4-}$ or $[Fe(CN)_6]^{3-}$ to a solution of Mg^{2+} containing Rb^+ or Cs^+ gives a white precipitate, better with some ethanol, in one of the most sensitive chemical tests known for Mg.

2.3 Calcium, $_{20}$Ca; Strontium, $_{38}$Sr; Barium, $_{56}$Ba and Radium, $_{88}$Ra

Oxidation number: (II), as in Ae^{2+}.

2.3.1 Reagents Derived from Hydrogen and Oxygen

Water. Water releases H_2 vigorously and forms $Ae(OH)_2$ from Ca, Sr, Ba or Ra. The hydrated ions are often $[Ae(H_2O)_n]^{2+}$, with $n = 6$ to 9.

In moist air CaO rapidly becomes $Ca(OH)_2$, with increase in volume and generation of much heat if sufficient water is present. The hydroxide (slaked lime) is commonly made by treating the oxide with water. Its usefulness combined with sand, to make mortar, is well known.

The hydroxide, $Ca(OH)_2$, is much less soluble than $Sr(OH)_2$ or $Ba(OH)_2$ in H_2O, 16 mM at 30 °C. It dissolves with evolution of heat, the solubility therefore decreasing with rising temperature, being about two-thirds of the quoted figure at the boiling point. A clear solution of the hydroxide in water is known as lime water, while a suspension of creamy consistency is called milk of lime.

Strontium hydroxide, $Sr(OH)_2$, is formed by the action (slaking) of water on the oxide or by heating the carbonate in steam:

$$SrCO_3 + H_2O \rightarrow Sr(OH)_2\downarrow + CO_2\uparrow$$

The slightly soluble $Sr(OH)_2$ shows an abnormal decrease in solubility in the presence of other bases.

Barium oxide reacts with water (slakes), releasing heat and forming $Ba(OH)_2$, which dissolves in its own weight of hot water.

Calcium acetate is efflorescent. The nitrate, chloride, bromide, iodide and chlorate are deliquescent. Calcium sulfite, $CaSO_3{\cdot}2H_2O$, is slightly soluble. The sulfate, $CaSO_4{\cdot}2H_2O$, gypsum, is slightly soluble at 25 °C, changing little up to 100 °C; from there to about 200 °C it decreases rather rapidly, an important factor in the problem of boiler scale.

The chromate, $CaCrO_4$, dissolves moderately in water, somewhat more in ethanol, and readily in acids including dichromic acid.

Strontium peroxide, $SrO_2{\cdot}8H_2O$, is only slightly soluble. Barium peroxide, BaO_2, is insoluble in H_2O.

Strontium acetate and nitrate are efflorescent. The hexafluorosilicate is soluble (distinction from Ba). The sulfate is practically insoluble, yet enough dissolves to allow its use as a reagent for traces of barium. The chloride is slightly deliquescent.

Most salts of Ba are stable in air, but the acetate is efflorescent. The acetate, cyanide, chloride, chlorate, perchlorate, bromide and iodide are readily soluble; the nitrate and the hexacyanoferrate(4–) moderately soluble; the fluoride slightly soluble; and the carbonate, oxalate, phosphate, sulfite, sulfate, iodate and chromate insoluble.

Table 2.1 lists the solubilities of some mostly less soluble compounds of Ca, Sr and Ba, for possible separations. The data from apparently reliable sources are so mutually discrepant that no more than one significant digit, if that many, is usually justified. We present molarities rather than millimolarities because , e.g., 200 mM wrongly suggests 100 times as much precision with three significant digits. Also, scientific symbolism (using E) would weaken the visual impact of differences. In any case, we do find some differences big enough to be tested for possible use in separations.

Table 2.1. Solubilities for certain difficultly soluble salts of Ca, Sr and Ba

	$[Ca^{2+}]$/M	$[Sr^{2+}]$/M	$[Ba^{2+}]$/M
$Ae(OH)_2$	0.02	0.07[a]	0.2[a]
$AeCO_3$	0.000 1	0.000 07	0.000 1
AeC_2O_4	0.000 06	0.000 3[a]	0.000 5
AeF_2	0.000 2	0.001	0.009
$AeSiF_6$	high	0.1[b]	0.000 9
$AeSO_4$	0.015	0.000 7	0.000 01
$AeSeO_4$	0.40	low	0.003
$Ae(IO_3)_2$	0.005[a]	0.000 7[c]	0.000 6[a]
$AeCrO_4$	0.15	0.005[c]	0.000 03[a]

All temperatures are 25 °C except a, 20; b, 18; and c, 15 °C.

The Ae phosphates are insoluble.

Seawater contains $CaCO_3$, $CaHCO_3^+$ and $CaSO_4$ complexes, and Ca^{2+}.

Oxonium. Metallic Ca, Sr and Ba react vigorously with acids, forming H_2 and Ae^{2+} or solid salts. The Ae oxides and hydroxides also combine with dilute acids to form H_2O and the same ions or salts, likewise the carbonates, cyanides etc. with not-too-weak acids.

Strontium chromate is soluble in many acids, including chromic.

Treating BaO_2 (formed when the oxide is heated to 600 °C in oxygen) with a non-reducing acid dissolves it and produces H_2O_2.

Hydroxide. Aqueous OH^- precipitates $Ae(OH)_2$ if Ae^{2+} is concentrated enough, and $Ca(OH)_2$ even from $CaSO_4$, especially with excess OH^-.

Peroxide. The peroxides, e.g., $CaO_2\cdot 8H_2O$ or $SrO_2\cdot 8H_2O$, are made by adding H_2O_2 or Na_2O_2 to (Ca,Sr,Ba)$(OH)_2$ (but acids release the H_2O_2):

$$Ca(OH)_2 + H_2O_2 + 6\,H_2O \rightarrow CaO_2\cdot 8H_2O\downarrow$$

Careful dehydration of $CaO_2\cdot 8H_2O$ by heating leaves CaO_2.

Heating BaO, but not CaO or SrO, in air forms the peroxide, BaO_2.

2.3.2 Reagents Derived from the Other 2nd-Period Non-Metals, Boron through Fluorine

Carbon oxide species. Calcium, strontium or barium oxide absorbs CO_2 from the air, becoming $AeCO_3$.

Alkali carbonates precipitate Ca^{2+}, Sr^{2+} and Ba^{2+} as white $AeCO_3$, insoluble in water free of CO_2, but decomposed by acids, including CH_3CO_2H. Calcium hydroxide may be used as a reagent to detect CO_2 but note that excess CO_2 (or of NH_4^+, also acidic by hydrolysis) dissolves precipitates of $AeCO_3$, although heat promotes precipitation:

$$AeCO_3 + CO_2 + H_2O \leftrightharpoons Ae^{2+} + 2\,HCO_3^-$$

Thus, although their carbonates are insoluble, their hydrogencarbonates dissolve readily, one of the important factors in "temporary" hardness.

Consuming CO_3^{2-} in leaching U from ores, a "parasitic reaction" converts $CaSO_4$ to $CaCO_3$. Boiling $SrSO_4$ in aqueous CO_3^{2-} transposes it to $SrCO_3$ (and SO_4^{2-}) rather readily. Boiling fresh $BaSO_4$ with at least 15 times its molar equivalent of 1–2 M Na_2CO_3 will convert 99 % of the $BaSO_4$ to $BaCO_3$ and Na_2SO_4 in an hour. Native barite requires about double the time. Filtration and digestion with H_2O will remove the SO_4^{2-} after which the $BaCO_3$ residue may be dissolved in HCl.

Some "simple" organic species. Aqueous $C_2O_4^{2-}$ quantitatively precipitates Ca^{2+} as $CaC_2O_4\cdot H_2O$. The precipitate is quite insoluble in CH_3CO_2H but readily soluble in HNO_3, H_2SO_4 and HCl. Precipitation is best effected by adding dilute NH_3 to a hot acidic solution containing both Ca^{2+} and $HC_2O_4^-$. If Sr^{2+} or Ba^{2+} is present in the solution to be tested (qualitatively), $(NH_4)_2SO_4$ should first be added (not K_2SO_4 as often suggested, because it forms an insoluble double salt with $CaSO_4$). After digesting, any precipitate that appears is removed and the oxalate test applied to the solute. Remarkably, in spite of the low solubility of MgC_2O_4 a quantitative separation can be effected, due to a great difference in the rate of precipitation of the two salts.

Aqueous $C_2O_4^{2-}$ also precipitates Sr^{2+} and Ba^{2+} as SrC_2O_4 and $BaC_2O_4 \cdot H_2O$, insoluble in H_2O, soluble in HCl or HNO_3. When first precipitated, $BaC_2O_4 \cdot H_2O$ may be dissolved in acetic or oxalic acid, but in a short time $H_2Ba(C_2O_4)_2 \cdot 2H_2O$ separates in the form of clear crystals. To explain the dissolution we note that CH_3CO_2H, although much less acidic than $H_2C_2O_4$, is almost as strong as $HC_2O_4^-$.

Reduced nitrogen. Ammonia free from CO_3^{2-} does not precipitate Ca^{2+}, Sr^{2+} or Ba^{2+}. Strontium peroxide, $SrO_2 \cdot 8H_2O$, is soluble in NH_4^+, and $SrCrO_4$ is more soluble in concentrated NH_4^+ than in water.

Oxidized nitrogen. Calcium is only slightly attacked by concentrated HNO_3 due perhaps to forming an insoluble coating of calcium nitrate.

The solubility of $Sr(NO_3)_2$ is diminished by HNO_3, but less so than with Ba^{2+}. Aqueous Ba^{2+} yields a fairly coarse, crystalline nitrate when treated with HNO_3, quite insoluble in concentrated HNO_3.

Fluorine species. The F^- ion precipitates Ae^{2+} as AeF_2.

2.3.3 Reagents Derived from the 3rd-to-5th-Period Non-Metals, Silicon through Xenon

Silicon species. Hexafluorosilicic acid, H_2SiF_6, does not precipitate Ca^{2+} even with an equal volume of ethanol (separation from Ba^{2+}). It does not precipitate Sr^{2+} even from quite concentrated solutions, especially in the presence of HCl. It does precipitate white crystalline $BaSiF_6$, slightly soluble in H_2O. Adding an equal volume of ethanol completes the precipitation of Ba^{2+}, with H_2SO_4 not giving a precipitate in the solute; Na^+ and K^+ interfere in this test.

Phosphorus species. Aqueous PHO_3^{2-}, but not $PH_2O_2^-$, precipitates Ba^{2+}.

A hydrated form of the mineral hydroxyapatite can be made, avoiding concomitant $Ca_4H(PO_4)_3 \cdot 2H_2O$, by slowly adding $Ca(NO_3)_2$ to $(NH_4)_2HPO_4$, first adjusting both to pH 12 with concentrated NH_3; using $CaCl_2$ would, on later calcination, give some chloroapatite:

$$5\,Ca^{2+} + 3\,HPO_4^{2-} + 4\,NH_3 + H_2O \rightarrow Ca_5OH(PO_4)_3 \cdot aq\downarrow + 4\,NH_4^+$$

Otherwise NH_3 and HPO_4^{2-}, or PO_4^{3-} can precipitate $Ca_3(PO_4)_2$. Aqueous HPO_4^{2-} alone precipitates $CaHPO_4 \cdot 2H_2O$, but, to avoid forming any $Ca_3(PO_4)_2$, a little $H_2PO_4^-$ is needed, so that $4 < pH < 5$. The more acidic salt—all of them are white—can be made, if it is washed with propanone (acetone); otherwise some H_3PO_4 in it causes some deliquescence; also:

$$CaCO_3 + 2\,H_3PO_4 \rightarrow Ca(H_2PO_4)_2 \cdot H_2O\downarrow + CO_2\uparrow$$

The HPO_4^{2-} and PO_4^{3-} ions precipitate Ba^{2+} as $BaHPO_4$ and $Ba_3(PO_4)_2$, respectively, similarly with Sr^{2+}.

Arsenic species. Neutral or ammoniacal solutions of arsenate(III) or arsenate(V) precipitate Ca^{2+}, e.g., as $[Ca(H_2O)_8]KAsO_4$. Neutral solutions of arsenates(III) do not precipitate Sr^{2+} or Ba^{2+}. Adding NH_3 precipitates part of the Sr^{2+} but not Ba^{2+} (distinction from Ca^{2+}). Aqueous arsenate(V) does not precipitate Sr^{2+} from a saturated (but still dilute) solution of $SrSO_4$ (distinction from Ca), but it does precipitate Ba^{2+} as $BaHAsO_4 \cdot H_2O$, white, slightly soluble in H_2O, soluble in acids. Strontium arsenate(V), precipitated from an alkali arsenate(V), resembles the corresponding Ba salt.

Reduced chalcogens. The sulfide ion, in moderately to strongly alkaline solution, precipitates Ca^{2+} as white, granular CaS [not $Ca(OH)_2$ as sometimes claimed]. Hydrogen sulfide dissolves CaS, forming Ca^{2+} and HS^-. Alkali sulfides precipitate Sr^{2+} possibly as $Sr(HS)_2$, white, from solutions not too dilute. Solutions of Ba^{2+} treated with an alkaline sulfide give a white precipitate, possibly $Ba(HS)_2$.

Oxidized chalcogens. Alkali sulfites precipitate Ca^{2+} as $CaSO_3 \cdot 2H_2O$, nearly insoluble in water, soluble in HCl, HNO_3, or aqueous SO_2.

Aqueous SO_3^{2-} precipitates Sr^{2+} as $SrSO_3$, white, from neutral or acetic acid solutions of Sr^{2+}. The precipitate is readily soluble in HCl.

Soluble sulfites precipitate Ba^{2+} as barium sulfite, $BaSO_3$, white, insoluble in water but soluble in HCl (distinction from the sulfate).

Concentrated H_2SO_4 is reduced by Ca, Sr or Ba to SO_2, S and H_2S.

Solutions of SO_4^{2-} precipitate Ca^{2+} as $CaSO_4$ from systems not too dilute. This compound is distinguished from $BaSO_4$ by its solubility in H_2O and HCl, and by the ease of conversion to the carbonate upon boiling with a solution of CO_3^{2-}. An aqueous solution of $CaSO_4$ is occasionally used to detect Sr^{2+} after the removal of Ba^{2+}.

The solubility of $CaSO_4 \cdot 2H_2O$ in most alkali salts is greater than in pure water; in fact it is readily soluble in hot $(NH_4)_2SO_4$ or in aqueous $S_2O_3^{2-}$. In ethanol it is almost insoluble but in acids (HNO_3 and HCl) its solubility is much greater than in H_2O. The double salt with K_2SO_4 is more insoluble with increasing K_2SO_4 concentration.

Sulfuric acid or SO_4^{2-} precipitates Sr^{2+} as $SrSO_4$ unless the solution is too dilute. A solution of $SrSO_4$ may be used to detect traces of Ba^{2+}. In dilute solutions

$SrSO_4$ separates very slowly. Precipitation is aided by boiling or by adding ethanol, prevented by HNO_3, HCl, and Ca^{2+} or other polyvalent metal ions in high concentration. The sulfate is less soluble in SO_4^{2-} salts or dilute H_2SO_4 than in H_2O; it is appreciably soluble in HNO_3 or HCl. It dissolves in concentrated H_2SO_4:

$$SrSO_4 + H_2SO_4 \rightarrow Sr^{2+} + 2\ HSO_4^-$$

Aqueous sulfate precipitates Ba^{2+} as barium sulfate, $BaSO_4$, white, slightly soluble in hot concentrated H_2SO_4. Immediate precipitation by a saturated solution of $CaSO_4$ distinguishes Ba^{2+} from Sr^{2+}, but precipitation by a solution of $SrSO_4$ (very dilute, due to its low solubility) offers a more certain distinction.

Reduced halogens. In concentrated HCl, barium nitrate is quite insoluble, the sulfate perceptibly soluble, the chloride almost insoluble.

Boiling $BaSO_4$ with HI forms soluble BaI_2 and volatile SO_2 and I_2.

Oxidized halogens. Iodate precipitates concentrated Sr^{2+} as $Sr(IO_3)_2$, also Ba^{2+} as $Ba(IO_3)_2 \cdot H_2O$, white, soluble at ambient *T* up to 6 dM.

2.3.4 Reagents Derived from the Metals Lithium through Uranium, plus Electrons and Photons

Reduction. Mercury, Ae^{2+} and e^- form amalgams, Ae_{Hg}.

Non-redox reactions. Normal chromates, CrO_4^{2-}, precipitate Ca^{2+} as yellow $CaCrO_4$ if not too dilute. This dissolves readily in acids. Aqueous molybdate precipitates Ca^{2+} from a slightly alkaline solution as $CaMoO_4$ (separation from Mg^{2+}). Aqueous tungstate completely precipitates Ca^{2+} as $CaWO_4$ (also separation from Mg^{2+}), but somewhat soluble in excess.

Aqueous CrO_4^{2-} precipitates strontium chromate, $SrCrO_4$, from solutions sufficiently concentrated. The precipitate is soluble in acids. In the absence of Ba^{2+}, Sr^{2+} may be separated from Ca^{2+} by adding CrO_4^{2-} to the nearly neutral solution containing one-third ethanol or propanone. At room temperature $CaCrO_4$ is about 20 times as soluble as $SrCrO_4$. Dichromates give no precipitate with Sr^{2+}.

Chromates or dichromate ions precipitate Ba^{2+} as barium chromate, $BaCrO_4$, yellow, insoluble in H_2O (separation from Sr and Ca except in concentrated solutions), sparingly soluble in acetic acid, readily soluble in HCl and HNO_3. If the solution is sufficiently buffered to absorb the H_3O^+ released, precipitation will be complete:

$$[Cr_2O_7]^{2-} + 3\ H_2O \leftrightarrows 2\ H_3O^+ + 2\ CrO_4^{2-}$$

$$CrO_4^{2-} + Ba^{2+} \leftrightarrows BaCrO_4\downarrow$$

Excess $K_4[Fe(CN)_6]$ precipitates Ca^{2+} as white $K_2Ca[Fe(CN)_6]$. An excess of NH_4^+ helps but then the composition of the precipitate varies. This test for Ca^{2+} seems to be more sensitive in the presence of Rb^+ or Cs^+, and most sensitive with added ethanol. Magnesium interferes.

Aqueous $[Fe(CN)_6]^{4-}$ does not precipitate Sr^{2+}.

Bibliography

See the general references in the Introduction, and some more-specialized books [1-.5]. Some articles in journals discuss: Be [6]; our neglect of Sr [7]; and Be complexes [8].

1. Lambert I, Clever HL (1992) Alkaline earth hydroxides in water and aqueous solutions. IUPAC, Blackwell, London
2. Brusset H (1976) Compléments au nouveau traité de chimie minéral: strontium. Masson, Paris
3. Novoselova AV, Batsanova LR (1966) Schmorak J (trans) (1968) Analytical chemistry of beryllium. Israel Program Scient Trans, Jerusalem
4. Everest DA (1964) The chemistry of beryllium. Elsevier, Amsterdam
5. Kirby HW, Salutsky ML (1964) The radiochemistry of radium. National Academy of Sciences, Washington
6. Alderighi L, Gans P, Midollini S, Vacca A (2000) Adv Inorg Chem 50:109
7. Nicholson JW, Pierce LR (1995 May) Chem Brit 31:74
8. Wong CY, Woollins JD (1994) Coord Chem Rev 130:243

3 The Rare-Earth and Actinoid Elements

First, some notes on nomenclature. Should one use the term "lanthanide", "lanthanon", "lanthanoid", or "rare earth"? The first uses the same ending, with a totally different meaning, as in "oxide" and so on. The second likewise shares its ending with the noble gasses. The third, albeit less-common, term is therefore preferred here, and by the IUPAC. All three focus attention nominally on the first member, hardly unique chemically, of the series. True, "lanthanoid" suggests "like lanthanum", thus conceivably excluding La itself, but La is also perfectly "like" La, and may therefore be included. The resemblances make this much more convenient and economical in expression than frequently saying "lanthanum and the lanthanoids".

Whether $_{57}La$ or $_{71}Lu$, and $_{89}Ac$ or $_{103}Lr$, are "the" proper (and therefore exclusive) congeners of $_{39}Y$ as members of the **d** block has been much disputed for decades, e.g. [1]. Of course the neighboring $_{71}Lu$, $_{72}Hf$, $_{73}Ta$, etc. are inherently likely to constitute a smoother series for all sorts of properties such as atomic radii than are the interrupted series $_{57}La$, $_{72}Hf$, $_{73}Ta$, etc. A more appropriate concern here, however, may be whether La or Lu is similar enough chemically to the intermediate elements to be classified with them. The question almost answers itself; clearly, both of them justify this classification. Regarding electronic structure, one can easily defend including the $\mathbf{f}^0$ at one end of the series for M^{3+}, say, as well as the $\mathbf{f}^{14}$ at the other end.

The fourth choice above, "rare earths", historically refers to the oxides rather than the elements, and their literal rarity is quite variable, but "rare earth" can include scandium and yttrium, which are very similar chemically although not in having a low-lying **4f**-electron subshell. Here then we prefer "lanthanoid" except when including scandium and yttrium as "rare earths".

For the 15 elements lanthanum through lutetium collectively we use the common symbol Ln. Actinium through lawrencium, the 15 "actinoids", are likewise represented as An. For the rare-earth elements collectively we propose and use the symbol Rth. This, like other symbols for elements, has just one upper-case letter, unlike R.E., and does not conflict with Re for rhenium.

We note also that $_{89}Ac$, plus $_{95}Am$ through $_{103}Lr$, resemble the Ln^{III}, although $_{90}Th$ through $_{93}Np$ sometimes resemble the **d**-block elements.

These elements show a kind of mini-periodicity [2] of characteristic extreme oxidation states, as seen in Table 3.1. A few of these known oxidation states, which represent exactly empty, half-full or full **f** subshells of electrons, are nevertheless not stable in water, as will be seen in the descriptions below. Table 3.2

shows the important oxidation states of An in water. Appendix C includes further oxidation states.

Table 3.1. Periodicity in the characteristic extreme oxidation states of Ln, An and their neighbors, not all in water

$n\mathbf{f}^m$	I	II	III	IV	V	VI	VII	VIII
$4\mathbf{f}^0$	${}_{55}$Cs	${}_{56}$Ba	${}_{57}$La	${}_{58}$Ce				
$4\mathbf{f}^7$		${}_{63}$Eu	${}_{64}$Gd	${}_{65}$Tb				
$4\mathbf{f}^{14}$		${}_{70}$Yb	${}_{71}$Lu	${}_{72}$Hf	${}_{73}$Ta	${}_{74}$W	${}_{75}$Re	${}_{76}$Os
$5\mathbf{f}^0$	${}_{87}$Fr	${}_{88}$Ra	${}_{89}$Ac	${}_{90}$Th	${}_{91}$Pa	${}_{92}$U	${}_{93}$Np	${}_{94}\underline{\text{Pu}}$
$5\mathbf{f}^7$		${}_{95}$Am	${}_{96}$Cm	${}_{97}$Bk	${}_{98}\underline{\text{Cf}}$			
$5\mathbf{f}^{14}$	${}_{101}\underline{\text{Md}}$	${}_{102}$No	${}_{103}$Lr	${}_{104}$Rf	${}_{105}$Db	${}_{106}$Sg	${}_{107}$Bh	${}_{108}$Hs

The underlined ones are doubtful.

Table 3.2. The important oxidation states of An in water

I	II	III	IV	V	VI	VII	VIII
		Ac					
		Th	**Th**				
			Pa	**Pa**			
		U	U	U	**U**		
		Np	Np	**Np**	Np	Np	
		Pu	**Pu**	Pu	Pu	Pu	?
	Am	**Am**	Am	Am	Am	Am	
		Cm	Cm				
		Bk	Bk				
		Cf	Cf	?			
	?	**Es**					
	Fm	**Fm**					
?	Md	**Md**					
	No	No					
		Lr					

The bold-faced ones are considered "the most stable".

The literature, incidentally, often does not clarify the meaning of "the most stable", but it normally means either the most thermodynamically or the most kinetically resistant to thermal decomposition, to dismutation, to oxidation by air and to hydrolysis, oxidation or reduction by water (liquid or vapor) at ambient *T*.

A similar, albeit less clear, mini-periodicity appears with the **d** subshells of the **d**-block elements and their neighbors, as in Table 3.3. Some of the congruences in *this* table are deceptive, however; the low-spin structure of $[Fe(CN)_6{}^{3-}]$, for example, is not that of a specially stabilized half-full, **3d**, high-spin, subshell. As is well known, incidentally, the **d**- and **f**-block elements do not add **d** and **f** electrons smoothly as *Z* rises, and Appendix B shows simple graphical explanations. We also note that many of the oxidation states in Table 3.3 are the ones that gave rise to Mendeleyev's arrangement.

Table 3.3. Periodicity in some oxidation states of **d**-block and near-by elements

	I	II	III	IV	V	VI	VII	VIII
$\mathbf{3d^0}$	$_{19}$K	$_{20}$Ca	$_{21}$Sc	$_{22}$Ti	$_{23}$V	$_{24}$Cr	$_{25}$Mn	
$\mathbf{4d^0}$	$_{37}$Rb	$_{38}$Sr	$_{39}$Y	$_{40}$Zr	$_{41}$Nb	$_{42}$Mo	$_{43}$Tc	$_{44}$Ru
$\mathbf{5d^0}$	$_{55}$Cs	$_{56}$Ba	La-Lu	$_{72}$Hf	$_{73}$Ta	$_{74}$W	$_{75}$Re	$_{76}$Os
$\mathbf{6d^0}$	$_{87}$Fr	$_{88}$Ra	Ac-Lr	$_{104}$Rf	$_{105}$Db	$_{106}$Sg	$_{107}$Bh	$_{108}$Hs
$\mathbf{3d^5}$		$_{25}$Mn	$_{26}$Fe	$_{27}$Co				
$\mathbf{4d^5}$			$_{44}$Ru	$_{45}$Rh				
$\mathbf{5d^5}$			$_{76}$Os	$_{77}$Ir				
$\mathbf{3d^{10}}$	$_{29}$Cu	$_{30}$Zn	$_{31}$Ga	$_{32}$Ge	$_{33}$As	$_{34}$Se	$_{35}$Br	
$\mathbf{4d^{10}}$	$_{47}$Ag	$_{48}$Cd	$_{49}$In	$_{50}$Sn	$_{51}$Sb	$_{52}$Te	$_{53}$I	$_{54}$Xe
$\mathbf{5d^{10}}$	$_{79}$Au	$_{80}$Hg	$_{81}$Tl	$_{82}$Pb	$_{83}$Bi	$_{84}$Po	$_{85}$At	

The common terms "lighter" and "heavier" for **f** and **d** metals can be identified less misleadingly as "low-*Z*" and "high-*Z*" (for atomic number) or "left-side" and "right-side" (of the most used periodic charts), or even "earlier" and "later" (despite connotations of time).

Before continuing, we note that mass as such has practically no effect on the chemistry of these metals, and some elements, e.g., Co, are famously heavier than their neighbors of next higher *Z*. We mention this partly because of the persistent myth that molecular mass is a major influence on boiling points [3].

Each Ln^{III} differs only slightly in non-redox reactivity from its immediate neighbors, but the earlier (lower-*Z*) ones differ enough from the mostly more distant later (higher-*Z*) ones, on the whole, that two subgroups have acquired special names, sometimes with several variations by the same authors. The earlier ones have thus been called the cerium Group, cerium subgroup, cerium earths, cerites and cerite Group, all for the most abundant member.

Yttrium(III) behaves most like Ho^{III} among the higher-*Z* lanthanoids, because of a near identity of radii due to the lanthanoid contraction and relativity, and the much greater abundance of Y has led to the names yttrium Group, yttrium subgroup, yttrium earths, ytter-earth Group, ytter earths and yttria Group for this subgroup, which may often be taken to begin at about Eu, just before the midpoint of the lanthanoids, but based on chemical behavior. Scandium, though with lower *Z*, acts like an extreme member beyond Lu, because of its smaller radius.

The ending "ite" in cerite of course has a different meaning, as does "earth" in spite of the established status of "rare earth". We propose "ceroid" and "yttroid", symbols Ced and Ytd, which avoid these problems, are brief, mutually consistent (unlike, say, cerites and ytter earths or yttria Group, used by the same authors) and analogous to the other collective terms, lanthanoid, actinoid and uranoid.

In the **5f** series some of the distinctively f-subshell chemical behavior arises later, with U, leading to the term "uranoid", especially for elements 92–95, which, unlike the other "actinoids" in the same period, have six (VI) as an important oxidation state in water. We may refer to the higher-*Z* actinoids, with III as the more characteristic oxidation state, as "post-uranoids". Nature is clever in complicating our task by precluding full consistency and simplicity in any periodic chart.

As to chemical behavior, the complexing abilities and acidities are, as expected: $An^{5+} > An^{4+} > An^{3+}$ and $AnO_2^{2+} > AnO_2^{+}$; together, most often: $An^{5+} > An^{4+} > AnO_2^{2+} > An^{3+} > AnO_2^{+}$ because O^{2-} does not completely quench 2+ charges on the cation; also complexing is stronger for actinoids than for lanthanoids. The acidities and hydrolytic tendencies are higher for the right-side (higher *Z*) than for the left-side (lower *Z*) members of each series. Stabilities with ligands mostly follow basicities: $F^- > H_2PO_4^- > NCS^- > NO_3^- > Cl^- > ClO_4^-$; $PO_4^{3-} > CO_3^{2-} > HPO_4^{2-} > SO_4^{2-}$.

The similarities of non-redox behavior within each of the seven aqueous oxidation states of Rth and An including M(0), combined with the size of the Group, 32 elements in all, make it convenient to examine each oxidation state separately, M^0 in sec. 3.1, M^{II} in 3.2, M^{III} in 3.3, M^{IV} in 3.4, M^{V} in 3.5, M^{VI} in 3.6, and M^{VII} in 3.7, with M for either Rth or An. This also facilitates comparisons within each oxidation state. The only generally important non-zero oxidation state for Rth in water is Rth^{III}, except for Eu^{II} and especially Ce^{IV}.

The actinoids are much more varied, as shown in Table 3.1 and the text below. The highest oxidation state is higher or more stable for many **5f** elements than for the **4f** because of the much greater relativistic destabilization of the **5f** electrons. A smaller amount of the same effect appears in comparing the **5d** and **4d** elements. Bases stabilize high oxidation states. Radiation, however, generates e^-, $H^•$, $OH^•$ and $HO_2^•$ radicals that reduce some of those states. We may compare the Gibbs energies of the aqua/hydroxo/oxo An species in Fig. 3.1 [4].

Mutual separations of the uranoid elements may be eased because solutions can simultaneously have UO_2^{2+}, NpO_2^{+}, Pu^{IV} and Am^{III} with the different complexation and extraction behaviors of the various oxidation states.

The higher-*Z* actinoids show some non-metallic behavior, beginning rather clearly with solid Pu, but can still form cations in water.

3.1 The Rare Earths Rth(0) and Actinoids An(0)

3.1.1 Reagents Derived from Hydrogen and Oxygen

Water. The metals Rth react slowly with cold H_2O, and An with hot H_2O, releasing H_2 and forming $Rth(OH)_3 \cdot aq$, $Eu(OH)_2$ or surface An_2O_3.

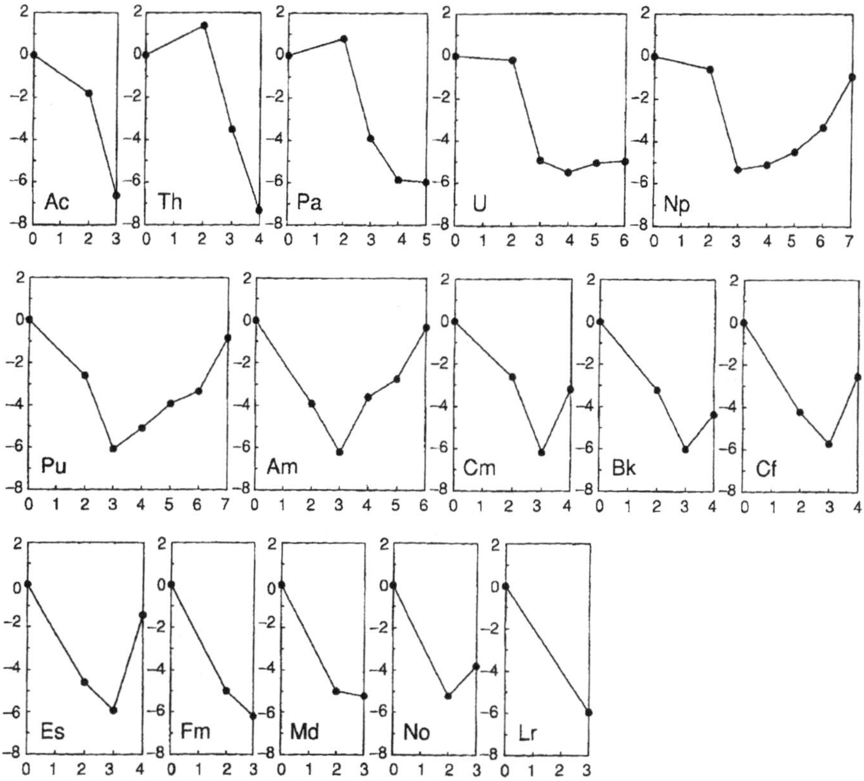

Figure 3.1. Gibbs energies (Frost-Ebsworth diagrams) of the aqua/hydroxo/oxo An^{n+} species at pH 0 (relative to the metal). The ordinate is $nE°/V$; the abscissa, n

Oxonium. The metals dissolve readily in acids (except concentrated H_2SO_4), resulting in, e.g., Eu^{2+}, Rth^{3+}, Th^{4+}, An^{3+} or No^{2+}; see Appendix C.

Hydroxide. Metallic Rth and An are not oxidized by OH^-, except as its H_2O attacks them; e.g., Eu and 10-M OH^- form pure $Eu(OH)_2 \cdot H_2O$, and Th, U and Pu are made passive by forming surface hydroxides.

Dioxygen. The Rth metals are attacked by moist air, mostly forming Rth^{III} hydroxides. However, it converts Eu to yellow $Eu(OH)_2 \cdot H_2O$, which slowly breaks down, even without O_2, to $Eu(OH)_3 \cdot aq$ and H_2.

3.1.2 Reagents Derived from the Other 2nd-Period Non-Metals, Boron through Fluorine

Reduced nitrogen. The Rth and An metals are not attacked by NH_3, except as its H_2O attacks them.

Oxidized nitrogen. Aqua regia readily dissolves Th as Th^{IV}, but in HNO_3 it soon becomes passive; Pa is insoluble in 8-M HNO_3 even with 1-cM HF. Uranium dissolves as U^{IV} in HNO_3. Concentrated HNO_3 passivates Th, U and Pu unless HF is present. The "Purex" process (Pu-U recovery by extraction, or other interpretations of the acronym) for used nuclear fuel begins with dissolution in 7-M HNO_3.

Fluorine species. Thorium and HF produce ThF_4, but not readily. Metallic Pa is attacked, but only briefly, by 12-M HF. Uranium dissolves slowly in HF.

3.1.3 Reagents Derived from the 3rd-to-5th-Period Non-Metals, Silicon through Xenon

Phosphorus species. Uranium dissolves slowly in H_3PO_4.

Oxidized chalcogens. Thorium dissolves as Th^{IV}, but not readily, in H_2SO_4. Metallic Pa is attacked, but only briefly, by 2.5-M H_2SO_4. Uranium dissolves slowly as U^{IV} in cold, dilute H_2SO_4.

Reduced halogens. Thorium and uranium dissolve as M^{IV} in HCl. Metallic Pa is attacked, but only briefly, by 8-M HCl, although 8-M HCl and 1-M HF, combined, may be the best of all solvents.

3.1.4 Reagents Derived from the Metals Lithium through Uranium, plus Electrons and Photons

Oxidation. Uranium is oxidized at least to U^{IV} when it reduces solutions of Pt, Cu, Ag, Au, Hg, Sn, Bi etc. to the metals.

3.2 The Rare Earths Rth(II) and Actinoids An(II)

We have only some lanthanoids(2+), not Sc^{2+} or Y^{2+}, in water. The ions Md^{2+} and No^{2+} are more stable than Eu^{2+} to oxidants; in metathesis they are like Ln^{2+} and Ba^{2+}. Electrode potentials have been determined for Am^{2+} etc. (see Appendix C), even without appreciable information to report here on their chemical behavior in water.

3.2.1 Reagents Derived from Hydrogen and Oxygen

Water. Water oxidizes Sm^{2+}, Tm^{2+} or Yb^{2+} to Ln^{III} in a few hours or minutes, but Eu^{2+} is mostly stable for weeks without platinum catalysts.

Hydroxide. The Eu^{2+} and OH^- ions precipitate a mixture of $Eu(OH)_2 \cdot H_2O$ and $Eu(OH)_3 \cdot aq$.

Dioxygen. Air gradually oxidizes Eu^{2+} and, in HCl, first forms H_2O_2:

$$Eu^{2+} + H_3O^+ + {}^1/_2\, O_2 \rightarrow Eu^{3+} + {}^1/_2\, H_2O_2 + H_2O$$

Moist air converts EuI_2 to EuOI.

3.2.2 Reagents Derived from the Other 2nd-Period Non-Metals, Boron through Fluorine

Oxidized nitrogen. The sulfate $SmSO_4$ dissolves in 2-M HNO_3, but $EuSO_4$ requires 6 M.

Fluorine species. Nobelium coprecipitates with BaF_2, revealing No^{2+}.

3.2.3 Reagents Derived from the 3rd-to-5th-Period Non-Metals, Silicon through Xenon

Oxidized chalcogens. Catalytic Ag^+, No^{2+} and $[S_2O_8]^{2-}$ give No^{III}.

Reduced halogens. Air-free Eu^{2+} and 10-M HCl precipitate $EuCl_2 \cdot 2H_2O$.

Oxidized halogens. Bromate and periodate oxidize No^{2+} to No^{III}.

3.2.4 Reagents Derived from the Metals Lithium through Uranium, plus Electrons and Photons

Oxidation. Aqueous Eu^{2+} is oxidized by and reduces $[CrX(H_2O)_5]^{2+}$ or $[CoX(NH_3)_5]^{2+}$; X = any halogen. The Yb^{2+} ion reduces $[CrX(H_2O)_5]^{2+}$ and $[CrX(NH_3)_5]^{2+}$ without catalysis by free Cl^-, but $[Co(NH_3)_6]^{3+}$ and $[Co(NH_3)_5(H_2O)]^{3+}$ with catalysis by Cl^-; X = N_3^-, NCS-*κN*, F^-, Cl^-, Br^- or I^-. All rates except for $[Cr(H_2O)_6]^{3+}$ are independent of $c(H_3O^+)$. Iron(III) and Ln^{2+} give Ln^{3+} and Fe^{2+}, which reduces $Cr_2O_7^{2-}$ in titrations.

To convert No^{2+} to No^{3+} requires strong oxidants, e.g., a peroxochromate, MnO_4^- or (apparently incompletely) Ce^{IV}. Then No^{3+} coprecipitates with LaF_3.

Light (UV), with catalytic Ni/Pd/Pt, energizes the oxidation of Eu^{2+}, but cf. **3.3.4 Reduction**:

$$Eu^{2+} + H_3O^+ + \gamma \rightarrow Eu^{3+} + {}^1/_2\, H_2\!\uparrow + H_2O$$

Other reactions. The coprecipitation of $MdSO_4$ along with $BaSO_4$ or $EuSO_4$ (perhaps after adding small amounts of the carrier cation and reducing the Md^{3+} with, say, Cr^{2+}, Yb^{2+} or Zn_{Hg}) may be used to separate very small quantities of Md^{n+} from An^{n+} such as Es^{3+} and Fm^{3+}. The $BaSO_4$ may then be made soluble by evaporating with aqueous HI.

The colors of Ln^{2+} are: Sm^{2+} **4f**^{6} blood-red; Eu^{2+} **4f**^{7} none; Tm^{2+} **4f**^{13} violet-red; Yb^{2+} **4f**^{14} pale yellow-green; not patterned as in Table 3.4 below.

3.3 The Rare Earths Rth(III) and Actinoids An(III)

The bonding in M^{III}-ligand is mainly ionic (with hard donors) and labile. The An^{III} ions bind a little more firmly than Ln^{III} to ligands containing the soft N, S or Cl.

3.3.1 Reagents Derived from Hydrogen and Oxygen

Water. Highly acidic solutions favor $[Sc(H_2O)_8]^{3+}$. Less-acidified, concentrated Sc^{III} gives $[\{Sc(H_2O)_5\}_2(\mu\text{-}OH)_2]^{4+}$ and so on. Hydrated Y^{3+} is $[Y(H_2O)_8]^{3+}$, and n in $[Ln(H_2O)_n]^{3+}$ is often 8 or 9. For Pu^{3+} one finds $[Pu(H_2O)_9]^{3+}$, and theoretical predictions of An^{3+} hydration give $[(Ac\text{–}Md)(H_2O)_9]^{3+}$ and $[(No,Lr)(H_2O)_8]^{3+}$ [5]. Moderate hydrolysis of Ln^{3+} yields $LnOH^{2+}$ and often $Ln_2(OH)_2{}^{4+}$ etc. The solids $YtdCl_3\cdot 6H_2O$ are $[YtdCl_2(H_2O)_6]Cl$, but we also have $[Ln(H_2O)_6](ClO_4)_3$.

The oxides Ln_2O_3 absorb H_2O from the air.

The ceroid hydroxides are somewhat soluble, and wet $La(OH)_3\cdot aq$ turns red litmus blue; the yttroid hydroxides are less soluble. For both Groups the carbonates are insoluble, and the oxalates (distinction from many M^{n+}), fluorides and phosphates are insoluble and more or less insoluble even in (cold) dilute H_3O^+. The nitrates, sulfides and sulfates are soluble; the basic nitrates are distinctly less so, but those of Ytd^{3+} are more soluble than those of Ced^{3+}. However, the oxalates, double nitrates and double sulfates of Ced^{3+} are more soluble than those of Ytd^{3+}. The stable chlorides $(La\text{-}Pr)Cl_3\cdot 7H_2O$ and $(Y,Nd\text{-}Lu)Cl_3\cdot 6H_2O$, and the bromides and iodides, are soluble. The perchlorates dissolve easily; the bromates $[Ln(H_2O)_9](BrO_3)_3$ dissolve in the molar range, but the solubilities of hydrated $Ln(IO_3)_3$ are only about 1 mM.

Hydrolysis rises from La to Lu and from Ac to Lr, and from their radii, the acidities should be $Ce^{3+} < Pu^{3+} < Pr^{3+}$ and $Eu^{3+} < Cf^{3+} < Gd^{3+}$. For most An^{3+} it is not appreciable below a pH of 4 or, for Am^{3+}, even in nearly neutral solutions, but Pu^{3+} goes about 70 % to $PuOH^{2+}$ at pH 3; it is hard to study further because of easy oxidation at pH 7 and higher.

The Th^{3+} and Pa^{3+} ions are oxidized by H_2O too fast to persist in it; U^{3+} also releases H_2 from H_2O but is stable in 1-M HCl for days; it hydrolyzes to polymers in rather less acidic solutions.

Thermodynamic evidence for Ln^{3+} and An^{3+} points to $NO_3{}^-$, Cl^-, Br^- and I^- complexes as being outer-sphere (beyond the hydration sphere) but to F^-, $SO_4{}^{2-}$ and $IO_3{}^-$ as replacing H_2O in the inner sphere.

In non-redox chemistry, ${}_{94}Pu^{3+}$ resembles ${}_{60}Nd^{3+}$.

Hot natural waters may contain $[Rth(CO_3)_4]^{5-}$, $[ScF_6]^{3-}$, $RthF^{2+}$, $RthCl^{2+}$ or $Rth(SO_4)_2{}^-$. Seawater contains $RthCO_3{}^+$, $Rth(CO_3)_2{}^-$ and less $RthOH^{2+}$.

Oxonium. The oxides and hydroxides are soluble in acids, and form, e.g., hydrated Ln^{3+} at a below pH 5.

In acid, Am is stable only as Am^{III} or AmO_2^{2+}.

Hydroxide. The OH^- ion and Sc^{III} produce $Sc(OH)_4^-$, $K_2Sc(OH)_5 \cdot 4H_2O$, $Na_3Sc(OH)_6 \cdot 2H_2O$, $(Ca,Ba)_3Sc_2(OH)_{12}$, $(Ca,Sr)Sc_2(OH)_8 \cdot 2H_2O$, etc.

The Ln^{3+} and OH^- ions in various solutions at pH $\approx$ 6 to 8 precipitate $Ln(OH)_{(3-x)}X_x$ where we have $0 < x \leq 1$ and X may be Cl^-, NO_3^-, $^1/_2\ SO_4^{2-}$, etc. Aging replaces more, or even all, X^- with OH^-. The precipitates are slightly soluble in concentrated OH^-, and solid $Na_3(Yb,Lu)(OH)_6 \cdot aq$ and $Na_4(Yb,Lu)(OH)_7 \cdot aq$ have been found. Using insufficient reagent leads to other basic salts. Reaction in the cold gives a slimy product, difficult to filter or wash. Acetate delays precipitation. Citrate or tartrate prevent it, although in some cases boiling promotes the separation of a complex tartrate, e.g., ammonium yttrium tartrate. All ceroid hydroxides are strong bases although only slightly soluble. The lower solubilities of the yttroids allow separations based on tedious fractional precipitations.

The precipitation of $La(OH)_3 \cdot aq$ or $Fe_2O_3 \cdot aq$ etc. by OH^- (perhaps after adding small amounts of the carrier cation) may be used for a preliminary separation of very small quantities of An species (from large amounts of the less-acidic cations) by coprecipitating them.

Digesting $Ln_3(PO_4)_4$ from ores with hot OH^- yields $Ln(OH)_3$.

Aqueous OH^- and Pu^{III} precipitate $Pu(OH)_3 \cdot aq$, which quickly goes to Pu^{IV}. The OH^- ion (or NH_3) precipitates Am^{III} as pink, gelatinous $Am(OH)_3 \cdot aq$, easily soluble in H_3O^+. Treating AmF_3 with 1-dM OH^- at 90 °C for 1 h produces $Am(OH)_3 \cdot aq$. In general, insoluble AnF_3 becomes acid-soluble $An(OH)_3 \cdot aq$ on treatment with concentrated OH^-. Aging the precipitated hydroxides gives rise to $An(OH)_3$.

Peroxide. A yellow color due to the oxidation of Ce^{III} to Ce^{IV} by means of NH_3 plus H_2O_2 is visible with as little as 0.3 mmol of Ce.

At fairly high pH, $Am(OH)_3 \cdot aq$ and O_2^{2-} form $Am(OH)_4 \cdot aq$. Concentrated OH^- with O_2^{2-} and $(Np,Pu,Am)^{<VII}$ form M^{VII}; see **3.6.1.**

Aqueous $Am(OH)_3 \cdot aq$ and H_2O_2 yield $Am(OH)_4 \cdot aq$. Ten min on a water bath with 6 to 7-M OH^- and 3-dM H_2O_2 forms black $Am(OH)_4 \cdot aq$.

Di- and trioxygen. Air oxidizes $U^{<VI}$, rapidly if only electron transfer is required (from UO_2^+ to form UO_2^{2+}), more slowly otherwise (with $U^{III,\ IV}$). This kinetic factor affects some other An ions with moderate oxidants.

Aqueous U^{III} is oxidized (first to U^{IV}) by both O_2 and H_2O, Np^{III} less readily by only O_2, and Pu^{III} with non-ligating anions by neither; Pu^{3+} still has a (less) favorable electrode potential but is inert to O_2 (because no simple electron transfer reduces O_2 to H_2O) which also does not oxidize any of the higher-Z An^{3+} to An^{IV}. The α rays, however, produce strong oxidants, and O_2 oxidizes Pu^{III} slowly in dilute SO_4^{2-} at pH 4, rapidly at higher pH and in HCO_3^- (thus with CO_3^{2-} complexes).

Aqueous U^{3+}, Np^{3+} or Pu^{3+}, all plus ozone, do form MO_2^{2+}.

Ozone with concentrated OH^- oxidizes Np^{III}, Pu^{III} or Am^{III} to M^{VII}, forming, for example, $Li_3(NpO_2)(OH)_6$, $(Na,K)_3[NpO_4(OH)_2]\cdot nH_2O$, $[Co(NH_3)_6][NpO_4(OH)_2]\cdot 2H_2O$ and $Li[Co(NH_3)_6]Np_2O_8(OH)_2\cdot 2H_2O$, the latter prepared from Np^{VII} in LiOH by adding $[Co(NH_3)_6]Cl_3$. Some Pu^{VII} salts are rather similar. Also see **Peroxide** above and **3.6.1** below.

Ozone can oxidize $Am(OH)_3\cdot aq$ completely in 1 h to Am^{VI}. It can also yield Am^{V} sulfate with some H_2SO_4 and evaporation.

Ozone plus Am^{III} [possibly from $Am(OH)_3\cdot aq$], dissolved variously in 3-cM $KHCO_3$ up to concentrated K_2CO_3, precipitate $KAmO_2CO_3$, $K_3AmO_2(CO_3)_2$ or $K_5[AmO_2(CO_3)_3]$ from hot solutions. Acids then give AmO_2^+. Washing with H_2O decomposes the carbonato complexes.

Somewhat likewise, passing O_3 for 1 h through Am^{III} in 2-M Na_2CO_3 at ambient T gives Am^{VI}, but heating this to 90 °C for 30–60 min precipitates an Am^{V} double carbonate free of Am^{III}. Treating Am^{III} and Rb_2CO_3 or "$(NH_4)_2CO_3$" with O_3 forms $RbAmO_2CO_3$ or $NH_4AmO_2CO_3$.

Ozone plus Am^{III} in HNO_3 form AmO_2^{2+}, but Cm^{III} in various media does not yield $Cm^{>III}$.

3.3.2 Reagents Derived from the Other 2nd-Period Non-Metals, Boron through Fluorine

Boron species. Saturated H_3BO_3 in 1-M to 6-M H_3O^+ dissolves AmF_3.

Carbon oxide species. The strongly basic Ln_2O_3 absorb aerial CO_2.

Aqueous CO_3^{2-} complexes Ln^{3+} in stages up to $[Ln(CO_3)_4]^{5-}$. Without a great excess of CO_3^{2-} it also precipitates normal (e.g., $Eu_2(CO_3)_3\cdot 3H_2O$) or basic carbonates of all rare earths. The yttroid carbonates, but not the ceroids, are fairly soluble in excess "$(NH_4)_2CO_3$". Barium carbonate gives no precipitate with Ytd^{3+} in the cold, and only partial precipitation in hot solution (distinction from Ced^{3+}, Al^{3+} and Th^{4+}).

Carbonate and Pu^{3+} form $PuCO_3^+$, $Pu(CO_3)_2^-$ and higher complexes.

Treating Am^{III} with $NaHCO_3$ saturated with CO_2 precipitates a pink $Am_2(CO_3)_3\cdot 4H_2O$, quite soluble in excess CO_3^{2-}. The addition of 5-dM $NaHCO_3$ to the solid forms $NaAm(CO_3)_2\cdot 4H_2O$, but 1.5-M Na_2CO_3 gives $Na_3Am(CO_3)_3\cdot 3H_2O$ instead. Carbonate solutions at $pH > 6$ may also form $AmCO_3^+$ and $Am(CO_3)_2^-$, but hydrolysis to $Am(OH)_2^+$ should appear at $pH \approx 11$. All Am from Am^{III} through Am^{VI} can be in equilibrium in 1.2 to 2.3 M HCO_3^- plus CO_3^{2-} (total). At only 1.2 to 6-dM CO_3^{2-}, Am^{III} is mainly $[AmOH(CO_3)_3]^{4-}$.

Carbonate precipitates white $Cm_2(CO_3)_2\cdot 4H_2O$ from Cm^{III} in weakly acidic solutions; air slowly darkens it, and it dissolves in 3-M CO_3^{2-}.

Cyanide species. Various $[Ln(NCS)_3(H_2O)_6]$ have a ligancy of nine.

Some “simple” organic species. Evaporating solutions of $Rth(OH)_3\cdot aq$ or $An(OH)_3\cdot aq$ in $HCHO_2$ yields formates, and $KCHO_2$ can produce, e.g., $KY(CHO_2)_4\cdot H_2O$ or $K_5Y(CHO_2)_8$. The acetates $Ln(CH_3CO_2)_3\cdot 2H_2O$ can be crystallized, and $CH_3CO_2^-$ complexes An^{3+} mildly, although $[Pu(CH_3CO_2)_5]^{2-}$, for example, is quite stable.

Oxalic acid and $C_2O_4^{2-}$ precipitate oxalates of the rare earths, e.g., $Sc_2(C_2O_4)_3\cdot 6H_2O$, practically insoluble in $H_2C_2O_4$ and other acids. The oxalates of the yttroids, but not the ceroids, dissolve in excess $C_2O_4^{2-}$. Some solids are $Ytd_2(C_2O_4)_3\cdot 3H_2O$ and $K_8Ytd_2(C_2O_4)_7\cdot 14H_2O$.

The oxalato complexes of the actinoids(III) are somewhat more stable than the acetato ones, up to $[An(C_2O_4)_4]^{5-}$, at pH 1 to 4, with $Pu(C_2O_4)_2^-$ as an important and stable example. The insolubility of the An^{III} oxalates in water is useful for separations.

Oxalate in acidic (especially for Bk^{III} or Cf^{III}) or neutral solutions of An^{III} precipitates $An_2(C_2O_4)_3\cdot nH_2O$, with n up to 11 depending on procedures, pink or rust-brown for Am^{III}. Its α rays decompose it in days, primarily to $Am_2(CO_3)_3\cdot 5H_2O$. However, 5-cM $H_2C_2O_4$ and pH 0–2 separate Am (with Ca, Sr, Rth^{III} etc.) well from much Cr, Fe, Ni, Al etc. Then 1-dM OH^- converts most of the oxalates to hydroxides. The Am oxalates yield $(Alk,NH_4)Am(C_2O_4)_2\cdot nH_2O$ in neutral solutions.

Curium(III) gives white $Cm_2(C_2O_4)_3\cdot 10H_2O$, gradually turning gray. Radiolysis quite quickly transforms it to $Cm_2(CO_3)_3$. The oxalate is converted by 5-dM OH^- to $Cm(OH)_3\cdot aq$.

The precipitation of $CaC_2O_4\cdot H_2O$ or $La_2(C_2O_4)_3\cdot 11H_2O$ may be used for the preliminary separation of small quantities of An^{III} species (from large amounts of Mg^{2+}, Cr^{III}, Mn^{2+}, Fe^{III}, Ni^{2+}, Al^{III} etc.) by coprecipitating them from high $c(NO_3^-)$. The oxalate may be formed slowly, to obtain larger crystals, by hydrolyzing $Me_2C_2O_4$ in dilute HNO_3.

Styrene-sulfonate cation-exchange resins, for example, preferentially bind the ceroids, while soluble organic chelators prefer the yttroids. The latter are therefore eluted first by, say, citrate at pH 5, enabling efficient separations of all of them.

Reduced nitrogen. The oxides and hydroxides are insoluble in NH_3, which, however, precipitates Sc^{3+} only partly, possibly due to forming ammines. Approaching $Ca(OH)_2$ in basicity is $La(OH)_3\cdot aq$, which liberates NH_3 from NH_4^+. Still, the precipitation of $La(OH)_3\cdot aq$ [or $Fe_2O_3\cdot aq$ etc.] by NH_3 (after adding a little, if needed, of the carrier, e.g., La^{3+} or Fe^{III}) may be used for the first separation of very small amounts of An species (from less-acidic cations) by coprecipitation.

Ammonia precipitates Cm^{III} as white, flocculent $Cm(OH)_3\cdot aq$, which darkens later and dissolves easily in H_3O^+. It precipitates Bk^{III} as white $Bk(OH)_3\cdot aq$, often greenish due to radiolytic oxidation to $Bk(OH)_4\cdot aq$.

Oxidized nitrogen. Aqueous Pu^{III} is stable in pure HNO_3 but not if contaminated by HNO_2 as usual. Autocatalytically it goes thus:

$$2\ NO_2^- + 2\ H_3O^+ \leftrightarrows 2\ HNO_2 + 2\ H_2O$$

$$HNO_2 + NO_3^- + H_3O^+ \leftrightarrows 2\ NO_2 + 2\ H_2O$$

$$2\ Pu^{3+} + 2\ NO_2 \leftrightarrows 2\ Pu^{4+} + 2\ NO_2^-\ \text{(slowest)}$$

$$2\ Pu^{3+} + NO_3^- + 3\ H_3O^+ \leftrightarrows 2\ Pu^{4+} + HNO_2 + 4\ H_2O$$

but its dismutation then to Pu^{3+} and PuO_2^{2+} leads finally to PuO_2^{2+}. Good to quench the HNO_2 to prevent this are $N_2H_5^+$, HSO_3NH_2 and Fe^{2+}.

Concentrated HNO_3 and Sc^{III} can give $(K,Rb,Cs)_2[Sc(NO_3)_5]$.

Nitric acid and La_2O_3 yield $[La(\eta^2\text{-}NO_3)_3(H_2O)_5]\cdot H_2O$, the first Ln^{III} species with ligancy (c. n.) 11, unstable in air at ambient *T*.

Evaporating HNO_3 solutions of Ln^{III} and An^{III} yields the nitrates. Stable nitrates include $Eu(NO_3)_3\cdot 6H_2O$. Complexes of NO_3^- with Ln^{3+} are stronger than those of Cl^- or ClO_4^- but weaker than those of SO_4^{2-}.

Historically important separations by fractional crystallization used, for example, the double nitrates $Rth_2Mg_3(NO_3)_{12}\cdot 24H_2O$ mixed with $Bi_2Mg_3(NO_3)_{12}\cdot 24H_2O$; the ceroids are the less soluble ones. More efficient but still difficult separations use the solvent extraction of the nitrates into Bu_3PO_4 as $[Rth(Bu_3PO_4)_3(NO_3)_3]$.

The NO_3^- ligand, unlike SO_4^{2-}, is usually didentate, as in at least some $[Ln(NO_3)_5]^{2-}$, $[(La,Ce)(NO_3)_6]^{3-}$, $[Ce(NO_3)_6]^{2-}$, $[Nd(NO_3)_3(H_2O)_4]\cdot 2H_2O$ and $[Gd(NO_3)_3(H_2O)_3]$; with these then, the ligancies can go up to 12.

Fluorine species. Aqueous F^- forms $RthF^{2+}$ etc., and HF and F^- precipitate Rth^{3+} as $RthF_3$, or, e.g., $EuF_3\cdot {}^1/_2H_2O$, all slightly soluble in H_2O and insoluble in excess F^-, but appreciably soluble in hot H_3O^+.

Fluoride and An^{3+} precipitate, e.g., $UF_3\cdot H_2O$; other An^{3+}, before precipitation, form complexes with stabilities: $Am^{3+} < Cm^{3+} < Bk^{3+} < Cf^{3+}$. In the mutual separations of uranoids, blue-violet $PuF_3\cdot aq$ (precipitated from HNO_3 or HCl) is luckily not gelatinous like $PuF_4\cdot aq$, but F(α-n) reactions release more neutrons than come from other precipitants.

Aqueous Am^{III} and HF give $AmF_3\cdot xH_2O$. Adding HF, made finally to be 1–2 M, to Cm^{III} in 1-M HNO_3 or 2-M HCl precipitates $CmF_3\cdot aq$. Non-aqueous and aqueous sources, with Alk^+ and Ae^{2+}, give numerous fluoro-complexes, as well as chloro- and bromo-complexes, e.g., $AlkAnX_4$, Alk_2AnX_5, Alk_3AnX_6, $AlkAn_2Cl_7$, $AeAnX_5$ and Ae_2AnX_7.

The precipitation of LaF_3 by HF (after adding small amounts of the carrier La^{3+}, if needed) may be used for the preliminary separation of very small quantities of An^{III} and An^{IV} species (from large amounts of most other elements) by coprecipitating them. Cerium(IV) can oxidize Np and Pu ions to An^{VI} so that U, Np and Pu stay dissolved as fluorides. One may also remove only Cm as CmF_3 if Am is kept as Am^{VI}.

3.3.3 Reagents Derived from the 3rd-to-5th-Period Non-Metals, Silicon through Xenon

Phosphorus species. Phosphoric acid complexes Ln^{3+} weakly (and Sc^{3+} more strongly), and it and phosphate ions, e.g., HPO_4^{2-}, precipitate rare-earth phosphates such as $CePO_4 \cdot 2H_2O$. A pH of 2.3 yields $Ln_2(HPO_4)_3$.

Acidified phosphate plus Ac^{III}, Pu^{III}, Am^{III} or Cm^{III} yield $AnPO_4 \cdot {}^1/_2H_2O$ (white for Ac, blue for Pu and light yellow for Cm, soluble in 4 to 6-M HCl) and gelatinous $Pu(HPO_4)_2 \cdot aq$. The phosphate complexes of An^{III} may include $An(H_2PO_4)_n^{(3-n)+}$, $1 \leq n \leq 4$.

The precipitation of $BiPO_4$ may be used for the preliminary separation of very small quantities of An^{III} species (from large amounts of Mg^{2+}, Ca^{2+}, Al^{III} etc.) by coprecipitating them, e.g., from 3-mM Bi^{III}, 9-cM H_3PO_4, 15-cM HNO_3 and 7-cM NH_3OHCl (to keep An^{III} reduced). Then the $BiPO_4$ is dissolved in 4 to 6-M HCl.

One may also separate Am and Cm from Pu rather similarly (with $BiPO_4$) after oxidizing the Pu to PuO_2^{2+}. Oxalate and Cl^- must first be removed, and Fe^{III} and Cr^{III} interfere if present at 3 dM.

Triphosphate forms hydrated $Ln[P_3O_{10}]_2^{7-}$ and other polyphosphates.

Reduced chalcogens. Sulfane, H_2S, does not react with Rth^{III}. Alkali sulfides precipitate the hydroxides by hydrolysis.

Oxidized chalcogens. Excess thiosulfate forms, e.g., $Ln(S_2O_3)_3^{3-}$ and $Ln(S_2O_3)_4^{5-}$ from Ln^{3+}.

Sulfate complexes include $Ln(SO_4)^+$, $Ln(SO_4)_2^-$ and $Ln(SO_4)_3^{3-}$. Some stable solids are $(La,Ce)_2(SO_4)_3 \cdot 9H_2O$ and $Ytd_2(SO_4)_3 \cdot 8H_2O$. The solubilities of Ln sulfates vary inversely, somewhat unusually, with T.

Aqueous SO_4^{2-} and Na^+ precipitate crystalline double salts, namely $NaCed(SO_4)_2 \cdot nH_2O$ with the ceroids(III) but less with the yttroids(III), thus separating the subgroups fairly well. However, $AlkGd(SO_4)_2 \cdot H_2O$ (Alk ≠ Li) can be crystallized. Also, $K_3Ytd(SO_4)_3$, but not $K_3Ced(SO_4)_3$, dissolve in saturated K_2SO_4. The Rb^+ ion gives $RbYtd(SO_4)_2 \cdot H_2O$, and $(Rb,Cs,NH_4)Gd(SO_4)_2 \cdot 4H_2O$ can be crystallized from water.

Aqueous U^{3+} with SO_4^{2-} yields $U_2(SO_4)_3 \cdot 5H_2O$. Other sulfates may be crystallized as $An_2(SO_4)_3 \cdot 8H_2O$. Such salts as $(Alk,Tl^I)Am(SO_4)_2 \cdot nH_2O$, $K_3Am(SO_4)_3 \cdot nH_2O$ or $(Alk,Tl^I)_8Am_2(SO_4)_7$ are found after adding M^I to Am^{III} in solutions with H_2SO_4 (HSO_4^-). Sulfate even at only 1 dM forms $AnSO_4^+$ and $An(SO_4)_2^-$ with Am^{III}, Cm^{III} and Cf^{III}.

Aqueous sulfate gives, e.g., $Pu(SO_4)^+$ and $Pu(SO_4)_2^-$, and we can crystallize $M(U,Np,Pu)(SO_4)_2 \cdot nH_2O$ with M = Na, K, Rb, Cs, Tl or NH_4, plus such further complexes as $K_5(Np,Pu)(SO_4)_4 \cdot 4H_2O$.

Alkali-metal An^{3+} double sulfates dissolve sparingly, somewhat like those of Ln^{3+}, but with formulas for Am (precipitated from great excesses of the M^+ ion and sometimes with ethanol added) as $(K,Rb,Cs,Tl)Am(SO_4)_2 \cdot nH_2O$, K_3Am-

$(SO_4)_3 \cdot H_2O$ and $(K,Cs,Tl)_8Am_2(SO_4)_7$. The $M^{I}Am(SO_4)_2 \cdot nH_2O$ salts and Tl salts are much more soluble than the others (but still not highly so).

The crystallization of $K_3La(SO_4)_3$ may be used to separate small quantities of Np^{IV}, Pu^{IV} and Am^{III} species from other elements (including $An^{>IV}$) and also from each other by coprecipitating them. Having the K_2SO_4 at 19 cM best promotes coprecipitating the Pu^{IV} but not the Am^{III}. One may also use $K_8Pu_2(SO_4)_7$ as a carrier for transplutonium ions.

Ignited (< 500 °C) Am_2O_3 dissolves easily in concentrated H_2SO_4.

Insoluble AnF_3 become soluble sulfates on evaporation with H_2SO_4.

Anomalous mixed crystals of Am^{III} coprecipitate with K_2SO_4.

With $[S_2O_8]^{2-}$ in acidic solution, Ce^{3+} is nearly alone among the Ln^{3+} in readily yielding the Ln^{IV}, i.e., yellow-orange Ce^{IV}.

Aqueous Am^{3+} plus cold $[S_2O_8]^{2-}$ form AmO_2^{+}, which dismutates in HNO_3 and $HClO_4$ (so that Am^{VI} can be made this way), and is reduced by its own α-decay products. It may be oxidized completely to AmO_2^{2+} in 5-cM to 2-dM HNO_3 (but incompletely at higher acidities), e.g., with > 5-mM $[S_2O_8]^{2-}$ and 3-cM Ag^{+} (catalyst) and heating to 85–100 °C for 5–10 min. Without Ag^{+} in 3-cM to 1-dM HNO_3 and with 2-dM $[S_2O_8]^{2-}$ it takes several h, or by heating at 85–95 °C for 15–20 min, it is more than 99 % complete. The Am^{III} hydroxide plus $[S_2O_8]^{2-}$ at high pH form $Am(OH)_4 \cdot aq$.

Heating $Am(OH)_3 \cdot aq$ with $S_2O_8^{2-}$ in 7-M OH^- at 90 °C may form black $Am(OH)_4$. Treating Am^{III} and Rb_2CO_3 or "$(NH_4)_2CO_3$" with $[S_2O_8]^{2-}$ produces Am^{V} in $RbAmO_2CO_3$ or $NH_4AmO_2CO_3$. A large excess of K_2CO_3 instead gives $K_3AmO_2(CO_3)_2$ or $K_5[AmO_2(CO_3)_3]$. More specifically, Am^{III} is oxidized and precipitated as $K_5[AmO_2(CO_3)_3]$ by treatment in concentrated K_2CO_3 with 1-dM $[S_2O_8]^{2-}$ for 2 h at 75–80 °C. In OH^- or < 5-dM H_3O^+, $S_2O_8^{2-}$ forms Am^{VI}. The Am^{III} and Cm^{III} can be separated by oxidizing Am^{III} to Am^{VI} and then precipitating CmF_3.

With $[S_2O_8]^{2-}$ and, e.g., $[P_2W_{17}O_{61}]^{10-}$ or a phosphotungstate, Cm^{III} may become an unstable (to reduction by H_2O), red Cm^{IV} complex. Also oxidized by $[S_2O_8]^{2-}$, more easily, is Bk^{III}.

Reduced halogens. The low stabilities of some Rth^{3+} and An^{3+} complexes are: $Cl^- > Br^- > I^- > ClO_4^-$, and usually An^{3+} (slightly) > Rth^{3+} (the **5f** orbitals are exposed more than the **4f**). The stronger complexing for An^{3+} in HCl-ethanol allows a group separation from Ln^{3+}. Ion exchange and organic extractants, with $c(HCl) > 8M$, form $AnCl_4^-$.

Concentrated HCl and Sc^{III} give $[ScCl_4(H_2O)_2]^-$.

Chlorides and bromides are crystallized with such formulas as $[(Ln,An)X_2(H_2O)_6]X$, $NH_4[UCl_4(H_2O)_4]$, $(Rb,NH_4)UCl_4 \cdot 5H_2O$.

Ignited (< 500 °C) Am_2O_3 (from the oxalate) dissolves slowly in dilute HCl. Concentrated HCl containing Am^{III} and Cs^{+}, cooled to 0 °C and saturated with HCl, yields yellow, deliquescent $CsAmCl_4 \cdot 4H_2O$.

Elemental and oxidized halogens. Chlorine, Np^{III} and 1-M HCl at 75 °C form NpO_2^{2+}; the same results from BrO_3^- or fuming $HClO_4$, and the latter likewise oxidizes $Pu^{<VI}$ to PuO_2^{2+}.

Lanthanum trihydroxide adsorbs iodine, somewhat as does starch. The blue color disappears on adding an acid or base.

Among the Ln^{3+} ions, HClO oxidizes only Ce^{3+} to Ln^{IV}; then one can precipitate and separate CeO_2 or $Ce(IO_3)_4$ from the others.

Treating $Am(OH)_3 \cdot aq$ with ClO^- at high pH gives $Am(OH)_4 \cdot aq$; in fact, heating with 2-dM OH^- and 2 to 6-dM ClO^- forms black $Am(OH)_4$. At 95 to 100 °C, Am^{III} precipitates as $K_5[Am^VO_2(CO_3)_3]$ on treatment in concentrated K_2CO_3 with 1-dM ClO^- for 10–15 min.

Treating Bk^{III} with BrO_3^- in 2-M H_2SO_4 at 90 °C for 3 min, or in 8–10 M HNO_3, converts Bk^{3+} to Bk^{4+}, which resembles Ce^{4+} in oxidizing power.

Aqueous IO_3^- precipitates Ln^{III} iodates, readily soluble in HNO_3.

Iodic acid oxidizes Am^{III} ions incompletely to Am^{VI}.

Xenon species. Aqueous XeO_3 oxidizes Pu^{3+} at least to Pu^{4+}.

The Am^{3+} ion, plus 3-cM hydronated (added H^+) XeO_6^{4-} in 2-dM HNO_3 or $HClO_4$ with some catalytic Ag^+, form AmO_2^{2+}, 99 % in 30 s; Bk^{3+}, but not Cm^{3+} ($\mathbf{5f^7}$) or Cf^{3+}, is oxidized, to Bk^{4+}.

The salt $Am^{III}{}_4(XeO_6)_3 \cdot 40H_2O$ has been isolated.

3.3.4 Reagents Derived from the Metals Lithium through Uranium, plus Electrons and Photons

Oxidation. With acids, Ce^{IV}, MnO_4^- or Ag^{II} oxidizes Np^{III} to NpO_2^{2+}. Aqueous Am^{3+} and Ce^{IV} or Ag^{II} in HNO_3 or $HClO_4$ forms AmO_2^{2+}.

Uranium(3+) reduces $[Cr(NCS)_6]^{3-}$ etc., and becomes U^{IV}.

Berkelium(3+) goes to Bk^{IV} when heated with 2-dM $[Cr_2O_7]^{2-}$ in 1-M HNO_3 or 5-dM H_2SO_4, or when treated with 8-cM CrO_3 in 4-M HNO_3.

Among the Ln^{3+} ions, MnO_4^- oxidizes only Ce^{3+} to the Ln^{IV}; then one can precipitate and separate CeO_2 or $Ce(IO_3)_4$ from the others.

Uranium(3+) goes at least to U^{4+} with, e.g., $[Co(NH_3)_4(H_2O)_2]^{3+}$.

Cerium(3+) is readily oxidized to Ce^{IV} by PbO_2 and acid.

Lead dioxide or $NaBiO_3$ oxidizes Bk^{3+} to Bk^{IV}.

In 2 to 5.5-M CO_3^{2-} and 1-M OH^-, anodes oxidize Ce, Pr or Tb trichlorides to stable, yellow Ce^{IV}, yellow Pr^{IV} or dark red-brown Tb^{IV}.

Anodes can oxidize Am^{III} (but apparently not Cm^{III}), to Am^{VI} in concentrated CO_3^{2-}, to Am^{IV} in 12-M H_3PO_4, or to Am^V in IO_3^-.

Reduction. The Ln^{3+} and An^{3+} ions that are reducible to Ln^{2+} and An^{2+} (even if not aqueous) also form amalgams easily (like Ae^{2+}), i.e., with M = Sm, Eu, Yb, Am, Md, No:

$$M^{3+} + 3\,Na_{Hg} \rightarrow M_{Hg} + 3\,Na^+$$

The other ceroids form amalgams much less easily, and the other yttroids still less. Also:

$$Eu^{3+} + Na_{Hg} + SO_4^{2-} \rightarrow EuSO_4\downarrow + Na^+$$

The amalgams are easily re-oxidized:

$$2\,Ln_{Hg} + {}^3/_2\,O_2 + 6\,H_3O^+ \rightarrow 2\,Ln^{3+} + 9\,H_2O$$

When Na_{Hg} reduces Eu^{3+} but not La^{3+}, presumably to $Ln^{2+}(e^-)_2$ in an amalgam, Md^{2+} goes along, evidence for that oxidation state; Eu^{2+}, Yb^{2+}, V^{2+}, Cr^{2+}, Zn and Zn_{Hg}, but not Ti^{3+}, also reduce Md^{3+} to Md^{2+}.

The Sm^{2+}, V^{2+} and Cr^{2+} ions do not reduce Lr^{3+}.

With Zn_{Hg}, only Eu^{3+} forms a rather stable Ln^{2+} in H_2O (and separable as $EuSO_4$) although the unstable Sm^{2+}, Tm^{2+} and Yb^{2+} are also known.

A mercury cathode reduces, e.g., Yb^{3+} to Yb^{2+}.

Light (UV) can reduce Eu^{3+}, forming radicals (cf. **3.2.4 Oxidation**):

$$Eu^{3+} + 2\,H_2O + \gamma \rightarrow Eu^{2+} + OH^{\bullet} + H_3O^+$$

Other reactions. Mixing the corresponding amounts of Y^{III} or Ln^{III} with $[cis\text{-}(Cr^{III},Co^{III})(OH)_2(NH_3)_4]^+$ and large excesses of Br^- or I^- may give $[Rth\{(Cr,Co)\,(NH_3)_4\}_2(\mu\text{-}OH)_2]X_7\cdot nH_2O$.

The octacoordinate "lacunary" (defect, monovacant) complexes of all lanthanoids, $(NH_4)_{11}[Ln(PMo_{11}O_{39})_2]\cdot nH_2O$, can be made by starting with $H_3[PMo_{12}O_{40}]$ as follows: Dissolve 1 mol in water, and add Li_2CO_3 to raise the pH to 4.3 and make $[PMo_{11}O_{39}]^{7-}$ the dominant species. Add half as much, 5 dmol, of Ln^{3+}, and enough Li_2CO_3 to restore the pH to 4.3. After 1h add EtOH slowly and store for a few days at 5 °C. Every final anion, for La through Lu, is isostructural! Complexes of $W_{10}O_{36}{}^{12-}$, $PW_{11}O_{39}{}^{7-}$, $SiW_{11}O_{39}{}^{8-}$ and so on with Am^{III} and Cm^{III} are rather stable.

Potentiometry uses $[Fe(CN)_6]^{4-}$ to precipitate Rth^{3+}. Also precipitated by $[Fe(CN)_6]^{4-}$ are $U^{III}H[Fe^{II}(CN)_6]\cdot(9\text{–}10)H_2O$ and black (due to charge transfer likely involving Pu^{IV} and Fe^{II}) $Pu^{III}[Fe^{III}(CN)_6]\cdot{\sim}7H_2O$. Titration of excess $[Fe(CN)_6]^{4-}$ by Ce^{IV} also determines Rth^{3+}.

Aqueous Ytd^{3+} and TcO_4^- or ReO_4^- crystallize as $Ytd(MO_4)_3\cdot 4H_2O$.

The ions $[(Ag,Au)(CN)_2]^-$ form $La[(Ag,Au)(CN)_2]_3\cdot 3H_2O$ crystals.

The main standard stepwise electrode potentials of Pu (Pu^{III}-Pu^{VI}) are nearly equal (see Appendix C, Table C.14) so that Pu^{III}, Pu^{IV}, Pu^{V} and Pu^{VI} easily occur together, albeit with little Pu^{V} at equilibrium, and it is often difficult to achieve a pure solution of any one of them. Natural waters low in organics, however, may have Pu^V as the dominant species.

The pale colors of Ln^{3+} show some similarity between those with electron structures [Xe]**4f**^{n} and those with [Xe]**4f**$^{14-n}$, but the explanation is beyond our scope. The **5f** electrons in An^{3+} are more exposed, making their colors more intense and less patterned; moreover, some are of course less well known. See Table 3.4.

Table 3.4. The colors of Ln^{3+} and An^{3+}, necessarily omitting nuances

n	**4f**^{n}			**4f**$^{14-n}$	**5f**^{n}			**5f**$^{14-n}$
0	$_{57}$La	none	none	$_{71}$Lu	$_{89}$Ac	none		$_{103}$Lr
1	$_{58}$Ce	none	none	$_{70}$Yb	$_{90}$Th	dp blue		$_{102}$No
2	$_{59}$Pr	yl-grn	lt grn	$_{69}$Tm	$_{91}$Pa	dk blue		$_{101}$Md
3	$_{60}$Nd	violet	pink	$_{68}$Er	$_{92}$U	red-prp		$_{100}$Fm
4	$_{61}$Pm	pink	yl-pink	$_{67}$Ho	$_{93}$Np	purple	lt pink	$_{99}$Es
5	$_{62}$Sm	dk yl	yellow	$_{66}$Dy	$_{94}$Pu	blue	green	$_{98}$Cf
6	$_{63}$Eu	lt pink	lt pink	$_{65}$Tb	$_{95}$Am	yl-pink	yl-grn	$_{97}$Bk
7	$_{64}$Gd	none	none	$_{64}$Gd	$_{96}$Cm	~ none	~ none	$_{96}$Cm

Abbreviations: dk, dark; dp, deep; grn, green; lt, light; prp, purple; yl, yellow; ~ none, pale yellow-green.

3.4 The Lanthanoids Ln(IV) and Actinoids An(IV)

The important ores pitchblende and/or uraninite are variously formulated as UO_2 or U_2O_5 up to the dark green oxide U_3O_8, also more realistically in some cases as $Ae^{2+}{}_jRth^{3+}{}_{2k}Th^{4+}{}_lU^{4+}{}_mU^{6+}{}_nO^{2-}{}_{j+3k+2l+2m+3n}$, $2m \geq n$, or $Ae^{2+}{}_jRth^{3+}{}_{2k}Th^{4+}{}_lU^{5+}{}_{2m}U^{6+}{}_nO^{2-}{}_{j+3k+2l+5m+3n}$, $m \geq 2n$. The instability of aqueous $UO_2{}^+$ (below) does not discredit any evidence for U_3O_8 as $U^V{}_2U^{VI}O_8$, but the ores may be mentioned here under both U^{IV} and U^V.

3.4.1 Reagents Derived from Hydrogen and Oxygen

Dihydrogen. Neptunium(>III) plus H_2 on Pt give Np^{III}, stable in H_2O.

Water. The most common thorium salt, $Th(NO_3)_4 \cdot 5H_2O$, and the chloride are soluble (forming $[ThCl_2(H_2O)_n]^{2+}$). Anhydrous $Th(SO_4)_2$ is soluble in ice water, but it separates as a hydrate on heating. If the solution is allowed to stand without boiling, a series of hydrates will separate, their compositions depending on conditions. This behavior can separate Th^{IV} quantitatively at 0 °C from the soluble Rth sulfates.

The ligancies of hydrated An^{4+} ions seem to vary from $[Th(H_2O)_{11}]^{4+}$ down to $[An(H_2O)_8]^{4+}$. Some only slightly soluble salts include $ThOCO_3 \cdot 8H_2O$, $ThF_4 \cdot 4H_2O$, $K_2[ThF_6] \cdot 4H_2O$, $K_4[Th(SO_4)_4] \cdot H_2O$ and $Th[Fe(CN)_6]$. The phosphates and iodates of M^{4+} are insoluble.

The ions Ce^{IV} (slowly but catalyzed by RuO_2 in 5-dM H_2SO_4, or by MnO_2 or $Co_2O_3 \cdot aq$), Pr^{IV}, Nd^{IV}, Tb^{IV} and Dy^{IV} oxidize H_2O to O_2, going to Ln^{III}. Water also reduces Am^{IV}, especially if warm and in high $c(H_3O^+)$, likewise Cm^{IV} in any case if not strongly complexed, giving An^{III}.

The $An(H_2O)_n{}^{4+}$ ions hydrolyze in the order $U^{4+} > Np^{4+} < Pu^{4+}$ even at a pH as low as 0 for the smaller (higher-Z) ions, or more than 1 for U^{4+}. Aqueous Th^{4+} goes to $Th(OH)_n{}^{(4-n)+}$ with $Th < 1$ mM, otherwise to polymers. Below pH 6 these include $Th_4(OH)_8{}^{8+}$, $Th_4(OH)_{12}{}^{4+}$, $Th_6(OH)_{14}{}^{10+}$ and $Th_6(OH)_{15}{}^{9+}$, also perhaps $U_6(OH)_{15}{}^{9+}$. The precipitation of hydroxides greatly complicates the study of the An^{IV} hydrolyses.

Natural waters may contain $Th(OH)_4$, $ThF_2{}^{2+}$, $Th(HPO_4)_2$, $Th(HPO_4)_3{}^{2-}$ or $Th(SO_4)_2$, and some hot natural waters may contain, e.g., $[Th(CO_3)_5]^{6-}$.

Uranium tetrafluoride, UF_4, can give hydrates such as $UF_4 \cdot H_2O$.

Water hydrolyzes U^{4+} only slightly in 1-M H_3O^+ and does not oxidize it. A pH not much over 3 forms $U(OH)_3{}^+$ etc. Fairly acidic media let Pu^{IV} hydrolyze to colloidal polymers, irreversibly on aging.

Above pH 1, Pu^{4+} tends, retarded by $UO_2{}^{2+}$, to hydrolyze to a colloid. Above pH 4 it precipitates $Pu(OH)_4$, becoming quite insoluble on aging.

Heat, $Pu(SO_4)_2$ and H_2O precipitate $Pu(OH)_2(SO_4) \cdot 4H_2O$.

Without complexants, Am^{4+} and H_2O form Am^{III} and O_2.

Oxonium. Thorium hydroxide, when freshly precipitated, is readily soluble in acids but after drying is more resistant. The oxide ThO_2, ,e.g., from igniting $Th(OH)_4$, dissolves only in HF/HNO_3.

The salts $Th(C_2O_4)_2 \cdot 6H_2O$, ThF_4 and $Th_3(PO_4)_4 \cdot 4H_2O$ are insoluble even in high $c(H_3O^+)$. The uranium(IV) salts are similarly insoluble.

Little affected by the anions, H_3O^+ and $Am(OH)_4$ react as Am^{4+}:

$$2\,Am^{4+} + 6\,H_2O \rightarrow Am^{3+} + AmO_2{}^+ + 4\,H_3O^+$$

Chemiluminescence occurs on dissolving $Li_xCm^{IV}O_y$ in H_3O^+, with its reduction to Cm^{III}.

Hydroxide. Raising the pH on Ce^{4+} gives $CeOH^{3+}$, then polymers and yellow, gelatinous $CeO_2 \cdot aq$.

A preliminary separation of very dilute An^{IV} (from large amounts of less-acidic cations) may coprecipitate them as $An(OH)_4 \cdot aq$, e.g., with $ZrO_2 \cdot aq$ by OH^- (after adding a little Zr^{IV} carrier if needed).

Aqueous Th^{IV} forms an insoluble, gelatinous, white $Th(OH)_4 \cdot aq$ with OH^- at about pH 6 after, for example, $ThOH^{3+}$ and especially $Th_6(OH)_{15}{}^{9+}$. The precipitate is insoluble in excess but is not formed in the presence of chelators like tartrate (separation from yttrium). The basic salts $Th(OH)_2CrO_4 \cdot H_2O$ and $Th(OH)_2$ $SO_4 \cdot H_2O$ are known. The oxalato complexes also give $Th(OH)_4 \cdot aq$ with OH^-. Digesting $Th_3(PO_4)_4$ from ores (e.g., monazite) with OH^- (e.g., several h at 150 °C) yields $Th(OH)_4$, insoluble in HCl at pH 3–4 but $Ln(OH)_3$ dissolve as Ln^{3+}.

Aqueous OH^- gives with U^{IV} a pale green precipitate, nearly insoluble in excess reagent but giving some $U(OH)_5^-$ above pH 6, rapidly oxidized in air to a brown color. No precipitate is obtained with chelating organic hydroxy-acid anions or excess CO_3^{2-}. Likewise OH^- does not precipitate Pu^{IV} from carbonates below a pH of 11 or 12 without reduction to Pu^{III}, e.g., as $Pu_3(OH)_3(CO_3)_3 \cdot H_2O$.

Concentrated OH^- converts insoluble AnF_4 to acid-soluble $An(OH)_4$.

The amorphous $An(OH)_4 \cdot aq$ ($AnO_2 \cdot aq$) structures are poorly known.

Peroxide. Cerium(IV) is readily reduced to Ce^{III} by H_2O_2 in acid.

Especially on warming neutral or slightly acidified Th^{IV}, H_2O_2 precipitates a variable hydrated peroxide, used to confirm thorium, soluble in excess H_2SO_4. One product is $Th_6(O_2)_{10}(NO_3)_4 \cdot 10H_2O$.

Carbonate and O_2^{2-} dissolve $U^{<VI}$ minerals as CO_3^{2-}-UO_2^{2+} complexes.

Aqueous H_2O_2 in dilute H_3O^+ precipitates Np^{IV} or Pu^{IV} as $MO_4 \cdot aq$, apparently really $M^{IV}(O_2)_2 \cdot aq$, but also reduces Pu^{IV} to Pu^{3+}. A low $c(H_2O_2)$ forms $Pu_2(\mu\text{-}O_2)_2^{4+}$.

Alkaline peroxide and $Np^{<VII}$ or $Pu^{<VII}$ form Np^{VII} or Pu^{VII}; see **3.6.1**.

Solids are known containing $[Pu^{IV}{}_2(CO_3)_6(\mu\text{-}O_2)_2]^{8-}$, or what might be elaborated as $[\{Pu(\eta^2\text{-}CO_3)_3\}_2\{\mu\text{-}(1,2\text{-}\eta\text{:}1,2\text{-}\eta)\text{-}O_2\}_2]^{8-}$, with ligancy 10, i.e., having side-by-side O_2 bridges (in two Pu_2O_2 rhombi, bent at O–O).

Aqueous H_2O_2 easily reduces Bk^{IV} to Bk^{III} even in concentrated HNO_3.

Di- and trioxygen. Air oxidizes Pa^{IV} rapidly to Pa^V. It also oxidizes U^{IV} or Np^{IV}, but not Pu^{IV}, slowly to UO_2^{2+} or NpO_2^{2+}, although all U^{IV} carbonato complexes go easily to U^{VI} in the air. In 2-M CO_3^{2-} at pH over 11.7, Np^{IV} becomes Np^V.

Hot air (or H_2O_2, faster) helps isolate uranium from some ores, and $[Cu(NH_3)_4]^{2+}$ exemplifies the many redox catalysts for it:

$$UO_2 + {}^1/_2\, O_2 + CO_3^{2-} + 2\, HCO_3^- \rightarrow [UO_2(CO_3)_3]^{4-} + H_2O$$

Adding NaOH up to pH 11 then recovers $Na_2U_2O_7$.

Aqueous U^{IV}, Np^{IV} or Pu^{IV} plus O_3 form MO_2^{2+} and even, with concentrated OH^-, Np^{VII} or Pu^{VII} (likewise Am^{VII}); see **3.6.1 trioxygen.**

Ozone in 1-dM OH^- converts $Am(OH)_4$ to soluble, yellow Am^{VI}.

3.4.2 Reagents Derived from the Other 2nd-Period Non-Metals, Boron through Fluorine

Carbon oxide species. Cerium(IV) and Na_2CO_3 can yield $Na_6[Ce(\eta^2\text{-}CO_3)_5] \cdot 12H_2O$ with the ligancy (c. n.) 10.

Thorium(IV) and CO_3^{2-} precipitate a basic carbonate, readily soluble in concentrated, difficultly in dilute, CO_3^{2-}. The complex is decomposed and precipitated by OH^- but not by NH_3, F^- or PO_4^{3-}. Treating $Th(OH)_4$ with CO_2 or "$(NH_4)_2CO_3$" yields $ThOCO_3$ or $(NH_4)_2Th(CO_3)_3$ in turn.

Alkali carbonates or bicarbonates give pale-green $U(OH)_4$ with U^{IV}. The precipitate is soluble in HCO_3^- or "$(NH_4)_2CO_3$" and reprecipitated on boiling and destroying the excess reagent. Barium carbonate completely precipitates both Th^{IV} and U^{IV} even in the cold.

Solutions of U^{IV} in $KHCO_3$ or "$(NH_4)_2CO_3$", treated with $C_2H_4(NH_3^+)_2$, are found to precipitate $C_2H_4(NH_3)_2[U(CO_3)_3(H_2O)]\cdot 2H_2O$. We may also form the guanidinium salt $[C(NH_2)_3]_4[U(CO_3)_4]$. Better known are the pentacarbonato salts of Th and U: $M_6[An(CO_3)_5]\cdot aq$, where $M_6 = Na_6$, K_6, Tl_6, $[C(NH_2)_3]_6$, $[Co(NH_3)_6]_2$, etc. E.g., dissolving fresh $U(OH)_4\cdot aq$ in $KHCO_3$ yields $K_6[U(CO_3)_5]\cdot 6H_2O$. Or one may treat warm $U(SO_4)_2$ with $[C(NH_2)_3]_2CO_3$ and cool to get the guanidinium salt. Complexes of Pu^{IV} and Am^{IV} also go up to $[M(CO_3)_5]^{6-}$. In $M_6[An(CO_3)_5]\cdot aq$ generally the carbonate is didentate. In natural waters the predominant Th species will often be $[Th(CO_3)_5]^{6-}$, but Pu^{IV} is more likely hydrolyzed to colloidal $Pu(OH)_4$. (Uranium will be U^{VI} in those waters.) Whether the mixed $An(OH)_n(CO_3)_2^{n-}$ complexes predominate is often unclear.

Even higher Pu^{IV} complexes can arise from dissolving the oxalate in Na, K or NH_4 carbonates, giving $M^I_{(2n-4)}Pu(CO_3)_n\cdot aq$ with $n = 4$, 5, 6 or 8, although some of the carbonate may be uncomplexed in at least the 8-salt, and we therefore omit the brackets, [], that would indicate definite complexes. Aqueous $[Pu(CO_3)_5]^{6-}$ has 10-coordination. One finds various greenish, amorphous, water-soluble powders after treating the initial ethanol-produced oils with more ethanol or drying by heat.

More reactions than we can mention here give $AnO(CO_3)\cdot nH_2O$, $Th(OH)_2CO_3\cdot 2H_2O$, $An(CO_3)_2\cdot aq$, $Na[Th(OH)(\eta^2\text{–}CO_3)_2(H_2O)_3]\cdot 3H_2O$, $Alk_2[Th(OH)_2(\eta^2\text{–}CO_3)_2(H_2O)_2]\cdot nH_2O$, $(Na,NH_4)_2[U(H\text{–}\eta^2\text{–}CO_3)_2F_4]$, $(NH_4)_2[Th(\eta^2\text{–}CO_3)_3]\cdot H_2O$, $K_3[Th(OH)(\eta^2\text{–}CO_3)_3(H_2O)_2]\cdot 3H_2O$, $[C(NH_2)_3]_5[Th(\eta^2\text{-}CO_3)_3F_3]$, $(Alk,NH_4)_4[An(\eta^2\text{–}CO_3)_4]\cdot nH_2O$, $Na_5[Th(OH)(\eta^2\text{–}CO_3)_4(H_2O)]\cdot 8H_2O$, $(Alk,Tl,NH_4)_6[An(\eta^2\text{–}CO_3)_5]\cdot nH_2O$, $Ae_3[Th(\eta^2\text{–}CO_3)_5]\cdot 7H_2O$, $[Co(NH_3)_6]_2[An(\eta^2\text{–}CO_3)_5]\cdot (4,5)H_2O$, and $(Alk,NH_4)_8[Pu(\eta^2\text{–}CO_3)_6]\cdot nH_2O$. Also, the mineral tuliokite is found to be $Na_6Ba[Th(\eta^2\text{–}CO_3)_6]\cdot 6H_2O$.

Cyanide species. We also find $[An(NCS)_4(H_2O)_4]$, $Rb[Th(NCS)_5(H_2O)_3]$, $Na_2[Th(NCS)_5(OH)\cdot aq]$, $(NH_4)_3Th(NCS)_7\cdot 5H_2O$, $M_4[(Th,U)(NCS)_8]\cdot aq$ (cubic coordination! with $M = Alk$ or NH_4) and $(Et_4N)_4(U,Np)(NCS)_8$.

Some "simple" organic species. Alcohols etc. dissolve $Th(NO_3)_4\cdot 5H_2O$.

Boiled with $CH_3CO_2^-$, Th^{IV} precipitates a basic acetate, but other An^{IV} formates and acetates are too numerous even to summarize here.

Cerium(IV) is readily reduced to Ce^{III} by $C_2O_4^{2-}$.

Oxalic acid precipitates An^{IV} from an inorganic-acid solution as $An(C_2O_4)_2\cdot 6H_2O$ (distinction from Al and Be but not the rare earths):

$$An^{4+} + 2\,H_2C_2O_4 + 10\,H_2O \rightarrow An(C_2O_4)_2\cdot 6H_2O\downarrow + 4\,H_3O^+$$

This is practically insoluble in an excess of the cold dilute reagent and only slightly soluble in dilute inorganic acids, but soluble in warm, concentrated HCl. For Th^{IV} at least, the oxalate is readily soluble in a mixture of acetate anion and acetic acid (distinction from the rare earths), also soluble in a hot concentrated oxalate solution, forming, e.g., $Th(C_2O_4)_4^{4-}$, reprecipitated by H_3O^+. Igniting the oxalate gives the dioxide. Cooling the solution and adding ethanol produce a white salt, which breaks down somewhat in water:

$$Th(C_2O_4)_2\cdot 6H_2O + 2\,C_2O_4^{2-} + 4\,K^+ \rightarrow K_4Th(C_2O_4)_4\cdot 4H_2O\downarrow + 2\,H_2O$$

Excess $C_2O_4^{2-}$ converts U^{IV} to $U(C_2O_4)_4^{4-}$ and to, for example, light-green $K_4U(C_2O_4)_4\cdot 5H_2O$, precipitated by ethanol, alternatively to $K_2(Ca,Sr)U(C_2O_4)_4\cdot 8H_2O$ or $Ba_2U(C_2O_4)_4\cdot 9H_2O$. They all reduce Ag^+.

At 98 °C, $H_2C_2O_4$ in 5-M HNO_3 or $NaNO_3$ reduces Pu^{IV} to Pu^{III}.

Cupferron, $C_6H_5N_2O_2^-$, is interesting in precipitating U^{IV} but not UO_2^{2+}. Thereby we can isolate uranium(VI) in a mixture by first precipitating Ti, Zr, V, Fe etc. with the $PhN_2O_2^-$, from H_2SO_4 solution. After that separation the UO_2^{2+} can be reduced to U^{IV}, precipitated with more cupferron, and thus separated also from Cr, Mn, Al, P and so on.

Reduced nitrogen. Thorium(IV) forms insoluble, gelatinous, white $Th(OH)_4$ with NH_3. The precipitate is insoluble in excess of the reagent but is not formed in the presence of chelating organic hydroxy-acid anions (separation from yttrium).

The precipitation of $Zr(OH)_4\cdot aq$ etc. by NH_3 (after adding small amounts of the carrier compound if needed) may be used for the preliminary separation of very small quantities of An species (from large amounts of the less-acidic cations) by coprecipitating them.

Ammonia gives with U^{IV} a pale green precipitate of $U(OH)_4$, insoluble in excess and rapidly oxidized in the air, changing to a brown color. No precipitate is obtained in the presence of, e.g., tartrate or excess CO_3^{2-}.

Ammonia and Bk^{IV} precipitate $Bk(OH)_4\cdot aq$.

Aqueous Ce^{4+} oxidizes $N_2H_5^+$ to the radical ion $N_2H_4^{\bullet +}$, which reduces Fe^{3+} and Cu^{2+} for example.

Plutonium(>III) and $N_2H_5^+$ or NH_3OH^+ form Pu^{3+}, with conveniently gaseous byproducts. This allows separation from the unaffected UO_2^{2+}.

Thorium(4+) and HN_3 yield a strongly reducing amber complex. However, boiling N_3^- with Th^{IV} precipitates $Th(OH)_4$. The test is distinctive in the absence of unreduced Ce^{IV} and any other M^{IV}.

Oxidized nitrogen. Nitrous acid readily reduces Ce^{IV} to Ce^{3+}. It quickly oxidizes U^{IV}, however, to UO_2^{2+}, releasing NO.

Cerium(IV) in $(NH_4)_2[Ce(\eta^2\text{-}NO_3)_6]$ is a standard volumetric oxidant.

Nitric acid does not dissolve $Ce(IO_3)_4$, and dissolves ThO_2 poorly.

Uraninite, $\sim UO_{\geq 2}$, dissolves in HNO_3 and aqua regia slowly, and U^{4+} reacts with HNO_3 forming HNO_2 slowly, both resulting in UO_2^{2+}.

Solids, from dissolving hydroxides or carbonates in higher or lower $c(HNO_3)$, include $[Th(\eta^2\text{-}NO_3)_4(H_2O)_4]$ and $[Th(\eta^2\text{-}NO_3)_6]^{2-}$ with ligancy (c.n.) 12 and $[Th(\eta^2\text{-}NO_3)_4(H_2O)_3]\cdot 2H_2O$ with ligancy 11 also in, e.g., $AlkTh(NO_3)_5\cdot nH_2O$ and $(Alk,Tl,NH_4)_2[An(NO_3)_6]$. The following hexanitrato-complexes are derived from 8 to 14-M HNO_3, with sulfamic acid added to prevent the oxidation of any uranium(IV) by any HNO_2, i.e., $[(Mg,$**3d**$)(H_2O)_6][An(NO_3)_6]\cdot 2H_2O$ plus $K_3H_3(Th,U)(NO_3)_{10}\cdot nH_2O$ etc. Partial hydrolysis gives $[\{Th(NO_3)_3(H_2O)_3\}_2(\mu\text{-}OH)_2]\cdot 2H_2O$, ligancy 11.

Concentrated HNO_3 oxidizes $U(C_2O_4)_2\cdot 6H_2O$ to UO_2^{2+}.

In 3-M HNO_3 Pu^{IV} is mainly $Pu(NO_3)_2^{2+}$; in $HNO_3 > 10$ M it has $[Pu(NO_3)_6]^{2-}$, like Ce^{IV}, Th^{IV} etc., hence $(Rb,Cs,NH_4)_2[Pu(NO_3)_6]\cdot 2H_2O$, also, from 16-M HNO_3, green $[Pu(\eta^2\text{-}NO_3)_4(H_2O)_3]\cdot 2H_2O$, ligancy 11.

Warm 3-dM HNO_3 and Pu^{4+} form PuO_2^{2+}, apparently via the Pu^{4+} dismutation to PuO_2^{2+} and Pu^{3+}, which is then oxidized back to Pu^{4+} etc.

Hot HNO_3 with Np^{IV} gives HNO_2 and NpO_2^+, which is stable in neutral solution but dismutates slowly at low pH.

Fluorine species. Adding F^- to Ce^{IV} precipitates $CeF_4\cdot H_2O$. Concentrated NH_4F yields $(NH_4)_4[CeF_8]$, or $(NH_4)_6[Ce_2F_{14}]\cdot 2H_2O$ at lower $c(NH_4F)$; this becomes $(NH_4)_2CeF_6$ and $(NH_4)_4[CeF_8]$ if dried.

Adding HF or F^- to dissolved Th^{IV} precipitates a bulky white ThF_4, insoluble separately in excess fluoride or strong acid (separation from Be, Ti, Zr and Al), but fluoride helps HNO_3 dissolve ThO_2, and a mixture containing 5-cM HF, 1-dM $Al(NO_3)_3$ (to buffer F^-) and 13-M HNO_3 can be used to dissolve ThO_2/UO_2 fuel, giving Th^{IV}, Pa^V and U^{VI}.

Uranium(IV) and HF or F^- precipitate green UF_4, $UF_4\cdot H_2O$ or UOF_2. In air slowly, or 16-M HNO_3 vigorously, this all dissolves as UO_2^{2+}. The tetrafluoride dissolves little in dilute H_3O^+; hot OH^- forms black UO_2.

Plutonium complexes such as PuF^{3+} and PuF_2^{2+} are quite stable, and other stabilities vary as $Th^{4+} < U^{4+} > Np^{4+} \geq Pu^{4+}$.

Concentrated alkali fluorides dissolve fresh $Am(OH)_4\cdot aq$ and CmF_4 as intensely colored An^{IV} complexes, and Am^{IV} in 13-M NH_4F or concentrated RbF precipitates $(NH_4)_4[AmF_8]$ or Rb_2AmF_6, although $Am(OH)_4$ dissolves in 13-M NH_4F at 25 °C only up to 2 cM. In solution, Am^{IV} persists generally only with strong complexers (see, e.g., phosphates just below and polytungstates in **3.4.4 Other reactions**). The fluoro-complexes do not dismutate even on heating to 90 °C. However, O_3 oxidizes them to Am^{VI}, and I^- reduces them to Am^{III}.

The fluoride CmF_4 in 15-M AlkF forms a Cm^{IV} complex. This is stable for 1 h at ambient *T*, but oxidizes H_2O. The An^{IV} ions form various further complexes, including $[An_6F_{31}]^{7-}$.

3.4.3 Reagents Derived from the 3rd-to-5th-Period Non-Metals, Silicon through Xenon

Phosphorus species. Phosphate precipitates Ce^{IV} as a phosphate. Adding phosphate or H_3PO_4 to Th^{IV} produces a gelatinous precipitate of $Th_3(PO_4)_4 \cdot 4H_2O$, insoluble even in strong acids except hot, concentrated H_2SO_4. Uranium(IV) or Pu^{IV}, and H_3PO_4, precipitate $An(HPO_4)_2 \cdot nH_2O$. Also precipitable are $Pu_3(PO_4)_4 \cdot nH_2O$, $Pu_2H(PO_4)_3 \cdot nH_2O$, $Cu^{I}Th_2(PO_4)_3$, $Pb_{1/2}Th_2(PO_4)_3$, $An[P_2O_7]$ such as $Th[P_2O_7]$ at pH 1, and $[An(PO_3)_4]_n$ (the metaphosphates). Ferroelectricity appears in (Na,K) $Th_2(PO_4)_3$.

The precipitation of $BiPO_4$ may be used for the preliminary concentration of small quantities of Np^{IV}, Pu^{IV} and Bk^{IV} by coprecipitation. Sulfuric acid keeps UO_2^{2+} complexed and dissolved. Zirconium phosphate can also carry Pu^{IV} etc. into its precipitate.

The many phosphate complexes of An^{IV} include, for example, $An(H_3PO_4)_x(H_2PO_4)_y(HPO_4)_z^{(y+2z-4)-}$, with high acidity naturally favoring high *x*, and high basicity high *z*. Concentrated H_3PO_4 stabilizes even Am^{IV} against the otherwise easy reduction to Am^{III}. Also, $[P_2O_7]^{4-}$ stabilizes Am^{IV}. Even Tb^{IV} is stabilized in $[P_3O_{10}]^{5-}$, but can then oxidize Ce^{3+} or Mn^{2+} in acid. Aqueous $[P_2O_7]^{4-}$ precipitates and separates Th^{IV} from all the Rth^{III} ions in approximately a 1.5-dM sulfate or 3-dM chloride solution. The insolubility of this $Th[P_2O_7]$ in dilute acid provides an excellent quantitative separation of Th^{IV} from cerium (reduced from Ce^{IV} to Ce^{III}) and determination of the thorium.

Reduced chalcogens. Sulfane, H_2S, reduces Ce^{IV} to Ce^{III}.

Sulfides in acid do not affect Th^{IV} or U^{IV}. Alkaline sulfides give $Th(OH)_4$. "Ammonium sulfide" forms, with U^{IV} in neutral solution, a pale green, rapidly darkened precipitate.

Oxidized chalcogens. Boiling $S_2O_3^{2-}$ and Th^{IV} precipitates $Th(OH)_4$ and sulfur (distinction from Ce).

Cerium(IV) is readily reduced to Ce^{III} by SO_2.

Plutonium(> III) and SO_2 form Pu^{III}, stable to H_2O and O_2. Sulfite solids include $Th(SO_3)_2 \cdot 4H_2O$, and $(Alk,NH_4)_2Th(SO_3)_3 \cdot nH_2O$, also $(Na,NH_4)_4Th(SO_3)_4 \cdot nH_2O$, $Na_{2n}U(SO_3)_{n+2} \cdot aq$ and mixed complexes.

From $Am^{>III}$, SO_2 yields stable Am^{3+}, a weak reductant.

Mixing the appropriate ions can precipitate, say, $An(SeO_3)_2 \cdot aq$, $Th(SeO_4)_2 \cdot 9H_2O$, or $An(TeO_3)_2 \cdot aq$, likewise either $ThO(TeO_4) \cdot nH_2O$ or $ThO(H_4TeO_6) \cdot (n\text{-}2)H_2O$.

Fairly concentrated, hot H_2SO_4 dissolves CeO_2 somewhat slowly, then concentrating this with alkali sulfates gives, e.g., $(NH_4)_4Ce(SO_4)_4 \cdot 2H_2O$.

Heating H_2SO_4 with aqueous Th^{IV} may precipitate a basic sulfate which will dissolve on cooling. Some simple sulfate hydrates, $Th(SO_4)_2 \cdot nH_2O$, $n = 8$ or 9, also crystallize at ambient *T*.

With Th^{IV} a saturated solution of K_2SO_4 forms an insoluble double salt that is not affected by an excess of the reagent but is dissolved by hot water (separation from the yttroids). The corresponding sodium and ammonium double salts are soluble in water and in SO_4^{2-} (distinction from the ceroids), and $Th(SO_4)_2$, $Th(SO_4)_3^{2-}$ and $Th(SO_4)_4^{4-}$ are known. The 0.03 mM solubility of $^7/_2K_2SO_4 \cdot Th(SO_4)_2$ in 3.5 dM K_2SO_4 separates it from the soluble Ln^{III} sulfates.

Thorium oxide, ThO_2, is insoluble in acids except hot, concentrated H_2SO_4. The thorium and other phosphates in monazite sand dissolve slowly in hot concentrated H_2SO_4. (Cold H_2O then allows removing residues of silica, rutile, zircon, etc., and H_2S can eliminate certain metals.) Careful neutralization reprecipitates the phosphate.

Uranium dioxide is difficultly soluble in H_2SO_4. In contrast to other U^{IV} salts, the sulfate, $U(SO_4)_2$, is fairly stable in air.

From Pu^{4+} and HSO_4^- arise $PuSO_4^{2+}$, $Pu(SO_4)_2 \cdot aq$ or $K_4Pu(SO_4)_4 \cdot 2H_2O$.

Sulfuric acid dissolves $Am(OH)_4$, very quickly forming the Am^{3+} and AmO_2^{2+} sulfato complexes; cf. **3.4.1 Oxonium.**

Insoluble AnF_4 become soluble sulfates on evaporation with H_2SO_4.

Anomalous mixed crystals of Pu^{IV} coprecipitate with K_2SO_4.

Sulfate ions complex An^{IV} firmly. Complexes include $An(SO_4)_n^{(2n-4)-}$, with the tetrasulfato dominating at $c(SO_4^{2-}) > 2$ dM.

Solid phases are found to include $[(Th,U,Pu)(SO_4)_2(H_2O)_4] \cdot 4H_2O$, $UOSO_4 \cdot 2H_2O$ from a pH of about 7, $(Alk,Tl^I,NH_4)_2[An(SO_4)_3] \cdot nH_2O$, $(Alk,NH_4)_4[An(SO_4)_4] \cdot nH_2O$ (green for Pu^{IV}), the pentasulfato $(Alk,NH_4)_6[An(SO_4)_5] \cdot nH_2O$, the hexasulfato $(NH_4)_8[An(SO_4)_6] \cdot nH_2O$, plus $Na_6[U_2(SO_4)_7] \cdot 4H_2O$ and mixed complexes with $C_2O_4^{2-}$ for example.

Reduced halogens. Complexes of Ce^{IV} include $[CeCl_6]^{2-}$.

Anion-exchange resins retain $AnCl_6^{2-}$ from 12-M HCl.

Thorium (mono)phosphates dissolve in HCl.

Uranium dioxide is difficultly soluble in HCl and HBr.

Adding CsCl in 6-M HCl to Pu^{IV} in 9-M HCl precipitates $Cs_2[PuCl_6]$, but Th^{IV} does not act similarly.

An interesting formula for a solid is $[UBr(H_2O)_8]Br_3 \cdot H_2O$.

Cerium(IV) is readily reduced to Ce^{III} by HI.

From Am(> III) I^- yields (chemically) stable Am^{3+}, a weak reductant.

Elemental and oxidized halogens. If ThO_2 is suspended in OH^- and the system saturated with Cl_2, no dissolution occurs (distinction from many other oxides but not cerium oxide). Chlorine, Np^{IV} and 1-M HCl at 75 °C form NpO_2^{2+}; BrO_3^- or fuming $HClO_4$ does the same.

Concentrated HCl with Cl_2 and a 20 % excess of Cs^+, saturated with HCl at –23 °C, dissolves $Bk(OH)_4·aq$ giving a red solution and, promptly, an orange-red precipitate of Cs_2BkCl_6.

Heating $Am(OH)_3·aq$ and ClO^- in 2-dM OH^-, 90 °C, forms $Am(OH)_4$.

Chlorate and H_2SO_4 dissolve U^{IV} ores as UO_2^{2+} sulfates.

Bromate, with $An^{<VI}$, gives UO_2^{2+}, NpO_2^{2+}, PuO_2^{2+} and AmO_2^{2+}.

The precipitation of $Ce(IO_3)_4$, with BrO_3^- as oxidant, may be used to separate small amounts of Bk^{IV} from other transplutonium elements by coprecipitation as $Bk(IO_3)_4$ in quite dilute HNO_3. This can also coprecipitate Th, Group 4, Mn, Ag, Sn^{IV}, Pb and Bi, but not Group 1, Group 2, Rth^{III}, An^{III}, U, Cr, Mo, Fe, Co, Ni, Cu, Group 12, etc.

Iodate precipitates $An(IO_3)_4$, even from 6-M HNO_3 but not H_2SO_4, thus separating them from other elements after reducing Ce^{IV}, perhaps by warming with H_2O_2 in acidic solution, to Ce^{III}.

Thorium perchlorate crystallizes as colorless $Th(ClO_4)_4·4H_2O$.

Periodate and Th^{IV} precipitate $ThHIO_6·5H_2O$.

The $Pu^{<VII}$ ions plus $H_2IO_6^{3-}$ and OH^- form Pu^{VII}; cf. **3.6.1 Peroxide.**

Xenon species. The $Pu^{<VII}$ ions react with XeO_3 and OH^-, or with XeO_6^{4-}, to form Pu^{VII}; again cf. **3.6.1 Peroxide.**

3.4.4 Reagents Derived from the Metals Lithium through Uranium, plus Electrons and Photons

Oxidation. With $An^{<VI}$, the oxidants Ce^{4+}, MnO_4^- or Ag^{II} give UO_2^{2+}, NpO_2^{2+}, PuO_2^{2+} and AmO_2^{2+}, for example:

$$2\,MnO_4^- + 5\,M^{4+} + 6\,H_2O \rightarrow 2\,Mn^{2+} + 4\,H_3O^+ + 5\,MO_2^{2+}$$

Leaching uranium from U^{IV} ores often best uses the concomitant, limited, Fe^{III}, with enough H_2SO_4 to prevent precipitation of phosphate, arsenate etc., and, say, ClO_3^- or MnO_2 to reoxidize the Fe^{II}, e.g.,:

$$UO_2 + 2\,FeSO_4^+ + (n - 2)\,HSO_4^- + (n - 2)\,H_2O \rightarrow$$

$$UO_2(SO_4)_n^{(2n-2)-} + 2\,Fe^{2+} + (n - 2)\,H_3O^+$$

Aqueous $[Fe(CN)_6]^{3-}$ reacts with U^{IV} to form $[Fe(CN)_6]^{4-}$ and UO_2^{2+}; these then give a red precipitate; see **3.6.4 Other reactions,** below.

Treating Np^{IV} with Fe^{III} yields Np^V and Fe^{2+}.

Uranium(IV) in acid precipitates metallic Ag, Au etc. from their solutions and goes to U^{VI}.

Reduction. Uranium(IV), e.g., from cathodic e^-, quickly reduces $Pu^{>III}$:

$$2\,Pu^{4+} + U^{4+} + 6\,H_2O \rightarrow 2\,Pu^{3+} + UO_2^{2+} + 4\,H_3O^+$$

Iron(2+) in acid readily reduces Ce^{IV} or Pu^{IV} to M^{3+}, thus allowing a separation of Pu^{3+}, e.g., by solvent extraction, from unaffected UO_2^{2+}.

In nearly 1-M HCF_3SO_3 Ru^{II} reduces Np^{IV} in an equilibrium, $n \geq 0$:

$$Np^{4+} + [Ru(NH_3)_{5+n}(H_2O)_{1-n}]^{2+} \leftrightarrows Np^{3+} + [Ru(NH_3)_{5+n}(H_2O)_{1-n}]^{3+}$$

Excess Ce^{IV} goes to Ce^{III}, oxidizing Ru^{IV} oxide in 5-dM H_2SO_4 completely to RuO_4, accompanied by some Ru-catalyzed release of O_2:

$$RuO_2 \cdot aq + 4\,Ce(SO_4)_x^{(4-2x)+} + 10\,H_2O \rightarrow$$

$$RuO_4 + 4\,CeSO_4^{+} + (4x - 4)\,HSO_4^{-} + 4x\,H_2O + (8 - 4x)\,H_3O^+$$

Neptunium(> III) plus Zn_{Hg} give Np^{3+}, stable in H_2O.

The very slow reduction of Ce^{IV} by Tl^+ is catalyzed by Ag^+; see **13.5.4 Oxidation.** Light (UV) also forms Ce^{III} and O_2 from Ce^{IV} and H_2O.

Other reactions. Traces of Np^{IV}, Pu^{III} and Pu^{IV} (and Ln^{III} and Th^{IV}) can be isolated by coprecipitation with LaF_3, MnO_2, $Fe(OH)_3 \cdot aq$ or $BiPO_4$.

Comparable to $M^ITh_2(PO_4)_3$ are $(K,Rb)Th_2(VO_4)_3$.

Dichromate or basic CrO_4^{2-} precipitates Th^{IV} as $Th(CrO_4)_2 \cdot (3,1)H_2O$ or $Th(OH)_2CrO_4 \cdot H_2O$ in turn. Here clarity with the name dichromate for both $K_2Cr_2O_7$ and a $Th(CrO_4)_2$ may require longer structural names such as μ-oxo-hexaoxodichromate(VI) and bis[tetraoxochromate(VI)].

A molybdate, Alk^+ and Th^{IV} yield $Alk_{2j}Th_k(MoO_4)_{j+2k}$, including a $K_8[Th(\eta^1\text{–}MoO_4)_4(\eta^2\text{–}MoO_4)_2]$ (ligancy 8), and a hydrated $Th(MoO_4)_2$.

The An^{4+} ions do complex, often strongly and as either 1:1 or 1:2, $[Nb_6O_{19}]^{8-}$, $[W_{10}O_{36}]^{12-}$, $[W_{12}O_{42}]^{12-}$, $[SiW_{12}O_{40}]^{4-}$, $[P_2W_{18}O_{62}]^{6-}$, $[P_5W_{30}O_{110}]^{15-}$ (encapsulating An^{4+}), $[NaP_5W_{30}O_{110}]^{14-}$, $[(B^{III},Si^{IV},P^V,As^V)W^{VI}{}_{11}O_{39}]^{n-}$ etc.

Americium(IV) is stabilized against reduction (except by its own radiolysis) in, e.g., $AmP_2W_{17}O_{61}{}^{6-}$. Related Cm^{IV} complexes are chemiluminescent during the reduction to Cm^{III} by H_2O. Even Cf^{IV} may perhaps be stabilized by phosphotungstates.

Treated with $[Fe(CN)_6]^{4-}$, Th^{IV} gives a white precipitate of $Th[Fe(CN)_6]$, a very sensitive test, in neutral or slightly acidic solutions. Aqueous $[Fe(CN)_6]^{4-}$ gives with U^{IV} a yellow-green precipitate, gradually being oxidized to red brown.

Uranium(IV) and $[Fe(CN)_6]^{4-}$ form $U[Fe(CN)_6] \cdot 6H_2O$; $[Ru(CN)_6]^{4-}$ or $[Os(CN)_6]^{4-}$ precipitate $U[M(CN)_6] \cdot 10H_2O$.

Low H_3O^+ favors dismutation for Pu^{4+} (catalyzed by UO_2^{2+} but stable in concentrated acid) or Am^{4+}:

$$3\,Pu^{4+} + 6\,H_2O \leftrightarrows 2\,Pu^{3+} + PuO_2^{2+} + 4\,H_3O^+$$

$$2\,Am^{4+} + 6\,H_2O \leftrightarrows Am^{3+} + AmO_2^{+} + 4\,H_3O^+$$

but AmO_2^+ then also dismutates to Am^{3+} and AmO_2^{2+}.

The colors of Ln^{4+} are: Ce $\mathbf{4f^0}$, yellow-orange; Pr $\mathbf{4f^1}$ yellow; Nd $\mathbf{4f^2}$, blue-violet; Tb $\mathbf{4f^7}$, red-brown; and Dy $\mathbf{4f^8}$, yellow-orange. The colors of An^{4+} are: Th $\mathbf{5f^0}$, none; Pa $\mathbf{5f^1}$, pale yellow; U $\mathbf{5f^2}$, green; Np $\mathbf{5f^3}$, yellow-green; Pu $\mathbf{5f^4}$, tan; Am $\mathbf{5f^5}$, orange; Cm $\mathbf{5f^6}$, pale yellow; Bk $\mathbf{5f^7}$, brown; Cf $\mathbf{5f^8}$, green; neither Ln^{4+} nor An^{4+} can show the pattern of Table 3.4, because H_2O is pulled in, interacting too strongly with the **f** electrons.

3.5 The Actinoids An(V)

3.5.1 Reagents Derived from Hydrogen and Oxygen

Water. The formulas for the simple An^V species, except Pa^V, in water are (linear) AnO_2^+, where An = U, Np, Pu or Am (unlike Nb or Ta). The apparent ionic charge of, e.g., PuO_2^+ felt by a ligated X^- is ~2.2+.

Protactinium(V) is hydrolyzed, much more strongly than the higher-Z, *smaller*, An^V ions; it forms $Pa_2O_5 \cdot aq$ and colloids that are adsorbed on containers and interfere greatly with its study; it may be $PaO(OH)^{2+}$, $Pa(OH)_3^{2+}$ or $PaO(OH)_2^+$, unlike the others (AnO_2^+) and with chemistry more like those of Nb^V and Ta^V than like those of other An^V.

Uranium pentafluoride reacts violently, giving UF_4 or UOF_2:

$$2\,UF_5 + 2\,H_2O \rightarrow UF_4\downarrow + UO_2^{2+} + 2\,F^- + 4\,HF$$

Hydrolysis makes UO_2^+, NpO_2^+, PuO_2^+ and AmO_2^+ especially as $[AnO_2(H_2O)_5]^+$ with linear AnO_2 from their AnX_5. The AnO_2^+ ions (excluding Pa^V) do not readily hydrolyze further at a pH < 7 or 9 (higher than for the other oxidation states). Then we get AnO_2OH, $AnO_2(OH)_2^-$ etc..

Some controversy may remain about Cf^V (~ stable $\mathbf{5f^7}$?) in OH^-.

Seawater is found now to contain Pu^V as ~ 10^{-14}-M PuO_2^+.

Oxonium. Uranium(V) dismutates rapidly but least at ~ pH 3:

$$2\,UO_2^+ + 3\,H_3O^+ \rightarrow UOH^{3+} + UO_2^{2+} + 4\,H_2O$$

High $c(H_3O^+)$ promote similar reactions of Np^V and Pu^V:

$$2\,MO_2^+ + 4\,H_3O^+ \leftrightarrows M^{4+} + MO_2^{2+} + 6\,H_2O$$

The NpO_2^+ ion, however, is stable from 1-dM to 2-M HNO_3 (but is fully oxidized >6 M HNO_3). Above ~0.01 μM, PuO_2^+ dismutates in neutral solution. The dismutation of AnO_2^+ is much faster for U and Pu (pH<2), with odd numbers of electrons, than for Np and Am, with even numbers. For Am it is complicated but ends as:

$$3\,AmO_2^+ + 4\,H_3O^+ \leftrightarrows 2\,AmO_2^{2+} + Am^{3+} + 6\,H_2O$$

This rate depends strongly on the pH, is also lowest at about 3, low in $HClO_4$ and high in H_2SO_4 (likewise for NpO_2^+ with H_2SO_4). Thus reductions of AmO_2^{2+} to AmO_2^+ finally give only Am^{3+}.

More easily than VO_2^+, NpO_2^+ in acid dimerizes between 1 dM and 1-M Np^V, but forms $Np_2O_4^{2+}$ and polymerizes further at >1 M.

Hydroxide. Aqueous OH^- precipitates a Pa^V hydroxide and apparently (Np,Pu, Am)$O_2OH\cdot aq$ from M^V. Yellow $AlkAmO_2(OH)_2\cdot aq$ is isolated from 1–5 dM AlkOH, and rose $Alk_2AmO_2(OH)_3\cdot aq$ from ~ 2-M AlkOH.

Peroxide. Alkaline peroxide and $Np^{<VII}$ or $Pu^{<VII}$ form Np^{VII} or Pu^{VII}; see **3.6.1**. Heating AmO_2^+ with acidified H_2O_2 yields Am^{III}.

Di- and trioxygen. Air oxidizes UO_2^+ and, with OH^-, Pu^V, to M^{VI}.

Ozone oxidizes: NpO_2^+ at pH 5 and 90 °C, or $NpO_2OH\cdot aq$, to $NpO_2(OH)_2\cdot aq$ or $NpO_3\cdot 2H_2O$; and AmO_2^+ in acids to AmO_2^{2+}, or, in HCO_3^- or CO_3^{2-}, to precipitable Am^{VI} complexes. The oxidations of UO_2^+, NpO_2^+ or PuO_2^+ by O_3 are rapid. Ozone with concentrated OH^- forms Np^{VII}, Pu^{VII} or Am^{VII} from $M^{<VII}$; see **3.6.1**.

3.5.2 Reagents Derived from the Other 2nd-Period Non-Metals, Boron through Fluorine

Carbon oxide species. Adding CO_3^{2-} or HCO_3^-, up to <1 dM, to NpO_2^+ (in dilute acid) precipitates $(Alk,NH_4)NpO_2CO_3\cdot nH_2O$ on standing. If 5 dM < c(carbonate) < 2 M, $(Alk,NH_4)_3NpO_2(CO_3)_2$ precipitates after some hours. [Other $M^I_3AnO_2(CO_3)_2$ are also known.] A great excess of carbonate produces $(Alk,NH_4)_5[NpO_2(CO_3)_3]$. Dissolving Np^V in basic carbonate produces, e.g., $[NpO_2(CO_3)_2(OH)]^{4-}$. Adding a solid alkali carbonate to PuO_2^+ (at pH=2), thus raising the pH to 7, forms $(K,NH_4)PuO_2CO_3\cdot nH_2O$. Another solid is $(NH_4)_2PuO_2(CO_3)(OH)\cdot nH_2O$.

Adding $KHCO_3$, to make it 1 dM, to AmO_2^+ and heating at 90 °C for 3–4h precipitates $KAmO_2CO_3$. The NH_4 and Rb salts are similar. A large excess of CO_3^{2-} produces, e.g., $K_3AmO_2(CO_3)_2$ or $K_5[AmO_2(CO_3)_3]$. These all dissolve in acids to give AmO_2^+. Small amounts of AmO_2^+ in CO_3^{2-} complexes may be separated from fission-product Rth^{III} and from Cm^{III} by coprecipitating the former with the UO_2^{2+} in $K_4[UO_2(CO_3)_3]$.

Groundwater with $E°$ below ~ 2 dV for Np and ~ 5 dV for Pu leaves them as insoluble An^{III} and An^{IV}; more oxidizing waters form mobile An^{V} and An^{VI} complexes of CO_3^{2-} and HCO_3^-.

The precipitation of tantalum(V) hydroxide may be used for the preliminary separation of very small quantities of Am^{V} (from large amounts of other An species) (formed, e.g., by ClO^-) by coprecipitating the Am^{V}. A mixture of the Am^{V} with Ta^{V} dissolved in carbonate is heated to precipitate the Ta^{V} hydroxide, entraining the Am^{V} also.

Cyanide species. One complex appears to be $Cs_4[NpO_2(NCS)_5]$.

Some "simple" organic species. Formates and acetates, both at least mostly didentate, appear to precipitate or crystallize $Cs_2[NpO_2(CHO_2)_3]$, $Cs_2[(Np,Pu,Am)O_2(CH_3CO_2)_3]$, $Na_4NpO_2(CH_3CO_2)_5$ and, for example, $Ba[NpO_2(CH_3CO_2)_3]\cdot 2H_2O$.

Such oxalates as, e.g., $Pa(C_2O_4)_2OH\cdot 6H_2O$, $NpO_2(HC_2O_4)\cdot 2H_2O$, $(NpO_2)_2C_2O_4\cdot H_2O$, $Alk(Np,Am)O_2C_2O_4\cdot nH_2O$, $Alk_3NpO_2(C_2O_4)_2\cdot nH_2O$ and even $Alk_5NpO_2(C_2O_4)_3\cdot nH_2O$ are precipitated from AnO_2^+ solutions.

Reduced nitrogen. Excess NH_3 precipitates Pa^{V} hydrous oxide even from 5-M HF. Aqueous Np^{V} gives a green hydroxide at pHs near 7, becoming grayish and less soluble on aging.

Treating NpO_2^+ in acid with $N_2H_5^+$, catalyzed by Fe^{III}, one gets Np^{4+} and N_2H_2, on to N_2 and $N_2H_5^+$, also to NH_4^+ and HN_3. With NH_3OH^+ one quickly finds Np^{4+} and N_2, or, with excess NpO_2^+, N_2O.

Useful reductions of PuO_2^+ also occur with $N_2H_5^+$ or NH_3OH^+, and of Am^{V} in OH^- with N_2H_5 or NH_2OH.

Oxidized nitrogen. Aqueous HNO_2 slowly reduces PuO_2^+ to $Pu^{\le IV}$.

Aqua regia (or HNO_3) easily oxidizes and dissolves the ~U_2O_5 ores as UO_2^{2+}, similarly with a mixture of concentrated CH_3CO_2H and HNO_3 (20v:1v) (distinction from ignited V_2O_5 and Fe_2O_3, which are insoluble).

In 1 to 12-M HNO_3, fresh Pa^{V} hydroxide dissolves as $Pa(NO_3)_i(OH)_j^{k+}$ with $1 \le i \le 4$, $1 \le j \le 3$, $-1 \le k \le 2$, and $k = 5 - i - j$. A high $c(HNO_3)$ gives solid $PaO(NO_3)_3\cdot nH_2O$ (from fuming HNO_3), $NpO_2NO_3\cdot nH_2O$ and $RbNpO_2(NO_3)_2\cdot H_2O$.

The NpO_2^+ ion is stable up to 2-M HNO_3, but becomes NpO_2^{2+} completely at > 6-M HNO_3, catalyzed by HNO_2.

Fluorine species. Some fluoro complexes are AnF_6^-, AnF_7^{2-} and AnF_8^{3-}, e.g., cubic PaF_8^{3-}. Also known are $PaF_5\cdot nH_2O$ and $NpOF_3\cdot 2H_2O$.

Acetone precipitates $(Alk,NH_4)_2PaF_7$ from 17-M HF and excess M^+.

Acidified (HNO_3) HF is often the best solvent for Pa^{V}, and it slowly dissolves most of the associated oxides in nature, e.g., of U, Ti, Zr, Nb, Ta, Mo, Fe and Si. Then, after one extracts U^{VI} with Bu_3PO_4, Al^{3+} can tie up all the F^- and precipitate Pa^{V} and other hydrous oxides and phosphates (or, Al^0 slowly precipitates Pa^0) and

OH^- then displaces phosphate and dissolves away nearly all but Pa, Zr and Fe; HF and further steps yield a Pa^V solution. Separating, e.g., fluoro-chloro complexes by anion exchange is effective even from the otherwise often especially troublesome Nb^V.

Although U^V is unstable in water, $Cs[UF_6]$ dissolves in liquid HF without change, but $Cs[NpF_6]$, from water-stable Np^V, dismutates in HF to NpF_4 and NpF_6, showing the importance of the solvent.

Dilute HF converts U_3O_8 to solid UF_4 and dissolved yellow UO_2F_2.

At pH 6, RbF or NH_4F plus PuO_2^+ give $(Rb,NH_4)PuO_2F_2$. Saturated KF or RbF plus AmO_2^+ in < 1-dM H_3O^+ precipitate white $AlkAmO_2F_2$. In acid, $RbAmO_2F_2$ is reduced after some hours, partly to Rb_2AmF_6.

3.5.3 Reagents Derived from the 3rd-to-5th-Period Non-Metals, Silicon through Xenon

Phosphorus species. Phosphates and M^V precipitate $PaO(H_2PO_4)_3 \cdot 2H_2O$, $NpO_2H_2PO_4$ and $NH_4PuO_2HPO_4 \cdot 4H_2O$, for example, and complexes such as $(Np,Pu)O_2(HPO_4)^-$ are found.

Oxidized chalcogens. Heating AmO_2^+ with $S_2O_3^{2-}$ yields Am^{III}. In OH^- at ambient *T*, $S_2O_4^{2-}$ or SO_3^{2-} slowly reduce Am^V at least to Am^{IV}.

With SO_2 in H_2SO_4, NpO_2^+ becomes Np^{IV}.

Some solids, crystallized by evaporation from M^V in HF with H_2SO_4, H_2SeO_4 or a sulfate salt, include $H_3PaO(SO_4)_3$, $H_3PaO(SeO_4)_3$, $[Co^{III}(NH_3)_6](Np,Am)O_2(SO_4)_2 \cdot nH_2O$ (without a PuO_2^+ salt either here or in the next examples), $(Na,K)_2[Co^{III}(NH_3)_6](Np,Am)O_2(SO_4)_3 \cdot nH_2O$, $[(Np,Am)O_2]_2SO_4 \cdot nH_2O$ and $CsAmO_2SO_4 \cdot nH_2O$.

Perhaps surprisingly, H_2SO_4/HF solutions precipitate $H_3PaO(SO_4)_3$ from Pa^V. Mixtures of H_2SeO_4 and HF precipitate $H_3PaO(SeO_4)_3$.

Aqueous $[S_2O_8]^{2-}$ oxidizes Am^V in $NaHCO_3$ to an Am^{VI} complex.

Reduced halogens. The oxide U_3O_8, i.e., $U^V{}_2U^{VI}O_8$, is difficultly soluble in HCl (distinction from V_2O_5).

The salt $NpO_2Cl \cdot H_2O$ may be formed from HCl.

Adding ethanol to AmO_2^+ and Cs^+ in 1-M or 6-M HCl gives light-yellow $Cs_3AmO_2Cl_4$, but Cl^- also reduces AmO_2^+ to Am^{III}.

At 100 °C, I^- and 5-M HCl reduce Np^V and Np^{VI} to Np^{IV}.

Elemental and oxidized halogens. Chlorine, NpO_2^+ and 1-M HCl at 75 °C form NpO_2^{2+}; BrO_3^- or fuming $HClO_4$ does the same.

One can crystallize $[NpO_2ClO_4(H_2O)_4]$ and $NpO_2IO_3 \cdot nH_2O$, for example, and one precipitated salt is $(NpO_2)_2[Co(NH_3)_6](IO_3)_5 \cdot 4H_2O$. Iodate dismutates Pu^V to Pu^{IV} and Pu^{VI}. Neptunium(< VII) reacts with $H_2IO_6^{3-}$ and OH^- to form Np^{VII}; see **3.6.1 Peroxide** for products.

Xenon species. The NpO_2^+ ion and other oxidation states react with XeO_3 and OH^- to form Np^{VII}; see **3.6.1 Peroxide** for some products.

3.5.4 Reagents Derived from the Metals Lithium through Uranium, plus Electrons and Photons

Oxidation. With acids, Ce^{IV}, MnO_4^- or Ag^{2+} oxidizes NpO_2^+ to NpO_2^{2+}, and PuO_2^+ to PuO_2^{2+}; likewise Ce^{IV} or Ag^{2+} oxidizes AmO_2^+ to AmO_2^{2+}.

Anodic treatment of Am^V in concentrated CO_3^{2-} can precipitate an Am^{VI} carbonate.

Reduction. Uranium(4+) reduces NpO_2^+ or PuO_2^+ to Np^{4+}, Pu^{4+} or Pu^{3+}.

In H_2SO_4, Fe^{2+} and Np^V or Np^{VI} forms Np^{IV}.

Heating AmO_2^+ with acidified Fe^{2+} yields Am^{III}.

Zinc, in Zn_{Hg}, reduces Pa^V to Pa^{IV}, a good reductant, stable in air-free H_2O, and possibly to Pa^{3+}.

Other reactions. Uranyl or M^{3+}, and NpO_2^+ but not AnO_2^{2+}, in $HClO_4$ yield $UNpO_4^{3+}$ or $MNpO_2^{4+}$ with $Fe^{3+} > In^{3+} > Sc^{3+} > Ga^{3+} > Al^{3+}$, also Cr^{3+} and Rh^{3+}, all found spectrophotometrically or isolated by ion exchange. For example, NpO_2^+ and $[Rh(H_2O)_6]^{3+}$ join weakly in an equilibrium favoring $[Rh(H_2O)_5(\mu\text{-}O)NpO]^{4+}$, $K = 3.3$ at 25 °C, catalyzed by F^-.

An incompletely formulated chromate appears on slowly evaporating Am^V with chromic acid.

Protactinium(V) occurs in $PaO(ReO_4)_3 \cdot nH_2O$.

The colors of AnO_2^+ are: Pa, none; U, pale purple; Np, green; Pu, red violet; and Am, yellow.

3.6 The Actinoids An(VI)

3.6.1 Reagents Derived from Hydrogen and Oxygen

Water. The cations are mainly $[AnO_2(H_2O)_5]^{2+}$ with linear AnO_2. Solids include $UO_3 \cdot 2H_2O$, $(\alpha,\beta,\gamma)UO_2(OH)_2$ and $U_3O_8(OH)_2$. The ordinary salts, e.g., $AnO_2(NO_3)_2 \cdot nH_2O$, are mostly soluble with little hydrolysis. The hydrolytic order $UO_2^{2+} > NpO_2^{2+} > PuO_2^{2+}$ opposes the order expected for decreasing size. The apparent ionic (central) charge of AnO_2^{2+} for a ligated X^- is ~3.3+. In 1-M $HClO_4$, the exchange of O between UO_2^{2+} and H_2O has a $t_{1/2}$ of 4 h.

Hydrolysis rapidly converts the readily soluble UF_6 and UCl_6 to UO_2F_2 and UO_2Cl_2, and PuF_6 to PuO_2F_2 and $PuOF_4$. Uranyl phosphate and ores such as carnotite, $KUO_2VO_4 \cdot {}^3/_2H_2O$, are practically insoluble.

Below a c of a few μM, UO_2^{2+} hydrolyzes first to $[UO_2(OH)(H_2O)_4]^+$. Above that, polymers predominate, especially $[(UO_2)_2(\mu\text{-}OH)_2]^{2+}$ with anions ligated, above a pH of 3, and $[(UO_2)_3(OH)_n]^{(n-6)-}$, especially with $n = 5, 7, 8$ or 10, above a pH of 5.

Mainly the α rays of the higher-Z AnO_2^{2+} form $H^{\bullet}$, $HO^{\bullet}$, $HO_2^{\bullet}$ and H_2O_2 from H_2O, mostly reducing them, e.g., to Am^{III}.

Natural waters may contain UO_2CO_3, $UO_2(CO_3)_2^{2-}$, $[UO_2(CO_3)_3]^{4-}$ (hot), U^{IV} fluorides, $UO_2(HPO_4)_2^{2-}$ and perhaps $UO_2HSiO_4^-$. Marine waters appear to contain Pu mainly as $PuO_2(OH)_2(HCO_3)^-$ and $PuO_2(OH)_2$ with some Pu^{III}, Pu^{IV}, Pu^{V} and $PuO_2(CO_3)_2^{2-}$.

Oxonium. If a solute (e.g., from some non-aqueous treatments of ores) containing sodium uranyl carbonate, is barely neutralized with acid, yellow $Na_2U_2O_7 \cdot 6H_2O$ separates. There are no discrete anions in, e.g., Na_2UO_4 or $Na_2U_2O_7$, unlike Na_2SO_4 or $Na_2[S_2O_7]$. Uranates are generally insoluble in H_2O but soluble in acids as UO_2^{2+} or complexes.

Hydroxide. Very many salts are known, formed from AlkOH, NH_3 or $Ae(OH)_2$, from or with anions like $AnO_2(OH)_3^-$, $AnO_2(OH)_4^{2-}$, $An_2O_7^{2-}$ or $An_8O_{25}^{2-}$, e.g., $Li_2U_3O_{10}$ [i.e., $Li_2(UO_2)_3O_4$], $Cs_2Np_3O_{10}$ and $Li_6(Np,Pu)O_6$. More specifically, NaOH, free from CO_3^{2-}, quantitatively precipitates UO_2^{2+} as the yellow salt, $\sim Na_2U_2O_7$, insoluble in excess reagent, readily soluble in "$(NH_4)_2CO_3$":

$$2\,UO_2^{2+} + 2\,Na^+ + 6\,OH^- \rightarrow Na_2U_2O_7\downarrow + 3\,H_2O$$

$$Na_2U_2O_7 + 6\,HCO_3^- \rightarrow 2\,[UO_2(CO_3)_3]^{4-} + 2\,Na^+ + 3\,H_2O$$

Tartrate and peroxide prevent precipitation. The other AnO_2^{2+} ions precipitate similar salts, e.g., at $pH > 13$ for Pu^{VI}. In leaching U^{VI} from ores with CO_3^{2-}, HCO_3^- lowers the pH to prevent this precipitation:

$$2\,[UO_2(CO_3)_3]^{4-} + 6\,OH^- + 2\,Na^+ \rightarrow Na_2U_2O_7\downarrow + 6\,CO_3^{2-} + 3\,H_2O$$

Uranyl hexacyanoferrate(II) dissolves in OH^- to a yellow solution.

In OH^-, Am^{VI} is slowly reduced to a light-tan product that dissolves in H_3O^+ as Am^V. In > 10-M OH^-, Am^{VI} may dismutate into Am^V and Am^{VII}.

Peroxide. Uranyl salts give with 10-M H_2O_2, a stable pale-yellow peroxide, $n = 2$ at > 70 °C, or = 4 at < 50 °C, soluble in excess reagent:

$$UO_2^{2+} + H_2O_2 + 4\,H_2O \rightarrow UO_2(O_2)\cdot nH_2O\downarrow + 2\,H_3O^+$$

Also known are $Na_4[UO_2(O_2)_3]$ and others.

In a solution of uranate with CO_3^{2-} or HCO_3^-, H_2O_2 forms a deep yellow to red solution of a peroxo-complex. This is a sensitive test for U, but Ti, V and Cr interfere.

Such salts as $Na_4An^{VI}O_2(O_2)_3 \cdot 9H_2O$ can be crystallized.

The $HO_2^{\bullet}$ and $H^{\bullet}$ from α rays reduce Np^{VI} and Pu^{VI} to lower states.

In 5-dM HNO_3, H_2O_2 and NpO_2^{2+} give NpO_2^+.

Concentrated OH^- with O_2^{2-} and NpO_2^{2+} or PuO_2^{2+} form Np^{VII} or Pu^{VII}; as, e.g., $[trans\text{-}NpO_4(OH)_2]^{3-}$ with nearly square-planar NpO_4 units, but no Pu^{VIII} as hoped. The more stable and easily obtained M^{VII} is Np^{VII} ([Rn]**5f⁰6d⁰7s⁰**). A sample of solids, some from non-aqueous sources, can be $Li_5(Np,Pu)O_6$, $K_3(Np,Pu)O_5$, $(K,Rb,Cs)NpO_4$ and $Ba_3(NpO_5)_2$. The Pu^{VII} and Am^{VII}, however, oxidize H_2O to O_2 in minutes.

Aqueous AmO_2^{2+} in HNO_3 at 85 °C with 1.8-M H_2O_2 is reduced completely in 5 min to Am^{III}, but to Am^V if the pH > 2.

Trioxygen. Ozone and Np^{VI}, Pu^{VI} or Am^{VI} (or lower) with concentrated OH^-, or Np^V hydroxide and O_3 alone, form M^{VII} and possibly Pu^{VIII}; also see **Peroxide** above.

3.6.2 Reagents Derived from the Other 2nd-Period Non-Metals, Boron through Fluorine

Carbon oxide species. A slurry of $UO_3 \cdot 2H_2O$, or aqueous UO_2^{2+}, precipitates with CO_2:

$$UO_3 \cdot 2H_2O + CO_2 \rightarrow UO_2CO_3\downarrow + 2H_2O$$

Alkali carbonates and UO_2^{2+} precipitate, e.g., $(Na,NH_4)_4[UO_2(CO_3)_3]$, yellow, readily soluble in excess, also formed from UF_6, CO_2 and NH_3.

Barium carbonate completely precipitates UO_2^{2+}, probably as $Ba_2[UO_2(CO_3)_3]$ (distinction from Mn^{2+}, Co^{2+}, Ni^{2+} and Zn^{2+}).

Uranyl sulfide, UO_2S, is insoluble in H_2O but readily soluble in "$(NH_4)_2CO_3$" (distinction from MnS, FeS, ZnS, etc.). Uranyl hexacyanoferrate(II) also dissolves in "$(NH_4)_2CO_3$".

Excess CO_3^{2-} converts UO_2^{2+} to stable complexes such as, at pH = 6, the cyclic trimer $[\{UO_2(CO_3)_2\}_3]^{6-}$. In general, high ionic strength and high AnO_2^{2+} concentrations especially favor $[\{AnO_2(CO_3)_2\}_3]^{6-}$, as well as such mixed ions as $[(UO_2)_2(Np,Pu)O_2(CO_3)_6]^{6-}$.

Other complexes in solution include UO_2CO_3, $UO_2(CO_3)_2^{2-}$ and $[UO_2(CO_3)_3]^{4-}$ in rapid equilibria, together with polymers and hydrolyzed forms, making the systems very difficult to disentangle.

Especially well studied are the solids $M^I_4[AnO_2(CO_3)_3] \cdot nH_2O$, prepared either by evaporating solutions of the components or by precipitating the AnO_2^{2+} with an excess of CO_3^{2-}:

$$AnO_2^{2+} + 3\,CO_3^{2-} + 4\,M^+ + n\,H_2O \rightarrow M_4[AnO_2(CO_3)_3] \cdot nH_2O\downarrow$$

Variably hydrated uranium minerals occur with $M_4 = K_3Na$, Na_2Ca, Mg_2, MgCa, Ca_2 or Pb_2. The CO_3^{2-} ion can leach U from some ores. A pH of ~ 11, however, favors $UO_2(OH)_3^-$ over carbonato complexes.

Wyartite is one of many minerals with a truly complex formula: $CaU^{V}O_2(U^{VI}O_2)_2(CO_3)O_2(OH)(H_2O)_7$.

Moderate acidification of $[NpO_2(CO_3)_3]^{4-}$ precipitates red-brown NpO_2CO_3. Adding CO_3^{2-} to PuO_2^{2+} and raising the pH from 4 to 7 is found to form pink or brown PuO_2CO_3.

The carbonato complexes of NpO_2^{2+} and PuO_2^{2+} are much less stable than those of UO_2^{2+}, and the hydrated solids decompose in days or weeks to Np^{V} carbonate or Pu^{IV} hydroxide in turn. Still, salts that may be $K_8[Pu(CO_3)_5]CO_3 \cdot nH_2O$ and $K_{12}[Pu(CO_3)_5](CO_3)_3 \cdot nH_2O$ are found.

Carbonate promotes the coexistence of Am^{III}, Am^{IV}, Am^{V} and Am^{VI}.

Various carbonato AmO_2^{2+} salts precipitate, e.g., from 0.2-mM Am^{VI} in saturated Na_2CO_3-$NaHCO_3$ or by adding methanol to Am^{VI} in 1-dM $NaHCO_3$ (giving a lemon-yellow Na salt), by adding Ba^{2+} (giving a red-brown Ba salt), etc. The similarly prepared $Na_4[UO_2(CO_3)_3]$ is not isostructural. One may also obtain $(Cs,NH_4)_4[AmO_2(CO_3)_3]$.

Heating Am^{VI} in 2-M Na_2CO_3 at 90 °C for 30–60 min reduces it to Am^{V} and precipitates a sodium americyl(V) carbonate.

Cyanide species. One may crystallize such salts (or complexes) as $UO_2(NCS)_2 \cdot 8H_2O$ and various complexes: $(Alk,NH_4)[UO_2(NCS)_3(H_2O)_2]$, $(NH_4)_2UO_2(NCS)_4 \cdot nH_2O$ and $Alk_3[UO_2(NSC)_5] \cdot nH_2O$. Aqueous NCS^- reduces AmO_2^{2+} to Am^{V}.

Some "simple" organic species. Much $CH_3CO_2^-$, with Alk^+ (or Ae^{2+}) and AnO_2^{2+}, precipitates $Alk[AnO_2(\eta^2\text{-}CH_3CO_2)_3]$, especially the less soluble $Na[AnO_2(CH_3CO_2)_3]$ (or, e.g., $Ae[UO_2(CH_3CO_2)_3]_2 \cdot nH_2O$).

When $2 < pH < 4$, a few cM UO_2^{2+} and 1-dM of a chelator L^{2-} (tartrate, malate or citrate) give $[(UO_2)_2L_2]$ in solution.

Oxalate precipitates, e.g., $AnO_2C_2O_4 \cdot 3H_2O$, with only one H_2O bonded to U in that salt. Oxalates favor $[UO_2(C_2O_4)]$ and $[UO_2(C_2O_4)_2]^{2-}$ in either strongly or weakly acidic solutions of UO_2^{2+}, also yielding, say, $(NH_4)_2(UO_2)_2(C_2O_4)_3$, but slowly reduce U^{VI} to U^{IV}. A high $c(C_2O_4^{2-})$ forms $[UO_2(C_2O_4)_3]^{4-}$ and $(NH_4)_4[UO_2(CO_3)_3]$, for example, but even $K_6[(UO_2)_2(C_2O_4)_5] \cdot H_2O$ and $K_6[(UO_2)_2(C_2O_4)_5] \cdot 10H_2O$ have been isolated, along with many other hydrates, plus mixed complexes containing OH^-, O_2^{2-}, CO_3^{2-}, NCS^-, SO_3^{2-}, SO_4^{2-}, SeO_3^{2-}, SeO_4^{2-} and halides. Oxalate reduces at least the carbonato complexes of AmO_2^{2+} to Am^{V}.

Evaporating $UO_2(NO_3)_2$ and urea forms lime-green, fluorescent crystals of $[trans\text{-}UO_2\{CO(NH_2)_2\text{-}\kappa O\}_5](NO_3)_2$.

Cupferron, $C_6H_5N_2O_2^-$ precipitates U^{IV} but not UO_2^{2+}. See **3.4.2** above.

Reduced nitrogen. The "yellow cake" for nuclear-fuel processing is a mixture averaging about $(NH_4)_2U_2O_7$, precipitable from NH_3 and UO_2^{2+} or UO_2F_2; cf. **Hydroxide** above. Certain ore solutions, after reduction and then precipitating out the vanadium with "$(NH_4)_2CO_3$", also yield ammonium diuranate on boiling off the excess reagent.

One can prepare $U^{IV}(U^{VI}O_2)(PO_4)_2$ from UO_2^{2+} and some $N_2H_5^+$, followed with concentrated H_3PO_4.

In 1-M H_3O^+, $N_2H_5^+$ or NH_3OH^+ plus NpO_2^{2+} give NpO_2^+.

Aqueous PuO_2^{2+} with $N_2H_5^+$ or NH_3OH^+ forms Pu^{III} with some complicated intermediate steps (but conveniently gaseous byproducts) and Am^{IV}, AmO_2^+ or AmO_2^{2+} quickly yields Am^{III}. The AmO_2^{2+} carbonato complexes plus N_2H_4 or NH_2OH, however, go to Am^V.

Oxidized nitrogen. With HNO_2 in 1-M HNO_3, NpO_2^{2+} forms NpO_2^+ and NO_3^-, although it can be re-oxidized to NpO_2^{2+} at high acidity and < 1-mM HNO_2; PuO_2^{2+} gives Pu^{4+}, stable in 6-M H_3O^+.

Aqueous NO_2^- quickly reduces AmO_2^{2+}, e.g., in HCO_3^-, to AmO_2^+. This dismutates to AmO_2^{2+} and Am^{3+}, thus finally giving Am^{3+}.

Uranium trioxide dissolves in HNO_3 and aqua regia as UO_2^{2+}, but uranyl phosphates and $NaUO_2VO_4$ in ores are not very soluble.

With much NO_3^- we appear to have $UO_2(NO_3)^+$, $UO_2(NO_3)_2$, $UO_2(NO_3)_3^-$, $NpO_2(NO_3)^+$, $NpO_2(NO_3)_2$ and $PuO_2(NO_3)^+$ in solution, as well as the salts $AnO_2(NO_3)_2 \cdot nH_2O$, easily obtained for further work, and $(Alk,NH_4)AnO_2(NO_3)_3 \cdot nH_2O$. Dilute HNO_3 gives us the commercial $UO_2(NO_3)_2 \cdot 6H_2O$; the concentrated acid forms $UO_2(NO_3)_2 \cdot 3H_2O$. Similar formulas are found for $NpO_2(NO_3)_2 \cdot 6H_2O$ and $PuO_2(NO_3)_2 \cdot 6H_2O$.

Fluorine species. Fluoride and AnO_2^{2+} give no precipitates, but form complexes AnO_2F^+ or $AnO_2F_n^{(n-2)-}$ ($n \geq 2$), with $UO_2^{2+} > NpO_2^{2+} > PuO_2^{2+}$.

Uranyl peroxide dissolves in HF:

$$UO_2(O_2) \cdot 2H_2O + 3\,HF \rightarrow [UO_2F_3]^- + H_2O_2 + H_3O^+ + H_2O$$

Aqueous $[UF_8]^{2-}$ shows a high ligancy (c. n.) of eight.

More reactions than we can describe here give $(Alk,NH_4)AnO_2F_3$, $(Alk,NH_4)_2AnO_2F_4 \cdot 2H_2O$, $(Alk,NH_4)(AnO_2)_2F_5$, $Ae_2UO_2F_4 \cdot 4H_2O$ and also AnO_2F_2 (aqueous), $(\mathbf{3d}^{II},Cd)UO_2F_4 \cdot 4H_2O$ and $(Alk,NH_4)PuO_2F_3 \cdot H_2O$.

3.6.3 Reagents Derived from the 3rd-to-5th-Period Non-Metals, Silicon through Xenon

Phosphorus species. The lower acids or their anions precipitate $UO_2(PH_2O_2)_2 \cdot nH_2O$ and $UO_2PHO_3 \cdot nH_2O$. Salts of Alk^+ or Ae^{2+} form $K_2(UO_2)_2(PHO_3)_3 \cdot 4H_2O$, $Ba(UO_2)_2(PHO_3)_3 \cdot 6H_2O$ and so on.

The $H_2PO_4^-$ ion, AnO_2^{2+} and K^+, Cs^+ or NH_4^+ can form hard-to-filter $AlkAnO_2PO_4 \cdot aq$. Beryllium and Al, but not V, interfere. Uranyl and HPO_4^{2-} precipitate a pale-yellow $UO_2HPO_4 \cdot 3H_2O$, inhibited by Fe^{III}, soluble in H_3O^+ but not CH_3CO_2H or $CH_3CO_2^-$, and apparently an ionic conductor of H^+.

One can prepare $U^{IV}(U^{VI}O_2)(PO_4)_2$ from UO_2^{2+} and some $N_2H_5^+$, followed with concentrated H_3PO_4.

The insolubility of these AnO_2^{2+} phosphates calls for studies at high $c(H_3O^+)$, but this destabilizes Np^{VI} and Pu^{VI}. Some known species are $UO_2PO_4^-$, $(UO_2)_3(PO_4)_2 \cdot 3H_2O$, $AnO_2(HPO_4) \cdot aq$, $UO_2(H_2PO_4)_2 \cdot 3H_2O$, $UO_2(H_2PO_4)^+$, $UO_2(H_3PO_4)^{2+}$, $UO_2(H_2PO_4)_2$, $UO_2(H_3PO_4)(H_2PO_4)^+$ and $(H,Alk,NH_4)AnO_2PO_4 \cdot nH_2O$ (at pH 3.5 to 4.0 for the last salts) as well as $(Ae,\mathbf{3d}^{II})(AnO_2)_2(PO_4)_2 \cdot nH_2O$ and $NpO_2(HPO_4)_2^{2-}$. The hydronated compounds easily exchange H^+ with Alk^+ or $^1/_2$ Ae^{2+}. The mineral autunite is $Ca(UO_2)_2(PO_4)_2 \cdot 11H_2O$.

Arsenic species. Aqueous Alk^+ or M^{2+}, AnO_2^{2+} and $H_2AsO_4^-$ precipitate $(H,Alk,NH_4)AnO_2AsO_4 \cdot nH_2O$ or $(Ae,\mathbf{3d}^{II})(AnO_2)_2(AsO_4)_2$, not all well studied, but including $UO_2HAsO_4 \cdot 4H_2O$ and $UO_2(H_2AsO_4)_2 \cdot H_2O$.

Reduced chalcogens. In acids, H_2S does not precipitate or reduce UO_2^{2+}. In the absence of tartrates etc., "$(NH_4)_2S$" precipitates a dark brown uranyl sulfide, UO_2S. This product is insoluble in excess of the reagent, but if the mixture is aerated, a red compound is obtained, apparently due to $S_2O_3^{2-}$. Uranyl sulfide is insoluble in NH_3, readily soluble in "$(NH_4)_2CO_3$" (distinction from, e.g., MnS, FeS or ZnS) and in acids.

Oxidized chalcogens. Yellow $UO_2S_2O_3$, alternately $UO_2SO_3 \cdot nH_2O$, precipitates when $S_2O_3^{2-}$ is added to a solution of UO_2^{2+}. A flocculent, yellow precipitate separates when SO_3^{2-} is the reagent.

Uranyl acetate, aqueous dithionite and HCl give dark-green U^{IV}:

$$UO_2^{2+} + S_2O_4^{2-} + 2\,H_3O^+ \rightarrow U^{4+} + 2\,HSO_3^- + 2\,H_2O$$

Aqueous UO_2^{2+} is not reduced by SO_2, but PuO_2^{2+} gives Pu^{4+}, stable in 6-M H_3O^+. Sulfite precipitates UO_2SO_3 and $UO_2SO_3 \cdot {}^9/_2H_2O$. However, U^{VI}, SO_2 and HF yield UF_4. In H_2SO_4, SO_2 reduces NpO_2^{2+} to NpO_2^+.

Sulfur dioxide and Am^{IV}, AmO_2^+ or AmO_2^{2+} yield Am^{III}.

Uranyl sulfate is an inert complex, and under various conditions Ba^{2+} gives a precipitate of $BaSO_4$ only after long standing. Hydroxosulfato complexes of UO_2^{2+} may also exist. Sulfate, like carbonate, can leach U from ore. Many uranyl sulfate double salts contain Alk^+ or **d**-block M^{II}.

Mixing $UO_2C_2O_4 \cdot 3H_2O$ with aqueous Rb_2SO_4 or Cs_2SO_4 yields $Alk_2UO_2SO_4 C_2O_4 \cdot H_2O$.

Acidified sulfate, AnO_2^{2+} and M^+ form AnO_2SO_4 or $M^I_2AnO_2(SO_4)_2$; $[Co(NH_3)_6]^{3+}$ gives $[Co(NH_3)_6]_2AnO_2(SO_4)_3(HSO_4)_2 \cdot nH_2O$.

Aqueous SeO_4^{2-} and UO_2^{2+} appear to form UO_2SeO_4, $UO_2SeO_4 \cdot 4H_2O$ and $(Mg,Co,Zn)UO_2(SeO_4)_2 \cdot 6H_2O$; tellurate can give UO_2TeO_4.

Treating NpO_2^{2+} or PuO_2^{2+} with $[S_2O_8]^{2-}$ generates Np^{VII} or Pu^{VII} only in strong alkali; see **3.6.1 Peroxide** for products.

Reduced halogens. Insoluble UO_2^{2+} salts generally dissolve in HCl.

Anion-exchange resins retain $[AnO_2Cl_4]^{2-}$ from 12-M HCl.

Aqueous HCl dissolves AnO_3, giving, e.g., $PuO_2Cl_2 \cdot 6H_2O$, slowly decomposing to Pu^{IV}. One can isolate $(Alk,NH_4)_2AnO_2(Cl,Br)_4 \cdot 2H_2O$.

In $Cs_7(Np^VO_2)(Np^{VI}O_2)_2Cl_{12}$ two oxidation states coexist.

Salts such as $Cs_2AmO_2Cl_4$ have been isolated for Am.

The carbonato complexes of AmO_2^{2+} are reduced by I^-, but not by Cl^- or Br^- even on heating in 1-dM $NaHCO_3$, to Am^V. In acids, AmO_2^{2+} plus Cl^- or Br^- go to AmO_2^+, even up to a pH of ~ 5; hot HCl gives Am^{III}. In HNO_3, AmO_2^{2+} is reduced only to AmO_2^+ by adding just enough I^- to form I_2, otherwise to Am^{III}.

Oxidized halogens. Treating $Np^{<VII}$ or $Pu^{<VII}$ with IO_6^{5-} in alkali generates Np^{VII} or Pu^{VII}; see **3.6.1 Peroxide.**

More reactions than we can describe here give $AnO_2(ClO_4)_2 \cdot nH_2O$ and $AnO_2(IO_3)_2 \cdot nH_2O$

Xenon species. Treating NpO_2^{2+} or PuO_2^{2+} etc. with XeO_3 or XeO_6^{4-} in alkali forms Np^{VII} or Pu^{VII}; see **3.6.1 Peroxide** for some products.

3.6.4 Reagents Derived from the Metals Lithium through Uranium, plus Electrons and Photons

Reduction. The uranyl ion, UO_2^{2+}, is readily reduced to U^{IV}, stable in air-free H_2O, by Na_{Hg}, Mg, Cr^{2+}, Fe, Co, Cu, Zn, Zn_{Hg}, Cd, Sn, Pb, etc., but not appreciably by $SnCl_2$, despite moderately favorable (formal) standard electrode potentials for the latter, even with heat in chloride media (i.e., absent the tightly ligating SO_4^{2-} for UO_2^{2+}), as verified by the author's experiments. The amalgams Na_{Hg} or Zn_{Hg} can then take it on to U^{3+}. The reduction of $[UO_2(CO_3)_3]^{4-}$ in mineral waters can yield UO_2:

$$[UO_2(CO_3)_3]^{4-} + 2\,Fe^{2+} \rightarrow UO_2\downarrow + Fe_2O_3 \cdot aq\downarrow + 3\,CO_2\uparrow$$

Iron(2+) reduces PuO_2^{2+} to PuO_2^+, Pu^{IV} and, incompletely, Pu^{III}.

The reducibilities of AnO_2^{2+} vary as: $U < Np < Pu < Am$. Aqueous AmO_2^{2+} is a bit more oxidizing (going to Am^{3+}) than MnO_4^- (going to Mn^{2+}); this is not obvious from the stepwise electrode potentials in Appendix C, but for Am $(2.62 + 0.82 + 1.60)/3 = 1.68$ V; then for Mn we arrive at $(1.51 + 0.95 + 2.90 + 1.28 + 0.90)/5 = 1.51$ V.

In HCl, $[SnCl_3]^-$ reduces Np^{VI} to Np^V.

Cathodic treatment of UO_2^{2+} gives U^{3+} or U^{4+}. For example, $Na_4[UO_2(CO_3)_3]$ precipitates $Na_6[U(CO_3)_5] \cdot aq$ after adding Na_2CO_3.

Likewise, Am^{VI} carbonato complexes can be reduced to Am^V or lower salts such as $KAmO_2CO_3$, $K_3AmO_2(CO_3)_2$ or $K_5[AmO_2(CO_3)_3]$.

Light and $UO_2(CHO_2)_2$ precipitate $UO(C_2O_4) \cdot 6H_2O$, or possibly $U(OH)_2(C_2O_4) \cdot 5H_2O$, oxidizing the formate and reducing the U.

Other reactions. An equilibrium in mineral waters favors U^{VI} vanadate:

$$Ca(UO_2)_2(PO_4)_2 \cdot aq + 2\,H_2VO_4^- + 2\,K^+ \leftrightarrows$$

$$2\,KUO_2VO_4 \cdot aq\downarrow + 2\,H_2PO_4^- + Ca^{2+}$$

The complex formation of AnO_2^{2+} with $V_{10}O_{28}^{6-}$, $Mo_7O_{24}^{6-}$ and heteropolymolybdates is weak.

The molybdate ion, MoO_4^{2-}, precipitates, for two examples, $(Alk,NH_4)_2(UO_2)(MoO_4)_2 \cdot nH_2O$ and $(Alk,NH_4)_6(UO_2)(MoO_4)_4 \cdot nH_2O$. The mineral iriginite is found to be $[UO_2Mo_2O_7(H_2O)_2] \cdot H_2O$.

Aqueous $[Fe(CN)_6]^{4-}$ forms with UO_2^{2+}, a deep red-brown precipitate of $(UO_2)_2[Fe(CN)_6]$ or $K_2UO_2[Fe(CN)_6].6H_2O$. (Small amounts of UO_2^{2+} give only a brown color.) This may be distinguished from $Cu_2Fe(CN)_6$ by treatment with OH^-, HCl, or "$(NH_4)_2CO_3$", which all dissolve the uranium compound, resulting in a yellow solution.

Basic salts such as $Zn(UO_2)_2SO_4(OH)_4 \cdot {}^3/_2H_2O$ are numerous.

Adding Tl^+ to a solution containing CO_3^{2-} and a little UO_2^{2+} gives a quite insoluble crystalline precipitate, a sensitive test for UO_2^{2+}.

Volumetrically, UO_2^{2+} may be reduced to U^{IV} with a Jones reductor (Zn_{Hg}), and then titrated with MnO_4^-, e.g.,:

$$UO_2^{2+} + Zn + 4\,H_3O^+ \rightarrow U^{4+} + Zn^{2+} + 6\,H_2O$$

$$2\,MnO_4^- + 5\,U^{4+} + 6\,H_2O \rightarrow 2\,Mn^{2+} + 4\,H_3O^+ + 5\,UO_2^{2+}$$

Partial reduction to a lower stage can give high results.

Light breaks up $[UO_2(C_2O_4)]$, as in a chemical actinometer, giving different products in acidic and neutral solutions:

$$[UO_2(C_2O_4)] + \gamma + 2\,H_3O^+ \rightarrow UO_2^{2+} + CO_2\uparrow + CO\uparrow + 3\,H_2O$$

A pH of 7 gives CHO_2^-, HCO_3^-, and homopoly uranium complexes.

The colors of AnO_2^{2+} are: U, yellow with a greenish fluorescence; Np, reddish pink; Pu, orangish; Am, yellow.

3.7 The Actinoids An(VII)

3.7.1 Reagents Derived from Hydrogen and Oxygen

Water. The Np^{VII} and Pu^{VII} species are barely stable, Am^{VII} even less so. With acidity (or even just moderate basicity for Pu^{VII} or Am^{VII}) they rapidly oxidize H_2O to become MO_2^{2+} (or other M^{VI}) plus O_2.

Hydroxide. At pH 10, Np^{VII} may precipitate as $NpO_3(OH)$. In solution the M^{VII} ions are $[PuO_4(OH)_2]^{3-}$ for example.

Trioxygen. Note **3.6.1** for the possibility of Pu^{VIII} in strong base.

3.7.2 Reagents Derived from the Other 2nd-Period Non-Metals, Boron through Fluorine

and

3.7.3 Reagents Derived from the Heavier Non-Metals, Silicon through Xenon

The author has scanned no data for these.

3.7.4 Reagents Derived from the Metals Lithium through Uranium, plus Electrons and Photons

Non-redox reactions. The colors of An^{VII} are: Np, deep green; Pu, deep blue; and Am, dark green.

References

1. Jensen WB (1982) J Chem Educ 59:634
2. Ternstrom T (1964) J Chem Educ 41:190
3. Rich RL (2003) U Chem Educ 7:35
4. Seaborg GT, Loveland WD (1990) The elements beyond uranium. Wiley, New York, p 88, modified, with permission
5. Wiebke J, Moritz A, Cao X, Dolg M (2007) Phys Chem Chem Phys 9:459

Bibliography

See the general references in the Introduction, and some more-specialized books [4, 6–58]. Some articles in journals discuss: actinide complexation and thermodynamics at elevated temperatures [59]; classifying lanthanoids by multivariate analysis, albeit with results that seem hard to defend [60]; designing sequestering agents for Pu and other actinoids [61]; lanthanoid compounds with complex inorganic anions, part of a thematic issue on lanthanoid chemistry [62]; Pm, discovery and chemistry [63]; recent Sc chemistry [64]; actinoid complexes [65]; the transuranium elements [66]; actinoid complexes with OH^- and CO_3^{2-} [67]; coordination numbers [68]; the aqueous chemistry and thermodynamics of Eu [69]; photo-oxidation-reduction of Np and Pu [70]; review of Pm [71]; Rth thermochemistry [72]; unusual oxidation states of Ln and An [73]; and Rth chemistry [74].

6. Cotton S (2006) Lanthanide and actinide chemistry, 2nd edn. Wiley, West Sussex
7. Morss LR, Edelstein NM, Fuger J, Katz JJ (honorary) (eds) (2006) The chemistry of the actinide and transactinide elements, 3rd edn, five vols. Springer, Berlin Heidelberg New York

8. Sastri VS, Bünzli JC, Rao VR, Rayudu GVS, Perumareddi JR (2003) Modern aspects of rare earths and their complexes. Elsevier, Amsterdam
9. Hoffman DC (ed) (2002) Advances in plutonium chemistry 1967–2000. American Nuclear Society, La Grange Park, IL
10. Aspinall HC (2001) Chemistry of the f-block elements. Gordon & Breach, London
11. Jones CJ (2001) D- and f-block chemistry. The Royal Society of Chemistry, Cambridge
12. Lemire RJ et al (2001) Chemical thermodynamics of neptunium and plutonium. Elsevier, Amsterdam
13. Kaltsoyannis N, Scott P (1999) The f elements. Oxford University Press, Oxford
14. Morss LR, Fuger J (eds) (1998) Transuranium elements–a half century. Oxford University Press, Oxford
15. Yoshida Z, Kimura T, Meguro Y (eds) (1997) Recent progress in actinides separation chemistry. World Scientific, Singapore
16. Nash KL, Choppin GR (1995) Separations of f elements. Plenum, New York
17. Siekierski S, Phillips SL (1994) Actinide nitrates. IUPAC, Blackwell, London
18. Baran V (1993) Uranium(VI)-oxigen [sic] chemistry. Hadronic, Palm Harbor
19. (a) Fuger J, Khodakovsky IL, Sergeeva EI, Medvedev EA, Navratil JD (1992) The chemical thermodynamics of actinide elements and compounds: part 12, the actinide aqueous inorganic complexes. (b) Fuger J, Oetting FL (1976) The actinide aqueous ions: part 2. International Atomic Energy Agency, Vienna
20. Grenthe I et al (1992) Chemical thermodynamic of uranium: vol 1. Elsevier, Amsterdam
21. Freeman AJ et al (eds) (1984–1991) Handbook on the physics and chemistry of the actinides, vols 1–6. North Holland, Amsterdam
22. Meyer G, Morss LR (eds) (1991) Synthesis of lanthanide and actinide compounds. Springer, Berlin Heidelberg New York
23. Seaborg GT, Loveland WD (1990) The elements beyond uranium. Wiley, New York
24. Katz JJ, Seaborg GT, Morris LR (eds) (1986) The chemistry of the actinide elements, 2 vol. Chapman Hall, London
25. Edelstein NM, Navratil JD, Schulz WW (1985) Americium and curium chemistry and technology. Reidel, Dordrecht
26. Carnall WT, Choppin GR (symposium eds) (1983) Plutonium chemistry. American Chemical Society, Washington
27. Sinha SP (1983) Systematics and properties of the lanthanides. Reidel, Dordrecht
28. Bickel L (1979) The deadly element: the story of uranium. Stein and Day, New York
29. Cleveland JM (1979) The chemistry of plutonium. American Nuclear Society, La Grange Park, IL
30. Gschneidner KA Jr, Eyring LE (eds) (1978 onward, series) Handbook on the physics and chemistry of rare earths. Elsevier, Amsterdam
31. Seaborg GT (ed) (1978) Transuranium elements: products of modern alchemy. Dowden, Hutchinson & Ross; distrib Academic, San Diego
32. Schulz WW (1976) The chemistry of americium. Technical Information Center, Oak Ridge
33. Horowitz CT (ed) (1975) Scandium. Academic, San Diego
34. Muxart R, Guillaumont R (1974) Compléments au nouveau traité de chimie minéral: protactinium. Masson, Paris
35. Myasoedov BF, Guseva LI, Lebedev IA, Milyukova MS, Chmutova MK (1974) Kaner N (trans) Slutzkin D (ed) Analytical chemistry of transplutonium elements. Wiley, New York
36. Horowitz CT (ed) (1973) Scandium. Academic, San Diego

37. Bagnall KW (ed) (1972) Lanthanides and actinides. Butterworth, London
38. Cleveland JM (1970) The chemistry of plutonium. Gordon & Breach, New York
39. Cordfunke EHP (1969) The chemistry of uranium. Elsevier, Amsterdam
40. Pal'shin ES, Myasoedov BF, Davydov AV (1968) Schmorak J (trans) (1970) Analytical chemistry of protactinium. Ann Arbor-Humphrey, Ann Arbor
41. Institut du radium (1966) Physico-chimie du protactinium. Centre National de la Recherche Scientifique, Paris
42. Ryabchikov DI, Ryabukhin VA (1966) Aladjem A (trans) (1970) Analytical chemistry of yttrium and the lanthanide elements. Ann Arbor-Humphrey, Ann Arbor
43. Sinha SP (1966) Complexes of the rare earths. Pergamon, Oxford
44. Milyukova MS, Gusev NI, Sentyurin IG, Sklyarenko, IS (1965) Schmorak J (trans) (1967) Analytical chemistry of plutonium. Ann Arbor-Humphrey, Ann Arbor
45. Topp NE (1965) The chemistry of the rare-earth elements. Elsevier, Amsterdam
46. Chernyaev II (ed) (1964) Mandel L (trans) Govreen M (ed) (1966) Complex compounds of uranium. Israel Program for Scientific Translations, Jerusalem
47. Coleman GH (1964) The radiochemistry of plutonium. National Academy of Sciences, Washington
48. Taube M (1974) Plutonium; a general survey. VCH, Weinheim
49. Gittus JH (1963) Uranium. Butterworth, London
50. Rand MH, Kubaschewski O (1963) The thermochemical properties of uranium compounds. Oliver & Boyd, Edinburgh
51. Vickery RC (1961) Analytical chemistry of the rare earths. Pergamon, Oxford
52. Gel'man AD, Moskvin AI, Zaitsev LM, Mefod'eva MP (1961) Schmorak J (trans) (1967) Complex compounds of transuranides. Ann Arbor-Humphrey, Ann Arbor
53. Ryabchikov DI, Gol'braikh EK (1960) Norris AD (trans) (1963) The analytical chemistry of thorium. Macmillan, New York
54. Gel'man AD (1962) Turton CN, Turton TI (trans) (1962) Complex compounds of transuranium elements. Consultants Bureau, New York
55. Spedding FH, Daane AH (eds) (1961) The rare earths. Wiley, New York
56. Vickery RC (1960) The chemistry of yttrium and scandium. Pergamon, Oxford
57. Vickery RC (1953) Chemistry of the lanthanons. Academic, San Diego
58. Yost DM, Russell H jr, Garner CS (1947) The rare-earth elements and their compounds. Wiley, New York
59. Rao L (2007) Chem Soc Rev 36:881
60. Horovitz O, Sârbu C (2005) J Chem Educ 82:473
61. Gorden AEV, Xu J, Raymond KN (2003) Chem Rev 103:4207
62. Wickleder MS (2002) Chem Rev 102:2011
63. Cotton SA (1999) Educ in Chem 36:96
64. Cotton SA (1999) Polyhedron 18:1691
65. Clark DL, Hobart DE, Neu MP (1995) Chem Rev 95:25
66. Seaborg GT (1995) Acc Chem Res 28:257
67. Fuger J (1992) Radiochim Acta 59:81
68. Marçalo J, Pires de Matos A (1989) Polyhedron 8:2431
69. Rard JA (1985) Chem Rev 85:555
70. Toth LM, Bell JT, Friedman HA (1980) ACS Symp Ser 177:253
71. Weigel F (1978) Chem Zeit 108:339
72. Morss LR (1976) Chem Rev 76:827
73. Asprey LB, Cunningham BB (1960) in Cotton AF (ed) Prog Inorg Chem 2:267 Interscience, New York
74. Pearce DW, Hanson RA, Butler JC, Russell RG, Barthauer GL, McCoy HN, Cooley RA, Yost DM (1946) Inorg Synth 2:29

4 Titanium through Rutherfordium

4.1 Titanium, $_{22}Ti$

Oxidation numbers: (II), (III) and (IV), as in Ti^{2+}, Ti^{3+} and TiO_2.

4.1.1 Reagents Derived from Hydrogen and Oxygen

Water. Titanium(III) salts are in general readily soluble in H_2O, forming a wine-red to violet solution, depending on the acidity.

Water, above pH 7, and $Ti_2O_3 \cdot aq$ form $TiO_2 \cdot aq$ and H_2, catalyzed by **3d**^{II}, Pt and Li^+ and Na^+.

Titanium dioxide is insoluble in H_2O. The hydrated oxide, $TiO_2 \cdot aq$, is slightly amphoteric, with both basic and acidic salts hydrolyzing readily to $TiO_2 \cdot aq$. Boiling makes it less hydrated and less soluble.

Among the titanium(IV) salts the hexacyanoferrate(II) and phosphate are insoluble. The soluble ones require acid to prevent hydrolysis, which may form $Ti(OH)_n^{(4-n)+}$, $[(TiO)_8(OH)_{12}]^{4+}$ etc.

Seawater and some freshwater contain Ti^{IV} as $TiO(OH)^+$ or $TiO(OH)_2$.

Oxonium. Metallic Ti is usually insoluble in cold H_3O^+ due to superficial passivation, but forms Ti^{III}, deep red to violet, in hot HCl.

Titanium dioxide is practically insoluble in the ordinary dilute acids; concentrated HCl, HNO_3 and aqua regia have only a slight effect. The hydrate $TiO_2 \cdot aq$, if precipitated from a cold solution, is readily soluble in dilute acids, otherwise not.

Hydroxide. Metallic Ti does not react even with hot OH^-.

The very reactive, dark-blue $[TiOH(H_2O)_5]^{2+}$, and then $Ti_2O_3 \cdot aq$ forms as a gelatinous, dark precipitate when Ti^{III} is treated with OH^-.

Aqueous OH^- has only a slight solvent effect on TiO_2. Treating Ti^{IV} salts with OH^-, in the absence of chelators (many organic compounds, diphosphates and so on), gives a white gelatinous $TiO_2 \cdot aq$, less hydrated at higher temperature. The more hydrated compound is slightly soluble in OH^-; the less hydrated form is not.

Peroxide. An acidic solution of Ti^{IV} when treated with H_2O_2 gives a yellow (at low $[Ti^{IV}]$ in colorimetry) to red color (at high $[Ti^{IV}]$) of $Ti(O_2)^{2+}$, which, with OH^-, precipitates $Ti(O_2)O \cdot 2H_2O$, also from H_2O_2 plus $TiO_2 \cdot H_2O$. This (first) reaction provides a very sensitive test (as little as 6 µmol/100 mL of solution gives a distinct color) for titanium in the absence of fluorides, phosphates and much Alk^+, which bleach the color. Iron(III) interferes due to its own color; Cr^{III} gives blue "CrO_5"; Mo gives a yellow, vanadium a red-brown, color:

$$[TiO(H_2O)_5]^{2+} + H_2O_2 \leftrightarrows [Ti(O_2)(H_2O)_5]^{2+} + H_2O$$

Dioxygen. Oxygen does not rapidly attack $[Ti(H_2O)_6]^{3+}$ but converts $[Ti(OH)(H_2O)_5]^{2+}$ to $[TiO(H_2O)_5]^{2+}$ without forming $[Ti(O_2)(H_2O)_5]^{3+}$.

4.1.2 Reagents Derived from the Other 2nd-Period Non-Metals, Boron through Fluorine

Carbon oxide species. Saturated Na_2CO_3 is without action on TiO_2. The gelatinous hydrate $TiO_2 \cdot aq$, if precipitated from a cold solution of Ti^{IV}, e.g., by $BaCO_3$, is readily soluble in alkali carbonates, especially "$(NH_4)_2CO_3$".

Some "simple" organic species. Boiled with $CH_3CO_2^-$, Ti^{IV} forms the less soluble form of $TiO_2 \cdot aq$.

Aqueous $C_2O_4^{2-}$ reacts with Ti^{3+}, thus forming $Ti_2(C_2O_4)_3 \cdot 10H_2O$ and, e.g., $(K,NH_4)Ti(\eta^2\text{-}C_2O_4)_2 \cdot 2H_2O$. The $Ti_2(C_2O_4)_3 \cdot 10H_2O$ has a ligancy (c.n.) of seven for the Ti^{III}, including two each from the (opposite) sides of a μ-oxalate, making five-membered rings, not from its ends (making four), as may be partly elucidated without more complicated symbolism: $[\{Ti(\eta^2\text{-}C_2O_4)(H_2O)_3\}_2(\mu\text{-}C_2O_4)] \cdot 4H_2O$. The Ti^{III} chelate $Cs[Ti(\eta^2\text{-}C_2O_4)_2(H_2O)_3] \cdot 2H_2O$ likewise has a ligancy of seven.

Aqueous $C_2O_4^{2-}$ precipitates Ti^{IV} as a white titanium oxalate.

A urea complex of Ti^{III}, stable in dry air for several weeks, is made by mixing $TiCl_3$ with excess urea, under N_2, e.g., adding oxygen-free water and $NaClO_4$, and cooling, to give blue $[Ti\{OC(NH_2)_2\}_6](ClO_4)_3$.

The separation of Ti from Nb and Ta is quite difficult. Reactions that are excellent with the individual metals are very poor with a typical mixture. One good method for removing Ti is to boil the precipitated hydrated hydroxides of the Ti, Zr, Hf, Nb, and Ta with a moderately dilute solution of 2-hydroxybenzoate [salicylate, $o\text{-}C_6H_4(OH)CO_2^-$] and 2-hydroxybenzoic acid, whereupon all of the Ti is dissolved and the other metals are left in the residue.

Many organic compounds form colored products with Ti^{IV}, especially in concentrated H_2SO_4. Cupferron, $C_6H_5N_2O_2^-$, forms a flocculent, canary-yellow precipitate, even in a strongly acidic solution. In general, organic compounds do not interfere. Thorium, U^{IV}, Zr, Hf, V, Nb, Ta, Fe^{III}, Cu^{2+} and Sn^{IV} also form precipitates, while Ce, W, Ag, Hg, Si, Pb and Bi are partly precipitated.

Reduced nitrogen. Treating Ti^{III} with NH_3 precipitates the very reactive, gelatinous, dark $Ti_2O_3 \cdot aq$.

Treating Ti^{IV} with NH_3 in the absence of chelators precipitates the white gelatinous $TiO_2 \cdot aq$. See **Hydroxide** above for more.

Oxidized nitrogen. Dilute HNO_3 attacks Ti slowly in the cold, forming $TiO_2 \cdot aq$; if the reagent is hot and concentrated, the less soluble form is obtained. Aqua regia dissolves some Ti, but a coating of the hydrated oxide soon stops the reaction. The trioxide Ti_2O_3 is insoluble in HNO_3.

Fluorine species. Under Ar, Ti wire, ~1-M HF and 2-M CF_3SO_3H yield Ti^{II}. Metallic titanium is readily soluble even in cold HF as colorless $[TiF_6]^{2-}$, but in the absence of H_2SO_4 much Ti may be lost due to the volatility of TiF_4; reductants can then give , e.g., $(NH_4)_3[TiF_6]$.

A good solvent for titanium compounds in general is a mixture of HF and H_2SO_4 containing a small amount of HNO_3.

Treating ilmenite, $FeTiO_3$ including more or less Fe_2O_3, on a steam bath with concentrated and solid NH_4F, then more H_2O, gives:

$$FeTiO_3 + 10\,F^- + 6\,NH_4^+ \rightarrow [TiF_6]^{2-} + [FeF_4]^{2-} + 6\,NH_3\uparrow + 3\,H_2O$$

$$Fe_2O_3 + 12\,F^- + 6\,NH_4^+ \rightarrow 2\,[FeF_6]^{3-} + 6\,NH_3\uparrow + 3\,H_2O$$

Then to precipitate and remove only the $[FeF_4]^{2-}$ and $[FeF_6{}^{3-}]$, as FeS and S or FeS_2, we may use HS^- or H_2S at pH 5.8 to 6.2, by adding NH_3 as needed. Following this, concentrated NH_3 gives a non-gelatinous product, easily soluble in acids:

$$[TiF_6]^{2-} + 4\,NH_3 + 2\,H_2O \rightarrow TiO_2 \cdot aq\downarrow + 6\,F^- + 4\,NH_4^+$$

4.1.3 Reagents Derived from the 3rd-to-5th-Period Non-Metals, Silicon through Xenon

Phosphorus species. Hot, concentrated H_3PO_4 dissolves metallic Ti.

When Ti^{IV} is treated with a phosphate, e.g., HPO_4^{2-}, a white precipitate of a basic phosphate, approximately $Ti(OH)PO_4$, is obtained even in a fairly strongly acidic solution (separation from Al). Tartrates do not interfere, but cold H_2O_2 prevents precipitation (distinction from Zr).

Reduced chalcogens. "Ammonium sulfide" and Ti^{IV} give $TiO_2 \cdot aq$.

Oxidized chalcogens. Boiling Ti^{IV} with $S_2O_3^{2-}$ or SO_3^{2-} quantitatively precipitates the Ti^{IV} as the hydrous oxide (distinction from Ln^{III} but similar to Sc, Ce^{IV}, Th, Zr, and Al). The dithionite ion, $S_2O_4^{2-}$, reacts with Ti^{IV} in dilute acid to give a red to violet solution of Ti^{III}. Unless protected by an inert atmosphere, the color quickly disappears.

Cold, dilute H_2SO_4 readily dissolves Ti to form Ti^{III}; the hot, concentrated acid gives Ti^{IV} and SO_2.

The trioxide Ti_2O_3 is soluble in H_2SO_4.

Alums such as $[Cs(H_2O)_6][Ti^{III}(H_2O)_6](SO_4)_2$ are easily formed.

Hot concentrated H_2SO_4 slowly converts TiO_2 to the sulfate, $Ti(SO_4)_2$, soluble in H_2O if sufficient acid is present to prevent hydrolysis.

Reduced halogens. In hot dilute HCl, Ti dissolves as Ti^{III} if oxidants are excluded, and one can isolate $[Ti(H_2O)_6]Cl_3$.

The trioxide Ti_2O_3 is insoluble in HCl.

Dissolving one mole of dry $TiCl_3$ and two of CsCl in minimal 2-M HCl under N_2, and evaporating over concentrated H_2SO_4, yields fairly stable, dichroic (red-violet and colorless) plates of $Cs_2TiCl_5 \cdot 4H_2O$.

Elemental and oxidized halogens. Aqueous I_2 (or I_3^-) oxidizes Ti^{II} or Ti^{III} to Ti^{IV}, catalyzed by Mo^{VI} but not Cr^{3+}, WO_4^{2-}, $[(Mo,W)^{IV/V}(CN)_8]^{n-}$, Mn^{2+}, Co^{2+}, Ni^{2+} or Cu^{2+}; cf. **4.1.4** for oxidation by VO^{2+}.

Elemental and oxidized halogens, including the often inert ClO_4^-, oxidize Ti^{III}, and sometimes surprisingly less rapidly, Ti^{II}.

4.1.4 Reagents Derived from the Metals Lithium through Uranium, plus Electrons and Photons

Oxidation. Vanadium(IV) oxidizes Ti^{II} and Ti^{III}, to Ti^{III} and Ti^{IV}, catalyzed by Mo^{VI} (mediated perhaps by monomeric Mo^V but not $Mo_2O_4^{2+}$) and Cu^{II} but not Cr^{3+}, WO_4^{2-}, $[(Mo,W)^{IV/V}(CN)_8]^{n-}$, Mn^{2+}, Co^{2+}, or Ni^{2+}, with similar catalytic rate constants for Ti^{II} and Ti^{III}.

The Ti^{III} species, powerful reductants, readily change Fe^{III} to Fe^{2+}, Cu^{2+} to Cu, and so on, being itself oxidized to Ti^{IV}, although O_2 quickly produces Ti^{IV}, limiting those reductions.

Aqueous Ti^{2+} reduces Co^{III} complexes. With excess oxidant, these reactions yield Ti^{IV}, but with excess reductant the main product is Ti^{III}. Despite rate-law differences, the ratios of rates for the two, $k(Ti^{II})/k(Ti^{III})$, clearly fall well below 10^4, which corresponds to estimated differences in formal potentials, so that the stronger Ti^{II} actually reacts the more slowly, catalyzed understandably then by Ti^{IV}, which forms Ti^{III}.

In oxidations of both Ti^{III} and Ti^{II} in the series $[Co(NH_3)_5X]^{2+}$ (X = F, Cl, Br, and I), the fluoro complex reacts much faster than its congeners, the iodo the most slowly, just as for reductions by Eu^{2+}, but opposite the order for Fe^{II}, Cr^{II}, Cu^I and In^I. The rates of the $[Co(NH_3)_5(Br,I)]^{2+}$ reactions with excess Ti^{II} are nearly independent of c(oxidant) during the first 80–90 % reaction, suggesting that $[Ti(H_2O)_6]^{2+}$ may first form an active ion, e.g., $[Ti(H_2O)_5]^{2+}$. Some other oxidants for Ti^{3+}, $TiOH^{2+}$ etc. in acids are UO_2^{2+}, Pu^{4+}, PuO_2^{2+}, VO^{2+}, V^V, $HCrO_4^-$ and Hg^{2+}.

Light and Ti^{2+} reduce H_2O to H_2, catalyzed by Ni, Pd and Pt.

Reduction. Titanium(IV) is reduced to Ti^{III} by Na_{Hg}, Mg, Zn, Zn_{Hg}, Al in acid, and Sn. With the Jones reductor (Zn_{Hg}), large amounts of the element may be determined volumetrically. The Ti^{III} is mixed with Fe^{III} sulfate, excluding air, and the Fe^{2+} produced is titrated with MnO_4^-.

Other reactions. Titanium(III), Cs^+ and SO_4^{2-} yield the sparingly soluble violet alum $[Cs(H_2O)_6][Ti(H_2O)_6](SO_4)_2$ if air is absent.

The following oxidation by Ce^{IV} is not, of course, of the Ti^{IV}, but rather of the O_2^{2-} to O_2^-. The first product is unstable and partly reverts to the $[Ti^{4+}(O_2^{2-})(H_2O)_5]$, i.e., $Ti(O_2)^{2+}$, but it is also partly oxidized further to O_2 and TiO^{2+}:

$$[Ti^{4+}(O_2^{2-})(H_2O)_5] + Ce^{4+} + H_2O \rightarrow$$
$$[Ti^{4+}(O_2^-)(OH^-)(H_2O)_4] + Ce^{3+} + H_3O^+$$

$$2\,[Ti^{4+}(O_2^-)(OH^-)(H_2O)_4] + H_2O \rightarrow$$
$$2\,[Ti^{4+}(O_2^{2-})(H_2O)_5] + {}^1/_2\,O_2\uparrow$$

$$[Ti^{4+}(O_2^-)(OH^-)(H_2O)_4] + Ce^{4+} + 2\,H_2O \rightarrow$$
$$[Ti^{4+}(O^{2-})(H_2O)_5] + Ce^{3+} + O_2\uparrow + H_3O^+$$

Aqueous $[Fe(CN)_6]^{4-}$ forms with Ti^{IV} a brown precipitate, and $[Fe(CN)_6]^{3-}$ yields a yellow product.

4.2 Zirconium, $_{40}$Zr; Hafnium, $_{72}$Hf; and Rutherfordium, $_{104}$Rf

Oxidation number: (IV), as in ZrO_2 or HfO_2. The oxidation states for Rf calculated relativistically to occur in water: (III) and (IV), especially (IV), and possibly (II). Zirconium and Hf are so similar, with the lanthanoid contraction and relativistic effects canceling the otherwise expected larger size of the latter in its ions, that we treat them together. The same goes for many subsequent elements. In what follows here, the rarer Hf can often be substituted for Zr, except in comparisons or separations.

4.2.1 Reagents Derived from Hydrogen and Oxygen

Water. The ions Zr^{4+}, Hf^{4+} and Rf^{4+} are large enough to form $[M(H_2O)_8]^{4+}$ but hydrolysis and polymerization yield mainly the inert [*quadro*-$\{M(\mu\text{-}OH)_2\text{-}(H_2O)_4\}_4]^{8+}$ or "$MO^{2+}\cdot 5H_2O$" (at least for Zr and Hf), even when $c(M^{IV})$ is < 1 cM or mM and $c(H_3O^+)$ is > 1 M. Lowering the acidity may lead to $[\{M(OH)(\mu\text{-}OH)_2(H_2O)_3\}_4]^{4+}$ by loss of H^+.

The alkali zirconates, $M^I_2ZrO_3$ or $M^I_4ZrO_4$, made by fusing ZrO_2 with the caustic alkalis, are insoluble, but largely hydrolyzed to hydroxides.

The fluorine compounds of zirconium are insoluble or difficultly soluble in H_2O. The other halide compounds are soluble but readily hydrolyzed. This hydrolysis goes so far that a solution of, e.g., $ZrOCl_2$ may be diluted with H_2O and the compound determined by titration of the H_3O^+ liberated. Zirconium nitrate is readily soluble in H_2O.

Some hot natural waters may contain $[ZrF_6]^{2-}$.

Aqueous Rf^{4+} may be somewhat less hydrolyzed than Zr^{4+}, Hf^{4+} (and $Hf^{4+} < Zr^{4+}$) or Pu^{4+}, but more than Th^{4+}, with $\log K_{11} = -2.6 \pm 0.7$, agreeing well with a relativistic quantum-mechanical prediction of ~ -4.

The extent of hydrolysis of $[MF_6^-]$ is both calculated and found to be $Rf \geq Zr > Hf$; the extent for $[MCl_6]^-$, on the other hand, is calculated to be $Zr > Hf > Rf$ but found to be $Rf > Zr > Hf$.

Oxonium. At moderate concentrations of M^{IV} the hydrolyzed polymer $[\{M\ (\mu\text{-}OH)_2(H_2O)_4\}_4]^{8+}$ resists hydronation (attachment of H^+) and a concomitant breakup even on refluxing in concentrated H_3O^+ for a week. With $HClO_4$ of > 5 dM and Zr^{IV} of < 1 cM, however, aqueous $[\{Zr(\mu\text{-}OH)_2(H_2O)_4\}_4]^{8+}$ begins to become $[Zr(H_2O)_8]^{4+}$.

Hydroxide. Aqueous OH^- reacts with Zr only very slightly.

Treating Zr^{IV} (cold) with OH^- precipitates white, impure $Zr(OH)_4 \cdot aq$ or $ZrO_2 \cdot aq$, bulky and gelatinous, readily soluble in H_3O^+ but insoluble in excess OH^-, any apparent solution being probably due to peptization. Slowly on standing, or faster with heating, this approaches the less soluble composition $ZrO(OH)_2$, more slowly soluble in the dilute acids, often requiring treatment for several days. In either form it is practically insoluble in H_2O, OH^-, CO_3^{2-} or NH_4^+ ions. It may be purified by solution in HCl and reprecipitation with NH_3. Sulfate gives a basic sulfate, which is converted to a normal hydroxide with difficulty.

Peroxide. Many peroxo complexes are white, rather stable at room temperature, and soluble in H_3O^+. Solid complexes such as these are known: $[Zr(O_2)(C_2O_4)(H_2O)_2] \cdot nH_2O$, $Zr(O_2)F_2 \cdot 2H_2O$, $(NH_4)_3[MF_6(HO_2)]$, $K_2MF_5(HO_2)$, $(NH_4)_3[Zr(O_2)F_5]$, $(K,Rb,Cs,NH_4)_3Zr_2(O_2)_2F_7 \cdot nH_2O$ and $[Zr_2(O_2)_3(SO_4)(H_2O)_4] \cdot nH_2O$.

4.2.2 Reagents Derived from the Other 2nd-Period Non-Metals, Boron through Fluorine

Carbon oxide species. The alkali carbonates precipitate Zr^{IV} as a basic carbonate, readily soluble in excess reagent and reprecipitated on boiling. Also isolated are $(Na,K)_4[M(\eta^2\text{-}CO_3)_4] \cdot nH_2O$, $(K,NH_4)_6[\{M(\eta^2\text{-}CO_3)_3\}_2(\mu\text{-}OH)_2] \cdot nH_2O$ and numerous mixed-ligand solids with $C_2O_4^{2-}$ or F^- for example, where the four-membered carbonate chelate rings are, perhaps surprisingly, more stable than the five-membered oxalate rings.

Some "simple" organic species. Aqueous $CH_3CO_2^-$ precipitates Zr^{IV} as a basic salt, which, on sufficient boiling, is fully converted to ~ $ZrO(OH)_2$.

Oxalates and tartrates precipitate Zr^{IV} as basic salts, soluble in excess reagent. From such solutions, OH^- and "$(NH_4)_2S$" do not precipitate $Zr(OH)_4 \cdot aq$. The high solubility of many zirconium compounds in oxalate shows the close resemblance between this element and thorium.

If some $ZrCl_2(OH)_2 \cdot 7H_2O$, often called $ZrOCl_2 \cdot 8H_2O$, structurally $[\{Zr(\mu\text{-}OH)_2(H_2O)_4\}_4]Cl_8 \cdot 12H_2O$, is dissolved and slowly added to a little excess of aqueous $K_2C_2O_4$ and $H_2C_2O_4$ we have (recalling that solid and dissolved species may differ) after adding ethanol (hafnium is similar):

$$^1/_4\,[Zr_4(OH)_8(H_2O)_{16}]^{8+} + 2\,C_2O_4^{\,2-} + 2\,HC_2O_4^{\,-} + 4\,K^+ \rightarrow$$

$$K_4[Zr(C_2O_4)_4] \cdot 5H_2O\downarrow + H_2O$$

Quite stable complexes of M^{IV} arise from α-hydroxycarboxylic acids, even up to 5-M H_3O^+. The stabilities of the 1:1 complexes are in the order Zr > Hf (as with the sulfates), and lactate > citrate > glycolate > malate > tartrate. Various 1:2, 1:3 and 1:4 complexes also are formed.

For separation on a cation-exchange resin, Zr^{IV} can be eluted before Hf^{IV} by 9-cM citric acid together with 4.5-dM HNO_3.

Reduced nitrogen. Treating Zr^{IV} with NH_3 precipitates a white flocculent $Zr(OH)_4 \cdot aq$. See **Hydroxide** above for more. For the gravimetric analysis of Zr^{IV}, precipitation as the hydroxide with NH_3 and subsequent ignition to ZrO_2 is very satisfactory.

Oxidized nitrogen. Compact Zr dissolves in aqua regia, giving $Zr(NO_3)_x(OH)_y^{(x+y-4)-}$ etc., but HNO_3 alone has almost no effect.

Fluorine species. In general, compact Zr is insoluble in all cold acids except HF and aqua regia. A mixture of HF and HNO_3 is very efficient.

Zirconium dioxide, after ignition, is insoluble in all acids except HF. One can isolate $ZrF_4 \cdot 3H_2O$. By repeated fuming with HF in the presence of only a little H_2SO_4, ZrO_2 may be almost completely volatilized.

Zirconium forms , e.g., $[ZrF_6]^{2-}$ and $[Zr_4F_{24}]^{8-}$, very little hydrolyzed although OH^- can replace some F^-. The first separation of Zr and Hf was by very many fractional crystallizations of $(NH_4)_2[ZrF_6]$ and $(NH_4)_2[HfF_6]$; their solubilities in water at 20 °C are 1.050 M and 1.425 M, respectively. Likewise, $K_2[HfF_6]$ is 1.7 times as soluble as (70 % more soluble than) $K_2[ZrF_6]$. In current English, regrettably, some writers confuse multiplication with addition and say that 1.425 is 1.357 times *more than* 1.050 instead of 0.357 times *more than* (multiplying and adding) or 1.357 times *as much as* (multiplying) 1.050. The original purpose of this in advertising must have been to exaggerate the difference but it may leave the reader uncertain.

Two of many further solid complexes are $NH_4ZrF_5 \cdot H_2O$ (with H_2O very weakly bound) and $[\{ZrF_3(H_2O)_3\}_2F_2]$. However, discrete $[ZrF_8]^{2-}$ can be found in $[Cu(H_2O)_6]_2[ZrF_8]$.

Stable complexes for Rf include $[RfF_6]^{2-}$ and lower Rf^{IV} species.

4.2.3 Reagents Derived from the 3rd-to-5th-Period Non-Metals, Silicon through Xenon

Phosphorus species. Phosphates completely precipitate Zr^{IV}, if strongly acidified with HCl or HNO_3, as $ZrO(H_2PO_4)_2$ or $Zr(HPO_4)_2 \cdot H_2O$. Upon warming, a gelatinous white precipitate will be obtained from as little as 0.06-mM Zr, reportedly soluble in pure water up to 1.4 mM. (Hafnium may differ a little.) The precipitate is not easily filtered. With H_2O_2 present, Zr and Hf are precipitated, and Rf (and, slightly, Nb and Ta) co-precipitated, but not Ti or other elements.

For gravimetry, the precipitate is washed, ignited to $Zr[P_2O_7]$ and weighed. With large amounts of Zr, this tends to give low results due to hydrolysis and consequent loss of phosphate.

Arsenic species. Arsenic acid and soluble arsenates precipitate Zr^{IV} from strongly acidic solutions.

Reduced chalcogens. Alkali sulfides precipitate Zr^{IV} as the hydroxide.

Hafnium thiocyanate is more soluble than, and thus separable from, zirconium thiocyanate in, for example, 4-methyl-2-pentanone (methyl isobutyl ketone, "hexone").

Oxidized chalcogens. Aqueous $S_2O_3^{2-}$ precipitates Zr^{IV} as the hydroxide or a basic thiosulfate, depending on conditions.

If SO_2 is passed into a neutral solution of Zr^{IV}, a slimy precipitate is obtained that dissolves on addition of excess SO_2.

Hot, concentrated H_2SO_4 acts energetically even on compact Zr.

Sulfuric acid, added to a solution of Zr^{IV}, gives a white, flocculent precipitate, readily soluble in excess reagent and other inorganic acids (separation from the rare earths and thorium). Aqueous SO_4^{2-} forms a basic sulfate, insoluble in excess SO_4^{2-}. Basic sulfates without discrete molecules or ions include $Zr(OH)_2SO_4$ and $Zr_2(OH)_2(SO_4)_3(H_2O)_4$.

From solutions of sulfato complexes, oxalates do not precipitate the Zr, and sulfites precipitate it very slowly in part. Sulfate solutions also can produce such solid complexes as $Na_2M(\eta^2\text{-}SO_4)_2(H_2O)_2(\mu\text{-}SO_4) \cdot H_2O$ (with an infinite chain structure), $[\{Zr(\eta^2\text{-}SO_4)(H_2O)_4\}_2(\mu\text{-}SO_4)_2] \cdot nH_2O$ and $K_4[\{Zr(\eta^2\text{-}SO_4)_2(H_2O)_2\}_2\text{-}(\mu\text{-}SO_4)_2]$. In addition, aqueous Hf^{IV} gives: $Na_4[Hf(\eta^2\text{-}SO_4)_2(\eta^1\text{-}SO_4)_2(H_2O)_2]$ and

$Na_6[Hf(\eta^2\text{-}SO_4)_2(\eta^1\text{-}SO_4)_3(H_2O)]$. The Hf complexes seem to be a little less stable than those of Zr. Many mixed complexes with CO_3^{2-}, $C_2O_4^{2-}$ and F^- also exist.

Sintering some ores with lime etc., and extracting with dilute HCl leaves calcium zirconate, which is treated with concentrated H_2SO_4 to form soluble $Zr(SO_4)_2$.

Solutions of H_2SeO_4, M^{IV}, F^- and Alk^+ form $K_2MF_2(SeO_4)_2 \cdot 3H_2O$ etc.

Reduced halogens. Chloride and $[\{Zr(\mu\text{-}OH)_2(H_2O)_4\}_4]^{8+}$, the main cation, yield $[Zr_4(OH)_8(H_2O)_{16}]Cl_8 \cdot 12H_2O$, but equivalent amounts of AlkCl (or NH_4Cl) and "$MOCl_2 \cdot 8H_2O$" or MCl_4 in HCl saturated with HCl gas do form white, moisture-sensitive $Alk_2[MCl_6]$. Similar white Br or yellow I complexes exist. In contrast to some other ligands, the Hf salts are more stable than the Zr salts.

The separation of Zr and Hf uses 6-M or 9-M HCl as eluents from a cation exchanger, e.g., Dowex-50, or an anion exchanger, e.g., Dowex-2. Liquid extraction is also used.

Moderately concentrated Cl^- does not so readily form $[ZrCl_6]^{2-}$ or $[HfCl_6]^{2-}$, but does form $[RfCl_6]^{2-}$, as suggested by its non-extraction into Bu_3PO_4. Lower Rf^{IV} complexes are also stable.

Oxidized halogens. If Zr^{IV} is treated with an iodate, a white precipitate of a basic zirconyl iodate is obtained, with a composition depending on conditions. Aqueous periodate reacts similarly.

4.2.4 Reagents Derived from the Metals Lithium through Uranium, plus Electrons and Photons

Non-redox reactions. Slowly evaporating solutions of MO_2 in aqueous HF at ambient T with **3d**-row difluorides at a 1:1 molar ratio yields $MnMF_6 \cdot 5H_2O$, $[(Fe,Co,Ni,Zn)(H_2O)_6]MF_6$ or $[Cu(H_2O)_4]MF_6$. Using half as much M^{IV} results in $(Co,Ni,Cu,Zn)_2[MF_8] \cdot 12H_2O$. Either ratio of CdF_2 and MO_2 in HF forms $Cd_2[MF_8] \cdot 6H_2O$. Equivalent amounts of KF, CuF_2 and ZrF_4 in HF give $K_2[Cu(H_2O)_6](ZrF_6)_2$.

Aqueous CrO_4^{2-} precipitates Zr^{IV} partly from weakly acidic solutions.

Aqueous $[Fe(CN)_6]^{4-}$ gives a green precipitate with Zr^{IV}.

Bibliography

See the general references in the Introduction, and some more-specialized books [1–6]. Some articles in journals discuss reductions by Ti^{II} [7], also Zr^{IV} and Hf^{IV} chemistry [8].

1. Hala J (1989) Halides, oxyhalides and salts of halogen complexes of titanium, zirconium, hafnium, vanadium, niobium and tantalum. IUPAC, Blackwell, London
2. Clark RJH, Bradley DC, Thornton P (1975) The chemistry of titanium, zirconium and hafnium. Oxford University Press, Oxford
3. Mukherji AK (1970) Analytical chemistry of zirconium and hafnium. Pergamon, Oxford
4. Clark RJH (1968) The chemistry of titanium and vanadium. Elsevier, Amsterdam
5. Elinson SV, Petrov KI (1965) Kaner N (trans) (1965) Analytical chemistry of zirconium and hafnium. Ann Arbor-Humphrey, Ann Arbor
6. Blumenthal WB (1958) The chemical behavior of zirconium. Van Nostrand, Princeton
7. Yang Z, Gould ES (2005) J Chem Soc Dalton Trans 2005:1781
8. Larsen EM (1970) Adv Inorg Chem Radiochem 13:1

5 Vanadium through Dubnium

The relative stabilities of the Group 5 species, plus the pseudoanalog Pa, in acidic solutions, at least without strong ligands, appear to be:

$V^{II} > Nb^{II} \geq Db^{II} > Ta^{II} > Pa^{II}$;
$V^{III} > Nb^{III} > Ta^{III} > Db^{III} > Pa^{III}$;
$V^{IV} >> Pa^{IV} > Nb^{IV} > Ta^{IV} > Db^{IV}$;
$Pa^{V} > Db^{V} > Ta^{V} > Nb^{V} > V^{V}$.

5.1 Vanadium, $_{23}V$

Oxidation numbers: (II), (III), (IV) and (V), as in V^{2+}, V^{3+}, VO^{2+} and V_2O_5. Aqueous V^{2+} is one of the strongest reductants known, being more active than Cr^{2+}, and V^{III} is more easily oxidized than Cr^{III}. The non-redox properties of V^{III} are similar to those of Cr^{III}, Fe^{III}, and Al^{III}.

5.1.1 Reagents Derived from Hydrogen and Oxygen

Water. Fresh $V(OH)_2$ releases H_2 from H_2O. The oxides VO, V_2O_3 and VO_2 are insoluble in water, but the dioxide is somewhat hygroscopic. The amphoteric hydrate $VO_2 \cdot H_2O$ occurs in a stable, green form, and an unstable, red one.

Vanadium halides are hygroscopic and tend to hydrolyze more readily, the higher the oxidation state.

Important hydrated cations include the lavender $[V(H_2O)_6]^{2+}$, green $[V(H_2O)_6]^{3+}$, green $[V(Cl,Br)_2(H_2O)_4]^+$, blue $[VO(H_2O)_5]^{2+}$, and the yellow or orange [*cis*-$VO_2(H_2O)_4]^+$ etc.

Among the V^{III} salts, VOCl and VOBr are slightly soluble in H_2O, readily soluble in HNO_3. Anhydrous $V_2(SO_4)_3$ is insoluble in water. The hydrolysis of V^{3+} to $[VOH(H_2O)_5]^{2+}$ and polymers occurs in low acidity.

The V^{IV} salts $VOSO_4$, $VOBr_2$ etc. dissolve to give $[VO(H_2O)_5]^{2+}$ and, up to pH 6, $[VO(OH)(H_2O)_4]^+$, $(VO)_2(OH)_2{}^{2+}$ etc. The blue salt $[VO(H_2O)_5](ClO_4)_2$ is very hygroscopic.

Vanadium pentoxide, V_2O_5, dark red to orange, and poisonous, turns moist blue litmus red. It dissolves to the extent of 4 mM.

The alkali vanadates are soluble, NH_4VO_3 least (a non-cyclic "metavanadate" quantitatively precipitated by NH_4Cl with ethanol). Many **d**- or **p**-block vanadates are insoluble in water or CH_3CO_2H, especially with a little excess precipitant. The products tend to become colloidal, the most with Fe, Cu, Zn and Al, the least with Ca, Hg and Pb.

Seawater and some freshwater contain traces of V^V complexes as H_3VO_4, $H_2VO_4^-$, HVO_4^{2-} and $NaHVO_4^-$.

Oxonium. Metallic V dissolves as V^{2+}.

The monoxide, VO, gray, dissolves in dilute acids without the evolution of H_2, to form V^{2+}, one of the most powerful reductants known in water. The trioxide, V_2O_3, black, is not easily soluble in acids. The dioxide VO_2 is readily soluble, yielding VO^{2+}.

Adding H_3O^+ to a fairly concentrated vanadate precipitates V_2O_5. This dissolves easily in strong acids, forming salts that hydrolyze readily.

The various V^V anions go to a "vanadyl", VO_2^+, i.e., $[cis\text{-}VO_2(H_2O)_4]^+$, below pH 3. In dilute solutions, say 10 μM to avoid polymerization, the acidities and basicities of $H_nVO_4^{(3-n)-}$ resemble those of $H_nPO_4^{(3-n)-}$, but the V species are more basic by 1 to 2 p*K* units.

Excess inorganic acids decompose all simple V^V compounds, forming *cis*-VO_2^+ salts. Rather concentrated (>9 M) H_2SO_4 or $HClO_4$ yields red $V_2O_3^{4+}$, i.e., $[cis\text{-}\{VO(H_2O)_4\}_2(\mu\text{-}O)]^{4+}$, and sulfato complexes.

Hydroxide. The V^{2+} and OH^- ions precipitate $V(OH)_2$, gelatinous, with various colors reported, one of the most powerful inorganic reductants; air changes it to greenish V(III) hydroxide.

The trioxide V_2O_3 is insoluble in the alkalis.

The basic, green $V_2O_3\cdot aq$ is obtained from V^{3+} plus OH^-, and is practically insoluble in the alkalis.

The dioxide VO_2 is readily soluble in alkalis, giving brown solutions containing $[V_{18}O_{42}]^{12-}$, stable if $9 < pH < 13$, easily oxidized or forming, e.g., $K_{12}[V_{18}O_{42}]\cdot aq$. Dilution seems to yield $[VO(OH)_3]^-$. In > 1-M OH^-, V^{IV} dismutates into V^{III} and V^V. Grayish $VO_2\cdot aq$ precipitates from OH^- and VO^{2+}. Slightly alkaline solutions yield $[VO(OH)_3]^-$ and perhaps $(VO)_2(OH)_5^-$ and other polymers, reverting to $[VO(OH)_3]^-$ at $pH > 12$.

Acidified vanadate(V) yields a brown precipitate with alkalis, soluble in excess. Excellent solvents for V_2O_5 are either OH^- or CO_3^{2-}, with a small (catalytic) amount of peroxide, forming ions having formulas like those of the mono- and linear polyphosphates as $[V_nO_{3n+1}]^{(n+2)-}$, depending on conditions. Cyclic $[(VO_3)_4]^{4-}$ and $[(VO_3)_5]^{5-}$ are especially prominent at $6.5 < pH < 8$. Other homopoly anions are also easily obtained. The alkaline species such as $[(VO_3)_n]^{n-}$ ("metavanadates"), paired with Alk^+ ions, are colorless; acidification forms the yellow or orange acidified polyvanadates, $H_m[V_nO_{3n+1}]^{(n+2-m)-}$.

The prominent cluster $[V_{10}O_{28}]^{6-}$ with overlapping VO_6 octahedra rearranges and dissociates slowly to HVO_4^{2-} and VO_4^{3-} in base, faster at $pH > 10$ and with the Alk^+ ion $K^+ > Na^+ > Li^+$.

Peroxide. A peroxide rapidly oxidizes V^{II} with a one-electron step, and V^{III} and V^{IV} are rapidly oxidized to V^{IV} or V^{V} by neutral or alkaline peroxide, but vanadates can also be reduced to V^{IV} by acidified H_2O_2.

Vanadium(V) readily forms a red complex, e.g., in cold, ~ 2-M H_2SO_4:

$$VO_2^+ + H_2O_2 \rightarrow VO(O_2)^+ + H_2O$$

and then a yellow complex, especially if the pH is at least 2:

$$VO(O_2)^+ + H_2O_2 + 2\,H_2O \leftrightarrows [VO(O_2)_2]^- + 2\,H_3O^+$$

The color is not extracted by ether (distinction from Cr), or affected by H_3PO_4 (distinction from Fe^{III}) or HF (distinction from Ti). It is especially stable under various conditions, and more so than "perchromic acid" (see **6.1.1 Peroxide**), but is destroyed by excess H_2O_2. The reaction readily reveals vanadium in a 4-mM solution. Some triperoxovanadate is also formed. Adding NH_4^+ and concentrated NH_3 gives, after standing, bright-yellow $(NH_4)[VO(O_2)_2(NH_3)]$.

Mixing (alkaline) vanadate(V) with H_2O_2 yields, depending on concentrations, $[VO_j(OH)_k(O_2)_l]^{(2j+k+2l-5)-}$. With $l=3$ it is pale yellow. Some solid salts are a blue $Na_3[V(O_2)_4]\cdot 14H_2O$ and a purple $K_3[V(O_2)_4]$.

Dioxygen. Oxygen and $[V(H_2O)_6]^{2+}$ seem first to form $[V(O_2)(H_2O)_6]^{2+}$; this then at $<$ 5-mM V^{2+} breaks up and the V^{2+} reacts with the H_2O_2:

$$[V(O_2)(H_2O)_6]^{2+} \leftrightarrows [VO(H_2O)_5]^{2+} + H_2O_2$$

$$[V(H_2O)_6]^{2+} + {}^1\!/_2\,H_2O_2 + H_3O^+ \rightarrow [V(H_2O)_6]^{3+} + 2\,H_2O$$

but a higher $c(V^{2+})$ gives:

$$[V(O_2)(H_2O)_6]^{2+} + [V(H_2O)_6]^{2+} \leftrightarrows [\{V(H_2O)_5O-\}_2]^{4+} + 2\,H_2O$$

$$[\{V(H_2O)_5O-\}_2]^{4+} \rightarrow 2\,[VO(H_2O)_5]^{2+}$$

so that more V^{II} actually yields more V^{IV} and less V^{III}.

Many V^{III} complexes go to V^{IV} or V^{V} in air.

In acid, VO^{2+} is stable, but aerial oxidation is fast at a $c(OH^-)$ over 6 mM, catalyzed by Fe^{III} but inhibited by Cr^{III}. Air oxidizes fresh VO_2.

5.1.2 Reagents Derived from the Other 2nd-Period Non-Metals, Boron through Fluorine

Carbon oxide species. Aqueous V^{2+} and CO_3^{2-} precipitate $VCO_3\cdot 2H_2O$, and CO_3^{2-} precipitates $VOSO_4$ or $VOCl_2$ as grayish $VO_2\cdot aq$.

One may prepare NH_4VO_3 by boiling V_2O_5 with CO_3^{2-} as in the following equation; then minimal MnO_4^- (to oxidize any $V^{<V}$) plus much NH_4^+ and cooling yield the "metavanadate" NH_4VO_3:

$$V_2O_5 + CO_3^{2-} \rightarrow 2\,VO_3^- + CO_2\uparrow$$

Cyanide species. Aqueous V^{2+} (from V^{III} and Zn) plus saturated KCN form yellow $K_4[V(CN)_6]$. Aqueous VCl_3 and excess concentrated KCN, with ethanol, give scarlet $K_4[V(CN)_7]\cdot 2H_2O$; this then decomposes and dismutates to $[V(CN)_6]^{4-}$ and $[VO(CN)_5]^{3-}$.

Cyanide (6 M) together with VO^{2+}, followed by methanol, precipitates $Alk_3[VO(CN)_5]$, with Alk = K, Cs or $[NMe_4]$, for example.

Thiocyanate and VO^{2+} yield $(K,NH_4)_2[\textit{trans}\text{-}VO(H_2O)(NCS)_4]\cdot 4H_2O$.

Some "simple" organic species. Vanadates can be reduced to blue V^{IV} by CH_2O or $H_2C_2O_4$.

Acetic acid yields the important orange decavanadate ion, isolated, e.g., as $Na_6[V_{10}O_{28}]\cdot 18H_2O$:

$$10\,VO_4^{3-} + 24\,CH_3CO_2H \rightarrow V_{10}O_{28}^{6-} + 24\,CH_3CO_2^- + 12\,H_2O$$

Ageing or warming $V_{10}O_{28}^{6-}$ with NH_4^+ or large Alk^+ can precipitate dark-red MV_3O_8, sometimes called "hexavanadates".

Oxalate ion or $H_2C_2O_4$, plus V^{2+}, form $VC_2O_4\cdot 2H_2O$. Aqueous V^{3+} and VO^{2+} form stable complexes such as $V(C_2O_4)^+$, $V(C_2O_4)_2^-$, $VO(C_2O_4)$ and $[VO(C_2O_4)_2(H_2O)]^{2-}$. An easily crystallized salt is $K_3[V(C_2O_4)_3]\cdot 3H_2O$.

In dilute acidic solution, vanadates give, with cupferron, $C_6H_5N_2O_2^-$, a deep red precipitate, sensitive to about 0.08 mM.

Reduced nitrogen. Ammonia does not attack V.

The green $V_2O_3\cdot aq$, is obtained from V^{3+} plus NH_3.

If to a solution of a vanadate, neutral or alkaline, solid NH_4Cl is added, the vanadium is completely precipitated as NH_4VO_3, ammonium metavanadate, colorless, crystalline, insoluble in NH_4Cl solution.

Ammonia, in only a slight excess, provides another preparation of decavanadates, precipitated as $(NH_4)_6[V_{10}O_{28}]\cdot 6H_2O$ by adding acetone:

$$5\,V_2O_5 + 6\,NH_3 + 3\,H_2O \rightarrow [V_{10}O_{28}]^{6-} + 6\,NH_4^+$$

Excess $N_2H_5^+$ with V_2O_5 in aqueous HF gives $N_2H_5VOF_3$.

Vanadium(IV) is rapidly oxidized by NH_2OH in alkaline solution. With KCN present, however, NH_2OH reduces V^V at 100°C to an orange, diamagnetic, $K_3[V(CN)_5NO]\cdot 2H_2O$ after being precipitated by ethanol. Excess NH_2OH can result in $[V(CN)_4(NO)_2]^{2-}$. Vanadate(V) plus KCN, NH_2OH, K_2S and KOH yield a yellow, diamagnetic product, $K_4[V(CN)_6NO]\cdot H_2O$. The oxidation states are subject to some dispute because of the variable charge on NO.

Vanadates can be reduced to blue V^{IV} by acidified NH_3OH^+.

A suspension of fresh $V(OH)_2$ and (required) $Mg(OH)_2$ yields some N_2H_4 and NH_3 from N_2 in a form of nitrogen fixation.

Oxidized nitrogen. Vanadates can be reduced to V^{IV} by NO_2, but also high c(V^{IV} nitrate) are unstable, and HNO_3 oxidizes $VOSO_4$ rapidly at 80°C after a long induction period perhaps for autocatalysis.

Vanadium dissolves slowly in HNO_3 or aqua regia.

The trioxide V_2O_3 is attacked by HNO_3.

Fluorine species. Vanadium dissolves slowly in HF.

The trioxide V_2O_3 is attacked by HF.

The F^- complexes of VO^{2+} are more stable than those of $\mathbf{3d}^{2+}$, less than those of UO_2^{2+}, and much more than with Cl^-, Br^- or I^-. One can isolate $Alk_3[VOF_5]$, $[M^{III}(NH_3)_6][VOF_5]$, $(K,NH_4)_2(VOF_4)$ etc., and hydrothermal treatment of V^{IV} with fluoride in aqueous $(CH_2OH)_2$ yields [*cis*-$VOF_4(H_2O)]^{2-}$, $[V_2O_2F_6(H_2O)_2]^{2-}$, $[V_2O_2F_8]^{4-}$ and $[V_4O_4F_{14}]^{6-}$.

5.1.3 Reagents Derived from the 3rd-to-5th-Period Non-Metals, Silicon through Xenon

Phosphorus species. Vanadates can be reduced to V^{IV} by H_2PHO_3.

Phosphate complexes VO^{2+} especially as $VO(\eta^2\text{–}HPO_4)$, but also as $VO(H_2PO_4)_2$ in 1-M $H_2PO_4^-$ at pH 2.

Diphosphate precipitates VO^{2+} as $(VO)_2[P_2O_7]$.

Phosphate and V^V form clusters including 14 V to one P, requiring months for equilibration in acids but only minutes otherwise.

Reduced chalcogens. An acidified vanadate gives no precipitate with H_2S, which reduces it to V^{IV}. "Ammonium sulfide" forms a thiovanadate. Treating an ammoniacal solution of vanadate with H_2S gives a violet-red color, a very good test in the absence of Mo. Aqueous VS_4^{3-} is deep violet but it decomposes to brown oxygenated species even in basic solution. Acidification then causes incomplete precipitation of V_2S_5, soluble in alkalis, alkali carbonates, and sulfides.

Oxidized chalcogens. Vanadates and the V^V oxide can easily be reduced to blue V^{IV} by $S_2O_3^{2-}$ or SO_2, e,g.:

$$V_2O_5 + SO_2 + 2\,H_3O^+ \rightarrow 2\,VO^{2+} + SO_4^{2-} + 3\,H_2O$$

Part of the peroxide in $[V(O_2)_3]^-$ can be reduced by SO_2 (no excess) while leaving the vanadium as vanadium(V):

$$[V(O_2)_3]^- + SO_2 + 3\,H_2O \rightarrow [VO(O_2)_2(H_2O)]^- + HSO_4^- + H_3O^+$$

Metallic vanadium dissolves slowly in hot, concentrated H_2SO_4.

Alums such as $[Cs(H_2O)_6][V(H_2O)_6](SO_4)_2$ arise from V^{3+}.

Sulfate and VO^{2+} form at least $VOSO_4$ and $VO(SO_4)_2^{2-}$ complexes.

Vanadium(IV) and (V) transfer e^- easily in H_2SO_4 but not in $HClO_4$.

Concentrated H_2SO_4 with V_2O_5 gives a blood-red solution that turns blue on dilution.

Vanadium(III) is oxidized slowly by $[S_2O_8]^{2-}$, but the following oxidation by $[S_2O_8]^{2-}$ is of O_2^{2-} to O_2, not of the V^V; it is highly catalyzed by VO^{2+}, i.e., $[VO(H_2O)_5]^{2+}$:

$$[VO(O_2)(H_2O)_4]^+ + HSO_3(O_2)^- \rightarrow [V(O)_2(H_2O)_4]^+ + O_2\uparrow + HSO_4^-$$

The same complex and peroxodisulfate, i.e., $[S_2O_8]^{2-}$, decompose each other with catalysis by Ag^+, with the final result, by way of VO^{2+}:

$$VO(O_2)^+ + [(SO_3)_2(O_2)]^{2-} + H_2O \rightarrow VO_2^+ + 2\,HSO_4^- + O_2\uparrow$$

Reduced halogens. Vanadium is insoluble in HCl.

Aqueous VCl_3 saturated with HCl gives green $AlkVCl_4 \cdot 6H_2O$ (Alk = K, Rb or Cs) at -10 to -30 °C, or red $Alk_2VCl_5 \cdot H_2O$ at 100 to 120 °C.

Vanadates and V_2O_5 dissolve in the acids and can be reduced to V^{IV} by HI (or even to V^{III}), HBr, or even HCl with H_3PO_4 present to stabilize the V^{IV}. One can remove the halides with $AgClO_4$ to get $VO(ClO_4)_2 \cdot aq$.

Heating V_2O_5 with 7-M HBr while Br_2 is released yields $VBr_3 \cdot 6H_2O$ or, with RbBr or CsBr and saturated HBr, $Alk_2VBr_5 \cdot 5H_2O$, all dark-green.

Elemental and oxidized halogens. Vanadium(III) is oxidized to V^{IV} or V^V, rapidly by Cl_2 or Br_2, and slowly by I_2.

Vanadium(IV) is rapidly oxidized by Cl_2.

The metal is oxidized by ClO_3^-, ClO_4^-, BrO_3^- or IO_3^-. An excellent volumetric method uses selective oxidation (to avoid interference from Cr, etc.) by BrO_3^- with SO_4^{2-} and a definite concentration of HCl. In this medium, $V^{<V}$ is oxidized to V^V, and, after removal of excess BrO_3^-, titrated with Fe^{2+}. The endpoint may be determined electrometrically.

Vanadium(III) is rapidly oxidized to V^{IV} or V^V by ClO_3^-.

To prepare, e.g., $VO(ClO_4)_2$ solutions, one may mix VO_2 and $HClO_4$, or $VOSO_4$ and $Ba(ClO_4)_2$.

5.1.4 Reagents Derived from the Metals Lithium through Uranium, plus Electrons and Photons

Oxidation. Aqueous $[Cr_2O_7]^{2-}$ or MnO_4^- rapidly oxidize V^{2+}, V^{III} or V^{IV} by steps to V^V. Many Co^{III} complexes are reduced by V^{2+}. A volumetric oxidant for VO^{2+} is MnO_4^-.

Metallic V dissolves as V^{2+} or higher while reducing $FeCl_3$, $CuCl_2$ or $HgCl_2$ to lower oxidation states or precipitating metallic Pt, Ag or Au.

Aqueous Fe^{3+} or $[Co(NH_3)_5(H_2O)]^{3+}$ oxidizes V^{2+} to $V^{>II}$. Likewise Fe^{III}, Cu^{2+} or Ag^+ converts V^{III} to V^{IV} or V^V; this can involve catalysis:

$$V^{III} + Cu^{II} \rightarrow V^{IV} + Cu^{I} \text{ (slow)}$$

$$Cu^{I} + Fe^{III} \rightarrow Cu^{II} + Fe^{II}$$

$$\overline{V^{III} + Fe^{III} \rightarrow V^{IV} + Fe^{II}}$$

Very acidified V^{2+}, catalyzed by Ni, Pd or Pt, faster with UV light (254 nm), reduces H_3O^+ to H_2, forming V^{III}, and V^{2+} reduces Pt^{IV} to Pt^{II}.

Reduction. Aqueous $V^{>II}$ plus Na_{Hg}, Zn or Zn_{Hg}, or at a cathode, can yield, e.g., V^{2+}; then evaporation over, say, P_4O_{10} can give the sulfate, $VSO_4 \cdot 7H_2O$, or $Alk_2[[V(H_2O)_6](SO_4)_2$, with Alk = K, Rb, Cs or NH_4:

$$VO^{2+} + 2\,e^- + H_3O^+ + HSO_4^- + 5\,H_2O \rightarrow [V(H_2O)_6]SO_4 \cdot H_2O\downarrow$$

This is a powerful reductant, forming mixed crystals, which are more stable in air with other $MSO_4 \cdot 7H_2O$ or double sulfates.

Aqueous V^{III} acetate plus K_{Hg} and CN^-, with ethanol but no air, go to $K_4[V(CN)_6] \cdot {\sim}3H_2O$, which is easily oxidized and hydrolyzed.

Magnesium and $V^{>III}$ can form V^{III}. Various metals under certain conditions give the lower oxidation states, often a mixture. Vanadates can be reduced to blue V^{IV} by Fe^{2+} or Hg.

In acidic solution, Zn, Cd and Al produce an interesting succession of colors from yellow to blue, green and violet due to the reduction of V^V to V^{IV}, then V^{III}, and finally V^{2+}.

A cathode with V_2O_5 dissolved in HCl, HBr or HI can produce $[V(H_2O)_6](Br,I)_2$ or $[VX_2(H_2O)_4]$ with X = Cl, Br or I. Vanadium(III) is often prepared from V^{IV} or V^V by electrolysis with air absent.

Violet and ultraviolet light reduce $[V^VO(O_2)]^+$ to $V^{IV}O^{2+}$:

$$2\,[VO(O_2)]^+ + 2\,H_3O^+ + \gamma \rightarrow 2\,VO^{2+} + {}^3/_2\,O_2\uparrow + 3\,H_2O$$

Other reactions. Aqueous Ba^{2+} and vanadate(V) precipitate a yellow $Ba(VO_3)_2$ (formula distinct from the PO_4^{3-} and AsO_4^{3-} cases), which becomes colorless on standing.

Vanadium species are both oxidized and reduced in the reactions of V^{2+} and V^{III} in acid with VO^{2+} and V^V, giving V^{III} and VO^{2+}.

Neutral or slightly acidified vanadates precipitate **d**- or **p**-block M^{n+}.

Complexes with $H_3[PW_{12}O_{40}]$ are deep purple (V^{IV}) and yellow (V^V).

Acidified vanadate solutions plus $[Fe(CN)_6]^{4-}$ form a green precipitate, insoluble in inorganic acids.

The following oxidation in, e.g., 1-M $HClO_4$ is formally of O_2^{2-} to O_2^-, not of the V^V; the product soon decomposes, actually reducing the V:

$$[VO(O_2)(H_2O)_4]^+ + Co^{3+} \rightarrow [VO(O_2)(H_2O)_4]^{2+} + Co^{2+}$$

$$[VO(O_2)(H_2O)_4]^{2+} + H_2O \rightarrow [VO(H_2O)_5]^{2+} + O_2\uparrow$$

Silver vanadate, yellow to orange, is obtained in a neutral solution.

Aqueous Hg_2^{2+} precipitates yellow mercurous vanadate, $Hg_2(VO_3)_2$. This has been used gravimetrically, with ignition to V_2O_5.

Lead acetate, $Pb(CH_3CO_2)_2$, forms a basic lead vanadate, yellow, turning to white on standing. Precipitation can be made quantitative.

Aqueous $V(C_2O_4)^+$ and $V(C_2O_4)_2^-$ sensitize the decomposition of $H_2C_2O_4$ by UV light at 254 nm:

$$H_2C_2O_4 \rightarrow CO_2\uparrow + CO\uparrow + H_2O$$

5.2 Niobium, $_{41}$Nb; Tantalum, $_{73}$Ta; and Dubnium, $_{105}$Db

The oxidation numbers of Nb and Ta: fractional, (III), (IV) and (V) (the most common one), as in $Ta_6Cl_{12}^{2+}$, $KNb(SO_4)_2 \cdot 4H_2O$, $K_4[Nb(CN)_8] \cdot 2H_2O$, and Ta_2O_5. In the lower states, Nb and Ta are more like Mo and W, but otherwise more like Group 4.

The oxidation states for dubnium calculated relativistically to be stable in water: (III), (IV) and (V), especially (V).

5.2.1 Reagents Derived from Hydrogen and Oxygen

Water. Solutions of niobates(V) or tantalates(V), upon boiling, readily hydrolyze to gelatinous precipitates of the hydrated oxides. (The alkali compounds are formed by fusion of the oxides.) Diluting and boiling sulfato complexes, perhaps from fusing Nb_2O_5 with $KHSO_4$, likewise with aqueous oxalato complexes, yields $Nb_2O_5 \cdot aq$ at pHs near 1, along with complex anions otherwise, including $[Nb_{12}O_{36}]^{12-}$ at a pH of, say, 4 to 7, or the dominant $[M_6O_{19}]^{8-}$ at high pH.

Quantum calculations on Nb, Ta, Pa and Db, and experiments on the first three, agree that the hydrolysis of M^V cations and the chlorides MCl_5 varies as Nb > Ta but reporters differ on Db and Pa.

Some hot natural waters may contain $[NbF_6]^{2-}$.

Oxonium. The metals resist HNO_3, H_3PO_4, H_2SO_4 (hot, dilute), HCl and HNO_3/HCl, but hot, concentrated HF and HNO_3/HF attack them.

The ignited oxides M_2O_5 are insoluble in all acids but HF.

Acidifying niobates gives similar results as in **Water** above.

Hydroxide. The metals react with fused alkali only at high *T*. Molten AlkOH or Alk_2CO_3 dissolve M_2O_5, and aqueous OH^- dissolves fresh $M_2O_5 \cdot aq$; then one may crystallize normal and hydrogen salts of $[M_6O_{19}]^{8-}$, i.e., [*octahedro*-$(MO)_6$-$(\mu\text{-}O)_{12}(\mu_6\text{-}O)]^{8-}$, but among the few soluble ones known is $K_8[Nb_6O_{19}] \cdot 16H_2O$. The insoluble ones are like mixed oxides, with no discrete anions in the solids. Even a pH near 14 with Nb^V still gives $[Nb_6O_{19}]^{8-}$, although concentrated OH^- does appear to yield $[MO_2(OH)_4]^{3-}$ ("$MO_4{}^{3-}$") for both Nb and Ta.

Peroxide. When precipitated hot, $Ta_2O_5 \cdot aq$ is almost insoluble, and when precipitated cold, only slightly soluble, in H_2O_2 (distinction from the more soluble Nb and Ti oxides).

Alkaline niobates and tantalates, when treated with H_2O_2, precipitate $Alk_3[M(\eta^2\text{-}O_2)_4]$. These release O_2 slowly, or explosively at 80°C. An acid, perhaps $[\{NbO(O_2)OH\}_2] \cdot aq$, arises on adding H_2SO_4.

Hydrogen peroxide forms a stable per-acid, i.e., $HTaO_4 \cdot aq$.

Niobium and H_2O_2 are found not to give a color as often claimed.

5.2.2 Reagents Derived from the Other 2nd-Period Non-Metals, Boron through Fluorine

Carbon oxide species. Alkali carbonates dissolve no Nb_2O_5 or Ta_2O_5.

Cyanide species. Aqueous NCS^- gives a bright yellow color with solutions of niobates (distinction from Ti and Ta).

Some "simple" organic species. Sulfato complexes, e.g., from fusing Nb_2O_5 with $AlkHSO_4$ or $Alk_2[S_2O_7]$, and $H_2C_2O_4$ form $Nb(C_2O_4)_n{}^{(2n-5)-}$.

Reduced nitrogen. The oxalato complexes plus NH_3 give $Nb_2O_5 \cdot aq$.

Oxidized nitrogen. Metallic Nb and Ta are not attacked even by hot, concentrated HNO_3 or aqua regia.

Fluorine species. The metals dissolve slowly in HF but readily in contact with Pt or in a mixture of HF and HNO_3.

Aqueous HF dissolves Nb_2O_5 and Ta_2O_5.

The complexation of M^V by F^- goes as $Pa > Nb \geq Db > Ta$.

The separation of Nb and Ta from each other is quite difficult. Reactions that seem excellent with the individual metals are very poor with a mixture. The fluorides, however, allow a rather good separation. The oxides Ta_2O_5 and Nb_2O_5 are dissolved in 1.0 to 1.5-M HF, the concentration of the solution is carefully adjusted and Ta is precipitated as K_2TaF_7 by adding the right amount of KF. Niobium favors the soluble $K_2NbOF_5 \cdot 2H_2O$. With proper control of conditions, four or five fractionations separate them fairly completely.

An easier, more modern, method extracts a Ta complex from dilute HF into 4-methyl-2-pentanone (methyl isobutyl ketone, "hexone"), followed by Nb at a higher acidity.

Aqueous HF, Nb_2O_5 and MCO_3 form $M[NbOF_5]$; M = **3d** or Cd. Fluorotantalates, however, arise from Ta_2O_5.

5.2.3 Reagents Derived from the 3rd-to-5th-Period Non-Metals, Silicon through Xenon

Oxidized chalcogens. Metallic Nb and Ta do not react with hot, dilute H_2SO_4, but boiling, concentrated H_2SO_4 slowly dissolves them.

The gelatinous $Ta_2O_5 \cdot aq$, precipitated from a cold, dilute solution of a tantalate by dilute H_2SO_4, dissolves in the hot, concentrated acid and reprecipitates when cold and diluted. Hydrated $Nb_2O_5 \cdot aq$ does not reprecipitate from a similar solution of Nb sulfate complexes.

Reduced halogens. Even hot, concentrated HCl does not attack Nb or Ta. The ignited oxides Nb_2O_5 and Ta_2O_5 are not attacked by HCl or HBr. Otherwise Nb_2O_5, $NbCl_5$ and $NbOCl_3$ do dissolve in concentrated HCl, and adding excess NH_4Cl yields $(NH_4)_2[NbOCl_5]$.

High-*T* syntheses of, e.g., M_6X_{14} followed by extraction with boiling H_2O give the octahedral, green $M_6(\mu\text{-}X)_{12}{}^{2+}$; then crystallization from aqueous HX, with X = Cl or Br, forms $[trans\text{-}(Ta_6Cl_{12})Cl_2(H_2O)_4] \cdot 3H_2O$. The bromides, but not the chlorides, dissolve well in water, where $Ta_6I_{12}{}^{2+}$ is unstable. The OH^- ion precipitates $M_6X_{12}(OH)_2 \cdot 8H_2O$, soluble in an excess as $M_6X_{12}(OH)_4{}^{2-}$, showing the inner X^- as inert. Therefore mixed-halide clusters such as $[Nb_6Cl_{12}(F,Br)_2(H_2O)_4] \cdot 4H_2O$ can also be made. All six M have a $^7/_3$ oxidation state.

The diamagnetic $M_6X_{12}{}^{2+}$ is oxidized to paramagnetic, yellow $M_6X_{12}{}^{3+}$ by O_2 (slow neutral, fast acidic or basic), I_2 or Hg^{II}, or to diamagnetic, red-brown $M_6X_{12}{}^{4+}$ by excess O_2 (for Ta), H_2O_2, Cl_2, BrO_3^-, Ce^{IV}, VO_2^+ or Fe^{III}. The chlorides are oxidized faster than the bromides. The 2+ ions may be restored with Cr^{2+}, V^{2+}, Cd or $SnCl_2$. The clusters with Nb, Br and higher charge are more hydrolyzed. All the (mixed) oxidation states also occur in $[(M_6X_{12})X_6]^{n-}$, with $4 \geq n \geq 2$, respectively.

Because of relativity, as in other Groups, periodic behavior such as complexation, ion exchange, and extractibility into organic solvents does not extrapolate Nb → Ta → Db. Accordingly, the tendency of M^V to favor chloro over hydroxo complexes in 4 to 12-M HCl is: Pa > Nb > Db > Ta. In ion-exchange media containing a small amount of HF, however, it is Pa > Db > Nb. Relativistic calculations on Nb, Ta, Pa and Db, and experiments on the first three, agree that the tendency of M^V to form, e.g., $[MCl_4(OH)_2]^-$ and to be extracted by anion exchangers varies as Pa >> Nb ≥ Db > Ta, and also with the sequence $[MF_6]^- > [MCl_6]^- > [MBr_6]^-$. Other complexes such as $[DbOCl_4]^-$ and $[DbCl_6]^-$ seem to exist at high HCl concentrations.

5.2.4 Reagents Derived from the Metals Lithium through Uranium, plus Electrons and Photons

Reduction. Zinc, H_2SO_4 and K^+ reduce Nb^V sulfato complexes to diamagnetic $KNb(SO_4)_2 \cdot 4H_2O$, sensitive to O_2. This solution, stable under CO_2, reduces NO to NH_3, and also reduces U^{VI}, V^V, Mo^{VI}, Fe^{III}, Cu^{II} and Tl^{III}. Furthermore it is a source of $Nb^VONb^{III}(PO_4)_2 \cdot 6H_2O$ etc.

In aqueous HCl Zn reduces $NbCl_5$ to $[NbOCl_4(H_2O)]^{2-}$. However, Zn and concentrated HCl produce a blue to brown color in Nb^V solutions even in the presence of F^- (distinction from Ta). Titanium in large amount interferes, forming a green color.

Cathodic treatment or Zn_{Hg}, and K^+ reduce Ta^V oxalato complexes to O_2-sensitive, diamagnetic, very labile $K_5[Ta(C_2O_4)_4]$.

Volumetrically, niobium can be determined by reduction to Nb^{III} in a Jones reductor (with Zn_{Hg}) and titration with MnO_4^-.

Treatment of methanolic $NbCl_5$ with a mercury cathode and then a concentrated aqueous solution of KCN yields an orange salt, $K_4[Nb(CN)_8] \cdot 2H_2O$, isomorphous with $K_4[Mo(CN)_8] \cdot 2H_2O$.

Other reactions. At pH 10.6, $K_7H[Nb_6O_{19}] \cdot 13H_2O$, $K[Co(CO_3)_2(NH_3)_2]$ and KOH form green $K_7[Co(\eta^3\text{-}Nb_6O_{19})(\eta^2\text{-}CO_3)(NH_3)]$. Other clusters include $Na_{12}Ni^{IV}Nb_{12}O_{38} \cdot (48\text{–}50)H_2O$ and $K_8Na_4Ni^{IV}Nb_{12}O_{38} \cdot 12H_2O$.

Aqueous $[FeCN)_6]^{4-}$ produces, with niobates, a pale yellow precipitate; with tantalates, only a yellow color in solution. The latter does not appear in the presence of oxalic, tartaric, citric or arsenic acids.

Relativity makes Tl_3TaSe_4 yellow green (absorbing higher-energy photons) in contrast to the deep violet of Tl_3NbSe_4.

Bibliography

See the general references in the Introduction, and some more-specialized books [1–4]. Journal articles include: the hydrothermal chemistry of vanadium oxyfluoride oligomers [5]; the hydronation and condensation of V^V etc. [6]; niobium compounds [7]; a thematic issue on polyoxometalates, especially of V, Nb, Ta, Mo and W; also see the references therein [8]; vanadium-peroxide complexes [9]; aqueous electron-transfer reductions by V^{IV} and Fe^{II} [10]; and thermochemistry and the oxidation potentials of V, Nb and Ta [11].

1. Hala J (1989) Halides, oxyhalides and salts of halogen complexes of titanium, zirconium, hafnium, vanadium, niobium and tantalum. IUPAC, Blackwell, London
2. Pope MT (1983) Heteropoly and isopoly oxometalates. Springer, Berlin Heidelberg New York
3. Clark RJH (1968) The chemistry of titanium and vanadium. Elsevier, Amsterdam
4. Fairbrother F (1967) The chemistry of niobium and tantalum. Elsevier, Amsterdam
5. Aldous DW, Stephens NF, Lightfoot P (2007) Dalton Trans 2271

6. Cruywagen JJ (2000) Adv Inorg Chem 49:127
7. Nowak I, Ziolek M (1999) Chem Rev 99:3603
8. Hill CL (ed) (1998) Chem Rev 98:1
9. Butler A, Clague MJ, Meister GE (1994) Chem Rev 94:625
10. Rosseinsky DR (1972) Chem Rev 72:215
11. Hill JO, Worsley IG, Hepler LG (1971) Chem Rev 71:127

6 Chromium through Seaborgium

6.1 Chromium, $_{24}Cr$

Oxidation numbers: mainly (II), (III) and (VI), as in Cr^{2+}, Cr_2O_3 and CrO_4^{2-}, plus (IV) and (V) in peroxo complexes etc.

6.1.1 Reagents Derived from Hydrogen and Oxygen

Dihydrogen. The Cu^{2+} ion catalyzes the reduction of Cr^{VI} (see **11.1.1**):

$$[Cr_2O_7]^{2-} + 3\,H_2 + 8\,H_3O^+ \rightarrow 2\,Cr^{3+} + 15\,H_2O$$

Water. Both Cr_2O_3 and $Cr_2O_3 \cdot aq$, also $CrPO_4$, are insoluble.

The Cr^{III} nitrate and even the basic nitrates dissolve readily.

Chromium(III) sulfide is hydrolyzed to $Cr_2O_3 \cdot aq$.

The common $[Cr^{III}Cl_2(H_2O)_4]Cl \cdot 2H_2O$ is green and easily soluble. The violet, sublimed, anhydrous $CrCl_3$ is insoluble in water or in dilute or concentrated acids. A tiny amount of Cr^{2+} or $SnCl_2$ catalyzes its dissolution; the $[Cr(H_2O_6]^{2+}$ ions reduce inert crystalline $CrCl_3$ units to labile Cr^{II}. This dissolves as $[Cr(H_2O_6]^{2+}$, continuing the cycle. Likewise Cr^{2+} catalyzes the equilibration of many Cr^{III} complexes.

The Cr^{III} bromide and sulfate also exist in soluble and insoluble modifications. All of these normal salts in solution react acidic by hydrolysis, which forms numerous mono- and polymeric products. The $[Cr(H_2O)_6]^{3+}$ ion is violet.

Water replaces NH_3 in $[Cr(NH_3)_6]^{3+}$, but not faster in high acidity.

Aqueous I_2 catalyzes the aquation of $[CrI(H_2O)_5]^{2+}$ to $[Cr(H_2O)_6]^{3+}$.

Chromium trioxide, CrO_3, is very soluble in H_2O, and yields H_2CrO_4, $H_2[Cr_2O_7]$ and their ions, depending on the concentrations. Yellow CrO_4^{2-} predominates at $pH > 8$, $HCrO_4^-$ and orange $[Cr_2O_7]^{2-}$ are in equlibrium when $2 < pH < 6$, and H_2CrO_4 predominates if $pH < 1$ at ordinary c:

$$HCrO_4^- + H_2O \leftrightarrows CrO_4^{2-} + H_3O^+$$

$$2\,CrO_4^{2-} + 2\,H_3O^+ \leftrightarrows Cr_2O_7^{2-} + 3\,H_2O$$

They are somewhat carcinogenic and mutagenic. Higher concentrations can provide, e.g., $K_2[Cr_3O_{10}]$ and $K_2[Cr_4O_{13}]$.

The chromates of the alkalis and Mg, Ca, Cu^{II} and Zn are soluble; those of Sr and Hg^{II} are slightly soluble; the insoluble salts include those of Ba, Mn, Ag, Hg^{I}, Pb and Bi—the neutral salt has not been prepared; bismuthyl dichromate, $(BiO)_2Cr_2O_7$, is often obtained.

The kinetics of hydrolysis of $[Cr^{III}X(H_2O)_5]^{n+}$ to $[Cr(H_2O)_6]^{3+}$ in paths of orders -1, 0 and 1 with respect to H_3O^+ all occur with $X = N_3^-$ and SO_4^{2-}, only for order 1 with $PH_2O_2^-$, for 0 with NCS-κN^-, for 0 and 1 with F^-, and for -1 and 0 with Cl^-, Br^- and I^-.

At pH 4 to 6, 25°C, in 30 min, $[Cr(H_2O)(NH_3)_5]^{3+}$ replaces much NH_3 with H_2O. Water and $[CrX(NH_3)_5]^{2+}$ ($X^- = NCS^-$, Cl^-, or Br^-) slowly form mainly $[Cr(H_2O)(NH_3)_5]^{3+}$; cf. **6.1.4 Other reactions** with light. The aquation of many Cr^{III} cyano-complexes is catalyzed by Cr^{2+} and Hg^{II}.

Seawater and some freshwater contain traces of Cr^{III} and Cr^{VI} complexes as $Cr(OH)_2^+$, CrO_4^{2-}, $HCrO_4^-$, $NaCrO_4^-$ and $KCrO_4^-$.

Oxonium. Metallic Cr dissolves in H_3O^+ as Cr^{2+}, better with heat when pure, readily when impure, but impurities catalyze further oxidation:

$$[Cr(H_2O)_6]^{2+} + H_3O^+ \rightarrow [Cr(H_2O)_6]^{3+} + {}^1\!/_2\, H_2\uparrow + H_2O$$

Chromium(III) oxide, Cr_2O_3, is slowly soluble in acids, best in HCl, unless previously ignited. The hydroxide is soluble in acids.

In reducing acids, CrO_3 forms Cr^{III}.

Hydroxide. Aqueous OH^- and Cr^{2+} precipitate brownish $Cr(OH)_2$. It slowly reduces H_2O to H_2, forming $Cr_2O_3 \cdot aq$.

Alkali hydroxides precipitate from Cr^{III}, $Cr_2O_3 \cdot aq$, gray green to gray blue (not precipitated in the presence of, e.g., glycerol or tartrates). The product retains traces of the alkali cation not easily removed by washing. and is soluble in acids and excess of OH^-, the latter yielding the green complex (or peptized precipitate):

$$Cr_2O_3 \cdot aq + 2\,OH^- \rightarrow 2\,Cr(OH)_4^-$$

The $Cr_2O_3 \cdot aq$ is completely reprecipitated on long boiling or standing (distinction from Al), or on heating with excess NH_4^+. Further addition of dilute alkali has little effect. The presence of some non-amphoteric hydrous oxides, e.g., $Fe_2O_3 \cdot aq$, greatly hinders the dissolution in OH^-, hence Cr cannot be separated from Fe by excess of OH^-.

Slowly adding limited OH^- to $Cr_2(SO_4)_3$ and refluxing 24 h free from CO_2 yields dark-green $[\{Cr(H_2O)_4\}_2(\mu\text{-}OH)(\mu\text{-}SO_4)]^{3+}$.

Concentrated NaOH solutions of $Cr_2O_3 \cdot aq$ or $CrCl_3$ yield the solids $Na_9[Cr(OH)_6]_2(OH)_3 \cdot 6H_2O$ or $Na_4[Cr(OH)_6]Cl \cdot H_2O$.

The well-known instability of some N_3^- compounds contrasts with the stability of Cr–N_3 in $[CrN_3(NH_3)_5]^{2+}$, whose hydrolysis by OH^- causes the replacement not only of N_3^- but also of NH_3.

Alkali hydroxides change dichromates to normal chromates:

$$[Cr_2O_7]^{2-} + 2\,OH^- \rightarrow 2\,CrO_4^{2-} + H_2O$$

Peroxide. The action of H_2O_2 or HO_2^- on ions of Cr depends on their oxidation state, the pH, the *T*, the amount of H_2O_2 present, etc. Heating alkaline, but not acidified, chromium(III) with H_2O_2 gives chromate:

$$2\,Cr(OH)_4^- + 3\,HO_2^- \rightarrow 2\,CrO_4^{2-} + OH^- + 5\,H_2O$$

From cold H_2O_2 and CrO_4^{2-} above pH 7 arises the unexpected red-brown Cr^V in $[Cr(\eta^2\text{-}O_2)_4]^{3-}$. This decomposes in base:

$$2\,[Cr(O_2)_4]^{3-} + H_2O \rightarrow 2\,CrO_4^{2-} + {}^7/_2\,O_2\uparrow + 2\,OH^-$$

A pH < 4 (from, say, HNO_3 or $HClO_4$) quickly gives temporary blue and violet intermediates, with the overall decomposition being:

$$[Cr^V(O_2)_4]^{3-} + 6\,H_3O^+ \rightarrow Cr^{3+} + {}^5/_2\,O_2\uparrow + 9\,H_2O$$

The deep-blue complex, probably $[Cr^{VI}O(O_2)_2(\text{solvent})]$, sometimes called CrO_5 or "perchromic acid", is often extracted (to detect chromium) and thus concentrated in oxygenated organic solvents such as diethyl ether, ethyl acetate or 1-pentanol, where it is much more stable. It is more stable below 0°C than above, decomposing fairly rapidly at room temperature, approximately as shown:

$$[CrO(O_2)_2(H_2O)] + 3\,H_3O^+ \rightarrow Cr^{3+} + {}^3/_2\,O_2\uparrow + {}^1/_2\,H_2O_2 + 5\,H_2O$$

This complex, or a dehydronated ("deprotonated") form may be an intermediate in various reactions. The formation of this "Chromium Blue" is an excellent test for Cr^{VI}; about 0.5 mM may be detected readily, especially if the ether is used, but this does not extract the products of neutral or alkaline solutions.

Moderate pHs give mixtures of products, and weak acids can yield more complicated behavior, such as oscillation. Neutral H_2O_2 plus $[Cr(O_2)_4]^{3-}$, or Cr^{VI} when $4 < pH < 7$, yield a violet $[CrO(O_2)_2(OH)]^-$ etc., even less stable than the blue "CrO_5".

Greenish $[Cr(O_2)_2(CN)_3]^{3-}$ and $[Cr(O_2)_2(NH_3)_3]$, surprisingly with Cr^{IV}, can be made by (1) adding the base, CN^- or NH_3, to $[Cr(O_2)_4]^{3-}$; (2) adding the base and H_2O_2 to CrO_3; and (3) treating CrO_5 with an excess of the base. In 1-M $HClO_4$, $[Cr(O_2)_2(NH_3)_3]$ decomposes to $[Cr(H_2O)_3(NH_3)_3]^{3+}$and O_2. With HCl, both H_2O and Cl^- replace the O_2^{2-}.

Solid peroxochromates mostly explode when struck or warmed, or even spontaneously at ambient *T*.

Peroxide helps separate and identify Cr mixed with similar metal species. Any Cr^{III} is precipitated along with Fe^{III} and Al^{III} by NH_3 in the presence of NH_4^+. Boiling with OH^- and HO_2^- oxidizes the chromium to CrO_4^{2-}, leaving the iron as $Fe_2O_3\cdot aq$ and the aluminum as $Al(OH)_4^-$. Boiling the separated solution with

NH_4Cl (or better, the sulfate) precipitates $Al_2O_3 \cdot aq$ and aids in removing excess peroxide. Chromium may be identified in the solution after acidifying (a) with acetic acid and adding Pb^{II} to precipitate yellow $PbCrO_4$; or (b) with H_2SO_4 and adding H_2O_2 to give the vanishing "Chromium Blue".

Di- and trioxygen. Chromium is inert in moist air up to 100 °C.

Exposing Cr^{2+} to air generates mainly $[\{Cr(H_2O)_4\}_2(\mu\text{-}OH)_2]^{4+}$, but injecting Cr^{2+} into 1-cM to 1-dM $HClO_4$, saturated with O_2 to ensure an excess, gives up to 0.5-mM of (hyperoxo) $Cr(O_2)^{2+}$, i.e., $Cr^{3+}(O_2^-)$ or $[Cr(O_2)(H_2O)_5]^{2+}$, inert enough, if stabilized with a little ethanol, to be followed further. Excess Cr^{2+} and the $Cr(O_2)^{2+}$ yield some $Cr^{IV}O^{2+}$, i.e., $[CrO(H_2O)_5]^{2+}$, with a decay $t_{1/2}$ of ~ 20 s at 25 °C and pH 1:

$$Cr^{2+} + CrO_2^{2+} \rightarrow 2\,CrO^{2+}$$

$$3\,CrO^{2+} + 2\,H_2O \rightarrow HCrO_4^- + 2\,Cr^{3+} + H_3O^+$$

and, more elaborately, but still in summary form:

$$[Cr(H_2O)_6]^{2+} + O_2 \rightarrow [Cr^{3+}(O_2^-)(H_2O)_5] + H_2O$$

$$[Cr(H_2O)_6]^{2+} + [Cr(O_2)(H_2O)_5]^{2+} \rightarrow [\{Cr^{III}(H_2O)_5O\text{–}\}_2]^{4+} + H_2O$$

$$[\{Cr(H_2O)_5O\text{–}\}_2]^{4+} + 2\,[Cr(H_2O)_6]^{2+} \rightarrow$$

$$2\,[\{Cr^{III}(H_2O)_5\}_2(\mu\text{-}OH)_2]^{4+} \text{ i.e. } 2\,Cr_2(OH)_2^{4+}$$

Over a 20-fold excess of O_2 first yields pure $[Cr(O_2)(H_2O)_5]^{2+}$, which then goes to other Cr^{VI} and Cr^{III} species, e.g.:

$$Cr(O_2)^{2+} + 5\,H_2O \rightarrow HCrO_4^- + 3\,H_3O^+$$

Both CrO^{2+} and $Cr(O_2)^{2+}$ oxidize I^- instantly to I_3^-; $HCrO_4^-$ or $[Cr_2O_7]^{2-}$ takes a few minutes.

The $Cr(O_2)^{2+}$ does not react directly with $HCrO_4^-$, but does react with the one-electron, outer-sphere reductants $[V(H_2O)_6]^{2+}$, $[Ru(NH_3)_6]^{2+}$ etc.:

$$[Cr^{III}(O_2)(H_2O)_5]^{2+} + e^- + H_3O^+ \rightarrow [Cr^{III}(O_2H)(H_2O)_5]^{2+} + H_2O$$

Then a one-electron oxidant, Ce^{IV}, reverses the reaction cleanly:

$$Cr(O_2H)^{2+} + Ce^{4+} + H_2O \rightarrow Cr(O_2)^{2+} + Ce^{3+} + H_3O^+$$

The $[Cr(O_2H)(H_2O)_5]^{2+}$ has a typical survival $t_{1/2}$ of about 15 min. Few such aqueous metal-hydroperoxo complexes are known.

Excess Fe^{II} reduces $Cr(O_2H)^{2+}$, with $1\text{ dM} < c(H_3O^+) < 5\text{ dM}$, thus:

$$Cr(O_2H)^{2+} + 2\,Fe^{2+} + 3\,H_3O^+ \rightarrow Cr^{3+} + 2\,Fe^{3+} + 5\,H_2O$$

Passing air through a cold mixture of aqueous $CrCl_2$, concentrated NH_3 and NH_4Cl yields a red complex:

$$2\,Cr^{2+} + 9\,NH_3 + {}^1/_2\,O_2 + NH_4{}^+ + 5\,Cl^- \rightarrow [\{Cr(NH_3)_5\}_2(\mu\text{-}OH)]Cl_5\downarrow$$

Adding this slowly to cold, concentrated HCl and heating gives a bright-red $[CrCl(NH_3)_5]Cl_2$, a good starter for other preparations:

$$[\{Cr(NH_3)_5\}_2(OH)]Cl_5 + H_3O^+ + Cl^- \rightarrow 2\,[CrCl(NH_3)_5]Cl_2\downarrow + 2\,H_2O$$

The product is somewhat sensitive to sunlight, but otherwise does not soon lose the inner Cl^- to aqueous Ag^+ when cold. Heating it with water forms $[Cr(H_2O)(NH_3)_5]Cl_3$; long boiling with OH^- gives $Cr_2O_3\cdot aq$.

Ozone oxidizes Cr^{II} or Cr^{III} to $CrO_4{}^{2-}$ or $[Cr_2O_7]^{2-}$.

6.1.2 Reagents Derived from the Other 2nd-Period Non-Metals, Boron through Fluorine

Carbon oxide species. Carbonates and suspensions of Cr^{II} acetate, $[\{Cr(H_2O)\}_2(\mu\text{-}CH_3CO_2)_4]$ (see below), give $M_4[\{Cr(H_2O)\}_2(\mu\text{-}CO_3)_4]$ with M as any Alk, NH_4 or ${}^1/_2\,Mg$.

From Cr^{III}, alkali carbonates precipitate $Cr_2O_3\cdot aq$ (in the absence of chelators). The precipitate is practically free from carbonate:

$$2\,Cr^{3+} + 3\,CO_3{}^{2-} \rightarrow Cr_2O_3\cdot aq\downarrow + 3\,CO_2\uparrow$$

Barium carbonate precipitates chromium from its solutions (better from $CrCl_3$) as a hydrous oxide with some basic salt, the precipitation being complete after long digestion in the cold.

Alkali carbonates change dichromates to normal chromates:

$$[Cr_2O_7]^{2-} + CO_3{}^{2-} \rightarrow 2\,CrO_4{}^{2-} + CO_2\uparrow$$

Cyanide species. The oxidation of $[Cr(CO)_5]^{2-}$ by NaCN and water at 10°C over three weeks yields a colorless, diamagnetic product, $Na[Cr(CO)_5CN]\cdot H_2O$. Further reaction at 75°C for 12 hours forms $Na_2[Cr(CO)_4(CN)_2]\cdot 2H_2O$ after crystallization.

Adding $[Cr_2(CH_3CO_2)_4(H_2O)_2]$ slowly to excess air-free KCN produces a red solution but then deep-green $K_4[Cr(CN)_6]\cdot 2H_2O$:

$$ {}^1/_2\,[Cr_2(CH_3CO_2)_4(H_2O)_2] + 6\,CN^- + 4\,K^+ + H_2O \rightarrow $$

$$K_4[Cr(CN)_6]\cdot 2H_2O\downarrow + 2\,CH_3CO_2{}^-$$

Then air and methanol give yellow, very H_2O-soluble, $K_3[Cr(CN)_6]$ (reducible again at cathodes):

$$2\,[Cr(CN)_6]^{4-} + {}^1/_2\,O_2 + H_2O \rightarrow 2\,[Cr(CN)_6]^{3-} + 2\,OH^-$$

Cold CN^- and Cr^{3+} precipitate $Cr_2O_3 \cdot aq$.

Chromium(III) acetate (after evaporating excess CH_3CO_2H), or the chloride, if poured into boiling KCN, then partly evaporated and cooled, yields the very soluble, pale-yellow $K_3[Cr(CN)_6]$. The complex hydrolyzes slowly in water, especially with light or heat, to $[Cr(CN)_n(H_2O)_{6-n}]^{(3-n)+}$, or with OH^-, especially hot, to $[Cr(CN)_n(OH)_{6-n}]^{3-}$ or $Cr_2O_3 \cdot aq$, and cyanide. Dilute H_3O^+ likewise dissociates it. It reacts with NH_2OH to give $[Cr(CN)_5NO)^{3-}$.

Boiling KCN and $[CrCl(NH_3)_5]^{2+}$ forms $K_3[Cr(CN)_5OH] \cdot H_2O$, yellow-orange. Many other mixed, as well as dinuclear, complexes are known.

Aqueous $[Cr(CN)_6]^{3-}$ precipitates , e.g., $[NBu_4]^+$ or $[PPh_4]^+$, and it precipitates M^{2+}, with M = **d**-block metals, as $M_3[Cr(CN)_6]_2 \cdot {\sim}14H_2O$.

At 60 °C KCN and $[Cr(O_2)_2(NH_3)_3]$ (see **Peroxide** above) form the explosive brown $K_3[Cr^{IV}(O_2)_2(CN)_3]$, precipitated by ethanol.

Heating alkaline CrO_4^{2-} with KCN and NH_2OH at 100°C, followed by cooling and adding ethanol, gives bright-green $K_3[Cr(CN)_5NO]$. Mild acidities hydrolyze this to $[Cr(CN)_2(H_2O)_3NO]$ and then to $[Cr(H_2O)_5NO]^{2+}$. See **8.1.2 Oxidized nitrogen** about oxidation states.

Thiocyanate and Cr^{2+} produce an unstable blue solution but can be crystallized with Na^+, not some other cations, as deep lilac-blue $Na_3Cr(NCS)_5 \cdot nH_2O$. On Cr^{III}, however, the *trans*-effect is seen:

$$[Cr(NCS)_6]^{3-} + CN^- \rightarrow [Cr(CN)(NCS)_5]^{3-} + NCS^-$$

$$[Cr(CN)(NCS)_5]^{3-} + CN^- \rightarrow [\textit{trans}\text{-}Cr(CN)_2(NCS)_4]^{3-} + NCS^-$$

$$[\textit{trans}\text{-}Cr(CN)_2(NCS)_4]^{3-} + CN^- \rightarrow [\textit{mer}\text{-}Cr(CN)_3(NCS)_3]^{3-} + NCS^-$$

$$[\textit{mer}\text{-}Cr(CN)_3(NCS)_3]^{3-} + CN^- \rightarrow [\textit{trans}\text{-}Cr(CN)_4(NCS)_2]^{3-} + NCS^-$$

$$[\textit{trans}\text{-}Cr(CN)_4(NCS)_2]^{3-} + CN^- \rightarrow [Cr(CN)_5(NCS)]^{3-} + NCS^-$$

$$[Cr(CN)_5(NCS)]^{3-} + CN^- \leftrightarrows [Cr(CN)_6]^{3-} + NCS^-$$

$$[Cr(CN)_5(NCS)]^{3-} + NCS^- \rightarrow [\textit{cis}\text{-}Cr(CN)_4(NCS)_2]^{3-} + CN^-$$

$$[\textit{cis}\text{-}Cr(CN)_4(NCS)_2]^{3-} + NCS^- \rightarrow [\textit{fac}\text{-}Cr(CN)_3(NCS)_3]^{3-} + CN^-$$

The hard Cr^{III} favors attachment to the harder N over the softer S of NCS^-, but $[Cr(SCN)(H_2O)_5]^{2+}$ can be made, along with other products, by the remote attack of Cr^{2+} on $[Fe(NCS)(H_2O)_5]^{2+}$ or by its adjacent attack on $[Co(SCN)(NH_3)_5]^{2+}$. Aqueous $[Cr(SCN)(H_2O)_5]^{2+}$ goes to $[Cr(H_2O)_6]^{3+}$ and $[Cr(NCS)(H_2O)_5]^{2+}$, both catalyzed by Cr^{2+}, Hg^{2+} etc.

Sephadex gel can separate $[\textit{cis-/trans}\text{-}Cr(NCS)_2(H_2O)_4]^+$, $[\textit{fac-/mer}\text{-}Cr(NCS)_3(H_2O)_3]$, and the other $[Cr(NCS)_n(H_2O)_{6-n}]^{(3-n)+}$.

Reinecke's salt, $NH_4[trans\text{-}Cr(NCS)_4(NH_3)_2] \cdot H_2O$, serves to precipitate large cations.

The NCO^- and $NCSe^-$ complexes are few, especially from water.

Some "simple" organic reagents. Chromium(VI) is reduced to Cr^{III} by CH_2O etc. plus H_3O^+.

Acetate and Cr^{2+} (e.g., from a Jones reductor) in cold water, under N_2, precipitate $[Cr_2(CH_3CO_2)_4]\cdot 2H_2O$, i.e., $[\{Cr(H_2O)\}_2(\mu\text{-}CH_3CO_2)_4]$, deep-red, a good source for other Cr^{II} compounds. Its slight solubility and the quadruple Cr–Cr bonding help make it a milder reductant than other Cr^{II} salts. Air oxidizes it slowly even when dry. Many carboxylates and other chelators behave rather similarly.

Fresh $Cr_2O_3\cdot aq$ dissolves in concentrated CH_3CO_2H as Cr^{III} acetate.

Chromium(III) precipitates no basic acetate when the corresponding compounds of Al^{III} and Fe^{III} are precipitated. A great excess of Fe^{III} or Al^{III}, however, co-precipitates Cr^{III} fairly completely.

From solutions of inert complexes such as $[CrCl_2(H_2O)_4]Cl$ or $[CrCl(H_2O)_5]Cl_2$, only $^1/_3$ or $^2/_3$ of the Cl^- can be precipitated promptly. The sulfates etc. are similar. In each case, however, adding acetate tends to displace the inner-sphere ligand and release the rest of it.

Oxalates do not precipitate the simple Cr(II or III) salts, but $[Cr(NH_3)_5(H_2O)]^{3+}$ at 50°C for 10 min forms $[Cr(NH_3)_5(C_2O_4)]^+$ and $[Cr(NH_3)_5(HC_2O_4)]^{2+}$. Oxalate reduces CrO^{2+}, but not at first to Cr^{3+}:

$$CrO^{2+} + HC_2O_4^- + H_3O^+ \rightarrow Cr^{2+} + 2\,CO_2\uparrow + 2\,H_2O$$

The bridged $[\{Cr(C_2O_4)_2\}_2(\mu\text{-}OH)_2]^{4-}$ and so on are well known.

Chromium(VI) can produce a green complex, with blue iridescence (chirally resolvable by organic and organo-metallic cations):

$$CrO_4^{2-} + 4\,H_2C_2O_4 + {}^1/_2\,C_2O_4^{2-} + 3\,K^+ \rightarrow$$

$$K_3[Cr(C_2O_4)_3]\cdot 3H_2O\downarrow + 3\,CO_2\uparrow + H_2O$$

Reduced nitrogen. Concentrated NH_3 and Cr^{2+} give a deep-blue color but then precipitate brown $Cr(OH)_2$.

Platinized asbestos catalyzes the formation of $[Cr(NH_3)_6]^{3+}$ from Cr^{2+} under N_2, and many mixed ammines are made from various sources:

$$Cr^{2+} + 6\,NH_3 + H_2O \rightarrow [Cr(NH_3)_6]^{3+} + {}^1/_2\,H_2\uparrow + OH^-$$

Ammonia precipitates gray-green Cr_2O_3.aq from Cr^{3+}:

$$2\,Cr^{3+} + 6\,NH_3 + 3\,H_2O \rightarrow Cr_2O_3\cdot aq\downarrow + 6\,NH_4^+$$

This dissolves slightly in excess cold NH_3 as violet ammines. The hydrous oxide is completely, but slowly, reprecipitated on boiling. Warming the original product with much NH_3 and NH_4^+ gives $[CrOH(NH_3)_5](OH)_2$ which, with much cold HNO_3, yields orange $NH_4[Cr(H_2O)(NH_3)_5](NO_3)_4$. This product is a good source, with NO_2^-, for $[Cr(NO_2)(NH_3)_5](NO_3)_2$, NO_2 linkage not given, also orange; with further treatment and heating, for $[Cr(NO_3)(NH_3)_5](NO_3)_2$, light tan; and with HBr, for $[CrBr(NH_3)_5]Br_2$. Many Cr^{III} ammines exist.

Ammonia precipitates a yellow-brown chromium(III) chromate(VI) from Cr^{3+} plus $[Cr_2O_7]^{2-}$.

Adding excess N_2H_4 to CrX_2 quickly gives pale-blue, probably polymeric, $CrX_2(\mu\text{-}N_2H_4)_2$, with X = F, Cl, Br or I, fairly stable in dry air except with F.

Diazanium ("hydrazinium"), $N_2H_5^+$, does not oxidize Cr^{2+} but reduces the O_2^-, not the Cr^{3+}, in CrO_2^{2+} approximately as:

$$CrO_2^{2+} + N_2H_5^+ + H_3O^+ \rightarrow Cr^{3+} + H_2O_2 + {}^1/_2\, N_2\uparrow + NH_4^+ + H_2O$$

Hydroxylamine and Cr^{2+} yield NH_3 and $[Cr(H_2O)_6]^{3+}$, not ammines.

Adding $H_2[Cr_2O_7]$ slowly to NH_3OH^+ gives red-brown $[Cr(H_2O)_5NO]^{2+}$ and many by-products; note this also in **Oxidized nitrogen** next and $[Fe(H_2O)_5(NO)]^{2+}$ under **8.1.2 Oxidized nitrogen.**

In base with CN^- and CrO_4^{2-}, NH_2OH forms a bright-green solid, $K_3[Cr(CN)_5NO]\cdot H_2O$. This is hydrolyzed as far as $[Cr(CN)_2NO(H_2O)_3]$ at $3 < pH < 5$, then at $pH < 2$ to $[CrNO(H_2O)_5]^{2+}$ and, quite slowly, to $[Cr(H_2O)_6]^{3+}$. Mercury(2+) accelerates the normal isomerization of Cr–CN in $[CrCN(NO)(H_2O)_4]^+$ to Cr–NC and it complexes the cation.

Aqueous Cr^{2+} and HN_3 give $[Cr(H_2O)_5NH_3]^{3+}$ and others.

In dilute CH_3CO_2H at 60 °C, N_3^- and $[Cr(NH_3)_5H_2O](ClO_4)_3$ form $[CrN_3(NH_3)_5](ClO_4)_2$. Refluxing N_3^- and $[Cr(NH_3)_6](ClO_4)_3$ produces $[Cr(N_3)_3(NH_3)_3]$.

Heating N_3^- with $CrCl_3\cdot 6H_2O$ at 50–60 °C 1 h gives $Cat_3[Cr(N_3)_6]$, violet and stable if precipitated with a large cation; e.g., $Cat^+ = [NBu_4]^+$.

Oxidized nitrogen. Treating Cr^{2+} with NO yields $[Cr(H_2O)_5NO]^{2+}$, to be considered (imperfectly) as $[Cr^{3+}(H_2O)_5(NO^-)]$, not $[Cr^+(H_2O)_5(NO^+)]$. Whether Cr-NO complexes are better classified as $Cr^+–NO^+$, $Cr^{2+}–NO^\bullet$ or $Cr^{3+}–NO^-$ is not always clear. See **8.1.2 Oxidized nitrogen**.

Cold, aqueous $[Cr(NH_3)_5(H_2O)]Cl_3$, NO_2^- and then HCl form $[Cr(NH_3)_5(NO_2\text{-}\kappa O)]Cl_2$. Such Cr^{III}–ONO species, unlike those of other metals, do not rearrange to $Cr^{III}–NO_2$.

Both dilute and concentrated HNO_3 tend to make metallic Cr passive.

Nitrites and nitrates do not react appreciably with Cr^{3+}, but NO_3^- labilizes the NH_3 in $[Cr(H_2O)(NH_3)_5]^{3+}$.

Fluorine species. Concentrated NH_4F and $[Cr(H_2O)_6]^{3+}$ give a violet salt.

Different reaction sequences can lead to different (octahedral) isomers (Py ≡ pyridine, C_5H_5N, and X = N_3, NCS or Br):

$$[Cr(H_2O)_6]^{3+} + 3\,F^- \rightarrow [Cr(H_2O)_6]F_3\downarrow$$

$$[Cr(H_2O)_6]F_3 + 3\,Py\ (\sim 100\ °C) \rightarrow [mer\text{-}CrF_3Py_3]\downarrow + 6\,H_2O$$

$$[mer\text{-}CrF_3Py_3] + 3\,NH_3\ (100\ °C) \rightarrow [mer\text{-}CrF_3(NH_3)_3]\downarrow + 3\,Py$$

We can keep these species meridional (with the 3 X on a great circle):

$$[mer\text{-}CrF_3(NH_3)_3] + 3\,H_3O^+ + 3\,ClO_4^- \text{ (12-M, 65 °C)} \rightarrow$$

$$[mer\text{-}Cr(NH_3)_3(H_2O)_3](ClO_4)_3\downarrow + 3\,HF$$

$$[mer\text{-}Cr(NH_3)_3(H_2O)_3](ClO_4)_3 + 3\,X^- \rightarrow$$

$$[mer\text{-}CrX_3(NH_3)_3]\downarrow + 3\,H_2O + 3\,ClO_4^-$$

Or we can change to the facial isomers (with the 3 X all adjacent):

$$[mer\text{-}CrF_3(NH_3)_3] + 3\,H_3O^+ + 3\,SO_3CF_3^- \text{ (70 °C)} \rightarrow$$

$$[fac\text{-}Cr(OSO_2CF_3)_3(NH_3)_3]\downarrow + 3\,HF + 3\,H_2O$$

$$[fac\text{-}Cr(OSO_2CF_3)_3(NH_3)_3] + 3\,H_2O + 3\,ClO_4^- \text{ (6 M, 70 °C)} \rightarrow$$

$$[fac\text{-}Cr(NH_3)_3(H_2O)_3](ClO_4)_3\downarrow + 3\,SO_3CF_3^-$$

$$[fac\text{-}Cr(NH_3)_3(H_2O)_3](ClO_4)_3 + 3\,X^- \rightarrow$$

$$[fac\text{-}CrX_3(NH_3)_3]\downarrow + 3\,H_2O + 3\,ClO_4^-$$

6.1.3 Reagents Derived from the 3rd-to-5th-Period Non-Metals, Silicon through Xenon

Phosphorus species. Aqueous HPH_2O_2 reduces Cr^{VI} to Cr^{III} and forms, e.g., $Cr(PH_2O_2)^{2+}$. Phosphate ions, such as HPO_4^{2-}, precipitate Cr^{2+} as $CrHPO_4\cdot 4H_2O$, and Cr^{III} as $CrPO_4$, the latter insoluble in CH_3CO_2H, decomposed by boiling with OH^-.

Arsenic species. Arsenite and arsenate ions form the corresponding salts with Cr^{III}. Arsenous acid instantly reduces $[Cr_2O_7]^{2-}$ to Cr^{III} and, on boiling, precipitates $CrAsO_4$.

Reduced chalcogens. Sulfane, H_2S, is without action on acidic or neutral solutions of Cr^{III}; $Cr(OH)_4^-$ is precipitated as $Cr_2O_3\cdot aq$:

$$2\,Cr(OH)_4^- + 2\,H_2S \rightarrow Cr_2O_3\cdot aq\downarrow + 2\,HS^- + 5\,H_2O$$

Similar precipitations of $Cr_2O_3\cdot aq$ occur with HS^- and NH_4^+ (forming S^{2-} and NH_3), also from Cr^{3+} with HS^- and NH_3 (forming H_2S and NH_4^+), all from the "$(NH_4)_2S$" mixture.

Aqueous Cr^{2+} and $S_2O_3^{2-}$, Ag_2S or PbS under N_2 give $[CrSH(H_2O)_5]^{2+}$, brownish-green, with a low yield; S_2^{2-} and H_3O^+ is better but still <20 %. It is aquated quite slowly, free of air, at 25 °C; further reactions include:

$$[CrSH(H_2O)_5]^{2+} + {}^1/_2\,O_2 + H_3O^+ \rightarrow [Cr(H_2O)_6]^{3+} + S\downarrow + H_2O$$

$$[CrSH(H_2O)_5]^{2+} + NCS^- \leftrightarrows [Cr(NCS)(SH)(H_2O)_4]^+ + H_2O$$

$$[Cr(NCS)(SH)(H_2O)_4]^+ + H_3O^+ \leftrightarrows [Cr(NCS)(H_2S)(H_2O)_4]^{2+} + H_2O$$

$$\text{Concd } [CrSH(H_2O)_5]^{2+} + HSO_4^- + H_2O \rightarrow [CrSH(H_2O)_5]SO_4\downarrow + H_3O^+$$

$$[CrSH(H_2O)_5]^{2+} + {}^1/_2\,I_2 + H_2O \rightarrow {}^1/_2\,[\{Cr(H_2O)_5\}_2(\mu\text{-}S_2)]^{4+} + I^- + H_3O^+$$

Chromium(VI) is reduced to Cr^{III} with liberation of sulfur; in neutral or alkaline solution $Cr_2O_3\cdot aq$ is again precipitated:

$$[Cr_2O_7]^{2-} + 3\,H_2S + 8\,H_3O^+ + 2n\,Cl^- \rightarrow$$

$$2\,[CrCl_n(H_2O)_{6-n}]^{(3-n)+} + 3\,S\downarrow + (2n+3)\,H_2O$$

$$2\,CrO_4^{2-} + 10\,HS^- \rightarrow Cr_2O_3\cdot aq\downarrow + 4\,S^{2-} + 3\,S_2^{2-} + 5\,H_2O$$

At times some $S_2O_3^{2-}$, SO_3^{2-} and SO_4^{2-} are obtained in addition.

The reaction of CrO_4^{2-} with "$(NH_4)_2S$" also gives the hydrous oxide.

Chromium(VI) is reduced to Cr^{III} by SCN^- in acidic solution.

Oxidized chalcogens. In acid, $S_2O_3^{2-}$ and $SO_2\cdot H_2O$ reduce Cr^{VI}, and we ignore the various resulting Cr^{III} complexes with SO_4^{2-} etc.:

$$[Cr_2O_7]^{2-} + 3\,SO_2\cdot H_2O + 2\,H_3O^+ \rightarrow 2\,Cr^{3+} + 3\,SO_4^{2-} + 6\,H_2O$$

Aqueous HSO_3^- and $[Cr(H_2O)_6]^{3+}$ quickly but reversibly form $[Cr(SO_3\text{-}\kappa O)(H_2O)_5]^+$, leaving the original Cr—O bond intact. Sulfite and selenite form various unstable complexes from $[Cr(NH_3)_5(H_2O)]^{3+}$.

Dilute H_2SO_4 with Cr forms Cr^{III} in the air, otherwise Cr^{2+}.

Concentrated H_2SO_4 induces passivity with Cr.

Concentrated sulfates and Cr^{2+} yield $(NH_4,Rb,Cs)_2[Cr(H_2O)_6](SO_4)_2$ or $(Na,K)_2Cr(SO_4)_2\cdot 2H_2O$, pale blue. The Cs salt, during easy dehydration, goes apparently to violet $Cs_4[\{Cr(H_2O)\}_2(\mu\text{-}SO_4)_4]$, like Cr^{II} acetate. There are many double sulfates of Cr^{III} and the alkali metals and NH_4 forming (violet) alums similar to those of Al or Fe. The inertness of the Cr^{III} equilibria with H_2O and SO_4^{2-}, however, allows a separation of Cr from Fe etc. Heating a mixture to 80 °C forms sulfato Cr ions which do not quickly revert on cooling to the $[Cr(H_2O)_6]^{3+}$ ions required for alums; the Fe^{III} alums can then be crystallized separately, but several days <30 °C are required to re-form $[Cr(H_2O)_6]^{3+}$ for crystallization.

Sulfate, HSO_4^- and $[Cr(NH_3)_5(H_2O)]^{3+}$ at 50°C, then quickly cooled to 0°C, form red-orange $[Cr(NH_3)_5(SO_4)]^+$, losing 1/3 of its NH_3 at 25°C.

Chromium trioxide, CrO_3, is formed as brown-red needles upon adding concentrated H_2SO_4 to a concentrated solution of $[Cr_2O_7]^{2-}$. To be freed from H_2SO_4 it must be recrystalized from water, in which it is readily soluble, or treated with the necessary amount of $BaCrO_4$. It is also prepared by transposing $BaCrO_4$ or $PbCrO_4$ with H_2SO_4.

The peroxo ion $HSO_3(O_2)^-$ oxidizes $[CrN_3(NH_3)_5]^{2+}$ nicely:

$$[CrN_3(NH_3)_5]^{2+} + HSO_3(O_2)^- \rightarrow [CrNO(NH_3)_5]^{2+} + HSO_4^- + N_2\uparrow$$

Reduced halogens. Chromium is soluble in HCl, yielding blue $CrCl_2 \cdot 4H_2O$ if the air is excluded, otherwise Cr^{III}.

From air-free Cr^{2+}, alkali salts and HCl or HBr one may crystallize light-blue $(Rb,Cs,NH_4)_2[$*trans*-$Cr^{II}(Cl,Br)_4(H_2O)_2]$.

Heating the calculated amounts of green $CrCl_3 \cdot 6H_2O$ and CsCl in 2-M HCl and evaporating slowly gives dark-green $Cs_2[$*trans*-$CrCl_2(H_2O)_4]Cl_3$.

The transposition of Ag_2CrO_4 with HCl yields CrO_3.

Concentrated HCl reduces CrO_3 mainly to the dark-green "hydrated chromic chloride" of commerce, $[$*trans*-$CrCl_2(H_2O)_4]Cl \cdot 2H_2O$, but water at ambient T for 24 h gives the light-green isomer $[CrCl(H_2O)_5]Cl_2 \cdot H_2O$.

A solution of CrO_3, or $Na_2[Cr_2O_7]$, plus concentrated HCl, with concentrated H_2SO_4 dropped into it slowly, keeping the aqueous solution below 10 °C, forms a dark-red, heavy separate liquid phase, chromyl chloride, CrO_2Cl_2, a powerful oxidant, e.g., for organics:

$$CrO_3 + 2\,Cl^- + 3\,H_2SO_4 \rightarrow CrO_2Cl_2\downarrow_{liq} + 3\,HSO_4^- + H_3O^+$$

It fumes in humid air, is hydrolyzed vigorously, and is better stored in the dark. Slowly adding this to hot aqueous K_2CrO_4 and cooling gives red-orange $KCrO_3Cl$.

Heating a dry chromate or dichromate with concentrated H_2SO_4 and a chloride (transposable by H_2SO_4) gives the brown fumes of CrO_2Cl_2:

$$K_2[Cr_2O_7] + 4\,KCl + 9\,H_2SO_4 \rightarrow$$

$$2\,CrO_2Cl_2\uparrow + 6\,K^+ + 3\,H_3O^+ + 9\,HSO_4^-$$

Boiling aqueous HCl reduces Cr^{VI}, e.g., to $CrCl^{2+}$:

$$[Cr_2O_7]^{2-} + 14\,H_3O^+ + 8\,Cl^- \rightarrow 2\,CrCl^{2+} + 3\,Cl_2\uparrow + 21\,H_2O$$

more readily and without releasing Cl_2 in the presence of faster reductants, such as ethanol or oxalic acid:

$$[Cr_2O_7]^{2-} + 8\,H_3O^+ + 3\,C_2H_5OH \rightarrow 2\,Cr^{3+} + 3\,CH_3CHO\uparrow + 15\,H_2O$$

The acids HBr and HI reduce Cr^{VI} to Cr^{III}, releasing Br_2 or I_2:

$$[Cr_2O_7]^{2-} + 12\,H_3O^+ + 6\,I^- + 2\,HSO_4^- \rightarrow$$

$$\text{e.g. } 2\,Cr(SO_4)^+ + 3\,I_2 + 19\,H_2O$$

Iodide and $Cr(HO_2)^{2+}$, catalyzed by H_3O^+, form Cr^{III}, HIO and I_3^- much faster than in the acid-catalyzed oxidation of I^- by uncoordinated H_2O_2.

The $[Cr(NH_3)_6]^{3+}$ ion is useful to precipitate large anions, e.g., I_7^-.

Elemental and oxidized halogens. Chlorine or bromine attacks Cr, forming Cr^{III}. Chromium(III) is oxidized to CrO_4^{2-} in alkalis by ClO^-, BrO^- etc. Boiling with ClO_3^- or BrO_3^- yields $[Cr_2O_7]^{2-}$:

$$5\,Cr_2O_3 \cdot aq + 6\,ClO_3^- + 6\,H_2O \rightarrow 5\,[Cr_2O_7]^{2-} + 3\,Cl_2\uparrow + 4\,H_3O^+$$

Saturated $NaIO_3$, plus $[Cr(NH_3)_5(H_2O)]^{3+}$ at 50 °C, and quickly cooled in ice, form red-violet $[Cr(NH_3)_5(IO_3)]^{2+}$, which reverts to the aqua form in water while keeping the Cr—O bond.

6.1.4 Reagents Derived from the Metals Lithium through Uranium, plus Electrons and Photons

Oxidation. Cerium(IV) quickly oxidizes Cr^{3+} in (perchloric) acid:

$$2\,Cr^{3+} + 6\,Ce^{4+} + 21\,H_2O \rightarrow [Cr_2O_7]^{2-} + 6\,Ce^{3+} + 14\,H_3O^+$$

Chromium(III) compounds are also oxidized to $[Cr_2O_7]^{2-}$ in acidic solution by $MnO_2 \cdot aq$, MnO_4^- and PbO_2; to CrO_4^{2-} in alkaline mixture by MnO_4^-, MnO_4^{2-}, CuO, Ag_2O, Hg_2O, HgO, PbO_2 etc.

Different pHs etc. give different results when $[Fe(CN)_6]^{3-}$ oxidizes Cr^{2+}, but $[(H_2O)_5Cr^{III}NCFe^{II}(CN)_5]^-$ may be one.

Aqueous Cr^{2+} and one-electron oxidants (Fe^{3+} or Cu^{2+}) form Cr^{3+}.

Slowly adding 5.0-mM Cr^{2+} to an equal volume of a 5.5-mM Fe^{III} and 4.5-mM NCS^- mixture gives some green $CrSCN^{2+}$, which, on standing, becomes Cr^{3+} and, somewhat less, purple $CrNCS^{2+}$. The $CrSCN^{2+}$ or $CrNCS^{2+}$, and Cl_2, form $CrCl^{2+}$ or Cr^{3+} respectively, but Hg^{2+} gives Cr^{3+} or $CrNCSHg^{4+}$ (Cl^- removes the Hg^{2+}) in turn. The $Cr'SCN^{2+}$ and Cr''^{2+} yield Cr'^{2+} and $Cr''NCS^{2+}$. Both isomers release NCS^- quickly at pH > 7. In general, the rates of aquation of CrX^{2+} are $I^- > SCN^- > Br^- > Cl^- > NCS^-$.

Aqueous $[Co(NH_3)_5(H_2O)]^{3+}$ oxidizes Cr^{2+} to Cr^{3+}.

The great reducing strength of $[Cr(CN)_6]^{4-}$, much more than that of $[Cr(H_2O)_6]^{2+}$, is seen in its generation even of the reactive $[Co(CN)_5]^{3-}$ in:

$$[Cr(CN)_6]^{4-} + [Co(CN)_5Br]^{3-} \rightarrow [Cr(CN)_6]^{3-} + [Co(CN)_5]^{3-} + Br^-$$

followed by (where the Co product can hydrogenate some organics):

$$[Cr(CN)_6]^{4-} + [Co(CN)_5]^{3-} + H_2O \rightarrow$$

$$[Cr(CN)_6]^{3-} + [Co(CN)_5H]^{3-} + OH^-$$

Aqueous Cr^{2+} and $[PtCl(NH_3)_5]^{3+}$ form $[Pt(NH_3)_4]^{2+}$ and $Cr^{IV}Cl^{3+}$, which quickly, with more Cr^{2+}, gives $CrCl^{2+}$ and Cr^{3+}.

The powerful Cr^{II} reductants convert Cu^{2+}, Hg^{2+}, Sn^{II} etc. to the metals.

The two-electron oxidant Tl^{III} (like O_2) oxidizes Cr^{2+} to CrO^{2+}, which joins more Cr^{2+} to form $[\{Cr(H_2O)_4\}_2(\mu\text{-}OH)_2]^{4+}$.

Light (254 nm), Cr^{2+}, HSO_4^- and a Ni/Pd/Pt catalyst give Cr^{III} and H_2.

Reduction. One-electron, outer-sphere reductants, such as $[V(H_2O)_6]^{2+}$, $[Cr(H_2O)_6]^{2+}$, $[Fe(H_2O)_6]^{2+}$ and $[Ru(NH_3)_6]^{2+}$, quite quickly reduce $[Cr(O_2)(H_2O)_5]^{2+}$ to $[Cr(O_2H)(H_2O)_5]^{2+}$ in acid. In some cases the retention of the peroxide structure is confirmed by the re-formation of $[Cr(O_2)(H_2O)_5]^{2+}$ by Ce^{IV}. At least $[Ti(H_2O)_6^{3+}$, $[V(H_2O)_6]^{2+}$, $[Fe(H_2O)_6]^{2+}$ and Cu^+, but not $[Ru(NH_3)_6]^{2+}$, reduce $[Cr(O_2H)(H_2O)_5]^{2+}$ further, e.g.:

$$CrO_2H^{2+} + 2\,Fe^{2+} + 3\,H_3O^+ \rightarrow Cr^{3+} + 2\,Fe^{3+} + 5\,H_2O$$

In acid, $H_2[Fe(CN)_6]^{2-}$ etc. reduce Cr^{VI} to Cr^{III}.

Aqueous Cr^{III} is reduced to the very air-sensitive blue Cr^{2+} ion by either Cr or Zn in H_3O^+, or by Zn_{Hg} or a cathode.

Other reactions. The CrO_4^{2-} and $[Cr_2O_7]^{2-}$ ions are precipitated mostly as normal chromates, not dichromates, when treated with Ba^{2+}, Ag^+, Hg_2^{2+} or Pb^{2+}, because of lower solubilities of the former and rapid equilibration, even when $[Cr_2O_7]^{2-}$ predominates in solution:

$$2\,Ba^{2+} + [Cr_2O_7]^{2-} + 3\,H_2O \leftrightarrows 2\,BaCrO_4\downarrow + 2\,H_3O^+$$

Barium chromate, $BaCrO_4$, yellow, is soluble in HCl, HNO_3, and slightly soluble even in chromic acid. Silver chromate, Ag_2CrO_4, is dark reddish brown, soluble in HNO_3 and NH_3; lead chromate, $PbCrO_4$, is yellow, soluble in 3-M HNO_3, insoluble in acetic acid;

The O_2^- in CrO_2^{2+} is reduced by V^{2+}, Fe^{2+}, $[Ru(NH_3)_6]^{2+}$ etc., leaving the Cr^{3+} unreduced.

Two Cr^{2+} and one Cr^{VI}, by successive one-electron transfers, first give Cr^{IV} and two Cr^{3+}; then the Cr^{IV} and a Cr^{2+} form $[\{Cr^{III}(H_2O)_4\}_2(\mu\text{-}OH)_2]^{4+}$.

It may be of interest that in the following reaction the equilibrium constant K_2 is only 2 pM for $n = 2$, but $K_3 \geq 1$ GM for $n = 3$:

$$[CrO_2(H_2O)_5]^{n+} + H_2O \leftrightarrows [Cr(H_2O)_6]^{n+} + O_2$$

so that we would have $K_3/K_2 \geq {\sim}5 \times 10^{20} \leq K$ for the Cr-Cr reaction:

$$[CrO_2(H_2O)_5]^{3+} + [Cr(H_2O)_6]^{2+} \leftrightarrows [CrO_2(H_2O)_5]^{2+} + [Cr(H_2O)_6]^{3+}$$

The $[Cr(NH_3)_6]^{3+}$ ion is useful to precipitate other large ions, especially with equivalent charge, such as $[Cr(CN)_6]^{3-}$, $[FeCl_6]^{3-}$ and the less common $[CuCl_5]^{3-}$. In $[Cr(NH_3)_6][Cr(CN)_6]$ the effective ionic charges are less than 3 (+ or −) because of CN·HN hydrogen bonds.

Aqueous Fe^{2+} precipitates $[Cr(CN)_6]^{3-}$ as $\sim Fe_3[Cr(CN)_6]_2$.

The $[Fe(CN)_6]^{4-}$ ion does not generally precipitate Cr^{III}.

Aqueous cyanocobalt(III) ions and Cr^{2+} form $[CrNC(H_2O)_5]^{2+}$ which changes, catalyzed by Cr^{2+}, to $[CrCN(H_2O)_5]^{2+}$.

Either Ag^+ or Hg^{2+} flips some CN Groups in $[Cr(CN)_n(H_2O)_{6-n}]^{(3-n)+}$ to NC, yielding complexes containing Cr–N≡C–M.

The alkali chromates are yellow and the dichromates orange.

Photoaquation often occurs, in general and as examples, respectively:

$$[Cr^{III}L_6] + \gamma + H_2O \to [Cr^{III}L_5(H_2O)] + L$$

Various ligands L are thus replaced by water, or sometimes by anions at high anionic concentrations.

Light and $[CrX(NH_3)_5]^{2+}$, with $X^- = CN^-$, Cl^- or Br^-, first form mainly [*cis*-$CrX(H_2O)(NH_3)_4]^{2+}$ in low quantum yields, but note the dark reaction in **6.1.1 Water**, and the different result here:

$$[Cr(NCS)(NH_3)_5]^{2+} + \gamma + H_2O \to$$

$$[\textit{trans}\text{-}Cr(NCS)(H_2O)(NH_3)_4]^{2+} + NH_3$$

Light and [*trans*-$CrCl_2(NH_3)_4]^+$, however, first replace a Cl^- and give primarily [*cis*-$CrCl(H_2O)(NH_3)_4]^{2+}$.

A chiral oxalato complex can be inverted:

$$[Cr(\eta^2\text{-}C_2O_4)_3]^{3-} + \gamma + H_2O \to [Cr(\eta^2\text{-}C_2O_4)_2(H_2O)(\eta^1\text{-}CO_2\text{–}CO_2)]^{3-} \to \text{racemic } [Cr(\eta^2\text{-}C_2O_4)_3]^{3-} + H_2O$$

Circularly polarized light can preferentially invert or decompose one chiral isomer.

6.2 Molybdenum, $_{42}$Mo; Tungsten, $_{74}$W and Seaborgium, $_{106}$Sg

Oxidation numbers in classical compounds of Mo and W: (II), (III), (IV), (V) and (VI), as in Mo_2^{4+} and $[W_2Cl_8]^{4-}$ (both quadruply bonded M to M), Mo^{3+}, $(Mo{\equiv}Mo)_2(\mu\text{-}OH)_2^{4+}$ and $[({\equiv}WCl_3)_2(\mu\text{-}Cl)_3]^{3-}$, MoO_2 and $[W(CN)_8]^{4-}$, $Mo_2O_4^{2+}$ or

$(-MoO)_2(\mu\text{-}O)_2^{2+}$ and $[(-WO)_2(\mu\text{-}O)_2F_6]^{4-}$, and MoO_4^{2-} and WO_4^{2-}. Many Mo^V and W^{III}, but not many W^V, are dinuclear.

For Sg calculated relativistically to be stable in water: (IV) and (VI), especially (VI). The stabilities of the highest oxidation states of the early **6d** elements go expectedly as: $Lr^{III} > Rf^{IV} > Db^V > Sg^{VI}$.

6.2.1 Reagents Derived from Hydrogen and Oxygen

Water. Aqueous Mo_2^{4+} may be written as $[Mo_2(H_2O)_8]^{4+}$, or actually as $[Mo_2(H_2O)_8(H_2O)_2]^{4+}$ with two weakly bound (axial) waters.

The salts of the lower oxidation states of Mo are nearly all soluble in water, but the anhydrous halides MoX_3 (like CrX_3) are insoluble.

Molybdenum trioxide is more soluble (but only slightly) cold than hot. Molybdates(VI) of Alk^+, Mg^{2+} and Tl^+ dissolve, the others do not.

Tungsten trioxide and the disulfide (tungstenite ore) are insoluble in H_2O. The trisulfide is slightly soluble. Normal tungstates of the alkali metals and Mg are soluble, those of the other metals, slightly soluble to insoluble in H_2O. The salt $[Co^{III}(NH_3)_6][W^V(CN)_8]$ is sparingly soluble.

Water quickly replaces three Cl^- in $[Mo_2Cl_8]^{4-}$ in non-complexing acids.

The complex $[MoCl_6]^{3-}$, e.g., 5 cM in air-free solutions of HSO_3CF_3, is hydrolyzed in days to very pale-yellow Mo^{3+}, i.e., $[Mo(H_2O)_6]^{3+}$.

The halides; e.g., WCl_4, WCl_5 WCl_6, $WOCl_4$ and WO_2Cl_2, are all more or less rapidly decomposed by H_2O.

The yellow K_3MoO_3N, made in liquid NH_3, is basic and hydrolyzed rapidly in water, slowly in air, to MoO_4^{2-}, NH_3 and OH^-.

Seaborgium(VI) seems to hydrolyze less than Mo^{VI} or W^{VI}.

Some natural waters may contain $H_mW^{VI}S_n^{(2n-m-6)-}$, and hot waters may contain W^{VI} carbonates or $[WF_8]^{2-}$.

Oxonium. This does not dissolve Mo or W without complexation.

Molybdenum(III) oxide is insoluble in H_3O^+; $Mo_2O_3 \cdot aq$ dissolves with difficulty. Molybdenum(V) oxide dissolves in warm acids. Fused MoO_3, also WO_3 and WS_2, are insoluble in most acids.

Molybdenum(VI) in concentrated H_3O^+ becomes *cis*-MoO_2^{2+} (unlike *trans*-AnO_2^{2+}), and [*cis,trans,cis*-$MoO_2(Cl,Br)_2(H_2O)_2$] can be isolated.

Weak acidification of MoO_4^{2-} generates especially $[Mo_7O_{24}]^{6-}$ and, with some acids, $[Mo_8O_{26}]^{4-}$ (with six different structures) but also , e.g., $[Mo_2O_7]^{2-}$, $[Mo_6O_{19}]^{2-}$, and $[Mo_{36}O_{112}(H_2O)_{16}]^{8-}$, omitting the hydronated ("protonated") forms. These equilibrate much faster than with W, and results depend of course on pH, concentration, etc. (Adding H^+ to MoO_4^{2-}, unlike WO_4^{2-}, perhaps then raises the ligancy to six, easing condensation, but some rapid tungstate condensations may involve only adding units.) Such polymolybdates have extremely weak basicity (i.e., are salts of very strong acids) but break up when attacked by either H_3O^+ or OH^-. They are colorless, except that $[Mo_6O_{19}]^{2-}$ is yellow.

Much H_3O^+ plus, say, 3-dM MoO_4^{2-}, with no other basic anions, form yellow $MoO_2O(H_2O)\cdot H_2O$ after a few weeks when cool, yellow $MoO_3\cdot H_2O$ quickly when hot. Much excess acid dissolves these, yielding colorless ions simplified, e.g., as $HMoO_3^+$, i.e., $[Mo(OH)_5(H_2O)]^+$, or as $H_2Mo_2O_6^{2+}$ and $H_3Mo_2O_6^{3+}$.

At $c(Mo^{VI}) < 0.1$ mM we may have $[HMoO_4]^-$, $[MoO_3(H_2O)_3]$, $[Mo(OH)_6]$ or $[MoO_2(OH)_2(H_2O)_2]$, with $pK_a \sim 4$.

If WO_4^{2-} is acidified (except by H_3PO_4), the trioxide precipitates. Precipitation from a hot solution with concentrated acid gives yellow $WO_3\cdot H_2O$ [structurally not $WO_2(OH)_2$]; from a cold solution, white $WO_3\cdot 2H_2O$ separates, turning yellow on boiling, insoluble in excess of the acid (distinction from MoO_3).

Acidifying $[MoS_4]^{2-}$ can apparently yield $[Mo^{IV}O(\eta^2\text{-}Mo^{VI}S_4)_2]^{2-}$, precipitable by Cs^+ etc.

The extent of hydrolysis of M^{6+} with $0 < pH < 1$ is both calculated and found to be Mo > W > Sg, but the basicity of SgO_4^{2-} is calculated to be between those of MoO_4^{2-} and WO_4^{2-} in:

$$MO_4^{2-} + H_3O^+ \leftrightarrows MO_3(OH)^- + H_2O \text{ and}$$

$$MO_3(OH)^- + H_3O^+ + H_2O \leftrightarrows [MO_2(OH)_2(H_2O)_2]$$

Experimentally, however, further acidification has Sg going farthest:

$$[MO_2(OH)_2(H_2O)_2] + H_3O^+ \leftrightarrows [MO(OH)_3(H_2O)_2]^+ + H_2O \text{ etc.}$$

Hydroxide. Aqueous OH^- does not attack Mo in the absence of oxidants such as ClO_3^-. Tungsten is slowly soluble in the alkalis.

Amorphous WO_2 dissolves in alkali hydroxides to form tungstates with the evolution of H_2. The crystalline dioxide is not affected by hot, concentrated, non-oxidizing alkalis.

Hydroxide added to Mo^V precipitates a brown hydroxide that then loses water to leave a brown-red $MoO(OH)_3$. Lower oxidation states also give precipitates with OH^-, forming the corresponding hydroxides or hydrous oxides.

The often-obtained Molybdenum Blue mixture of Mo^V and Mo^{VI}, when suspended in OH^-, dismutates to the brown-red $MoO(OH)_3$ and MoO_4^{2-}. Stable blue compounds, $Mo_4O_{10}(OH)_2$ and $Mo_8O_{15}(OH)_{16}$, have been obtained, however, along with "Blues" soluble in water.

Molybdenum(VI) oxide dissolves readily in OH^- to form MoO_4^{2-}, which combines with excess MoO_3 to produce very complex ions.

Amorphous MoS_3 plus KOH give the salt $K_2[Mo_3S_{13}]\cdot aq$, containing $[\{Mo(\eta^2\text{-}S_2)\}_3(\mu\text{-}\eta^2\text{-}S_2)_3(\mu_3\text{-}S)]^{2-}$, with a ligancy (c. n.) of seven.

Tungsten(VI) oxide reacts with bases, in excess or not, to form normal or polytungstates respectively.

The trisulfide WS_3 is easily soluble in OH^-.

Peroxide. Cold 10-M H_2O_2 dissolves Mo powder, then forms a yellow solid, perhaps $MoO_2(O_2)\cdot H_2O$. It also dissolves W powder, perhaps as $[W_2O_3(O_2)_4(H_2O)_2]^{2-}$.

Adding excess H_2O_2 to MoO_4^{2-} at pH 7 to 9 gives red $[Mo^{VI}(\eta^2\text{-}O_2)_4]^{2-}$, slowly releasing O_2 at higher pH, thus catalyzing the decomposition of H_2O_2, and crystallized as red-brown $[Zn(NH_3)_4][Mo(O_2)_4]$. The dark-red salts of M^+ and M^{2+} are explosive when heated or struck. Also, WO_4^{2-} forms yellow $[W^{VI}(\eta^2\text{-}O_2)_4]^{2-}$. We note that $[M(\eta^2\text{-}O_2)_4]^{n-}$ occurs with Cr^V but with Mo^{VI} and W^{VI}. The Mo and W anions below pH 5 become $[\{M^{VI}O(\eta^2\text{-}O_2)_2(H_2O)\}_2(\mu\text{-}O)]^{2-}$, again with a ligancy of seven, i.e.:

$$2\,[M(O_2)_4]^{2-} + 3\,H_2O + 2\,H_3O^+ \leftrightharpoons [M_2O_3(O_2)_4(H_2O)_2]^{2-} + 4\,H_2O_2$$

Acidified Mo^{VI} and H_2O_2 produce a yellow color, not extracted by ether (good for detection, but vanadates and titanates interfere).

In 1-dM H_3O^+ and <40-mM Mo species, or in 1-M H_3O^+ and <20-mM W species, $[MoO(O_2)_2(H_2O)_2]$ or $[WO(O_2)_2(H_2O)_2]$ appears to be formed without dimerization. Less acidified solutions lead to either $[MoO(O_2)_2(OH)(H_2O)]^-$ or $[WO(O_2)_2(OH)(H_2O)]^-$. Moreover, various concentrations of peroxide react with W^{VI} in rather acidic through slightly alkaline solutions to give, e.g., $[WO_3(HO_2)]^-$, $[WO(O_2)_2(OH)]^-$, $[WO(O_2)_2(H_2O)]$, $[W_2O_3(O_2)_4(OH)]^{3-}$, $[W_2O_2(O_2)_4(HO_2)_2]^{2-}$, $[W_4O_{12}(O_2)_2]^{4-}$, $[W_7O_{23}(O_2)]^{6-}$ and $[W_7O_{22}(O_2)_2]^{6-}$.

Many other peroxo complexes are known, e.g., $[WO(O_2)_2(C_2O_4)]^{2-}$, as well as polynuclear species especially from low concentrations of H_2O_2; with O_2^{2-} treated as didentate the ligancy (c. n.) is often seven.

Hydrogen peroxide reduces $[W(CN)_8]^{3-}$ to $[W(CN)_8]^{4-}$.

Dioxygen. Neither air nor water oxidizes pure Mo or W at ambient *T*. Tungsten dissolves slowly in OH^- and O_2 (or NO_2^-, NO_3^-, ClO_3^- or PbO_2 as oxidant). Tungsten dioxide, WO_2, brown, is stable in air.

Oxygen converts Mo^{3+} first to $Mo(O_2)^{3+}$; this then with excess Mo^{3+} forms a bright-yellow $Mo_2O_2^{6+}$, which proceeds further to $Mo_2O_4^{2+}$:

$$[Mo(O_2)(H_2O)_5]^{3+} + [Mo(H_2O)_6]^{3+} \rightarrow [\{Mo(H_2O)_5O\text{–}\}_2]^{6+} + H_2O$$

$$[\{Mo(H_2O)_5O\text{–}\}_2]^{6+} + 2\,H_2O \rightarrow [Mo_2O_4(H_2O)_6]^{2+} + 4\,H_3O^+$$

Excess O_2, however, yields the $Mo_2O_4^{2+}$ without the $Mo_2O_2^{6+}$, and then, much more slowly, Mo^{VI}. Air oxidizes all the less-oxidized solid hydrous oxides to Molybdenum Blue, $(Mo^VO_2OH)_x(Mo^{VI}O_3)_{1-x}$.

In hot 6-M HCl, however, oxygen may oxidize Mo^{II} only to Mo^{III}:

$$[Mo_2(CH_3CO_2)_4] + {}^1\!/_2\,O_2 + 6\,H_3O^+ + 10\,Cl^- \rightarrow$$

$$2\,[MoCl_5(H_2O)]^{2-} + 4\,CH_3CO_2H + 5\,H_2O$$

Treatment of aqueous $MoCl_5$ with CN^- and O_2 produces a $[\{(CN)_5Mo^{V}(O)Cl\}_2O]^{8-}$. Including $CoCl_2$ produces a green peroxo complex, $[(CN)_5Co^{III}–O_2–Mo^{VI}(O)Cl(CN)_5]^{6-}$, isolated as a K^+ salt.

An air stream with $K_3[W_2Cl_9]$ and excess KCN on a steam bath yields:

$$[W_2Cl_9]^{3-} + 16\,CN^- + {}^1/_2\,O_2 + H_2O \rightarrow$$

$$2\,[W(CN)_8]^{4-} + 9\,Cl^- + 2\,OH^-$$

Decolorizing charcoal, removal of KCl and KCN on cooling, and then ethanol give orange-yellow $K_4[W(CN)_8]\cdot 2H_2O$, stable in darkness.

6.2.2 Reagents Derived from the Other 2nd-Period Non-Metals, Boron through Fluorine

Boron species. Mixing K_2MoO_4 and $K[BH_4]$ gives MoO_2, sometimes mixed with bronzes K_xMoO_3 ($x = 0.26$, red; 0.30, blue) and others.

Cyanide, MoO_4^{2-} and $[BH_4]^-$, together with CH_3CO_2H, followed by ethanol, can yield $K_4[Mo(CN)_8]\cdot 2H_2O$.

Mixing $[MS_4]^{2-}$, $[BH_4]^-$ and H_3O^+ slowly at ambient T, then passing air at > 90°C for 20 h if M is Mo or 5 h if M is W, followed by ion-exchange separations, gives a green $Mo_3S_4^{4+}$ or blue-violet $W_3S_4^{4+}$, cathodically reducible in stages, both stable in air and acid, but not in HNO_3, which forms colorless solutions:

$$6\,[MS_4]^{2-} + 3\,[BH_4]^- + 23\,H_3O^+ + 4\,H_2O \rightarrow$$

$$2\,[M_3S_4(H_2O)_9]^{4+} + 3\,H_3BO_3 + 16\,H_2S\uparrow + 6\,H_2\uparrow$$

Also arising are some $M_3(O_nS_{4-n})^{4+}$, e.g purplish-red $[W_3OS_3]^{4+}$. These persistent incomplete-cuboidal structures are like that of cubane, C_8H_8, but with alternate M and S atoms at the corners and with one M missing. They can be written with increasing information as $M_3S_4^{4+}$, $[M_3S_4(H_2O)_9]^{4+}$, $[\{M(H_2O)_3\}_3S_4]^{4+}$ or $[\{M(H_2O)_3\}_3(\mu\text{-}S)_3(\mu_3\text{-}S)]^{4+}$, still with no detailed geometric data, although some of these data, such as an approximately octahedral structure involving M and S, may be inferred.

Reduction of $Mo_2O_3S^{2+}$ by $[BH_4]^-$ and HCl, both added slowly, and on heating 2 h with an air stream, followed by column chromatography, gives mainly red-purple $[\{Mo^{IV}(H_2O)_3\}_3(\mu\text{-}O)_3(\mu_3\text{-}S)]^{4+}$, or gray-green $[\{Mo^{IV}(H_2O)_3\}_3O_2S_2]^{4+}$ etc.

Starting the preceding treatment with $Mo_2O_2S_2^{2+}$ produces cubane-type green $[\{Mo(H_2O)_3\}_4(\mu_3\text{-}S)_4]^{5+}$, slightly air-sensitive and a little more acidic than H_3PO_4, plus green $[\{Mo(H_2O)_3\}_3OS_3]^{4+}$ etc., all related to molybdenum-enzyme structures. In each of these $Mo_3(O,S)$ species, an S (i.e., S^{2-}) is the "cap" in the μ_3 position. Further oxidation of the $[\{Mo(H_2O)_3\}_4S_4]^{5+}$ to $[\{Mo(H_2O)_3\}_4S_4]^{6+}$, and treatment with NCS^-, yield $[\{Mo(NCS)_3\}_4S_4]^{6-}$. A similar yellow-brown

$[\{Mo(H_2O)_3\}_3Se_4]^{4+}$ releases red selenium in a matter of days, but NCS^- easily forms $[\{Mo^{IV}(NCS)_3\}_3Se_4]^{5-}$. See **cuboidal clusters** in **6.2.5** below.

Treating a molybdate(VI) or tungstate(VI) with $K[BH_4]$, KCN and CH_3CO_2H, followed by ethanol, produces a golden-yellow, diamagnetic $K_4[Mo(CN)_8]\cdot 2H_2O$ or orange, diamagnetic $K_4[W(CN)_8]\cdot 2H_2O$.

Carbon oxide species. The molybdenum salts of lower oxidation states give precipitates with CO_3^{2-}, forming the corresponding hydrous oxides. With MoO_3 it forms MoO_4^{2-}.

Soluble tungstates are formed slowly by boiling WO_3 with CO_3^{2-}. Excess CO_3^{2-} and tungstic acid react at ambient *T*:

$$H_2WO_4 + 2\,CO_3^{2-} \rightarrow WO_4^{2-} + 2\,HCO_3^- + H_2O$$

The trisulfide WS_3 is easily soluble in alkali carbonates.

Cyanide species. Adding $K_3[MoCl_6]$ or $K_2[MoCl_5]\cdot H_2O$ to KCN in air-free water, then adding ethanol, yields very dark-green $K_4[Mo(CN)_7]\cdot 2H_2O$, easily oxidized to $[Mo(CN)_8]^{4-}$. Aqueous KCN and $[Mo_2(CH_3CO_2)_4]$ readily form yellow $K_5[Mo(CN)_7]\cdot H_2O$. Treating this with CH_3CO_2H, H_2S or HCl forms the hydride $K_4[MoH(CN)_7]\cdot 2H_2O$.

Ethanol precipitates $K_4[Mo(CN)_8]\cdot 2H_2O$ from Mo^{III} plus KCN and air, for example, or from Mo^V reduced by excess KCN. The aqueous anion slowly decomposes in light or hot, dilute H_3O^+, but HCl or ion exchange gives $H_4[Mo\ (CN)_8]\cdot 6H_2O$, apparently a strong tetrabasic acid.

Treating MoS_3 with CN^- as reductant and ligand yields the green incomplete cuboidal $[\{Mo(CN)_3\}_3(\mu\text{-}S)_3(\mu_3\text{-}S)]^{5-}$, sometimes isolated as $K_5[\{Mo(CN)_3\}_3S_4]\cdot 7H_2O$ (with the 3 Mo and 4 S at the cube's corners).

Cyanide also changes $(NH_4)_2[W^VOCl_5]$ to $[W^{IV}(CN)_8]^{4-}$ and tungstate(VI). The Ag, Cd and Pb salts of the W^{IV} dissolve slightly. The K salt is neutral and inactive to dilute H_3O^+ and OH^- in darkness. Fuming HCl with the saturated W^{IV} salt forms $H_4[W(CN)_8]\cdot 4HCl\cdot 12H_2O$, then yellow $H_4[W(CN)_8]\cdot 6H_2O$. This strong acid also arises from ion exchange or from the Ag salt plus HCl. Ionization constants are: $K_1 > K_2 > K_3 > 1$ dM; $K_4 = 2.5 \pm 0.8$ cM. Aqueous MnO_4^- or Ce^{IV} easily forms $[W(CN)_8]^{3-}$.

Aqueous $[WH(CN)_7]^{4-}$, treated with either NO, H_2S or SO_2, forms, either $[W(CN)_8]^{4-}$, no change, or $K_4[W(CN)_7(SO_3H\text{-}\kappa O)]$, respectively.

The NCS^- ion, Mo^{II}_2 and NH_4^+ form $(NH_4)_4[Mo_2(NCS)_8]\cdot nH_2O$.

Heating $[MoCl_6]^{3-}$ in 7-M KNCS at 60 °C for 2 h yields a red-orange, inert $K_3[Mo(NCS)_6]\cdot 4H_2O$, rapidly oxidized by $[IrCl_6]^{2-}$ to Mo^{IV}. Further oxidation forms $[MoO(NCS)_5]^{2-}$ and dimers.

Thiocyanate can replace several H_2O in $[Mo^{IV}{}_3O_4(H_2O)_9]^{4+}$.

Aqueous $[MoOCl_5]^{2-}$ and HNCS or HNCSe form $[MoO(NCS)_5]^{2-}$ or $[MoO(NCSe)_5]^{2-}$, respectively.

Some structures of $[M(CN)_8]^{4-}$ and $[M(CN)_8]^{3-}$ are often described as dodecahedral but without explanation. Most structures are outside the scope of this book, but this dodecahedron is not the regular, pentagonal, Platonic solid; it has often puzzled students and staff members even at leading universities, and diagrams have not always sufficed. A different description may help.

A regular pentagon with one corner missing has four vertices like a symmetrical trapezoid; the two identical ones at the ends of the open pentagon may be called "outer", the other two identical ones, "inner". Now we imagine the "outer" vertices as being near the tip of the forefinger and the tip of the thumb of a partly open hand, with the "inner" vertices being near the base of the forefinger and the base of the thumb. When we prepare to shake hands then, two open pentagons approach each other coaxially (on a line bisecting them) but with each rotated 90° from the plane of the other. In a complex a ligand is located at each vertex; "outer" and "inner" refer only to positions on the perimeters of the open pentagons, not to distances of the ligands from the metal atom, although these distances may differ for the two sets. The four equivalent "inner" vertices of a complex ion form a tetrahedron (stretched along the fourfold inversion axis, the axis of approaching hands above), and the four equivalent "outer" ones form another tetrahedron (squeezed along the same axis). The eight vertices form a dodecahedron of four isosceles and eight scalene triangles. The symmetry, D_{2d}, is as in allene, $C(=CH_2)_2$ (with no dodecahedron).

The other frequent structure for $[M(CN)_8]^{n-}$ is square antiprismatic, as in $K_3[M(CN)_8]\cdot H_2O$, made by turning one face of a cube 45° around its perpendicular, leaving the antiprism bounded, ideally, by two opposed squares and eight isosceles triangles. The energy differences between the dodecahedral and square antiprismatic structures are small, so preferences are hard to predict, and $(NH_4)_4$-$[Mo(CN)_8]\cdot {}^1/_2H_2O$ has its $[Mo(CN)_8]^{4-}$ in both the dodecahedral and square antiprismatic geometries in the solid, but predominantly dodecahedral in solution.

Some "simple" organic reagents. Dissolving $(NH_4)_2[MoCl_5(H_2O)]$ in saturated (9-M) $NaCHO_2$ and 1-dM $HCHO_2$ gives, after a day, a light-green $Na_3[Mo(CHO_2)_6]$. This reacts usefully with H_3O^+ in a few minutes to yield, say, 5-dM $[Mo(H_2O)_6]^{3+}$.

Molybdenum(3+) gives a dark gray precipitate with acetates, but no precipitate with oxalic acid. Acetate added to Mo^V precipitates a brown hydroxide that then loses water to leave perhaps $MoO(OH)_3$, red-brown.

Acetic and organic chelating acids (e.g., oxalic, tartaric or citric) complex Mo^{VI} and W^{VI} so that H_3O^+ does not precipitate MoO_3 or WO_3.

Aqueous $HC_2O_4^-$ complexes Mo^{II}_2, bridging the Mo–Mo 4-fold bond with both O on the same C of each $HC_2O_4^-$, completing 5-membered rings of –Mo–O–C–O–Mo–.

Reduced nitrogen. The molybdenum salts of oxidation states below (VI) give precipitates with NH_3, forming the corresponding hydroxides or hydrous oxides.

Molybdenum(VI) oxide dissolves readily—less so if fused—in NH_3, and $(NH_4)_2MoO_4$ can be crystallized from solutions having excess NH_3. Keeping the solution at 100°C for an hour produces the well-known ammonium "paramolybdate", $(NH_4)_6[Mo_7O_{24}]\cdot 4H_2O$. Then if we add acid it slowly rearranges to $(NH_4)_4[Mo_8O_{26}]\cdot 5H_2O$. Some other solids include $(NH_4)_2MoO_4$ and $(NH_4)_2[Mo_2O_7]$.

Tungsten(VI) oxide dissolves in NH_3, forming WO_4^{2-}. Metatungstates and NH_3 form normal tungstates:

$$W_4O_{13}^{2-} + 6\,NH_3 + 3\,H_2O \rightarrow 4\,WO_4^{2-} + 6\,NH_4^+$$

Molybdate(VI), $N_2H_5^+$ and H_3O^+ give $Mo_2O_4^{2+}$. Using HCl or HBr produces $[MoOX_5]^{2-}$. At least 7-M HCl stabilizes $[MoOCl_5]^{2-}$.

Treating MoO_3 with KCN, $N_2H_5^+$ and HCl to lower the pH to 8, with heating, cooling and adding methanol, yields $[Mo(CN)_8]^{4-}$:

$$MoO_3 + 6\,HCN + 2\,CN^- + 2\,N_2H_4 + 4\,K^+ \rightarrow$$

$$K_4[Mo(CN)_8]\cdot 2H_2O\downarrow + N_2\uparrow + 2\,NH_4^+ + H_2O$$

yellow, very soluble, oxidizable by Ce^{IV} to $[Mo(CN)_8]^{3-}$. Long exposure to sunlight changes solutions to red, then pale green, releasing HCN.

Diazane (N_2H_4), $K_4[Mo(CN)_4O_2]\cdot 6H_2O$ and KCN, heated at 60°C 2 h, later deposit some yellow $K_5[Mo(CN)_7]\cdot H_2O$.

A fine test for Mo(VI) (using $N_2H_5^+$) or reductants (using Mo(VI) depends on the production of the dark-blue, often colloidal $(Mo^VO_2OH)_n(Mo^{VI}O_3)_{1-n}$, Molybdenum Blue, in HCl or CH_3CO_2H mixture. However, W(VI) in HCl gives a similar dark-blue, colloidal $(W^VO_2OH)_x(W^{VI}O_3)_{1-x}$, "tungsten blue", an excellent test also for tungsten or reductants (H_2S, Zn, etc.) Digestion with $N_2H_6^{2+}$ in concentrated HCl on a water bath for 2 h yields green $[MoCl_5(H_2O)]$, or a greenish- or reddish-brown solution in < 10-M HCl:

$$\tfrac{1}{2}\,[Mo_2O_4Cl_4] + N_2H_6^{2+} + 2\,H_3O^+ + 4\,Cl^- \rightarrow$$

$$[MoCl_5(H_2O)] + \tfrac{1}{2}\,N_2\uparrow + NH_4Cl\downarrow + 3\,H_2O$$

Then evaporating the HCl and adding O_2-free water yields yellow $Mo^V{}_2O_4^{2+}$, i.e., $[Mo_2O_4(H_2O)_6]^{2+}$, reasonably stable under N_2.

In 3-M HCl we find, along with chloro complexes:

$$[Mo_2O_4Cl_4] + \tfrac{1}{2}\,N_2H_6^{2+} + 3\,H_2O \rightarrow$$

$$Mo_2O_4^{2+} + \tfrac{1}{2}\,N_2\uparrow + 4\,Cl^- + 3\,H_3O^+$$

Adding a stoichiometric amount of S^{2-} to the acidic solution gives yellow $Mo_2O_3S^{2+}$ and a little yellow $Mo_2O_2S_2^{2+}$, both air-stable.

Oxidized nitrogen. Aqueous HNO_2 in H_3O^+ yields $[Mo(CN)_8]^{3-}$, perhaps from both $[Mo(CN)_8]^{4-}$ and $H[Mo(CN)_8]^{3-}$ via NO^+, just as $[W(CN)_8]^{3-}$ arises from both $[W(CN)_8]^{4-}$ and $H[W(CN)_8]^{3-}$.

Molybdenum, but not tungsten, dissolves in HNO_3, with oxidation to MoO_2^{2+}, but soon becomes passive, especially in the concentrated acid, probably due to a protective coating of MoO_3.

Molybdenum dissolves in a mixture of HF and HNO_3, faster on heating. Tungsten dissolves quickly. This mixture, concentrated, is the best solvent for W. Molybdenum dissolves slowly in cold aqua regia (HCl/HNO_3), but tungsten dissolves, and rapidly, only on heating.

Nitric acid oxidizes all lower oxidation states of Mo to Mo^{VI}, and precipitates, from molybdates, $MoO_3 \cdot aq$ (see **Oxonium** above), soluble in excess of the reagent.

Aqua regia has slight effect on WS_2 (as in the ore tungstenite), but a mixture of HNO_3 and HF dissolves it readily.

Fluorine species. Aqueous HF does not attack Mo or W, but it dissolves MoO_3 and WO_3 even if fused. Theoretical calculations and elution from cation-exchange resins by 1-dM HNO_3 and 0.5-mM HF (combined) show Sg^{VI} forming various complexes such as perhaps $[SgO_2F_3(H_2O)]^-$. At around 1-M HF, Mo^{VI}, W^{VI} and Sg^{VI} appear to form $MO_2F_3^-$, or, with more HF, $[MOF_5]^-$. Whether Sg^{VI} is more or less complexed than W^{VI} depends on the c(HF) and pH.

6.2.3 Reagents Derived from the 3rd-to-5th-Period Non-Metals, Silicon through Xenon

Silicon species. Acidification of mixtures of MoO_4^{2-} and SiO_3^{2-}, SiO_4^{4-}, etc. at 60 °C forms $[SiMo_{12}O_{40}]^{4-}$ and $H_4[SiMo_{12}O_{40}] \cdot aq$ (prepared by ion exchange or by extracting "etherates"), rather resembling the phosphomolybdates. The acid is found to neutralize four equiv of base at pH 5 to 6, but eight equiv at pH 8 to 10 cold, and 24 equiv at 100°C, with breakup of the complex to MoO_4^{2-} and a form of silicic acid.

Similar treatment of WO_4^{2-} at 100°C forms $[SiW_{12}O_{40}]^{4-}$ and the very stable, extremely soluble, white $H_4[SiW_{12}O_{40}] \cdot aq$, resembling the phosphotungstates. Hydrogen sulfate, H_2SO_4, is a poor choice for the acid because of low volatility in later purification, and CH_3CO_2H acts as a reductant, but HCl works well. The product acid again neutralizes four equiv of base at pH 5 to 6, but 24 equiv at pH 8 to 10 and 100°C, decomposing the complex to WO_4^{2-} and a form of silicic acid.

Light slowly reduces the aqueous solution and turns it blue, but chlorine reverses this. The complex is used to precipitate proteins, alkaloids and some amino acids. The ammonium and potassium salts are much less soluble than that of sodium. The acid is completely extractable by ether, with which it forms a dense third liquid layer.

Phosphorus species. Acidified molybdate and HPH_2O_2 give a deep-blue precipitate or solution, Molybdenum Blue, depending on the amount of Mo present. Phosphinate, $PH_2O_2^-$, added to tungstate containing excess H_2SO_4 reduces it similarly, on heating, to "Tungsten Blue". Both are mixed-valence M^V-M^{VI} compounds.

Heating $W_3S_7Br_4$ with HPH_2O_2 and concentrated HCl at 90°C under N_2 for 15 h forms purple $[W_3S_4(H_2O)_9]^{4+}$ with a 20 % yield. See **6.2.5 … cuboidal clusters** for more on hydrated $W_3S_4^{4+}$, i.e., $[W_3S_4(H_2O)_9]^{4+}$, etc.

Heating $W_3Se_7Br_4$ and HPH_2O_2 with concentrated HCl as a catalyst at 90°C for 8 h or more under N_2 forms the green $[W_3Se_4(H_2O)_9]^{4+}$, stable for months in air at 5°C but somewhat sensitive to light and requiring (dilute) acid to prevent the polymerization concomitant with losing H^+. Red Se_8 appears slowly. **Beware** the toxic H_2Se:

$$W_3Se_7Br_4 + 3\,HPH_2O_2 + 12\,H_2O \rightarrow$$

$$[W_3Se_4(H_2O)_9]^{4+} + 4\,Br^- + 3\,H_2Se\uparrow + 3\,H_2PHO_3$$

Cesium ion, 6 mM, plus 4-mM $[Mo_2Cl_8]^{4-}$ in 2-M phosphoric acid and air for a day or two, precipitate purple $Cs_2[\{Mo^{III}(H_2O)\}_2(\mu\text{-}HPO_4)_4]$ with Mo≡Mo triple bonding.

Phosphoric acid and its salts precipitate rather slowly from solutions of ammonium molybdate with much HNO_3 (faster on warming) yellow triammonium phosphododecamolybdate, or, to give a name completely approved by the IUPAC for this example: triammonium hexatriacontaoxo(tetraoxophosphato)dodecamolybdate(3−). This salt is soluble in either NH_3 or OH^-, and slightly soluble in excess PO_4^{3-}:

$$H_3PO_4 + 12\,MoO_2^{2+} + 3\,NH_4^+ + 39\,H_2O \rightarrow$$

$$(NH_4)_3[PMo_{12}O_{40}]\downarrow + 27\,H_3O^+$$

(We note in passing that retaining "ammonium" instead of "azanium" in the name here avoids confusing "triazanium" with a possible $N_3H_6^+$, in analogy with "diazanium" for $N_2H_5^+$.) Some solutions on standing yield $[P_2Mo_{18}O_{62}]^{6-}$. The major species in phosphate-molybdate mixtures, however, are often $[PMo_9O_{31}(H_2O)_3]^{3-}$, pale yellow, or $[P_2Mo_5O_{23}]^{6-}$, colorless, or their acid forms. The acids of various heteropolymetalates like these, and their salts with small Alk^+, are quite soluble, but large cations such as Cs^+, Ba^{2+}, Tl^+ and Pb^{2+} usually precipitate them, with NH_4^+, K^+ and Rb^+ salts being in between.

Concentrated HCl, added slowly to a boiling solution of a tungstate and some excess of a phosphate over the calculated amount, forms a white, extremely soluble, strong heteropoly acid as follows:

$$12\,WO_4^{2-} + HPO_4^{2-} + 26\,H_3O^+ \rightarrow H_7[PW_{12}O_{42}]\cdot aq\downarrow + 36\,H_2O$$

In solution, light slowly reduces this and turns it blue; heating with Cl_2 reverses that. The acid is completely extractable by ether, which forms a dense, third liquid layer.

The acids of various heteropolymetalates like these, from numerous other manipulations, and their salts with small cations, such as Na^+, are quite soluble, but large cations such as Cs^+, Ba^{2+}, Tl^+ and Pb^{2+} usually precipitate them, with K^+, Rb^+ and NH_4^+ salts having low solubilities.

The hydrated H^+ in heteropoly acids is generally $[H(H_2O)_2]^+$. The unsolvated acids, from vacuum or heat treatment of the hydrates, are superacids, even when ions like Cs^+ replace some of the H^+.

The one-electron reducibility of the anions and acids is about as expected from the uncomplexed-species behavior: $V^V > Mo^{VI} > W^{VI}$. The reducibility of the Ti^{IV}, Nb^V and Ta^V units in related complexes is correspondingly limited.

More-complex species from many other elements are well known, taking **3d** elements M as examples:

$$[M(H_2O)_m]^{n+} + [PW_{11}O_{39}]^{7-} \rightarrow [PW_{11}O_{39}M(H_2O)]^{(7-n)-} + (m - 1)\ H_2O$$

Reducibility again resembles the normal $E°$ in the following series: $Co(H_2O)_6^{3+} > Mn(H_2O)_6^{3+} > Fe(H_2O)_6^{3+} > Cu(H_2O)_4^{2+}$.

Many homopoly and heteropoly species catalyze the oxidation, sometimes helped by light, of numerous organic and other substances.

Arsenic species. Acidified molybdate and As^{III} give a yellow $[As_2Mo_{12}O_{42}]^{6-}$. Arsenate forms with Mo^{2+} a gray precipitate; with Mo^{3+}, molybdenum(III) arsenate.

Arsenic acid precipitates, from HNO_3 solutions of ammonium molybdate, on warming to 60–70°C but not when cold, $(NH_4)_3[AsMo_{12}O_{40}]$, yellow, soluble in NH_3 and OH^-, e.g.:

$$H_3AsO_4 + 12\ [MoO_2(H_2O)_4]^{2+} + 3\ NH_4^+ \rightarrow$$

$$(NH_4)_3[AsMo_{12}O_{40}]\downarrow + 27\ H_3O^+ + 9\ H_2O$$

Reductants, Cl^- and tartaric acid hinder this test for arsenic.

If $3 < pH < 5$, molybdate and arsenate form $[AsMo_9O_{31}(H_2O)_3]^{3-}$ and more; cf. the phosphates. A lower pH gives $H_4[As_4Mo_{12}O_{50}]^{4-}$ etc.

Reduced chalcogens. Like As_2S_5, MoS_3, although insoluble even in concentrated HCl, dissolves in HNO_3. After precipitation by H_2S and dissolution in HNO_3, the As^V in an unknown mixture may be removed with magnesia mixture, and Mo^{VI} may be detected in the filtrate as $(NH_4)_3[PMo_{12}O_{40}]$ (unless in small amount) or by the SCN^- test.

Neutral and alkaline solutions of MoO_4^{2-} are colored deep yellow, brown or red by S^{2-}, forming $[MoO_{4-n}S_n]^{2-}$; then H_3O^+ precipitates MoS_3.

Bubbling H_2S into MoO_4^{2-} with much NH_3, then warming to 60°C with more H_2S for 30 min and cooling to 0°C, yields $(NH_4)_2[MoS_4]$. A somewhat similar treatment of H_2WO_4, but with passing H_2S for 8 h at 60°C, yields $(NH_4)_2[WS_4]$, although contaminated with $(NH_4)_2[WOS_3]$ if done without the long t, high T or continued stream of H_2S.

A sensitive test for molybdenum(VI) or reductants calls for the production of $(Mo^{V}O_2OH)_n(Mo^{VI}O_3)_{1-n}$, Molybdenum Blue, in an HCl mixture, by H_2S (or Zn, $N_2H_5^+$, or SO_2, for example):

$$2\,MoO_3\cdot aq + n\,H_2S \rightarrow 2\,(MoO_2OH)_n(MoO_3)_{1-n} + n\,S\downarrow$$

Acidified solutions of molybdate, treated with a small amount of H_2S, give a blue, possibly colloidal, solution; treatment with more H_2S, however, slowly restores Mo^{VI} as a brown precipitate of MoS_3, the reaction being complete only under pressure or at 100°C. The precipitate is soluble in S_x^{2-}, especially when warm and not too concentrated, yielding red $[MoS_4]^{2-}$, from which acids reprecipitate MoS_3, insoluble in boiling $H_2C_2O_4$ (distinction from W species, and separation from SnS_2).

Molybdate heated at 90°C ~20 min with a slight excess of HS^- (from Na_2S and HCl) is reduced at a pH of 11–13 and forms a red-orange complex related to molybdenum enzymes, $[Mo^{V}{}_2O_2S_2]^{2+}$.

Passing H_2S into concentrated K_2MoO_4 and excess KCN, yields, after purifications, green $K_6[\{Mo^{IV}(CN)_6\}_2(\mu\text{-}S)]\cdot 4H_2O$.

Sulfane, H_2S, does not precipitate WS_3 from acidic or alkaline solutions of W^{VI}, or from phosphotungstates. Alkalis give $[WS_4]^{2-}$, which precipitates WS_3 incompletely upon acidification. The trisulfide dissolves easily in alkaline sulfides, as $[WS_4]^{2-}$. Relativity makes $[WS_4]^{2-}$ yellow (absorbing higher-energy photons) although $[MoS_4]^{2-}$ is red.

A mutual separation of Mo and W, from low-quality $CaWO_4$, scheelite, uses high-pressure leaching with Na_2CO_3, then HS^-, which forms $[MoS_4]^{2-}$ much faster than $[WS_4]^{2-}$. Slowly adding H_2SO_4 until $2 < pH < 3$ can precipitate MoS_3 completely with very little WS_3.

Blue $K_6[\{Mo^{III}(CN)_4\}_2(\mu\text{-}S)_2]\cdot 4H_2O$ arises from MoO_4^{2-}, KCN and HS^- (from H_2S) under N_2, after the intermediate $[Mo^{IV}O_2(CN)_4]^{4-}$, along with some of the more reduced product $K_5[Mo(CN)_7]\cdot H_2O$; oxygen gives $K_{14}[\{Mo^{IV}(CN)_6\}_2\text{-}(\mu\text{-}S)]_2Mo^{VI}O_4\cdot 10H_2O$.

Purple $[NEt_4]_2[MoSe_4]$ and blue $[PPh_4]_2[MoSe_4]$ are precipitated from non-aqueous media.

Excess $(NH_4)_2S_3$, $[MoS_4]^{2-}$ and NEt_4^+ form brown $(NEt_4)_2[Mo^{IV}S(S_4)_2]$.

In $CHONMe_2$, WCl_6 reacts easily with Te_3^{2-} up to Te_5^{2-} to form $[W^{IV}O(\eta^2\text{-}Te_4)_2]^{2-}$. Lower yields arise from WCl_4 or $WOCl_4$.

Molybdenum(3+) gives complexes with SCN^-, e.g., $Mo(SCN)_6^{3-}$. The orange-red complex of Mo^{V} with SCN^- in acid is suitable for colorimetry. Thiocyanate in molybdate(VI) solution acidified with HCl produces a yellow color, changing to a deep carmine red on addition of a reducing agent, e.g., Zn, $SnCl_2$, etc. The color is not affected by H_3PO_4 (distinction from Fe^{III}, which will not interfere, however, if reduced completely to Fe^{2+}). Tartaric and other organic chelating acids interfere. Using $SnCl_2$ gives a sensitivity of 20 μM.

Acidified NCS^- reduces WO_4^{2-}, yielding various complexes of WO^{3+} and NCS^-, with colors depending on pH.

A sensitive reagent for MoO_4^{2-}, $C_2H_5OCS_2^-$ (ethylxanthate) in CH_3CO_2H, gives a deep-red color when added dropwise to a solution containing as little as 6 μM Mo. The intensity of the color is unaffected by Ti, V, or W species. Any CrO_4^{2-}, however, should first be reduced to Cr^{3+}. The $C_2O_4^{2-}$ and UO_2^{2+} ions, and Fe, Co, Ni and Cu species, interfere.

Oxidized chalcogens. Adding $S_2O_3^{2-}$ to MoO_4^{2-}, slightly acidified, gives a blue precipitate and blue solution. If the acidity is greater, a red-brown precipitate forms.

If WO_4^{2-} is heated with $S_2O_3^{2-}$ no action is noted. On adding HCl, a white precipitate and a blue liquid result, the latter from the reduction of the WO_4^{2-}. Nitric acid in place of HCl gives a brown liquid.

Treating molybdate and acid with SO_2 gives an intense bluish-green precipitate or color, or Mo^{3+}. depending on the amounts of reactants.

Alkaline SO_3^{2-} reduces $[Mo(CN)_8]^{3-}$ to $[Mo(CN)_8]^{4-}$.

Tungsten(VI) is reduced to W^V with excess SO_2 in acidic solution.

The dark-blue, colloidal $(W^VO_2OH)_x(W^{VI}O_3)_{1-x}$, "Tungsten Blue", arising in an HCl mixture of W^{VI} with a little H_2S or SO_2 provides an excellent test for tungsten or reductants:

$$2\,WO_3{\cdot}aq + x\,SO_2 + 4x\,H_2O \rightarrow$$

$$2\,(WO_2OH)_x(WO_3)_{1-x}{\downarrow} + x\,SO_4^{2-} + 2x\,H_3O^+$$

Metallic Mo and W are not attacked by H_2SO_4, which precipitates no MoO_3 from Mo^{VI}, but even fused MoO_3 dissolves in concentrated H_2SO_4.

Sulfate displaces chloride from $[Mo_2Cl_8]^{4-}$ to form a pink $[Mo_2(SO_4)_4]^{4-}$ and $K_4[Mo_2(SO_4)_4]$, and a lavender $[Mo_2(SO_4)_4]^{3-}$ (likely due to O_2). From the former, aqueous $Ba(SO_3CF_3)_2$ removes the SO_4^{2-} under N_2, giving red Mo_2^{4+}, i.e., $[Mo_2(H_2O)_8(H_2O)_2]^{4+}$ with two weakly bound axial waters, stable for up to 3 h in 1-dM HSO_3CF_3.

The bright-yellow alum, $[Cs(H_2O)_6][Mo(H_2O)_6](SO_4)_2$, from Mo^{3+}, turns brown in air in a few hours.

Treating $K_4[Mo_2Cl_8]$ with H_2SO_4 gives red $K_4[Mo_2(\mu\text{-}SO_4)_4]\cdot 2H_2O$. Air converts a saturated solution to blue $K_3[Mo_2(H_2O)_2(\mu\text{-}SO_4)_4]\cdot {}^3/_2H_2O$. Light (UV) in 5-M H_2SO_4 also forms this, as does H_2O_2 in 2-M H_2SO_4, this time acting on $K_4[Mo_2Cl_8]$ with some KCl. Electrochemistry reveals a one-electron oxidative process.

The amorphous dioxide, WO_2, is readily soluble in warm H_2SO_4.

Tungsten(VI) oxide is insoluble even in hot concentrated H_2SO_4.

Reduced halogens. Metallic Mo and W are passivated by cold HCl, but Mo dissolves slowly in hot, dilute HCl.

Molybdenum(VI) is mainly $[MoO_2Cl_2(H_2O)_2]$ with *cis*-O_2 from 2- to 6-M HCl, $MoO_2Cl_3^-$ from 6- to 12-M, and $[cis\text{-}MoO_2Cl_4]^{2-}$ > 12-M HCl.

Concentrated HCl, K^+ and $[Mo^{II}_2(CH_3CO_2)_4]$ (from a non-aqueous reaction with $[Mo(CO)_6]$), at 0°C give red $K_4[Mo_2Cl_8]\cdot 2H_2O$. Similarly, HBr can yield $(NH_4)_4[Mo_2Br_8]$. Saturation of the former with HCl at 0°C, plus NH_4Cl, followed by warming to ambient T under , e.g., N_2, yield wine-red $(NH_4)_5[Mo^{II}_2Cl_9]\cdot H_2O$. Alternately, the acetate with ordinary concentrated HCl or HBr (not saturating the solution) at 60°C for 1 h under N_2, followed by CsCl or CsBr and cooling, produces insoluble, oxidized (by H^+) and previously unexpected, monohydrido complexes, yellow or brown respectively, and surprisingly, rather stable in dry air:

$$[Mo_2(CH_3CO_2)_4] + 5\,H_3O^+ + 8\,X^- + 3\,Cs^+ \rightarrow$$

$$Cs_3[(Mo^{III}X_3)_2(\mu\text{-}H^{-I})(\mu\text{-}X)_2]\downarrow + 5\,H_2O + 4\,CH_3CO_2H$$

This may result also from electrolyzing $[Mo_2Cl_8]^{4-}$, or from treating it with 6 to 12-M HCl. Some procedural variations and HI form the corresponding $[(Mo^{III}I_3)_2HI_2]^{3-}$. The $[(Mo^{III}Cl_3)_2HCl_2]^{3-}$ in < 3-M HCl decomposes to H_2 and chloro-complexes of $Mo_2(OH)_2{}^{4+}$. In < 5-dM HCl it quickly forms an intermediate, $[Mo_2HCl_7(OH)]^{3-}$.

The anhydrous Mo^{II} or W^{II} salts M_6X_{12} (from non-aqueous sources) of the 3rd-to-5th-period halogens Cl, Br or I, are cluster compounds, [*octahedro*-$M_6(\mu_3\text{-}X)_8X_2]X_2$, most familiar with Cl and Br. The four outer ions can be replaced quickly, the inner eight slowly, in various ways by other halides, including F^-. Two more X^- (halide), H_2O, etc. can be added, yielding, say, $[(MoX)_6(\mu_3\text{-}X)_8]^{2-}$. The Mo clusters are weak reductants but are unstable to, e.g., OH^-, CN^- or SH^-. The oxonium solid $(H_3O)_2[Mo_6Cl_{14}]\cdot 6H_2O$ is soluble but (slowly) unstable; the potassium salt, recrystallized from 6-M HCl, is stable.

The W clusters may reduce H_2O. Aqueous $[W_6Cl_{14}]^{2-}$ is easily oxidized at an anode to $[W_6Cl_{14}]^-$, a good oxidant. At least $[Mo_6Cl_{14}]^{2-}$, $[Mo_6Br_{14}]^{2-}$ and $[W_6Cl_{14}]^{2-}$ are ordinarily luminescent.

Saturated, i.e., 12-M, HCl can give $[MoCl_6]^{3-}$, whose pink or red salts of K, Rb, Cs and NH_4 are stable in air, but 6-M HCl reverses this:

$$[MoCl_5(H_2O)]^{2-} + Cl^- \leftrightarrows 2\,[MoCl_6]^{3-} + H_2O$$

Molybdenum dioxide is insoluble in HCl.

Amorphous WO_2 dissolves readily in warm HCl to a red solution, which, on standing, loses its color with oxidation of the W. Crystalline WO_2, however, is not affected by hot, concentrated, non-oxidizing acids.

Molybdenum(V) tends strongly to dimerize below 2-M HCl, but not above 10-M HCl. Concentrated HBr forms, e.g., $[(MoBr_4)_2(\mu\text{-}O)_2]^{2-}$.

Concentrated HCl allows the following slow equilibrium at 0°C, favoring the left, but 4-cM $[NHMe_3]^+$ isolates the blue product in 2–3 d (air and the W^V give the same result), i.e., $[NHMe_3]_2[W_4O_8Cl_8(H_2O)_4]$:

$$[W^VOCl_5]^{2-} + [W^{VI}O_2Cl_4]^{2-} + 5\,H_2O \leftrightarrows$$

$${}^1/_2\,[\textit{quadro-}\{WOCl_2(H_2O)\}_4(\text{linear-}\mu\text{-}O)_4]^{2-} + 5\,Cl^- + 2\,H_3O^+$$

From this and NCS^- arises a similar mixed-valence ion, $[W_4O_8(NCS)_{12}]^{6-}$.

Molybdate(VI) in HCl over 6 M becomes, for example, $[MoCl_2O_2]$ or $[Mo_2Cl_4O_4]$. Cooling K_2WO_4 in ice, and then adding the slurry slowly to much concentrated, cold HCl forms complexes:

$$WO_4^{2-} + 8\,H_3O^+ + n\,Cl^- \rightarrow WCl_n^{(n\text{-}6)-} + 12\,H_2O$$

but we also find $[cis\text{-}WO_2Cl_4]^{2-}$ even at concentrations over 12-M HCl.

Dilute HCl and $[WS_4]^{2-}$ give $[W^{IV}O(\eta^2\text{-}W^{VI}S_4)_2]^{2-}$.

Iodide reduces $[W(CN)_8]^{3-}$ to $[W(CN)_8]^{4-}$.

Aqueous HI reduces $[MoCl_2O_2]$ in concentrated HCl to $[MoCl_5H_2O]$.

Elemental and oxidized halogens. Aqueous Cl_2 oxidizes $[W(CN)_8]^{4-}$ to $[W(CN)_8]^{3-}$.

Aqueous $[W_2Cl_9]^{3-}$ and Cl_2, Br_2 or I_2 form, e.g., violet $[W_2Cl_9]^{2-}$, i.e., $[(WCl_3)_2(\mu\text{-}Cl)_3]^{2-}$ with a $^5/_2$ W–W bond.

The calculated amounts of I_3^- plus Mo_2^{4+} appear to give yellow MoI^{2+}.

The Mo^{II} and Mo^{III} aqua ions, and even some trimeric Mo^{IV} ions overnight, reduce ClO_4^-. For a weakly coordinating and non-oxidizing anion then, one must choose, e.g., $CF_3SO_3^-$.

6.2.4 Reagents Derived from the Metals Lithium through Uranium, plus Electrons and Photons

Oxidation. Molybdenum(3+) and Mo^V can be titrated to Mo^{VI} with Ce^{4+}, VO_2^+, $[Cr_2O_7]^{2-}$ and MnO_4^-. This (aqua) Mo^{3+} with VO^{2+}, Fe^{3+}, $[Co(C_2O_4)_3]^{3-}$ or $[IrCl_6]^{2-}$ soon goes to $Mo^V{}_2O_4^{2+}$, then to Mo^{VI}. The $[Co(C_2O_4)_3]^{3-}$, however, does not oxidize $Mo^{IV}{}_3$, so the (aqua) Mo^{III} and $Mo^{III}{}_2$, which it does attack, apparently generate more oxidizable Mo^{IV} and $Mo^{IV}{}_2$ intermediates, as supported also by electrochemistry, although the stronger oxidant $[IrCl_6]^{2-}$ can oxidize $Mo^{IV}{}_3$.

The calculated amount of $[Ag(NH_3)_2]^+$, plus $[Mo_2(\mu\text{-}OH)_3(CO)_6]^{3-}$ in concentrated NH_3, followed by HSO_3CF_3 to a pH of 2, appear to give a small yield of the formato complex, $[Mo^{III}(CHO_2)(NH_3)_4H_2O]^{2+}$. A MnO_4^- titration takes 5 eq, presumably including CHO_2^- to CO_3^{2-}.

Electron transfer (outer sphere) is very fast from $[W(CN)_8]^{4-}$ to $[Mo(CN)_8]^{3-}$, $[W(CN)_8]^{3-}$, $[Fe(CN)_6]^{3-}$ and $[IrCl_6]^{2-}$.

The $[W(CN)_8]^{4-}$ is more easily oxidized than $[Mo(CN)_8]^{4-}$, and reacts with Ce^{IV} and MnO_4^-, going to $[W(CN)_8]^{3-}$. Precipitation with Ag^+, transposition with KCl, and then evaporation, yield the pale-yellow $K_3[W(CN)_8]\cdot 2H_2O$. The Ag salt plus HCl likewise give violet-brown $H_3[W(CN)_8]\cdot H_2O$. The ionization constants are: $K_1 > 1$ dM; $K_2 = 2 \pm 2$ cM; $K_3 = 4.5 \pm 1.5$ mM.

Molybdenum(II), Mo_2^{4+}, in 1-M HCl at an anode gives unstable $Mo^{III}{}_2Cl_4^{2+}$. Anodic treatment, Ce^{IV}, Cr^{VI}, MnO_4^- etc. change $[Mo(CN)_8]^{4-}$ to yellow $[Mo(CN)_8]^{3-}$, very light sensitive (turning red-brown) and easily reduced by, say, SO_2 or I^-. It can be precipitated by Ag^+ and then converted to various salts by Cl^-.

Light (254 nm) and Mo_2^{4+} with 1-M CF_3SO_3H, or $[Mo_2(SO_4)_4]^{4-}$ with 2.5-M H_2SO_4, or $[Mo_2Br_8]^{4-}$ with 3-M HBr, form H_2 and $Mo_2(OH)_2^{4+}$, with low, modest, or low quantum yields respectively.

Reduction with metallic species. At low pH, Eu^{2+}, Ti^{2+}, Ti^{III}, V^{2+} and Ge^{II} reduce H_2MoO_4 to $Mo_2O_4^{2+}$. Excess Cr^{2+}, in the inner sphere, reduces $Mo^{IV}{}_3O_4^{4+}$ and $Mo^{V}{}_2O_4^{2+}$ in 1.9-M H_3O^+ as far as $Mo^{III}{}_2(OH)_2^{4+}$ in 24 h.

Molybdenum(VI) may be titrated to Mo^V with Ti^{3+}, Cr^{2+} and $SnCl_2$. It can also be reduced to Mo^V by Cu, Ag, Hg, Bi or a controlled Hg-pool cathode for titration, or to Mo^{3+} by Mg, Zn, Zn_{Hg}, Cd, Cd_{Hg}, Al, Sn_{Hg}, Pb_{Hg} or Bi_{Hg}. Treating ammonium molybdate(VI) with metallic tin in hot, concentrated HCl, followed by NH_4Cl and cooling, can be used to make a red Mo^{III} complex, albeit in a low yield, without excluding air (although air slowly oxidizes the solution, but not the solid):

$$2\,[Mo_7O_{24}]^{6-} + 21\,Sn + 96\,H_3O^+ + 147\,Cl^- + 42\,NH_4^+ \rightarrow$$

$$14\,(NH_4)_3[MoCl_6]\downarrow + 21\,SnCl_3^- + 144\,H_2O$$

An intensely blue species arises from Fe^{2+} and $[Mo(CN)_8]^{3-}$.

The trioxide WO_3 will give a characteristic blue color when rubbed on a bright surface of Fe, Cu, Zn or Al; the test is facilitated by slight moistening. The best way is to put the possible WO_3 on Al, moisten with H_2O, then add 1–2 drops of dilute HCl.

Iron(2+) gives a brown precipitate with WO_4^{2-}. On adding an acid, no blue color is obtained (distinction from MoO_4^{2-}).

Zinc(Hg) with molybdate and HSO_3CF_3 under N_2 give the blue-green $Mo^{III}{}_2(OH)_2^{4+}$, stable at 0 °C in 2-M acid under N_2 for two weeks.

Zinc(Hg) reduces air-free $[Mo_2Cl_8H]^{3-}$ and $[Mo_2Cl_9]^{3-}$ to $Mo^{II}{}_2$, it reduces $[Mo^{IV}{}_3O_4(H_2O)_9]^{4+}$ to a green $Mo^{III}{}_3$ ion, but not the also known green $[Mo^{III}{}_2(\mu\text{-}OH)_2(H_2O)_8]^{4+}$, and it reduces $[Mo^{IV}{}_3O_4(C_2O_4)_3(H_2O)_3]^{2-}$.

Numerous reductants, e.g., $N_2H_5^+$ and H_2S, depending greatly on the reagent and conditions, often react with Mo^{VI}, as oxidants react with Mo^V or lower, to give a range of Molybdenum Blue mixtures, $(Mo^VO_2OH)_n(Mo^{VI}O_3)_{1-n}$. In fact, a sensitive test for reductants or Mo^{VI} uses the production of this in HCl mixture by Zn (avoiding excess; also HF interferes), Al, Sn^{II} or Hg_2^{2+} plus I^-:

$$2\,MoO_3{\cdot}aq + x\,Zn + 2x\,H_3O^+ + 3x\,Cl^- \rightarrow$$

$$2\,(MoO_2OH)_x(MoO_3)_{1-x}\downarrow + x\,[ZnCl_3]^- + 2x\,H_2O$$

Some of the deep-blue Mo^V-Mo^{VI} materials from the mild reduction of heterododecamolybdates and others are useful for their color.

If excess concentrated HCl is added to a dilute solution of WO_4^{2-} until any precipitate first formed dissolves, the resulting solution, upon successive additions of small pieces of Zn, will develop various colors, especially a brilliant red. This detects about 4 μmol at the lower limit.

Thiocyanate and metallic zinc, added to a concentrated HCl mixture with W^{VI}, give a deep green color. If the SCN^- is added to WO_4^{2-}, then HCl, and finally Zn, a beautiful amethyst color results.

The commonly two-electron reductants In^+, $[GeCl_4]^{2-}$ and $[SnCl_3]^-$ are oxidized in two one-electron steps to reduce $[Mo(CN)_8]^{3-}$ and $[W(CN)_8]^{3-}$ to $[Mo(CN)_8]^{4-}$ and $[W(CN)_8]^{4-}$. The Mo oxidant is much faster than the W, and the In reductant is much faster than the others.

Tin, WO_4^{2-} and 12-M HCl give a deep-purple $[(WCl_5)_2(\mu\text{-}O)]^{4-}$.

Tin dichloride and WO_4^{2-} give a yellow precipitate which becomes the blue $(W^VO_2OH)_x(W^{VI}O_3)_{1-x}$ upon warming with HCl or H_2SO_4. This is a sensitive test if no interfering substance is present. Metallic Sn or Zn, plus acid, give the blue color with WO_4^{2-}. Acetic acid does not interfere with the test with Zn, but $SnCl_2$ forms a brown precipitate. A slight variation begins with tungsten(VI), HCl and a little Sn^{II}:

$$2\,WO_3{\cdot}aq + x\,[SnCl_3]^- + 3x\,Cl^- + 2x\,H_3O^+ \rightarrow$$

$$2\,(WO_2OH)_x(WO_3)_{1-x}\downarrow + x\,[SnCl_6]^{2-} + 2x\,H_2O$$

The blue $W^{V/VI}$ may remain in solution if a complex salt is tested. Excess reductant however, may lead to a brown color.

Reduction by excess Sn in saturated HCl at 40 °C, or cathodic e^- with W^{VI} in much cold HCl, makes the solution deep purple:

$$2\,WCl_n^{(n\text{-}6)-} + 3\,Sn + (21 - 2n)\,Cl^- \rightarrow [W_2Cl_9]^{3-} + 3\,[SnCl_4]^{2-}$$

$$3\,[W_2Cl_9]^{3-} + Cl^- \leftrightarrows 2\,[W_3Cl_{14}]^{5-}$$

Two volumes of ethanol, with rapid filtration by vacuum, give the greenish $K_3W_2Cl_9$, i.e., $K_3[(WCl_3)_2(\mu\text{-}Cl)_3]$ with a W≡W (triple) bond:

$$W_2Cl_9^{3-} + 3\,K^+ \rightarrow K_3W_2Cl_9\downarrow$$

much less soluble in concentrated HCl or ethanol than in water, stable in concentrated HCl, but in water oxidized slowly by air. With more KCl we get the more soluble, more oxidizable, red $K_5W_3Cl_{14}$.

Concentrated HBr likewise gives $[W_2Br_9]^{3-}$, or $[(WBr_3)_2(\mu\text{-}Br)_3]^{3-}$, which can also be prepared by an exchange:

$$[W_2Cl_9]^{3-} + 9\,Br^- \leftrightarrows [W_2Br_9]^{3-} + 9\,Cl^-$$

Reduction with electrons and photons. Controlled cathodic electrolysis of MoO_3 in acids is a good route to Mo^V or Mo^{III}. In 12-M HCl it can give green, brown and then red complexes down to $[MoCl_6]^{3-}$ or $(Alk,NH_4)_3[MoCl_6]$, $[MoCl_5(H_2O)]^{2-}$ or $(NH_4)_2[MoCl_5(H_2O)]$, and $[Mo_2Cl_9]^{3-}$. Aqueous 9-M HBr forms similar complexes. The $[Mo_2X_9]^{3-}$ are bridged, i.e., $[(MoX_3)_2(\mu\text{-}X)_3]^{3-}$. Aqueous K^+ and 11-M

HCl yield pink $K_3[MoCl_6]$, stable in dry air but sensitive to moist O_2, going in H_2O to $[Mo(H_2O)_6]^{3+}$.

Cathodic e^- in 1-dM H_3O^+ reduce $[\{Mo^{IV}(H_2O)\}_3(\mu\text{-}CH_3CO_2)_6(\mu_3\text{-}O)_2]^{2+}$ (non-aqueous source) to an $Mo^{III}{}_3$ species, oxidizable to $Mo^{(III+III+IV)/3}{}_3$ (all Mo in the same fractional oxidation state) and the original $Mo^{IV}{}_3$. The related $[\{W^{(III+III+IV)/3}(H_2O)\}_3(\mu\text{-}CH_3CO_2)_6(\mu_3\text{-}O)]^{2+}$, dark blue, is similarly reduced to $[\{W^{III}(H_2O)\}_3(\mu\text{-}CH_3CO_2)_6(\mu_3\text{-}O)]^+$, unstable to H_2O and O_2, oxidized to $[\{W(H_2O)\}_3(\mu\text{-}CH_3CO_2)_6(\mu_3\text{-}O)_2]^+$ much more easily, like many other W species, than Mo^{III}. The acetato $Mo^{(III+III+IV)/3}{}_3$ ion, unlike $Mo^{(III+III+IV)/3}{}_3(OH)_4{}^{6+}$, does not dismutate in acid into Mo^{III} and Mo^{IV}.

Many chemists prefer not to speak of fractional oxidation states; clearly we do not find fractional electrons. Sometimes, however, this formalism may promote clarity with oxidation and reduction if we keep in mind that an integral number of electron clouds can be spread over several atoms. Cf. **7.2.4 Reduction** for $[Tc^{(II+III)/2}{}_2Cl_8]^{3-}$.

Cathodic e^- and WO_3 in ≤ 6-dM HNO_3 or 1-M $LiNO_3$ form H_xWO_3 ($x \le 0.84$) or Li_xWO_3 ($x \le 0.36$); Li_xWO_3 is the much more stable one in air.

Light (UV), $K_3[W_2Cl_9]$ and KCN in the absence of air give a good yield of bright-yellow, diamagnetic $K_5[W(CN)_7]\cdot H_2O$, quickly oxidized by moist air, stable if dissolved only in excess air-free OH^- or CN^-, otherwise apparently in equilibrium with $[WH(CN)_7]^{4-}$.

Other reactions. Normal tungstates and homopolytungstates generally precipitate Ca^{2+} and **d**- or **p**-block M^{2+}.

Aqueous $UO_2{}^{2+}$ and $[Mo(CN)_8]^{4-}$ give a red-brown 1:1 complex in water, but acetone precipitates $(UO_2)_2[Mo(CN)_8]\cdot aq$. The Cr^{3+} and Fe^{3+} ions also form 1:1 complexes and, with M = K, Rb, Cs or $NH_4{}^+$, precipitate $MFe[Mo(CN)_8]$. Not too dilute solutions can give $Fe_4[Mo(CN)_8]_3\cdot 12H_2O$, reminiscent of "Prussian Blue".

Solutions of $[Mo(CN)_8]^{4-}$ and $\mathbf{3d}^{2+}$ yield $\mathbf{3d}_2[Mo(CN)_8]\cdot aq$.

Manganese(II) produces the surprising and magnetically interesting $[Mn^{II}{}_6(H_2O)_9][W^V(CN)_8]_4\cdot 13H_2O$ [1].

Aqueous $(NH_4)_2[MoS_4]$ and Cu^{2+} precipitate $NH_4Cu^IMoS_4$. Adding one $[MS_4]^{2-}$ to one or two $[Mc(CN)_n]^{(n-1)-}$, where M is Mo or W, and Mc is a coinage metal, Cu or Ag, yields $[MS_2(S_2McCN)]^{2-}$ or, in turn, $[M(S_2McCN)_2]^{2-}$. Many complexes $[MoS_4(CuX)_n]^{2-}$, X = CN, Cl, Br etc., and $n \le 4$, are known.

Thiometalates, **d**-block ions $\mathbf{d}^{2+}$ (**d** = Fe, Co, Ni, Pd, Pt, Zn, Cd or Hg), and $PPh_4{}^+$ or $AsPh_4{}^+$ precipitate $[(P,As)Ph_4]_2[\mathbf{d}(MO_{2-n}S_{2+n})_2]$, $n \ge 0$, but also $[PPh_4](Fe,Co)MoS_4$; Cu^{2+} or Ag^+ gives $[PPh_4]Mc^IMS_4$.

The $[\mathbf{d}\{(Mo,W)S_4\}_2]^{2-}$ are reduced reversibly by one or two cathodic e^-, easier for Fe > Co > Ni. The $[Fe(MoS_4)_2]^{2-}$ is otherwise unstable.

Molybdenum(3+) with $[Fe(CN)_6]^{4-}$ produces a dark brown, with $[Fe(CN)_6]^{3-}$ a red-brown, precipitate. Aqueous $[Fe(CN)_6]^{4-}$ also forms a red-brown precipitate from molybdates(VI) acidified with HCl.

If to WO_4^{2-} a slight excess of H_3O^+ is added, followed by $[Fe(CN)_6]^{4-}$, the solution will become deep reddish brown. On standing, a precipitate of the same color appears.

Molybdate(VI) is unidentate in $[Co(\eta^1\text{-}MoO_4)(NH_3)_5]^+$ and didentate in $[Co(\eta^2\text{-}MoO_4)(NH_3)_4]^+$. Coordination isomerism occurs, for example, in $[CoCl(NH_3)_5]MoO_4$ and $[Co(MoO_4)(NH_3)_5]Cl$.

The alkali molybdates precipitate most other M^+ and M^{2+}, e.g.:

$$Hg_2^{2+} + MoO_4^{2-} \rightarrow Hg_2MoO_4\downarrow$$

Molybdenum thus may be precipitated as $PbMoO_4$, separately, with sufficient acid, from the common elements except V and W.

Light (254 nm), $[Mo_2H^{-I}Cl_8]^{3-}$ and 3-M HCl yield H_2 and $Mo_2(OH)_2^{4+}$.

Intense photolysis of aqueous $[W^V(CN)_8]^{3-}$ has apparently yielded the oxygen complexes $[W(CN)_7(O_2)]^{3-}$, which may show the first W to $\eta^1\text{-}O_2$ bond, and $[\{W(CN)_7\}_2(\mu\text{-}O_2)]^{6-}$, based on Raman bands.

Ultraviolet light causes $K_4[Mo(CN)_6]\cdot 2H_2O$ to decompose to $K_3[Mo(CN)_6]\cdot H_2O$, KOH and H_2. Near-UV light removes up to four CN^- ions from $[Mo(CN)_8]^{4-}$, giving a blue solution, from which ethanol can subsequently precipitate blue $K_3[\textit{trans}\text{-}MoO(OH)(CN)_4]\cdot aq$, easily acidified to produce $[\textit{trans}\text{-}MoO(CN)_4(H_2O)]^{2-}$; solid KOH, on the other hand, precipitates red $K_4[\textit{trans}\text{-}MoO_2(CN)_4]\cdot aq$, which reverts to the blue in water. Salts of $[MoO(CN)_5]^{3-}$ are also isolated.

Light quickly changes yellow $[W(CN)_8]^{4-}$ to a reddish-brown $[\textit{trans}\text{-}WO_2(CN)_4]^{4-}$, and then to purple $[\textit{trans}\text{-}WO(OH)(CN)_4]^{3-}$. Ethanol precipitates purple $K_3[\textit{trans}\text{-}WO(OH)(CN)_4]$; alternately, solid KOH precipitates a brownish-yellow $K_4[\textit{trans}\text{-}WO_2(CN)_4]\cdot 6H_2O$.

Light and $[(Mo,W)(CN)_8]^{4-}$ plus OH^- form $[(Mo,W)(CN)_7(OH)]^{4-}$; then without light, further CN^- ions are released.

Sulfur and $[MoO_2S_2]^{2-}$ arise from the photolysis of $[MoS_4]^{2-}$ in air.

Some colors for Mo are: Mo_2^{4+}, red; Mo^{3+}, pale yellow; $Mo_2(OH)_2^{4+}$, green; $Mo_2O_4^{2+}$, yellow; and MoO_4^{2-}, colorless.

6.2.5 Reactions Involving Chalcogeno Mo and W Clusters

Polyoxohomopolymetalates. The brilliant "tungsten bronzes", with $x < 1$ in $(M^n)_{x/n}(WO_3)$, from high-temperature reductions of the Alk^I, Ae^{II}, or Ln^{III} tungstates and WO_3, are insoluble even in hot, concentrated, strong acids and bases. Similar compounds of Mo are less stable.

At a pH of 3 to ~5.5, MoO_4^{2-} becomes mainly $[H_nMo_7O_{24}]^{(6-n)-}$, with n up to 3. A pH of 2 to 3 leads to salts of $[Mo_8O_{26}]^{4-}$. Some solids that are isolated include: $(NH_4)_6[Mo_7O_{24}]\cdot 4H_2O$, $(NH_4)_4[Mo_8O_{26}]\cdot 5H_2O$, and even $K_8[Mo_{36}O_{112}(H_2O)_{16}]\cdot 36H_2O$.

Aqueous Mo^{VI} in 2-M HCl reacts with $[Mo^{III}Cl_6]^{3-}$ or $[Mo^{III}Cl_5(H_2O)]^{2-}$, ratio 1:2, under N_2, giving the red $Mo^{IV}{}_3O_4{}^{4+}$, i.e., $[Mo_3O_4(H_2O)_9]^{4+}$ or $[\{Mo(H_2O)_3\}_3(\mu\text{-}O)_3(\mu_3\text{-}O)]^{4+}$. When freed from Cl^- by ion exchange, this reacts with air only slowly. The H_2O is replaceable by NCS^- for example. The ratio 2:1 (Mo^{VI}:Mo^{III}) yields the yellow $Mo^{V}{}_2O_4{}^{2+}$, i.e., $[Mo_2O_4(H_2O)_6]^{2+}$.

The weak acidification of $WO_4{}^{2-}$, depending on pH, concentrations and *T*, generates especially $[W_7O_{24}]^{6-}$ ("paratungstate A", formed quickly), pale-yellow $[W_{10}O_{32}]^{4-}$ (metastable, reducible by e^- and photons) and $[H_2W_{12}O_{42}]^{10-}$ ("paratungstate B"), particularly the W_7 and W_{12} above pH 6, but also $[W_4O_{16}]^{8-}$ and $[H_2W_{12}O_{40}]^{6-}$. The two hydrons (H nuclei) in the dodecatungstates are buried in the center, like the Si in $[SiW_{12}O_{42}]^{4-}$, and are not ionizable. Equilibration is quite slow below pH 5.5; it is much faster with Mo. Consider the "paratungstates":

$$12\,[W_7O_{24}]^{6-} + 2\,H_3O^+ + 4\,H_2O \leftrightarrows 7\,[H_2W_{12}O_{42}]^{10-}$$

Diluting these to <2 cM W^{VI} seems to lead to $WO_4{}^{2-}$ and $[H_2W_{12}O_{40}]^{6-}$.

We may note in passing that the common statement that "the rate is slow" confuses the (low) rate, an abstraction of the process, with the (slow) physical process itself. Either "the rate is low" or "the process is slow" would be unambiguous.

Such polytungstates have extremely weak basicity (i.e., are salts of very strong acids) but break up when attacked by either H_3O^+ or OH^-. Most are metastable, except the "Keggin"-type anion, $\alpha\text{-}[H_2W_{12}O_{40}]^{6-}$, "metatungstate". Concentrated solutions can produce such solids as $K_7[HW_5O_{19}]$ or $Na_5[H_3W_6O_{22}]$ for example. Some additional isolated solids are: $Na_5[H_3W_6O_{22}]\cdot 18H_2O$, $Na_6[W_7O_{24}]\cdot 21H_2O$, $Li_{14}[W_7O_{28}]\cdot 4H_2O$, $K_4(W_{10}O_{32})\cdot 4H_2O$, $Na_6[H_2W_{12}O_{40}]\cdot 29H_2O$ and $(NH_4)_{10}\,[H_2W_{12}O_{42}]\cdot 10H_2O$.

A large, ball-shaped cluster can be made from 4.5 mmol (5.6 g) of $(NH_4)_6[Mo_7O_{24}]\cdot 4H_2O$, 6.1 mmol (79 cg) of $N_2H_6SO_4$ and 162 mmol (12.5g) of $NH_4CH_3CO_2$, stirring in 250 mL H_2O 10 min (becoming blue green), then adding 42 mL of 17.5-M ("glacial") CH_3CO_2H and 42 mL of H_2O. This is stored 4 d at 20 °C in an open flask without stirring, changing slowly to dark brown and forming reddish-brown crystals of water-soluble $(NH_4)_{42}[Mo^{V}{}_{60}Mo^{VI}{}_{72}O_{372}(CH_3CO_2)_{30}(H_2O)_{72}]\cdot aq$.

A true equation for this, starting with the heptamolybdate, is formidable (albeit presentable) because the 7 does not divide evenly into the other subscripts. Let us note the mathematical equivalences and near equivalence in the following, and the fact that chemical reaction does provide small amounts of $HMoO_4{}^-$ at any instant:

$$Mo_7O_{24}{}^{6-} + 4\,H_2O = 7\,MoO_4{}^{2-} + 8\,H^+ \approx$$

$$7\,MoO_4{}^{2-} + 7\,H^+ = 7\,HMoO_4{}^-$$

Then, although Nature can handle large numbers that we may choose to cut down, we can write the still complicated equation, similar to, but more faithful to the species actually present than, the original [2]:

$$44\,HMoO_4^- + 5\,N_2H_5^+ + 35\,CH_3CO_2H \rightarrow 25\,CH_3CO_2^- + 5\,N_2\uparrow +$$

$$^1/_3\,[\{Mo^{VI}(Mo^{VI}{}_5O_{21})(H_2O)_6\}_{12}\{Mo^V{}_2O_4(\mu\text{-}CH_3CO_2)\}_{30}]^{42-} + 28\,H_2O$$

From this and HPH_2O_2 can be made very similar dark-brown crystals, $(NH_4)_{42}[\{Mo^{VI}(Mo^{VI}{}_5O_{21})(H_2O)_6\}_{12}\{Mo^V{}_2O_4(\mu\text{-}PH_2O_2)\}_{30}]\cdot aq$.

Polyoxoheteropolymetalates. Acidification of mixtures of MoO_4^{2-} and M^{IV} with M = Ti, Zr or Ge, forms $[MMo_{12}O_{40}]^{4-}$. For M = Ce or Th we find $[MMo_{12}O_{42}]^{8-}$. Aqueous H_2O_2 or $S_2O_8^{2-}$ with Mn^{II} or Ni^{II} gives $[M^{IV}Mo_9O_{32}]^{6-}$. With M = Cr, Fe, Co, Rh or Al in M^{3+} we have $[MMo_6O_{24}]^{9-}$. However, Co^{2+}, $Mo^{VI}{}_7O_{24}^{6-}$, Br_2 and NH_4^+ yield a green $[Co^{III}\text{-}\eta^6\text{-}cyclo\text{-}\{MoO(OH)(\mu\text{-}O)_2\}_6]^{3-}$: $(NH_4)_3[CoMo_6O_{18}\text{-}(OH)_6]\cdot 12H_2O$. Oxidizing $[Co^{II}W_{12}O_{40}]^{6-}$ yields $[Co^{III}W_{12}O_{40}]^{5-}$, a good outer-sphere oxidant ($E° = 1.00$ V), with the unusual tetrahedral Co^{III} at the center.

Also, Te^{VI} and I^{VII} produce $[TeMo_6O_{24}]^{6-}$ and $[IMo_6O_{24}]^{5-}$ and many more. These heteropolymolybdates(VI), and heteropolytungstates(VI) (similar clusters) break up with OH^- but often not with H_3O^+, and then can be made, e.g., by ion exchange, into strong acids, unlike the homopolymolybdates and homopolytungstates. Colorless hetero-atoms produce more-or-less yellow species. Similarly well known are homo- and heteropolymolybdates(V).

The mild reduction of hetero-dodecatungstates and others results in deep-blue W^V-W^{VI} materials, useful for their color. Homo- and heteropolytungstates(V) are also available.

Slowly mixing $[WO_4]^{2-}$, HCl to pH 7.7, and some extra Al^{3+}, with refluxing over an hour, leads to $[Al\{Al(H_2O)\}W_{11}O_{39}]^{6-}$:

$$11\,[WO_4]^{2-} + 2\,Al^{3+} + 10\,H_3O^+ \rightarrow [Al\{Al(H_2O)\}W_{11}O_{39}]^{6-} + 14\,H_2O$$

Slowly adding concentrated H_2SO_4 to this at 0 °C to below pH 0 and then refluxing 6 d yields $[AlW_{12}O_{40}]^{5-}$ and its acidified forms:

$$12\,[Al\{Al(H_2O)\}W_{11}O_{39}]^{6-} + 56\,H_3O^+ \rightarrow$$

$$11\,[AlW_{12}O_{40}]^{5-} + 13\,Al^{3+} + 96\,H_2O$$

Extraction by ether gives two yellowish isomers of $H_5[AlW_{12}O_{40}]$. Gradually adding K_2CO_3 to this at 60 °C and cooling produces a white, "lacunary" (deficit structure) salt:

$$2\,H_5[AlW_{12}O_{40}] + 15\,K_2CO_3 + H_2O \rightarrow$$

$$2\,K_9[AlW_{11}O_{39}]\downarrow + 2\,HWO_4^- + 10\,HCO_3^- + 5\,CO_2 + 12\,K^+$$

The anion of the K salt adds VO^{2+}, as in $VOSO_4$, at ambient *T*, forming, after cooling to 5 °C for 2 h, dark-purple $K_7[AlV^{IV}W_{11}O_{40}]$. In acid this is oxidized by O_3, HClO, Br_2 etc. to yellow $K_6[AlV^{V}W_{11}O_{40}]$.

The present work cannot describe many interesting structures, but we note that WO_4^{2-} mixed with $SnCl_2$ at a pH of 3.5 has been found [3] to produce an orange crystalline product with the complex structures $[Na_{12}(OH)_4(H_2O)_{28}][Sn^{II}{}_8W^{VI}{}_{18}O_{66}]\cdot 18H_2O$, where one might expect, say, $[Na_{12}(H_2O)_{30}][Sn^{IV}{}_8W^{V}{}_{16}W^{VI}{}_2O_{68}]\cdot 18H_2O$ with no "NaOH" (at the pH of 3.5) even though $[Na_{12}(H_2O)_{28}]^{12+}$ may indeed have enough charge field to hold some OH^- in the crystal even at that pH.

Chalcogeno (S, Se) cuboidal clusters, general. One source of hydrated $Mo_3S_4^{4+}$, often a reactant below, is MoS_4^{2-} plus Mo^{3+}. The hydrated $Mo_3S_4^{4+}$ is stable for years in air with dilute acid to prevent hydrolytic polymerization. The somewhat similar hydrated $Mo_3Se_4^{4+}$, $W_3S_4^{4+}$ and $W_3Se_4^{4+}$, details omitted here, also all require acid but are stable in air only for months at 5 °C. The Se complexes are somewhat sensitive to light, depositing red Se_8.

The reaction of aqueous thiometalates with **d**-block M^{2+} gives various $[M\{\eta^2\text{-(Mo,W)}S_4\}_2]^{2-}$ where M = Fe, Co, Ni, Pd, Pt, Zn, Cd or Hg; some similar complexes of $[(Mo,W)O_2S_2]^{2-}$ etc., again coordinate to $(\mathbf{d})^{2+}$ via S. Yet further products are $[\mathit{trans}\text{-}Fe(H_2O)_2(\eta^2\text{-}WS_4)_2]^{2-}$ and the reduced, more stable $[Fe(\eta^2\text{-}MoS_4)_2]^{3-}$. The Co and Ni complexes are also reducible, although less easily.

The M–M bonding in the clusters described hereafter is fascinating and important but beyond our scope. The bridging by other atoms will be recognized briefly. We abbreviate (Mo,W) as M and (S,Se) as Q, when either choice would fit. Some of the following formulas are tentative. This is a small sample of a large field, but often involving organic moieties omitted here. See especially [4].

Many of these formulas, more fully expressed, would appear as, e.g., $[\{W(H_2O)_2\}_3(\mu\text{-}Se_2)_3(\mu_3\text{-}Se)]^{4+}$ or, omitting some of the structural data, $[Mo\{\{Mo(H_2O)_3\}_3S_4\}_2]^{8+}$, $[(RhCl_3)\{Mo(H_2O)_3\}_3S_4]^{4+}$ and, for one more example, $[\{Fe(H_2O)\}\{Mo(H_2O)_3\}_3S_4]^{4+}$. When still less structural information is to be given, especially on repetition, clarity may often be served by dropping one level of enclosing and other marks as follows: $[(WAq_2)_3(Se_2)_3Se]^{4+}$, $[Mo\{(MoAq_3)_3S_4\}_2]^{8+}$, $[(RhCl_3)(MoAq_3)_3S_4]^{4+}$ or $[(FeAq)(MoAq_3)_3S_4]^{4+}$. We may also write $Mo(Mo_3S_4)_2^{8+}$, $(RhCl_3)Mo_3S_4^{4+}$ or $(FeAq)Mo_3S_4^{4+}$, where consistent with related formulas. These omit some H_2O, as is common when discussing aqua ions, partly because the hydrated $M_3Q_4^{4+}$ (incomplete-cube) occurs so frequently below. The first and last examples, however, may need to retain some Aq to convey new information about the coordination of the hetero-atom. In this discussion of clusters, then, Aq denotes one H_2O or (H_2O), not the indefinite number written as aq elsewhere in this book. Additional abbreviation, as in $Mo_7S_8^{8+}$, is used often elsewhere but seldom here.

Chalcogeno (S, Se) cuboidal clusters, homometallic. Reductants including Mg, V, $[MoCl_6]^{3-}$, $[Mo_2Cl_8]^{4-}$ and HPH_2O_2 in various conditions convert $Mo_3S_4^{4+}$

to mixtures, sometimes in low yields, of a green $Mo_4S_4^{4+}$ together with $Mo(Mo_3S_4)_2^{8+}$, where one Mo vertex or corner is shared by the two otherwise distinct $Mo_4S_4^{4+}$ cubes, making a double cube.

In one oxidation on the other hand, VO_2^+ (no excess allowed) oxidizes $Mo_4S_4^{5+}$ to red $Mo_4S_4^{6+}$, which decomposes to $Mo_3S_4^{4+}$, although anodes (instead of VO_2^+) break the $Mo_4S_4^{5+}$ down mainly to this $Mo_3S_4^{4+}$ and other byproducts. The green $Mo_4S_4^{5+}$ ion is quantitatively reduced to the orange, air-sensitive $Mo_4S_4^{4+}$ by $[BH_4]^-$, V^{2+}, Cr^{2+} and cathodic e^-.

Oxalic acid replaces some of the H_2O in $[\{Mo(H_2O)_3\}_3S_4]^{4+}$, and Cs^+ then precipitates $Cs_2[\{Mo(C_2O_4)(H_2O)\}_3S_4]\cdot 3H_2O$. All the O of the H_2O are *trans* to the μ_3-S cap, while the oxalate O are *trans* to the three μ-S atoms. Likewise, $[\{W(H_2O)_3\}_3S_4]^{4+}$ yields $[\{W(C_2O_4)(H_2O)\}_3S_4]^{2-}$.

Concentrated NH_3 forms $[\{Mo(NH_3)_3\}_4S_4]^{4+}$ from $[(MoAq_3)_4S_4]^{4+}$. Air and H_3O^+ then produce $[(MoAq_3)_4S_4]^{5+}$. Oxalate breaks down the $[(MoAq_3)_4S_4]^{5+}$ cluster and forms $[\{Mo(C_2O_4)Aq\}_3S_4]^{2-}$.

Molybdate heated with $(NH_4)_2S_x^{2-}$, depending on x, T, t etc., can form $(NH_4)_2[Mo^V{}_2S_{12}]\cdot 2H_2O$, i.e., $(NH_4)_2[\{Mo^V(\eta^2\text{-}S_2)_2\}_2(\mu\text{-}S_2)_2]\cdot 2H_2O$. Heating $(NH_4)_6[Mo_7O_{24}]\cdot 4H_2O$ with $(NH_4)_2S_x$ for 5 d gives high yields of a dark-red, $(NH_4)_2[Mo^{IV}{}_3S_{13}]\cdot{}^1/_2H_2O$, i.e., $[\{Mo^{IV}(\eta^2\text{-}S_2)\}_3(\mu\text{-}1^2,2^2\text{-}S_2)_3(\mu_3\text{-}S)]^{2-}$, somewhat cuboidal, stable in air, not very soluble, and inert in HCl, perhaps roughly as in the following equation (simplified with S_2^{2-}); heating for 3–4 h with added NH_3OH^+ gives lower yields. The S_2 bridges are not in Mo–S–S–Mo chains; the S–S pairs are the short diagonals in bent (at S–S) Mo_2S_2 rhombi:

$$3\,[Mo_7O_{24}]^{6-} + 119\,S_2^{2-} + 144\,NH_4^+ \rightarrow$$

$$7\,[Mo_3(S_2)_6S]^{2-} + 49\,S_3^{2-} + 144\,NH_3 + 72\,H_2O$$

The $[Mo_3S_7Br_6]^{2-}$ ion, i.e., $[(MoBr_2)_3(S_2)_3S]^{2-}$ is a good starter to introduce other ligands such as SCN^- to give $[\{Mo(NCS)_2\}_3(S_2)_3S]^{2-}$. Also, air and NCS^- convert $[(MoAq_3)_4S_4]^{4+}$ to purple $[\{Mo(NCS)_3\}_4S_4]^{6-}$.

In $[\{W(H_2O)_3\}_3S_4]^{4+}$, NCS^- can replace H_2O, apparently changing from the initial κS to the final κN isomer, giving a green $[\{W(NCS)_3\}_3S_4]^{5-}$, with a high yield of a Cs^+ salt. A similar green $[\{W(H_2O)_3\}_3Se_4]^{4+}$ plus NCS^- easily form $[\{W(NCS)_3\}_3Se_4]^{5-}$, still containing W^{IV}.

In $[\{Mo(H_2O)_3\}_3S_4]^{4+}$ or $[\{W(H_2O)_3\}_3S_4]^{4+}$, Cl^- can replace some but not all H_2O weakly, even in concentrated HCl; 1 to 3-M HCl forms $[\{MoCl(H_2O)_2\}_3S_4]^+$ or $[\{WCl(H_2O)_2\}_3S_4]^+$. In this case all the H_2O are *trans* to the μ-S bridges, while the Cl atoms (or Cl^- ions) are *trans* to the μ_3-S cap. Aqueous NCS^- replaces H_2O more firmly.

The reaction of $[W^{VI}S_4]^{2-}$ with $[W^{III}{}_2Cl_9]^{3-}$ yields W^{IV} in $[W_3OS_3(H_2O)_9]^{4+}$, $[W_3O_2S_2(H_2O)_9]^{4+}$, and $[W_3O_3S(H_2O)_9]^{4+}$.

A non-aqueous direct union of the elements provides $(M_3Q_7Br_4)_x$, which, by way of Br^- and $M_3Q_7Br_6^{2-}$, gives, for the Se complexes here, $M_3Se_7^{4+}$ or $[(MAq_2)_3(Se_2)_3Se]^{4+}$, i.e., $[(MAq_2)_3(\mu\text{-}\eta^2\text{-}Se_2)_3(\mu_3\text{-}Se)]^{4+}$. As in some other cases above, an electron count including the M–M bonding shows nine bonds to the metallic atom with an (outer) 18-electron or noble-gas structure. Then we may, however, substitute CN^- for the H_2O and Se_2^{2-}, or treat $(M_3Se_7Br_4)_x$ directly with hot CN^-, and revert to the also stabilized and familiar type of formula, and presumably $SeCN^-$:

$$^1/_x\,(M_3Se_7Br_4)_x + 12\,CN^- \rightarrow [\{M(CN)_3\}_3Se_4]^{5-} + 3\,SeCN^- + 4\,Br^-$$

The Mo cluster is brown, the W green, and one may isolate $Cs_6[\{M(CN)_3\}_3Se_4]Cl\cdot4H_2O$ for example.

The $[Mo_2Cl_8]^{4-}$ dimeric complex reduces and partly combines with $[(MAq_3)_3Q_4]^{4+}$ (M = W or a mixture with Mo) by, in effect, adding $MoAq_3^+$, and the resulting $[(MAq_3)_4Q_4]^{5+}$ has a random positioning of Mo and W with almost equal radii. Electrodes can then add or remove one electron reversibly. Oxidation of the 5+ ion by O_2 or Fe^{3+} simply gives, first, $[(MAq_3)_4S_4]^{6+}$, which then expels one WAq_3^{2+} only (i.e., not the added Mo), forming $[(MAq_3)_3S_4]^{4+}$.

Chalcogeno (S, Se) cuboidal clusters, heterometallic. Many metallic elements M^0, i.e., Fe, Co, Rh, Ni, Pd, Pt, Cu, Cd, Hg, Ga, In, Tl, Ge, Sn, Pb, As, Sb and Bi, can substitute for one Mo in $Mo_4S_4^{4+}$ and/or for the unique Mo in $Mo(Mo_3S_4)_2^{8+}$. Sometimes bond lengths etc. suggest oxidation states of (IV) for the three Mo in each cube, and (0) for the unique or heteroatom. In what follows, one reagent is always $Mo_3S_4^{4+}$ (or $W_3S_4^{4+}$ where appropriate) unless stated otherwise. Many of these heterometallic complexes are quite sensitive to air.

For chromium, however, Cr^{2+}, but not Cr^0, forms brown $CrMo_3S_4^{4+}$ and presumably Cr^{3+}. Air restores the $Mo_3S_4^{4+}$ and releases Cr^{III}.

Iron wire and H_3O^+ give reddish purple $(FeAq)Mo_3S_4^{4+}$, with a tetrahedral Fe. Oxygen then forms Fe^{II} and the original $Mo_3S_4^{4+}$. Alternatively, Cl^- quickly yields $(FeCl)Mo_3S_4^{3+}$. From this, concentrated NH_3 gives dark purple $[(FeAq)\{Mo(NH_3)_3\}_3S_4]Cl_4$.

Metallic Co forms at least $(CoMo_3S_4)_2^{8+}$, brown, with two cubes bonded on their edges, not sharing corners (or edges).

Heating with $RhCl_3$ in 4-M HCl a few hours gives brown $(RhCl_3)Mo_3S_4^{4+}$; after some days in 5-dM HCl this yields the aqua complex $(RhAq_3)Mo_3S_4^{7+}$.

Metallic Ni and H_3O^+ give blue-green $(NiAq)Mo_3S_4^{4+}$ or green $(NiAq)W_3S_4^{4+}$ after some hours or days. Excess Ni^{2+}, excess $[BH_4]^-$ and $Mo_3S_4^{4+}$ in 5-dM HCl yield the same in < 1 min. It is stable in air for about an hour; heating it in air returns $Mo_3S_4^{4+}$ and Ni^{II}, but HCl soon forms green $(NiCl)Mo_3S_4^{3+}$, also from the simpler $[MoS_4]^{2-}$, Ni powder and 2-M HCl, and NCS^- results in $Ni(NCS)Mo_3S_4^{3+}$.

In 2-M HCl, Pd (sometimes $PdCl_2$ plus HPH_2O_2) forms dark-blue $(PdCl)Mo_3S_4^{3+}$, stable in air for several weeks. The analogs with Se and W can react similarly. The $(PdCl)Mo_3S_4^{3+}$ ion reacts with H_2PHO_3, or $PH(O)(OH)_2$, as the unusual tautomeric ligand $P(OH)_3$ to form $[PdP(OH)_3]Mo_3S_4^{4+}$. In a similar way, H_3AsO_3 or $As(OH)_3$ yields $[PdAs(OH)_3]Mo_3S_4^{4+}$. Moreover, excess $SnCl_3^-$ gives $(PdSnCl_3)Mo_3S_4^{3+}$, similarly with the Se complex.

The $[PtCl_4]^{2-}$ ion (but not Pt black) and HPH_2O_2 form a brown $[\{PtMo_3S_4\}_2]^{8+}$, stable in air, after a few days; $W_3S_4^{4+}$ reacts similarly.

Metallic Cu yields a brown product, $(CuAq)Mo_3S_4^{4+}$, oxidizable to $(CuAq)Mo_3S_4^{5+}$. Air extracts Cu^{II} from this. The Cu^{2+} ion with $[BH_4]^-$, also CuCl alone, give $(CuAq)Mo_3S_4^{5+}$, which dismutates into $(CuAq)Mo_3S_4^{4+}$ and Cu^{2+}. However, HCl forms a more stable $(CuCl)Mo_3S_4^{4+}$. Metallic Cu forms green $(CuAq)W_3S_4^{5+}$, air-sensitive, but apparently not the 4+ ion.

Heating Cd in 5-dM H_3O^+ at 70 °C for 1 h yields orange-brown $(CdAq_3)Mo_3S_4^{4+}$; Cd^{2+}, HPH_2O_2 and 2-M HCl also give this. Air or H_3O^+ returns the $Mo_3S_4^{4+}$, and 1-M HCl releases H_2 with a $t_{1/2}$ of about 5 min.

Metallic Hg and 2-M $HClO_4$ over many days form deep-purple $Hg(Mo_3S_4)_2^{8+}$; the Mo-Se and W-Se, but not W-S analogs, give corresponding products; 4-M HCl forms $Hg[Mo_3Cl_2Aq_7S_4]_2^{4+}$, blue, with Cl^- rather randomly replacing certain of the 9 H_2O normally on the Mo.

Metallic Ga in 2-M HCl at 90 °C goes to dark-brown $(GaAq_3)Mo_3S_4^{5+}$, with Ga^I as a reasonable assignment of oxidation state; Ga^{III} in 4-M HCl plus $[BH_4]^-$ give the same; $W_3S_4^{4+}$, however, does not react these ways.

The In^+ ion quickly and quantitatively yields $(InAq_3)M_3S_4^{5+}$. Metallic In with $Mo_3S_4^{4+}$ in 4-M *p*-$MeC_6H_4SO_3H$ forms a similar red-brown product but with the acid anion replacing two H_2O on the In in this case. Largely correspondingly, one can prepare purple $(InAq_3)W_3S_4^{5+}$ and blue-green $(InAq_3)W_3Se_4^{5+}$.

The In^{3+} ion with HPH_2O_2 gives red-orange $In(Mo_3S_4)_2^{8+}$, stable only with excess HPH_2O_2, but $W_3S_4^{4+}$ forms $(InAq_3)W_3S_4^{5+}$. These indium products, like the related aqueous In^+, all reduce H_3O^+ to H_2.

Mixing $Mo_3S_4^{4+}$ with $InW_3S_4^{5+}$ transfers the In and gives $InMo_3S_4^{5+}$ and $W_3S_4^{4+}$ quickly and completely.

Metallic Tl with 2-M H_3O^+, or better, TlCl and $[BH_4]^-$ or HPH_2O_2, form blue-green, air-sensitive $Tl(Mo_3S_4)_2^{8+}$. Then H_3O^+ gives $Tl^+ + Mo_3S_4^{4+}$.

At 90 °C, GeO, or GeO_2 and HPH_2O_2, form $Ge(Mo_3S_4)_2^{8+}$ or the Se analog. With $W_3Q_4^{4+}$ we find $(GeAq_3)W_3Q_4^{4+}$; then $[BH_4]^-$ and more $W_3Q_4^{4+}$ yield $Ge[W_3Q_4]_2^{8+}$. Oxidation of the S complex gives $(GeAq_3)W_3S_4^{6+}$; finally 2-M Cl^- forms $(GeCl_3)W_3S_4^{3+}$.

Metallic Sn and $Mo_3Q_4^{4+}$ produce either purple $Sn(Mo_3S_4)_2^{8+}$ or brown $Sn(Mo_3Se_4)_2^{8+}$, but $W_3S_4^{4+}$ forms $SnW_3S_4^{4+}$. Oxygen or Fe^{3+} converts $Sn(Mo_3Q_4)_2^{8+}$ first to $(SnAq_3)Mo_3Q_4^{6+}$, then, with more oxidant, back to $Mo_3Q_4^{4+}$ plus Sn^{IV}. Tin(II) also yields $(SnAq_3)Mo_3S_4^{6+}$, or, with 5-cM Cl^-, $(SnCl_3)Mo_3S_4^{3+}$. The NCS^- ion replaces H_2O only at the Mo.

Reactions transferring $SnCl_3^-$ show these interesting preferences: $Mo_3S_4^{4+} > Mo_3Se_4^{4+} > W_3Se_4^{4+} > W_3S_4^{4+}$, as with the In complexes above.

Metallic Pb in 2-M H_3O^+ forms blue-green $Pb(Mo_3S_4)_2^{8+}$, very sensitive to O_2. With $Mo_3Se_4^{4+}$ we get dark-green $Pb(Mo_3Se_4)_2^{8+}$, and $W_3Se_4^{4+}$ gives wine-red $Pb(W_3Se_4)_2^{8+}$, but $W_3S_4^{4+}$ is inert, as also to many other metals. Oxygen or Fe^{3+} quickly oxidizes $[Pb(M_3Q_4)_2]^{8+}$ to Pb^{II} and $Mo_3Q_4^{4+}$.

Gray As does not join $M_3Q_4^{4+}$, but As^{III}, HPH_2O_2 and $M_3Q_4^{4+}$ go to blue-green $As(Mo_3S_4)_2^{8+}$, green $As(Mo_3Se_4)_2^{8+}$, or red $As(W_3Se_4)_2^{8+}$ in high yields, exposing the partly metallic nature of As. Again, O_2 or Fe^{3+} converts $As(M_3Q_4)_2^{8+}$ to As^{III} and $M_3Q_4^{4+}$.

Metallic Sb and $Mo_3Q_4^{4+}$ produce green $Sb(Mo_3S_4)_2^{8+}$ or dark-green $Sb(Mo_3Se_4)_2^{8+}$ in one week. With $SbCl_3$, HPH_2O_2 and $W_3Se_4^{4+}$ one finds blue-green $Sb(W_3Se_4)_2^{8+}$. Oxygen or Fe^{3+}, as expected, releases Sb^{III} and $M_3Q_4^{4+}$. Without reductant we have, tentatively, $(SbCl_3)W_3Se_4^{4+}$, yellow-brown. The isoelectronic $SnCl_3^-$ easily replaces the $SbCl_3$.

Bismuth(III) and $[BH_4]^-$ quickly, or Bi^0 slowly, form $Bi(Mo_3S_4)_2^{8+}$, blue. Bismuth(III) citrate, HPH_2O_2 and $M_3Se_4^{4+}$ go to green $Bi(Mo_3Se_4)_2^{8+}$ or blue-green $Bi(W_3Se_4)_2^{8+}$. Oxygen or Fe^{3+} then yields Bi^{III} and $M_3Q_4^{4+}$.

We see that many hetero-atoms occupy the unique vertices in these clusters with $W_3S_4^{4+}$, but many more are known to do so with $Mo_3S_4^{4+}$.

The reactions of $[\{Mo(H_2O)_3\}_3S_4]^{4+}$ with "lacunary" (deficit) anions, for example $[SiW_{11}O_{39}]^{8-}$, to form $[\{(SiW_{11}O_{39})Mo_3S_4(H_2O)_3(\mu\text{-}OH)\}_2]^{10-}$ at pH 1.8, are interesting but largely beyond the scope of this book.

References

1. Zhong ZJ et al (2000) Inorg Chem 39:5095
2. Müller A et al (2004) in Shapley JR (ed) Inorg Synth 34:195
3. Sokolov MN et al (2003) J Chem Soc Dalton Trans 2003:4389
4. Sokolov MN, Fedin VP, Sykes AG in McCleverty JA, Meyer TJ (eds) (2004) Comprehensive coordination chemistry II, vol 4. Elsevier, Amsterdam, p 761

Bibliography

See the general references in the Introduction, and some more-specialized books [5–16]. Some articles in journals discuss: Sg solution chemistry and predictions of hydrolysis of Mo, W and Sg [17]; the hydronation ("protonation") and condensation of Mo^{VI}, W^{VI} etc. [18]; the photochemistry of Cr^{III} complexes [19]; polyoxometalates, especially Mo and W, plus V, Nb and Ta; also see the references therein [20]; recent Cr chemistry [21]; peroxo and hyperoxo complexes of Cr, Mo and W [22]; an oxochromium(IV) intermediate arising from the hydroperoxochromium(III) ion [23]; additional molybdenum-oxygen chemistry [24]; the photochemistry of polyoxometalates of Mo, W and V, including much general information on reactions [25]; tungsten complexes [26]; the thermodynamic properties and standard potentials of Cr, Mo and W [27]; 7- and 8-coordinate and cyanide and related molybdenum complexes [28]; and Cr ammines [29].

5. Lassner E, Schubert WD (1999) Tungsten. Springer, Berlin Heidelberg New York
6. Braithwaite ER, Haber J (eds) (1994) Molybdenum: an outline of its chemistry and uses. Elsevier, Amsterdam
7. Parker GA (1983) Analytical chemistry of molybdenum. Springer, Berlin Heidelberg New York

8. Pope MT (1983) Heteropoly and isopoly oxometalates. Springer, Berlin Heidelberg New York
9. Aubry J, Burnel D, Gleitzer C (1976) Compléments au nouveau traité de chimie minéral: molybdène. Masson, Paris
10. Mitchell PCH (ed) (1973) Chemistry and uses of molybdenum. Climax Molybdenum, London
11. Rollinson CL (1973) The chemistry of chromium, molybdenum and tungsten. Pergamon, Oxford
12. Elwell WT, Wood DF (1971) Analytical chemistry of molybdenum and tungsten. Pergamon, Oxford
13. Rieck GD (1967) Tungsten and its compouds. Pergamon, Oxford
14. Busev AI (1962) Schmorak J (trans) (1964) Analytical chemistry of molybdenum. Ann Arbor-Humphrey, Ann Arbor
15. Udy MJ (1956) Chromium. Reinhold, New York
16. Agte C, Vacek J (1954) NASA (trans) (1963) Tungsten and molybdenum. NASA, Washington
17. Pershina V, Kratz JV (2001) Inorg Chem 40:776
18. Cruywagen JJ (2000) Adv Inorg Chem 49:127
19. Kirk AD (1999) Chem Rev 99:1607
20. Hill CL (ed) (1998) Chem Rev 98:1
21. House DA (1997) Adv Inorg Chem 44:341
22. Dickman MH, Pope MT (1994) Chem Rev 94:569
23. Wang WD, Bakac A, Espenson JH (1993) Inorg Chem 32:2005
24. Pope MT (1991) in Lippard SJ (ed) Prog Inorg Chem 39:181
25. Papaconstantinou E (1989) Chem Soc Rev 18:1
26. Dori Z (1981) in Lippard SJ (ed) Prog Inorg Chem 28:239
27. Dellien I, Hall FM, Hepler LG (1976) Chem Rev 76:283
28. Lippard SJ (1976) in Lippard SJ (ed) Prog Inorg Chem 21:91
29. Garner CS, House DA (1970) Transition Met Chem 6:59

7 Manganese through Bohrium

7.1 Manganese, $_{25}Mn$

Oxidation numbers: (II), (III), (IV), (V), (VI) and (VII), as in Mn^{2+}, Mn_2O_3, MnO_2, MnO_4^{3-} ("hypomanganate"), MnO_4^{2-} (manganate) and MnO_4^- (permanganate). Remarkably, all six oxidation states can be found, rarely or often, in a tetrahedral oxoanion, MnO_4^{n-}.

7.1.1 Reagents Derived from Hydrogen and Oxygen

Dihydrogen. Hydrogen reduces acidic or alkaline $Mn^{>IV}$ slowly to $MnO_2 \cdot aq$ at ambient *T*.

Water. Manganese reacts with warm water to give $Mn(OH)_2$ and H_2.

Aquated Mn^{2+} is faint violet $[Mn(H_2O)_6]^{2+}$, and Mn^{3+} occurs in red, rather unstable, alums such as $[Cs(H_2O)_6][Mn(H_2O)_6](SO_4)_2$, but some dissolved, green, dismutating species may be hydrolyzed even in acids.

All oxides and hydroxides of Mn except Mn_2O_7 are insoluble, as are Mn^{II} borate, carbonate, oxalate, phosphate, sulfide and sulfite, but the nitrate, sulfate and chloride are deliquescent. Seawater and some freshwater contain traces of $MnCl^+$, $MnCl_2$, $MnHCO_3^+$ and $MnSO_4$.

A very soluble salt is $Na_5[Mn(CN)_6]$; sparingly soluble $K_5[Mn(CN)_6]$ is colorless and diamagnetic. Hot water with $[Mn(CN)_6]^{5-}$ releases H_2.

Warm 7.5-M H_2SO_4 and MnO_4^- oxidize H_2O to O_2, leaving Mn^{III} and Mn^{IV}. The MnO_4^- alone also decomposes slowly but autocatalytically to $MnO_2 \cdot aq$ and O_2.

At pH 7 or a little higher, MnO_4^- is stable, but light or, e.g., warm 8-M H_2SO_4 forms O_2 and Mn^{III} or Mn^{IV}, autocatalyzed by $MnO_2 \cdot aq$.

Alkali manganates and permanganates are soluble in H_2O, but the former (like the latter) decompose, faster on warming, dilution with H_2O or acidification, slower with free alkali:

$$3\ MnO_4^{2-} + 2\ H_2O \rightarrow 2\ MnO_4^- + MnO_2 \cdot aq\downarrow + 4\ OH^-$$

Oxonium. Dilute acids readily dissolve Mn to form Mn^{2+} and H_2.

The manganese(II) oxide, hydroxide, borate, carbonate, oxalate, phosphate, sulfide and sulfite dissolve readily as Mn^{2+} in dilute acids.

The mixed oxide Mn_3O_4 breaks up in boiling, dilute HNO_3 or H_2SO_4:

$$Mn_3O_4 + 4\ H_3O^+ \rightarrow 2\ Mn^{2+} + MnO_2 \cdot aq\downarrow + 6\ H_2O$$

Evaporating $HMnO_4$ gives, inter alia, $(H_3O)_2[Mn^{IV}(Mn^{VII}O_4)_6] \cdot 5H_2O$, unstable above −4 °C.

Hydroxide. Alkalis precipitate from Mn^{2+} (in the absence of air and tartrates, etc.), after forming complexes, white $Mn(OH)_2$. Air quickly oxidizes it to brown ~$MnO(OH)$:

$$2\ Mn(OH)_2 + {}^1/_2\ O_2 \rightarrow 2\ MnO(OH)\downarrow + H_2O$$

The dihydroxide is soluble only in quite concentrated OH^-, giving, e.g., yellow $Na_2[Mn(OH)_4]$. Manganese(III) in concentrated OH^- yields green ions and solids, perhaps of $[Mn(OH)_6]^{3-}$, with Na^+, Sr^{2+} or Ba^{2+}.

The rare $[Mn^{IV}(OH)_6]^{2-}$ ion occurs in the yellowish mineral jouravskite, $Ca_3[Mn(OH)_6](CO_3)(SO_4) \cdot 12H_2O$.

Aqueous MnO_4^- is reduced to green MnO_4^{2-} on boiling with OH^-:

$$2\ MnO_4^- + 2\ OH^- \rightarrow 2\ MnO_4^{2-} + H_2O + {}^1/_2\ O_2\uparrow$$

The K_2MnO_4 salt crystallizes at 0 °C; even CO_2 or H_2O is acidic enough to convert it to $MnO_2 \cdot aq$ and MnO_4^-. See **Reduced halogens** for MnO_4^{3-}.

Peroxide. Alkaline or neutral manganese(II) is oxidized to $MnO_2 \cdot aq$ by HO_2^-. Cold, concentrated KOH, plus Mn^{II} and H_2O_2, give brownish, slightly soluble solids written variously as $K_4[Mn(O_2)_4]$ etc. These products explode above 0 °C.

Manganese dioxide and its hydrates are insoluble in HNO_3, dilute or concentrated, but adding some H_2O_2 causes rapid dissolution with the formation of Mn^{2+} and O_2.

Fresh $MnCO_3$, together with H_2O_2 and KCN, produce dark-red $K_3[Mn(CN)_6]$, to be recrystallized from KCN to avoid hydrolysis.

The reduction of acidified MnO_4^- by H_2O_2 to produce Mn^{2+} and O_2 has complicated kinetics, including autocatalysis by Mn^{2+}:

$$MnO_4^- + {}^5/_2\ H_2O_2 + 3\ H_3O^+ \rightarrow Mn^{2+} + {}^5/_2\ O_2\uparrow + 7\ H_2O$$

We note that infinitely many equations can be written for this, e.g.:

$$MnO_4^- + {}^7/_2\ H_2O_2 + 3\ H_3O^+ \rightarrow Mn^{2+} + 3\ O_2\uparrow + 8\ H_2O$$

because beside the balanced reduction of (one) Mn^{VII} to Mn^{II} and the oxidation of (five) O^{-I} to O^0, one may imply the additional dismutation of any number of H_2O_2 molecules to H_2O and ${}^1/_2$ O_2. We must therefore write such reactions separately when they are indeed distinct.

Di- and trioxygen. Air oxidizes $[Mn(CN)_6]^{5-}$ to $[Mn(CN)_6]^{4-}$ and then $[Mn(CN)_6]^{3-}$. Ozone and neutral Mn^{2+} precipitate brown $MnO_2 \cdot aq$. About 1-M H_2SO_4 forms MnO_4^- but ≥ 4-M H_2SO_4 yields Mn^{III} sulfate, and HNO_3 or even HCl gives similar results.

7.1.2 Reagents Derived from the Other 2nd-Period Non-Metals, Boron through Fluorine

Boron species. Borates and Mn^{2+} can give $MnB_2O_4 \cdot 3H_2O$ (sussexite), $MnB_4O_7 \cdot 9H_2O$, $MnB_6O_{10} \cdot 8H_2O$ or $CaMnB_2O_5 \cdot H_2O$ (roweite).

"Perborate", $[B_2(OH)_4(O_2)_2]^{2-}$, reduces violet MnO_4^- in sequence, during 1–2 min, to green MnO_4^{2-}, blue MnO_4^{3-} and brownish MnO_4^{4-}.

Carbon oxide species. Permanganate oxidizes CO to CO_2:

$$2\,MnO_4^- + 3\,CO + 2\,H_3O^+ \rightarrow 2\,MnO_2 \cdot aq\downarrow + 3\,CO_2\uparrow + 3\,H_2O$$

with the uncatalyzed mechanism apparently starting with:

$$MnO_4^- + CO \rightarrow [:(C{=}O){-}O{-}MnO_3]^-$$

$$[:(C{=}O){-}O{-}MnO_3]^- + 3\,H_2O \rightarrow MnO_4^{3-} + CO_2\uparrow + 2\,H_3O^+$$

Strong catalysts for this are Ag^+ and Hg^{2+}, which may first form, e.g., $[Ag{-}(CO){-}O{-}MnO_3]$; this then gives CO_2, MnO_3^- etc., and Ag^+ again.

Alkali carbonates precipitate manganese(II) carbonate, $MnCO_3$, white, oxidized by the air to form manganese(III) oxide-hydroxide. Before oxidation, precipitation is incomplete if NH_4^+ is present. The mineral sidorenkite, with η^2-CO_3, is $Na_3MnCO_3PO_4$.

Cyanide species. Soluble cyanides, as CN^-, precipitate manganese(II) cyanide, $Mn(CN)_2$, white, darkening in the air, soluble in excess reagent, forming $[Mn(CN)_6]^{4-}$ which, in air, becomes red $[Mn(CN)_6]^{3-}$. With H_3O^+ this dismutates to $Mn^{II}[Mn^{IV}(CN)_6 \cdot nH_2O$, green, but it precipitates $\mathbf{3d^{2+}}$ at pH ≤ 7. Heating rearranges one result, $Fe_3[Mn(CN)_6]_2$, to $[Fe(CN)_6]^{4-}$. Iron(III) and Mn^{2+} may be separated by treating them with an excess of CN^- and then with I_2. The Mn is precipitated as $MnO_2 \cdot aq$ while the Fe remains in solution.

Limited KCN, with Mn^{2+}, precipitates a rose-colored product turning green and very insoluble, apparently $K_2Mn[Mn(CN)_6]$. Excess CN^-, with Mn^{2+}, under, e.g., N_2, forms a yellow solution and soluble, yellow $Na_4[Mn(CN)_6] \cdot nH_2O$, or a less soluble, blue-violet $K_4[Mn(CN)_6] \cdot 3H_2O$. The CN^- must be at least 1.5 M to avoid depositing $K_2Mn[Mn(CN)_6]$. Solutions of $[Mn(CN)_6]^{4-}$ also contain aquated and dinuclear species. Adding OH^- gives $[Mn(CN)_5OH]^{4-}$ and dimers. Hydrogen sulfide and $Pb_2[Mn(CN)_6]$ yield the acid $[Mn(CN)_2(CNH)_4]$ or $H_4[Mn(CN)_6]$.

Slowly adding $MnPO_4{\cdot}H_2O$ to KCN at 80 °C yields red $K_3Mn(CN)_6$:

$$MnPO_4{\cdot}H_2O + 6\ CN^- \rightarrow Mn(CN)_6{}^{3-} + HPO_4{}^{2-} + OH^-$$

Boiling this with water gives $Mn_2O_3{\cdot}aq$. Potassium amalgam reduces dissolved $K_3Mn(CN)_6$ to dark-blue $K_4Mn(CN)_6$.

Aqueous KCN reduces $KMnO_4$ to a yellow-brown crystalline $K_7[\{Mn^{III}(CN)_5\}_2(\mu\text{-}O)]{\cdot}CN$.

Manganese(2+) and NCO^- or NCS^- can form $Mn(NCS)_2{\cdot}3H_2O$ and yellowish $[Mn(NCO)_4]^{2-}$ or $[Mn(NCS)_4]^{2-}$, or even $K_4[Mn(NCS)_6]{\cdot}3H_2O$ or $Cs_4[Mn(NCS)_6]$.

Some "simple" organic species. Aqueous MnO_4^- is reduced to MnO_4^{2-} in base by adding dilute CH_2O, avoiding excess.

Formate and acetate, with Mn^{2+}, yield $(Na,K,NH_4)_2Mn(RCO_2)_4{\cdot}nH_2O$.

Adding $C_2O_4^{2-}$ to Mn^{2+} precipitates $MnC_2O_4{\cdot}nH_2O$, soluble in H_3O^+ not too dilute. Other salts are $Alk_2Mn(C_2O_4)_2{\cdot}nH_2O$, $K_2Mn(C_2O_4)(NO_2)_2{\cdot}H_2O$, $K_2Mn(C_2O_4)S_2O_3{\cdot}2H_2O$ and even a green $K_4[\{Mn^{IV}(C_2O_4)_2\}_2(\mu\text{-}O)_2]{\cdot}aq$. This Mn^{IV} dimer is stable for days in darkness at –6 °C.

All $Mn^{>II}$ are reduced to Mn^{2+} on warming with $H_2C_2O_4$ and H_3O^+. The (cold) volumetric oxidation of oxalate by MnO_4^- is important:

$$MnO_4^- + {}^5/_2\ H_2C_2O_4 + 3\ H_3O^+ \rightarrow Mn^{2+} + 5\ CO_2\uparrow + 7\ H_2O$$

Reduced nitrogen. Manganese(II) hydroxide is insoluble in NH_3, but soluble in NH_4^+. As this suggests, NH_3 precipitates the Mn incompletely from solutions of Mn^{2+}, as the hydroxide. If sufficient NH_4^+ is initially present, no precipitate is obtained (separation of Mn from M^{III}), due to the common-ion effect, and slightly stable Mn^{II} ammines can be detected. However, air readily oxidizes the alkaline Mn^{II} to a brown MnO(OH) precipitate.

Manganese(>IV) is reduced to Mn^{IV} by NH_3.

Concentrated NH_3 gradually reduces MnO_4^- to $MnO_2{\cdot}aq$.

Adding NH_2OH to Mn^{II} and excess CN^- gives $K_3[Mn(CN)_5NO]{\cdot}2H_2O$, purple and diamagnetic. Poor yields come from NO plus $[Mn(CN)_6]^{4-}$. The K salt and H_3O^+ form carmine $[Mn(CN)_2(CNH)_3]$. Bromine or HNO_3 oxidizes $[Mn(CN)_5NO]^{3-}$ to yellow $[Mn(CN)_5NO]^{2-}$; $E° = 6$ dV. Oxygen and UV light convert $[Mn(CN)_5NO]^{2-}$ to $[Mn(CN)_5NO_2]^{3-}$. Whether we write NO as NO^+, NO or NO^- affects our choice of oxidation state for the former Mn; see **8.1.2 Oxidized nitrogen**.

Manganese(2+) and N_3^- yield the explosive $Mn(N_3)_2$ and the anion $[Mn(N_3)_4]^{2-}$. Large-organic-cation, shock-insensitive, salts of the latter turn brown in light.

Oxidized nitrogen. If an excess of NO_2^- is added to a neutral solution of Mn^{2+} at room temperature, a yellow liquid is obtained which, on adding oxalic acid, becomes a deep cherry red, due to forming perhaps $[Mn(C_2O_4)_3]^{3-}$. The color is quite permanent and the reaction has been suggested to detect small amounts of Mn in the presence of much Fe.

Nitrate and Mn^{2+} form $[Mn(H_2O)_6](NO_3)_2$, $[cis\text{-}Mn(\eta^1\text{-}NO_3)_2(H_2O)_4]$, $Mn(NO_3)_2 \cdot 2H_2O$, $Mn(NO_3)_2 \cdot H_2O$ and $(Na,K)_2Mn(NO_3)_4$, for example.

Warm, concentrated HNO_3 and concentrated H_3PO_4 together oxidize Mn^{II} to a gray-green Mn^{III} precipitate:

$$3\,Mn^{2+} + NO_3^- + 3\,H_3PO_4 + 6\,H_2O \rightarrow$$

$$3\,MnPO_4 \cdot H_2O\downarrow + NO\uparrow + 5\,H_3O^+$$

Fluorine species. The most stable of all manganese(III) salts is MnF_3. Some solids are $(K,Rb,Cs)_2MnF_5 \cdot H_2O$.

Aqueous HF dissolves MnO_2 very slightly.

7.1.3 Reagents Derived from the 3rd-to-5th-Period Non-Metals, Silicon through Xenon

Silicon species. Some hydrated salts and minerals are $Zn_2Mn^{II}(OH)_2SiO_4$ (hodgkinsonite), $K_2Mn^{II}{}_5Si_{12}O_{30} \cdot H_2O$, $CaMn^{II}{}_4OH(Si_5O_{14}OH) \cdot H_2O$, $(Li,Na)Mn^{II}{}_4Si_5O_{14}OH$ (nambulite), and $Ca_2Mn^{II}{}_7(OH)_2Si_{10}O_{28} \cdot 5H_2O$.

Phosphorus species. Phosphane, PH_3, reduces $Mn^{>IV}$ at least to Mn^{IV}.

Manganese(>II), with HPH_2O_2, goes to Mn^{II}. Aqueous HPO_4^{2-} precipitates, from neutral Mn^{2+}, the "normal" phosphate, $Mn_3(PO_4)_2$, white, slightly soluble in H_2O, soluble in dilute acids. It turns brown in the air. Slowly adding dilute NH_3 to hot, acidified Mn^{2+}, NH_4^+ and $H_2PO_4^-$ will precipitate $MnNH_4PO_4$ quantitatively. Various anions under other conditions may also precipitate $MnPHO_3 \cdot H_2O$, $MnHPO_4 \cdot 3H_2O$, $Mn(H_2PO_4)_2 \cdot {}^5/_2H_2O$, $Mn_2[P_2O_7]$, $Mn_5(OH)_4(PO_4)_2$ and so on.

Manganese(III), with a mono- or diphosphate, yields a red, soluble $AlkMn(HPO_4)_2 \cdot nH_2O$ or a red or violet $AlkMn[P_2O_7] \cdot nH_2O$. The "Manganese Violet" pigment is $NH_4Mn[P_2O_7]$.

Concentrated H_3PO_4 dissolves Mn^{III} acetate as violet $[Mn(PO_4)_2]^{3-}$ etc.

Treating MnO_4^- with a very high $c(H_3PO_4)$ gives a reddish-brown phosphato-Mn^{IV}, which precipitates $MnPO_4 \cdot H_2O$ on dilution or standing.

Diphosphate and Mn^{2+} plus HNO_3 or MnO_4^- form violet Mn^{III} as a useful titrimetric oxidant, which is rather stable if $4 \le pH \le 6$, and may be $[Mn(H_2P_2O_7)_3]^{3-}$ at $pH \le 4$.

Arsenic species. Arsane, AsH_3, reduces $Mn^{>IV}$ to Mn^{IV} or possibly lower.

Soluble arsenites or arsenates precipitate manganese(II) arsenite or arsenate, soluble in acids. Solutions of As^{III} reduce MnO_4^{2-} and MnO_4^- to $MnO_2 \cdot aq$ or Mn^{2+}, depending upon the conditions. One method for determining Mn involves oxidizing Mn^{2+} with $[S_2O_8]^{2-}$, using Ag^+ as a catalyst, and titrating the MnO_4^- with arsenite.

An arsenate and Mn^{III} can yield brown-violet $MnAsO_4 \cdot H_2O$, dark-violet $Mn(H_2AsO_4)_3 \cdot 3H_2O$ or even $Mn^{II}{}_2Mn^{III}(OH)_4AsO_4$ (flinkite).

Reduced chalcogens. Sulfane, H_2S, precipitates pink MnS (metastable, with MnS_4 tetrahedra) from an NH_3 solution containing Mn^{2+}, incompletely from a neutral acetate solution, and not in the presence of weakly acidic CH_3CO_2H. Acetic acid acting on the precipitated sulfides, MnS, "CoS", "NiS" and ZnS, separates Mn from Co and Ni, and from most of the Zn. Sulfane forms stable, green MnS in a hot ammoniacal solution of Mn^{2+}, with the same result when S^{2-} reduces $MnO_4{}^{2-}$ and $MnO_4{}^{-}$ in more-alkaline solutions.

In testing, Mn may be precipitated together with Co, Ni and Zn by HS^- from an ammoniacal solution. Delayed digestion in cold 1-M HCl dissolves only MnS and ZnS. After boiling out the H_2S, the solution is treated with an excess of OH^- and an oxidant (H_2O_2, ClO^-, Br_2, etc.). The Mn is precipitated as $MnO_2 \cdot aq$, while Zn remains in solution as $[Zn(OH)_4]^{2-}$. After filtration and washing, the $MnO_2 \cdot aq$ is dissolved in HNO_3 and H_2O_2 and the solution tested for Mn by either the Pb_3O_4 test (some find that 7 μM Mn can be detected, but that excess Fe seriously interferes), or the $[S_2O_8]^{2-}$ or $H_3IO_6{}^{2-}$ test. In the absence of reducing agents, or after their removal, these tests may be applied to portions of an original unknown solution.

Air oxidizes MnS, giving a mixture such as:

$$2\ MnS + {}^3/_2\ O_2 + H_2O \rightarrow 2\ MnO(OH) + 2\ S\downarrow$$

$$MnS + 2\ O_2 \rightarrow MnSO_4$$

All the higher oxidized forms of Mn (in solution or freshly precipitated) are reduced to Mn^{II} by soluble sulfides, forming S to $SO_4{}^{2-}$, depending on temperature, concentration, etc:

$$2\ MnO_4{}^- + 7\ HS^- + 9\ NH_4{}^+ \rightarrow 2\ MnS\downarrow + 9\ NH_3 + 5\ S\downarrow + 8\ H_2O$$

The SCN^- ion reduces $Mn^{>II}$ to Mn^{II}, also forming HCN and $SO_4{}^{2-}$.

Oxidized chalcogens. Thiosulfate quickly reduces $Mn^{>IV}$ to Mn^{2+} in acidic solution; alkaline solutions give $MnO_2 \cdot aq$, e.g.:

$$8\ MnO_4{}^- + 3\ S_2O_3{}^{2-} + H_2O \rightarrow 6\ SO_4{}^{2-} + 8\ MnO_2 \cdot aq\downarrow + 2\ OH^-$$

Aqueous sulfite precipitates from solutions of Mn^{2+}, $MnSO_3 \cdot nH_2O$, white, insoluble in H_2O, soluble in acids. It also rapidly reduces $MnO_4{}^-$ even at high pH, generating short-lived Mn^V and Mn^{VI}.

Incidentally, "short-lived" is better pronounced with a "long i" to rhyme with "arrived", not a "short i" as in "lived". It is not a past-tense verb (and may refer to the present or future), but rather an adjective derived from the noun "life", just as "broad-leaved" is derived from "leaf" with the same phonetic change of a consonant as in going from "wife" to "wives".

Alkaline SO_3^{2-} reduces $[Mn(P_2O_7)_3]^{9-}$ to Mn^{II}.

Selenite and Mn^{2+} precipitate $MnSeO_3 \cdot nH_2O$—tellurite similarly forms $MnTeO_3 \cdot nH_2O$—but heat, with H_2SeO_3, surprisingly yields an insoluble, reddish-orange $Mn^{IV}(SeO_3)_2$, also made from SeO_2 and MnO_2 in boiling water, from Se, MnO_2 and H_2SO_4, or from SeO_2 and MnO_4^-. In $(H,Alk,NH_4)Mn(SeO_3)_2 \cdot nH_2O$ and $Mn_2(SeO_3)_3 \cdot 3H_2O$ the oxidant Mn^{III}, like Mn^{IV}, coexists with the reductant SeO_3^{2-}.

Manganese dissolves in concentrated H_2SO_4 if warm, releasing SO_2.

Hot, concentrated H_2SO_4 and Mn_2O_3 give dark-green, hygroscopic $Mn_2(SO_4)_3$. With 11.5-M acid we have red $[H(H_2O)_n][Mn(SO_4)_2]$ or $(Alk,NH_4)[Mn(SO_4)_2]$. The $Mn_2(SO_4)_3$ is soluble in dilute H_2SO_4, but forms Mn^{2+} and O_2 if treated with H_2O alone; some discrepancies remain, but hot, 6 to 15-M H_2SO_4 and $MnO_2 \cdot aq$ give Mn^{III} and O_2. Hot, 18-M (concentrated) H_2SO_4 decomposes $MnO_2 \cdot aq$, yielding Mn^{II} and O_2:

$$MnO_2 \cdot aq + 2\,H_2SO_4 \rightarrow MnSO_4 + {}^1/_2\,O_2\uparrow + H_3O^+ + HSO_4^-$$

The reddish purple permanganate $KMnO_4$ is at once decomposed by adding hot concentrated H_2SO_4 to the solid salt:

$$2\,KMnO_4 + 3\,H_2SO_4 \rightarrow 2\,MnSO_4 + {}^5/_2\,O_2\uparrow + K_2SO_4 + 3\,H_2O$$

Aqueous 8.1-M H_2SO_4, Mn^{2+} and MnO_4^- form a brown solution and unstable, black solid, perhaps $Mn(SO_4)_2$.

Selenate and Mn^{2+} can give crystals of $MnSeO_4 \cdot nH_2O$.

Tellurate yields, e.g., Mn_3TeO_6 but not $MnTeO_4$.

If a Mn^{2+} solution or mixture free from halides is treated with $[S_2O_8]^{2-}$, the Mn^{2+} is oxidized to $MnO_2 \cdot aq$:

$$Mn^{2+} + [S_2O_8]^{2-} + 4\,H_2O \rightarrow MnO_2 \cdot aq\downarrow + 2\,HSO_4^- + 2\,H_3O^+$$

With a little Ag^+ as a catalyst, however, the $[S_2O_8]^{2-}$ will oxidize this to Ag^{2+}, which will then change the Mn^{2+} to MnO_4^-, giving the net result:

$$Mn^{2+} + {}^5/_2\,[S_2O_8]^{2-} + 7\,H_2O \rightarrow MnO_4^- + 5\,HSO_4^- + 3\,H_3O^+$$

The test is very sensitive but fails with moderate or large amounts of Mn due to precipitation of $MnO_2 \cdot aq$, although H_3PO_4 protects 1-cM Mn. An excess of Mn^{2+} or Ag^+ precipitates $MnO_2 \cdot aq$ or Ag_2O_2, respectively.

Mn(<VII) is oxidized to MnO_4^- by warming with an excess (to avoid forming $MnO_2 \cdot aq$) of $[S_2O_8]^{2-}$, plus Ag^+, in either HNO_3 or H_2SO_4 (aq).

Reduced halogens. Aqueous Mn^{2+} and Cl^- or Br^- form $MnCl_2 \cdot 4H_2O$, $MnBr_2 \cdot nH_2O$, $(Rb,Cs)MnCl_3 \cdot 2H_2O$, $(Alk,NH_4)_2MnCl_4 \cdot 2H_2O$ etc.

In solution or below –40 °C, $MnCl_3$ [from MnO(OH) and HCl] can persist; the Mn^{III} goes on to Mn^{2+} quite slowly without a catalyst like Cu^{2+} or Ag^+. Boiling and evaporation to the solid also reduce it to Mn^{2+}.

Halide ions are oxidized to X_2 by MnO_2 with Cl^- and only H_3O^+, with Br^- and weak acids (CH_3CO_2H), or with I^- and even $CO_2 \cdot aq$. Cold, concentrated HCl dissolves $MnO_2 \cdot aq$ into a greenish-brown solution, depositing $MnO_2 \cdot aq$ on great dilution, but forming Mn^{2+} and Cl_2 on warming. Generally $Mn^{>II}$ is thus reduced to Mn^{2+}.

When HCl reacts with MnO_4^- or MnO_4^{2-}, the products depend on the proportions. Excess MnO_4^- or MnO_4^{2-} yields $MnO_2 \cdot aq$, or, if HCl is in excess, Mn^{2+} as follows, e.g.:

$$MnO_4^- + 4\ H_3O^+ + 3\ Cl^- \rightarrow MnO_2 \cdot aq\downarrow + {}^3/_2\ Cl_2\uparrow + 6\ H_2O$$

$$MnO_4^{2-} + 8\ H_3O^+ + 4\ Cl^- \rightarrow Mn^{2+} + 2\ Cl_2\uparrow + 12\ H_2O$$

However, $Ca(MnO_4)_2$ and 13-M HCl plus KCl can form dark-red, unstable $K_2[MnCl_6]$. Aqueous Br^- and I^- reduce MnO_4^- more readily.

Alkaline I^- readily reduces MnO_4^- to MnO_4^{2-} (distinction from Cl^- and Br^-). An excess quickly reduces it further:

$$2\ MnO_4^- + I^- + H_2O \rightarrow 2\ MnO_2 \cdot aq + IO_3^- + 2\ OH^-$$

Iodide can be used to obtain Mn^{VI} as dark-green $BaMnO_4$ without appreciable $MnO_2 \cdot aq$:

$$8\ MnO_4^- + I^- + 8\ Ba^{2+} + 8\ OH^- \rightarrow 8\ BaMnO_4\downarrow + IO_4^- + 4\ H_2O$$

Mostly MnO_4^{3-}, and , e.g., $H_2IO_6^{3-}$ at high pH, arise without the Ba^{2+} to precipitate the MnO_4^{2-}. To prevent the dismutation of MnO_4^{3-} into MnO_4^{2-} and $MnO_2 \cdot aq$ requires at least 8-M OH^-. The blue salts of MnO_4^{3-} are quite sensitive to moisture. Then to prevent the rapid dismutation of MnO_4^{2-} into MnO_4^- and $MnO_2 \cdot aq$ requires $> \sim$ 1-M OH^-.

Elemental and oxidized halogens. Chlorine or Br_2 and Mn form Mn^{2+}.

Hot OH^- plus Cl_2 and $MnO_2 \cdot aq$ (MnO_2 more slowly) form MnO_4^-.

Iron(III) and Mn^{2+} may be separated by treating them with an excess of CN^- and then with I_2. The $[Mn(CN)_6]^{4-}$, but not the Fe^{III}, precipitates (as $MnO_2 \cdot aq$). Chlorine or Br_2 oxidizes alkaline Mn^{II} similarly.

Alkaline manganese(II) is oxidized to $MnO_2 \cdot aq$ by ClO^- (to MnO_4^- if Ag^I or Cu^{II} is present) or BrO^- (to MnO_4^- if Cu^{II} is present).

Aqueous ClO_2^- and $HClO_2$ reduce MnO_4^-.

Manganese(2+) and ClO_3^- or ClO_4^- crystallize as $[Mn(H_2O)_6](ClO_n)_2$.

A chlorate or bromate, when boiled with 12-M H_2SO_4 or concentrated HNO_3, and Mn^{2+}, precipitates $MnO_2 \cdot aq$ quantitatively. Reducing agents (Cl^-, Br^- etc.) should be absent:

$$5\ Mn^{2+} + 2\ ClO_3^- + 12\ H_2O \rightarrow 5\ MnO_2 \cdot aq\downarrow + Cl_2\uparrow + 8\ H_3O^+$$

Iodate and Mn^{II} precipitate insoluble, white $Mn^{II}(IO_3)_2$, structurally similar to the insoluble $Mn^{IV}(SeO_3)_2$ (above). The $Mn(IO_3)_2$ eventually releases I_2. Dissolving MnO_2 in HIO_3 with some periodate gives brownish solutions that may deposit red $K_2Mn^{IV}(IO_3)_6$ etc.

A periodate, $Mn_3(IO_5)_2$, decomposes above 15 °C.

Mn(<VII) is oxidized quickly and quantitatively to MnO_4^- by warming with an excess (to avoid producing $MnO_2 \cdot aq$) of H_5IO_6 in either HNO_3 or H_2SO_4 solution. This is more dependable than the $S_2O_8^{2-}$ method. Other conditions can yield red $(Na,K)_7H_4[Mn^{IV}(\eta^2\text{-}IO_6)_3] \cdot nH_2O$.

Xenon species. Aqueous XeO_3 oxidizes Mn^{2+} to $MnO_2 \cdot aq$ and MnO_4^-.

7.1.4 Reagents Derived from the Metals Lithium through Uranium, plus Electrons and Photons

Oxidation. If OH^- or CO_3^{2-} is present, $[Fe(CN)_6]^{3-}$ oxidizes Mn^{II} to $MnO_2 \cdot aq$, the $[Fe(CN)_6]^{3-}$ becoming $[Fe(CN)_6]^{4-}$.

Mn(<VII) is oxidized to MnO_4^- by warming with excess (to avoid producing $MnO_2 \cdot aq$) PbO_2 or Pb_3O_4 in either HNO_3 or H_2SO_4 solution. Reducing agents (Cl^-, Br^- etc.) should be absent:

$$2\ Mn^{2+} + 5\ PbO_2 + 4\ H_3O^+ \rightarrow 2\ MnO_4^- + 5\ Pb^{2+} + 6\ H_2O$$

The bismuthate method of analyzing Mn involves oxidizing Mn^{2+} to MnO_4^- with sodium bismuthate and HNO_3, removing the excess solid oxidant, adding excess $FeSO_4$, and titrating the excess with $KMnO_4$.

Anodes and Mn^{2+} form unstable, easily hydrolyzed Mn^{3+}.

Light (UV, 254 nm), Mn^{2+}, and a pH < 1 release H_2.

Reduction. All of the common metals reduce MnO_4^- in acidic solution; in dilute, neutral solution, even finely divided Pt and Au react. Aqueous VO^{2+}, $Mo_2O_4^{2+}$, $[Mo(CN)_8]^{4-}$, $[Fe(CN)_6]^{4-}$ and $[Ru(CN)_6]^{4-}$ etc. reduce MnO_4^-. Rapid reductants include U^{IV}, Ti^{III}, V^{IV} and $[PtCl_4]^{2-}$, but Cr^{III} is slow. Manganese(>II) is reduced to Mn^{II} also by Cr^{2+}, Cu^+, Hg_2^{2+}, Tl^+, Sn^{II} and Sb^{III}. Manganese(>IV) is reduced at least to black $MnO_2 \cdot aq$ by SbH_3. The reaction of MnO_4^- and Mn^{2+} forms the same:

$$2\ MnO_4^- + 3\ Mn^{2+} + 6\ H_2O \rightarrow 5\ MnO_2 \cdot aq\downarrow + 4\ H_3O^+$$

The volumetric reduction of MnO_4^- by Fe^{2+} is important:

$$MnO_4^- + 5\ Fe^{2+} + 5\ SO_4^{2-} + 8\ H_3O^+ \rightarrow$$

$$Mn^{2+} + \text{(e.g.)}\ 5\ FeSO_4^+ + 12\ H_2O$$

Both the Leclanché cells in ordinary flashlight batteries and the alkaline cells (or batteries) reduce MnO_2 to $MnO(OH)$. The rechargeable types, essentially by definition, reverse this. Leclanché:

$$8\ MnO_2 + 4\ Zn + ZnCl_2 + 9\ H_2O \rightarrow$$

$$8\ MnO(OH) + \sim Zn_5Cl_2(OH)_8 \cdot H_2O\downarrow$$

Cathodic e^-, Alk_{Hg}, Ae_{Hg} or Al reduce $[Mn(CN)_6]^{4-}$ to yellow Mn^I with $E°$ at ~ -1.06 V, but at -0.24 V for $[Mn(CN)_6]^{3-}$ to Mn^{II}:

$$[Mn(CN)_6]^{4-} + e^- \leftrightarrows [Mn(CN)_6]^{5-}$$

$$[Mn(CN)_6]^{3-} + e^- \leftrightarrows [Mn(CN)_6]^{4-}$$

Moist air oxidizes both $Na_5[Mn(CN)_6]$ and $K_5[Mn(CN)_6]$ (slower), but dry air does not. Water alone also slowly oxidizes the $[Mn(CN)_6]^{5-}$.

If $7 < pH < 14$, light and MnO_4^- oxidize H_2O to O_2, leaving MnO_2.

Other reactions. In processing used nuclear fuel, MnO_2 adsorbs (highly radioactive) Zr and Nb species but not Ce^{3+}, Am^{3+}, Pu^{4+} or UO_2^{2+}.

Mixing CrO_4^{2-} but not $[Cr_2O_7]^{2-}$, with Mn^{2+}, soon forms a dark brown precipitate, although not $MnCrO_4$, soluble in acids and NH_3.

Both oxidation and reduction of Mn occur when MnO_4^{2-} or MnO_4^- precipitates Mn^{2+} from neutral solution as $MnO_2 \cdot aq$:

$$3\ Mn^{2+} + 2\ MnO_4^- + 11\ H_2O \rightarrow 5\ MnO_2 \cdot aq\downarrow + 4\ H_3O^+$$

Treating $Mn(CH_3CO_2)_2 \cdot 4H_2O$ with MnO_4^- in a mixed solvent of H_2O and CH_3CO_2H forms the interesting $[Mn_{12}O_{12}(CH_3CO_2)_{16}(H_2O)_4]$, whose structure can be partly stated as cuboidal $[Mn^{IV}{}_4(\mu_3\text{-}O)_4]$, surrounded by and attached to $[Mn^{III}(\mu\text{-}CH_3CO_2)_2(\mu\text{-}O)Mn^{III}(H_2O)(\mu\text{-}CH_3CO_2)_2(\mu\text{-}O)]_4$, a non-planar ring, so that all the μ_2-O (μ-O) become μ_3-O by the additional bonding to the central Mn.

Substituting $Fe(CH_3CO_2)_2$ for the Mn^{II} salt in the above procedure produces $[Mn_8Fe_4O_{12}(CH_3CO_2)_{16}(H_2O)_4]$, with Mn^{III} and Fe^{III} alternating in the ring, and the Fe^{III} holding the H_2O.

Aqueous $[Fe(CN)_6]^{4-}$ and Mn^{2+} precipitate white $Mn_2[Fe(CN)_6]$ (soluble in HCl), even with tartrate and with NH_3 recently added (distinction from Fe). Aqueous $[Fe(CN)_6]^{3-}$ precipitates brown $Mn_3[Fe(CN)_6]_2$ or $KMn[Fe(CN)_6]$, insoluble in acids (separation from Co, Ni and Zn), but decomposed by hot, concentrated HCl.

D-block M^{2+} or Ag^+, or Pb^{2+} precipitates neutral or acidic $[Mn(CN)_6]^{3-}$. The deep-blue iron(II) product appears to rearrange on standing, or on oxidation to Fe^{III}, or on reduction to Mn^{II} salts of $[Fe(CN)_6]^{4-}$.

The positive electrode in rechargeable flashlight batteries may have MnO(OH) or a less stable, more oxidized mixture when charged, or a partly reduced mixture with $Mn(OH)_2$ when discharged.

Some colors are: Mn^{2+}, pale pink; Mn^{3+}, red; MnO_4^{3-}, blue; MnO_4^{2-}, green; MnO_4^-, deep purple.

7.2 Technetium, $_{43}$Tc; Rhenium, $_{75}$Re and Bohrium, $_{107}$Bh

Oxidation numbers for Tc and Re in classical compounds: (I), (III), (IV), (V), (VI) and (VII), as in $[M(CN)_6]^{5-}$, $[M_2Cl_8]^{2-}$, MO_2, $[MOX_4]^-$, MO_4^{2-} and MO_4^- or $[MH_9]^{2-}$. Relativistic calculations for Bh to be stable in water: (III), (IV), (V) and (VII), especially (III).

7.2.1 Reagents Derived from Hydrogen and Oxygen

Dihydrogen. Temperatures ≤ 290 °C and 50-atm pressure produce greenish-yellow $[ReCl_6]^{2-}$ and modest yields of a Re–Re quadruple bond in blue $[Re_2Cl_8]^{2-}$. Large cations, e.g., $[NBu_4]^+$, precipitate this:

$$ReO_4^- + 2\,H_2 + 4\,Cl^- + 4\,H_3O^+ \rightarrow {}^1/_2\,[Re_2Cl_8]^{2-} + 8\,H_2O$$

Water. Water and $TcCl_4$ quickly give dark-brown $TcO_2\cdot aq$, and $[TcX_6]^{2-}$ yields $TcO_2\cdot H_2O$; X = Cl, Br or I.

The dismutation of $[TcOCl_4]^-$ to $TcO_2\cdot aq$ and colorless TcO_4^- is fast in water, although quite slow in HCl > 2 M.

The hydrolysis of Re^V forms double bonds with $Re{=}O^{3+}$, $O{=}Re{=}O^+$ and $O{=}Re{-}O{-}Re{=}O^{4+}$ in many complexes. Water and $[ReOCl_5]^{2-}$ in HCl equilibrate with [*trans*-$ReO(H_2O)Cl_4]^-$.

The aquation of $[ReCl_6]^{2-}$ is very slow.

The $[Re(CN)_8]^{3-}$ ion, although stabilized toward redox by its 18-electron or noble-gas electron structure, quickly breaks down in water to $[ReO_2(CN)_4]^{3-}$. This plus HCl give, with other products, red-brown $[ReO(OH)(CN)_4]^{2-}$ (quickly), and purple $[\{ReO(CN)_4\}_2(\mu\text{-}O)]^{4-}$ (slowly); both K and other salts are known, and alkalis reverse these reactions.

Yellow-red ReO_4^{2-} dismutates to very stable, colorless ReO_4^-:

$$3\,ReO_4^{2-} + 2\,H_2O \rightarrow 2\,ReO_4^- + ReO_2\!\downarrow + 4\,OH^-$$

The partial hydrolysis of $[ReOCl_5]^-$ gives pale purple $[Re_2O_3Cl_8]^{2-}$.

Water very slowly hydrolyzes TcH_9^{2-} to Tc, H_2 and OH^-.

From non-aqueous sources, Tc_2O_7, is very hygroscopic, forming aqueous $HTcO_4$. The commercial salt, $NH_4[TcO_4]$, is moderately soluble, and only

$Li[TcO_4]$ and $Na[TcO_4]$ of the $Alk[TcO_4]$ dissolve easily. Large organic cations precipitate or extract this colorless anion.

The higher oxides of Re dissolve readily. Water and Re_2O_7 form $[ReO_3(\mu\text{-}O)\{ReO_3(H_2O)_2\}]$, with tetrahedral and octahedral Re, a strong acid that dissolves $(Fe,Al)_2O_3 \cdot aq$ etc., and releases H_2 with Fe or Zn.

The solubilities of some rhenates(VII) at 20 °C are: $NaReO_4$ ~1 M; $KReO_4$ 3 cM; $RbReO_4$ 1 cM; $CsReO_4$ 2 cM; $AgReO_4$ 1 cM; $TlReO_4$ 4 mM.

The pale-yellow nitridotrioxorhenate(VII), K_2ReO_3N, made in liquid NH_3, is weakly basic and hydrolyzed in water to ReO_4^-, NH_3 and OH^-, but fairly stable in air.

Oxonium. Dilute H_3O^+ does not attack Tc.

Acids and $[TcH_9]^{2-}$ or $[ReH_9]^{2-}$ yield Tc or Re, and H_2.

Acids and $[Re(CN)_7]^{4-}$ form $H[Re(CN)_7]^{3-}$, pK_a 1.3.

Strong acids hydronate ("protonate") bright-yellow $[TcO_2(CN)_4]^{3-}$ to blue $[TcO(H_2O)(CN)_4]^-$, pK_a 2.9, by way of $[TcO(OH)(CN)_4]^{2-}$ or purple $[\{TcO(CN)_4\}_2(\mu\text{-}O)]^{4-}$.

Aqueous ReO_4^-, is stable in acids, more basic toward H_3O^+ than is ClO_4^-, but still weak.

Hydroxide. Aqueous OH^- does not attack Tc.

The chloride, $ReCl_4$, and $[ReCl_6]^{2-}$ give a black precipitate of $ReO_2{\cdot}H_2O$ when treated with a slight excess of OH^-.

Solutions of the rhenates, ReO_4^{2-}, are stable in alkaline media.

Alkalis easily convert Re_2O_7 to ReO_4^-.

Peroxide. Alkaline HO_2^- does not attack Tc.

Powdered Re dissolves in 10-M H_2O_2 to a colorless solution.

Hydrogen dioxide readily oxidizes Re^{IV} to Re^{VII}.

An aqueous solution of Re_2O_8 reacts like a peroxide.

Aqueous KCN and H_2O_2 plus ReO_2 or $[ReCl_6]^{2-}$ give orange, diamagnetic $K_3[ReO_2(CN)_4]$, separable also as $[Co^{III}(NH_3)_6][Re^VO_2(CN)_4]$, obtainable additionally from the partial hydrolysis of $K_3[Re(CN)_8]$.

Peroxide in NH_3 dissolves Re_2S_7 at 40°C as ReO_4^-.

Dioxygen. Moist air slowly tarnishes Tc.

Air quickly oxidizes acidified but not alkaline Re^{IV} to Re^{VII}.

Air and $K_3[Re(CN)_8]$ (from a non-aqueous source), below pH 5, give a purple, paramagnetic solution from which one can, if not delayed, use large cations to precipitate, e.g., purple $[AsPh_4]_2[Re^{VI}(CN)_8]$ or $[Co^{III}(NH_3)_6]_2[Re^{VI}(CN)_8]_3$.

7.2.2 Reagents Derived from the Other 2nd-Period Non-Metals, Boron through Fluorine

Boron species. Mixtures of $K[BH_4]$ either with $K_4[ReO_2(CN)_4]$, KCN and KCl, or with $K_2[ReCl_6]$ and KCN, under N_2, give blue-green, diamagnetic $K_5[Re(CN)_6]$.

Reducing $K_3[Re(CN)_8]$ (from a non-aqueous source) with $[BH_4]^-$ and adding $[Co(NH_3)_6]^{3+}$ yields green $[Co(NH_3)_6][Re^{III}(CN)_6]$.

Treating TcO_4^- with $ThfBH_3$ [Thf = $(CH_2)_4O$] and CO gas results first in a hydrogen-bridged $[\{Tc(CO)_4\}_3(\mu\text{-}H)_3]$ and then, with Cl^-, colorless $[fac\text{-}TcCl_3(CO)_3]^{2-}$ and, with H_2O, $[fac\text{-}Tc(CO)_3(H_2O)_3]^+$. A high pH then gives $[\{Tc(CO)_3\}_2(\mu\text{-}OH)_3]^-$ and $[\{Tc(CO)_3\}_3(\mu\text{-}OH)_3(\mu_3\text{-}OH)]^-$, with multiple bridging, and even $[\{Tc(CO)_3\}_4(\mu_3\text{-}OH)_4]$.

Cyanide species. Refluxing $[TcI_6]^{2-}$ with air-free CN^- 24 h gives yellow-orange $[Tc(CN)_7]^{4-}$ (reducible by I^-), with an 18-e^- structure due to the ligancy of 7.

Aqueous CN^- dissolves TcO_2; addition of Tl^+ then precipitates dark-brown $Tl_3[Tc(OH)_3(CN)_4]$ or $Tl_3[TcO(OH)(CN)_4]$.

Cyanide and $ReCl_4$ give a yellow solution. Aqueous KCN and $K_2[ReCl_6]$ yield $K_3[ReO_2(CN)_4]$, $K_5[Re(CN)_6]$ etc. with excess KCN, and, with ethanol, a dark-gray $K_4[ReO_2(CN)_4]$. Acids and $[ReO_2(CN)_4]^{3-}$ form $[ReO(OH)(CN)_4]^{2-}$ and then $[Re_2O_3(CN)_8]^{4-}$.

Excess CN^- plus TcO_4^- and K_{Hg} form green $K_5[Tc(CN)_6]$, slightly soluble. This precipitates a brick-red salt with Tl^+. Dry air does not attack the salts, but the aqueous potassium salt is very prone to oxidation to Tc^{IV}.

With X = Cl or Br, $[TcX_4N]^-$ and CN^- in acetonitrile/water give $[Tc(CN)_4N(H_2O)]^{2-}$. A high *c* of CN^- or N_3^- then replaces the H_2O and yields $[Tc(CN)_5N]^{3-}$ or $[Tc(CN)_4(N_3)N]^{3-}$.

Excess CN^- is inactive to ReO_4^- but with H_2S (SH^-) forms $[\{Re(CN)_4\}_2(\mu\text{-}S)_2]^{4-}$, blue green and diamagnetic.

Excess NCS^- with $[Tc(NO)Br_4]^-$ gives deep-blue $[Tc(NO)(NCS)_5]^{2-}$, reducible by N_2H_4 or a cathode to red-brown $[Tc(NO)(NCS)_5]^{3-}$.

Thiocyanate, ReO_4^- and, e.g., Sn^{II} in HCl yield a stable, reddish-brown $[Re(NCS)_6]^{2-}$, useful in spectrophotometry, and $[ReO(NCS)_5]^{2-}$. Aqueous $[Re(NCS)_6]^{2-}$ and Tl^+ precipitate $Tl_2[Re(NCS)_6]$.

Acidified TcO_4^- and NCS^- produce deep-red-violet $[Tc(NCS)_6]^{2-}$. Then $N_2H_5^+$ quickly gives yellow-orange, air-sensitive $[Tc(NCS)_6]^{3-}$, re-oxidizable reversibly at SCE + 18 cV.

Some "simple" organic reagents. Formaldehyde, CH_2O, does not reduce ReO_4^- to Re. In hot, concentrated HCl or HBr, CH_2O reduces ReO_4^- to ReX_6^{2-}. Evaporation of ReO_4^- with formic acid causes the appearance of a blue color.

The treatment of $[TcF_6]^{2-}$ with $H_2C_2O_4$ at 80°C for 3 d forms dark-red $[Tc_2(C_2O_4)_4O_2]^{4-}$, i.e., $[\{Tc(\eta^2\text{-}C_2O_4)_2\}_2(\mu\text{-}O)_2]^{4-}$. Oxalic acid and $[TcBr_6]^{2-}$ produce pale-yellow $[Tc(C_2O_4)_3]^{2-}$. Using $C_2O_4^{2-}$ gives TcO_2. Oxalic acid and

$[ReCl_6]^{2-}$ give olive-green $[\{Re(C_2O_4)_2\}_2(\mu\text{-}O)_2]^{4-}$, forming ReO_4^- slowly in acids but stable for days at pH 7.

Reduced nitrogen. The reaction of Re_3Cl_9 with NH_3, and evaporation, give a purple $Re_3Cl_6(NH_2)_3(NH_3)_3$.

Ammonia and $ReCl_4$ precipitate black $ReO_2 \cdot H_2O$.

Rhenium dioxide, $ReO_2 \cdot aq$, black, is obtained by reducing ReO_4^- (rapidly if hot and concentrated) in 12-M OH^- by N_2H_4 from N_2H_5Cl.

Aqueous KCN, $KReO_4$ and N_2H_4 give $[\textit{trans}\text{-}ReO_2(CN)_4]^{3-}$ or $K_3[ReN(CN)_5]$ depending on conditions.

In hot, concentrated HCl or HBr, $N_2H_5^+$ reduces ReO_4^- to ReX_6^{2-}.

Mixing $(NH_4)_2[TcCl_6]$ with 2-M NH_3OH^+ gives, after evaporation, a pink $[\textit{trans}\text{-}Tc(NO)(H_2O)(NH_3)_4]^{2+}$, p$K_a$ 7.3, oxidizable by Ce^{IV} in 2-M H_3O^+ to green $[\textit{trans}\text{-}Tc(NO)(H_2O)(NH_3)_4]^{3+}$, stable only in acid. The nearly straight Tc–N–O moiety points to a $Tc^+{=}N^+{=}O$ structure.

Aqueous NH_2OH reduces ReO_4^-, primarily to green $[Re(OH)_4(NO)]^-$, apparently. Heating in HX (X = Cl or Br) converts this to $[ReX_5(NO)]^-$ and $[ReX_4(NO)_2]^-$. However, HI gives only $[ReI_6]^{2-}$. Alternately, NCS^- or N_3^- (as X^-) in alkali produces $[ReX_3(NO)(H_2O)]^-$.

Alkaline NH_2OH, $KReO_4$ and KCN yield a red, diamagnetic $K_3[Re(CN)_5(NO)] \cdot 3H_2O$.

Refluxing TcO_4^- and HN_3 (from N_3^-) with HCl forms air-stable $[TcCl_4N]^-$, obviously resistant to H_3O^+, precipitated by large cations; CsCl can give $Cs_2[TcCl_5N]$. A Tc≡N triple bond persists through many changes (like Re≡N) but can be easily reduced. Another frequent feature is a central $Tc_2(\mu\text{-}O)_2$, roughly square, as follows. The $[TcCl_4N]^-$ ion with excess H_2O yields $[\{TcN(H_2O)_2\}_2(\mu\text{-}O)_2]^{2+}$, among others, but some HCl or CN^- forms $[(TcCl_2N)_2(\mu\text{-}O)_2]^{2-}$ or $[\{Tc(CN)_2N\}_2(\mu\text{-}O)_2]^{2-}$, respectively. Treating TcO_4^- or $[TcOCl_4]^-$, plus HN_3 (from N_3^- as reductant or oxidant respectively), with concentrated HBr, and the addition of $[NEt_4]^+$, precipitate $[NEt_4][TcBr_4N(H_2O)]$ in high yields.

Oxidized nitrogen. Gaseous NO and $TcO_2 \cdot H_2O$ in 4-M HBr produce a blood-red $[Tc(NO)Br_4]^-$. Anion exchange with Cl^- or I^- gives the green Cl^- complex or the I^- complex.

Six-M HNO_3 readily and completely dissolves Tc or Re, leading to the strong acids dark-red $HTcO_4$ and H_3OReO_4, and yellow $HReO_4$:

$$3\,M + 4\,H_3O^+ + 7\,NO_3^- \rightarrow 3\,MO_4^- + 7\,NO\uparrow + 6\,H_2O$$

Adding $CsNO_3$ to Re_3Cl_9 in ice-cold HNO_3 forms $Cs_3[Re_3Cl_9(NO_3)_3]$ and, with some oxidation, Cs_2ReCl_6.

Generally, HNO_3 readily oxidizes Re^{IV} also to Re^{VII}.

Fluorine species. Technetium and Re are practically insoluble in HF.

A high yield of colorless, hydrolysis-resistant $[TcF_6]^{2-}$ comes from $[TcBr_6]^{2-}$ and AgF in 24-M (40 %) HF:

$$[TcBr_6]^{2-} + 6\ HF + 6\ Ag^+ + 6\ H_2O \rightarrow [TcF_6]^{2-} + 6\ AgBr\downarrow + 6\ H_3O^+$$

A pale-red intermediate may be $[TcF_5(H_2O)]^-$. Any such metathesis can be shown nicely as a concerted process if we keep in mind that the real mechanism is often not so pretty (cf. **11.1.4 Reduction.**):

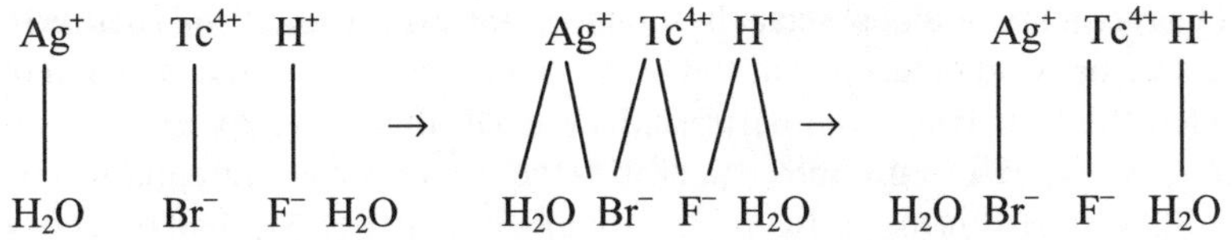

Water does not hydrolyze $[ReF_6]^{2-}$ appreciably.

7.2.3 Reagents Derived from the 3rd-to-5th-Period Non-Metals, Silicon through Xenon

Phosphorus species. Reducing TcO_4^- with HPH_2O_2 in HCl forms dark-green $[TcOCl_4]^-$, precipitable by large cations.

Heating ReO_4^- and HPH_2O_2 in about 6-M HCl or HBr gives ReX_6^{2-} plus a little quadruply-bonded Re^{III}–Re^{III} (2nd equation below), stable in acid:

$$2\ ReO_4^- + 3\ HPH_2O_2 + 12\ X^- + 10\ H_3O^+ + 4\ K^+ \rightarrow$$

$$2\ K_2[ReX_6]\downarrow + 3\ H_2PHO_3 + 15\ H_2O$$

With HCl the color changes during 2 h at 95 C, from pale yellow through very dark green to a pale emerald green, and cooling yields pale-green crystals of $K_2[ReCl_6]$. With HBr the color changes during 2 h at 110 C, from yellow to a very dark red, then fades noticeably during 14 h of further heating; cooling yields dark-red crystals of $K_2[ReBr_6]$. These are hydrolyzed by water, faster with the bromide. The chloride in air gives some ReO_4^-, and alkalis cause a complicated dismutation. The original reduction, as stated above, also forms a further product:

$$ReO_4^- + 2\ HPH_2O_2 + 4\ X^- + 4\ H_3O^+ \rightarrow$$

$$\tfrac{1}{2}\ [Re_2X_8]^{2-} + 2\ H_2PHO_3\ (?) + 6\ H_2O$$

Large cations such as $[NBu_4]^+$ precipitate a blue $[NBu_4]_2[Re_2Cl_8]$. Similarly, 6-M HBr gives greenish brown $[NBu_4]_2[Re_2Br_8]$, also from heating the $[Re_2Cl_8]^{2-}$ 5 min with concentrated HBr and methanol.

Phosphates have no effect on ReO_4^- and thus have been suggested for distinguishing Re from Mo.

Reduced chalcogens. Treating a neutral solution of ReO_4^- with H_2S first forms yellow $[ReO_3S]^-$, which soon begins to dismutate:

$$4\,[ReO_3S]^- \rightarrow [ReS_4]^- + 3\,ReO_4^-$$

Acidification precipitates a black sulfide, faster at high acidity:

$$2\,H_3O^+ + 7\,[ReO_3S]^- \rightarrow Re_2S_7\downarrow + 5\,ReO_4^- + 3\,H_2O$$

The precipitate is not appreciably soluble in the alkalis, alkali sulfides or H_3O^+. Nitric acid or H_2O_2 oxidizes it to ReO_4^-. If saturating a very dilute solution of this rare element with H_2S gives no precipitate even after some time, one may add MoO_4^{2-}, which will form MoS_3 and adsorb the Re, thus concentrating it. Treating $[ReO_3S]^-$ with H_2S leads on to orange $[ReO_2S_2]^-$, red $[ReOS_3]^-$ and reddish-violet $[ReS_4^-]$.

An alkaline ReO_4^- treated with H_2S first gives a pink solution, then slowly deposits Re_2S_7. However, S_x^{2-} shows no quick visible action at ambient *T*, but letting it stand, with darkening, completely precipitates Re_2S_7. Or, acidification gives a rose-red color, then, slowly, a gray, incomplete mixture of Re_2S_7 and S. Boiling the acidic solution with excess $S_2O_3^{2-}$ forms black, amorphous Re_2S_7 completely.

At 60–65 °C, ReO_4^- and S_n^{2-} form cuboidal $[Re^{IV}{}_4(\mu\text{-}S_3)_6(\mu_3\text{-}S)_4]^{4-}$.

Treating ReO_4^- in acids with H_2Se precipitates black Re_2Se_7.

Dilute solutions of ReO_4^- give, with thiocyanates, a yellow or yellow-red color, suggested for distinguishing Re from Mo; concentrated perrhenates give, in the cold, a dark-red solution that turns black when heated. Ether extracts the complex as a rose or dark-red solution.

Much KSeCN (reductant) plus $K_2[ReCl_6]$ and KCN under N_2 yield $K_4[Re(CN)_7]\cdot H_2O$.

Oxidized chalcogens. Aqueous ReO_4^- is easily reduced by SO_2, forming a yellow color, which disappears on standing.

In hot, concentrated HCl or HBr, SO_2 reduces ReO_4^- to $[ReX_6]^{2-}$.

Rhenium dissolves slowly in H_2SO_4. Up to 1–10 M H_2SO_4 complexes ReO_4^-, one to one and also apparently as $[ReO_2(SO_4)_2]^-$.

Reduced halogens. Rhenium is practically insoluble in HCl.

Aqueous $[Re_3Cl_8]^{3-}$ is quite short lived, but the blue $[Re_3Cl_8]^{2-}$, from a non-aqueous source, is stable.

Concentrated HCl converts red $ReCl_3$ and CsCl to $Cs_3[(ReCl_3)_3(\mu\text{-}Cl)_3]$ with three Re=Re bonds; black $ReBr_3$ and $Cs_3[(ReBr_3)_3(\mu\text{-}Br)_3]$ are similar. The bridging halides are, as expected, much less labile than the others to exchange for CN^-, NCS^- or N_3^-. The three equatorial ones are most easily replaced, e.g., by neutral ligands. The $[Re_3Br_{12}]^{3-}$ in air and HBr form $[ReBr_6]^{2-}$ and $[ReOBr_4]^-$. Mixtures of Cl^- and Br^- lead to various mixed complexes.

In concentrated Cl^-, $[TcOCl_4]^-$ gives olive-green $(Alk,NH_4)_2[TcOCl_5]$.

Concentrated HCl quickly reduces TcO_4^- to $[TcOCl_4]^-$, and then quantitatively, with prolonged refluxing, to olive-green $[TcCl_6]^{2-}$, but yellow $La_2[TcCl_5(OH)]_3$ or $Zn[TcCl_5(OH)]$ can also be isolated.

Concentrated HCl, plus TcO_4^- with concentrated H_2SO_4 produce the volatile oxidant $[TcO_3Cl]$. Concentrated HCl plus ReO_4^-, saturated with HCl, yield $[ReCl_3O_3]^{2-}$ and, with Cs^+, Cs_2[*fac*-$ReCl_3O_3$].

Aqueous HCl slowly reduces ReO_4^-:

$$ReO_4^- + 9\,Cl^- + 8\,H_3O^+ \rightarrow [ReCl_6]^{2-} + {}^3/_2\,Cl_2\uparrow + 12\,H_2O$$

In 10-M HCl, however, Cs^+ gives:

$$ReO_4^- + 3\,Cl^- + 2\,H_3O^+ + 2\,Cs^+ \rightarrow Cs_2[fac\text{-}ReCl_3O_3]\downarrow + 3\,H_2O$$

Concentrated HBr reduces TcO_4^- to a reddish-golden $[TcOBr_4]^-$ with a weakly coordinated H_2O, and then, faster than with HCl (the weaker reductant), red $[TcBr_6]^{2-}$. Large cations variously precipitate the former as $[TcOBr_4]^-$ or [*trans*-$TcO(H_2O)Br_4]^-$. Ligand exchange of $[TcCl_6]^{2-}$ with concentrated HBr also forms $[TcBr_6]^{2-}$. In the same way, deep-purple $[TcI_6]^{2-}$ is made from either TcO_4^- plus dilute HI, or ligand exchange of $[TcX_6]^{2-}$ with concentrated HI, yielding, e.g., $Rb_2[TcI_6]$.

If $[TcOI_4]^-$ is desired, HI and TcO_4^- form mixtures with Tc^{IV} and I_n^-, but ligand exchange between $[TcOCl_4]^-$ and NaI in acetone gives $[TcOI_4]^-$. Either 11-M or concentrated HCl, plus TcO_4^- and KI (with I^- as reductant and ligand, or only as reductant), yield either red $K_2[TcI_6]$ or yellow $K_2[TcCl_6]$ respectively. One can also precipitate $K_2[TcCl_5(OH)]$.

Iodide reduces ReO_4^- in hot concentrated HCl in several hours, giving at least green $K_2[ReCl_6]$ or $(NH_4)_2[ReCl_6]$ after cooling:

$$ReO_4^- + 3\,I^- + 6\,Cl^- + 8\,H_3O^+ + 2\,K^+ \rightarrow$$

$$K_2[ReCl_6]\downarrow + {}^3/_2\,I_2\uparrow + 12\,H_2O$$

Preparations of $[ReBr_6]^{2-}$ and $[ReI_6]^{2-}$ are broadly similar, also with HPH_2O_2 or Sn^{II} as reductant.

7.2.4 Reagents Derived from the Metals Lithium through Uranium, plus Electrons and Photons

Oxidation. In 1-M KCl, 25 °C, $[Re(CN)_7]^{4-}$ loses e^- reversibly at 64 cV.

Reduction. Cyanide with $[ReCl_6]^{2-}$ and K_{Hg} precipitates $K_5[Re(CN)_6]$ or its trihydrate. Reducing TcO_4^- with K_{Hg} and KCN yields bright-olive-green $K_5[Tc(CN)_6]$, stable in dry air but easily oxidized when wet. Cyanide together with ReO_4^- and Na_{Hg} also give Re^I. This, with air, acid and ethanol, forms a brown, hygroscopic $Na_3[Re(CN)_5(H_2O)]$.

In aqueous ethylenediamine, Na or K reduces ReO_4^- to colorless, strongly reducing $[ReH_9]^{2-}$; with acids this gives Re and H_2.

In hot, concentrated HCl or HBr, Cr^{2+} reduces ReO_4^- to $[ReX_6]^{2-}$.

Metallic Ni, Cu, Zn, Sn or Pb, and Fe^{2+} or Hg_2^{2+} reduce TcO_4^- to Tc and perhaps some TcO_2.

Concentrated HCl and Zn at 75 °C reduce $[TcCl_6]^{2-}$ (possibly releasing a little of the harmful Tc_2O_7) to green $[Tc_2Cl_8]^{2-}$ and to blue-gray $[Tc_2Cl_8]^{3-}$. The latter is $[Tc^{(II+III)/2}{}_2Cl_8]^{3-}$, with a $^7/_2$-order Tc-Tc bond, if we may write $Tc^{(II+III)/2}{}_2$, to distinguish it, with a mixed oxidation state for two equivalent atoms sharing an unpaired electron, from a hypothetical $Tc^{II}Tc^{III}$ pair having non-equivalent atoms. Or could we write $Tc^{V/2}{}_2$? [Cf. **6.2.4 Reduction with electrons** for $Mo^{(III+III+IV)/3}{}_3$.] It hydrolyzes quickly in H_2O, and is found in black $(NH_4)_3[Tc_2Cl_8]\cdot 2H_2O$ and in blue-green $Y[Tc_2Cl_8]\cdot 9H_2O$. Air oxidizes it further to $[Tc_2Cl_8]^{2-}$ and more slowly to $[TcCl_6]^{2-}$. Titration with Ce^{IV} confirms the $^5/_2$-average oxidation state.

Aqueous ReO_4^- is easily reduced by Zn, forming a yellow color. Upon standing, the color disappears. If the rhenium solution is concentrated, a black precipitate is formed. Zinc, ReO_4^- and HCl, HBr or HI in concentrated H_2SO_4 give $[ReOX_4]^-$.

Tin dichloride reduces ReO_4^- only as far as Re^{IV}.

Cathodes reduce $[Re(Cl,Br,I)_6]^{2-}$ to $[Re(Cl,Br,I)_6]^{3-}$ with complications. Metallic Re is formed cathodically from ReO_4^- at $pH \le 7$.

Cathodic treatment of orange $[TcCl_4N]^-$ gives $[TcCl_4N]^{2-}$. The less stable bromide is similar, and both are colorless. The SCE potentials are +21 cV (Cl^-) and +32 cV (Br^-).

A Pt cathode at –63 cV reduces TcO_4^- to air-sensitive TcO_4^{2-}, but this easily dismutates to TcO_4^- and TcO_4^{3-}.

Other reactions. The very slow aquation of $[ReCl_6]^{2-}$ is catalyzed by Cd^{II}, Hg^{II}, In^{III} and Tl^{III}. Thallium(III) catalyzes the more labile $[ReBr_6]^{2-}$.

Aqueous TcO_4^- is pale yellow; ReO_4^- is colorless and forms soluble salts that are more stable than the permanganates. The addition of K^+, Rb^+, Cs^+ or NH_4^+, not too dilute, or of Ag^+, Hg_2^{2+} or Tl^+ to a solution of ReO_4^- gives a precipitate of the corresponding salt.

Dissolved and solid species often differ, but $[Mg(H_2O)_6]^{2+}$ and $[ReCl_6]^{2-}$ form $[Mg(H_2O)_6][ReCl_6]$ with the same constituents in both.

Iron(III) hydroxide and $HReO_4$ form $[mer\text{-}Fe(ReO_4)_3(H_2O)_3]$.

With $ReCl_4$, $[Fe(CN)_6]^{4-}$ forms a blood-red solution. Aqueous $[Fe(CN)_6]^{4-}$ has no effect on ReO_4^- and thus has been suggested for distinguishing Re from Mo.

Aqueous Ag^+ precipitates orange $Ag_2[ReCl_6]$ from pale-green $[ReCl_6]^{2-}$, and $[ReBr_6]^{2-}$ precipitates brown $Ag_2[ReBr_6]$.

Bibliography

See the general references in the Introduction, and some more-specialized books [1–5]. Some articles in journals discuss: Re^{VII} oxo and imido complexes [6]; Tc complexes [7]; a bromate and diphosphate clock reaction for Mn^{III} [8]; and "recent" Re chemistry [9].

1. Yoshihara K, Omori T (eds) (1996) Technetium and Rhenium. Springer, Berlin Heidelberg New York
2. Peacock RD (1966) The chemistry of technetium and rhenium. Elsevier, Amsterdam
3. Colton R (1965) The chemistry of rhenium and technetium. Wiley, New York
4. Lebedev KB (1962) The chemistry of rhenium. Butterworth, London
5. Tribalat S (1957) rhénium et technétium. Gauthier-Villars, Paris
6. Romão CC, Kühn FE, Hermann WA (1997) Chem Rev 97:3197
7. Baldas J (1994) Adv Inorg Chem 41:1
8. Rich RL, Noyes RM (1990) J Chem Educ 67:606
9. Rouschias G (1974) Chem Rev 74:531

8 Iron through Hassium

8.1 Iron, $_{26}Fe$

Oxidation numbers in simple species: (−II), (0), (II), (III) and (VI), as in $[Fe(CO)_4]^{2-}$, $[Fe(CO)_5]$, Fe^{2+} ("ferrous" ion), $Fe_2O_3 \cdot aq$ ("ferric" oxide), Fe_3O_4, i.e., $Fe^{II}Fe^{III}{}_2O_4$, and $FeO_4{}^{2-}$ ("ferrate").

8.1.1 Reagents Derived from Hydrogen and Oxygen

Dihydrogen. The reduction of Fe^{3+} is catalyzed by finely divided, or especially, colloidal, Pt, also by way of hydrides formed from Cu^{2+} and $[RuCl_6]^{3-}$, as described in sections **11.1.1** and **8.2.1**:

$$Fe^{3+} + {}^1/_2\,H_2 + H_2O \rightarrow Fe^{2+} + H_3O^+$$

Water. The sulfate, $FeSO_4 \cdot 7H_2O$, is efflorescent. This formula, actually $[Fe(H_2O)_6]SO_4 \cdot H_2O$, shows us that the overall hydration (and other contents) of solids may mislead us about that of the constituent or dissolved ions; i.e., we do not have $[Fe(H_2O)_7]^{2+}$. Also, solid $FeCl_2 \cdot 4H_2O$, [*trans*-$FeCl_2(H_2O)_4$], dissolves to give $[Fe(H_2O)_6]^{2+}$, and $FeCl_3 \cdot 6H_2O$ is [*trans*-$FeCl_2(H_2O)_4$]$Cl \cdot 2H_2O$.

Iron and H_2O very slowly form H_2 and $Fe(OH)_2$, some of whose OH^- ions precipitate $Ca(HCO_3)_2$ in tap water as $CaCO_3$; this and the rust (see **Di- and trioxygen** below) may cover the Fe, slowing the rusting.

Water and Fe^{3+} form $[Fe(OH)_n(H_2O)_{6-n}]^{(3-n)+}$ with $0 \le n \le 4$ and especially $[\{Fe(H_2O)_4\}_2(\mu\text{-}OH)_2]^{4+}$ etc. The latter and various anions form transients, $Fe^{III}{}_2(OH)Q^{m+}$, where Q = $PH_2O_2{}^-$, $PHO_3{}^{2-}$, $HAsO_4{}^{2-}$, ${}^1/_2\,AsO_4{}^{3-}$, $SO_4{}^{2-}$ and $SeO_3{}^{2-}$, then giving $Fe^{III}Q^{m+}$, with Q = $PH_2O_2{}^-$, $HPHO_3{}^-$, $H_2AsO_4{}^-$, $SO_4{}^{2-}$ and $HSeO_3{}^-$. Arsenic(III) catalyzes this.

Iron(III) nitrate, sulfate, chloride and bromide are deliquescent. Many Fe^{III} compounds in solution have a brownish-yellow color, redden litmus and color the skin yellow. The free $[Fe(H_2O)_6]^{3+}$ ion, however, when protected from hydrolysis by excess HNO_3 or $HClO_4$, shows its true pale-violet color. Aqueous Fe^{2+} is much less hydrolyzed.

Boiling aqueous Fe^{III} frequently precipitates much of the Fe^{III} as a basic compound, especially if other soluble compounds are present.

A solution of fresh $Fe_2O_3 \cdot aq$ in $FeCl_3$ forms a residue on gentle evaporation to dryness which will redissolve in water if not more than 10 moles of $Fe_2O_3 \cdot aq$ are present to one of $FeCl_3$.

Water reacts with $NaFeO_2$ (from the fusion of Fe_2O_3 with NaOH) to form approximately FeO(OH).

Anhydrous $H_4[Fe(CN)_6]$ is a white, crystalline, non-volatile solid, stable in dry air, readily soluble in water, ionizing two H^+ strongly and two weakly. At 100 °C partial decomposition occurs:

$$3\ H_4[Fe(CN)_6] \rightarrow 12\ HCN\uparrow + Fe_2[Fe(CN)_6]\downarrow$$

The normal hexacyanoferrates(II) of Alk^+ and Ae^{2+} (except Ba^{2+}) are readily soluble in water; those of the other metals are insoluble. There are many double salts, some soluble and others with these small aqueous solubilities: $K_2Mg[Fe(CN)_6]$, 6.0 mM; $(NH_4)_2Mg[Fe(CN)_6]$, 9.2 mM; $(NH_4)_2Ca[Fe(CN)_6]$, 9.0 mM; all at 17°C; $K_2Sr[Fe(CN)_6]$ and $K_2Ba[Fe(CN)_6]$ are both only slightly soluble; the same is true of many other such salts having a higher-*Z* metal and an alkali for cations.

The soluble hexacyanoferrates(II) are yellow if hydrated, white when anhydrous. The salts $Alk_2Ae[Fe(CN)_6]$ are white. Several of the other insoluble salts have the following colors: $Cr_4[Fe(CN)_6]_3$, gray green; $K_2Mn[Fe(CN)_6]$, pink; $Fe_2[Fe(CN)_6]$, white if pure, usually pale blue; $Fe_4[Fe(CN)_6]_3 \cdot aq$, blue; $Co_2[Fe(CN)_6]$, green; $Ni_2[Fe(CN)_6]$, gray green; $Cu_2[Fe(CN)_6]$, red brown; $Ag_4[Fe(CN)_6]$, white, slowly turns blue; $K_2Zn_3[Fe(CN)_6]_2$, white; $K_2Cd_3[Fe(CN)_6]_2$, white; $(Hg_2)_2[Fe(CN)_6]$, yellow becoming blue, then gray, slowly decomposing; $Hg_2[Fe(CN)_6]$, white, quickly decomposing; Sn^{II} salt, white; Sn^{IV} salt, white gel; $Pb_2[Fe(CN)_6]$, white; $KSb[Fe(CN)_6]$, white; $KBi[Fe(CN)_6]$, white.

Long boiling of $[Fe(CN)_6]^{4-}$ and suspensions of its insoluble salts releases HCN and forms $[Fe(CN)_5H_2O]^{3-}$ etc. The aquation of $[Fe(CN)_6]^{3-}$ also forms $[Fe(CN)_6]^{4-}$ and $(CN)_2$ or NCO^-.

Solid $Fe(OH)_2$ can reduce H_2O to H_2 if catalyzed by colloidal Pt.

Ferrate(VI) slowly decomposes on standing in water:

$$2\ FeO_4^{2-} + 2\ H_2O \rightarrow Fe_2O_3 \cdot aq\downarrow + 4\ OH^- + {}^3/_2\ O_2\uparrow$$

Anhydrous $H_3[Fe(CN)_6]$ is a non-volatile, crystalline solid, readily soluble as a brown, strongly acidic solution, easily decomposed by heat.

For $[Fe(CN)_6]^{3-}$ the salts of Alk^+ and Ae^{2+} are readily soluble (distinction from $[Fe(CN)_6]^{4-}$ for Ae^{2+}). Those of most $\mathbf{d}^{2+}$, except Hg^{2+}, plus Bi^{3+}, are insoluble or slightly soluble. Most of the other **d**-block aquacations M(>II) (in the absence of oxidation and reduction) and of the **p**-block cations do not form precipitates. The soluble hexacyanoferrates(III) have reddish colors. The insoluble ones have the following somewhat pronounced colors: $Mn_3[Fe(CN)_6]_2$, brown; $Co_3[Fe(CN)_6]_2$, red; $Ni_3[Fe(CN)_6]_2$, yellow; $Cu_3[Fe(CN)_6]_2$, yellow-green; $Ag_3[Fe(CN)_6]$, brick red; $Zn_3[Fe(CN)_6]_2$, yellow; $Cd_3[Fe(CN)_6]_2$, orange, distinction and separation from SCN^-, a more sensitive test than S^{2-} for Cd^{2+}; $(Hg_2)_3[Fe(CN)_6]_2$, yellow becoming gray; $Bi[Fe(CN)_6]$, brown.

The $K_3[Fe(CN)_6]$ salt, from the oxidation of $[Fe(CN)_6]^{4-}$, is the usual source for other hexacyanoferrates(III). Its large red crystals are readily soluble in H_2O and ethanol (distinction from $K_4[Fe(CN)_6]$).

Water is oxidized to O_2 by FeO_4^{2-}, leaving Fe^{III}, if the pH < 8.8.

Seawater and some fresh and hot natural waters contain $FeOH^+$, $Fe(OH)_3$, $FeSO_4$, $FeCl^+$, $FeCl_2$ and so on.

Oxonium. Pure metallic iron dissolves in aqueous HCl and dilute H_2SO_4 for example, forming Fe^{2+}, with liberation of H_2, thus:

$$Fe + 2\,H_3O^+ \rightarrow Fe^{2+} + H_2\uparrow + 2\,H_2O$$

When commercial iron dissolves in H_3O^+, the carbon that it contains as carbide, Fe_3C, escapes as gaseous hydrocarbons, and the graphitic carbon remains undissolved.

Iron(II) oxide and hydroxide react with acids, forming iron(II) compounds, usually mixed with more or less iron(III). The iron(II) compounds are perhaps more readily prepared by the action of dilute acids on the metal, or on $FeCO_3$ or FeS. Acids dissolve Fe_2O_3, but extremely slowly if it has been ignited. Aqueous HCl is the best common solvent; warm 13-M H_2SO_4 has also been recommended. If the oxide is heated with alkalis or alkali carbonates, it then dissolves more readily in acids. Iron(III) hydroxide is readily soluble in acids. Magnetite, Fe_3O_4, treated with a small amount of HCl, yields Fe^{2+} and $Fe_2O_3 \cdot aq$; on treatment with excess HCl, a mixture of Fe^{2+} and Fe^{III} is obtained which, when treated with excess NH_3 and dried at 100 °C, again exhibits the magnetic properties of the original.

The acid $H_4[Fe(CN)_6]$, [*trans*-$Fe(CN)_2(CNH)_4$], is formed on adding , e.g., HCl to $[Fe(CN)_6]^{4-}$, or by ion exchange, and may be extracted with ether; $K_1 > K_2 > 1$ dM; $K_3 = 6 \pm 2$ mM; $K_4 = 67 \pm 3$ μM. The usual source for this acid or any of its salts is $K_4[Fe(CN)_6]$.

Cold, dilute acids do not decompose $[Fe(CN)_6]^{4-}$ greatly; warm solutions, however, e.g., 13.5-M H_2SO_4, liberate HCN:

$$2\,[Fe(CN)_6]^{4-} + 2\,K^+ + 6\,H_3O^+ \rightarrow$$

$$K_2Fe[Fe(CN)_6]\downarrow + 6\,HCN\uparrow + 6\,H_2O$$

The strong acid $H_3[Fe(CN)_6]$, consisting of lustrous, brownish-green needles, very soluble in water and ethanol, may be made by adding concentrated HCl to cold, saturated $K_3[Fe(CN)_6]$ and, before appreciable decomposition, drying in a vacuum the precipitate that forms.

In the reaction of most hexacyanoferrates(III) with acids the $[Fe(CN)_6]^{3-}$ is destroyed, but alkalis are distinctly more effective in this respect. Hot, dilute H_3O^+

with $[Fe(CN)_6]^{3-}$ gives HCN and complex products; concentrated H_2SO_4 produces CO and a little CO_2.

Hydroxide. Iron is not affected by OH^-. The hydroxide $Fe(OH)_2$ is formed on treating Fe^{2+} with OH^- or NH_3; it is white when pure, but seldom obtained sufficiently free from Fe^{III} to be white. It quickly changes in the air to a mixed hydroxide of a dirty green to black color, then to $Fe_2O_3 \cdot aq$, often apparently FeO(OH), reddish brown. The "fixed" alkalis, e.g., NaOH, are adsorbed by $Fe(OH)_2$. Sugar, many organic acids, and NH_4^+ to a slight extent, dissolve $Fe(OH)_2$ or prevent its formation. These organic chelators, but not NH_4^+, hold Fe^{III} in solution much more effectively.

Solutions of Fe^{III}, when treated with bases such as OH^- or NH_3, yield iron(III) hydrous oxide, $Fe_2O_3 \cdot aq$, perhaps ~FeO(OH), reddish brown, insoluble in modest excess (distinction from the Al, Cr and Zn compounds, which are soluble in excess of OH^-, and from the Co, Ni, Cu and Zn ones, which are soluble in NH_3).

Precipitation from a cold solution often gives a positively charged colloid, which may coagulate on boiling or treating with a doubly or triply negative anion. Aqueous CO_3^{2-} however, does not work thus unless in excess, because of its destruction, for example as follows:

$$2\,Fe^{3+} + 3\,CO_3^{2-} \rightarrow Fe_2O_3 \cdot aq\downarrow + 3\,CO_2\uparrow \text{ or}$$

$$2\,Fe^{3+} + 6\,CO_3^{2-} + 3\,H_2O \rightarrow Fe_2O_3 \cdot aq\downarrow + 6\,HCO_3^-$$

Salts of the alkali metals are adsorbed by this precipitate and held tenaciously. Concentrated OH^-, however, yields some $[Fe(OH)_6]^{3-}$.

Insoluble hexacyanoferrates(II) are transposed by alkalis.

The action of $[Fe(CN)_6]^{3-}$ upon heating in alkaline solution is somewhat complex and depends on conditions. Some of the possible products are CO_3^{2-}, CN^-, NCO^-, NH_3, $Fe_2O_3 \cdot aq$ and $[Fe(CN)_6]^{4-}$. In similar conditions CO_3^{2-}, HCO_3^- and NH_3 give the same results as OH^-.

Another complicated reaction with a complex starts with $[Fe(CO)_5]$ and OH^- in water or ethanol and produces a pale-yellow product:

$$[Fe(CO)_5] + OH^- \rightarrow [Fe(CO)_4CO_2H]^-$$

$$[Fe(CO)_4CO_2H]^- + OH^- \rightarrow [Fe(CO)_4CO_2]^{2-} + H_2O$$

$$[Fe(CO)_4CO_2]^{2-} \rightarrow [Fe(CO)_4]^{2-} + CO_2$$

$$[Fe(CO)_4]^{2-} + H_2O \leftrightarrows [FeH(CO)_4]^- + OH^-$$

$$CO_2 + 2\,OH^- \rightarrow CO_3^{2-} + H_2O$$

$$[Fe(CO)_5] + 3\,OH^- \rightarrow [FeH(CO)_4]^- + CO_3^{2-} + H_2O$$

This can be precipitated by large cations such as $[NR_4]^+$ or $[PPh_4]^+$, or isolated on an ion-exchange resin. The Fe anion is stable in water at high pH but not in the air. Then dilute HCl and a current of CO give a stream of vapor of the very unstable colorless liquid $[FeH_2(CO)_4]$. The $[FeH(CO)_4]^-$ ion is an extremely weak acid.

Another interesting reaction (reversed by strong acids) resembles those of Ru and Os and goes thus:

$$[Fe(CN)_5(NO)]^{2-} + 2\ OH^- \leftrightarrows [Fe(CN)_5(NO_2)]^{4-} + H_2O$$

Peroxide. Iron is passivated by H_2O_2 until the latter decomposes enough to allow corrosion to begin.

In "Fenton's reagent", Fe^{2+} plus H_2O_2, we find partly:

$$Fe^{2+} + {}^1/_2\ H_2O_2 \rightarrow FeOH^{2+}$$

However, the destruction of the strong oxidant H_2O_2 (also by an acid-catalyzed path) can accompany, perhaps surprisingly, the oxidation of H_2O_2-resistant substances such as H_2 and certain organic compounds. A mechanism here (even though still not showing the truly elementary steps) in spite of our general de-emphasis on mechanisms, gives a perhaps especially interesting and useful example of such results:

$$Fe^{2+} + H_2O_2 \rightarrow FeOH^{2+} + OH$$

$$OH + H_2O_2 \rightarrow H_2O + HO_2$$

$$HO_2 + H_2O_2 \rightarrow O_2\uparrow + H_2O + OH$$

$$FeOH^{2+} + H_2O_2 \rightarrow Fe^{2+} + HO_2 + H_2O$$

$$Fe^{2+} + OH \rightarrow FeOH^{2+}$$

Some of the OH radicals then can cause further otherwise-difficult oxidations of many species. Energy is still conserved; the other reactions are inhibited only by kinetic factors. The $FeOH^{2+}$ may then undergo further hydrolysis, condensation and polymerization, ligation with the anion, and precipitation.

The one-electron oxidations of Fe^{2+} and other **d**-block complexes by H_2O_2 are only slightly faster than those by $Cr(HO_2)^{2+}$, where the HO_2^- ligand oxidizes both the Fe^{II} and the Cr^{III} to Fe^{III} and the less stable Cr^{IV}:

$$Fe^{2+} + Cr(HO_2)^{2+} \rightarrow FeOH^{2+} + CrO^{2+}$$

Aqueous H_2O_2 oxidizes $H_n[Fe(CN)_6]^{(4-n)-}$ in acid to $[Fe(CN)_6]^{3-}$.

Di- and trioxygen. Iron is attacked slowly by moist air, faster with a more ionized electrolyte, forming brown rust, chiefly a hydrated oxide, written variously as $Fe_2O_3{\cdot}aq$, $Fe_2O_3{\cdot}xH_2O$, $Fe_2O_3(H_2O)_x$ etc., or Fe_3O_4, magnetite, with less air. The

x slowly decreases on standing. In some contexts the short formulas $Fe(OH)_3$ or $FeO(OH)$ may be acceptable.

Dissolved Fe^{2+} is unstable in contact with air, catalyzed by Cu^{II} and especially $PdSO_4$ (perhaps 1 mM), changing to Fe^{III} (as with H_2O_2 above), which is precipitated partly as a basic compound.

Pyrite or marcasite reacts slowly at first this way:

$$FeS_2 + {}^{7}/_{2}\,O_2 + 3\,H_2O \rightarrow Fe^{2+} + 2\,SO_4^{2-} + 2\,H_3O^+$$

sometimes forming a little S, $S_2O_3^{2-}$, SO_3^{2-} or $S_nO_6^{2-}$, then Fe^{III}.

Consuming CO_3^{2-} in leaching U from ores, a "parasitic reaction" is:

$$2\,FeS_2 + 8\,CO_3^{2-} + {}^{15}/_{2}\,O_2 + 5\,H_2O \rightarrow$$

$$2\,FeO(OH){\cdot}aq\downarrow + 4\,SO_4^{2-} + 8\,HCO_3^-$$

Hexacyanoferrates(II) are less easily oxidized than simple Fe^{II} salts, yet are moderately strong reductants, being themselves converted to hexacyanoferrates(III). Air oxidizes the Fe^{II} in $Ag_4[Fe(CN)_6]$ slightly to $AgFe[Fe(CN)_6]$, which colors the salt bluish.

Heating $[Fe(CN)_6]^{4-}$ with OH^- and air 60 h at 90 °C gives some Fe_2O_3.

Ozone appears to form FeO^{2+} transiently from Fe^{2+} at pH 0 to 2, then clearly Fe^{III}, or $Fe_2O_3{\cdot}aq$ at higher pH.

Purple or blue FeO_4^{2-}, "ferrate", may be made by treating a fresh suspension of $Fe_2O_3{\cdot}aq$ in OH^- with O_3 below 50 °C.

8.1.2 Reagents Derived from the Other 2nd-Period Non-Metals, Boron through Fluorine

Boron species. Soluble borates precipitate iron(II and III) borates, often with complex compositions, elaborated in references on borates.

Carbon oxide species. Four moles of CN^- and one of Fe^{2+}, stirred under CO for 48 h, yield $[trans\text{-}Fe(CN)_4(CO)_2]^{2-}$, stable in air for months.

Gaseous CO can displace NH_3:

$$[Fe(CN)_5(NH_3)]^{3-} + CO + CH_3CO_2H \rightarrow$$

$$[Fe(CN)_5(CO)]^{3-} + NH_4^+ + CH_3CO_2^-$$

Yellow-green $K_3[Fe(CN)_5(CO)]$ is light-sensitive and diamagnetic, but the anion resists oxidation. With CN^- it gives $[Fe(CN)_6]^{4-}$ and CO.

The alkali hexacyanoferrates(II) are not transposed in the cold by CO_2 (distinction from ionic cyanides, which do become carbonates).

Iron(II) carbonate precipitates from Fe^{2+} and CO_3^{2-}, but Fe^{III} gives $Fe_2O_3{\cdot}aq$ and either CO_2 or HCO_3^- (with excess CO_3^{2-}).

Cyanide species. The yellow $K_4[Fe(CN)_6]\cdot 3H_2O$ can be prepared by dissolving many Fe^{II} salts in hot KCN and then cooling; the last CN^- joins the complex much more slowly than the first five; i.e., low-spin $[Fe(CN)_5(H_2O)]^{3-}$ is rather inert. The product is so stable that its toxicity is low, in spite of the warning labels on some containers due to the cyanide. One Fe^{2+} and five CN^- also yield $[Fe_2(CN)_{10}]^{6-}$. Aqueous Fe^{2+} with more limited CN^- precipitates a yellowish-red iron(II) cyanide.

Solutions of Fe^{III} with CN^- yield $Fe_2O_3\cdot aq$ with the release of HCN and some $(CN)_2$ and CNO^- (in base) by reducing some Fe^{III}, e.g.:

$$2\,[FeCl_4]^- + 6\,CN^- + 3\,H_2O \rightarrow Fe_2O_3\cdot aq\downarrow + 8\,Cl^- + 6\,HCN\uparrow$$

A small amount of the $Fe_2O_3\cdot aq$ dissolves in excess CN^-, forming the toxic $[Fe(CN)_6]^{3-}$, "ferricyanide", but a better preparation is to oxidize $[Fe(CN)_6]^{4-}$. A possible interpretation of Mössbauer evidence, incidentally, could show the former as $[Fe^{2+}(CN)_6{}^{5-}]$ instead of the usually implied $[Fe^{3+}(CN)_6{}^{6-}]$. The $[Fe(CN)_6]^{3-}$ ion, in turn, oxidizes $[BH_4]^-$, CN^-, N_2H_4, NH_2OH, NO_2^-, NO_2, AsO_3^{3-}, SO_3^{2-}, I^- etc. Ion exchange forms $H_3[Fe(CN)_6]$, a strong tribasic acid.

Substituting ligands on $[Fe(CN)_6]^{4-}$ is difficult, but catalyzed by Hg^{2+}.

Some "simple" organic reagents. The $K_3[Fe(CN)_6]$ and $Na_3[Fe(CN)_6]$ salts are precipitated only slightly, or not at all, from water by ethanol (separation from $[Fe(CN)_6]^{4-}$). Aliphatic alcohols, ROH, react with $[H_4Fe(CN)_6]$ to form $[Fe(CN)_2(CNR)_4]$ and others; HCN then displaces RNC, making the Fe^{II} a catalyst for $ROH + HCN \rightarrow RNC + H_2O$.

Much excess $CH_3CO_2^-$ with Fe^{III} quite quickly forms a dull-red solution of basic acetates, especially $[\{Fe(H_2O)\}_3(\mu\text{-}CH_3CO_2)_6(\mu_3\text{-}O)]^+$. The red color is not affected by $HgCl_2$ (unlike the thiocyanates below).

This solution on standing, or faster on boiling, precipitates an oxide acetate, approximately $Fe_3O_3(OH)_2(CH_3CO_2)$. The reaction is complete at the boiling point, but tends to reverse on cooling. For complete precipitation, the solution of Fe^{III} is neutralized with NH_3 until a precipitate forms that does not disappear on stirring. The solution is then cleared with a minimum of HCl. After adding excess $CH_3CO_2^-$ and some dilution, the solution is boiled 3–5 minutes and the flocculent precipitate is quickly removed. This separates Fe^{III} and Al^{III} from various M^{2+}. Phosphate is precipitated as $FePO_4$ in the basic acetate precipitation; also, while Cr^{III} by itself will not precipitate a basic acetate, it precipitates fairly completely with excess Fe^{III} or Al^{III}. With excess Cr^{III}, the Fe^{III} and Al^{III} precipitate incompletely. The basic precipitate is soluble in H_3O^+, and is transposed by OH^- to the hydrous oxide.

Double oxalates, with M = Mg, Mn, Co, Ni or Zn, may be prepared as follows under N_2:

$$Fe + 2\,CH_3CO_2H \rightarrow Fe^{2+} + H_2\uparrow + 2\,CH_3CO_2^-$$

$$M^{2+} + 2\,Fe^{2+} + 3\,H_2C_2O_4 + 6\,CH_3CO_2^- + 6\,H_2O \rightarrow$$

$$MFe_2(C_2O_4)_3\cdot 6H_2O\downarrow + 6\,CH_3CO_2H$$

Igniting these in air yields iron(III) spinels, MFe_2O_4.

Oxalic acid and soluble oxalates precipitate from solutions of Fe^{2+}, $FeC_2O_4 \cdot 2H_2O$, yellowish white, crystalline, insoluble in H_2O, soluble in strong acids. Iron(III) is complexed and not precipitated by oxalates, except as reduction to Fe^{2+} occurs. The light-green complex may, however, be isolated with a sequence such as:

$$3\ BaC_2O_4 + FeSO_4^+ + Fe(SO_4)_2^- \rightarrow$$

$$FeC_2O_4^+ + Fe(C_2O_4)_2^- + 3\ BaSO_4\downarrow$$

$$FeC_2O_4^+ + Fe(C_2O_4)_2^- + 3\ C_2O_4^{2-} + 6\ K^+ + 6\ H_2O \rightarrow$$

$$2\ K_3[Fe(C_2O_4)_3] \cdot 3H_2O\downarrow$$

Hydrated $FeSO_4^+$ and $Fe(SO_4)_2^-$ are some of the actual constituents of aqueous $Fe_2(SO_4)_3$. The stability of $FeC_2O_4^+$ makes Fe^{III} chloride dissolve even CaC_2O_4. Another ready source of $K_3[Fe(C_2O_4)_3] \cdot 3H_2O$ is a solution of $FeC_2O_4 \cdot 2H_2O$, H_2O_2, $H_2C_2O_4$ and excess $K_2C_2O_4$.

Ethanol, ether, glycerol, tartrates etc. reduce FeO_4^{2-} to Fe^{III}.

Reduced nitrogen. The actions of the base, NH_3, on Fe^{2+} and Fe^{III} were given above, along with those of OH^-.

Saturating aqueous $Na_2[Fe(CN)_5(NO)]$ with NH_3 for 48 h at 0 °C, or still better, treating $Na_2[Fe(CN)_5(NO)]$ with concentrated NH_3 and $NaCH_3CO_2$ overnight, forms pale-yellow $Na_3[Fe(CN)_5(NH_3)] \cdot 6H_2O$:

$$[Fe(CN)_5(NO)]^{2-} + 3\ NH_3 \rightarrow [Fe(CN)_5(NH_3)]^{3-} + N_2\uparrow + NH_4^+ + H_2O$$

One may then dissolve the crude product in minimal cold water and precipitate it with ethanol; however, H_2O replaces the NH_3 in minutes, but reversibly, a good route, with ascorbic acid, to $[Fe(CN)_5(H_2O)]^{3-}$.

The reductant N_2H_4 in base, and $N_2H_5^+$ in acid, respectively, give:

$$[Fe(CN)_5(NO)]^{2-} + N_2H_4 + H_2O \rightarrow [Fe(CN)_5(H_2O)]^{3-} + N_2O\uparrow + NH_4^+$$

$$Fe^{3+} + N_2H_5^+ + H_2O \rightarrow Fe^{2+} + {}^1/_2\ N_2\uparrow + NH_4^+ + H_3O^+$$

Hydroxylamine in acid is another reductant:

$$4\ Fe^{3+} + 2\ NH_3OH^+ + 5\ H_2O \rightarrow 4\ Fe^{2+} + N_2O\uparrow + 6\ H_3O^+$$

or in base an oxidant in the first reaction following, or a reductant in the second, which also yields some $[Fe_2(CN)_{10}]^{6-}$:

$$2\ Fe(OH)_2 + NH_2OH \rightarrow Fe_2O_3 \cdot aq\downarrow + NH_3 + 2\ H_2O$$

$$[Fe(CN)_5(NO)]^{2-} + NH_2OH + CO_3^{2-} \rightarrow \text{brown } [Fe(CN)_5(H_2O)]^{3-} + N_2O + HCO_3^-$$

The ferrates(VI) are strongly reduced to Fe^{III} by NH_3, NO_2^-, etc.

"Azide" or triazide (N_3^-) and Fe^{III} yield the very reddish $Fe(N_3)_3$ and so on in solution, rather like the better-known thiocyanate complexes.

Oxidized nitrogen. With Fe in cold, dilute HNO_2 and HNO_3, Fe^{2+} and NH_4^+, N_2O and/or H_2 arise, although various conditions yield other nitrogen species:

$$4\ Fe + 10\ H_3O^+ + NO_3^- \rightarrow 4\ Fe^{2+} + NH_4^+ + 13\ H_2O$$

$$4\ Fe + 10\ H_3O^+ + 2\ NO_3^- \rightarrow 4\ Fe^{2+} + N_2O\uparrow + 15\ H_2O$$

$$Fe + 2\ H_3O^+ \rightarrow Fe^{2+} + H_2\uparrow + 2\ H_2O$$

Moderately dilute HNO_3 and heat give mainly Fe^{3+} and NO:

$$Fe + 4\ H_3O^+ + NO_3^- \rightarrow Fe^{3+} + NO\uparrow + 6\ H_2O$$

Nitrous acid and $[Fe(CN)_6]^{4-}$ or $H[Fe(CN)_6]^{3-}$, but not $H_2[Fe(CN)_6]^{2-}$ etc., form $[Fe(CN)_6]^{3-}$.

Strong oxidants, e.g., cold 16-M HNO_3 or $H_2[Cr_2O_7]$, induce passivity from a superficial oxide film, which may be destroyed by immersion in reducing agents or HCl, or by scratching the surface of the metal.

Nitric acid and all Fe^{II} compounds form Fe^{III}, faster with heat:

$$FeS + 4\ H_3O^+ + NO_3^- \rightarrow Fe^{3+} + S\downarrow + NO\uparrow + 6\ H_2O$$

In the cold and with a lower layer of 18-M H_2SO_4, the "brown-ring" complex, $[Fe(H_2O)_5(NO)]^{2+}$, is obtained (a common test for nitrates or nitrites). More-concentrated HNO_3 yields NO_2 and N_2O_4 more than NO.

Aqueous NO_2^- converts the $[Fe(CN)_6]^{4-}$ ion into the well-known "nitroprusside", as in red $Na_2[Fe(CN)_5(NO)]\cdot 2H_2O$, that is, the pentacyanonitrosylferrate(2−), which is stable, but not in water in light:

$$[Fe(CN)_6]^{4-} + NO_2^- \leftrightarrows [Fe(CN)_5(NO_2)]^{4-} + CN^-$$

$$[Fe(CN)_5(NO_2)]^{4-} + H_2O \leftrightarrows [Fe(CN)_5(NO)]^{2-} + 2\ OH^-$$

We may drive these equilibria to the right with the mild acidity of a current of CO_2 plus Ba^{2+}, along with heating and removing the HCN:

$$2\ [Fe(CN)_6]^{4-} + 2\ NO_2^- + 3\ CO_2 + 3\ Ba^{2+} + H_2O \rightarrow$$

$$2\ [Fe(CN)_5(NO)]^{2-} + 2\ HCN\uparrow + 3\ BaCO_3\downarrow$$

With HNO_3 more of the CN^- is oxidized:

$$H_2[Fe(CN)_6]^{2-} + NO_3^- + 2\ H_3O^+ \rightarrow$$

$$[Fe(CN)_5(NO)]^{2-} + CO_2\uparrow + NH_4^+ + 2\ H_2O$$

Ion exchange and evaporation yield the acid, $H_2[Fe(CN)_5(NO)]$.

Ammonia, $[Fe(CN)_5(NO)]^{2-}$ and $CH_3CO_2^-$ give $[Fe(CN)_5(NH_3)]^{3-}$ and NO_2^-, but excess NH_3 yields $[Fe(CN)_5(NH_3)]^{3-}$ and N_2.

Alkaline NO_2^- with $[Fe(CN)_5(NH_3)]^{3-}$ forms $[Fe(CN)_5(NO_2)]^{4-}$. The $Na_3[Fe(CN)_5(NH_3)]\cdot 6H_2O$ salt plus $NaNO_2$ and CH_3CO_2H, form dark-yellow $Na_2[Fe(CN)_5(NO)]\cdot 2H_2O$, precipitated by ethanol and ether.

Aqueous $[Fe(CN)_5NO]^{2-}$ plus HS^-, $S_2O_4^{2-}$ or $[BH_4]^-$ at pH 7–10 yield $[Fe(CN)_5NO]^{3-}$, or with H_2S, $S_2O_4^{2-}$ or $[BH_4]^-$ at pH 4, $[Fe(CN)_4NO]^{2-}$.

Similar to the ability of $[Fe(CN)_5(NO)]^{2-}$ to add oxide in base, thereby forming $[Fe(CN)_5(NO_2)]^{4-}$, is its ability to add sulfide in alkali, yielding the purple $[Fe(CN)_5(NOS)]^{4-}$. This provides a sensitive test for sulfides, even many insoluble and organic ones. The ion soon breaks down in water to $[Fe(CN)_6]^{4-}$, $Fe_2O_3\cdot aq$, Prussian Blue, NO, N_2, etc. The solid Na and K salts, however, i.e., $Alk_4[Fe(CN)_5(NOS)]$, are stable.

The generally helpful IUPAC names and formulas are problematic with the above complexes of NO. The recommendations [1] treat the "cyano" ligand as CN^- but the "nitrosyl" ligand as neutral NO. Taken alone, this points us to the formula $[Fe^{2+}(H_2O)_5(NO)]$ instead of the $[Fe^+(H_2O)_5(NO^+)]$ indicated by its magnetic moment or the $[Fe^{3+}(H_2O)_5(NO^-)]$ suggested by spectral data. This can remind us that oxidation numbers, although so useful in classification and in writing (balanced) equations, are based on simple assignments of electrons and may be misleading. Electron clouds are spread out, and the "real" charges on bonded atoms are hardly integral. See **6.1.2 Oxidized nitrogen** also. The name sometimes given to the $[Fe(CN)_5(NO)]^{2-}$ ion then, pentacyanonitrosylferrate(II), with the oxidation (Stock) number for the metal instead of the indisputable charge for the whole ion (Ewens-Bassett number), i.e., as in pentacyanonitrosylferrate(2–), points to $[Fe^{2+}(CN^-)_5(NO)]$, with an incorrect total charge (3–). Of course, a ligand with a variable charge is inherently problematic when we want to assign separate charges to the components of the whole species.

Aqueous $[Fe(CN)_5(NO)]^{2-}$ plus HS^-, $S_2O_4^{2-}$ or $[BH_4]^-$ at pH 7–10 yield $[Fe(CN)_5(NO)]^{3-}$, or with H_2S, $S_2O_4^{2-}$ or $[BH_4]^-$ at pH 4, $[Fe(CN)_4(NO)]^{2-}$.

Fluorine species. Iron dissolves in warm HF; ethanol then yields white $FeF_2\cdot 4H_2O$, turning brown in air. Boiling concentrated HF with much excess Fe yields red $Fe_2F_5\cdot 2H_2O$ while hot, or yellow $Fe_2F_5\cdot 7H_2O$, i.e., $[Fe^{II}(H_2O)_6][Fe^{III}F_5(H_2O)]$, when cool. Saturated $FeF_2\cdot 4H_2O$ after several days in the air also precipitates $Fe_2F_5\cdot 7H_2O$.

Iron(III) and F^- make colorless $[FeF_6]^{3-}$ etc. that are stable enough to interfere with some characteristic reactions, such as with NCS^- or $[Fe(CN)_6]^{4-}$. The stability order is $F^- > Cl^- > Br^-$.

8.1.3 Reagents Derived from the 3rd-to-5th-Period Non-Metals, Silicon through Xenon

Silicon species. Soluble silicates precipitate iron(II and III) silicates, often with complex compositions not elaborated here.

Phosphorus species. Phosphinic acid, HPH_2O_2, and phosphonic acid, H_2PHO_3, reduce Fe^{III} to Fe^{2+}, as in:

$$4\ FeCl_2^+ + HPH_2O_2 + 6\ H_2O \rightarrow 4\ Fe^{2+} + H_3PO_4 + 4\ H_3O^+ + 8\ Cl^-$$

In alkaline solutions PHO_3^{2-} and $PH_2O_2^-$ are oxidized to PO_4^{3-} while $[Fe(CN)_6]^{3-}$ is reduced to $[Fe(CN)_6]^{4-}$.

Solutions of Fe^{2+} with HPO_4^{2-} precipitate a mixture of $FeHPO_4$ and $Fe_3(PO_4)_2$, white to bluish white, soluble in H_3O^+:

$$Fe^{2+} + HPO_4^{2-} \rightarrow FeHPO_4\downarrow$$

$$3\ Fe^{2+} + 4\ HPO_4^{2-} \rightarrow Fe_3(PO_4)_2\downarrow + 2\ H_2PO_4^-$$

If acetate ion is present, only the tertiary phosphate precipitates:

$$3\ Fe^{2+} + 2\ HPO_4^{2-} + 2\ CH_3CO_2^- \rightarrow Fe_3(PO_4)_2\downarrow + 2\ CH_3CO_2H$$

With Fe^{III}, phosphates form $FePO_4$, insoluble in acetic acid, readily soluble in strong acids. Hence Fe^{III} is precipitated by H_3PO_4 and the anion it forms in the presence of acetate:

$$FeBr_3 + H_2PO_4^- + 2\ CH_3CO_2^- \rightarrow FePO_4\downarrow + 3\ Br^- + 2\ CH_3CO_2H$$

Phosphoric acid dissolves $FePO_4$, giving colorless (unlike the usual Fe^{III}) complexes, mainly $FeHPO_4^+$, perhaps $Fe(\eta^2\text{-}HPO_4)^+$, in dilute H_3O^+. Aqueous OH^- transposes freshly precipitated $FePO_4$, forming $Fe_2O_3\cdot aq$ and PO_4^{3-}. The transposition is incomplete in the cold.

Arsenic species. Arsenic reduces Fe^{III} to Fe^{2+}. Soluble arsenites and arsenates precipitate the corresponding arsenites and arsenates from neutral or faintly acid solutions of Fe^{2+} and Fe^{III}. Basic Fe^{III} arsenite, approx. $4Fe_2O_3\cdot As_2O_3\cdot 5H_2O$, is formed when an excess of fresh $Fe_2O_3\cdot aq$ is added to H_3AsO_3. The product is insoluble in acetic acid. This is also formed when moist $Fe_2O_3\cdot aq$ is given as an antidote in arsenic poisoning. A mixture of milk of magnesia, $Mg(OH)_2$, and $Fe_2(SO_4)_3$ has been generally indicated to make the oxide fresh.

Arsenite reacts with $[Fe(CN)_5(NO)]^{2-}$ and Na^+, giving an orange product, $Na_4[Fe(CN)_5(AsO_2)]\cdot 4H_2O$. Warm CN^- then forms $[Fe(CN)_6]^{4-}$.

Reduced chalcogens. Sulfane, H_2S, is without action on Fe^{2+} in acidic solution. Alkali sulfides form FeS, brown or dark gray, insoluble in excess reagent, readily soluble in dilute strong acids while releasing H_2S in the absence of strong oxidants. The moist precipitate is slowly converted, in the air, to $FeSO_4$, and finally to a basic sulfate, $Fe_2O(SO_4)_2$. Iron(III) salts are reduced by H_2S, e.g.:

$$2\,FeCl^{2+} + H_2S + 2\,H_2O \rightarrow 2\,Fe^{2+} + 2\,Cl^- + S\downarrow + 2\,H_3O^+$$

Alkali sulfides, with most iron compounds, give FeS or Fe_2S_3, the latter quickly changing to FeS and S. An example is $FePO_4$ warmed with "$(NH_4)_2S$":

$$2\,FePO_4 + NH_3 + 3\,HS^- \rightarrow 2\,FeS\downarrow + S\downarrow + NH_4^+ + 2\,HPO_4^{2-}$$

The FeS lattice, however, tends to have many Fe vacancies and can often be represented with greater precision, when required, as ~ $Fe_{0.9}S$. Solid structures are not normally perfect, but the imperfections here are just extra great. Polysulfides, S_x^{2-}, give similar results but with the deposition of additional S or an increase in the value of x in any unconsumed S_x^{2-}.

Sulfides including H_2S reduce $[Fe(CN)_6]^{3-}$ to $[Fe(CN)_6]^{4-}$ and, with K^+, some $K_2Fe^{II}[Fe^{II}(CN)_6]$.

Sulfane plus the rather insoluble Cu^{II} salt of $[Fe(CN)_5(CO)]^{3-}$ give colorless, light-sensitive $H_3[Fe(CN)_5(CO)]$ after evaporation.

The reactions of Se^{2-} and Te^{2-} are rather similar to those of S^{2-}, but with FeSe and FeTe being even less soluble than FeS in water.

Thiocyanate ion, SCN^-, gives no reaction with Fe^{2+}; Fe^{III} yields blood-red complexes:

$$FeCl_3 + n\,SCN^- \rightarrow Fe(SCN)_n^{(3-n)+} + 3\,Cl^-$$

This is a very sensitive test for Fe^{III}, 0.02 μM or less giving a perceptible pink. Because the reaction is reversible, excess of the reagent is an important factor, as is also the acidity. Basic anions (from the salts of weak acids) give considerable interference. The red compounds are very soluble in water, ethanol and ether; more so in ether than in water; they may be extracted and concentrated in that solvent, thus increasing the sensitivity of the test. The red color is decreased or destroyed by Hg^{II} (due to the formation of colorless and less dissociated thiocyanato complexes), phosphates, borates, acetates, oxalates, tartrates, citrates, etc., and acids of these salts. Strongly oxidizing acids such as HNO_3 or $HClO_3$ also interfere by forming red "perthiocyanogen", $H(SCN)_3$.

Oxidized chalcogens. Iron(II) sulfite, $FeSO_3$, from $Fe(OH)_2$ and SO_2, is moderately soluble in water, readily soluble in excess SO_2 solution. The moist compound is oxidized rapidly by air. The iron(III) salt is known only as a red solution formed, e.g., by the action of SO_2 on freshly precipitated $Fe_2O_3\cdot aq$, rapidly dismutated and reduced to iron(II):

$$Fe_2(SO_3)_3 \rightarrow FeSO_3 + Fe[S_2O_6]$$

Iron(II) thiosulfate is formed, along with some FeS and $FeSO_3$, by the action of aqueous SO_2 on Fe or FeS. Iron(III) is reduced to Fe^{2+} by $S_2O_3^{2-}$, SO_2 or SO_3^{2-}:

$$FeCl_2^+ + S_2O_3^{2-} \rightarrow Fe^{2+} + 2\ Cl^- + {}^1/_2\ [S_4O_6]^{2-}$$

In acidic solution H_2SO_4 and S are formed:

$$2\ FeCl_3 + S_2O_3^{2-} + 3\ H_2O \rightarrow 2\ Fe^{2+} + 6\ Cl^- + 2\ H_3O^+ + SO_4^{2-} + S\downarrow$$

Cold $S_2O_3^{2-}$ has no effect on $[Fe(CN)_6]^{3-}$, but in hot alkali it reduces the iron. Sulfite (SO_3^{2-}), $S_2O_4^{2-}$ and alkaline CN^- all produce a similar result:

$$[Fe(CN)_6]^{3-} + {}^1/_2\ S_2O_4^{2-} \rightarrow [Fe(CN)_6]^{4-} + SO_2\uparrow$$

$$2\ [Fe(CN)_6]^{3-} + CN^- + 2\ OH^- \rightarrow$$

$$2\ [Fe(CN)_6]^{4-} + NCO^- + H_2O$$

The SO_3^{2-} ion forms $[Fe(CN)_5CNSO_3]^{5-}$ and $[Fe(CN)_5CNSO_3]^{4-}$, then by hydrolysis $[Fe(CN)_6]^{4-}$ and SO_4^{2-}.

Cold $Fe_2O_3\cdot aq$, suspended in NH_3, plus SO_2, produce an orange $(NH_4)_9[Fe(SO_3\text{-}\kappa O)_6]$.

Alkaline SO_3^{2-}, $[Fe(CN)_5(NO)]^{2-}$ and Na^+, after 24 h and the addition of ethanol, form a red oil; repeated treatment with water and ethanol give pale-yellow $Na_5[Fe(CN)_5(SO_3\text{-}\kappa S)]\cdot 2H_2O$.

A basic Fe^{III} sulfate, $[\{Fe(H_2O)\}_3(\mu\text{-}SO_4)_6(\mu_3\text{-}O)]^{5-}$, is like the acetate.

Acidified Fe^{III} and sulfate contain $FeSO_4^+$, $Fe(SO_4)_2^-$, $FeHSO_4^{2+}$ etc.

Concentrated H_2SO_4 and $[Fe(CN)_6]^{4-}$ yield $H_3[Fe(CN)_5(CO)]$ and:

$$H_4[Fe(CN)_6] + H_2SO_4 + 6\ H_3O^+ \rightarrow FeSO_4 + 6\ CO\uparrow + 6\ NH_4^+$$

The acid $H_3[Fe(CN)_6]$ may be obtained by treating its lead salt with (cool) H_2SO_4 but not with H_2S, which would reduce the Fe^{III}.

Iron(II) and $HSO_3(O_2)^-$ yield Fe^{III} and HSO_4^-. Peroxosulfates decompose $[Fe(CN)_6]^{3-}$.

Reduced halogens. Dissolving Fe in HCl and crystallizing yields $[FeCl_2(H_2O)_4]$. Iron(II) and large cations in Cl^- solutions give salts of $[FeCl_4]^{2-}$. Solutions of Fe in HBr form different $FeBr_2\cdot nH_2O$ at various temperatures. Likewise different $FeI_2\cdot nH_2O$ arise with I^-, including a green $FeI_2\cdot 4H_2O$ by evaporation at ambient *T*.

In < 1-M HCl, Fe^{III} predominates as $[Fe(H_2O)_6]^{3+}$, but in 3–4 M HCl as $[FeCl_2(H_2O)_4]^+$. Aqueous Fe^{3+} and Cl^- or Br^- form several yellow or brown complexes. Slowly evaporating $FeCl_3$ and AlkCl or NH_4Cl yields salts of $[FeCl_5(H_2O)]^{2-}$. Large cations are effective to crystallize or precipitate, e.g., $[NMe_4][FeCl_4]$ or $[PPh_4][FeBr_4]$. Only Cl^-, not the larger and softer Br^-, also occurs as $[FeX_6]^{3-}$, in $[Co(NH_3)_6][FeCl_6]$ for example. Iodide reduces acidified Fe^{III} to Fe^{2+}, forming I_2 or I_3^-:

$$2\ Fe^{3+} + 2\ I^- \rightarrow 2\ Fe^{2+} + I_2$$

but base reverses this:

$$2\ Fe(OH)_2 + I_2 + 2\ OH^- \rightarrow Fe_2O_3 \cdot aq\downarrow + 2\ I^- + 3\ H_2O$$

The Cl^- ion, under ordinary conditions, apparently does not affect $[Fe(CN)_6]^{3-}$; Br^- tends to reduce the iron, while I^- first forms an addition species which decomposes to give $[Fe(CN)_6]^{4-}$ and I_2. Aqueous $[Fe(CN)_6]^{3-}$ may in fact be determined by reduction to $[Fe(CN)_6]^{4-}$ with I^- in acidic solution and titration of the I_2 with $S_2O_3^{2-}$.

Elemental and oxidized halogens. Iron(II) is oxidized to iron(III) by Cl_2, Br_2, HClO, HBrO, $HClO_2$, and acidified ClO_3^-, BrO_3^-, or IO_3^-.

Aqueous $[Fe(CN)_6]^{4-}$ in acidic solution, e.g., $H_2[Fe(CN)_6]^{2-}$, is oxidized to $[Fe(CN)_6]^{3-}$ by Cl_2, Br_2, HClO, HBrO, $HClO_2$, $HClO_3$, $HBrO_3$ and HIO_3.

Bromine water oxidizes $[Fe(CN)_5H_2O]^{3-}$ and $[Fe_2(CN)_{10}]^{6-}$ to yellow-green $[Fe(CN)_5H_2O]^{2-}$ (often containing $[Fe_2(CN)_{10}]^{4-}$) and intensely blue $[Fe_2(CN)_{10}]^{4-}$ respectively.

The $Na_3[Fe(CN)_5(NH_3)]\cdot 6H_2O$ salt reacts with BrO^-, forming dark-yellow $Na_2[Fe(CN)_5(NH_3)]\cdot 2H_2O$, precipitated by ethanol and ether.

Even in base $[Fe(CN)_6]^{4-}$ is oxidized by bromite, for example:

$$4\ [Fe(CN)_6]^{4-} + BrO_2^- + 2\ H_2O \rightarrow 4\ [Fe(CN)_6]^{3-} + Br^- + 4\ OH^-$$

The purple or blue FeO_4^{2-} may be made by treating fresh $Fe_2O_3\cdot aq$ suspended in 1–2 M OH^- with Cl_2 (becoming ClO^-) or Br_2 below 50 °C:

$$2\ FeO(OH) + 3\ ClO^- + 4\ OH^- \rightarrow 2\ FeO_4^{2-} + 3\ Cl^- + 3\ H_2O$$

Neutral or acidified FeO_4^{2-} solutions, stronger oxidants than MnO_4^-, quickly form Fe^{III} and O_2.

At least the ClO^-, ClO_3^- and BrO^- ions decompose $[Fe(CN)_6]^{3-}$, usually forming some $[Fe(CN)_5NO]^{2-}$, less when warm or hot.

The orthoperiodate complex $[Fe^{III}{}_2(OH)(H_3I^{VII}O_6)]^{3+}$ exemplifies a series formed with Fe^{III} dimers. This one is relatively (but of course not absolutely) inert kinetically and stable thermodynamically. Other such complexes have phosphinate, phosphonate, phosphate, arsenite, arsenate, sulfite, sulfate, or selenite instead of orthoperiodate.

8.1.4 Reagents Derived from the Metals Lithium through Uranium, plus Electrons and Photons

Oxidation. Solutions of easily reduced metal ions, such as those of Pt, Cu, Ag, Au and Hg, but also of Sn, Pb and Bi, dissolve Fe as Fe^{2+} at low pH. With a dilute non-reducing acid present, such as H_3PO_4, H_2SO_4 or $HClO_4$, Fe^{2+} is oxidized further, more or less rapidly, to Fe^{III} by various metallic ions and complexes, such as Cr^{VI}, $Mn^{>II}$, Ag^+, $Au^{I\text{ or }III}$, or, in some conditions, $Pd^{II\text{ or }IV}$ or $Pt^{II\text{ or }IV}$:

$$6\,Fe^{2+} + [Cr_2O_7]^{2-} + 14\,H_3O^+ \rightarrow 6\,Fe^{3+} + 2\,Cr^{3+} + 21\,H_2O$$

The $[Fe(CN)_6]^{4-}$ ion, usually at low pH, e.g., as $H_2[Fe(CN)_6]^{2-}$, is oxidized to $[Fe(CN)_6]^{3-}$ by VO_2^+, $[Cr_2O_7]^{2-}$, $[Mo(CN)_8]^{3-}$ (fast by simple e^- transfer), $Mn^{>II}$, $Co_2O_3 \cdot aq$, $[IrCl_6{}^{2-}]$ (very fast by e^- transfer), NiO_2, Pb_3O_4, PbO_2 and $NaBiO_3$. As a complex example, MnO_4^- acts faster at pH 2 (the first equation) than at pH 6 (the second equation):

$$5\,H_2[Fe(CN)_6]^{2-} + MnO_4^- \rightarrow$$

$$5\,[Fe(CN)_6]^{3-} + Mn^{2+} + 2\,H_3O^+ + 2\,H_2O$$

$$5\,K[Fe(CN)_6]^{3-} + MnO_4^- + 8\,H_3O^+ \rightarrow$$

$$5\,[Fe(CN)_6]^{3-} + Mn^{2+} + 5\,K^+ + 12\,H_2O$$

Hexacyanoferrates(II) may be determined in H_2SO_4 solution by titration with MnO_4^-. Gravimetric methods are unsatisfactory because practically all of the insoluble salts are amorphous and hard to filter.

Ferrocene, $[Fe(\eta^5\text{-}C_5H_5)_2]$, is oxidized by $FeCl_3$ etc. to blue ferrocene(1+), $[Fe(\eta^5\text{-}C_5H_5)_2]^+$, "ferricenium". Rather large anions such as I_3^-, $[Cr(NCS)_4(NH_3)_2]^-$ or $[SiW_{12}O_{40}]^{4-}$ precipitate this. (Note that the the names "ferricenium" and ferrocene are not analogous to ammonium and ammonia, implying the addition of H^+ to the latter).

Without O_2, Fe^{2+} and Pd^{II} precipitate Pd^0 immediately. Oxygen delays this until all Fe^{II} becomes Fe^{III}, but not with 0.2 mM Cl^- included. Oxygen may oxidize a Pd^I intermediate faster than Fe^{II} reduces it.

The negative electrode in Ni-Fe rechargeable batteries uses:

$$Fe + 2\,OH^- \leftrightarrows Fe(OH)_2\downarrow + 2e^-$$

Yellow $K_4[Fe(CN)_6] \cdot 3H_2O$ arises from anodic treatment of Fe in KCN. Further electrolytic oxidation of the $[Fe(CN)_6]^{4-}$ gives $[Fe(CN)_6]^{3-}$.

Light (UV) and at least $Fe(SO_4)_n{}^{(2n-2)-}$ in acid form a little Fe^{III} and H_2.

Air and UV light slowly convert $[Fe(CN)_6]^{4-}$ to $Fe_2O_3 \cdot aq$ and some NCO^- with OH^- present, but to $Fe_4[Fe(CN)_6]_3 \cdot aq$, with H_3O^+. Light first releases electrons (to be solvated), e.g., from cyano-complexes:

$$[Fe(CN)_6]^{4-} + \gamma \rightarrow [Fe(CN)_6]^{3-} + e^-$$

Reduction. Aqueous $Na_2[Fe(CN)_5(NO)]$ and Na_{Hg} (amalgam) form pale-yellow $Na_3[Fe(CN)_5(NH_3)] \cdot 6H_2O$.

Iron(III) is reduced to Fe^{2+} by V^{III} and Cu^+, also by various metals, including Mg, Fe, Co, Ni, Cu, Zn, Zn_{Hg}, Cd, Al, Sn, Pb, Sb, and Bi. Excess Mg reduces Fe^{2+} further to the metal. Many other metals do not easily carry out this last step. At least the reduction by V^{III} is catalyzed by Cu^{II}:

$$V^{III} + Cu^{II} \rightarrow V^{IV} + Cu^{I} \text{ slowly}$$

$$Cu^{I} + Fe^{III} \rightarrow Cu^{II} + Fe^{II} \text{ faster}$$

Aqueous $[Fe(CN)_6]^{3-}$ is reduced to $[Fe(CN)_6]^{4-}$ by the metals Mg, Th, Pd and As, but not, according to various reports, by Mn, Fe, Co, Pt, Cu, Ag, Au, Zn, Cd, Hg, Al, Pb, Sb, Bi or Te. However, when a piece of a metal except Pt or Au is placed in contact with a mixture of $[Fe(CN)_6]^{3-}$ and Fe^{III}, a coating of Prussian Blue is soon formed. In alkaline solution Cr^{III}, Mn^{II}, Sn^{II} and Pb^{II} all reduce $[Fe(CN)_6]^{3-}$ to $[Fe(CN)_6]^{4-}$.

Aqueous Cr^{2+} reduces Fe^{III} rapidly, less with Fe^{3+} than with $FeCl^{2+}$, producing $[CrCl(H_2O)_5]^{2+}$, or with $FeOH^{2+}$.

Aqueous $[Fe(CN)_6]^{3-}$ does not precipitate Cr^{3+}, but they become $[Fe(CN)_6]^{4-}$ and CrO_4^{2-} in base.

Aqueous $[Fe(CN)_6]^{3-}$ is reduced by and combined with $[Co(CN)_5]^{3-}$ to form the reddish-yellow, dinuclear $[(NC)_5Co^{III}NCFe^{II}(CN)_5]^{6-}$. Iodine (and partly, $[Fe(CN)_6]^{3-}$) oxidizes this to $[(NC)_5CoNCFe(CN)_5]^{5-}$, but SO_3^{2-} reduces it back to the (6–) ion. Water at 80 °C for 8 h hydrolyzes the (6–) ion to $[Fe(CN)_6]^{4-}$ and $[Co(CN)_5(H_2O)]^{2-}$. Somewhat similar is:

$$[Ru(CN)_6]^{4-} + [Fe(CN)_5(H_2O)]^{2-} \rightarrow [(NC)_5RuCNFe(CN)_5]^{6-} + H_2O$$

Ferrocene(1+), $[Fe(\eta^5\text{-}C_5H_5)_2]^+$, is reduced to $[Fe(C_5H_5)_2]$ by $SnCl_2$ etc.

The $[Fe(CN)_6]^{3-}$ ion and acidified Sb^{III} go to $H_2[Fe(CN)_6]^{2-}$ and Sb^V.

Cathodic e^- with $[Fe(CN)_6]^{4-}$ and excess CN^- form perhaps $[Fe(CN)_5]^{4-}$ or $[FeH(CN)_5]^{3-}$, colorless, both seen in pulse radiolysis.

Light (UV, 254 nm) and Fe^{III} (including $[Fe(CN)_6]^{3-}$) in acid form a little Fe^{II} and O_2, catalyzed by TiO_2, WO_3, RuO_2 on WO_3 and ZnO.

Acidified $[Fe(C_2O_4)_3]^{3-}$ and light of 254 to 500 nm release Fe^{II} and CO_2 in chemical actinometers.

Other reactions. The K^+ ion strongly catalyzes the $[Fe(CN)_6]^{4-}$-$[Fe(CN)_6]^{3-}$ electron exchange.

Aqueous Ba^{2+} plus FeO_4^{2-} precipitate the comparatively stable $BaFeO_4$. This compound is very slightly soluble and is not decomposed by H_2O or by cold, dilute H_2SO_4.

In analysis $[Fe(CN)_6]^{4-}$ is recognized by its precipitation from neutral solution as $KCe[Fe(CN)_6]$, from dilute acidic solution as $Th[Fe(CN)_6]$, and from dilute HCl solution by precipitation with $(NH_4)_2MoO_4$. All of these methods have been used as separations from $[Fe(CN)_6]^{3-}$ and SCN^-. It is also recognized by its reactions with Fe^{III} and Cu^{2+}.

Aqueous $[Fe(CN)_6]^{4-}$ precipitates various salts as $K_2M^{II}[Fe(CN)_6]$ and $KM^{III}[Fe(CN)_6]$, especially of **d**-block ions. Many of these are good ion exchangers; e.g., $K_2Zn[Fe(CN)_6]$ takes in Cs^+ as $Cs_2Zn[Fe(CN)_6]$, decomposable by HNO_3 or NH_4NO_3. The Ti^{IV} salt adsorbs larger ions well, with the following preferences:

$$Cs^+ > Rb^+ > K^+ > NH_4^+ > Na^+ > Li^+ \text{ and}$$

$$Ba^{2+} > Sr^{2+} > Ca^{2+} > Mg^{2+} > Be^{2+}$$

Complexes of $Fe(OH)_3$ with ions of V, Cr, Mn, Co, Ni and Cu may be important in natural waters and in forming ores.

Iron(III) hydroxide, freshly precipitated, readily dissolves in solutions of $CrCl_3$ or $FeCl_3$, but not of $AlCl_3$.

Mixing the basic acetate $[\{Fe(H_2O)\}_3(CH_3CO_2)_6O]^+$ or $Fe_2O_3{\cdot}aq$ with WO_4^{2-} and $[H_2P_2W_{12}O_{48}]^{12-}$ forms $[\{H_6P_2W_{14}O_{54}\}_2(Fe_8O_{12})]^{16-}$, with the unusual nearly cubic $[Fe_8(\mu\text{-}O)_{12}]$ center and new $P_2W_{14}O_{54}$ moiety [2].

Aqueous $[Fe(CN)_6]^{3-}$ may be detected by its reactions with Fe^{2+} or Fe^{III}. It may be separated (a) from $[Fe(CN)_6]^{4-}$ by precipitating the latter as $Th[Fe(CN)_6]$ or as a double salt of Tl^+ and Ca^{2+}; (b) from thiocyanates by precipitating the $[Fe(CN)_6]^{3-}$ as $Cd_3[Fe(CN)_6]_2$.

Treating the fulminato complex $[Fe(CNO)_6]^{4-}$ with a suspension of $Fe(OH)_2$ reduces the CNO^- (but not the Fe^{II}, of course) to $[Fe(CN)_6]^{4-}$.

Aqueous $[Fe(H_2O)_6]^{3+}$ and $[Fe(CN)_6]^{3-}$ equilibrate with the ion pair $[Fe(H_2O)_6][Fe(CN)_6]$ and the complex $[Fe(CN)_5{-}CN{-}Fe(H_2O)_5]$.

Other compounds are: $Ln[Fe(CN)_6]{\cdot}5H_2O$ and similar salts of Y^{3+} and Bi^{3+}; and $Zn_3[Fe(CN)_6]_2$ and $CsZn[Fe(CN)_6]$. Also $Cs_2Li[Fe^{III}(CN)_6]$ and $Cs_2Mg[Fe^{II}(CN)_6]$ with similar sizes for Li^+ and Mg^{2+} are isostructural.

Fresh $MnO_2{\cdot}aq$ and $HFe(CO)_4^-$ buffered with NH_3 and NH_4^+ produce a black trinuclear complex (under N_2) after making a dark-red mixture:

$$HFe(CO)_4^- + MnO_2{\cdot}aq + 3\ NH_4^+ \rightarrow$$

$$^1/_3\ Fe_3(CO)_{12}\downarrow + Mn^{2+} + 3\ NH_3 + 2\ H_2O$$

Aqueous $[Fe(CN)_5(NH_3)]^{3-}$ catalyzes the formation of $[Fe(CN)_5Q]^{3-}$ from $[Fe(CN)_5(NH_3)]^{2-}$ where $Q^- = OH^-$, N_3^- or SCN^-. The substitution on the catalyst then determines the rate, followed by rapid redox.

Evaporating a solution of the sodium salts of $[Fe(CN)_5(NH_3)]^{2-}$ and $[Fe(CN)_6]^{4-}$, or of $[Fe(CN)_5(NH_3)]^{3-}$ and $[Fe(CN)_6]^{3-}$, yields the dinuclear $Na_6[(NC)_5Fe^{II}NCFe^{III}(CN)_5]\cdot 2H_2O$, easily oxidized and reduced.

Aqueous $[Fe(CN)_5(NO)]^{2-}$ precipitates most **d**-block cations, and we find, for example, $Fe[Fe(CN)_5(NO)]Cl\cdot {}^1/_2H_2O$. The $[Fe(CN)_5(CO)]^{3-}$ ion forms, e.g., green $Co_3[Fe(CN)_5(CO)]_2\cdot 6H_2O$.

Aqueous $[Fe(CN)_5(H_2O)]^{3-}$ reacts incompletely with $[Co(CN)_6]^{3-}$ to form the dinuclear complex $[(NC)_5Co^{III}CNFe^{II}(CN)_5]^{6-}$, and mixing the $[Fe(CN)_5(H_2O)]^{2-}$ ion with $[Co(CN)_6]^{3-}$ yields $[(NC)_5Co^{III}CNFe^{III}(CN)_5]^{5-}$. Likewise, letting $[Fe(CN)_5(H_2O)]^{2-}$ react with $[Ru(CN)_6]^{4-}$ produces $[(NC)_5RuCNFe(CN)_5]^{6-}$.

Iron(III) chloride and $[Co(NH_3)_6]^{3+}$ in warm 3-M HCl form orange $[Co(NH_3)_6][FeCl_6]$.

Salts of $[Fe(H_2O)_6]^{2+}$, in crystals and in solution, have a pale-green color. Light acts on $[Fe(CN)_6]^{3-}$ to give $Fe_2O_3\cdot aq$, $[Fe(CN)_5H_2O]^{2-}$ and $[Fe(CN)_5H_2O]^{3-}$, also Prussian Blue, i.e., $Fe_4[Fe(CN)_6]_3\cdot aq$, etc. Heating $[Fe(CN)_5H_2O]^{3-}$ without air forms $[Fe(CN)_6]^{4-}$ and $[Fe(H_2O)_6]^{2+}$.

Light without air slowly breaks $[Fe(CN)_6]^{4-}$ down to HCN, CN^- and Fe^{II}, and raises the pH; darkness reverses this somewhat. This slow photochemical, thermal or acidic aquation of $[Fe(CN)_6]^{4-}$ is catalyzed by, e.g., Pt^{IV}, warm Ag^+, Au^{III}, Hg_2^{2+} and Hg^{2+}, with at least Ag^+ and Hg^{2+} forming complexed intermediates:

$$[Fe(CN)_6]^{4-} + \gamma + H_3O^+ + H_2O \leftrightarrows [Fe(CN)_5(H_2O)]^{3-} + HCN + H_2O$$

Light and $[Fe(CN)_6]^{4-}$ also form $[Fe_2(CN)_{10}]^{6-}$ or $[Fe_2(CN)_{10}(H_2O)]^{6-}$ and:

$$(H^+)_j[Fe(CN)_6]^{4-} + (H^+)_k[Fe(CN)_5(H_2O)]^{3-} + i\,H_3O^+ + (j+k)\,H_2O \leftrightarrows$$

$$(H^+)_i[Fe_2(CN)_{11}]^{7-} + (j+k)\,H_3O^+ + (i+1)\,H_2O$$

Aqueous $[Fe(CN)_6]^{3-}$ is fairly stable in the dark but not in the light.

8.1.5 Reactions Involving the Prussian Blues

The acid $H_4[Fe(CN)_6]$ absorbs O_2 from the air, especially when warmed, releasing HCN and depositing one kind of Prussian Blue:

$$7\,H_4[Fe(CN)_6] + O_2 \rightarrow Fe_4[Fe(CN)_6]_3\cdot aq\downarrow + 24\,HCN\uparrow + 2\,H_2O$$

Oxygen oxidizes $H[Fe(CN)_6]^{3-}$ and $H_2[Fe(CN)_6]^{2-}$ to $[Fe(CN)_6]^{3-}$ or to a Prussian Blue (see below), quite slowly and only with acids. Often-faster oxidants include H_2O_2, O_3, $[S_2O_8]^{2-}$, Cl_2 and Br_2. Thus:

$$4\,H[Fe(CN)_6]^{3-} + {}^1/_2\,O_2 + 2\,K^+ + 10\,H_3O^+ \rightarrow$$

$$2\,KFe[Fe(CN)_6]\cdot H_2O\downarrow + 12\,HCN + 9\,H_2O$$

We note that HCN is not much released from a cold solution.

Iron(III) plus $[Fe(CN)_6]^{4-}$ give (1) Prussian Blue, and Fe^{2+} plus $[Fe(CN)_6]^{3-}$ form (2) Turnbull's Blue, but these semiconducting products have identical structures, with some variability of composition. One could expect (1) $Fe^{III}{}_4[Fe^{II}(CN)_6]_3$:

$$4\ FeCl_2{}^+ + 3\ [Fe^{II}(CN)_6]^{4-} \rightarrow Fe^{III}{}_4[Fe^{II}(CN)_6]_3 \cdot aq\downarrow + 8\ Cl^-$$

(where the brackets show the complex with direct Fe–C bonds, while the other Fe atoms are coordinated to the N and in some cases to H_2O), and (2) $Fe^{II}{}_3[Fe^{III}(CN)_6]_2 \cdot aq$, and (1), "Insoluble Prussian Blue", does arise with excess Fe^{III}, but with equimolar reagents we get the same colloidal form of "Soluble [peptizable] Prussian Blue" in each case, e.g.:

$$FeCl_2{}^+ + [Fe^{II}(CN)_6]^{4-} + K^+ \rightarrow KFe^{III}[Fe^{II}(CN)_6] \cdot aq\downarrow + 2\ Cl^-$$

$$Fe^{2+} + [Fe^{III}(CN)_6]^{3-} + K^+ \rightarrow KFe^{III}[Fe^{II}(CN)_6] \cdot aq\downarrow$$

An excess of Fe^{2+} over $[Fe^{III}(CN)_6]^{3-}$ yields, not $Fe^{II}{}_3[Fe^{III}(CN)_6]_2$, but:

$$6\ Fe^{2+} + 4\ [Fe^{III}(CN)_6]^{3-} + 14\ H_2O \rightarrow$$

$$Fe^{III}{}_4[Fe^{II}(CN)_6]_3 \cdot 14H_2O\downarrow + Fe^{II}{}_2[Fe^{II}(CN)_6]\downarrow$$

In each case we can treat the Fe^{2+} and $[Fe^{III}(CN)_6]^{3-}$ as first becoming Fe^{3+} and $[Fe^{II}(CN)_6]^{4-}$ because of the extra stability of the $[Fe^{II}(CN)_6]^{4-}$ and Fe^{3+} electron structures. Many other **d**-block cations M^{2+} (M = Mn, Co, Ni, Cu, Zn, Cd) are not so easily oxidized, however, and they do precipitate $M_3[Fe^{III}(CN)_6]_2$, similarly $Ag_3[Fe^{III}(CN)_6]$ and $Bi[Fe^{III}(CN)_6]$.

Analogues of $Fe^{III}{}_4[Fe^{II}(CN)_6]_3 \cdot \sim 14H_2O$, the "Insoluble Prussian Blue", contain $[(Ru^{II},Os^{II})(CN)_6]^{4-}$; one also finds numerous related complex compounds such as $(Mn,Fe,Co,Ni,Cd)_3[(Cr^{III},Co^{III})(CN)_6]_2 \cdot aq$, $Cu_2[(Fe^{II},Ru^{II},Os^{II})(CN)_6] \cdot aq$, $Cd[Pd^{IV}(CN)_6]$ etc.

The intense color of the Prussian Blues, however, is from rapid electron exchange between Fe^{II} and Fe^{III}, so that the oxidation states cannot be distinguished well, and we may well write the formulas simply as $Fe_4[Fe(CN)_6]_3 \cdot aq$ and $KFe[Fe(CN)_6] \cdot aq$.

The products are insoluble in acids, but transposed by alkalis:

$$Fe_4[Fe(CN)_6]_3 \cdot aq + 12\ OH^- \rightarrow$$

$$2\ Fe_2O_3 \cdot aq\downarrow + 3\ [Fe(CN)_6]^{4-} + 6\ H_2O$$

$$2\ KFe[Fe(CN)_6] + 6\ OH^- \rightarrow$$

$$Fe_2O_3 \cdot aq\downarrow + 2\ [Fe(CN)_6]^{4-} + 2\ K^+ + 3\ H_2O$$

Iron(2+) precipitates $K_4[Fe(CN)_6]$ as white $K_2Fe[Fe(CN)_6]$ or $Fe_2[Fe(CN)_6]$ (with no colorful electron transfer between the Fe^{II}s):

$$Fe^{2+} + 2\,K^+ + [Fe(CN)_6]^{4-} \rightarrow K_2Fe[Fe(CN)_6]\downarrow$$

insoluble in dilute acids, transposed by the alkalis:

$$K_2Fe[Fe(CN)_6] + 2\,OH^- \rightarrow Fe(OH)_2\downarrow + [Fe(CN)_6]^{4-} + 2\,K^+$$

The original precipitates are converted into Prussian Blue gradually by exposure to the air, or immediately by dissolved oxidants:

$$2\,K_2Fe[Fe(CN)_6] + {}^1\!/_2\,O_2 + CO_2 \rightarrow$$

$$2\,KFe[Fe(CN)_6]\cdot aq\downarrow + 2\,K^+ + CO_3^{2-}$$

If $[Fe(CN)_6]^{4-}$ is added in large excess to Fe^{III}, the precipitate is partly dissolved or peptized, forming a blue liquid. In this way 0.04-μM Fe^{III} may be detected. Iron(III) and $[Fe(CN)_6]^{3-}$ give no precipitate, but the solution is colored brown (with fresh reagent) or green (with an old solution). Nearly black $Fe_4[Fe(CN)_6]_3\cdot 14H_2O$, free of Alk^+, arises from $[FeCl_4]^-$ and $[H_4Fe(CN)_6]$ in 10-M HCl over some weeks.

The $[Fe(CN)_6]^{3-}$ ion is useful for the detection of Fe^{2+} in the presence of Fe^{III}. The solution should be diluted enough to permit the detection of the "Prussian-blue" precipitate in the presence of the dark-colored liquid due to any Fe^{III} present. If no precipitate is obtained (indicating the absence of Fe^{2+}), a drop of $SnCl_2$ or other strong reductant constitutes a sensitive test for Fe^{III} (now reduced to Fe^{2+}) and confirms the negative result for original Fe^{2+}.

Prussian Blue, $Fe_4[Fe(CN)_6]_3\cdot aq$, and KOH give rise to $Fe_2O_3\cdot aq$ and yellow $K_4[Fe(CN)_6]\cdot 3H_2O$.

8.2 Ruthenium, $_{44}Ru$; Osmium, $_{76}Os$ and Hassium, $_{108}Hs$

Oxidation numbers in simple species: (−II), (0), (II), (III), (IV), (V), (VI), (VII) and (VIII), as in $[M(CO)_4]^{2-}$, $[M(CO)_5]$, Ru^{2+} and $[Os(CN)_6]^{4-}$, Ru^{3+} and $[OsCl_6]^{3-}$, $MO_2\cdot aq$, $[MCl_6]^-$, RuO_4^{2-} and $[OsO_2(OH)_4]^{2-}$, MO_4^- and the volatile, explosive RuO_4 and volatile OsO_4.

The stable oxidation states for Hs in water, calculated relativistically: (III), (IV), (VI) and (VIII), especially (III) and (IV). Experiments show stability for HsO_4, as predicted.

8.2.1 Reagents Derived from Hydrogen and Oxygen

Dihydrogen. Only *powdered* Os absorbs 1600 volumes H_2 at room *T*.

Hydrogen reduces $Ru_2O_3\cdot aq$ imperfectly at ambient *T*, but H_2 and $[RuCl_6]^{3-}$, for example, first form $[RuCl_5H]^{3-}$, which, with an oxidant, may revert to $[RuCl_6]^{3-}$, thus catalyzing a reduction:

$$[RuCl_6]^{3-} + H_2 + H_2O \rightarrow [RuCl_5H]^{3-} + Cl^- + H_3O^+$$

$$[RuCl_5H]^{3-} + 2\ FeCl_3 + H_2O \rightarrow [RuCl_6]^{3-} + 2\ Fe^{2+} + 5\ Cl^- + H_3O^+$$

$$2\ FeCl_3 + H_2 + 2\ H_2O \rightarrow 2\ Fe^{2+} + 6\ Cl^- + 2\ H_3O^+$$

Platinum black and H_2 reduce airless red-brown $[RuCl_5(H_2O)]^{2-}$ in 1-dM CF_3SO_3H in hours to blue $[Ru(H_2O)_6]^{2+}$. Longer treatment gives Ru.

Aqueous $[RuCl_5CO]^{2-}$ and H_2 (80 °C, 5 h) form $[RuCl_4CO(H_2O)]^{2-}$.

Hydrogen reduces $[RuCl_6]^{2-}$ to $[RuCl_6]^{3-}$–then see above–and some Ru ions (such as RuO_4^{2-}, slowly) to Ru; it precipitates RuO_4 first as $RuO_2\cdot aq$, then as Ru. It reduces $OsO_2\cdot aq$ to Os at ambient *T*.

Water. In H_2O, $[Ru(H_2O)(NH_3)_5]^{2+}$ and $[Ru(H_2O)(NH_3)_5]^{3+}$ are a bit more acidic than H_2O and CH_3CO_2H, respectively.

The expected $[Os(H_2O)_6]^{n+}$, $n = 2$ to 4, seem to be unknown.

Aqueous $[Os(NH_3)_6]^{3+}$ is less acidic than H_2O, i.e., $[OsNH_2(NH_3)_5]^{2+}$ is a strong base, but $[Os(NH_3)_6]^{4+}$, a strong acid, is more acidic than H_3O^+. The known di-, tri- and some other halo Os^{IV} ammines are intermediate. The *cis*-dihalo Os^{IV} ions are more acidic and oxidizing than the *trans*.

Of the oxides of osmium, OsO, Os_2O_3, OsO_2 and OsO_4, the first three are basic in water, the last one nearly neutral or slightly acidic.

Chloro-Ru^{II} species in acid, but not $[Ru(H_2O)_n(NH_3)_{6-n}]^{2+}$, reduce H_2O, but various $[Ru^{II}Q(NH_3)_5]^{n+}$ also quickly become $[Ru(H_2O)(NH_3)_5]^{2+}$.

The salts of Os^{2+} and Os^{III} appear to be unstable in aqueous solution.

The aquation of $[RuCl(NH_3)_5]^+$, i.e., to $[Ru(H_2O)(NH_3)_5]^{2+}$, is about 10^6 times as fast as that of $[RuCl(NH_3)_5]^{2+}$. Other saturated ligands resemble Cl^- on this point. The aquation of $[Ru(Cl,Br,I)(NH_3)_5]^{2+}$ is faster in base than in acid, and is promoted by Hg^{2+}. With X = Cl, Br or I, the hydrolysis of $[cis\text{-}RuX_2(NH_3)_4]^+$ retains the *cis* configuration.

Aqueous $RuCl_3$ is easily hydrolyzed to $[RuCl_n(H_2O)_{6-n}]^{(3-n)+}$, and then polymerized. Heating pushes it farther to $Ru_2O_3\cdot aq$. Aquation of $[mer\text{-}RuCl_3(H_2O)_3]$ is faster than that of $[fac\text{-}RuCl_3(H_2O)_3]$, and it thus generates $[cis\text{-}RuCl_2(H_2O)_4]^+$. Heating blue Ru^{II} chlorides under N_2 gives the *trans* Ru^{III}; ion exchange separates these. The aquation of $[RuCl(H_2O)_5]^{2+}$ takes many months, the $[RuCl_6]^{3-}$ only seconds.

Osmium disulfide is slightly soluble, and $OsCl_4$ is slightly soluble and hydrolyzed to OsO_2; OsS_4 is insoluble in water.

At pH 1 and more for *trans* isomers, or pH 2 or above but quickly for *cis*, $[Os^{IV}X_2(NH_3)_4]^{2+}$ dismutates–note the *trans* result–e.g.:

$$3\ [OsCl_2NH_2(NH_3)_3]^+ + 3\ H_2O \rightarrow$$

$$2\ [OsCl_2(NH_3)_4]^+ + [\textit{trans}\text{-}OsO_2(NH_3)_4]^{2+} + H_3O^+ + 2\ Cl^-$$

similarly for any isomer of $[OsX_{3+n}(NH_3)_{3-n}]^{(n-1)-}$, but the Os^{VI} products from $[OsX_{1-n}(NH_3)_{5+n}]^{(3+n)+}$ get an Os≡N moiety (quickly losing three H^+).

In H_2O, OsF_6 (from Os and F_2) dismutates to $[OsF_6]^{2-}$ and OsO_4.

Water reduces RuO_4^- to Ru^{VI}, more slowly than in base (below):

$$2\ RuO_4^- + H_2O \rightarrow 2\ HRuO_4^- + {}^1/_2\ O_2\uparrow$$

Ruthenium tetraoxide is slightly soluble and weakly acidic in water. Osmium tetraoxide dissolves up to 2 dM at 15 °C.

Oxonium. Metallic Ru is inert to cold single acids.

Acid and $[Ru(NH_3)_6]^{2+}$ give $[Ru(NH_3)_5(H_2O)]^{2+}$, faster at high $c(H_3O^+)$, replacing the second and third NH_3 more slowly. However, 4-M HCl at 0°C for 30 min under N_2 forms a mixed or fractional blue Ru^{II}-Ru^{III}:

$$2\ [Ru(NH_3)_6]^{2+} + 7\ H_3O^+ + 3\ Cl^- \rightarrow$$

$$[\{Ru(NH_3)_3\}_2(\mu\text{-}Cl)_3]^{2+} + 6\ NH_4^+ + {}^1/_2\ H_2\uparrow + 7\ H_2O$$

precipitable by $[MCl_4]^{2-}$, M = Zn, Cd or Hg, or by 2 $[SnCl_3]^-$. Titrating this consumes 1/2 equiv of Ce^{IV} per Ru; the resulting Ru^{III} (or from exposure to O_2), on standing, deposits red *fac*-$[RuCl_3(NH_3)_3]$.

Acids dissolve $Ru_2O_3 \cdot aq$ as brown Ru^{III}. They do not attack RuO_2, but $RuO_2 \cdot aq$ and H_3O^+ produce $Ru_4O_6^{4+}$ etc.

Ammines of Os^{IV} are slightly more acidic than similar ones of Ir^{IV} but 10^5 to 10^6 times as acidic as Pt^{IV}, thus requiring high $c(H_3O^+)$ to persist.

Nitric acid and even CO_2 dismutate RuO_4^{2-}:

$$2\ RuO_4^{2-} + 4\ CO_2 + 4\ H_2O \rightarrow RuO_4 + Ru(OH)_4 \cdot aq\downarrow + 4\ HCO_3^-$$

The dismutation of RuO_4^- in acid is favored; $K = 2.5 \times 10^{27}\ M^{-5}$:

$$4\ RuO_4^- + 4\ H_3O^+ \leftrightarrows 3\ RuO_4 + RuO_2 \cdot aq\downarrow + 6\ H_2O$$

This RuO_4 melts at ~ 25°C, is a very strong oxidant, poisonous and can explode with reductants or at high *T*. It oxidizes or catalyzes the oxidation of various organic substances.

Osmium is not attacked by non-oxidizing acids.

The ionization quotients Q for OsO_4 are 1×10^{-12}M and ~ 3×10^{-15}M:

$$OsO_4 \cdot aq + 3\ H_2O \leftrightarrows [OsO_3(OH)_3]^- + H_3O^+$$

$$[OsO_3(OH)_3]^- + H_2O \leftrightarrows [OsO_4(OH)_2]^{2-} + H_3O^+$$

Hydroxide. Aqueous OH^- and $RuCl_3$ precipitate a dark-yellow or black $Ru_2O_3 \cdot aq$, contaminated by the alkali, soluble when fresh in acids, insoluble in excess OH^-; $[Os(OH)_6]^{3-}$ and $[Os(OH)_6]^{2-}$ seem unknown.

Base and $[Ru(NH_3)_6]^{2+}$ produce $[Ru(NH_3)_5(H_2O)]^{2+}$.

The OH^- ion turns $[Ru(NH_3)_6]^{3+}$ yellow with $[RuNH_2(NH_3)_5]^{2+}$.

The $[Ru(NO)(NH_3)_5]^{3+}$ ion and OH^- under various conditions give $[Ru(NH_2)(NO)(NH_3)_4]^{2+}$, $[Ru(NO_2)(NH_3)_5]^+$, $[Ru(OH)(NO)(NH_3)_4]^{2+}$ or even $[Ru(NH_3)_5N_2]^{2+}$.

Aqueous OH^- precipitates $RuO_2 \cdot aq$ or $Ru(OH)_4 \cdot aq$ from $RuCl_4$; this dissolves in excess OH^- as a yellow anion. Alkaline oxidation of most $Ru^{<VI}$ forms orange RuO_4^{2-}. From $[OsX_6]^{2-}$, OH^- precipitates brown $Os(OH)_4$, which retains some of the alkali firmly.

Distilled or swept by a non-reactive gas from oxidized solutions, RuO_4 is readily soluble in and may be collected in cold, dilute OH^-, which reduces it in stages to "perruthenate" and "ruthenate":

$$2\ RuO_4 + 2\ OH^- \rightarrow 2\ RuO_4^-\ \text{(yellow-green)} + {}^1\!/_2\ O_2\uparrow + H_2O$$

$$2\ RuO_4^- + 2\ OH^- \rightarrow 2\ RuO_4^{2-}\ \text{(orange to red)} + {}^1\!/_2\ O_2\uparrow + H_2O$$

The RuO_4^{2-} is a two-electron oxidant for various organic compounds. Dilute RuO_4^{2-} becomes greenish by dismutation although base favors the reverse reaction; $K = 6 \times 10^{-9}\ \text{M}^3$:

$$3\ RuO_4^{2-} + 2\ H_2O \leftrightarrows 2\ RuO_4^- + RuO_2 \cdot aq\downarrow + 4\ OH^-$$

Various ratios of AlkOH and OsO_4 give $Na_2[OsO_4(OH)_2] \cdot 2H_2O$, $K_2[OsO_4(OH)_2]$, $M_2[OsO_4(OH)_2]$, yellow-orange $M[OsO_4OH]$ and yellow $M[(OsO_4)_2(OH)]$; M = Rb or Cs. The $Alk_2[cis\text{-}OsO_4(OH)_2] \cdot nH_2O$, deep-red, and $(Sr,Ba)[OsO_4(OH)_2] \cdot nH_2O$, red-brown, are the "perosmates".

Peroxide. Metallic Ru or Os or compounds plus HO_2^- yield (little) RuO_4 or OsO_4. Fusing Na_2O_2 and Ru gives RuO_4^{2-} (separation from Ir in H_2O).

Hydrogen peroxide oxidizes $Ru_2O_3 \cdot aq$ to $Ru(OH)_4 \cdot aq$, $K_3[Ru(C_2O_4)_3]$ to black $K_2[Ru(C_2O_4)_3]$, and powdered Os to (toxic) OsO_4.

The $OH^\bullet$ radical oxidizes $[Ru(NH_3)_6]^{3+}$ to $[Ru(NH_3)_6]^{4+}$, which quickly becomes $[Ru(NH_3)_6]^{3+}$ and $[Ru(NH_3)_6]^{5+}$.

Solid Na_2O_2 and RuO_2 form RuO_4^{2-} and some higher states in water.

Aqueous OsO_4 catalyzes the decomposition of H_2O_2, and we have no inorganic peroxo Os complexes.

Dioxygen. Oxygen and HCl attack Ru at ambient *T*.

Air with powdered Os quickly yields OsO_4, and even massive Os slowly produces an odor of OsO_4.

Air oxidizes $[Ru(H_2O)_6]^{2+}$ to yellow $[Ru(H_2O)_6]^{3+}$, and $Ru_2O_3 \cdot aq$ to ~$RuO_2 \cdot aq$. Also, $[Ru(NH_3)_6]^{3+}$ at pH 13 yields $[RuNO(NH_3)_5]^{3+}$.

8.2.2 Reagents Derived from the Other 2nd-Period Non-Metals, Boron through Fluorine

Boron species. Aqueous RuO_4^{2-}, RuO_4^- and RuO_4, plus $[BH_4]^-$, form dark-blue $RuO_2 \cdot aq$; $Ru^{>II}$ and $[BH_4]^-$ with $[BF_4]^-$ yield blue Ru^{2+}.

Carbon oxide species. Aqueous $[Ru(CO)(NH_3)_5]^{2+}$ results from treating $[Ru(NH_3)_5(H_2O)]^{2+}$ with CO and H_2SO_4 for 48 h, or with CO_2 and Zn_{Hg}, or from $[RuCl(NH_3)_5]^{2+}$, CO, Zn_{Hg} and H_3O^+.

Treating $RuCl_3$ with CO and HCl at 80°C for 16 h forms $[RuCl_5CO]^{2-}$, but $OsO_2 \cdot aq$ and CO at room *T* yield Os.

A solution of $OsCl_3$, when treated with an excess of CO_3^{2-}, forms Os_2O_3, black. From Os^{IV}, CO_3^{2-} precipitates $Os(OH)_4 \cdot aq$, which holds some alkali tenaciously.

Cyanide species. Aqueous CN^- and Ru^{2+} form a gray-green product, possibly $Ru(CN)_2$. This with KCN, and heating with the higher oxidation states plus cooling, finally give a colorless, diamagnetic $K_4[Ru(CN)_6] \cdot 3H_2O$. Treatment with HCl and extraction with ether yield the acid $H_4[Ru(CN)_6]$.

Apparently $[Ru(CN)_6]^{4-}$ becomes yellow $[Ru(CN)_6]^{3-}$ by the action of acidified H_2O_2, neutral O_3, or acidified Ce^{IV} or BiO_3^-. The $[Ru(CN)_6]^{3-}$ precipitates $\mathbf{d}^{2+}$ and Ag^+; cf. $[Fe(CN)_6]^{3-}$.

Chlorine causes several color changes with $[Ru(CN)_6]^{4-}$; warm H_2SO_4 then precipitates dark-green $Ru(CN)_3 \cdot 5H_2O$ which, with concentrated NH_3, becomes also-insoluble $Ru(CN)_3 \cdot 2NH_3 \cdot H_2O$. We also have:

$$[Ru(CN)_6]^{4-} + Br_2 + H_2O \rightarrow [Ru(CN)_5(H_2O)]^{3-} + BrCN + Br^-$$

Aqueous HCN and either $[Ru(NH_3)_5(H_2O)]^{2+}$, or $[RuCl(NH_3)_5]^{2+}$ and Zn_{Hg}, form $[Ru(HCN)(NH_3)_5]^{2+}$. Dissolved potassium cyanide and $[\{RuCl_4(H_2O)\}_2(\mu\text{-}N)]^{3-}$ give a quite stable $K_5[\{Ru(CN)_5\}_2(\mu\text{-}N)] \cdot 3H_2O$.

Heating RuO_4^{2-} or $[OsO_2(OH)_4]^{2-}$ with CN^-, or boiling "$RuCl_3 \cdot aq$" with excess CN^-, forms colorless $[M(CN)_6]^{4-}$. These are like $[Fe(CN)_6]^{4-}$ in precipitating, for example, $\mathbf{d}^{II}{}_2[M(CN)_6] \cdot nH_2O$ and $\mathbf{d}^{III}{}_4[M(CN)_6]_3$. They can likewise be converted to the tetrabasic acids.

Refluxing $[RuCl(NH_3)_5]^{2+}$ with NCO^-, NCS^- or $NCSe^-$ ("NCQ^-") forms $[Ru(NCQ)(NH_3)_5]^{2+}$. Excess NCS^- yields $[Ru(NCS)_2(NH_3)_4]^+$ etc., possibly including $[Ru(NCS\text{-}\kappa N)_2(NCS\text{-}\kappa S)_2(NH_3)_2]^-$.

The ions $[Ru(NCS\text{-}\kappa N)_n(NCS\text{-}\kappa S)_{6\text{-}n}]^{3-}$ are formed from NCS^- and "$RuCl_3 \cdot aq$" or $[RuCl_6]^{3-}$; $[RuCl_5(NO)]^{2-}$ gives $[Ru(NCS)_5(NO)]^{2-}$. All the isomers $[Os(NCS\text{-}\kappa N)_n(NCS\text{-}\kappa S)_{6\text{-}n}]^{3-}$ except for $n = 0$ are also known; Ru^{III} and Os^{III} thus lie on the "hard-soft" boundary. Refluxing $[OsCl_6]^{2-}$ with NCS^- forms those *cis* and *fac* complexes; 60 °C gives the *trans* and *mer* types. A short treatment favors the κS, a long one the κN isomers.

Distilling Os metal or compounds with HNO_3 forms OsO_4 as a vapor.

Nitric acid, KCN and $K_2[trans\text{-}OsO_2(OH)_4]$ yield red, slightly soluble $K_2[Os(CN)_5(NO)] \cdot 2H_2O$, but $[trans\text{-}Os(CN)_4(NO)(H_2O)]^-$, among others, arises with additional HNO_3. The slow action of KCN on OsO_4 yields $K_2[trans\text{-}OsO_2(CN)_4]$, reducible by excess CN^- but stable to hot H_3O^+; at least Cu^{2+} and Ag^+ precipitate it.

Slowly adding $(CN)_2$ to $[Ru(NH_3)_5(H_2O)]^{2+}$ at pH 4 under Ar, plus Br^-, yield $[\{Ru(NH_3)_5NC-\}_2]Br_4$. An anode gives $[\{Ru(NH_3)_5NC-\}_2]^{5+}$ with the Ru^{II}-Ru^{III} charges delocalized even over the N≡C–C≡N distance.

Some "simple" organic species. Ethene rapidly reduces RuO_4^{2-} to Ru.

Ethanol reduces RuO_4^{2-} to a black oxide or hydroxide and then to finely divided Ru. In HCl it reduces RuO_4 to "$RuCl_3 \cdot aq$".

Ethanol reduces $[OsO_4(OH)_2]^{2-}$, and with KOH precipitates it as purple $K_2[trans\text{-}OsO_2(OH)_4]$, or with NH_4Cl as $[OsO_2(NH_3)_4]Cl_2$; with excess KCl and OsO_4 it forms $K_2[OsCl_6]$; all are used in further studies.

Methanal, CH_2O, converts $[RuCl_5NO]^{2-}$ to $[\{RuCl_4(H_2O)\}_2(\mu\text{-}N)]^{3-}$. This $[Ru_2NCl_8(H_2O)_2]^{3-}$ and CN^- become $[Ru_2N(CN)_{10}]^{5-}$, quite stable, probably with Ru=N=Ru bonds and a noble-gas Ru structure, with Ru–CN bonds between single and double.

Refluxing HCO_2H and HCl with "$RuCl_3 \cdot aq$" for up to 30 h and adding CsCl at various times yields, successively, red $Cs_2[RuCl_5(CO)]$, green $Cs_2[trans\text{-}RuCl_4(CO)(H_2O)]$, orange $Cs_2[cis\text{-}RuCl_4(CO)_2]$ and yellow $Cs[fac\text{-}RuCl_3(CO)_3]$. The bromides and iodides give similar salts.

Zinc, HCO_2H and $[OsCl(NH_3)_5]^{2+}$ form $[Os(CO)(NH_3)_5]^{2+}$, oxidizable by $[IrCl_6]^{2-}$ to $[Os(CO)(NH_3)_5]^{3+}$ ($pK_a \sim 2.5$). The $[Os(CO)(NH_3)_5]^{2+}$, with HNO_2 and HCl, becomes $[cis\text{-}Os(CO)(NH_3)_4(N_2)]^{2+}$, but with Ce^{IV}, MnO_4^-, $S_2O_8^{2-}$ or an anode it yields $[\{Os(CO)(NH_3)_4\}_2(\mu\text{-}N_2)]^{4+}$.

Oxalic acid and either $[RuOH(NH_3)_5]^{2+}$ or $[RuCl(NH_3)_5]^{2+}$ form $[Ru(C_2O_4)(NH_3)_4]^+$, reducible by $[Ru(NH_3)_6]^{2+}$ or $[Ru(NH_3)_5(H_2O)]^{2+}$.

Aqueous $C_2O_4^{2-}$ and $[RuCl_5(H_2O)]^{2-}$ yield $[Ru(C_2O_4)_3]^{3-}$.

Excess $NH_4HC_2O_4$ converts $RuO_2 \cdot aq$ to $(NH_4)_3[Ru(C_2O_4)_3] \cdot {}^3/_2H_2O$.

Ice-cold $H_2C_2O_4$, $Cs_2C_2O_4$ and RuO_4 become $Cs_2[trans\text{-}RuO_2(C_2O_4)_2]$. Aqueous $H_2C_2O_4$ and OsO_4 similarly form $[trans\text{-}OsO_2(C_2O_4)_2]^{2-}$.

Reduced nitrogen. Refluxing $[Ru(NH_3)_5N_2]^{2+}$ with concentrated NH_3 yields $[Ru(NH_3)_6]^{2+}$.

Ammonia precipitates from $RuCl_3$ solutions a dark-yellow hydroxide, $Ru_2O_3\cdot aq$, soluble in excess NH_3, giving a greenish-brown liquid.

Ammonia and some Ru^{IV} chlorides give $[Ru_3O_2(NH_3)_{14}]^{n+}$, with $n = 6$ or 7 for "Ruthenium Red" or "Ruthenium Brown" and with so-called fractional oxidation states for all three Ru, of 10/3 and 11/3 respectively, in linear $[trans\text{-}Ru(NH_3)_4\{Ru(NH_3)_5(\mu\text{-}O)\}_2]^{n+}$.

Ammonia and $[Os(Cl,Br)_6]^{2-}$ at 0°C form $[\{OsX(NH_3)_4\}_2(\mu\text{-}N)]^{3+}$. Refluxing with Q^- (NCS^-, N_3^-, Cl^-, Br^-, I^-) gives $[\{OsQ(NH_3)_4\}_2(\mu\text{-}N)]^{3+}$.

Concentrated NH_3 plus RuO_4^{2-} give $[RuO_2(OH)_2(NH_3)_2]$, at times mistaken for $(NH_4)_2RuO_4$ with the same empirical formula.

Aqueous $[OsO_2(OH)_4]^{2-}$ and NH_4Cl yield pale-yellow, slightly soluble $[OsO_2(NH_3)_4]Cl_2$.

Concentrated NH_3 and concentrated RuO_4 (changing the color from yellow to gray-brown) form $(NH_4)_2RuO_5$ by evaporation.

Osmium(III) ammines in acidified, and Os^{IV} ammines in strongly acidic solutions, are extremely inert to substitution.

From Os^{IV}, NH_3 gives $Os(OH)_4\cdot aq$, which strongly retains some alkali.

Concentrated KOH plus $[OsO_4(OH)_2^{2-}]$ and NH_3 produce a yellow, non-basic, slightly soluble "osmiamate":

$$[OsO_4(OH)_2]^{2-} + NH_3 + K^+ \rightarrow K[OsO_3N]\downarrow + OH^- + 2\ H_2O$$

This is not hydrolyzed in water, but a bit sensitive to light, reducible by HCN or $H_2C_2O_4$ to, e.g., $K[trans\text{-}Os(C_2O_4)_2N(H_2O)]$:

$$K[OsO_3N] + 6\ HCN \rightarrow K[Os(CN)_4N(H_2O)]\downarrow + (CN)_2\uparrow + 2\ H_2O$$

Hot, concentrated OH^- does not release NH_3 from $K[OsO_3N]$. Concentrated HCl or HBr, however, reduces it to $K_2[OsNX_5]$, both purple and soluble, and H_2O slowly replaces the *trans*-X:

$$K[OsO_3N] + 7\ X^- + 6\ H_3O^+ + K^+ \rightarrow K_2[OsX_5N]\downarrow + X_2\uparrow + 9\ H_2O$$

Aqueous OH^- decomposes this without releasing NH_3. Ozone restores the $K[OsO_3N]$, and $[SnCl_3]^-$ and HCl yield $K_2[OsCl_5NH_2]$.

Treating "$RuCl_3\cdot aq$", $[RuCl_6]^{2-}$, $[RuCl_5(H_2O)]^{2-}$ or $[RuCl(NH_3)_5]^{2+}$ with N_2H_4 and NH_4^+ reduces them to the synthetically useful $[Ru(NH_3)_6]^{2+}$, but $N_2H_5^+$ or HNO_2 also oxidizes $[Ru(NH_3)_6]^{2+}$ to colorless $[Ru(NH_3)_6]^{3+}$. In contrast, $[Os(NH_3)_6]^{2+}$, from cathodic e^- and $[Os(NH_3)_6]^{3+}$, is unstable.

Adding neat diazane hydrate to solid $(NH_4)_2OsCl_6$, with refluxing, i.e., starting at 119 °C, results in a very stable dinuclear form, tentatively:

$$8\ (NH_4)_2OsCl_6 + 49\ N_2H_4\cdot H_2O \rightarrow$$

$$4\ [\{Os^{IV}(NH_3)_5\}_2N^{-III}]Cl_5\cdot H_2O\downarrow + 7\ N_2\uparrow + 28\ Cl^- + 28\ N_2H_5^+ + 45\ H_2O$$

Contrarily, adding the osmium salt over 15 minutes to a large excess of $N_2H_4 \cdot H_2O$ avoids large local excesses of the former. Refluxing for 30 hours changes the color from brown to yellow, and intermediate washings with ethanol and ether remove [*cis*-$Os(N_2)_2(NH_3)_4]Cl_2$. Further treatments produce a good yield of $[Os^{II}(NH_3)_5(N_2)]Cl_2$ by way of:

$$4\ (NH_4)_2[OsCl_6] + 27\ N_2H_4 \cdot H_2O \rightarrow$$

$$4\ [Os(NH_3)_5(N_2)]Cl_2\downarrow + N_2\uparrow + 16\ Cl^- + 16\ N_2H_5^+ + 27\ H_2O$$

$$(NH_4)_2[OsCl_6] + 5\ N_2H_4 \cdot H_2O \rightarrow$$

$$[cis\text{-}Os(NH_3)_4(N_2)_2]Cl_2\downarrow + 4\ Cl^- + 4\ NH_4^+ + 5\ H_2O$$

$$4\ [cis\text{-}Os(NH_3)_4(N_2)_2]Cl_2 + 3\ N_2H_4 \cdot H_2O \rightarrow$$

$$4\ [Os(NH_3)_5(N_2)]Cl_2\downarrow + 5\ N_2\uparrow + 3\ H_2O$$

Then Ce^{IV} at 50°C for 10 min yields $[Os(NH_3)_5(N_2)]^{3+}$. Such Os^{III}–N_2 species are more labile than those of Os^{II} but much less than those of Ru^{III}. Base causes dismutation to Os^{II} and Os^{VI}. The [*cis*-$Os(NH_3)_4(N_2)_2]^{2+}$ also arises from HNO_2 and $[Os(NH_3)_5(N_2)]^{2+}$.

The $[Ru(NO)(NH_3)_5]^{3+}$ ion and NH_2OH in basic solutions yield $[Ru(N_2O)(NH_3)_5]^{2+}$. Hydroxylamine reduces $RuCl_4$ to "$RuCl_3 \cdot aq$".

One may prepare $[Os(N_3)_5(NO)]^{2-}$ by stirring 25 mL H_2O with 0.39 mmol OsO_4, 5.7 mmol NH_3OHCl and 5.4 mmol NaN_3 at 60–70 °C for 15 min, adding 5.4 mmol NaN_3 again and heating at 80 °C for 30 min, then cooling the deep reddish-brown solution to ambient *T* and acidifying to pH 5 with 6-M HCl. Adding $[NEt_4]^+$ or hot aqueous $[NBu_4]^+$ or $[PPh_4]^+$ precipitates the solids.

Aqueous HN_3 and $[Ru(NH_3)_5(H_2O)]^{2+}$ form $[Ru(NH_3)_6)]^{3+}$, perhaps via a nitrene; [*trans*-$RuO_2(Cl,Br)_4]^{2-}$ and HCl/HBr plus Cs^+ give $Cs_2[RuNX_5]$.

In dilute CH_3CO_2H, [*trans*-$OsO_2(OH)_4]^{2-}$, N_3^- and Cs^+ give $Cs[OsO_3N]$.

Ice-cold concentrated HCl or HBr, [*trans*-$(Ru,Os)O_2X_4]^{2-}$, excess N_3^- and Cs^+ give diamagnetic $Cs_2[M^{VI}NX_5]$, both purple, plus N_2 and X_2.

Elemental nitrogen and nitrogen-fixation related. The hope to "fix" (convert to compounds) N_2 under mild conditions (but without Nature's enzymes) created interest in reactions such as:

$$[Ru(NH_3)_5(H_2O)]^{2+} + N_2 \rightarrow [Ru(NH_3)_5(N_2)]^{2+} + H_2O$$

$$[RuCl(NH_3)_5]^{3+} + Zn_{Hg} + N_2 \rightarrow [Ru(NH_3)_5(N_2)]^{2+} + ZnCl^+$$

$$[Ru(NH_3)_5(N_2)]^{2+} + [Ru(NH_3)_5(H_2O)]^{2+} \rightarrow$$

$$[Ru(NH_3)_5(\mu\text{-}N_2)Ru(NH_3)_5]^{4+} + H_2O$$

Dinitrogen at 500 kPa (5 atm) and 5 to 10-cM $[Ru(H_2O)_6]^{2+}$ in darkness for 72 h also form $[\{Ru(H_2O)_5\}_2(\mu\text{-}N_2)]^{4+}$, slower than above. This dimer does not yield $N^{\neq 0}$ with strong acids, bases, or oxidants.

The following sources (some poor) of $[Ru(NH_3)_5N_2]^{2+}$ from other reagents are not even first steps toward fixing N_2 but are listed here for comparison: $RuCl_3$ and N_2H_4; $[Ru(NH_3)_6]^{3+}$ and N_2H_4; $[Ru(NH_3)_5(H_2O)]^{3+}$ and N_2H_4 for 1 h; $[Ru(NO)(NH_3)_5]^{3+}$ and N_2H_4; $[RuCl_6]^{2-}$ or $[RuCl_5(H_2O)]^{2-}$ and N_2H_4 for 12 h; $[Ru(NH_3)_5(H_2O)]^{3+}$, N_3^-, NH_3 and H_3O^+; $[cis\text{-}RuCl_2(NH_3)_4]^+$, HN_3 and H_3O^+, then NH_3; $[Ru(NH_3)_6]^{3+}$ and NO at pH 8.45; $[Ru(NH_3)_6]^{2+}$ and Cl_2 at 0 °C; $[Ru(NH_3)_5(N_2O)]^{2+}$ and V^{2+} or Cr^{2+}; $RuCl_3$, Zn and NH_3; and $[RuCl(NH_3)_5]^{3+}$, Zn and N_2O.

The internal dismutation of, e.g., $[(Os^{III}L_5)_2(\mu\text{-}N_2\text{-}N,N')]$ to 2 $Os^{VI}L_5N$ with (undetermined) ligands L would "fix" N_2, but known $Os^{III}\text{–}(\mu\text{-}N_2)$ bonds are unstable. Another unsuccessful example of splitting N_2 has been the Os^{II}, Os^{III} complex $[\{Os(NH_3)_5\}_2(\mu\text{-}N_2)]^{5+}$, made from a treatment of $[Os(NH_3)_5(H_2O)]^{3+}$ and $[Os(NH_3)_5N_2]^{2+}$.

The oxidized ion, $[Ru(NH_3)_5N_2]^{3+}$, quickly aquates to $[RuOH(NH_3)_5]^{2+}$.

Adding N_2O to $[Ru(NH_3)_5(H_2O)]^{2+}$, from $[Ru(NH_3)_5Cl]^{2+}$ plus $H_2(Pt)$, or Cr^{2+} or Zn_{Hg}, equilibrates $[Ru(NH_3)_5(N_2O)]^{2+}$ and $[Ru(NH_3)_5(H_2O)]^{2+}$. Then Fe^{3+} slowly liberates N_2. After some days, $[Ru(NH_3)_5(H_2O)]^{2+}$ and N_2O form $[Ru(NH_3)_5(N_2)]^{2+}$, $[\{Ru(NH_3)_5\}_2\mu\text{-}N_2]^{4+}$ and $[RuCl(NH_3)_5]^{2+}$ (in Cl^-).

The $(Os^{III}XNH_3)^{2+}$ moiety in haloammines, plus HNO_2, give NO and $(Os^{IV}XNH_3)^{3+}$, ionizing to $(Os^{IV}XNH_2)^{2+}$. This and the NO then become $\{Os^{III}X(N_2)\}^{2+}$. Similarly $[Os^{II}(NH_3)_5(N_2)]^{2+}$ and NO_2^- at pH > 3 (the transitory Os^{III} is less acidic than the Os^{IV}) form $[Os^{II}(NH_3)_4(N_2)_2]^{2+}$. The acidities of $[Os(NH_3)_6]^{2+}$ and $[Os(NH_3)_5(N_2)]^{2+}$ are strikingly different, less than that of H_2O, and similar to that of $HPHO_3^-$, in turn. The Os^{III} N_2 ions dismutate quickly in base to the Os^{II} N_2 species and Os^{VI}.

Oxidized nitrogen. Nitrogen(II) oxide converts $[Ru^{II}(H_2O)_6]^{2+}$ or $[Ru^{III}(NH_3)_6]^{3+}$, in turn, to $[Ru(H_2O)_5(NO)]^{3+}$ or $[Ru(NH_3)_5(NO)]^{3+}$, and Cr^{2+} (in the absence of Cl^- for the ammine) may then yield:

$$[Ru(H_2O,NH_3)_5(NO)]^{3+} + 4\ Cr^{2+} + 3\ H_3O^+ + H_2O \rightarrow$$

$$[Ru(H_2O,NH_3)_5(\mu\text{-}NH)Cr(H_2O)_5]^{5+} + 3\ Cr^{3+}$$

Treating "$RuCl_3 \cdot aq$" with NO and HCl for 48 h gives $[RuCl_5(NO)]^{2+}$; then Br^- or I^- can replace the Cl^-.

Nitrogen oxide and an acid can replace NH_3 in Ru ammines much faster than H_2O replaces it, and diamagnetism etc. point to the structure of a product as, e.g., $[Ru^{2+}(NH_3)_5(NO^+)]$:

$$[Ru(NH_3)_6]^{3+} + NO + H_3O^+ \rightarrow [Ru(NH_3)_5(NO)]^{3+} + NH_4^+ + H_2O$$

Above pH 8.3, however, only $[Ru(NH_3)_5(N_2)]^{2+}$ is produced.

Aqueous $[Os(NH_3)_6]^{3+}$ and NO produce $[Os(NH_3)_5(NO)]^{3+}$; then a base gives [*trans*-$Os(NH_2)(NH_3)_4(NO)]^{2+}$ and [*trans*-$Os(OH)(NH_3)_4(NO)]^{2+}$, which, with HX, forms [*trans*-$OsX(NH_3)_4(NO)]^{2+}$. In contrast, NO and $[Os^{IV}X_2(NH_2)(NH_3)_3]^+$ or $[Os^{IV}X_3(NH_2)(NH_3)_2]$ rapidly convert an NH_3 to an N_2 attached to Os^{III}. Quickly if pH $\geq$ 1, very slowly in 6-M HCl, [*cis*- or [*trans*-$Os^{IV}X_2(NH_3)_4]^{2+}$ and NO form $[Os^{III}X_2(N_2)(NH_3)_3]^+$.

Also, HNO_2 and Os^{III} haloammines (with X, X_2 or X_3) form Os^{III}–N_2 species, the essential moieties giving:

$$Os^{4+}NH_2^- + NO \rightarrow Os^{3+}N_2 \text{ (\textit{trans} with } OsXN_2) + H_2O$$

$$Os^{3+}NH_3 + HNO_2 \rightarrow (\text{via } Os^{IV}) \rightarrow Os^{3+}N_2 + 2\ H_2O$$

$$Os^{3+}N_2 + H_2O \rightarrow Os^{3+}(H_2O) + N_2$$

Sometimes an NO displaces an X^-, but [*cis*-$OsX_2(NH_3)_4]^+$ and NO slowly form mainly $[OsX_2(NH_3)_3(NO)]^+$; many further reactions are known.

A weakly alkaline NO_2^- solution (with CO_3^{2-}) turns aqueous $RuCl_3$ yellow orange, containing some [*trans*-$Ru(OH)(NO_2)_4(NO)]^{2-}$; a little "$(NH_4)_2S$" makes it carmine red, and still more precipitates sulfides.

Aqueous $[Ru(NH_3)_6]^{3+}$ and HNO_2 yield $[Ru(NH_3)_5NO]^{3+}$.

The acids HX and HNO_2, plus $[OsX_6]^{2-}$, form $[OsX_5(NO)]^{2-}$ with I^- for X^- being the least stable; hydrolysis gives [*trans*-$OsX_4(H_2O)(NO)]^-$ and then [*trans*-$OsX_4(OH)(NO)]^{2-}$; X is a halogen.

Excess KNO_2 with $[OsCl_6]^{2-}$ forms K_2[*trans*-$Os(NO_2)_4(OH)(NO)]$.

The NO_2^- ion in base reduces RuO_4^- to RuO_4^{2-}:

$$2\ RuO_4^- + NO_2^- + 2\ OH^- \rightarrow 2\ RuO_4^{2-} + NO_3^- + H_2O$$

Depending on conditions, NO_2^- reduces $[OsO_4(OH)_2]^{2-}$ to a reddish $K_2[OsO_2(OH)_4]$ or, slowly, to brown $K_4[\{OsO_2(NO_2)_2\}_2(\mu\text{-}O)_2]\cdot 6H_2O$, a good precursor to Os^{IV} and Os^{VI} species because acids remove NO_2^-:

$$[OsO_4(OH)_2]^{2-} + NO_2^- + 2\ K^+ + H_2O \rightarrow K_2[OsO_2(OH)_4] + NO_3^-$$

Aqua regia (HNO_3 and HCl) dissolves Ru quite slowly.

Nitric acid or NO changes $K_4[Ru(CN)_6]$ to $K_2[Ru(CN)_5(NO)]\cdot 2H_2O$, red-brown. The anion precipitates Co^{2+}, Ni^{2+}, Cu^{2+}, Ag^+ etc. Aqueous S^{2-} produces red, unstable $[Ru(CN)_5(NOS)]^{4-}$, and 4-M KOH precipitates yellow $K_4[Ru(CN)_5(NO)]\cdot 2H_2O$, both reminiscent of the iron salts.

Some ruthenium compounds, if once treated with HNO_3 or NO_2^-, tend to retain one NO ligand very firmly through many later changes, so that Ru nitrosyls are the most numerous ones. However, HNO_3 and Ru^{III} nitrosyls generate a red-brown Ru^{IV}, perhaps $[Ru_4(OH)_{12}]^{4+}$. From this are formed $[Ru_4(OH)_4]^{8+}$ and intermediates.

Used nuclear fuel in HNO_3 has many problematic Ru species, which may be represented as $[Ru(NO)(NO_2)_x(NO_3)_y(OH)_z(H_2O)_{5-x-y-z}{}^{(x+y+z-3)-}]$. Typical aquations in 45-cM HNO_3 and nitrations in 10-M HNO_3 at 0°C take hours or days. One salt is Na_2[*trans*-Ru(NO)(OH)$(NO_2)_4$]·$2H_2O$.

Compact Os is scarcely attacked by acids. The precipitated metal, or finely divided "osmiridium" or other material containing Os, is slowly dissolved by aqua regia, hot concentrated HNO_3, or fuming HNO_3. When distilled from such solutions OsO_4 may be absorbed in OH^-, forming $[OsO_4(OH)_2]^{2-}$. The tetraoxide exists as colorless, glistening needles, melting at 40.5 °C. Its solutions have a penetrating odor, resembling that of chlorine. The fumes are very poisonous and inflame the eyes; H_2S has been recommended as an antidote.

Fluorine species. The complex $K_3[RuF_6]$ (from fusing $RuCl_3$ with KHF_2) dissolves in dilute H_3O^+ to give mainly $[RuF_6]^{3-}$.

Non-aqueous products Alk$[OsF_6]$, plus AlkOH, form $Alk_2[OsF_6]$, white and stable in water.

Ruthenium tetraoxide does not react with HF.

8.2.3 Reagents Derived from the 3rd-to-5th-Period Non-Metals, Silicon through Xenon

Reduced chalcogens. Sulfane (H_2S) and $[Ru(NH_3)_5(H_2O)]^{2+}$ yield pale-yellow $[Ru(NH_3)_5(H_2S)]^{2+}$ reversibly but needing an excess of Eu^{2+} even under Ar to prevent forming (probably) $[Ru(NH_3)_5SH]^{2+}$, orange. Even solid $[Ru(NH_3)_5(H_2S)][BF_4]_2$ releases H_2 but gives no RuS_x. Air yields $[\{Ru(NH_3)_5\}_2(\mu\text{-}S_2)]^{4+}$, green, after an orange intermediate, at pH 1. The $[Ru(NH_3)_5(H_2S)]^{2+}$ ion is a little more acidic than CH_3CO_2H, and $[Ru(NH_3)_5(H_2S)]^{3+}$ is a strong acid. The $[\{Ru(NH_3)_5\}_2(\mu\text{-}S_2)]^{4+}$, found in $[\{Ru(NH_3)_5\}_2(\mu\text{-}S_2)]Cl_4\cdot 2H_2O$, has *trans* Ru–S–S–Ru; the S–S is mainly hyperthio S_2^- (little S_2^{2-}), leaving mixed Ru_2^{5+}. The $[Ru(NH_3)_5SH]^+$ is a weaker reductant than $[Ru(NH_3)_5OH]^+$ but releases H_2 faster.

From "$RuCl_3$·aq", H_2S slowly forms a soluble, reduced "Ruthenium blue", and precipitates a brown or black pyrite-like $Ru^{II}S_2$, insoluble in S^{2-}. The latter is formed at once by "$(NH_4)_2S$" and is difficultly soluble in excess. From that solution, OH^- precipitates a black hydroxide, soluble in acids but not OH^-.

If "$(NH_4)_2S$" is added to Ru nitrito-complexes, a characteristic crimson liquid is obtained. On standing, a brown precipitate appears.

Sulfane (H_2S) reduces OsO_2·aq to Os at ambient *T*.

Alkaline S^{2-} and $[OsCl_6]^{2-}$ may form Os^{IV} sulfido or sulfanido anions.

From OsO_4, H_2S precipitates brown OsS_2 with an inorganic acid, but OsS_4 if neutral. Both compounds are insoluble in "$(NH_4)_2S$".

Thiocyanate forms with $RuCl_3$, after some time in the cold, a red color which, on heating, becomes a beautiful violet and finally black.

Traces of Os may be detected by the distinct blue color obtained when SCN^- is added to an acidic solution of OsO_4. Extraction with ether gives a sensitivity of 5 μM in the original solution.

Ruthenium solutions, heated with thiourea, i.e., $CS(NH_2)_2$, and HCl, become blue in a test sensitive to 30 μM Ru, or 3 μM on extraction with ether. More specifically, Ru^{III} in HCl with excess $CS(NH_2)_2$ at 100°C forms $[Ru\{CS(NH_2)_2\text{-}\kappa S)\}_6]^{3+}$, isolated as the $[HgI_4]^{2-}$ salt.

Thiourea (Tu), KOH and $[Os(CN)_5(NO)]^{2-}$ yield $K_2[Os(CN)_5Tu]$.

An intense red color appears when $[OsCl_6]^{2-}$ is boiled with thiourea, $CS(NH_2)_2$, in the presence of a little HC1. The test will detect 50-μM Os.

Thiourea and OsO_4 become $[OsO_2Tu_4]^{2+}$ and then $[OsTu_6]^{3+}$, isolated with [*trans*-$Cr(NH_3)_2(NCS)_4]^-$ ("Reinecke's salt") or $[Cr(NCS)_6]^{3-}$ in HCl; from H_2SO_4 arises brown $[OsO_2Tu_4]SO_4$.

Selenourea, $CSe(NH_2)_2$, gives a blue-green complex with Ru^{IV} in HCl.

Oxidized chalcogens. Thiosulfate and $[Ru(NH_3)_5(H_2O)]^{3+}$ under Ar forms a red, somewhat unstable, $[Ru(NH_3)_5(S_2O_3)]^+$. However, treating $[Ru(NH_3)_6]Cl_3$ with SSO_3^{2-} or SPO_3^{3-}, and O_2, in darkness, gives high yields of yellow $[Ru(NHSO_3\text{-}\kappa N)(NH_3)_5]Cl$, probably by transferring (to NH_2^-) and then oxidizing an S atom. With concentrated HBr this leads to $[Ru(NH_2SO_3)(NH_3)_5]Br_2$. Sulfite also, more slowly, converts $[Ru(NH_3)_6]^{3+}$ to $[Ru(NHSO_3\text{-}\kappa N)(NH_3)_5]^+$.

Aqueous SO_2 or $[S_2O_5]^{2-}$, and $[RuCl(NH_3)_5]^{2+}$ form $[Ru(HSO_3)_2(NH_3)_4]$.

Complexes of $Ru(NH_3)_4{}^{2+}$ or $Ru(NH_3)_5{}^{2+}$ with SO_2, HSO_3^- or SO_3^{2-} arise from ligand substitution or reducing Ru^{III} by Zn_{Hg} followed by S^{IV} etc.

At 70°C for 2 h, $K_2[S_2O_5]$ and $K_2[OsCl_6]$, after cooling, form light-brown $K_4[Os(SO_3\text{-}\kappa S)_3(H_2O)_3]$. A solution of SO_2, HSO_3^- and $[Os(NH_3)_5(H_2O)]Cl_3$ at 80 °C yields $[Os(NH_3)_5(SO_2)]Cl_2$.

Tan $Na_6[OsO_2(SO_3)_4]\cdot 2H_2O$ arises from $Na_2[OsO_4(OH)_2]$ and SO_2.

Mixed $[OsO_4(OH)_2]^{2-}$, SO_3^{2-} and CO_3^{2-} may give $[Os(SO_3)_6]^{8-}$; and $[OsCl_6]^{2-}$ and $NaHSO_3$ may form $Na_8[Os(SO_3)_6]\cdot 3H_2O$.

Aqueous SO_3^{2-} or SO_2 reduces OsO_4 to Os^{II}. The solution changes through violet to blue, and finally precipitates $OsSO_3$. With NaOH, Na_2SO_3 and $[OsO_4(OH)_2]^{2-}$ yield $Na_6[OsO_2(SO_3\text{-}\kappa S)_4]\cdot 5H_2O$.

Osmium is practically insoluble even in fused $KHSO_4$, but hot, concentrated H_2SO_4 volatilizes powdered Os or $OsO_2\cdot aq$ as OsO_4.

Alkaline $[S_2O_8]^{2-}$ and $RuCl_3$ form $[RuO_4]^{2-}$.

A gas stream sweeps OsO_4 out of acidified $[S_2O_8]^{2-}$ and Os solutions.

Reduced halogens. Ruthenium and even powdered Os are not attacked by HCl without air; with it HCl attacks Os only slightly. Concentrated HCl at 150 °C slowly oxidizes spongy Os to a yellow-green solution.

Both Cl^- and Br^- quickly equilibrate between the ions $[Ru(NH_3)_5(H_2O)]^{2+}$ and $[RuX(NH_3)_5]^+$.

Aqueous $[Ru(H_2O)_6]^{2+}$ after some hours in HCl yields a Ru mirror.

Anhydrous and rather inert $RuCl_3$ can be made active by dissolving in concentrated HCl and evaporating dry at ~100 °C. Aqueous $RuCl_3$ slowly equilibrates all the various $[RuCl_n(H_2O)_{6-n}]^{(n-3)-}$.

Aqueous HCl or HBr, and $K_3[Ru(C_2O_4)_3]$, form $K_2[RuX_5(H_2O)]$. Concentrated HCl and $[RuC_2O_4(NH_3)_4]I$ give [*cis*-$RuCl_2(NH_3)_4[Cl\cdot H_2O$.

Refluxing $[Ru(NH_3)_6]^{3+}$ with 6–12 M HCl for four hours yields $[RuCl(NH_3)_5]^{2+}$; longer heating gives some $[RuCl_2(NH_3)_4]^+$, and HBr or HI forms $[RuX(NH_3)_5]^{2+}$. Refluxing $[Ru(C_2O_4)(NH_3)_4]^+$ with HCl, HBr or HI generates $[RuX_2(NH_3)_4]^+$.

Aqueous HCl, HBr or HI, and $[Os(H_2O)(NH_3)_5]^{3+}$ form $[OsX(NH_3)_5]^{2+}$.

Recrystallizing $[RuCl_5(H_2O)]^{2-}$ salts from 12-M HCl gives salts of $[RuCl_6]^{3-}$. Concentrated HBr, "$RuCl_3\cdot aq$", ethanol and KBr form $K_3[Ru_2Br_9]$. Iodide ion, with hot aqueous $RuCl_3$, precipitates black RuI_3.

The acids HCl and HBr dissolve $RuO_2\cdot aq$ or $OsO_2\cdot aq$ as $[MCl_6]^{2-}$ and $[MBr_6]^{2-}$, but not as $[RuI_6]^{2-}$ from HI, due to reduction.

Water and $[OsCl_6]^{2-}$ equilibrate with $[OsCl_5(H_2O)]^-$ and Cl^-, but also give various other species. The $[OsCl_6]^{2-}$ precipitates Ag^+, Tl^+ etc.

With AlkCl in 5-M HCl (Alk = K, Rb or Cs), either $RuO_2\cdot aq$, Alk_2RuO_4 or RuO_4 produces $Alk_4[Ru_2OCl_{10}]$; Br^- and RuO_4 give $K_4[Ru_2OBr_{10}]$. Excess RbCl or CsCl plus RuO_4 in dilute HCl form Alk_2[*trans*-RuO_2Cl_4]. The bromides yield Cs_2[*trans*-RuO_2Br_4]. Boiling 10-M HCl and $Cs_2[RuO_2Cl_4]$ give $Cs_2[RuCl_6]$, but water forms $RuO_2\cdot aq$ and RuO_4.

Cold, dilute HCl with K_2RuO_4 yields black $K_2[RuCl_6]$ etc., but I^- at ambient T releases $^3/_2$ I_2, pointing to a product Ru^{III}. Water slowly reduces $[RuCl_6]^{2-}$ to $[RuCl_5(H_2O)]^{2-}$, and HCl can separate $[RuO_4]^{2-}$ and $[OsO_4(OH)_2]^{2-}$ (from fusing some mixed metals with Na_2O_2) as volatile OsO_4 and $[RuCl_6]^{3-}$, precipitable as $(NH_3)_3[RuCl_6]$.

Aqueous HCl and $[OsO_2(OH)_4]^{2-}$ form red [*trans*-OsO_2Cl_4]$^{2-}$, an "osmyl" (i.e., OsO_2, usually *trans*) ion. The bromide is also red.

Aqueous HCl, HBr or HI reduces the oxidants RuO_4 and HNO_3 to $[RuX_3(NO)(H_2O)_2]$, good starters to synthesize Ru nitrosyls.

Ruthenium tetraoxide, plus concentrated CsCl (similarly with RbCl) and a little HCl, slowly crystallize $Cs_2[RuO_2Cl_4]$. Treated with warm HCl however, it becomes $RuCl_4\cdot 5H_2O$, $Ru(OH)Cl_3$, $[RuCl_6]^{2-}$ etc. (plus Cl_2). Repeated evaporation can produce pure $RuCl_3\cdot 3H_2O$, a good organic oxidation catalyst, but the commercial "hydrated trichloride", "$RuCl_3\cdot aq$", may contain much of Ru^{IV}, hydrolysates and polymers.

Evaporating RuO_4, HCl, and C_2H_5OH yields the acid of [*cis*-$RuCl_4(H_2O)_2$]$^-$. Ethanol, HCl or HBr, plus RuO_4, plus K^+, Rb^+, Cs^+ or NH_4^+, and Cl^- or Br^-, give $(Alk,NH_4)_2[RuX_5(H_2O)]$, red with Cl^-.

Concentrated HCl, HBr or HI with OsO_4 yield $[OsX_6]^{2-}$, but with a separate reductant, e.g., $FeCl_2$, for the chloride. The $[OsCl_6]^{2-}$ is a good precursor for many species. Refluxing $[OsCl_6]^{2-}$ with HI forms $[OsI_6]^{2-}$. Similar procedures, with mixed halides X^- and Y^- but not F^-, produce $[OsX_nY_{6-n})^{2-}$, and all the isomers are separable by, e.g., chromatography.

Specifically, refluxing OsO_4 with concentrated HBr for 2 h gives:

$$OsO_4 + 10\ Br^- + 8\ H_3O^+ \rightarrow [OsBr_6]^{2-} + 2\ Br_2\uparrow + 12\ H_2O$$

Then NH_4Br, cooling and adding ethanol yield black $(NH_4)_2[OsBr_6]$, slightly soluble and red, in cold water; hot water produces black OsO_2. Cathodic e^- and $K_2[OsBr_6]$ form $K_3[OsBr_6]$ at $E° = 31$ cV in 4-M HBr.

Adding HBr or HI to RuO_4 immediately forms soluble $RuBr_3·aq$ or black, slightly soluble RuI_3, and Br_3^- or I_3^-. Adding HBr to $Ru_2O_3·aq$, or I^- to "$RuCl_3·aq$", also gives these Ru compounds.

Refluxing [*trans*-$OsCl_2(NH_3)_4]^+$ with HI gives [*trans*-$OsI_2(NH_3)_4]^+$. Mercury(II) catalyzes the hydrolysis of the iodides.

Elemental and oxidized halogens. A stream of Cl_2 sweeps RuO_4 (explosive as a solid) or OsO_4 (stable), both toxic, out of acidified $M^{<VIII}$.

Chlorine oxidizes $[RuCl_5(H_2O)]^{2-}$ to $[RuCl_6]^{2-}$, and RuO_2 or RuO_4^{2-} to RuO_4, but $K_2[RuBr_6]$ results from Br_2 and $K_2[RuBr_5(H_2O)]$, or from Br_2, HBr and either $RuCl_3·aq$ or $K_3[RuCl_6]$, or from HBr and $K_2[RuCl_6]$, e.g.:

$$RuO_2·aq + 2\ Cl_2 + 6\ H_2O \rightarrow RuO_4 + 4\ Cl^- + 4\ H_3O^+$$

Excess Br^-, Ru^{II} ammine and Br_2 form a stable, yellow Ru^{III} powder:

$$[Ru(NH_3)_6]Cl_2 + {}^1/_2\ Br_2 + 2\ Br^- \rightarrow [Ru(NH_3)_6]Br_3\downarrow + 2\ Cl^-$$

Neutral ClO^- and powdered Ru or Os easily form $RuO_2·aq$ or OsO_4.

Alkaline ClO^- readily dissolves finely divided Ru or Os, or $M^{<VI}$ as RuO_4^{2-} or OsO_4^{2-} (separation from Rh, Ir, Pd and Pt).

Ruthenate(VI) and Cl_2 or ClO^- give RuO_4^-, or RuO_4 with excess oxidant. From the former green solutions the quite dark-green $KRuO_4$ or more soluble $NaRuO_4·H_2O$ can be crystallized.

Metallic Ru may be dissolved after fusion with KOH plus oxidants such as KNO_3 or $KClO_3$. Fusing Os with, e.g., $KClO_3$ and KOH yields K_2OsO_4, an "osmate", which in water may become [*trans*-$OsO_2(OH)_4]^{2-}$.

Aqueous $HClO_4$ and $[Ru(H_2O)_6]^{2+}$, catalyzed by halides, first form $[Ru(H_2O)_6]^{3+}$ and ClO_3^-, but ClO^-, ClO_3^-, ClO_4^- (hot and concentrated, distilling), BrO_3^- or H_5IO_6, all with H_3O^+, go on to oxidize Ru or Os compounds to MO_4. The RuO_4 from the safe and convenient oxidation of RuO_2 by H_5IO_6 at 0°C, like OsO_4, may be distilled, swept out by a gas stream or extracted by CCl_4. The hot oxidants dissolve both metals.

8.2.4 Reagents Derived from the Metals Lithium through Uranium, plus Electrons and Photons

Oxidation. A gas stream or distillation sweeps MO_4 out of acidic Ru or Os solutions or the metals with Ce^{IV}, $[Cr_2O_7]^{2-}$, $[AuCl_4]^-$, MnO_4^- or BiO_3^-.

The $[Ru(CN)_6]^{4-}$ ion and Ce^{IV} give $[Ru(CN)_6]^{3-}$, which then appears to go partly to $[Ru(CN)_5(H_2O)]^{2-}$ and a dimer, and with CN^- as reductant back to $[Ru(CN)_6]^{4-}$, but it may also dismutate:

$$2\ [Ru(CN)_6]^{3-} + 2\ OH^- \rightarrow [Ru(CN)_6]^{4-} + [Ru(CN)_5(CNO)]^{4-} + H_2O$$

Aqueous Ce^{IV}, H_3O^+ and $[Os(N_2)(NH_3)_5]^{2+}$ form $[Os(H_2O)(NH_3)_5]^{3+}$, ionized by base to $[Os(OH)(NH_3)_5]^{2+}$. Cerium(IV) also quickly and fully oxidizes $[Os(NH_3)_5CO]^{2+}$ to $[Os(NH_3)_5CO]^{3+}$; this dismutates with the Os reduced and one NH_3 oxidized:

$$3\ [Os(NH_3)_5CO]^{3+} + 3\ H_2O \rightarrow$$

$$^1/_2\ [\{Os(NH_3)_4CO\}_2(\mu\text{-}N_2)]^{4+} + 2\ [Os(NH_3)_5CO]^{2+} + 3\ H_3O^+$$

Anodes, $S_2O_8{}^{2-}$ and, less cleanly, $MnO_4{}^-$ in acid, do the same.

Excess Ce^{IV} oxidizes Ru^{IV} oxide in 0.5-M H_2SO_4 completely to RuO_4, accompanied by some Ru-catalyzed release of O_2:

$$RuO_2{\cdot}aq + 4\ Ce(SO_4)_n{}^{(4-2n)+} + 10\ H_2O \rightarrow$$

$$RuO_4 + 4\ CeSO_4{}^+ + (4n - 4)\ HSO_4{}^- + 4n\ H_2O + (8 - 4n)\ H_3O^+$$

Aqueous $[OsCl(NH_3)_5]^{2+}$ with excess Ce^{IV} followed by 6-M HCl and methanol forms a nitride, perhaps $[OsN(NH_3)_4Cl]Cl_2$, bright-yellow. Cerium(IV) and $[Os(NH_3)_6]^{3+}$ may give $[OsN(NH_3)_4(H_2O)]^{3+}$ as a soluble perchlorate but, from HCl, a rather insoluble chloride.

Various complexes oxidize ruthenate(vi); cf. **Reduction** below:

$$RuO_4{}^{2-} + [Mo(CN)_8]^{3-} \rightarrow RuO_4{}^- + [Mo(CN)_8]^{4-}$$

$$RuO_4{}^{2-} + [Ru(CN)_6]^{3-} \rightarrow RuO_4{}^- + [Ru(CN)_6]^{4-}$$

Most $Ru^{<VI}$, with $MnO_4{}^-$ etc. oxidants in OH^- at ambient T, give $RuO_4{}^{2-}$.

In $HClO_4$, $MnO_4{}^-$ oxidizes $[Ru(CN)_6]^{4-}$ completely to $[Ru(CN)_6]^{3-}$ by way of $[RuH(CN)_6]^{3-}$ and $[RuH_2(CN)_6]^{2-}$.

Aqueous $[Fe(H_2O)_6]^{3+}$ oxidizes $[Ru(H_2O)_6]^{2+}$ to $[Ru(H_2O)_6]^{3+}$.

In 1-dM HCl, the *cis*- or *trans*- isomers of $[OsX_2(NH_3)_4]^+$ (X = Cl, Br or I) and excess $FeCl_3$ generate the same isomers of $[OsX_2(NH_3)_4]^{2+}$. The [*mer*-$OsX_3(NH_3)_3$] complexes are oxidized similarly. Also in 1-dM HCl, Fe^{III}, O_2 or an anode converts $[OsCl_5NH_3]^{2-}$ to $[OsCl_5NH_3]^-$.

Aqueous $[PtCl_6]^{2-}$ or $[AuCl_4]^-$ oxidizes $[Ru(NH_3)_6]^{2+}$ to $[Ru(NH_3)_6]^{3+}$.

Solid $NaBiO_3$ and $RuO_2{\cdot}aq$ form $RuO_4{}^{2-}$ in water.

Anodes can convert $[Ru_2(Cl,Br)_9]^{3-}$ to $[Ru_2X_9]^{2-}$ or unstable $[Ru_2X_9]^-$, and $[Ru(H_2O)_6]^{2+}$ to [*tetrahedro*-$\{Ru(H_2O)_3\}_4)(\mu\text{-}O)_6]^{4+}$, but they can also be used to oxidize Ru and Os metal or compounds in general to MO_4.

Reduction. Metallic Mg, Zn, Zn_{Hg} and Al reduce Ru^{III} to Ru^{II}, and finally to Ru. Boiling a solution of "$RuCl_3{\cdot}aq$" in concentrated NH_3 with zinc dust for a few minutes yields a yellow ammine:

$$2\ RuCl_3 + Zn + 16\ NH_3 \rightarrow 2\ [Ru(NH_3)_6]^{2+} + [Zn(NH_3)_4]^{2+} + 6\ Cl^-$$

Adding NH_4Cl and cooling give crystals of $[Ru(NH_3)_6]Cl_2$. Or the solution may be barely neutralized with concentrated HCl and treated with more $ZnCl_2$ instead of NH_4Cl to produce $[Ru(NH_3)_6][ZnCl_4]$ with a higher yield. The dry ammines are stable for weeks, especially if cold.

Aqueous Sm^{2+}, Eu^{2+}, Yb^{2+}, U^{3+}, $TiOH^{2+}$, V^{2+} or Cr^{2+} reduces both $[Ru(NH_3)_6]^{3+}$ and $[Ru(H_2O)(NH_3)_5]^{3+}$ to Ru^{II}. Aqueous Eu^{2+}, V^{2+} or Cr^{2+} also reduces $[RuQ(NH_3)_5]^{2+}$ (Q = OH, RCO_2, Cl, Br or I) to Ru^{II}, some by way of the inner-sphere, and Cr^{2+} attacks the uncoordinated O of RCO_2. At least the halo ammines give $[Ru(H_2O)(NH_3)_5]^{2+}$ and , e.g., $CrCl^{2+}$. The V^{2+} and Cr^{2+} ions reduce $[Ru(NH_3)_6]^{3+}$ 10^6 times as fast as they do $[Co(NH_3)_6]^{3+}$, and V^{2+} reduces $[Ru(NH_3)_6]^{2+}$ further to Ru.

In 1-M HCl, excess Eu^{2+}, V^{2+} or Sn^{II} reduces $[OsCl_5N]^{2-}$ to $[OsCl_5NH_3]^{2-}$, somewhat air sensitive.

Titanium(III) and $[RuCl(NH_3)_5]^{2+}$ form $[RuCl(NH_3)_5]^{+}$.

Chromium(2+) reduces $[Os\{CS(NH_2)_2\}_6]^{3+}$ to $[Os\{CS(NH_2)_2\}_6]^{2+}$.

Powdered Ag appears to reduce $[OsX_6]^{2-}$ to $[OsX_6]^{3-}$; X = Cl, Br or I.

The $[Ru(NH_3)_6]^{3+}$ or $[RuCl(NH_3)_5]^{2+}$ ion and Zn_{Hg} yield $[Ru(NH_3)_6]^{2+}$ or $[Ru(H_2O)(NH_3)_5]^{2+}$ respectively.

Aqueous HF, $RuCl_3 \cdot aq$, SnF_2 and (Na,K,NH_4)F form $Alk_4[Ru(SnF_3)_6]$.

Aqueous $[SnCl_3]^-$ reduces $[RuCl_6]^{2-}$ to Ru^{III}. In dilute HCl, $[SnCl_3]^-$ reduces "$RuCl_3 \cdot aq$", $[RuCl_5NO]^{2-}$ and RuO_4 all to yellow-orange $[RuCl(SnCl_3\text{-}\kappa Sn)_5]^{4-}$, or, with much excess $[SnCl_3]^-$, to $[Ru(SnCl_3)_6]^{4-}$.

At 70 °C [*trans*-$OsO_2(NH_3)_4]^{2+}$ is reduced by Sn^{II} in 6-M HCl for 12 h, in 6-M HBr for 5 h, or by Fe wire in 4-M HI for 12 h, to [*trans*-$OsX_2(NH_3)_4]^{+}$. Refluxing for 2–3 days converts *cis* Cl_2 and Br_2 Os^{III} ions to *trans* isomers, likely via the Os^{IV} amides from some oxidation or dismutation. The *cis* Cl_2 and Br_2, but not I_2, Os^{IV} ions quickly isomerize to [*trans*-$OsX_2(NH_3)_4]^{2+}$.

Zinc dust, $[OsCl_6]^{2-}$ and NH_3 form $[Os(NH_3)_6]^{3+}$. So do Zn_{Hg}, $HClO_4$ and $[Os(NH_3)_5(NO)]^{3+}$ for a higher yield. Isolated salts include $[Os(NH_3)_6]Br_3$, $[Os(NH_3)_6][Os^{III}Br_6]$ and $[Os(NH_3)_6][Os^{IV}Br_6]Br \cdot H_2O$.

The $[OsCl_6]^{2-}$ ion, treated with HF, HCl or HBr, plus SnX_2, yields colorless $[Os(SnF_3)_6]^{4-}$, $[Os(SnCl_3)_6]^{4-}$, pale-yellow $[OsCl(SnCl_3)_5]^{4-}$ or red $[OsBr(SnBr_3)_5]^{4-}$. Hydrolyzing $[OsCl(SnCl_3)_5]^{4-}$ generates the hydroxo complex $[OsCl\{Sn(OH)_3\}_5]^{4-}$, all κSn (with Os–Sn bonding). Some complexes catalyze the isomerization of alkenes.

Certain complexes reduce ruthenate(vii); cf. **Oxidation** above:

$$RuO_4^- + [W(CN)_8]^{4-} \rightarrow RuO_4^{2-} + [W(CN)_8]^{3-}$$

$$RuO_4^- + [Fe(CN)_6]^{4-} \rightarrow RuO_4^{2-} + [Fe(CN)_6]^{3-}$$

Various common metals, Mg, Zn, Hg, etc., react with OsO_4, especially in the presence of H_3O^+, yielding Os^0. The OsO_4 also is reduced to OsO_2 by Fe^{2+}. Tin dichloride produces a brown precipitate, soluble in HCl.

Heating OsO_4 with concentrated HCl and excess $FeCl_2$ on a water bath for two hours, while the deep-green solution becomes orange-red, followed with cooling and adding NH_4Cl, results in deep-red crystals, slightly soluble and greenish-yellow in cold water:

$$OsO_4 + 4\ Fe^{2+} + 22\ Cl^- + 8\ H_3O^+ + 2\ NH_4^+ \rightarrow$$

$$(NH_4)_2[OsCl_6]\downarrow + 4\ [FeCl_4]^- + 12\ H_2O$$

Less drastic conditions yield $[(OsCl_5)_2(\mu\text{-}O)]^{4-}$.

Aqueous HCl and either Eu^{2+}, V^{2+} or $[SnCl_3]^-$ reduce $K[OsO_3N]$ to $K_2[OsCl_5(NH_3)]$.

Lead (activated for 15 min in 5.6-M HNO_3) reduces RuO_4 in 1-M $H_2[SiF_6]$ to pink $[Ru(H_2O)_6]^{2+}$.

Cathodic e^- and various chloro-Ru^{III} complexes give Ru^{II}, quickly becoming deep-blue $[Ru(H_2O)_6]^{2+}$ if oxidants and most ligands other than, say, $[BF_4]^-$ are removed by ion exchange, but certain conditions and reductants also produce other "Ruthenium blue" Ru^{II} or $Ru^{II,III}$ (mixed) complexes. Some blue-violet bromides are similar.

Cathodic electrons reduce $[OsBr_6]^{2-}$ to $[OsBr_6]^{3-}$, and they reduce $[Os(NH_3)_5(Cl,I)]^{2+}$, likely to unstable $[Os(NH_3)_5X]^+$.

A cathode and RuO_4 form Ru^{IV}, perhaps $[Ru_4(OH)_{12}]^{4+}$. Further electro-reduction can generate $[Ru_4(OH)_4]^{8+}$, which decomposes. Hydroxo-ruthenium complexes catalyze hydrogenations.

Photons (UV) aquate $[Ru(NH_3)_6]^{2+}$ but also yield Ru^{III} and H_2.

Other reactions. The alkaline-earth ions precipitate RuO_4^{2-} as black $MgRuO_4\cdot aq$, black $CaRuO_4\cdot aq$, red $SrRuO_4\cdot aq$ and red $BaRuO_4\cdot H_2O$, actually $Ba[\textit{trans}\text{-}RuO_3(OH)_2]$, a good way to isolate Ru^{VI}.

Aqueous $[Ru(CN)_6]^{4-}$ precipitates Ba^{2+}, Ln^{3+}, Fe^{3+}, Cu^{2+}, Ag^+ etc. The iron(III) salt, $Fe_4[Ru(CN)_6]_3$, is a semiconductor.

An example of forming a dinuclear complex is:

$$[Ru(CN)_6]^{4-} + [Fe(CN)_5(H_2O)]^{2-} \rightarrow [(NC)_5RuCNFe(CN)_5]^{6-} + H_2O$$

Aqueous $[Ru(H_2O)(NH_3)_5]^{2+}$ and $[RuCl(NH_3)_5]^{2+}$ reduce and oxidize each other to $[Ru(H_2O)(NH_3)_5]^{3+}$ and $[RuCl(NH_3)_5]^+$; the latter then goes to $[Ru(H_2O)(NH_3)_5]^{2+}$, autocatalyzing the aquation. Other $[RuX(NH_3)_5]^{2+}$, H_2O_2 etc., are also reduced by $[Ru(NH_3)_5(H_2O)]^{2+}$. The substitution of Cl^-, Br^- or I^- onto $[Ru(H_2O)_6]^{3+}$ is catalyzed by $[Ru(H_2O)_6]^{2+}$.

Very alkaline solutions promote the following:

$$OsO_2\cdot aq + [OsO_4(OH)_2]]^{2-} + 2\ OH^- + 2\ H_2O \leftrightarrows 2\ [OsO_2(OH)_4]^{2-}$$

Silver(1+) completely precipitates RuO_4^{2-} as black Ag_2RuO_4.

Yellow $[OsCl_6]^{2-}$ precipitates brown $Ag_2[OsCl_6]$ and olive-green $Tl_2[OsCl_6]$. Red $[OsBr_6]^{2-}$ gives black $Tl_2[OsBr_6]$.

Hot $[RuCl(NH_3)_5]^{2+}$ and Ag_2O yield $[Ru(H_2O)(NH_3)_5]^{3+}$ and AgCl.

A little Zn_{Hg} with $[Os(H_2O)(NH_3)_5]^{3+}$ and $[Os(N_2)(NH_3)_5]^{2+}$ produce $[\{Os(NH_3)_5\}_2(\mu\text{-}N_2)]^{5+}$, more stable than $[\{Os(NH_3)_5\}_2(\mu\text{-}N_2)]^{4+}$, but oxidizable by Cl_2 to blue $[\{Os(NH_3)_5\}_2(\mu\text{-}N_2)]^{6+}$.

Tin(II) chloride converts $[RuCl_5NO]^{2-}$ to $[\{RuCl_4(H_2O)\}_2(\mu\text{-}N)]^{3-}$, i.e., $[Ru_2NCl_8(H_2O)_2]^{3-}$. Whether or not the Ru is considered "oxidized" while the NO is reduced depends on the charge assigned to the NO.

Ruthenium is both oxidized and reduced in pseudocapacitor layers:

$$RuO(OH) + x\ H_3O^+ + x\ e^- \leftrightarrows RuO_{1-x}(OH)_{1+x} + x\ H_2O$$

Anodic oxidation and cathodic reduction of $[Ru(NH_3)_5(H_2O)]^{3+}$ and its basic derivatives give Ru^{IV} and Ru^{II} (a better known reaction) in turn.

Light, $[Ru(NH_3)_6]^{2+}$ and H_3O^+ yield $[Ru(NH_3)_5(H_2O)]^{2+}$. Photons, $[Ru(NH_3)_5(H_2O)]^{2+}$ and Cl^- form $[RuCl(NH_3)_5]^+$; $[cis\text{-}Ru(Br,I)_2(NH_3)_4]^+$ and H_2O give $[cis\text{-}Ru(Br,I)(H_2O)(NH_3)_4]^{2+}$; the chloride does not react.

Light (185 nm) reduces $[Ru(NH_3)_6]^{3+}$; alcohols scavenge the radicals.

Light slowly breaks aqueous RuO_4 down into a black oxide and O_2.

Postscript. "It is an interesting sign of the times that when a new element is discovered, there is a rush from many sides to torture the baby by oxidation, chlorination, fractionation, and so many other appliances which the chemist has at his disposal, yet here, in ruthenium, there is an element of an age exceeding four score years and ten, which is treated with so much respect that it yet awaits the severe ordeal it must inevitably undergo before it can occupy a worthy place in our records. We have read the properties of ruthenium so frequently that we are inclined to give the stereotyped records far more confidence than the evidence justifies [3]."

References

1. Leigh GJ (ed) (1990) Nomenclature of inorganic chemistry. IUPAC, Blackwell, London, pp 156, 157
2. Godin B et al (2005) Chem Commun 5624
3. Mellor JW (ed) (1936) Mellor's comprehensive treatise on inorganic and theoretical chemistry, vol XV. Longman, London, p 513

Bibliography

See the general references in the Introduction, specifically [116], [121] and [313], and some more-specialized books [4–9]. Some articles in journals include: aqueous iron chemistry, condensation etc. [10]; Fe^{III} photochemistry [11]; $[FeH(CO)_4]^-$ [12]; oxo- and hydroxo-bridged Fe_2 complexes [13]; ruthenium chemistry and thermodynamics [14]; the hydrolysis of iron(III) [15]; the catalyzed oxidation of FeS_2 by O_2 in water etc. [16]; reactions of Os ammines and N_2 species [17]; and electron-transfer reductions by V^{IV} and Fe^{II} [18].

4. Mielczarek EV, McGrayne SB (2000) Iron, nature's universal element. Rutgers University, New Brunswick
5. Silver J (ed) (1993) Chemistry of iron. Blackie, London
6. Seddon EA, Seddon KR (1984) The chemistry of ruthenium. Elsevier, Amsterdam
7. Griffith WP (1967) The chemistry of the rarer platinum metals (Os, Ru, Ir and Rh). Interscience, London
8. Avtokratova TD (1962) Analytical chemistry of ruthenium. Ann Arbor-Humphrey, Ann Arbor
9. American Cyanamid Company (1953) The chemistry of the ferrocyanides. American Cyanamid Company, New York
10. Jolivet JP, Chanéac C, Tronc E (2004) Chem Commun 2004:477
11. Sima J, Makanova J (1997) Coord Chem Rev 160:161 [Editor: Sima has an upside-down ^ over the S.]??
12. Brunet JJ (1990) Chem Rev 90:1041
13. Kurtz DM Jr (1990) Chem Rev 90:585
14. Rard JA (1985) Chem Rev 85:1
15. Flynn CM Jr (1984) Chem Rev 84:31
16. Lowson RT (1982) Chem Rev 82:461
17. Buhr JD, Taube H (1979) Inorg Chem 18:2208, (1980) Inorg Chem 19:2425; Buhr JD, Winkler JR, Taube H (1980) Inorg Chem 19:2416
18. Rosseinsky DR (1972) Chem Rev 72:215

9 Cobalt through Meitnerium

9.1 Cobalt, $_{27}$Co

Oxidation numbers: (–I) in $[Co(CO)_4]^-$ (II) in Co^{2+}, (III) in Co^{3+} and (IV) in CoO_2. In $[Co(CO)_3(NO)]$ (from a non-aqueous source) we could, without further structural information, classify Co as Co^0 but might also well see it as $[Co^-(CO)_3(NO^+)]$ in spite of the electronegativities (because NO^+ is isoelectronic with the very stable N_2 and CO), or as $[Co^+(NO^-)(CO)_3]$, assigning the metal as usual a positive oxidation state. (These assignments, within molecules, are partly but not entirely arbitrary.) Various experiments and calculations, not described here, also reveal NO both as a primarily neutral radical, e.g., in $[Ir^{3+}(Cl^-)_5(NO^\bullet)]$ and as NO^+ in $[Ru^{3+}(Cl^-)_5(NO^+)]$. See **6.1.2** and **8.1.2 Oxidized nitrogen** also.

9.1.1 Reagents Derived from Hydrogen and Oxygen

Dihydrogen. This can act as an oxidant, changing Co^{II} to Co^{III}. The forward reaction, the first one written below, is favored, $K = 10^5\ M^{-1}$ at 25 °C, showing the Co^{II} moiety's strong preference for 18 outer electrons, and that the more electronegative H can become H^-. Attempts to precipitate this complex often decompose it, but colorless $Cs_2Na[Co(CN)_5H]$ etc. are known. Ion-exchange gives $H_3[Co(CN)_5H]\cdot aq$, an unstable acid. The anion is a strong catalyst for hydrogenation (often organic) , e.g.:

$$[Co(CN)_5]^{3-} + {}^1/_2\ H_2 \leftrightarrows [Co(CN)_5H]^{3-}$$

$$[Co(CN)_5]^{3-} + {}^1/_2\ H_2O_2 \rightarrow [Co(CN)_5OH]^{3-}$$

$$\underline{[Co(CN)_5]^{3-} + [Co(CN)_5OH]^{3-} \rightarrow 2\ [Co(CN)_5]^{3-} + H_2O}$$

$${}^1/_2\ H_2O_2 + {}^1/_2\ H_2 \rightarrow H_2O$$

Compare similar reductions of NH_2OH, NCI (to HCN and HI), Br_2 etc.:

$$2\ [Co(CN)_5]^{3-} + YZ \rightarrow [Co(CN)_5Y]^{3-} + [Co(CN)_5Z]^{3-}$$

with YZ = H_2, $(HO)_2$, $(NH_2)(OH)$, halogen$_2$, I(CN) etc. But we also have:

$$2\ [Co(CN)_5]^{3-} + Y \rightarrow [\{Co(CN)_5\}_2(\mu\text{-}Y)]^{6-}$$

with Y = O_2, C_2H_2, SO_2, $SnCl_2$ etc. For Y = O_2 and with K^+ a brown product is $K_6[\{Co(CN)_5\}_2(\mu\text{-}O_2)]\cdot 4H_2O$, but $[\{Co(NH_3)_5\}_2(\mu\text{-}O_2)]^{5+}$ and KCN give a $K_5[\{Co\ (CN)_5\}_2(\mu\text{-}O_2)]\cdot 5H_2O$, magenta.

Water. Cobalt becomes passive in water.

Cobalt(II) oxide and hydroxide are insoluble, the acetate and nitrate deliquescent; the sulfate, efflorescent; the chloride, hygroscopic. The borate, carbonate, cyanide, oxalate, phosphates, sulfide and hexacyanoferrate(II and III) are insoluble. The ordinary cobalt(II) ammines and the hexacyanocobaltate(III) salts of Alk^+ and Ae^{2+} are soluble, those of the **d**-block M^{2+} and Ag^+ ions insoluble.

Most acidopentaamminecobalt(III) nitrates are slightly soluble.

Dissolved and solid species often differ, but $[Co(H_2O)_6]SO_4$ has the same constituents in both media. Solid $CoCl_2\cdot 6H_2O$, however, turns out to be [*trans*-$CoCl_2(H_2O)_4]\cdot 2H_2O$ rather than $[Co(H_2O)_6]Cl_2$, although the cobalt dissolves, as in other salts (at least when dilute), as $[Co(H_2O)_6]^{2+}$.

Less acidic media than H_2O can support the basic $[Co(CN)_5]^{4-}$, but:

$$[Co(CN)_5]^{4-} + H_2O \rightarrow [Co(CN)_5H]^{3-} + OH^-$$

and oxidation numbers based on electronegativities lead to the assignments $[Co^+(CN^-)_5]$ and $[Co^{3+}(CN^-)_5H^-]$ in this oxidation by H^+.

The aging of not-too-dilute $[Co(CN)_5]^{3-}$ proceeds primarily thus:

$$2\ [Co(CN)_5]^{3-} + 2\ H_2O \rightarrow [Co(CN)_5H]^{3-} + [Co(CN)_5(H_2O)]^{2-} + OH^-$$

Water replaces SO_3^{2-} much faster from [*trans*-$Co(CN)_4(SO_3)_2]^{5-}$ than from $[Co(CN)_5(SO_3)]^{4-}$; this may show a *trans* effect (commonly noted in square-planar complexes) in octahedral complexes, although CN^- is at or near the top of the usual lists of *trans*-effect ligands. See **10.2.2 Reduced nitrogen.** Of course other factors also intrude.

The rate of replacing X by H_2O in $[Co(NH_3)_5X]^{(3-n)+}$ covers a wide range for X as $NH_3 \ll PO_4^{3-} < NO_2^- < CH_3CO_2^- < CF_3CO_2^- < SO_4^{2-} < Cl^- < H_2O < Br^- < NO_3^- < CH_3SO_3^- \ll SO_3F^- < CF_3SO_3^- < NH_2SO_2NH_2 < ClO_4^- \ll N_4O\ (N_3^- + NO^+)$. The "triflate" ($CF_3SO_3^-$) complex, much safer than ClO_4^-, then provides rapid access to other ligands.

Many Co–H species are too firmly bound to be titrated, but $[CoH(PF_3)_4]$ (non-aqueous source) ionizes strongly in water.

Water and Co^{3+} in H_3O^+, especially warm, give $[Co(H_2O)_6]^{2+}$ and O_2.

Seawater and some freshwater contain traces of Co^{II} complexes as $CoOH^+$, $Co(OH)_2$, $CoCO_3$, $CoSO_4$ and $CoCl^+$.

Oxonium. Warm, dilute HCl or H_2SO_4 dissolves Co slowly as Co^{2+} and H_2, although pure Co is hardly attacked except in contact with, e.g., Pt.

Cobalt(II) oxide and hydroxide are soluble in acids.

Higher oxides and hydroxides, but not Co_3O_4 from high-temperature treatments, release O_2 with non-reducing acids, forming Co^{2+}.

With $[Co(CN)_5]^{3-}$, H_3O^+ seems to split into H_2O^+, forming only $[Co(CN)_5\,H_2O]^{2-}$, and H, forming both $^1/_2\,H_2$ and $[Co(CN)_5H]^{3-}$.

Hydroxide. Cobalt is not affected by OH^-. Cobalt(II) hydroxide, $Co(OH)_2$, blue, is precipitated from Co^{II} with OH^-, but dissolves in hot, concentrated OH^- as $[Co(OH)_4]^{2-}$. It turns pink if warmed. It absorbs O_2 from the air and turns gray green, as cobalt(II-III) hydroxide, and then slowly forms $Co_2O_3\cdot aq$.

Cobalt(II) oxide and hydroxide dissolve slightly in hot, concentrated OH^-, giving a blue colored solution (distinction from Ni).

If Co^{2+} is precipitated as $Co(OH)_2$ with a slight excess of OH^-, and the precipitate dissolved in the minimum amount of CN^-, adding an oxidant (H_2O_2, $[S_2O_8]^{2-}$ or ClO^-) to the cold solution causes no precipitation (distinction from Ni), but boiling completely precipitates $Co_2O_3\cdot aq$.

Brown $Co_2O_3\cdot aq$ is precipitated on treating Co^{3+} with OH^-.

Peroxide. In a mixture with OH^- (not NH_3), $Co(OH)_2$ is readily oxidized by HO_2^- to brown $Co_2O_3\cdot aq$, and may go to black $CoO_2\cdot aq$. These do not dissolve in NH_3 plus NH_4^+, or in CN^-.

Saturating ice-cold Co^{2+} with excess $NaHCO_3$ and adding 10-M H_2O_2 yields an apple-green cobalt(III) product (distinction from Ni). As a test this will detect 0.04 mM Co in 10 mL:

$$Co^{2+} + 5\,HCO_3^- + {}^1/_2\,H_2O_2 + 3\,Na^+ \rightarrow$$

$$Na_3[Co(CO_3)_3]\cdot 3H_2O\downarrow + 2\,CO_2\uparrow$$

The green product, stable when dry, is moderately stable in solution with excess HCO_3^-, although the chirally resolved form racemizes quickly. This complex, with various amounts of CN^-, NH_3 (and NH_4^+) and NO_2^-, and often with catalytic charcoal and heat, can be converted conveniently to numerous corresponding mixed complexes, e.g., [*mer*-$Co(CN)_3(NH_3)_3$], slightly soluble and yellow. The carbonate ions are especially safely displaced by otherwise easily oxidized ligands.

At 55 °C, $Co(CH_3CO_2)_2$, H_2O_2 and $K_2C_2O_4$ give light-, heat- and base-sensitive, green $K_4[\{Co(C_2O_4)_2\}_2(\mu\text{-}OH)_2]$.

In partial sequence, Co^{2+}, $K_2C_2O_4$, $KHCO_3$ and H_2O_2, then KNO_2 for 2 h, all at 40–50 °C, yield a red, unstable $K_3[Co(C_2O_4)_2(NO_2)_2]\cdot H_2O$.

Mixtures of Co^{II} and H_2O_2 react with various weak bases to form Co^{III} complexes. A rose-colored aqua ammine and a lavender bromo ammine arise, with two hours heating in the second case, from:

$$Co^{2+} + 3\,Br^- + 4\,NH_3 + NH_4^+ + {}^1/_2\,H_2O_2 \rightarrow [Co(NH_3)_5(H_2O)]Br_3\downarrow$$

$$[Co(NH_3)_5(H_2O)]^{3+} + 3\,Br^- \rightarrow [CoBr(NH_3)_5]Br_2\downarrow + H_2O$$

Or a charcoal catalyst and more NH_3 can produce $[Co(NH_3)_6]Br_3$.

Aqueous $[Co(CN)_5]^{3-}$ and H_2O_2 form $[Co(CN)_5(OH)]^{3-}$, followed by $[Co(CN)_5(H_2O)]^{2-}$ except at high pH. Adding I^- gives $[Co(CN)_5I]^{3-}$.

Oxidizing $[Co(NH_3)_5(NCS)]^{2+}$ in acid gives an interesting mixture:

$$[Co(NH_3)_5(NCS)]^{2+} + 4\ H_2O_2 \rightarrow$$

$$[Co(NH_3)_6]^{3+} + CO_2\uparrow + HSO_4^- + 2\ H_2O$$

retaining Co^{III}, especially in cool, more acidic solutions, along with:

$$[Co(NH_3)_5(NCS)]^{2+} + {}^7/_2\ H_2O_2 + 4\ HSO_4^- \rightarrow$$

$$Co^{2+} + CO_2\uparrow + 5\ SO_4^{2-} + 6\ NH_4^+ + H_2O$$

giving Co^{II} together with the CO_2, plus:

$$[Co(NH_3)_5(NCS)]^{2+} + {}^5/_2\ H_2O_2 + 3\ HSO_4^- \rightarrow$$

$$Co^{2+} + HCN\uparrow + 4\ SO_4^{2-} + 5\ NH_4^+ + H_2O$$

This is interesting even more because of the way in which some of the Co^{III} and external oxidant, especially in warm, less acidic solutions, both finally take one electron (first apparently the peroxide, then the Co^{III}) in that part of the process, as in the last two equations, leading to Co^{II}.

One-electron and sometimes two-electron oxidants can show different results with $[Co(NH_3)_5(HC_2O_4)]^{2+}$; Mo^{VI} slowly catalyzes this:

$$[Co(NH_3)_5(HC_2O_4)]^{2+} + H_2O_2 + H_3O^+ \rightarrow$$

$$[Co(NH_3)_5(H_2O)]^{3+} + 2\ CO_2\uparrow + 2\ H_2O$$

Compare with related reactions under **oxidized halogens** below.

Di- and trioxygen. The metal is not oxidized on exposure to air or when heated in contact with alkalis unless in powder form.

Air, Co^{2+}, "$(NH_4)_2CO_3$" and NH_3 at ambient T for 24 h produce red $[Co(\eta^1\text{-}CO_3)(NH_3)_5]NO_3\cdot{}^3/_2H_2O$.

Cold aqueous Co^{2+} plus CN^- and rapid oxidation with O_2 give:

$$2\ [Co(CN)_5]^{3-} + O_2 \rightarrow [(NC)_5Co\text{–}O\text{–}O\text{–}Co(CN)_5]^{6-}$$

precipitable by ethanol as brown $K_6[\{Co(CN)_5\}_2(\mu\text{-}O_2)]\cdot 4H_2O$, or as the less-soluble orange tribarium trihydrate. Slow oxidation allows:

$${}^1/_2\ [(NC)_5CoO_2Co(CN)_5]^{6-} + [Co(CN)_5]^{3-} + 3\ H_2O \rightarrow$$

$$2\ [Co(CN)_5(H_2O)]^{2-} + 2\ OH^-$$

before going with more CN^- to $[Co(CN)_6]^{3-}$. A product in low yield is magenta (orange-brown in solution) $K_5[(NC)_5CoO_2Co(CN)_5]\cdot 5H_2O$, which appears to be an O_2^- (i.e., hyperoxo) complex of Co^{III}, not an O_2^{2-} complex of Co(III and IV). Other sources of this salt are $[(NC)_5CoO_2Co(CN)_5]^{6-}$ plus KBrO at 0 °C followed by ethanol, or better, $[(NH_3)_5CoO_2Co(NH_3)_5]^{5+}$ plus KCN.

Aqueous $[Co(CN)_5(H_2O)]^{2-}$ does not react with CO, H_2S or Cl^-; but NH_3, N_2H_4, N_3^- and SCN^- do replace the H_2O.

Aqueous $[(NC)_5CoO_2Co(CN)_5]^{6-}$ in acidic solution quickly becomes $[(NC)_5Co(O_2H)Co(CN)_5]^{5-}$, which then splits into $[Co(CN)_5(O_2H)]^{3-}$ and $[Co(CN)_5(H_2O)]^{2-}$, and finally into two $[Co(CN)_5(H_2O)]^{2-}$ and H_2O_2.

The $[Co(CN)_5(O_2H)]^{3-}$ ion also comes from $[Co(CN)_5H]^{3-}$ plus O_2; then K^+ with methanol and acetone give an impure salt.

Cobalt(II) in NH_3 is easily oxidized by air to red $[Co(NH_3)_5(H_2O)]^{3+}$, but catalysts such as activated charcoal or $[Ag(NH_3)_2]^+$ with excess NH_3 can yield the yellow-orange $[Co(NH_3)_6]^{3+}$ ion, reddish-brown in large crystals, e.g., (crystallized with much HCl and cooling):

$$2\ Co^{2+} + {}^1/_2\ O_2 + 2\ NH_3 + 10\ NH_4^+ + 6\ Cl^- \rightarrow$$

$$2\ [Co(NH_3)_6]Cl_3\downarrow + H_2O$$

Substitution of the nitrate and HNO_3 yields $[Co(NH_3)_6](NO_3)_3$, also obtainable from $[Co(NH_3)_6]Cl_3$ plus HNO_3. Common anions, and N_3^-, $H_2AsO_4^-$, $S_2O_3^{2-}$-κO, $S_2O_3^{2-}$-κS, $HSeO_3^-$, ClO_2^-, ClO_3^-, CrO_4^{2-}, MoO_4^{2-}, ReO_4^- and neutral ligands, some starting with Co^{2+}, can replace the H_2O in $[Co(NH_3)_5(H_2O)]^{3+}$. Even the pink $[Co(NH_3)_5(ClO_4)](ClO_4)_2$ (quickly aquated) arises by nitrosating $[CoN_3(NH_3)_5](ClO_4)_2$ in concentrated $HClO_4$. Some reactions of $[Co(ClO_2$-$\kappa O)(NH_3)_5]^{2+}$, for example, include reduction by SO_2, VO^{2+} or Fe^{2+}:

$$[Co(ClO_2)(NH_3)_5]^{2+} + 2\ SO_2 + 5\ H_2O \rightarrow$$

$$[Co(H_2O)(NH_3)_5]^{3+} + 2\ HSO_4^- + Cl^- + 2\ H_3O^+$$

Oxygen, NO_2^- and NH_3 give more ammines:

$$2\ Co^{2+} + 6\ NO_2^- + {}^1/_2\ O_2 + 4\ NH_3 + 2\ NH_4^+ \rightarrow$$

$$2\ [mer\text{-}Co^{III}(NO_2)_3(NH_3)_3]\downarrow + H_2O\text{, but also} \rightarrow$$

$$[trans\text{-}Co^{III}(NO_2)_2(NH_3)_4][trans\text{-}Co^{III}(NO_2)_4(NH_3)_2]\downarrow + H_2O$$

Treating aqueous $CoCl_2$ with $MoCl_5$, CN^- and O_2 gives the unexpected green peroxo complex, $[(NC)_5Co^{III}–O_2–Mo^{VI}(O)Cl(CN)_5]^{6-}$.

With a base, O_3 and $Co(OH)_2$ or CoS readily form $Co_2O_3\cdot aq$, dark-brown, and even $CoO_2\cdot aq$, black. Neutral Co^{2+} yields some $Co_2O_3\cdot aq$.

9.1.2 Reagents Derived from the Other 2nd-Period Non-Metals, Boron through Fluorine

Boron species. Aqueous $[Co(CN)_5(H_2O)]^{3-}$ and $[BH_4]^-$ at pH 9, but not much higher, slowly yield $[Co(CN)_5H]^{3-}$ and $[Co(CN)_6]^{3-}$.

Aqueous $[CoBr(CN)_5]^{3-}$ and $[BH_4]^-$ also give $[Co(CN)_5H]^{3-}$.

Carbon oxide species. One can synthesize some carbonyls in water during several hours, if air is kept out; colors in the following go through blue and pink to yellow, while part of the CO reduces the cobalt to Co^{-I} [but also with some apparent dismutation of the Co^{II} (at < 0.1 M) to Co^{I} and Co^{III}]:

$$2\,[Co(CN)_5]^{3-} + 11\,CO + 12\,OH^- \rightarrow$$

$$2\,[Co(CO)_4]^- + 3\,CO_3^{2-} + 10\,CN^- + 6\,H_2O$$

$$2\,[Co(CN)_5]^{3-} + 2\,CO \rightarrow [Co(CN)_3(CO)_2]^{2-} + [Co(CN)_6]^{3-} + CN^-$$

The Co(-I) ion reacts with dilute HCl to produce the strongly acidic, very malodorous, poisonous, volatile, light-yellow, liquid, $[Co(CO)_4H]$, which decomposes to the dark-brown solid $[Co_2(CO)_8]$ and H_2.

Another source provides one similar result:

$$[Co(CN)_5H]^{3-} + 2\,CO + OH^- \rightarrow [Co(CN)_3(CO)_2]^{2-} + 2\,CN^- + H_2O$$

Aqueous Co^{2+} and HCO_3^- containing free CO_2, precipitate red $CoCO_3 \cdot 6H_2O$ at room temperature, but the blue anhydrous salt near the boiling point. Aqueous CO_3^{2-} precipitates a basic cobalt(II) carbonate. The precipitate is soluble in "$(NH_4)_2CO_3$" or NH_4^+, but only very slightly soluble in CO_3^{2-}. Carbonates of Mg, Ca, Sr or Ba do not precipitate Co^{2+} in the cold (separation from Cr^{III}, Fe^{III} and Al^{III}), but prolonged boiling in the air completely oxidizes and precipitates it as $Co_2O_3 \cdot aq$.

Warm Co^{2+} and K_2CO_3 give rose-pink $K_2[\textit{trans}\text{-}Co(\eta^1\text{-}CO_3)_2(H_2O)_4]$.

Carbonate under CO_2 can produce a deep-red bridged complex:

$$2\,[Co(NH_3)_5(H_2O)]I_3 + 3\,Ag_2CO_3 + 2\,(NH_4)_2SO_4 + 2\,CO_2 + 4\,H_2O$$

$$\rightarrow [\{Co(NH_3)_5\}_2(\mu\text{-}CO_3)](SO_4)_2 \cdot 4H_2O + 6\,AgI\downarrow + 4\,NH_4^+ + 4\,HCO_3^-$$

Cyanide species. Aqueous CN^- precipitates Co^{2+} as light-brown or red-brown $Co(CN)_2 \cdot {\sim}2H_2O$, soluble in HCl, not in acetic acid or HCN, soluble in excess CN^- as green $[Co(CN)_5]^{3-}$ or $[Co(CN)_5(H_2O)]^{3-}$ with the H_2O weakly bound:

$$Co(CN)_2 + 3\,CN^- \rightarrow [Co(CN)_5]^{3-}$$

(Some Co^{II} reduces some CN^- to CH_3NH_2, and at pH ~10, H_2O to H_2.) Dilute acids soon reprecipitate $Co(CN)_2$ (as with Ni):

$$[Co(CN)_5]^{3-} + 3\ H_3O^+ \rightarrow Co(CN)_2\downarrow + 3\ HCN + 3\ H_2O$$

The green complex can be crystallized as a brown paramagnetic $K_3[Co(CN)_5]$, as red-violet, air-sensitive $K_6[\{-Co(CN)_5\}_2]\cdot 4H_2O$ or air-stable $Ba_3[\{-Co(CN)_5\}_2]\cdot 13H_2O$, each with a weak Co—Co single bond.

Oxidative addition to $[Co(CN)_5]^{3-}$ by H_2, O_2, Br_2 etc. readily forms $[Co(CN)_5H]^{3-}$, $[Co(CN)_5(O_2)]^{3-}$, $[Co(CN)_5Br]^{3-}$ and so on. We also find brown $K_6[\{Co^{III}(CN)_5\}_2(\mu\text{-}O_2)]$, oxidizable by Br_2 to a red, paramagnetic $K_5[\{Co^{III}(CN)_5\}_2(\mu\text{-superoxo-}O_2)]$, and $Zn_5[\{Co^{III}(CN)_5\}_2(\mu\text{-}O_2)]_2\cdot 2H_2O$. With some liberated or excess CN^-, it reduces even boiling water to H_2 in a few minutes, and, as expected therefore, and more readily, O_2 to OH^-:

$$[Co(CN)_5]^{3-} + H_2O + CN^- \rightarrow [Co(CN)_6]^{3-} + {}^1/_2\ H_2\uparrow + OH^-$$

$$2\ [Co(CN)_5]^{3-} + {}^1/_2\ O_2 + 2\ CN^- + H_2O \rightarrow 2\ [Co(CN)_6]^{3-} + 2\ OH^-$$

This Co^{III} corresponds to $[Fe(CN)_6]^{3-}$, but to no such nickel complex. The oxidation is also faster with oxidants like ClO^- or CrO_4^{2-}. The $[Co(CN)_6]^{3-}$ ion is pale yellow and stable, and acids cause no immediate precipitation (important distinction from Ni, whose unoxidized solutions do precipitate with acids), but concentrated strong acids slowly decompose it. It is unreactive to OH^-, H_2O_2, Cl_2 and H_2S. Ion exchange and evaporation, or HCl and extraction by ether, yield the strong, tribasic acid $H_3[Co(CN)_6]\cdot {}^1/_2H_2O$. We force fed a mouse with a dose of $K_3[Co(CN)_6]$ equivalent to > 40 g for a man, without obvious stress.

Aqueous $[Co(CN)_5(H_2O)]^{2-}$ and CN^- react very slowly at ambient T and negligibly at very high pH; at 90°C however, and without added OH^-, $[Co(CN)_6]^{3-}$ is quickly formed.

A suspension of $[CoCl(NH_3)_5]Cl_2$ plus KCN and traces of catalytic $[Co(CN)_5]^{3-}$ at 0°C yield $K_3[CoCl(CN)_5]$.

Without a catalyst, e.g., charcoal or $[Co(CN)_5]^{3-}$, dilute $[Co(NH_3)_6]^{3+}$ is inert to CN^- at ambient T; otherwise at 0°C we can get both [*cis*- and [*trans*-$Co(CN)_2(NH_3)_4]NO_3\cdot H_2O$ (and both yellow-orange) for example.

At 25°C, KCN converts $[Co(NH_3)_5(N_3)]^{2+}$, after precipitation by methanol, to $K_3[Co(CN)_5(N_3)]\cdot 2H_2O$, yellow. At 100°C KCN and $[Co(NH_3)_5(NO_2)]^{2+}$ form a light-yellow $K_3[Co(CN)_5(NO_2)]\cdot 2H_2O$ after cooling and precipitating by methanol. Cyanide also converts $[Co(NH_3)_5(S_2O_3\text{-}\kappa S)]^+$ to yellow $K_4[Co(CN)_5(S_2O_3\text{-}\kappa S)]$.

Again at ambient T, KCN, $[Co(NH_3)_5NCS]^{2+}$ and traces of catalytic $[Co(CN)_5]^{3-}$ first form $[Co^{III}(NH_3)_5\text{–}NCS\text{–}Co^{II}(CN)_5]^-$, and then (after adding a little methanol) yellow $K_3[Co^{III}(CN)_5SCN]$ plus Co^{II}. However, $K_2[Co(CN)_5(H_2O)]$ and KSCN at 40 °C, after adding ethanol, give $K_3[Co^{III}(CN)_5NCS]$, the isomer; this rearranges extremely slowly in water to the other (κS) structure. Other (pseudo)halides, N_3^-, Br^- and I^-, also replace the H_2O.

Rather similarly, $[Co(NH_3)_5NCSe]^{2+}$ yields brown $K_3[Co^{III}(CN)_5SeCN]$, which isomerizes slowly and, with NH_4^+, yields $(NH_4)_3[Co^{III}(CN)_5NCSe]$.

The ions $[Co^{3+}(CN^-)_5X]$, with X = H_2O, OH^-, N_3^-, SCN^-, Cl^-, Br^- or I^-, plus CN^- and catalytic $[Co(CN)_5]^{3-}$, become $[Co(CN)_6]^{3-}$, although much more slowly if X = (isotopic) CN^-. Various substitutions of NH_3 by CN^- also conclude as follows when $Y^{n-} = NH_3$, CO_3^{2-}, RCO_2^-, PO_4^{3-} or SO_4^{2-}:

$$[CoY(NH_3)_5]^{(3-n)+} + 6\ CN^- \rightarrow [Co(CN)_6]^{3-} + Y^{n-} + 5\ NH_3$$

However, Y^- as OH^-, NCS^-, N_3^-, NO_2^-, Cl^-, Br^- or I^- gives $[CoY(CN)_5]^{3-}$, apparently via $[Co(CN)_5{-}Y{-}Co(NH_3)_5]^-$. More specifically, KCN, with $[CoCl(NH_3)_5]^{2+}$ at 0 °C, yields light-yellow $K_3[CoCl(CN)_5]$. Cyanide and $[Co(S_2O_3)(NH_3)_5]^+$ likewise give $K_4[Co(S_2O_3)(CN)_5]$.

Aqueous KCN, with $[CoBr(NH_3)_5]^{2+}$ and then ethanol as precipitant, forms cream-colored $K_3[CoBr(CN)_5]$.

Another example of complexes with CN^- is an unstable red ion:

$$[Co(CO_3)_3]^{3-} + 2\ CN^- \rightarrow [\mathit{cis}\text{-}Co(CN)_2(CO_3)_2]^{3-} + CO_3^{2-}$$

from which ammonia etc. can produce , e.g., an orange *cis-cis-* ion:

$$[\mathit{cis}\text{-}Co(CN)_2(CO_3)_2]^{3-} + 2\ NH_3 \rightarrow$$

$$[\mathit{cis}\text{-}\mathit{cis}\text{-}Co(CN)_2(NH_3)_2(CO_3)]^- + CO_3^{2-}$$

and perchloric acid at 0 °C can then yield:

$$[\mathit{cis}\text{-}\mathit{cis}\text{-}Co(CN)_2(NH_3)_2(CO_3)]^- + 2\ H_3O^+ \rightarrow$$

$$[\mathit{cis}\text{-}\mathit{cis}\text{-}\mathit{cis}\text{-}Co(CN)_2(NH_3)_2(H_2O)_2]^+ + CO_2\uparrow + H_2O$$

Heating $[Co(SO_3)_3]^{3-}$ with CN^- gives $[\mathit{cis}\text{-}Co^{III}(CN)_4(SO_3)_2]^{5-}$ in a small yield if the heating is stopped before it all goes to $[Co(CN)_6]^{3-}$. The mixed complex can be changed further to $[Co^{III}(CN)_5(SO_3)]^{4-}$.

Cyanide, SO_2, O_2 and $Co(CH_3CO_2)_2$ yield $[\mathit{trans}\text{-}Co^{III}(CN)_4(SO_3)_2]^{5-}$.

Alkaline fulminate (CNO^-), Co^{II} and air form $[Co(CNO)_6]^{3-}$.

Cyanate (NCO^-) and $[Co(NH_3)_5(H_2O)]^{3+}$ give not the cyanate but the carbamate, $[Co(NH_3)_5(CO_2NH_2\text{-}\kappa O)]^{2+}$, keeping the original Co–O bond. This and NO^+ yield $[Co(NH_3)_5(H_2O)]^{3+}$, CO_2 and N_2.

Cyanate and others can substitute, however, for "triflate", $CF_3SO_3^-$:

$$[Co(NH_3)_5(CF_3SO_3)]^{2+} + NCO^- \rightarrow [Co(NH_3)_5(NCO)]^{2+} + CF_3SO_3^-$$

Concentrated NCS^-, with Co^{2+}, forms a deep-blue $[Co(NCS)_4]^{2-}$; extraction by pentanol gives a quite sensitive test (distinction from Ni^{2+}). The red color of Fe^{III} thiocyanate complexes interferes, but Sn^{II} reduces, or CO_3^{2-} precipitates, the Fe^{III}. Acetate and Hg^{II} also interfere some.

Heat, NCS^- and Co^{III}–OH_2 species yield Co^{III}–NCS, usually stable for long times in both acid and alkali, and reversibly binding with Fe^{3+}, Cu^{2+}, Ag^+, Hg^2, Tl^{3+} etc. as Co^{III}–NCS–M^{n+}, e.g., yellow $[Co(NH_3)_5NCSHg]^{4+}$ from orange $[Co(NH_3)_5NCS]^{2+}$.

Base, $[Co(NO_3)(NH_3)_5]^{2+}$ and much excess NCS^- quickly give the less stable but rather inert $[CoSCN(NH_3)_5]^{2+}$, less of the more stable $[CoNCS(NH_3)_5]^{2+}$, and considerable $[CoOH(NH_3)_5]^{2+}$. However, base catalyzes $[Co(S_2O_3\text{-}\kappa O)(NH_3)_5]^+$ to become $[Co(S_2O_3\text{-}\kappa S)(NH_3)_5]^+$.

Concentrated NaNCS and $[Co(NH_3)_5(NO_3,I)](ClO_4)_2$, and a little NaOH, then HCl, give (separable by ion exchange) $[Co(H_2O)(NH_3)_5]^{3+}$, violet $[Co(SCN)(NH_3)_5]Cl_2\cdot{}^3/_2H_2O$ and orange $[Co(NCS)(NH_3)_5]Cl_2$. Salts of N_3^- and NO_2^- may replace the NaNCS. If not cold and dark, SCN-*κS* becomes NCS-*κN*. The S of both SCN-*κS* and NCS-*κN* joins any added $[Co(CN)_5]^{3-}$, each giving $[Co(CN)_5(SCN\text{-}\kappa S)]^{3-}$. Similarly, $[Co(CN)_5]^{3-}$ and $[Hg(SeCN)_4]^{2-}$ yield $[Co(CN)_5(SeCN\text{-}\kappa Se)]^{3-}$ and $Hg_2(SeCN)_2$.

Some "simple" organic reagents. Ethene, ethyne etc. (at 0 °C) can join cobalt atoms:

$$2\,[Co(CN)_5]^{3-} + C_2H_4 \rightarrow [(NC)_5Co\text{–}CH_2\text{–}CH_2\text{–}Co(CN)_5]^{6-}$$

$$2\,[Co(CN)_5]^{3-} + C_2H_2 \rightarrow [\textit{trans}\text{-}(NC)_5Co\text{–}CH\text{=}CH\text{–}Co(CN)_5]^{6-}$$

with ethanol to precipitate, e.g., yellow $K_6[\{Co(CN)_5(CH\text{=})\}_2]\cdot 4H_2O$.

The presence of chelating organic acids or sugars prevents the precipitation of Co^{2+} by alkalis.

Oxalic acid and oxalates precipitate Co^{2+} as reddish cobalt(II) oxalate, CoC_2O_4. At first only a cloudiness is obtained, then finally complete precipitation. The salt is soluble in strong acids and NH_3. A green oxalato complex, sensitive to both light and heat, can be made as follows for example, with ethanol as a final precipitant:

$$2\,CoCO_3 + 6\,HC_2O_4^- + PbO_2 + 2\,CH_3CO_2H + 6\,K^+ \rightarrow$$

$$2\,K_3[Co(C_2O_4)_3]\downarrow + Pb(CH_3CO_2)_2 + 2\,CO_2\uparrow + 4\,H_2O$$

In acid, Co^{3+} is reduced to Co^{2+} by $H_2C_2O_4$.

Oxalate can be used to precipitate $[Co(NH_3)_6]^{3+}$ quantitatively as $[Co(NH_3)_6]_2(C_2O_4)_3\cdot 4H_2O$.

Warming $[Co(NH_3)_5(H_2O)]^{3+}$ with $H_2C_2O_4$ gives $[Co(NH_3)_5(HC_2O_4)]^{2+}$.

If a slightly acidic solution of Co^{2+} is treated with 1-M $CH_3CO_2^-$ and butanedionedioxime (dimethylglyoxime), adding an alkaline sulfide makes the solution wine red.

From a non-aqueous source, $[Co(\eta^5\text{-}C_5H_5)_2]^+$, cobaltocene(1+) or "cobalticenium", resembles Cs^+ in size, salt solubilities and ability to be reduced to a neutral molecule or metal respectively.

Reduced nitrogen. Cobalt(II) oxide and hydroxide, and most of the Co^{II} salts insoluble in water, dissolve in (aqueous) NH_3. The presence of NH_4^+ prevents the precipitation of Co^{2+} by the alkalis. Ammonia without NH_4^+ produces the same precipitate as OH^-; incomplete, even at first, due to the NH_4^+ formed in the reaction; soluble in excess of NH_3 to give a solution that turns brown due to oxidation and is not affected by OH^-.

The higher oxides and hydroxides are insoluble in NH_3 or NH_4^+ [separation from $Ni(OH)_2$ after oxidation of Co^{II} but not Ni^{II} with IO^-].

Treating $CoCl_2$ in cool, 7-M NH_3 under N_2 with NO gives a 10-% yield of lustrous black, diamagnetic $[Co(NH_3)_5(NO)]Cl_2$, stable if quite dry. It is decomposed even by cold water (forming basic Co^{II} chlorides), by air (forming Co^{III} ammines), by 15-M NH_3 (forming $[Co(NH_3)_6]^{2+}$ under N_2) and by 12-M HCl (forming $[CoCl_4]^{2-}$). Thiocyanate, H_2O and acetone give blue $[Co(NCS)_4]^{2-}$, suggesting Co^{II}, but diamagnetism points to either $[Co^+(NH_3)_5(NO^+)](Cl^-)_2$ or $[Co^{3+}(NH_3)_5(NO^-)](Cl^-)_2$. The lustrous appearance suggests electron exchange.

Air or H_2O_2 oxidizes Co^{II} in cold NH_3 and "$(NH_4)_2CO_3$":

$$[Co(NH_3)_6]^{2+} + {}^1/_2\, H_2O_2 + HCO_3^- \rightarrow$$

$$[Co(\eta^2\text{-}CO_3)(NH_3)_4]^+ + 2\, NH_3 + H_2O$$

which can be crystallized to red products with $C_2O_4^{2-}$, NO_3^-, SO_4^{2-}, SeO_4^{2-}, Cl^-, Br^-, I^- etc. Dilute acids change this to $[cis\text{-}Co(H_2O)_2(NH_3)_4]^{3+}$; concentrated HX give $[cis\text{-}CoX(H_2O)(NH_3)_4]^{2+}$ or $[cis\text{-}CoX_2(NH_3)_4]^+$, with X = Cl^-, Br^- etc. Treating the $[cis\text{-}Co(H_2O)_2(NH_3)_4]^{3+}$ with excess hot, dilute NH_3 yields the very stable $[Co(H_2O)(NH_3)_5]^{3+}$. These tetraammines and pentaammines serve well for further syntheses.

Bubbling air for 48 h through $Co(NO_3)_2$, concentrated NH_3, and "$(NH_4)_2CO_3$", while adding more NH_3 at times, then storing at 5 °C, can yield pink $[Co(\eta^1\text{-}CO_3)(NH_3)_5]NO_3$.

The following equation leads to a purely inorganic, chirally resolvable, lustrous violet-brown complex, "hexol", slightly soluble in water and completely precipitated by CrO_4^{2-}, $[Cr_2O_7]^{2-}$ or $[PtCl_6]^{2-}$; it recalls the historic [1] proof that chirality does not require organic ligands, thus erasing the long-held organic/inorganic distinction:

$$4\, [cis\text{-}CoCl(H_2O)(NH_3)_4]^{2+} + 2\, NH_3 + 3\, SO_4^{2-} + 6\, H_2O \rightarrow$$

$$[Co\{(\mu\text{-}OH)_2Co(NH_3)_4\}_3](SO_4)_3\cdot 4H_2O\downarrow + 6\, NH_4^+ + 4\, Cl^-$$

Catalysis by charcoal yields a yellow-brown source for triammines:

$$Co(NH_3)_6^{2+} + {}^1/_2\, H_2O_2 + 3\, NO_2^- + NH_4^+ \rightarrow$$

$$[Co(NH_3)_3(NO_2)_3]\downarrow + 4\, NH_3 + H_2O$$

Concentrated HNO_3 forms $[Co(H_2O)_3(NH_3)_3](NO_3)_3$, very hygroscopic.

Oxidation of an ammoniacal solution of $CoCl_2$ yields $[Co(NH_3)_5Cl]Cl_2$. This is only slightly soluble in concentrated HCl and, with enough Cl^- present, may be used to separate Co^{2+} from Ni^{2+}. Only the outer two-thirds of the chlorine is precipitated quickly by Ag^+.

Aqueous $[Co(NH_3)_6]^{3+}$ in acid and darkness is extremely inert.

Ammonia, concentrated in this case, yields many ammines from other Co^{III} complexes, e.g., this deep-blue, chiral product:

$$[Co(CO_3)_3]^{3-} + 2\ NH_3 \rightarrow [\textit{cis}\text{-}Co(NH_3)_2(CO_3)_2]^- + CO_3^{2-}$$

This is reasonably stable in solution with excess HCO_3^-, although the resolved form racemizes with a half time of 3 min.

Diazane, N_2H_4, at 40 °C replaces H_2O in $[Co(CN)_5(H_2O)]^{2-}$, producing $[Co(CN)_5(N_2H_4)]^{2-}$, which precipitates, e.g., Ag^+ as $Ag_2[Co(CN)_5(N_2H_4)]$.

Aqueous $[Co(CN)_5]^{3-}$ splits NH_2OH into two parts (like H_2O_2), forming $[Co(CN)_5(NH_3)]^{2-}$ and $[Co(CN)_5(H_2O)]^{2-}$ after hydronation.

Aqueous Co^{2+} or $[Co(CN)_5(H_2O)]^{2-}$ and N_3^- give $[Co(N_3)_4]^{2-}$ or $[Co(CN)_5(N_3)]^{3-}$ respectively.

Oxidized nitrogen. Treatment of $[Co(CO)_4]^-$ with NO for several hours yields the red liquid $[Co(CO)_3NO]$. The drive of the **d**-block metal with its unsaturated ligands for the 18-electron configuration actually uses the potential oxidant NO to release the reductant H_2:

$$[Co(CO)_4]^- + NO + H_2O \rightarrow$$

$$[Co(CO)_3(NO)]_{liq}\downarrow + CO\uparrow + {}^1/_2\ H_2\uparrow + OH^-$$

Nitrosyl Co complexes are not made by removing O from attached NO_2^-, ONO^- or NO_3^- nor by oxidizing attached NH_3 or NH_2OH, but NO, NH_3 and CoX_2 (X = Cl or NO_3) at 0 °C for 45 min give diamagnetic, black $[Co(NH_3)_5(NO)]X_2$; at ambient T in 2 h they appear to yield a "hyponitrite", red, diamagnetic $[Co(NH_3)_5\text{–}ON{=}NO\text{–}Co(NH_3)_5]X_4$. At 0°C, $[Co(NH_3)_5(NO)]^{2+}$ and KCN form a dimer or yellow, diamagnetic $K_3[Co(CN)_5(NO)]\cdot nH_2O$; more CN^- and H_3O^+ give $[Co(CN)_6]^{3-}$.

A neutral or acetic-acid solution of Co^{2+} and KNO_2 saturated with KCl precipitates golden-yellow $K_3[Co(NO_2\text{-}\kappa N)_6]$, faster with shaking and nearly complete in about ten minutes (separation from Ni):

$$Co^{2+} + 7\ NO_2^- + 3\ K^+ + 2\ CH_3CO_2H \rightarrow$$

$$K_3[Co(NO_2)_6]\downarrow + NO\uparrow + 2\ CH_3CO_2^- + H_2O$$

Air, Co^{2+}, NO_2^-, NH_3 and K^+ give K[*trans*-$Co(NO_2)_4(NH_3)_2$], yellow- brown. Aqueous Co^{2+}, CH_3CO_2H, NO_2^-, NH_3, H_2O_2, charcoal and heat, partly in sequence, yield yellow-brown $[Co(NO_2)_3(NH_3)_3]$ isomers etc.

Cold, aqueous $[Co(NH_3)_5(H_2O)]Cl_3$, NO_2^- and then HCl mainly attach NO^+ to the H_2O to form $[Co(NH_3)_5(NO_2\text{-}\kappa O)]Cl_2$, isomerizing warm to $NO_2\text{-}\kappa N$, but with some direct attack of $NO_2^-\text{-}\kappa N$ on $[Co(NH_3)_5]^{3+}$.

Nitric acid, $[Co(\eta^2\text{-}CO_3)(NH_3)_4]^+$, NO_2^- and heat can give a yellow isomer, [*cis*-$Co(NO_2)_2(NH_3)_4]NO_3$ but Co^{2+}, NO_2^-, NH_3, NH_4^+ and air produce yellow-brown [*trans*-$Co(NO_2)_2(NH_3)_4]NO_3$.

Nitrite can form, as further examples of complexes:

$$[Co(NH_3)_5N_3]^{2+} + HNO_2 + H_3O^+ \rightarrow$$

$$[Co(NH_3)_5H_2O]^{3+} + N_2O\uparrow + N_2\uparrow + H_2O$$

$$[cis\text{-}Co(NH_3)_2(\eta^2\text{-}CO_3)_2]^- + 2\,NO_2^- + 2\,CH_3CO_2H \rightarrow$$

$$[cis\text{-}cis\text{-}Co(NO_2)_2(NH_3)_2(\eta^2\text{-}CO_3)]^- + CO_2\uparrow + 2\,CH_3CO_2^- + H_2O$$

Aqueous N_2O_3 and Co–OH complexes form Co–ONO rather rapidly at ambient T if $3 < pH < 5$, with retention of the Co–O oxygen isotope.

Nitrite, $[Co(NH_3)_5(H_2O)]^{3+}$ and HCl or $HClO_4$ (HX) yield orange $[Co(\text{–}ONO)(NH_3)_5]X_2$. Red [*cis*-$Co(\text{–}ONO)_2(NH_3)_4]Q$ arises from H_3O^+, $[Co(CO_3)(NH_3)_4]^+$, then NO_2^- at 5 °C for 10 min; Q = NO_3 or ClO_4.

Excess H_3O^+ plus NO_2^- (giving NO^+) and $[Co(N_3)(NH_3)_5]^{2+}$ may form $[Co\text{–}N\text{=}N\text{=}N\text{–}N\text{=}O(NH_3)_5]^{3+}$, promptly yielding $[Co(H_2O)(NH_3)_5]^{3+}$, N_2 and N_2O.

Heating a Co–OH_2 complex with NO_2^- in 1-cM H_3O^+ at 60–80 °C for 20 min usually converts Co–ONO to Co–NO_2, reversible by light, which thus promotes the acidic removal of the $NO_2\text{-}\kappa O$ Groups, catalyzed by NCS^-, Cl^-, Br^- and I^-. Otherwise the hydrolysis of Co–NO_2 is quite slow, although faster with hot H_3O^+ or OH^-.

The metal dissolves quickly as Co^{2+} on warming in dilute HNO_3, but concentrated HNO_3 passivates it. Concentrated NO_3^- and Co^{2+}, however, form $[Co\,(\eta^2\text{-}NO_3)_4]^{2-}$, with the two O atoms of NO_3^- bound unequally.

9.1.3 Reagents Derived from the 3rd-to-5th-Period Non-Metals, Silicon through Xenon

Phosphorus species. Aqueous Co^{3+} is reduced to Co^{2+} by HPH_2O_2, which also slowly reduces $[Co(CN)_6]^{3-}$ to impure $Co(CN)_2$.

Phosphates, e.g., HPO_4^{2-}, precipitate Co^{2+} as a red $CoHPO_4$, soluble in acids and NH_3. Diphosphate forms a gelatinous precipitate with Co^{2+}, soluble in excess $[P_2O_7]^{4-}$. Adding CH_3CO_2H then causes reprecipitation even in the presence of tartrates (separation from Ni^{II}, but not from Fe^{II} or Mn^{II}). If Co^{2+} is treated with saturated $(NH_4)_2HPO_4$, and the precipitate dissolved in HCl, when heated the addition of an excess of NH_3 precipitates blue $CoNH_4PO_4$ (separation from Ni).

Heating $[Co(H_2O)(NH_3)_5]^{3+}$ with H_3PO_4 and $H_2PO_4^-$ (mixed) at 70–80°C for 1 h, followed by ion-exchange separation in dilute OH^-, allows one to isolate $[CoPO_4(NH_3)_5]\cdot 2H_2O$. This retains the phosphate at least for hours in solution, where it is $[Co(H_nPO_4)(NH_3)_5]^{n+}$ and $n = 0, 1, 2$ or 3 for pH > 9, 4 to 8, < 3 or very low (concentrated $HClO_4$), in turn, but Cr^{2+} reduces these quickly, and OH^- gives $[CoOH(NH_3)_5]^{2+}$ and an oxide.

From various cobalt(III) ammines and polyphosphates one can get red $[Co(NH_3)_5(\eta^1\text{-}HP_2O_7)]\cdot H_2O$, red-violet $[Co(NH_3)_4(\eta^2\text{-}HP_2O_7)]\cdot 2H_2O$ and red $[Co(NH_3)_4(\eta^2\text{-}H_2P_3O_{10})]\cdot H_2O$, each with a –Co–O–P–O–P–O– ring, and lavender $[Co(NH_3)_3(\eta^3\text{-}H_2P_3O_{10})]$ with two 6-rings.

Arsenic species. Soluble arsenites and arsenates precipitate Co^{2+}, forming the corresponding cobalt arsenite or arsenate, bluish white, soluble in NH_3 or in acids, including arsenic acid.

At pH 6.5 and 40 °C for 35 min, $H_2AsO_4^-$ and $[Co(NH_3)_5(H_2O)]^{3+}$ give a red $[Co(AsO_4)(NH_3)_5]$. At 40 °C for 30 min, $HAsO_4^{2-}$ and $[Co(NH_3)_4(H_2O)_2]^{3+}$ yield a violet $[Co(\eta^2\text{-}AsO_4)(NH_3)_4]$.

Reduced chalcogens. Sulfane, H_2S, gradually and incompletely precipitates black cobalt(II) sulfide, "CoS", from solutions of Co^{2+}; from acetate solution the separation is fairly prompt and complete, but in the presence of strong acids, no precipitate forms. If, however, the cobalt is in NH_3 solution, the reaction is rapid and complete. Alkali sulfides immediately and completely precipitate "CoS", which is insoluble in excess of the reagent. At first the precipitate is distinctly soluble in dilute HCl, but upon standing 10 to 15 minutes, quite insoluble. The simple formula "CoS" represents the $Co(OH)_x(SH)_{2-x}$ produced free of air. Air rapidly forms Co(OH)S, still acid-soluble, which, with more sulfide reagent, becomes a less soluble Co^{III} sulfide.

In some unknown mixtures, Co is thus precipitated along with Mn, Ni and Zn. The sulfides are digested with cold, 1-M HCl, which dissolves any MnS and ZnS (also traces of "CoS" and "NiS"). The residue is dissolved readily in HNO_3 or even more easily in aqua regia, and any Co^{2+} detected by means of tests applicable in the presence of Ni: the HCO_3^- plus H_2O_2 test, or the production of $[Co(NCS)_4]^{2-}$ or $[Co(NO_2)_6]^{3-}$.

Upon continued exposure to the air, freshly precipitated "CoS" is gradually oxidized to the sulfate, as occurs with FeS.

In acid the Co^{3+} ion is reduced to Co^{2+} by H_2S.

Cyanide, $Co(CH_3CO_2)_2$ and $K_2Hg(SeCN)_4$ form $K_3[Co(CN)_5(NCSe)]$.

Aqueous $K_3[Co(CN)_5]$ and CS_2, followed by ethanol and purification, appear to yield $K_6[(NC)_5Co\text{–}S\text{–}C(=S)\text{–}Co(CN)_5]$, light-yellow.

The "ethylxanthate" ion, $EtOCS_2^-$, forms a green precipitate in neutral or slightly acidic solutions of Co^{2+}. The Ni compound precipitates too, but dissolves alone in NH_3 as a blue solution.

Elemental and oxidized chalcogens. Sulfur and $K_3[Co(CN)_5]$ yield brown $K_6[\{Co(CN)_5\}_2(\mu\text{-}S_2)]\cdot 4H_2O$; also $[\{Co(CN)_5\}_2(\mu\text{-}O_2)]^{6-}$, or even $[Co(CN)_5\,(OH)]^{3-}$ plus H_2S, does the same. Air slowly oxidizes one of the S atoms to SO_2 in $-S-S(O_2)-$ without breaking the other bonds.

Selenium and $K_3[Co(CN)_5]$ yield brown $K_6[\{Co(CN)_5\}_2(\mu\text{-}Se_2)]\cdot 5H_2O$.

When Co^{2+} is boiled with $S_2O_3^{2-}$ in neutral solution, "CoS" is partly precipitated. Cobalt(II) acetate, $K_2S_2O_3$ and HCN at 0°C give rise to yellow $K_4[Co^{III}(CN)_5(S_2O_3\text{-}\kappa S)]$.

A brown Co^{II} "sulfoxylate", i.e., dioxosulfate(2−), $CoSO_2\cdot{\sim}2H_2O$, is hygroscopic but unusually stable, formally derived from unstable "sulfoxylic" acid, H_2SO_2, but not convertible to other metal salts of that acid, and with the S easily oxidized by HNO_3, Cl_2, Br_2 etc.:

$$Co^{2+} + S_2O_4^{2-} + HCO_3^- \rightarrow CoSO_2\cdot aq\downarrow + HSO_3^- + CO_2\uparrow$$

From $K_3[CoBr(CN)_5]$ and K_2SO_3, followed by methanol, arises yellow $K_4[Co(CN)_5(SO_3)]\cdot 4H_2O$. Aqueous $K_3[Co(CN)_5]$ and SO_2, followed by methanol, may yield orange $K_6[\{Co(CN)_5\}_2(\mu\text{-}SO_2\text{-}1\kappa S{:}2\kappa O)]\cdot 4H_2O$.

Cobalt(II) acetate plus much SO_2 and concentrated CN^- form yellow $Na_5[\textit{trans}\text{-}Co(CN)_4(SO_3)_2]\cdot 3H_2O$ or $K_5[\textit{trans}\text{-}Co(CN)_4(SO_3)_2]\cdot H_2O$. Or, one may obtain the isomeric $Na_2K_3[\textit{cis}\text{-}Co(CN)_4(SO_3)_2]\cdot{}^5/_2H_2O$ from $Na_3[Co(SO_3)_3]$ plus KCN, and then precipitation by methanol.

In acid, SO_2 reduces Co^{3+} to Co^{2+}.

Sulfite and $[Co(NO_2)_6]^{3-}$ give $[Co(SO_3)_3]^{3-}$.

Aqueous $[Co(NH_3)_5OH]^{2+}$ and $SO_2\cdot H_2O$ yield $[Co(NH_3)_5(SO_3\text{-}\kappa O)]^+$ immediately, soon reverting in acid to $[Co(NH_3)_5(H_2O)]^{3+}$, but in less acid slowly giving Co^{2+} and SO_4^{2-}, 2:1.

At a pH 1 to 3, $[Co(NH_3)_5(H_2O)]^{3+}$ and H_2SeO_3 quickly form stable $[Co(NH_3)_5(HSeO_3\text{-}\kappa O)]^{2+}$, much faster than HSO_4^- or $HSeO_4^-$ reacts. Rather similarly, $SO_2\cdot H_2O$ gives a κO ion, but rapidly goes to the κS.

At pH 8, $[Co(NH_3)_5(OH)]^{2+}$ and $HSeO_3^-$ give $[Co(NH_3)_5(SeO_3\text{-}\kappa O)]^+$ immediately, but less quickly with SeO_3^{2-} at pH 10.

At pH 5.5, $S_2O_5^{2-}$ and $[Co(NH_3)_5(H_2O)]^{3+}$ give an unstable, red $[Co(SO_3\text{-}\kappa O)(NH_3)_5]^+$, but they form brown $[Co(SO_3\text{-}\kappa S)(NH_3)_5]Cl\cdot H_2O$ from NH_3 and Cl^- at 40–60 °C. This and concentrated HCl for 30 min yield yellow-brown $[\textit{trans}\text{-}Co(SO_3\text{-}\kappa S)(H_2O)(NH_3)_4]Cl$. Then more Na_2SO_3, then methanol, give yellow $Na[\textit{trans}\text{-}Co(SO_3\text{-}\kappa S)_2(NH_3)_4]\cdot 2H_2O$. The brown isomeric ion in $NH_4[\textit{cis}\text{-}Co(SO_3\text{-}\kappa S)_2(NH_3)_4]\cdot 3H_2O$ arises quickly from HSO_3^-, $[\textit{cis}\text{-}Co(NH_3)_4(H_2O)_2]^{3+}$ and NH_4^+.

Heating $[Co(NH_3)_5(H_2O)](Br,ClO_4)_3$ with SeO_3^{2-} at 70 °C for a few minutes and cooling give $[Co(NH_3)_5(SeO_3\text{-}\kappa O)](Br,ClO_4)\cdot H_2O$, bright pink or red respectively, reversible in acid. At pH 1 to 3, $HSeO_3^-$ reacts $\geq 10^3$ times as fast with $[Co(NH_3)_5(H_2O)]^{3+}$ as the exchange with solvent H_2O. At pH 6 to 10, $[(Co,Rh)(NH_3)_5OH]^{2+}$ quickly gives $[M(NH_3)_5SeO_3]^+$.

Dissolving $Co_2O_3 \cdot aq$ in H_2SO_4 releases O_2, forming Co^{2+}, but the very reactive Co^{3+} ion can be isolated (cold) as the deep-blue alum, $[Cs(H_2O)_6][Co(H_2O)_6](SO_4)_2$, easily dehydrated and decomposed.

Concentrated H_2SO_4 liberates CO from $[Co(CN)_6]^{3-}$.

Treating Co^{2+} with OH^- (not NH_3), and $[S_2O_8]^{2-}$, yields a dark-brown precipitate of $Co_2O_3 \cdot aq$, soluble neither in NH_3 plus NH_4^+ nor in CN^-.

Aqueous $[Co(CN)_5]^{3-}$ and $[S_2O_8]^{2-}$ form $[Co(CN)_5(H_2O)]^{2-}$.

Aqueous $[S_2O_8]^{2-}$ can act as a (slow) one-electron oxidant, when catalyzed by Ag^+, via Ag^{2+}, in this example (cf. $\mathbf{H_2O_2}$ or $\mathbf{Cl_2}$), although reducing the Co one step while oxidizing the oxalate by two:

$$[Co(NH_3)_5(HC_2O_4)]^{2+} + {}^1/_2\,[S_2O_8]^{2-} + 4\,H_3O^+ \rightarrow$$

$$Co^{2+} + SO_4^{2-} + 2\,CO_2\uparrow + 5\,NH_4^+ + 4\,H_2O$$

Peroxo becomes hyperoxo with $[\{Co(NH_3)_5\}_2(\mu\text{-}O_2)]^{4+}$ and $[S_2O_8]^{2-}$ forming $[\{Co(NH_3)_5\}_2(\mu\text{-}O_2)]^{5+}$. Then $[Ru(NH_3)_6]^{2+}$ reverses this.

Reduced halogens. The higher oxides and hydroxides, also Co^{3+}, release X_2 and possibly Cl_2O from HX, with warming if need be, forming Co^{2+}, but HCl does not reduce Co^{III} ammines or $[Co(CN)_6]^{3-}$.

In dilute solution $CoCl_2$ is pink, as $[Co(H_2O)_6]^{2+}$; adding concentrated HCl changes it to deep blue $[CoCl_4^{2-}]$, known as, e.g., $Cs_3[CoCl_4]Cl$, not $Cs_3[CoCl_5]$. This serves to detect 1 μmol of Co. Nickel and iron(III) interfere, giving a green and a yellow color, in turn; Mn^{2+} does not interfere. Also known are $[CoBr_4]^{2-}$ and $[CoI_4]^{2-}$.

The halides replace H_2O in $[Co(CN)_5(H_2O)]^{2-}$ with rates for I_3^- (which releases I_2) > I^- > Br^- > Cl^-, opposite the order for $[Co(NH_3)_5(H_2O)]^{3+}$. The I_3^- or I_2 then also catalyzes the (reverse) aquation of $[Co(CN)_5I]^{3-}$.

Excess KBr, with $K_2[Co(CN)_5(H_2O)]$, yields $K_3[CoBr(CN)_5]$.

Aqueous KI, with $[Co(CN)_5(H_2O)]^{2-}$ and then ethanol as precipitant, yields red-brown $K_3[Co(CN)_5I]$.

Elemental and oxidized halogens. The halogens and Co form Co^{2+}.

If Co^{2+} is treated with Cl_2 and digested cold with $BaCO_3$, $Co_2O_3 \cdot aq$ precipitates (distinction from Ni). Aqueous Cl_2 and $Co(OH)_2$ also yield the less soluble Co_2O_3, thus removing cobalt from some mixtures.

An example of the (slow) oxidation of a ligand is:

$$[Co^{III}(NH_3)_5(HC_2O_4)]^{2+} + Cl_2 + 2\,H_2O \rightarrow$$

$$[Co(NH_3)_5(H_2O)]^{3+} + 2\,Cl^- + 2\,CO_2\uparrow + H_3O^+$$

Chlorine and $K_3[Co(CN)_5]$ form $K_3[Co(CN)_6]$ and light-yellow $K_3[CoCl(CN)_5]$. Bromine and then ethanol give $K_3[CoBr(CN)_5]$.

Aqueous KI_3, with $[Co(CN)_5]^{3-}$ and then ethanol as precipitant, yields red-brown $K_3[Co(CN)_5I]$.

Iodine or ClO^-, like O_2 or H_2O_2, q.v., oxidizes Co^{II} ammines to Co^{III}.

If ClO^- is added to a slightly acidic Co^{2+} solution, a precipitate of $Co_2O_3 \cdot aq$ forms in a short time. The brown $Co_2O_3 \cdot aq$ is precipitated also on treating $Co(OH)_2$ with ClO^-, BrO^- or IO^- (from Cl_2, Br_2 or I_2) in the presence of OH^- or CO_3^{2-} (not NH_3), and it may go to black $CoO_2 \cdot aq$. It does not dissolve in NH_3 plus NH_4^+, or in CN^-.

Aqueous Co^{2+}, ClO^-, IO_4^-, OH^- and acidic ion exchange, alternately $[Co(CO_3)_3]^{3-}$, $NaIO_4$ and $HClO_4$, form an interesting, diamagnetic, stable, dark-green acid, $(H_3O)_3[Co^{III}\text{-}\eta^6\text{-}cyclo\text{-}\{Co^{III}(H_2O)_2(\mu\text{-}O)_2IO_2(\mu\text{-}O)_2\}_3]$, that is, $(H_3O)_3[Co\{Co(IO_6)(H_2O)_2\}_3]$, soluble in water, precipitated by K^+, Cs^+, Ag^+ and large cations, and reduced by acidified SO_2, I^- and Fe^{2+}:

$$4\,[Co(CO_3)_3]^{3-} + 3\,H_5IO_6 + 12\,H_3O^+ \rightarrow$$

$$(H_3O)_3[Co_4(IO_6)_3(H_2O)_6]\downarrow + 12\,CO_2\uparrow + 15\,H_2O$$

9.1.4 Reagents Derived from the Metals Lithium through Uranium, plus Electrons and Photons

Oxidation. Reagents MnO_4^- and PbO_2 oxidize Co^{II} ammines to Co^{III}, somewhat similarly to the action of O_2 or H_2O_2, which see. This prevents cobalt's precipitation by OH^- (separation from nickel).

An interesting redox result is:

$$[Co^{II}(CN)_5]^{3-} + [Fe^{III}(CN)_6]^{3-} \rightarrow [Co^{III}(CN)_5NCFe^{II}(CN)_5]^{6-}$$

The $[Co(CN)_5H]^{3-}$ ion is rather inert for electron transfer, e.g., with $[Fe(CN)_6]^{3-}$, but high pH forms $[Co(CN)_5]^{4-}$ and then:

$$[Co(CN)_5]^{4-} + 2\,[Fe(CN)_6]^{3-} \rightarrow$$

$$Co^{III}(CN)_5NCFe^{II}(CN)_5]^{6-} + [Fe(CN)_6]^{4-}$$

and such salts as $Ba_3[Co(CN)_5NCFe(CN)_5] \cdot 16H_2O$.

In water, $NiO(OH)$ and $Co(OH)_2$ yield $Ni(OH)_2$ and the less soluble Co_2O_3, thus making cobalt removable from some mixtures.

Cathodic e^-, surprisingly, can actually oxidize Co^{II} to Co^{III}:

$$[Co^{II}(CN)_5]^{3-} + e^- + H_2O \leftrightarrows [Co^{III}(CN)_5H^{-I}]^{3-} + OH^-$$

Charging the positive electrode in "lithium-ion" batteries, using a solid electrolyte, oxidizes the cobalt (reversed during discharge):

$$LiCoO_2 \rightarrow CoO_2 + Li^+ + e^-$$

Light (UV, 254 nm), Co^{2+}, and a pH < 1 release H_2 from H_3O^+.

Reduction. At least Mg, Zn and Cd precipitate Co from Co^{2+}.

Acidified Cr^{2+} reduces $[Co(NH_3)_5(H_2O)]^{3+}$ or $[Co(NH_3)_5(S_2O_3\text{-}\kappa S)]^+$ for example, forming Cr^{3+} or $[Cr(H_2O)_5(S_2O_3\text{-}\kappa O)]^+$ (which goes on to Cr^{3+} and $S_2O_3^{2-}$), and Co^{2+} and NH_4^+ in each case.

Aqueous Eu^{2+}, V^{2+}, Cr^{2+} or Fe^{2+} reduces $[\{Co(NH_3)_5\}_2(\mu\text{-}O_2)]^{5+}$ (hyperoxo) with one e^- to Co^{2+} and O_2, which immediately consumes four more moles of the Eu^{2+}, V^{2+} or Cr^{2+}; otherwise:

$$[\{Co(NH_3)_5O\text{–}\}_2]^{5+} + Fe^{2+} + 10\ H_3O^+ \rightarrow$$

$$2\ Co^{2+} + Fe^{3+} + O_2 + 10\ NH_4^+ + 10\ H_2O$$

The reduction of $[Co^{III}Cl(NH_3)_5]^{2+}$ by Cr^{2+}, i.e., $[Cr(H_2O)_6]^{2+}$, has special interest for the early showing of direct atom transfer in forming the activated complex $[(NH_3)_5CoClCr(H_2O)_5]^{4+}$ before becoming the inert $[Cr^{III}Cl(H_2O)_5]^{2+}$ and labile Co^{II}. Some relative rates are I > Br > Cl > F, also with V^{2+}, $[Co(CN)_5]^{3-}$ or (outer-sphere) $[Ru(NH_3)_6]^{2+}$ as reductants. The (much slower) order is F > Cl > Br > I with Eu^{2+} or Fe^{2+}, paralleling the thermodynamic stabilities of Eu^{III} and Fe^{III}.

In a ClO_4^- medium, catalyzed by Ti^{IV}, Ti^{2+} also reduces $[CoBr(NH_3)_5]^{2+}$ and $[CoI(NH_3)_5]^{2+}$ to Co^{II}. Aqueous $[Co(C_2O_4)_3]^{3-}$ in acid is reduced by Ti^{2+} (slower than by Ti^{III}!), Ti^{III}, V^{III}, Fe^{2+}, Ga^{I}, In^{I}, Ge^{II} and Sn^{II}.

Aqueous Cr^{2+} reduces $[Co^{III}Q(NH_3)_5]$ to Co^{II}, with Q = CHO_2^-, N_3^-, $NO_2^-\text{-}\kappa O$, NO_3^-, $H_2PO_4^-$, $SCN^-\text{-}\kappa S$, $S_2O_3^{2-}\text{-}\kappa S$, SO_4^{2-} etc. (especially rapidly for N_3^- and NO_2^-, both easily reducible); for example:

$$[Co(S_2O_3\text{-}\kappa S)(NH_3)_5]^+ + Cr^{2+} + 5\ H_3O^+ \rightarrow$$

$$Co^{2+} + Cr(S_2O_3\text{-}\kappa O)^+ + 5\ NH_4^+ + 5\ H_2O$$

The reduction of cobalt(III) is actually induced by some one-electron *oxidants*, e.g., Ce^{IV} or $[Co(H_2O)_6]^{3+}$, acting on complexes containing reductants, such as oxalate:

$$[Co(NH_3)_5(HC_2O_4)]^{2+} + 4\ H_3O^+ \rightarrow$$

$$Co^{2+} + e^- + 2\ CO_2\uparrow + 5\ NH_4^+ + 4\ H_2O$$

Excess Cr^{2+} mixed with $[Co(NH_3)_5SCN]^{2+}$ in rapid flow quickly yields Co^{2+} and both $CrSCN^{2+}$ from an adjacent attack, and $CrNCS^{2+}$ from the remote. Then

Cr^{2+} more slowly catalyzes the change of $CrSCN^{2+}$ to $CrNCS^{2+}$. Still quick but less so is the reduction of $[Co(NH_3)_5NCS]^{2+}$.

Vanadium(2+) reduces $[Co(CN)_5X]^{3-}$, for X = Cl, Br or I, also N_3 and NCS, forming the intermediate $[Co(CN)_5]^{3-}$ and some VX^{2+} product:

$$[Co(CN)_5X]^{3-} + V^{2+} + 5\ H_3O^+ \rightarrow Co^{2+} + V^{3+} + X^- + 5\ HCN + 5\ H_2O$$

Dissolved Co^{III} with Mn^{2+} or Fe^{2+} in H_3O^+ forms Co^{2+} and Mn^{III} or Fe^{III}.

Aqueous $[Ru(NH_3)_6]^{2+}$ reduces many Co^{III} to Co^{II}.

Aqueous $[Co(CN)_5]^{3-}$ quickly reduces $[Co(NH_3)_5NCS]^{2+}$ and, even faster, $[Co(NH_3)_5SCN]^{2+}$, to Co^{II}, both also giving $[Co(CN)_5SCN]^{3-}$.

Copper(1+) in, e.g., 2-dM $HClO_4$ reduces (to Co^{II}) $[CoX(NH_3)_5]^{2+}$ with X = OH, CN, NCS, N_3, F, Cl and Br, but also some $[CoX_2(NH_3)_4]^+$ etc.

Tin(II) does not reduce $[CoCl(NH_3)_5]^{2+}$, $[CoBr(NH_3)_5]^{2+}$ and so on.

Light can cause redox changes, producing transient X, when X = Br or I for example (Cl gives more nitrogenous radicals) as in:

$$[CoX(NH_3)_5]^{2+} + \gamma + 5\ H_3O^+ \rightarrow Co^{2+} + X + 5\ NH_4^+ + 5\ H_2O$$

Other reactions. Acidified Cr^{2+} and $[CoBr(CN)_5]^{3-}$ give $[Co(CN)_5H^{-I}]^{3-}$, but the H^+ in the H_3O^+, not the Co^{3+}, is seen to be the reduced moiety.

Aqueous CrO_4^{2-} precipitates basic cobalt(II) chromate, reddish brown, from Co^{2+} in a neutral solution. The product is soluble in NH_3 or in acids. No precipitate is obtained with $[Cr_2O_7]^{2-}$.

Cobalt(II) in NH_3, warmed with H_2O_2 and then acidified with CH_3CO_2H, is precipitated by $(NH_4)_2MoO_4$. The $[Co(NH_3)_5(H_2O)]^{3+}$ ion and MoO_4^{2-} or WO_4^{2-} quickly generate $[Co(NH_3)_5(MO_4)]^+$.

Aqueous $[Co(CN)_5]^{3-}$ forms characteristic, insoluble precipitates with many **d**- or **p**-block metal ions. The $[Co(CN)_6]^{3-}$ ion, however, selectively precipitates Ln^{3+} and the M^I and M^{II} **d**-block ions, including Mn^{2+}, Fe^{2+}, Co^{2+}, Ni^{2+}, Cu^{2+}, Ag^+, Zn^{2+}, Cd^{2+}, Hg_2^{2+} and Hg^{2+}, ∴ not Pb^{2+}, e.g., as $M^{II}_3[Co(CN)_6]_2 \cdot 12\text{–}14H_2O$, $Ag_3[Co(CN)_6]$ and $(Hg_2)_3[Co(CN)_6]_2$; this can serve to distinguish experimentally between the low-valent IUPAC **d**-block and "main" Groups, with Group 12 in the former. A system of qualitative analysis has used this property; see Appendix A, ref 4.

The $[Fe(CN)_6]^{4-}$ ion, with Co^{2+}, precipitates $Co_2[Fe(CN)_6]$.aq, gray green, insoluble in acids. Aqueous $[Fe(CN)_6]^{3-}$ precipitates $Co_3[Fe(CN)_6]_2$, brownish red, insoluble in acids. A fairly distinctive test for Co^{2+} is obtained by adding $[Fe(CN)_6]^{4-}$ to an ammoniacal solution, whereupon a blood-red color (and precipitate, if sufficient Co is present) appears (distinction from Ni).

Heating 6 h at 50 °C joins $[Co(NH_3)_5(CN)]^{2+}$ and $[Co(CN)_5(H_2O)]^{2-}$ to form, after evaporation at 40 °C, $[Co(NH_3)_5\text{–}CN\text{–}Co(CN)_5] \cdot H_2O$, yellow or orange. Base hydrolyzes this slowly, with isomerization of the CNCo group, and yields $[Co(NH_3)_5(H_2O)]^{3+}$ and $[Co(CN)_6]^{3-}$.

We also find, with immediate aquation of the Co^{II} pentaammine:

$$[Co(CN)_5]^{3-} + [Co(NCS)(NH_3)_5]^{2+} \rightarrow [Co(SCN)(CN)_5]^{3-} + [Co(NH_3)_5]^{2+}$$

With Co^{III}, however, $[Co(CN)_5(H_2O)]^{2-}$ and $[Co(NCS)(NH_3)_5]^{2+}$ yield $[Co(NH_3)_5{-}NCS{-}Co(CN)_5]\cdot H_2O$, orange; but $[Co(SCN)(NH_3)_5]^{2+}$ forms $[Co(NH_3)_5{-}SCN{-}Co(CN)_5]\cdot H_2O$, pink; also $[Co(H_2O)_6]^{2+}$, KCN and $K_2[Hg(SCN)_4]$ give brown $K_4[\{Co^{III}(CN)_4\}_2(\mu\text{-}NCS)(\mu\text{-}SCN)]\cdot 5H_2O$ (with oxidation).

The reaction of $[Co(CN)_5]^{3-}$ and $[Co(NH_3)_5{-}NO_2]^{2+}$ appears to form $[Co(CN)_5{-}ON({=}O){-}Co(NH_3)_5]^{-}$, hydrolyzing this to $[Co(CN)_5{-}ONO]^{3-}$, then rearranging the CoONO group, yielding $[Co(CN)_5{-}NO_2]^{3-}$.

Mixtures of $[Co(NH_3)_5(H_2O)](ClO_4)_3$ and $[Co(NH_3)_5(C_2O_4)]ClO_4$ form red $[\{Co(NH_3)_5\}_2(\mu\text{-}C_2O_4)](ClO_4)_4$ at pH 4, 70–75 °C, in 2–3 h.

Freshly precipitated $Zn(OH)_2$, HgO and $Pb(OH)_2$ precipitate $Co(OH)_2$ from solutions of various Co^{2+} salts at 100 °C.

A standard calibrant for magnetic susceptibility measurements is insoluble $Hg[Co(NCS\text{-}\kappa N)_4]$ or $Co[Hg(SCN\text{-}\kappa S)_4]$.

Aqueous $K_3[Co(CN)_5]$ complexes the Hg from aqueous $Hg(CN)_2$, forming a yellow, diamagnetic $K_6[(NC)_5Co{-}Hg{-}Co(CN)_5]$ after adding ethanol. For the anion we might assign the shared electrons to either Co or Hg with only a small electronegativity difference, as in (respectively) $[(NC)_5Co^{I}{-}Hg^{II}{-}Co^{I}(CN)_5]^{6-}$ or $[(NC)_5Co^{III}{-}Hg^{-II}{-}Co^{III}(CN)_5]^{6-}$, or even a combination, and Cd seems to react similarly.

One Tl^{I} and two $[Co(CN)_5]^{3-}$ produce a diamagnetic ion, possibly $[(NC)_5Co{-}Tl{-}Co(CN)_5]^{5-}$. Likewise $SnCl_2$ gives a species that may be $[(NC)_5Co{-}(SnCl_2){-}Co(CN)_5]^{6-}$.

Light substitutes H_2O for CN^- well in $[Co(CN)_6]^{3-}$, and H_2O for I^- in $[Co(CN)_5I]^{3-}$, but with a poor quantum yield in aquating $[Co(NH_3)_6]^{3+}$.

Light (248 nm) and $[Co^{III}H^{-I}(CN)_5]^{3-}$ give H_2 and $[Co(CN)_5]^{3-}$.

Light can cause linkage isomerization of nitro to nitrito (mixed with aquation and decomposition):

$$[Co(NH_3)_5(NO_2)]^{2+} + \gamma \rightarrow [Co(NH_3)_5(ONO)]^{2+}$$

Light (UV) hydrolyzes $[Co(CN)_6]^{3-}$ to $[Co(CN)_5(H_2O)]^{2-}$ and CN^-. Blue light works with UO_2^{2+} as sensitizer, and this promotes substitution of the H_2O by N_3^-, SCN^- and I^-. Light also hydrolyzes $[Co(CN)_5(OH)]^{3-}$ further, mainly to [*cis*-$Co(CN)_4(OH)_2]^{3-}$; $[Co(CN)_5(H_2O)]^{2-}$ goes much less readily. Visible and near-UV light hydrolyze $[Co(CN)_5X]^{3-}$, most efficiently for X = I, least for X = Cl, to $[Co(CN)_5H_2O]^{2-}$. Light also replaces H_2O (isotopic), NCS^-, NH_3 and N_3^- in these complexes by H_2O.

Cobalt(II) varies nicely in color. The hydrated salts and dilute $[Co(H_2O)_6]^{2+}$, are pink; the anhydrous salts and $[CoX_4]^{2-}$ from concentrated weakly basic ligands, tend to be blue.

The acidopentaamminecobalt(III) salts normally have the following colors: Co–O, pink to red; Co–(NO_2-*κN*), orange; Co–F, pink; Co–Cl, red; Co–Br, purple; Co–I, olive green.

9.2 Rhodium, $_{45}$Rh; Iridium, $_{77}$Ir and Meitnerium, $_{109}$Mt

Oxidation numbers in classical compounds in water: (II), (III) and (IV), as in $[Rh_2(H_2O)_{10}]^{4+}$ (in equilibrium with Rh^{2+}), M_2O_3 and MO_2. The oxidation states for Mt, calculated relativistically to be stable in water: (I), (III), and (VI), especially (I).

9.2.1 Reagents Derived from Hydrogen and Oxygen

Dihydrogen. Hydrogen reduces $Rh_2{}^{4+}$ to Rh. High-pressure H_2 reduces $[RhCl_6]^{3-}$ to Rh quantitatively.

Water. The nitrate and sulfate of Rh^{III} are readily soluble.

The $[Rh(NO_2)_6]^{3-}$ salts, except of Na^+, are insoluble to slightly soluble.

Rhodium trichloride, $RhCl_3$, obtained by chlorination of the metal, is insoluble in H_2O; $RhCl_3 \cdot 3H_2O$ readily gives reddish, yellow or brownish solutions and, like $[RhCl_6]^{3-}$, yields various $[RhCl_n(H_2O)_{6-n}]^{(3-n)+}$.

Dry $IrCl_3$ may be brown or red; both are insoluble, but the several hydrates dissolve in H_2O, making it acidic. Salts such as $(NH_4)_3[IrCl_6]$ are soluble in water but insoluble in ethanol. Some higher-valence salts, e.g., $(NH_4)_2[IrCl_6]$, are only slightly soluble in water, but some other $M^I{}_2[IrX_6]$, with X = F, Cl or Br, are soluble.

Aqueous *cis*- and [*trans*-$Rh^{III}Q(H_2O)(NH_3)_4]^{n+}$, with Q = H_2O, OH^-, CN^-, NH_3, Cl^- or Br^-, are acidic in water with pK_1 from about 5 to 8, and with *trans* the more acidic ones, except for CN^-.

Dark-red $K_3[RhCl_6]$ (from a high-*T* chlorination) and H_2O give a wine-red $[RhCl_5(H_2O)]^{2-}$, whose further aquation (in strong acids because OH^- is *trans*-labilizing too) illustrates the *trans*-effect of Cl^-; it yields only [*cis*-$RhCl_4(H_2O)_2]^-$ and [*fac*-$RhCl_3(H_2O)_3$], then essentially no more. As expected, Hg^{2+} or $HgCl^+$, but not $HgCl_2$, accelerates the removal of Cl^-.

The aquation of $[IrCl_6]^{3-}$ produces $[IrCl_{6-n}(H_2O)_n]^{(3-n)-}$, $1 \le n \le 3$.

Some of the H_2O in $[Rh_2(H_2O)_{10}]^{4+}$ is easily replaced by $CH_3CO_2{}^-$, $SO_4{}^{2-}$ (which see) etc., and in $[Rh(H_2O)_6]^{3+}$ by many ligands, but we have no $[Ir(H_2O)_6]^{3+}$. In $[IrX(NH_3)_5]^{2+}$, H_2O replaces the X increasingly readily in the order Cl < Br < I < NO_3, but more slowly than for Rh^{III}.

Water and IrF_6 give $IrO_2 \cdot aq$ and O_2, but $M^{II}[IrF_6]_2$ form $[IrF_6]^{2-}$ and O_2.

Oxonium. Oxonium ion dissolves $Rh_2O_3 \cdot H_2O$ as $[Rh(H_2O)_6]^{3+}$.

Hydrated $Rh_2O_3 \cdot 5H_2O$ dissolves in HNO_3, H_2SO_4, HCl and CH_3CO_2H.

Solid Rh_2O_3, RhO_2, IrO_2 and $Ir(OH)_4$ are insoluble in H_3O^+.

Anhydrous rhodium trichloride is insoluble in acids.

One of many examples of **d**-block species promoting the hydrolysis of ligands is that $[Rh(NCO\text{-}\kappa N)(NH_3)_5]^{2+}$ and H_3O^+ first produce $[Rh(NH_2CO_2H\text{-}\kappa N)(NH_3)_5]^{3+}$ and then $[Rh(NH_3)_6]^{3+}$ and CO_2.

Hydroxide. Aqueous OH^- slowly precipitates Rh^{III} as the yellow hydroxide, $Rh_2O_3 \cdot H_2O$, soluble in excess OH^-. From a hot solution the product is darker, and the separation is faster from a sulfate solution than from a chloride one. The precipitate dissolves in excess NH_3. The oxide, Rh_2O_3, is slightly soluble in concentrated OH^-. The dioxide, RhO_2, is insoluble in OH^-, but $RhO_2 \cdot aq$ dissolves as a green anion.

Alkalis change $[Rh(CN)_5H]^{3-}$ to $[Rh(CN)_4]^{3-}$, which can add HCN to restore $[Rh(CN)_5H]^{3-}$:

$$[Rh(CN)_5H]^{3-} + OH^- \rightarrow [Rh(CN)_4]^{3-} + CN^- + H_2O$$

$$[Rh(CN)_4]^{3-} + HCN \rightarrow [Rh(CN)_5H]^{3-}$$

The stable complex $[RhH(NH_3)_5]^{2+}$ is not acidic toward OH^-.

Under Ar, adding an alkali to $IrCl_3$, evaporating dry, and extracting impurities with OH^- and H_2SO_4 alternately, gives impure $Ir_2O_3 \cdot aq$.

Aqueous OH^- and $[IrCl_6]^{3-}$ give $[Ir(OH)_6]^{3-}$ or $[Ir(OH)_5(H_2O)]^{2-}$, thence, e.g., $K_2[Ir(OH)_6]$, $Zn[Ir(OH)_6]$ and $Cd[Ir(OH)_6]$.

Base helps isomerize $[Ir(NO_2\text{-}\kappa O)(NH_3)_5]^{2+}$ to $[Ir(NO_2\text{-}\kappa N)(NH_3)_5]^{2+}$.

Dioxygen. Air and HCl slowly dissolve finely divided "Rh black".

Air and acidified $[\{Rh(H_2O)_5\}_2]^{4+}$ form $[\{Rh(H_2O)_5\}_2(\mu\text{-}O_2)]^{4+}$ briefly, then slowly produce a violet, hyperoxo $[Rh^{III}{}_2(O_2)(OH)_2(H_2O)_n]^{3+}$ and $[Rh(H_2O)_6]^{3+}$, but treating the $Rh_2{}^{4+}$ slowly with O_2 by diffusion yields yellow Rh^{III} cationic polymers. Aqueous Cr^{2+} and the first dimer, while fresh, regenerate the $Rh_2{}^{4+}$.

Dioxygen converts $[RhH(CN)_4(H_2O)]^{2-}$ to $[Rh(O_2H)(CN)_4(H_2O)]^{2-}$.

Air partly oxidizes $Ir_2O_3 \cdot aq$ to $IrO_2 \cdot aq$. Boiling Ir^{III} and Ir^{IV} chlorides and chlorocomplexes in air, with either OH^- or $CO_3{}^{2-}$, precipitates dark-blue $Ir(OH)_4 \cdot aq$, insoluble in base.

Acidified $[IrCl_6]^{3-}$ solutions, O_2 and Na^+ form black, soluble $Na_2[IrCl_6]$, a starter for other Ir(IV) preparations.

9.2.2 Reagents Derived from the Other 2nd-Period Non-Metals, Boron through Fluorine

Boron species. Even cold, concentrated $[B_{10}H_{10}]^{2-}$, $[BF_4]^-$, $[BPh_4]^-$ or $[B_{12}Cl_{12}]^{2-}$ does not precipitate $[Rh_2(H_2O)_{10}]^{4+}$.

The $[BH_4]^-$ ion reduces $[RhCl(NH_3)_5]^{2+}$, apparently to $[RhH(NH_3)_5]^{2+}$; the H^- has a strong labilizing *trans*-effect, promoting various syntheses.

Tetraborate gives a precipitate with Rh^{III}.

Carbon oxide species. The four equatorial pairs of H_2O on the Rh–Rh pair in $[\{-Rh(H_2O)(H_2O)_4\}_2]^{4+}$, i.e., $[Rh_2(H_2O)_{10}]^{4+}$ or Rh_2^{4+}, can be replaced by bridges such as CO_3^{2-}, HCO_3^-, $C_2O_4^{2-}$, SO_4^{2-} and $H_2PO_4^-$, e.g.:

$$[Rh_2(H_2O)_{10}]^{4+} + 4\ CO_3^{2-} \rightarrow [\{Rh(H_2O)\}_2(\mu\text{-}CO_3)_4]^{4-} + 8\ H_2O$$

The wintergreen color, under an inert gas, quickly becomes dark blue, and Na^+ or K^+ precipitates a purple solid. More convenient is to suspend $Rh_2(\mu\text{-}CH_3CO_2)_4$ in 2-M Alk_2CO_3 at 100 °C for 10–15 min. Strong acids, which turn it green, release CO_2 only slowly from the $Rh_2(CO_3)_4^{4-}$, unlike $[Co(CO_3)(NH_3)_4]^+$, but H_2O also soon aquates it reversibly. Two equiv of Ce^{IV} in H_2SO_4 oxidize it to two Rh^{III}. Raising the pH to 4-to-5, with more treatment plus acid, give $Rh_2(HCO_3)_2^{2+}$, oxidized by O_2 in days but stable even in air-free concentrated H_2SO_4 at 100 °C for a week!

Aqueous CO_3^{2-} slowly precipitates Rh^{III} as $Rh_2O_3\cdot aq$. Rhodium(III) solutions may be treated with $BaCO_3$ to precipitate Rh hydroxide (distinction from Pt) without making the solution highly alkaline, thus:

$$2\ [RhCl_6]^{3-} + 3\ BaCO_3 \rightarrow$$

$$Rh_2O_3\cdot aq\downarrow + 3\ CO_2\uparrow + 3\ Ba^{2+} + 12\ Cl^-$$

Carbon dioxide and $[Ir(OH)(NH_3)_5]^{2+}$ form $[Ir(HCO_3\text{-}\kappa O)(NH_3)_5]^{2+}$ reversibly, and this ionizes reversibly to $[Ir(CO_3\text{-}\kappa O)(NH_3)_5]^+$. Aqueous CO_3^{2-} and $[IrCl_6]^{3-}$ form greenish or bluish, often colloidal, $Ir_2O_3\cdot aq$.

Iridium solutions may be treated with $BaCO_3$ to precipitate the hydroxide (distinction from Pt).

Cyanide species. Heating aqueous CN^- and $RhCl_3.3H_2O$ yields yellow $Rh(CN)_3\cdot 3H_2O$, soluble in excess KCN, finally giving, after considerable processing, pale-yellow $K_3[Rh(CN)_6]$. This plus HCl and extraction with ether form white $H_3[Rh(CN)_6]$. Treating $(NH_4)_3[IrCl_6]$ with KCN yields $K_3[Ir(CN)_6]$; Mn^{2+} through Zn^{2+} can then precipitate $\mathbf{3d}_3[Ir(CN)_6]_2\cdot aq$.

Excess dry KCN and $[Rh(CO)_2Cl]_2$ in methanol form white, diamagnetic and very hygroscopic $K_3[Rh(CN)_5H]$. We mention this non-aqueous process for the parallel between the Rh and Co products.

Fusing KCN with $(NH_4)_3[IrCl_6]$ or $K_3[IrCl_6]$, and crystallizing from water, forms a very pale-yellow $K_3[Ir(CN)_6]$. The acid, $H_3[Ir(CN)_6]$, is much like $H_3[Rh(CN)_6]$.

Hydrated $IrCl_3$ and $Hg(CNO\text{-}\kappa C)_2$ (fulminate) yield $[Ir(CNO)_6]^{3-}$.

Some "simple" organic species. An oxidative addition is:

$$[Rh^{I}(CN)_4]^{3-} + CH_3I \rightarrow [Rh^{III}(CH_3)(CN)_4I]^{3-}$$

Formic acid (in hot solution) reduces Ir compounds to the metal.

Acetic acid dissolves $RhO_2 \cdot aq$ as a blue complex. Refluxing, reducing and complexing $RhCl_3 \cdot 3H_2O$ or $[RhCl_6]^{3-}$ with RCO_2H or RCO_2^- and other reagents in water plus C_2H_5OH (reductant) form the remarkably stable series $[\{-Rh(H_2O)\}_2(\mu\text{-}CH_3CO_2)_4]$, $Li_2[(-RhCl)_2(\mu\text{-}CH_3CO_2)_4]$ etc., and even $[\{Rh(NO\text{-}\kappa N)\text{–}Rh(NO_2\text{-}\kappa N)\}(\mu\text{-}CH_3CO_2)_4]$, all with $Rh^{II}\text{–}Rh^{II}$ bonds and H_2O, NO^+ or NO_2^- on the ends of the Rh–Rh axis. Axial ligands-κO give green to blue species; ligands-κN or -κS, red to orange.

The $[\{Rh(H_2O)\}_2(CH_3CO_2)_4]$ is inert to O_2, but reacts with O_3 to give $[\{Rh^{III}(H_2O)\}_3(\mu\text{-}CH_3CO_2)_6(\mu_3\text{-}O)]^+$, p$K_a$ 8.3, (made from $RhCl_3 \cdot 3H_2O$ and $AgCH_3CO_2$ also). However, Cl_2, Ce^{IV} or PbO_2, but not H_2O_2 or Ag^+, appears to form $Rh_2(CH_3CO_2)_4^+$, slowly (or with Zn_{Hg}) reverting to the original blue-green, although O_2^{2-} and $S_2O_8^{2-}$ yield bright-yellow Rh^{III}.

At 100 °C under N_2, 5 mM $[\{Rh(H_2O)\}_2(CH_3CO_2)_4]$ in 1-M $HClO_4$ slowly appears to form $Rh_2(CH_3CO_2)_3^+$ and $Rh_2(CH_3CO_2)_2^{2+}$, both reverting to $Rh_2(CH_3CO_2)_4$ with excess acetate. Air oxidizes the stable $Rh^{II}_2(CH_3CO_2)_2^{2+}$, only slowly even at 60 °C in 1-M H_3O^+. Heating it with 1-M H_2SO_4 for 20 min appears to coordinate SO_4^{2-}.

Oxalic acid does not precipitate Rh^{III}, but $C_2O_4^{2-}$ forms $[Rh(C_2O_4)_3]^{3-}$. Refluxing this with $HClO_4$ gives [*cis*-$Rh(C_2O_4)_2(H_2O)_2]^-$. In 1-M H_2SO_4, Ce^{IV} oxidizes $[Rh(C_2O_4)_3]^{3-}$ slowly to $[Rh(C_2O_4)_2(H_2O)_2]^-$ and CO_2.

Oxalic acid reduces Ir^{IV} to Ir^{III} (separation from Au). Oxalates reduce $[IrCl_6]^{2-}$ to $[IrCl_6]^{3-}$, and $C_2O_4^{2-}$ substitutes for H_2O and other ligands in various Ir^{III} species, generally giving chelates like $[IrY_4(C_2O_4)]^{3-}$, $[IrY_2(C_2O_4)_2]^{3-}$ or $[Ir(C_2O_4)_3]^{3-}$, with Y_n as some mixture of NO_2, Cl etc.

Reduced nitrogen. One of many ammines, somewhat analogous to those of Co^{III}, is made by dissolving $RhCl_3 \cdot 3H_2O$ in water and ethanol (a catalyst), 5v:1v, at 30 °C. Concentrated NH_3 is added until the resulting suspension dissolves. This is boiled, giving a pale-yellow color, and then cooled in ice:

$$RhCl_3 \cdot 3H_2O + 5\ NH_3 \rightarrow [RhCl(NH_3)_5]Cl_2\downarrow + 3\ H_2O$$

One may also treat $RhCl_3 \cdot 3H_2O$, $[RhCl_6]^{3-}$ or $[IrCl_6]^{3-}$ with NH_4Cl (the pH buffer) and "$(NH_4)_2CO_3$" (the source of NH_3) on a steam bath for 3 h (Rh) or 6 h (Ir), followed by cooling, giving $[MCl(NH_3)_5]Cl_2$ plus [*trans*-$MCl_2(NH_3)_4]Cl\downarrow$; M = Rh or Ir. The $[MCl(NH_3)_5]Cl_2$ is insoluble in cold 3-M HCl; the $[MCl_2(NH_3)_4]Cl$ is much more soluble.

Further reactions of the first product with various reagents in sequence are cited here in order to illustrate efficiently and briefly the wide range of applicable procedures and products for rhodium(III) ammines. The NH_3 in the pentaammines is often inert, and simply heating $[RhCl(NH_3)_5]Cl_2$ with other ligands often replaces the Cl^-.

Boiling the pentaammine with $AgClO_4$ readily gives the perchlorate of $[Rh(H_2O)(NH_3)_5]^{3+}$, with a pK_a of 6 to 7. Refluxing $[MCl(NH_3)_5]Cl_2$ with OH^- (8 h for Ir, less for Rh) followed by cold HCl yields $[M(H_2O)(NH_3)_5]Cl_3$. Replacement of the H_2O is easier than that of Cl^-, and CO_2 and SO_2 attach directly to form CO_3^{2-} and SO_3^{2-} ligands-κO. Anyway, the $[RhCl(NH_3)_5]^{2+}$ does lose NH_3 slowly at 120 °C with aqueous $HC_2O_4^-$ in an autoclave in 24 h:

$$[RhCl(NH_3)_5]^{2+} + HC_2O_4^- \rightarrow [Rh(C_2O_4)(NH_3)_4]^+ + NH_4^+ + Cl^-$$

Cooling and adding $HClO_4$ yield crude $[Rh(C_2O_4)(NH_3)_4]ClO_4$. Boiling the product one minute with 6-M HCl, then cooling and adding methanol, give yellow $[cis\text{-}RhCl_2(NH_3)_4]Cl\cdot{}^1/_2H_2O$. Now refluxing for several hours with Ag^+ and away from light causes aquation:

$$[cis\text{-}RhCl_2(NH_3)_4]^+ + 2\,Ag^+ + 2\,H_2O \rightarrow$$

$$[cis\text{-}Rh(H_2O)_2(NH_3)_4]^{3+} + 2\,AgCl\downarrow$$

Then pyridine can distinguish the two stages of acidity:

$$[cis\text{-}Rh(H_2O)_2(NH_3)_4]^{3+} + C_5H_5N \rightarrow$$

$$[cis\text{-}Rh(H_2O)(OH)(NH_3)_4]^{2+} + C_5H_5NH^+$$

This is useful partly because of forming with $[S_2O_6]^{2-}$ a sparingly soluble, and therefore easily isolable, dithionate, a pale-yellow complex, $[cis\text{-}Rh(H_2O)(OH)(NH_3)_4][S_2O_6]$. Then the dithionate, heated at 120 °C for 20h, loses some of its water and becomes a bridged complex, $[\{(NH_3)_4Rh\}_2(\mu\text{-}OH)_2][S_2O_6]_2$. Stirring this 1 h with saturated NH_4Br forms $[\{(NH_3)_4Rh\}_2(\mu\text{-}OH)_2]Br_4\cdot 4H_2O$, pale yellow.

Excess NH_3 with $Na_7[IrCl_2(SO_3)_4]\cdot 7H_2O$ forms $Na_3[Ir(SO_3)_3(NH_3)_3]$, white. Refluxing $IrCl_3\cdot aq$ with $CO(NH_2)_2$ (source of NH_3) and CH_3CO_2H 5h, further work and ion-exchange separations yield $[Ir(NH_3)_6]Cl_3$, $[IrCl(NH_3)_5]Cl_2$, $[cis\text{-}IrCl_2(NH_3)_4]Cl\cdot{}^1/_2H_2O$, $[trans\text{-}IrCl_2(NH_3)_4]Cl\cdot H_2O$, $[mer\text{-}IrCl_3(NH_3)_3]Cl_3$ and other salts.

The dioxide, RhO_2, is insoluble in NH_3. Heating $[RhCl_6]^{2-}$ with concentrated NH_3 yields the same yellow complex as above, $[RhCl(NH_3)_5]Cl_2$, insoluble in 6-M HCl.

In acid, $N_2H_5^+$ reduces at least $[IrCl_6]^{2-}$ through $[IrCl_4(H_2O)_2]$, and $[IrBr_6]^{2-}$, to iridium(III).

Hydroxylamine reduces Ir^{IV} to Ir^{III} (separation from Au).

Refluxing N_3^- and $[Rh(H_2O)(NH_3)_5]^{3+}$ for 1 h gives $[Rh(N_3)(NH_3)_5]^{2+}$.

Aqueous N_3^-, $[IrCl_6]^{3-}$, $[NBu_4]^+$ and ethanol yield $[NBu_4]_3[Ir(N_3)_6]$.

Oxidized nitrogen. Cold, aqueous $[M(H_2O)(NH_3)_5]Cl_3$ (M = Rh or Ir) but not $[IrCl(NH_3)_5]^{2+}$ or $[IrCl_5(NH_3)]^{2-}$, plus NO_2^- and then HCl form, e.g., white $[M(NO_2\text{-}\kappa O)(NH_3)_5]Cl_2$, isomerizing, when warmed, to $NO_2\text{-}\kappa N$. However, (anodic) $[IrCl_5(NH_3)]^-$ and NO return only $[IrCl_5(NH_3)]^{2-}$.

A complicated mechanism yields:

$$2\,[RhN_3(NH_3)_5]^{2+} + 6\,HNO_2 \rightarrow$$

$$2\,[Rh(H_2O)(NH_3)_5]^{3+} + 5\,N_2O + 2\,NO_2^- + H_2O$$

Metal sulfides precipitated from dilute HCl may be dissolved in aqua regia, evaporated just to dryness, treated with NO_2^- and CH_3CO_2H (aiding Ir^{IV} to Ir^{III}), heated 5 min at 60–70 °C to form $[(Rh,Ir)(NO_2)_6]^{3-}$, yellow, and then treated with OH^-, no excess, to separate all those other metals as solid hydroxides fairly completely from the Rh and Ir ions:

$$[IrCl_6]^{3-} + 6\,NO_2^- \rightarrow [Ir(NO_2)]^{3-} + 6\,Cl^-$$

The $[Ir(NO_2)]^{3-}$ is not precipitated by Ag^+, as is $[Ir(CN)_6]^{3-}$.

Fusion of Rh with KNO_3 and KOH appears to form RhO_2.

Metallic rhodium or an alloy with Au is almost insoluble, unless very finely divided, in HNO_3/HCl. Alloyed with Cu, Pt, Pb or Bi, rhodium is soluble in HNO_3. Freshly precipitated Ir may be dissolved in aqua regia. Compact or ignited Ir is insoluble in all aqueous acids. A Pt-Ir alloy containing 25 to 30 % Ir is not attacked by aqua regia.

Anhydrous Rh_2O_3 is insoluble in aqua regia.

Concentrated HNO_3 and $IrCl_3$ form $[Ir(NO_3)_6]^{3-}$, not $(NO_2)_6$, at 100 °C.

Aqua regia dissolves $(NH_4)_2[IrCl_6]$ as $[IrCl_6]^{2-}$ and can yield $IrCl_4{\cdot}aq$.

Fluorine species. The F^- ion decolorizes Rh_2^{4+} but gives no solid.

Rhodium(III) in HF forms $[RhF_n(H_2O)_{6\text{-}n}]^{(3\text{-}n)+}$.

9.2.3 Reagents Derived from the 3rd-to-5th-Period Non-Metals, Silicon through Xenon

Phosphorus species. Aqueous $[Rh_2(H_2O)_{10}]^{4+}$ and $H_2PO_4^{2-}$ give the complex $[\{-Rh(H_2O)\}_2(\mu\text{-}H_2PO_4)_4]$; cf. the acetate above.

Phosphates and Rh^{3+} form $[Rh(PO_4)_2(H_2O)_4]^{3-}$ and a precipitate as well as $[Rh(HP_2O_7)(H_2O)_4]$ etc. with $[P_2O_7]^{4-}$.

Even cold, concentrated $[PF_6]^-$ does not precipitate $[Rh_2(H_2O)_{10}]^{4+}$.

Reduced chalcogens. Sulfane (H_2S) reduces Ir^{IV} to Ir^{III} and then precipitates Ir_2S_3, brown, soluble in alkali sulfides.

Rhodium(III) unites with SCN^- or $SeCN^-$ in $[Rh\{(S,Se)CN\}_6]^{3-}$.

Thiocyanate reduces $[IrCl_6]^{2-}$ by 1st- and 2nd-order paths in SCN^-:

$$6\,[IrCl_6]^{2-} + SCN^- + 11\,H_2O \rightarrow 6\,[IrCl_6]^{3-} + SO_4^{2-} + HCN + 7\,H_3O^+$$

Oxidized chalcogens. Aqueous SO_2 precipitates Ir, black, from hot solutions of various Ir compounds.

Warming $[RhCl_6]^{3-}$ (from $RhCl_3$ in hot 5-M HCl) with excess concentrated NH_4HSO_3 (NH_3 plus SO_2) until colorless, and cooling, first yields $(NH_4)_2SO_3$, then white $(NH_4)_3[Rh(SO_3)_3(NH_3)_3]\cdot{}^3/_2H_2O$. Aqueous $[RhCl_6]^{3-}$ in HCl, and $K_2S_2O_5$ form yellow $K_3[Rh(SO_3)_3]\cdot 2H_2O$.

Warming $K_3[IrCl_6]$ with K_2SO_3 (from K_2CO_3 plus SO_2) for 2 h gives a light-orange $K_5[\textit{trans}\text{-}IrCl_4(SO_3)_2]\cdot 6H_2O$. However, warming $Na_3[IrCl_6]$ with excess $NaHSO_3$ ($NaHCO_3$ saturated with SO_2) at 75 °C for 2 h yields yellow $Na_7[IrCl_2(SO_3)_4]$.

At pH 8, $[Rh(NH_3)_5(OH)]^{2+}$ and $HSeO_3^-$ give $[Rh(NH_3)_5(SeO_3\text{-}\kappa O)]^+$ immediately, less quickly with SeO_3^{2-} at pH 10.

Hot, concentrated H_2SO_4 dissolves only very finely divided "Rh black", and fusion with $KHSO_4$ yields $Rh_2(SO_4)_3$.

The $[Rh_2(H_2O)_{10}]^{4+}$ ion and $(NH_4)_2SO_4$ under Ar form probably $(NH_4)_4[\{Rh(H_2O)\}_2(\mu\text{-}SO_4)_4]\cdot{}^5/_2H_2O$. Aqueous $[\{Rh(H_2O)\}_2(\mu\text{-}CH_3CO_2)_4]$, HSO_4^-, heat and Cs^+ give $Cs_4[\{Rh(H_2O)\}_2(\mu\text{-}SO_4)_4]$.

Reduced halogens. Chloride, Br^- and I^- dismutate Rh_2^{4+} to Rh and Rh^{III}, and the I^- slowly reduces it completely to Rh, although I^-, from a hot solution of Rh^{III}, first precipitates a dark-brown RhI_3.

Treating $Rh_2O_3\cdot$aq with minimal concentrated HCl gives a wine-red solution, and evaporation at ~ 100 °C yields $[RhCl_3(H_2O)_3]$ (with small impurities of HCl etc.), often a source for other syntheses. This precipitates Ag^+ quite slowly. Heating Rh^{3+} with less, or more, HCl yields $[RhCl_n(H_2O)_{6-n}]^{(3-n)+}$, with $1 \le n \le 6$. These catalyze the hydration of C_2H_2, especially with the Cl_5 but not the Cl_6 ion. Chlorine or Ce^{IV} converts a suspension of $Cs_3[RhCl_6]$ at 0 °C to cold-insoluble $Cs_2[RhCl_6]$, which soon loses Cl_2 in solution. Other unknown $[RhX_6]^{2-}$ may be unstable.

The successive chloridation of $[Rh(H_2O)_6]^{3+}$, when done in hot HCl (taking two to four days) because the OH^- even in neutral solution is trans labilizing too, illustrates the trans effect of Cl^-, after it forms $[RhCl(H_2O)_5]^{2+}$. This quickly takes the second step, thereby yielding $[\textit{trans}\text{-}RhCl_2(H_2O)_4]^+$ (stable for 30 days at 5°C), $[\textit{mer}\text{-}RhCl_3(H_2O)_3]$, $[\textit{trans}\text{-}RhCl_4(H_2O)_2]^-$, $[RhCl_5(H_2O)]^{2-}$ and $[RhCl_6]^{3-}$, separable, e.g., by ion exchange.

The trans-effect order, also for Rh^{III} ammines, is $Cl^- \approx OH^- \gg H_2O$. Boiling 8-M HCl dissolves the golden-yellow dichloro Rh^{III} chloride but scarcely the pale-yellow monochloro complex, which can be recrystallized from boiling water (a small amount to prevent aquation).

Substitutions of the H_2O in $[RhCl_5(H_2O)]^{2-}$ by Cl^-, Br^- and I^-, also by NCS^-, N_3^- and NO_2^-, have similar rates, as in a dissociative mechanism.

Dissolving $RhCl_3 \cdot aq$ in mixtures of HCl and HBr can give all 10 of the isomeric ions $[RhCl_nBr_{6-n}]^{3-}$.

Iridium trichloride forms complexes, e.g., $[IrCl_6]^{3-}$ with Cl^-:

$$[IrCl_6]^{3-} + H_2O \leftrightarrows [IrCl_5(H_2O)]^{2-} + Cl^-$$

The dioxide, RhO_2, dissolves in HCl, releasing Cl_2 and forming red $[RhCl_6]^{3-}$. Acetone and NaCl give $Na_3[RhCl_6] \cdot 2H_2O$. Cooling and KCl yield $K_3[RhCl_6] \cdot H_2O$. Water and NH_4^+ yield, e.g., $(NH_4)_2[RhCl_5(H_2O)]$:

$$[RhCl_6]^{3-} + H_2O \leftrightarrows [RhCl_5(H_2O)]^{2-} + Cl^-$$

The solutions turn brown on standing or heating.

Adding AlkCl to $IrO_2 \cdot aq$ suspended in HCl gives $Alk_2[IrCl_6] \cdot nH_2O$.

Excess HCl dissolves IrO_2 as $[IrCl_6]^{2-}$, but $[IrCl_3(H_2O)_3]^+$ and $[IrCl_4(H_2O)_2]$ also exist. Iridium(IV) and KCl and NH_4Cl precipitate the dark-colored, slightly soluble $K_2[IrCl_6]$ and $(NH_4)_2[IrCl_6]$, respectively. We can prepare $IrBr_3 \cdot 4H_2O$ and $IrI_3 \cdot 3H_2O$ from $Ir_2O_3 \cdot aq$ with HBr or HI, and we may reduce Ir^{IV} complexes to $M^I{}_3[IrX_6]$ and $M^I{}_2[IrX_5(H_2O)]$.

Aqueous HBr converts Ir_2O_3 and $[IrCl_6]^{3-}$ (with repeated treatment) to $[IrBr_6]^{3-}$, crystallized as $[Co(NH_3)_6][IrBr_6]$, $Alk_3[IrBr_6] \cdot nH_2O$ and/or $(H_3O)K_8[IrBr_6]_3 \cdot 9H_2O$. Also, treating $Alk_3[IrCl_6]$ or $Alk_2[IrCl_5(H_2O)]$ with HBr can form $Alk_2[IrBr_6]$ or $Alk_2[Ir_2Br_9]$, depending on conditions.

Aqueous HI and Ir_2O_3 can form $[IrI_6]^{3-}$.

Iodide ion and $[IrBr_6]^{2-}$ yield $[IrBr_6]^{3-}$ and I_2.

Elemental and oxidized halogens. The best solvent for Ir may be $Cl_{2\,aq}$.

Aqueous Rh_2^{4+} and Cl_2, Br_2 or I_2 form $[RhX(H_2O)_5]^{2+}$, and Cl_2 or Br_2 can yield $[RhX(H_2O)_5]^{3+}$.

From Rh^{III} either ClO^- or BrO^- forms $RhO_2 \cdot 2H_2O$ or ill-defined species of Rh^V or Rh^{VI}: purple cations (pH 2), blue anions (pH 6), green (pH 8), and yellow-orange (pH 11), also from BrO_4^- (pH 11), but blue K_2RhO_4 is found from concentrated KOH and RhO_2. Excess BrO_4^- with $[RhCl_6]^{3-}$ may give a violet Rh^{VI}, slowly going to a blue, better-known Rh^V, which dismutates in mild acid:

$$6\ HRhO_4^{2-} + 4\ H_3O^+ \rightarrow 4\ RhO_4^{2-} + Rh_2O_3 \cdot aq\downarrow + 9\ H_2O$$

Concentrated $HClO_4$ converts $RhCl_3 \cdot 3H_2O$ to $[Rh(H_2O)_6](ClO_4)_3$, yellow, with a pK_a of 3.3 at 25 °C. In aqueous $HClO_4$ at $T \geq 130$ °C, $[Ir(NH_3)_5(H_2O)](ClO_4)_3$ decomposes to metallic Ir.

9.2.4 Reagents Derived from the Metals Lithium through Uranium, plus Electrons and Photons

Oxidation. Cerium(IV) and Rh_2^{4+} quickly form $[Rh(H_2O)_6]^{3+}$.

Aqueous Ce^{IV}, $[Ir(C_2O_4)_3]^{3-}$ and H_3O^+ give $[Ir(C_2O_4)_3]^{2-}$.

Anodes convert $[Rh^{II}Q(NH_3)_5]^{n+}$, with Q = H_2O, OH^-, NCO^-, CHO_2^-, NH_3, NO_2^- or halides, to Rh^{III}; $n = 2, 1, 1, 1, 2, 1$ or 1 in turn.

Aqueous $[\{Rh^{II}(H_2O)\}_2(\mu\text{-}Q)_4]^{n-}$, with Q = CO_3^{2-}, $CH_3CO_2^-$ or SO_4^{2-}, loses one electron reversibly at $E°$ a little over 1 V each; $n = 4, 0$ or 4.

Anodes passivate Rh in 1-dM H_2SO_4 or NaOH at 10 μA/cm² or less.

Anodic treatment of Rh^{3+} in H_3O^+ yields green Rh^{IV}; RhO_2 in dilute $HClO_4$ appears to give RhO_4^{2-} and hence a Ba^{2+} salt.

Light (254 nm), $[IrCl_6]^{3-}$ and 12-M HCl yield H_2 and $[IrCl_6]^{2-}$.

Reduction. With HCl, Mg or Zn precipitates Rh from many Rh species.

The Eu^{2+}, Ti^{III} or V^{2+} ion reduces $[RhCl(H_2O)_5]^{2+}$ to Rh. The Ti^{III} largely separates Rh from the less reducible Ir^{III}.

Aqueous Cr^{2+} and $[RhCl(H_2O)_5]^{2+}$, unlike other $[Rh^{III}L_5Cl]^{n\pm}$, quickly and completely form $[Rh_2(H_2O)_{10}]^{4+}$, not isolable with $[B_{10}H_{10}]^{2-}$, $[BF_4]^-$, $[BPh_4]^-$, $[PF_6]^-$ or $[Fe(CN)_6]^{4-}$, and halides (any not complexed by the resulting Cr^{III}) catalyze dismutation to Rh and Rh^{III}. A large excess of Cr^{2+} gives Rh slowly, but Eu^{2+} or V^{2+}, with $RhCl^{2+}$, produces only Rh.

The $[Cr(H_2O)_6]^{2+}$ ion reduces $[IrCl_6]^{2-}$ and $[IrBr_6]^{2-}$ by both outer- and inner-sphere paths, via $[IrX_5(\mu\text{-}X)Cr(H_2O)_5]$ inner, resulting in $[IrX_6]^{3-}$, $[Cr(H_2O)_6]^{3+}$, $[IrX_5(H_2O)]^{2-}$ and $[CrX(H_2O)_5]^{2+}$.

Iron(2+) or $SnCl_2$ reduces Ir^{IV} only to Ir^{III} (separation from Au).

Zinc reduces $Rh_2O_3 \cdot aq$ in alkaline CN^- to a square-planar $[Rh(CN)_4]^{3-}$. Zinc, with H_3O^+ or heat, precipitates Ir from its compounds.

In 3-M HCl, Rh^{III} and $[SnCl_3]^-$ form $[\{RhCl(SnCl_3)_2\}_2]^{4-}$. As a test (detecting 6 μM Rh), if 2 to 3-M $SnCl_3^-$ in concentrated HCl is added to a very acidic solution of a rhodium salt, with heating to boiling, a brown color develops that changes to raspberry-red on cooling.

Other reactions. Even cold, concentrated $[Fe(CN)_6]^{4-}$ does not precipitate $[Rh_2(H_2O)_{10}]^{4+}$. The $[Fe(CN)_6]^{4-}$ and $[Fe(CN)_6]^{3-}$ ions, when heated with Rh^{III}, give a greenish-brown color.

Aqueous $[IrCl_6]^{2-}$ and $[Co(CN)_5]^{3-}$ form $[IrCl_6]^{3-}$ and $[Co(CN)_5OH]^{3-}$.

Orange $[IrCl_6]^{2-}$ precipitates blue $Ag_2[IrCl_6]$ or dark-green $Tl_2[IrCl_6]$.

A cream-colored hydrido complex is formed with $(NH_4)_2SO_4$ and $[RhCl(NH_3)_5]Cl_2$ suspended in 8-M NH_3, by heating to 60 °C, adding Zn dust, keeping it warm for 2 min, then making it ice-cold, stirring more and saturating with gaseous NH_3:

$$[RhCl(NH_3)_5]Cl_2 + Zn + SO_4^{2-} + 3\ NH_3 + NH_4^+ \rightarrow$$

$$[Rh^{III}H^{-I}(NH_3)_5]SO_4\downarrow + [Zn(NH_3)_4]^{2+} + 3\ Cl^-$$

The solution is air-sensitive, giving a blue peroxo complex, but the solid sulfate is quite stable. The overall complex has clearly been reduced, but the hydrogen in such species is more electronegative than the metal and is thus treated as anionic, so the Rh may still be called Rh^{III}, i.e., not reduced. Water establishes the equilibrium:

$$[RhH(NH_3)_5]^{2+} + H_2O \leftrightarrows [RhH(H_2O)(NH_3)_4]^{2+} + NH_3$$

Acidified $[SnBr_3]^-$ and $[IrBr_6]^{2-}$ form $[IrBr_4(SnBr_3\text{-}\kappa Sn)_2]^{2-}$.

Acidified photolysis (254 nm), even if extended, changes $[Rh(CN)_6]^{3-}$ only to $[Rh(CN)_5(H_2O)]^{2-}$ and $[Ir(CN)_6]^{3-}$ to $[Ir(CN)_5(H_2O)]^{2-}$. Then OH^- or warm Cl^-, Br^- or I^- produces $[M(CN)_5OH]^{3-}$ or $[M(CN)_5X]^{3-}$, precipitable as $[Co(NH_3)_6][M(CN)_5X]$ etc.

Light and $[(Rh,Ir)(NH_3)_6]^{3+}$ first yield $[(Rh,Ir)(H_2O)(NH_3)_5]^{3+}$.

Light and $[RhX(NH_3)_5]^{2+}$, with $X^- = Cl^-$, Br^- or I^-, form both $[Rh(H_2O)(NH_3)_5]^{3+}$ and [*trans*-$RhX(H_2O)(NH_3)_4]^{2+}$.

Either *cis*- or [*trans*-$RhCl_2(NH_3)_4]^+$ gives [*trans*-$RhCl(H_2O)(NH_3)_4]^{2+}$ when irradiated. However, either *cis*- or [*trans*-$RhCl(OH)(NH_3)_4]^+$, treated with OH^- and light, forms [*cis*-$Rh(OH)_2(NH_3)_4]^+$.

Light (254 nm) and HCl can oxidize and reduce $[IrCl_6]^{3-}$ and $[IrCl_6]^{2+}$ cyclically, yielding H_2 and Cl_2.

Ultraviolet light changes $[Ir(CN)_6]^{3-}$ to $[Ir(CN)_5(H_2O)]^{2-}$, and warm Cl^-, Br^- or I^- can form $[Ir(CN)_5X]^{3-}$, precipitable as $[Co(NH_3)_6][Ir(CN)_5X]$.

Photolyzing $[Rh(N_3)(NH_3)_5]^{2+}$ in 1-M HCl releases N_2, likely via a nitrene, $[RhNH(NH_3)_5]^{3+}$, and producing both $[Rh(NH_2Cl)(NH_3)_5]^{3+}$, which is reducible by I^- to $[Rh(NH_3)_6]^{3+}$ (at the N–Cl bond), and smaller amounts of $[RhQ(NH_3)_5]^{3+}$, with Q = NH_3, NH_2OH or H_2O. Light and $[RhI(NH_3)_5]^{2+}$ form $[RhI_2(NH_3)_4]^+$ and $[RhI(H_2O)(NH_3)_4]^{2+}$.

At 518 nm the quantum yield to aquate $[RhCl_6]^{3-}$ is only 0.02.

Photons replace the substituent and/or NH_3 in substituted Rh^{III} polyammines with H_2O, among other reactions, often at rates in the order $H_2O > Cl^- > Br^- > I^-$; the loss of NH_3, but not that of Br^-, is greatly slowed by OH^-. However, light also replaces the H_2O in $[Rh(H_2O)(NH_3)_5]^{3+}$ with Cl^- or Br^-.

Photons (UV) and HCl efficiently convert $[Ir(N_3)(NH_3)_5]^{2+}$ to N_2 and $[Ir(NH_2Cl)(NH_3)_5]^{3+}$, and HSO_4^- leads to $[Ir(NH_2\text{–}OSO_3)(NH_3)_5]^{2+}$. This with HCl or H_2O goes to $[Ir(NH_2Cl)(NH_3)_5]^{3+}$ or $[Ir(NH_2OH)(NH_3)_5]^{3+}$, and the $[Ir(NH_2Cl)(NH_3)_5]^{3+}$ with OH^- or HI forms $[Ir(NH_2OH)(NH_3)_5]^{3+}$ or $[Ir(NH_3)_6]^{3+}$ and I_2 respectively.

Light (UV), H_3O^+ and $[Ir(NH_3)_6]^{3+}$ or $[IrCl(NH_3)_5]^{2+}$ form the rather inert $[Ir(H_2O)(NH_3)_5]^{3+}$ and NH_4^+ or Cl^-.

Reference

1. (a) Werner A (1920) Neuere Anschauungen auf dem Gebiete der anorganischen Chemie, 4th ed. Friedrich Vieweg, Brunswick (b) Werner A (1914) Ber Deutsch Chem Ges 47:3087

Bibliography

See the general references in the Introduction, specifically [116], [121] and [313], and some more-specialized books [2–5]. Some articles in journals discuss: DF theory for $[Rh_6(PH_3)_6H_m]_n$, m = 12, 14 or 16 [6]; reductions of Co^{III} by metallic ions [7]; iridium [8]; mononuclear cyanocobalt(III) complexes [9]; Ir^{III} chloro and bromo species [10] and metal-metal bonding in Rh^{II} [11].

2. Griffith WP (1967) The chemistry of the rarer platinum metals (Os, Ru, Ir and Rh). Wiley, New York
3. Young RS (1966) The analytical chemistry of cobalt. Pergamon, London
4. Pyatnitskii IV (1965) Kaner N (trans) Slutzkin D (ed) (1966) Analytical chemistry of cobalt. Ann Arbor-Humphrey, Ann Arbor
5. Young RS (1960) Cobalt. Reinhold, New York
6. Brayshaw SK, Green JC, Hazari N, Weller AS (2007) Dalton Trans 2007: 1781
7. Yang Zh, Gould ES (2004) J Chem Soc Dalton Trans 2004:3601
8. Housecroft CE (1992) Coord Chem Rev 115:163
9. Burnett MG (1983) Chem Soc Rev 12:267
10. Fergusson JE, Rankin DA (1983) Aust J Chem 36:863; Rankin DA, Penfold BR, Fergusson JE (1983) ibid:871
11. Felthouse TR (1982) in Lippard SJ (ed) Prog Inorg Chem 29:73

10 Nickel through Darmstadtium

10.1 Nickel, $_{28}Ni$

Oxidation numbers: (I), (II), (III) and (IV), as in $[Ni_2(CN)_6]^{4-}$, Ni^{2+}, and hydrated Ni_2O_3 and NiO_2.

10.1.1 Reagents Derived from Hydrogen and Oxygen

Dihydrogen. Finely divided nickel will dissolve about 17 times its own volume of H_2. Nickel-hydride (less toxic than Ni-Cd) storage-battery negative electrodes use alloys of Ni, often with much Ln, to absorb H, with an $E°$ near that of the standard H_2 electrode:

$$H_2O + e^- \leftrightarrows H_{Ni,Ln} + OH^-$$

At high pressure, H_2 reduces Ni ammines autocatalytically, e.g.:

$$Ni(NH_3)_2{}^{2+} + H_2 \leftrightarrows Ni\downarrow + 2\ NH_4{}^+$$

Water. Nickel(II) oxide and hydroxide are insoluble in water.

Nickel(II) borate, carbonate, cyanide, oxalate, phosphate, sulfide, hexacyanoferrate(II and III) and hexacyanocobaltate(III) are insoluble.

Hydrated Ni^{II} acetate is efflorescent, as are the nitrate and chloride in dry air, but both are deliquescent in moist air, giving green $[Ni(H_2O)_6]^{2+}$.

Solid $NiCl_2{\cdot}6H_2O$ is [*trans*-$NiCl_2(H_2O)_4]{\cdot}2H_2O$, not $[Ni(H_2O)_6]Cl_2$.

The salts of $[Ni(CN)_4]^{2+}$ and alkali metals are soluble in water.

Nickel(0) in $K_4[Ni(CN)_4]$ (from $K_2[Ni(CN)_4]$ and K in liquid NH_3) reacts with H_2O to release H_2 and form $[Ni_2(CN)_6]^{4-}$, with a Ni–Ni bond and two planar $Ni(CN)_3$ units almost mutually perpendicular in salts.

Water and NiO_2 form hydrated Ni_2O_3 and Ni_3O_4, and release O_2.

Seawater and some freshwater contain traces of Ni^{II} complexes as $NiOH^+$, $Ni(OH)_2$, $NiCO_3$, $NiSO_4$ and $NiCl^+$.

Oxonium. Dilute or concentrated HCl or H_2SO_4 and Ni slowly give Ni^{II}.

Nickel(II) oxide and hydroxide are soluble in acids. Non-reducing acids dissolve the higher oxides while producing O_2 and Ni^{II}.

Slowly adding dilute H_2SO_4 to boiling $[Ni(CN)_4]^{2-}$ precipitates a pale-violet $Ni(CN)_2 \cdot 2H_2O$ and releases HCN.

Hydroxide. Nickel is not affected by OH^-. Alkalis and Ni^{2+} first form mainly $Ni_4(OH)_4{}^{4+}$, and precipitate from Ni^{2+} (absent organic chelators) pale green $Ni(OH)_2$, not oxidized by air, dilute H_2O_2 or I_2 [distinction from $Co(OH)_2$]; oxidized by $[S_2O_8]^{2-}$, ClO^-, BrO^- or $[Fe(CN)_6]^{3-}$ even in the presence of minimal CN^-, to black NiO(OH) or $NiO_2 \cdot H_2O$, soluble in H_3O^+ as Ni^{2+} and O_2 (or X_2 with HX). Concentrated NaOH and $Ni(OH)_2$ form $Na_2[Ni(OH)_4]$.

Di- and trioxygen. Alkaline suspensions of $Ni(OH)_2$ or NiS give hydrated Ni_3O_4, Ni_2O_3 or possibly NiO_2 with O_3 but not dilute $HO_2{}^-$.

10.1.2 Reagents Derived from the Other 2nd-Period Non-Metals, Boron through Fluorine

Boron species. The $[Ni(NH_3)_6]^{2+}$ ion and $[BH_4]^-$ precipitate Ni.

Carbon oxide species. The action of CO on metallic Ni at 50 °C forms nickel carbonyl, $Ni(CO)_4$. The non-aqueous syntheses may be well known, but the following summarize some aqueous syntheses of the volatile and highly poisonous liquid, using various reductants:

$$Ni^{2+} + 4\ CO + 2\ OH^- + CN^- \rightarrow [Ni(CO)_4]_{liq}\downarrow + NCO^- + H_2O$$

$$3\ Ni^{2+} + 12\ CO + 6\ OH^- + S^{2-} \rightarrow 3\ [Ni(CO)_4]_{liq}\downarrow + SO_3{}^{2-} + 3\ H_2O$$

Aqueous $[Ni_2(CN)_6]^{4-}$ reacts with CO, apparently giving $[Ni(CN)_4]^{2-}$ and $[Ni(CN)_2(CO)_2]^{2-}$.

Alkali carbonates precipitate from Ni^{2+} a green, basic carbonate, $Ni_5(CO_3)_2(OH)_6 \cdot 4H_2O$; the normal carbonate, $NiCO_3 \cdot 6H_2O$, is obtained if an excess of CO_2 is present. The precipitate separating upon adding "$(NH_4)_2CO_3$" to Ni^{2+} dissolves in excess of the reagent. The carbonates of Mg, Ca, Sr and Ba do not react with Ni^{2+} in the cold (distinction from Cr^{III}, Fe^{III} and Al^{III}), but on boiling, they completely precipitate the Ni.

Cyanide species. Aqueous CN^-, without excess, precipitates Ni^{2+} as greenish $Ni(CN)_2 \cdot aq$, or, from boiling water, blue-gray $Ni(CN)_2 \cdot {}^3/_2H_2O$, i.e., polymeric [*trans*-$Ni(H_2O)_2(N\equiv)_4$][*quadro*-$(\equiv C)_4Ni$]$\cdot H_2O$. In NH_3 the $Ni(CN)_2$ forms [*trans*-$Ni(NH_3)_2(N\equiv)_4$][*quadro*-$(\equiv C)_4Ni$]$\cdot {}^1/_2H_2O$, if we may thus also suggest the structure of $Ni(CN)_2NH_3 \cdot {}^1/_4H_2O$, again writing N≡C for the bridges. The $Ni(CN)_2$ is insoluble in cold, dilute HCl, readily soluble in excess CN^-, crystallized as K_2[*quadro*-$Ni(CN)_4$]$\cdot H_2O$, yellow, very soluble. The overall pK_{dissoc} for $[Ni(CN)_4]^{2-}$ is ~30. Acid reprecipitates $[Ni(CN)_4]^{2-}$ as $Ni(CN)_2 \cdot aq$.

Concentrated CN^- forms red $[Ni(CN)_5]^{3-}$, isolated, with slow cooling to −5 °C, as orange-red $[Cr(NH_3)_6][Ni(CN)_5]\cdot 2H_2O$. This $[Ni(CN)_5]^{3-}$ salt is ~square pyramidal, but its other salts are trigonal bipyramidal or both. The pK_{dissoc} for $[Ni(CN)_5]^{3-}$ into $[Ni(CN)_4]^{2-}$ and CN^- is near 0.

Excess CN^- decomposes $[Ni_2(CN)_6]^{4-}$ (see **Water** above):

$$\tfrac{1}{2}\,[Ni_2(CN)_6]^{4-} + CN^- + H_2O \rightarrow [Ni(CN)_4]^{2-} + \tfrac{1}{2}\,H_2\uparrow + OH^-$$

Volumetrically, Ni may be determined by titrating $[Ni(NH_3)_4]^{2+}$ with CN^-, back titrating with Ag^+ and using I^- as an internal indicator for Ag^+, with citrate to keep any Fe^{III} from precipitating as the hydroxide:

$$[Ni(NH_3)_4]^{2+} + 4\,CN^- \rightarrow [Ni(CN)_4]^{2-} + 4\,NH_3$$

$$2\,CN^- + Ag^+ \rightarrow [Ag(CN)_2]^-$$

Anions $Ni(NCY\text{-}\kappa N)_4{}^{2-}$ and $[Ni(NCY\text{-}\kappa N)_6]^{4-}$ with Y = O, S or Se, are known, as in $Na_4[Ni(NCS\text{-}\kappa N)_6]\cdot 8H_2O$, also $Hg[Ni(NCS\text{-}\kappa N)_4(H_2O)_2]$.

Some "simple" organic reagents. An interesting clathrate is $Ni(NH_3)_2[Ni(CN)_4]\cdot 2C_6H_6$ or $C_6H_6\cdot NH_3\cdot Ni(CN)_2$ (following the IUPAC nomenclature recommendations), from shaking benzene with $Ni(CN)_2$ dissolved in (aqueous) NH_3. In related clathrates, planar Pd or Pt, or tetrahedral Cd or Hg, replaces the cyano (planar) Ni, or Mn, Fe, Co, Cu, Zn or Cd replaces the other (octahedral) Ni. Also, 1,2-diaminoethane, ethylenediamine, $(-CH_2NH_2)_2$, may replace the two NH_3, and furan, pyridine, pyrrole, thiophene etc. may replace the benzene.

Acetic acid and $NiCO_3$ yield [*trans*-$Ni(\eta^1\text{-}CH_3CO_2)_2(H_2O)_4$] with some $Ni(CH_3CO_2)^+$ in solution.

A sensitive test for Ni^{2+}, with precise claims differing widely, depends on the characteristic red precipitates that certain dioximes form in ammoniacal or buffered acetic-acid solutions. The most common example, 2,3-butanedionedioxime, "dimethylglyoxime", abbreviated as "H_2Dmg", that is $[Me\text{–}C(=NOH)\text{–}]_2$, gives $[Ni(HDmg)_2]$ or $[Ni(C_4H_7N_2O_2)_2]$, soluble in CN^-. Other **3d**$^{2+}$ ions form similar, but soluble complexes. Cobalt interferes if present in an excess of more than 10 Co to 1 Ni. Iron(2+) gives a red color but also no precipitate.

Oxalic acid and $C_2O_4{}^{2-}$ precipitate nickel oxalate, green, from Ni^{2+}. The separation is slow, being almost complete after about 24 hours.

Nickel dioxide is reduced to Ni^{II} by $H_2C_2O_4$, releasing CO_2:

$$NiO_2\cdot H_2O + 2\,H_2C_2O_4 \rightarrow NiC_2O_4\cdot 2H_2O\downarrow + 2\,CO_2\uparrow + H_2O$$

Reduced nitrogen. A small amount of NH_3 precipitates $Ni(OH)_2$ from solutions of Ni^{2+}, soluble in excess, as is NiO also, to give complexes up to $[Ni(NH_3)_6]^{2+}$, in various shades of blue or violet. No precipitate is formed if considerable $NH_4{}^+$ is present. Excess of OH^- will slowly (rapidly if boiled) precipitate $Ni(OH)_2$ from ammoniacal solutions (distinction from Co^{III}). The violet complex, $[Ni(NH_3)_6]Br_2$, is precipitated upon adding concentrated NH_3 to a hot solution of $NiBr_2$ (separation from $[Co(NH_3)_6]^{3+}$ etc.). The similar iodide is less soluble than the bromide. These are converted by boiling with OH^- to the hydroxide. Many salts of Ni^{II} form soluble ammines:

$$Ni^{2+} + 2\,X^- + 6\,NH_3 \rightarrow Ni(NH_3)_6X_2\downarrow$$

Aqueous $K_2[Ni(CN)_4]$ and N_2H_4 with much OH^- give apparently an extremely reactive $K_3[Ni(CN)_4]$. Also in base, $K_2[Ni(CN)_4]$, NH_2OH and O_2 form a violet, diamagnetic product:

$$[Ni(CN)_4]^{2-} + NH_2OH + {}^1/_2\,O_2 + OH^- + 2\,K^+ \rightarrow$$

$$K_2[Ni(CN)_3(NO)]\downarrow + CN^- + 2\,H_2O$$

A large excess of KN_3 yields $K_4[Ni(N_3)_6]\cdot 2H_2O$ from Ni^{2+}.

Interesting bridging structures arise between N_3^- and Ni^{2+} or Cu^{2+}, including μ–1,1 (end-on), μ–1,3 (end-to-end), μ_3–1,1,1, μ_3–1,1,3, μ_4–1,1,1,1 and μ_4–1,1,3,3, but these also involve large organic ligands and can barely be mentioned here [1].

Oxidized nitrogen. Nitrite ion, in presence of acetic acid, does not oxidize Ni^{2+} (distinction from Co).

Nickel dioxide is reduced to Ni^{II} by HNO_2, forming nitrate:

$$NiO_2\cdot H_2O + 2\,HNO_2 \rightarrow Ni^{2+} + NO_3^- + NO_2^- + 2\,H_2O$$

Dilute HNO_3 dissolves Ni readily; concentrated HNO_3 passivates it.

Fluorine species. Fluoride complexes Ni^{2+} weakly to form NiF^+.

10.1.3 Reagents Derived from the 3rd-to-5th-Period Non-Metals, Silicon through Xenon

Phosphorus species. Nickel dioxide is reduced to Ni^{II} by HPH_2O_2, possibly this way:

$$4\,NiO_2\cdot H_2O + 2\,HPH_2O_2 + 2\,H_2O \rightarrow Ni_3(PO_4)_2\cdot 8H_2O\downarrow + Ni(OH)_2\downarrow$$

Hydrogenphosphate, HPO_4^{2-}, precipitates the green nickel phosphate, $Ni_3(PO_4)_2\cdot 8H_2O$, soluble in acids, including CH_3CO_2H.

Arsenic species. Nickel(2+) is precipitated by either AsO_3^{3-} or AsO_4^{3-}, as a pale green salt, soluble in acids.

Reduced chalcogens. Sulfane, H_2S, precipitates from neutral solutions of nickel salts, a part of the Ni as "NiS", black. Precipitation takes place slowly, but, from a solution containing sufficient acetate ion, is complete. In the presence of high $c(H_3O^+)$, no precipitation takes place.

Sulfane passed into an ammoniacal solution of similar metals for analysis, precipitates black "NiS" along with "CoS", MnS and ZnS. Separation may be delayed somewhat, permitting "NiS" to change to the less soluble form. Then the precipitate is digested with cold 1-M HCl. The residue of "NiS" and "CoS" is dissolved in HNO_3 or aqua regia and the Ni^{2+} detected in the presence of Co^{2+} or after its removal.

Alkali sulfides precipitate all of the Ni as the black sulfide. Although a high $c(H_3O^+)$ prevents precipitation, the precipitate, once formed, becomes quite insoluble in dilute HCl, slowly soluble in concentrated HCl (less with aging), but readily in HNO_3 and aqua regia. The situation with "NiS" is like that with "CoS"; see **9.1.3 Reduced chalcogens**.

Nickel sulfide, "NiS", is partially soluble (or peptized) in $(NH_4)_2S_x$, from which brown-colored system it is precipitated on adding acetic acid and boiling (distinction from Co). Freshly precipitated "NiS" is soluble in KCN; then H_3O^+ precipitates $Ni(CN)_2$ (separation from Co).

Nickel dioxide is reduced to a nickel sulfide by H_2S, forming S.

Aqueous $K_2[Ni(CN)_4]$ and K_2S, over 24 h, yield $K_3[Ni(CN)_3S]\cdot H_2O$ after using ethanol as precipitant.

Nickel(2+) is not visibly affected by SCN^- (distinction from Co). Nickel dioxide is reduced to Ni^{II} by HSCN, forming sulfate ions and a cyanide product.

Aqueous CS_3^{2-} and $[Ni(NH_3)_6]^{2+}$ give a red-brown solution.

Ethoxydithiocarbonate ion, "ethylxanthate", $EtOCS_2^-$, prepared by the action of ethanolic KOH on CS_2, precipitates Ni^{2+} (and Co^{2+}) from neutral or slightly acidic solution. The precipitate is soluble in NH_3 to give a blue solution (distinction from Co), and reprecipitated by "$(NH_4)_2S$". The $EtOCS_2^-$ also precipitates Ni^{II} from alkaline solutions in the presence of $[P_2O_7]^{4-}$ (separation from Fe^{III}).

Oxidized chalcogens. When Ni^{2+} is boiled with $S_2O_3^{2-}$, a portion of the Ni^{2+} is precipitated as "NiS". If a nitrite is added along with the thiosulfate, a permanganate-colored liquid is obtained, from which dark-purple crystals soon separate (distinction from large amounts of Co).

Nickel dioxide is reduced to Ni^{II} by SO_2, forming the sulfate:

$$NiO_2\cdot H_2O + SO_2 \rightarrow Ni^{2+} + SO_4^{2-} + H_2O$$

One way of distinguishing Ni and Co begins with a hot ammoniacal solution of them as M^{II}. Adding $[S_2O_8]^{2-}$ to the hot solution oxidizes any cobalt to form a red cobalt(III) ammine. (There should be no precipitate at this stage.) After removal of any large excess of NH_3 by boiling, the solution is cooled. Upon adding OH^- and shaking, nickel is oxidized to dark brown to black $NiO_2\cdot aq$, which slowly precipitates. If Co is present the filtrate will be pink to red. No Co is precipitated unless the solution is warm. The amount of OH^- required to precipitate the Ni depends on the excess of NH_3 and NH_4^+ present.

Reduced halogens. Halides complex Ni^{2+} extremely weakly to form NiX^+, but more strongly when hot, and to form, e.g., $[NiBr_4]^{2-}$ in solids.

The higher oxides of Ni dissolve in HX, giving Ni^{2+} and the halogen.

Fresh $NiO_2 \cdot H_2O$ is also reduced by neutral I^- (distinction from Co).

Elemental and oxidized halogens. With OH^- and ClO^- or BrO^- but not IO^- (distinction from Co^{II}), $Ni(OH)_2$ becomes NiO(OH) or $NiO_2 \cdot H_2O$.

Alkaline $[Ni(CN)_4]^{2-}$ and Cl_2, ClO^-, Br_2 or BrO^-, e.g., in a test for Ni, yield the brown or black NiO(OH) or $NiO_2 \cdot aq$, not $[Ni(CN)_6]^{3-}$ (distinction from Fe and Co). The test is affected by the excess of cyanide, in that, to avoid failure, a large amount of oxidant must be used when too much CN^- has been added, whereupon, due to dilution, only a brown coloration will appear. For example, Br_2 oxidizes CN^- to CNBr before attacking the cyano-complex:

$$CN^- + Br_2 \rightarrow Br^- + CNBr$$

$$[Ni(CN)_4]^{2-} + 5\,Br_2 + 4\,OH^- \rightarrow$$

$$NiO_2 \cdot aq\downarrow + 6\,Br^- + 4\,CNBr + 2\,H_2O$$

The $[Ni(NH_3)_6]^{2+}$ ion plus ClO_4^- precipitate blue $[Ni(NH_3)_6](ClO_4)_2$. Aqueous $[Co(NH_3)_6]^{2+}$ gives a yellowish-red precipitate, but not if boiled with H_2O_2 before adding the ClO_4^- (separation of Ni from Co).

Boiling Ni^{2+} with $[S_2O_8]^{2-}$ and either $Na_3H_2IO_6$ or KIO_4 changes it from a green solution through red and yields dark-purple, unusual Ni^{IV} orthoperiodates with a metallic sheen, practically insoluble in cold water; they seem to emit some O_3 in air:

$$Ni^{2+} + H_2IO_6^{3-} + [S_2O_8]^{2-} + Na^+ + H_2O \rightarrow NaNiIO_6 \cdot H_2O\downarrow + 2\,HSO_4^-$$

$$Ni^{2+} + IO_4^- + [S_2O_8]^{2-} + K^+ + 4^1/_2\,H_2O \rightarrow$$

$$KNiIO_6 \cdot {}^1/_2H_2O\downarrow + 2\,HSO_4^- + 2\,H_3O^+$$

10.1.4 Reagents Derived from the Metals Lithium through Uranium, plus Electrons and Photons

Oxidation. Anodes in base convert $Ni(OH)_2$ to ~NiO(OH), or a less stable (releasing O_2) mixture with Ni^{IV}. The positive electrode in some rechargeable flashlight batteries contains NiO(OH) when charged, or a partly reduced mixture with $Ni(OH)_2$ when discharged.

Reduction. Nickel(II) is reduced to Ni by finely divided Zn, Cd, Sn etc.

Aqueous $K_2[Ni(CN)_4]$ and K_{Hg}, or cathodic e^-, yield under H_2, after adding ethanol slowly, a red, diamagnetic $K_4[\{-Ni(CN)_3\}_2]$, whose two planar $-Ni(CN)_3^{2-}$ units are nearly perpendicular. This absorbs CO or NO, forming, e.g., $K_2[Ni(CN)_3NO]$.

Nickel dioxide is reduced to Ni^{II} by acidic solutions of hexacyanoferrate(II), forming possible products:

$$NiO_2\cdot H_2O + 2\ H_2[Fe(CN)_6]^{2-} + K^+ \rightarrow$$

$$KNi[Fe(CN)_6]\downarrow + [Fe(CN)_6]^{3-} + 3\ H_2O$$

$$3\ NiO_2\cdot H_2O + 6\ H_2[Fe(CN)_6]^{2-} \rightarrow$$

$$Ni_3[Fe(CN)_6]_2\downarrow + 4\ [Fe(CN)_6]^{3-} + 9\ H_2O$$

Discharging one kind of rechargeable "lithium-ion" batteries causes, at the positive electrode (with reversal during charging):

$$Li_xNiO_2 + y\ Li^+ + y\ e^- \rightarrow Li_{x+y}NiO_2$$

Gamma rays reduce cyano nickel(II) to nickel(I) complexes.

Other reactions. Aqueous CrO_4^{2-} precipitates from neutral solutions of Ni^{2+} a yellow to brown basic chromate, $NiCrO_4\cdot 2NiO$, soluble in acids; $K_2[Cr_2O_7]$ forms no precipitate. A saturated solution of $(NH_4)_2MoO_4$ slowly forms, in neutral or slightly acidic solutions of Ni^{2+} at about 70°C, a greenish-white precipitate (distinction from Co).

Mixing Ni^{II} with $[MoS_4]^{2-}$ or $[WS_4]^{2-}$, better in aqueous CH_3CN, yields $[Ni\{\eta^2\text{-}(Mo,W)S_4\}_2]^{2-}$.

Aqueous $[Fe(CN)_6]^{4-}$ precipitates a greenish-white $Ni_2[Fe(CN)_6]$, insoluble in acids, soluble in NH_3, transposed by OH^-. Aqueous $[Fe(CN)_6]^{3-}$ precipitates greenish-yellow $Ni_3[Fe(CN)_6]_2$, insoluble in acids, soluble in NH_3 to give a green solution.

Aqueous $[Ni(H_2O)_6]^{2+}$ has a pale-green color in crystals and in solution; the ordinary anhydrous salts are yellow. A solution containing $[Ni(H_2O)_6]^{2+}$ and $[Co(H_2O)_6]^{2+}$ at about 3:1 is colorless.

10.2 Palladium, $_{46}$Pd; Platinum, $_{78}$Pt and Darmstadtium, $_{110}$Ds

Oxidation numbers of Pd: (II) and (IV), as in PdO and PdO_2. Oxidation numbers of Pt: (II), (IV) and (VI), as in PtO ("platinous" oxide), PtO_2 ("platinic" oxide) and PtO_3 (unstable). The oxidation states for Ds calculated relativistically to be stable in water: (0), (II), (IV) and (VI).

10.2.1 Reagents Derived from Hydrogen and Oxygen

Dihydrogen. Finely divided spongy Pd absorbs thousands of times its volume of H_2, up to ~ $PdH_{0.7}$, retaining most of it even at 100 °C.

Hydrogen reduces $Pd(OH)_2$ and $Pd(OH)_4$ to Pd even incandescently.

Acidified $[PdCl_4]^{2-}$ oxidizes H_2 to H_3O^+, leaving Pd and Cl^-.

Platinum black catalyzes very many reactions of hydrogen.

Water. Aqueous Pd^{2+} is red-brown, acidic $[Pd(H_2O)_4]^{2+}$.

Palladium(II) oxide and hydroxide are insoluble. The chloride, bromide and iodide are moderately soluble (the chloride) to insoluble (the iodide). The sulfate dissolves easily but decomposes on standing.

The $[Pt(CN)_4]^{2-}$ salts of the Group-1 and Group-2 metals, but not the late **d**-block metals, and all of the $[Pt(CN)_6]^{2-}$ salts, are soluble.

Platinum(II) and platinum(IV) nitrate are soluble in H_2O but easily decomposed, precipitating basic salts. Platinum difluoride is soluble.

Platinum(II) sulfide may be even less soluble than HgS. The disulfide, PtS_2, is insoluble.

Platinum dichloride (e.g., from heating $H_2[PtCl_4]$), dibromide and diiodide are insoluble. The complex $[PtCl_4]^{2-}$ is hydrolyzed slowly to $[PtCl_3(H_2O)]^-$ and [*cis*-$PtCl_2(H_2O)_2$].

The $[PtCl_4]^{2-}$ salts of Na and Ba are soluble; of NH_4, K and Zn sparingly soluble; of Ag and Pb, insoluble in water.

The [*cis*-$PtCl_2(NH_3)_4]^{2+}$ ion is more acidic in water than the *trans*.

The platinum(II) salts of oxoacids, i.e., containing $[Pt(H_2O)_n]^{2+}$, are unstable; however, $[PtCl_3(NH_3)]^-$ and H_2O form $[PtCl_2(NH_3)(H_2O)]$, which ionizes to $[PtCl_2(NH_3)OH]^-$.

Palladium dioxide, PdO_2, is unstable; when freshly prepared, boiling with H_2O will cause loss of O_2.

Water hydrolyzes PtF_4 violently, but thermodynamically unstable $K_2[PtF_6]$ can be crystallized from boiling water, although moist air hydrolyzes $K_2[PdF_6]$.

The salts $K_2[PdCl_6]$ and $(NH_4)_2[PdCl_6]$, are slightly soluble in water, insoluble in ethanol, and partly decomposed by both solvents. Such alkaline-earth salts as $Ca[PdCl_6]$ are soluble in water and ethanol.

Platinum(IV) chloride and sulfate are soluble, the bromide (and iodide with decomposition) only slightly so. Many salts of $[PtCl_6]^{2-}$ are soluble, including those of Na, Group 2, Cu, Zn and Al; those of NH_4 and K are slightly soluble in H_2O but not ethanol. Water hydrolyzes $H_2[PtCl_6]$ to $H[PtCl_5(H_2O)]$ and $[PtCl_4(H_2O)_2]$, reversible by adding HCl.

Oxonium. Fresh $Pd(OH)_2$ is soluble in dilute H_3O^+ as $[Pd(H_2O)_4]^{2+}$ and even in CH_3CO_2H, but rather insoluble, dried, even in HNO_3 and H_2SO_4.

Platinum(IV) oxide and hydroxide are soluble in acids except acetic.

Oxonium ion precipitates unstable PtO_3 from (electrolytic) PtO_4^{2-}.

Hydroxide. Aqueous OH^- precipitates Pd^{II} as a brown, basic salt or as brown $Pd(OH)_2$, soluble in excess of hot OH^-. Boiling $[PtCl_4]^{2-}$ with limited OH^- produces $Pt(OH)_2$ or, with excess, Pt plus $[Pt(OH)_6]^{2-}$. Gentle heating may convert $Pt(OH)_2$ to PtO, easily dismutated.

From $[PdCl_6]^{2-}$ arises dark-red $PdO_2 \cdot aq$, which gives Pd^{2+} with H_3O^+. It dissolves slightly in concentrated OH^- as $[Pd(OH)_6]^{2-}$, slowly forms PdO and O_2 at ambient *T*, and is easily reduced by H_2, H_2O_2 and organics.

Heating $[PtCl_4]$ or $[PtCl_6]^{2-}$ with limited OH^- produces $Pt(OH)_4 \cdot aq$, soluble, if fresh, in excess OH^- as $[Pt(OH)_6]^{2-}$, also soluble, when fresh, in various acids. Gentle heating converts the hydrate to PtO_2. Adding CH_3CO_2H to $[Pt(OH)_6]^{2-}$ precipitates $Pt(OH)_4 \cdot aq$.

Peroxide. Its catalytic destruction of H_2O_2 (to O_2 and H_2O) enables 4-nM, colloidal Pd to be detected by using OH^- and H_2O_2.

Hydrogen peroxide easily reduces $Pd(OH)_2$ and $Pd(OH)_4$ to Pd.

Aqueous HCl containing H_2O_2 dissolves platinum (slowly for the massive metal). Peroxide does not reduce Pt^{IV}.

A small excess of H_2O_2, with dilute H_2SO_4 and K^+, oxidizes $[Pt(CN)_4]^{2-}$ partly on warming, cooling and evaporating, to make the interesting, bronze-colored, electrically conducting, ionic solid, a linear polymer (one of several with other cations); HNO_3 and PbO_2 act similarly:

$$4\,[Pt(CN)_4]^{2-} + {}^1/_2\,H_2O_2 + 7\,K^+ + H_3O^+ + 4\,H_2O \rightarrow$$

$$4\,{\sim}K_{7/4}[Pt(CN)_4] \cdot {}^3/_2 H_2O\downarrow$$

Further action by H_2O_2 gives a deep-blue substance with the empirical formula ${\sim}KPt(CN)_4$. The acid $H_2[Pt(CN)_4]$ and H_2O_2 precipitate, after drying, yellow ${\sim}Pt(CN)_3$, soluble in hot aqueous CN^-.

Similar partial oxidation of $[Pt(C_2O_4)_2]^{2-}$ leads to similar conducting products, such as ${\sim}(Rb,NH_4)_{5/3}[Pt(C_2O_4)_2]$ or ${\sim}(Mg,Co)_{5/6}[Pt(C_2O_4)_2]$.

Mixing 60 μmol $[Pt(NH_3)_4]Cl_2$ with 10 mL 7-M H_2SO_4 which is made 5–10 cM in H_2O_2 yields long, dichroic, orange-pink or almost colorless crystals of $[Pt^{II}(NH_3)_4]Pt^{IV}(NH_3)_4Cl_2(HSO_4)_4$ after some weeks.

Aqueous H_2O_2 and $[Pt(NH_3)_4]^{2+}$ form [*trans*-$Pt(OH)_2(NH_3)_4]^{2+}$. With $[PtCl_2(NH_3)_2]$ the peroxide gives a $[PtCl_2(OH)_2(NH_3)_2]$.

Di- and trioxygen. Air at ambient *T* tarnishes Pd only slightly.

Platinum shows no change in air (or H_2O) at any temperature. Nevertheless platinum black has marked catalytic power; e.g., it unites O_2 with SO_2 to form SO_3 (the "contact process" for making H_2SO_4); with it air oxidizes C_2H_5OH to CH_3CO_2H, but HCO_2H and $H_2C_2O_4$ to CO_2; As^{III} becomes As^V, and a stream of air mixed with hydrogen ignites when passed over it. (Washing the precipitated metal with methanol in air surprised the author by cracking the sintered filter, red hot from the catalyzed oxidation of the methanol.)

Platinum(II) oxide and hydroxide are subject to oxidation by air.

Ozone precipitates $Pd(OH)_4$ from Pd^{II}, or forms a Pd^{IV} anion in alkali.

10.2.2 Reagents Derived from the Other 2nd-Period Non-Metals, Boron through Fluorine

Boron species. Tetrahydroborate can be used to produce very fine Pt:

$$[PtCl_6]^{2-} + [BH_4]^- + 6\,H_2O \rightarrow Pt\downarrow + H_3BO_3 + 2\,H_2\uparrow + 3\,H_3O^+ + 6\,Cl^-$$

Carbon oxide species. Passing CO into a solution of Pd^{2+} reduces MoO_4^{2-}, or a phosphomolybdate, to Molybdenum Blue. This is specific for Pd among the "platinum metals" (Ru, Rh, Pd, Os, Ir and Pt). Acidified $[PdCl_4]^{2-}$ oxidizes CO to CO_2. A detector of CO shows the dark Pd^0 formed from $PdCl_2 \cdot 2H_2O$, also forming CO_2 and HCl.

At a P of 107 or 120 kPa respectively, CO and $[PtCl_6]^{2-}$ form $[Pt_{38}(CO)_{44}]^{2-}$ or $[\{(PtCO)_3(\mu\text{-}CO)_3\}_{n>6}]^{2-}$ in one step at high yields [2].

Aqueous CO_3^{2-} precipitates brown $Pd(OH)_2$ from $PdCl_2$; boiling CO_3^{2-} and $[PtCl_6]^{2-}$ gives $PtO_2 \cdot aq$.

Cyanide species. Hot, concentrated CN^- and Pt sponge form $[Pt(CN)_4]^{2-}$ and H_2. Acids then precipitate yellow $Pt(CN)_2$, soluble in NH_3 or CN^-.

Aqueous CN^- and Pd^{II} precipitate $Pd(CN)_2.aq$, white or yellow, soluble in excess of the reagent to form $[Pd(CN)_4]^{2-}$; one can crystallize, e.g., white $K_2[Pd(CN)_4] \cdot 3H_2O$, efflorescent, soluble, reverting with warm, dilute acids to $Pd(CN)_2$, completely decomposed by boiling with concentrated H_2SO_4. Saturated $K_2[Pd(CN)_4]$ and cold, concentrated HCl saturated with ether yield $H_2[Pd(CN)_4]$ after evaporation of the ether layer over H_2SO_4.

Cyanide ion reduces $[PdCl_6]^{2-}$ to $[Pd(CN)_4]^{2-}$ and Pd. However, $K_2[S_2O_8]$ prevents reduction and yields a little $K_2[Pd(CN)_6]$.

Aqueous CN^- reacts with $[PtCl_4]^{2-}$ and many other compounds of Pt^{II} and Pt^{IV} (mostly with reduction by CN^-) to give $[Pt(CN)_4]^{2-}$ and, depending on T, $K_2[Pt(CN)_4] \cdot nH_2O$. The isolated ion is colorless but many solid hydrated salts show various colors due to Pt–Pt bonding. The strong acid, red $H_2[Pt(CN)_4] \cdot 5H_2O$ can be made by ion-exchange, or from the Ba salt plus H_2SO_4, and extracted by ether, also, e.g., from HCl solution. The potassium salt is made especially conveniently from the bromo complex in warm solution, followed by a salt-ice bath:

$$K_2[PtBr_4] + 4\,CN^- \rightarrow K_2[Pt(CN)_4] \cdot aq\downarrow + 4\,Br^-$$

Aqueous $[Pt(CN)_4]^{2-}$ is not oxidized by $(CN)_2$, thus precluding one route to $[Pt(CN)_6]^{2-}$. However, $K_2[Pt(CN)_4]$ and ICN form $K_2[Pt(CN)_5I]$; added CN^- begins after a few minutes to deposit $K_2[Pt(CN)_6]$. This precipitates, e.g., a silver salt which, with HI, yields the strong acid $H_2[Pt(CN)_6] \cdot 2H_2O$, and this dissolves Zn, yet without reducing the Pt^{IV}.

Aqueous KCN and $K_2[PtI_6]$ form the stable $K_2[Pt(CN)_6]$ and some Pt^{II}, although a dry mixture favors $K_2[Pt(CN)_6]$.

However, Cl_2, Br_2 or I_2 oxidizes $K_2[Pt(CN)_4]$ to $K_2[Pt(CN)_4X_2]$, which becomes $[Pt(CN)_4(NH_3)_2]$ on refluxing with NH_3.

Cyanide, no excess, and heat convert [*trans*-$Pt(Cl,NO_2)_2(NH_3)_2$] to white [*trans*-$Pt(CN)_2(NH_3)_2$]. Heating $Pt(CN)_2$ with concentrated NH_3, however, forms white [*cis*-$Pt(CN)_2(NH_3)_2$].

Among various further complexes we just mention $[Pt(CN)_4(OH)_2]^{2-}$ and finally $[Pt(CN)_n(X)_{6-n}]^{2-}$ (where X may be a halogen).

Some "simple" organic reagents. Ethene, "ethylene", reacts slowly with $[PdCl_4]^{2-}$, yielding the forerunner of many Pd-olefin π-complexes:

$$[PdCl_4]^{2-} + C_2H_4 \rightarrow [PdCl_3(\eta^2\text{-}C_2H_4)]^- + Cl^-$$

It likewise reacts slowly with $[PtCl_4]^{2-}$, catalyzed by $SnCl_2$, yielding, e.g., "Zeise's salt", $Na[PtCl_3(\eta^2\text{-}C_2H_4)]$, the first known olefin complex. Refluxing $K_2[PtCl_6]$ in ethanol produces the same anion; then OH^- forms [*trans*-$PtCl_2(C_2H_4OH)(OH)]^{2-}$, or CN^- gives $[Pt(CN)_4]^{2-}$:

$$K_2[PtCl_6] + 2\ C_2H_5OH + H_2O \rightarrow$$

$$K[PtCl_3(\eta^2\text{-}C_2H_4)]\downarrow + CH_3CHO\uparrow + 2\ H_3O^+ + 3\ Cl^- + K^+$$

Finely divided palladium sponge absorbs over 1000 times its volume of ethyne, "acetylene", C_2H_2. Ethyne passed into an acidic solution of Pd^{2+} forms a brown precipitate (quantitative separation from Cu). Caution! Metal "acetylides" are in general very explosive.

Ethanol at the boiling point, and formate, CHO_2^-, reduce Pd^{2+} to Pd.

Boiling Pt residues with C_2H_5OH and OH^- or CO_3^{2-} recovers the Pt as a fine powder, "platinum black", similarly with glycerol and OH^-, or with HCO_2H, which may be used to detemine Pt gravimetrically.

Acetate and Pd^{2+} form solid $[\{Pd(\mu\text{-}CH_3CO_2)_2\}_3]$ or, with $c(H_3O^+)$ and $c(CH_3CO_2H)$ near 1 M:

$$[Pd(H_2O)_4]^{2+} + CH_3CO_2H \leftrightarrows [Pd(CH_3CO_2)(H_2O)_3]^+ + H_3O^+$$

Alkali tartrates or citrates give yellow precipitates in neutral Pd^{2+}.

Acetic acid, CH_3CO_2H, added to a solution of Pt^{IV} in nitric acid, reduces it to Pt^{II} acetate, with some danger of explosion.

Oxalic acid does not reduce Pd^{2+} (distinction from Au). Warming concentrated $K_2C_2O_4$ in mixtures with $Pd(OH)_2$, $Pd(CH_3CO_2)_2$, $Pd(NO_3)_2$ or $PdCl_2$ yields $K_2[Pd(C_2O_4)_2]$. Other dicarboxylates are similar.

Limited $H_2C_2O_4$ or $C_2O_4^{2-}$ and $[PtCl_6]^{2-}$ give the reddish, light-stable $[PtCl_4]^{2-}$. Further $K_2C_2O_4$ and $K_2[PtCl_4]$ yield $K_2[Pt(C_2O_4)_2]$. Also $H_2C_2O_4$ and $K_2[Pt(NO_2)_4]$ produce $K_2[Pt(C_2O_4)(NO_2\text{-}\kappa N)_2]$ and $K_2[Pt(C_2O_4)_2]$, all chelated. Oxalic acid and [*cis*-$Pt(OH)_2(NH_3)_2$] form $[Pt(C_2O_4)(NH_3)_2]$.

"Dimethylglyoxime", 2,3-butanedionedioxime, $[Me–C(=NOH)–]_2$, or "H_2Dmg", provides a characteristic test by precipitating a yellow palladium(II) dioximate, $[Pd(HDmg)_2]$, i.e., $[Pd(C_4H_7N_2O_2)_2]$, with Pd^{II} even in acidic solutions (distinction from Ni^{II}).

Reduced nitrogen. Palladium(II) oxide and hydroxide dissolve in concentrated NH_3 or "ammonium carbonate".

Boiling $PdCl_2$ or $[PdCl_4]^{2-}$ with excess NH_3, added slowly, gives:

$$PdCl_2 + 4\ NH_3 \rightarrow \text{light-yellow } [Pd(NH_3)_4]^{2+} + 2\ Cl^-$$

Cold HCl plus $[Pd(NH_3)_4]^{2+}$ precipitate yellow $[\textit{trans}\text{-}PdCl_2(NH_3)_2]$.

Excess $HClO_4$ with $[Pd(NH_3)_4]^{2+}$ generates $[\textit{cis}\text{-}Pd(NH_3)_2(H_2O)_2]^{2+}$. Then adding concentrated NaX precipitates $[\textit{cis}\text{-}PdX_2(NH_3)_2]$, stable if X = Cl or Br, but becoming *trans* over many months if X = I.

Mixing $[Pd(NH_3)_4]^{2+}$ and $[PdCl_4]^{2-}$ precipitates $[Pd(NH_3)_4][PdCl_4]$, rose-red. Boiling and cooling give $[\textit{trans}\text{-}PdCl_2(NH_3)_2]$.

Ammonia dissolves $Pd(CN)_2$, perhaps resulting in $[Pd(CN)_2(NH_3)_2]$. Dilute acetic acid then forms a monoammine.

Limited NH_3 with $Pt(NO_2)_4{}^{2-}$ gives:

$$[Pt(NO_2)_4]^{2-} + 2\ NH_3 \rightarrow [\textit{cis}\text{-}Pt(NO_2)_2(NH_3)_2]\downarrow + 2\ NO_2{}^-$$

showing the trans effect, whereby certain ligands in **d**-Group complexes accelerate the replacement of ligands trans to themselves. The trans effect decreases generally thus: $C_2H_4 \sim CO \sim NO \sim (CN\text{-}\kappa C)^- > H^- > (SO_3\text{-}\kappa S)^{2-} > [SC(NH_2)_2\text{-}\kappa S]^{2-} > (SO_3H\text{-}\kappa S)^- > (NO_2\text{-}\kappa N)^- \sim (SCN\text{-}\kappa S)^{2-} \sim I^- > Br^- > Cl^- > NH_3 > OH^- > H_2O \sim F^-$.

Platinum(II) chloride dissolves in (aqueous) NH_3 as $[Pt(NH_3)_4]Cl_2$.

Ammonia with $[PtCl_4]^{2-}$ produces $[PtCl_3NH_3]^-$ and precipitates green $[Pt(NH_3)_4][PtCl_4]$ and yellow $[\textit{cis}\text{-}PtCl_2(NH_3)_2]$, "cisplatin", an antitumor agent, which is sparingly soluble and slowly isomerized to the *trans* and hydrolyzed in water. This $[\textit{cis}\text{-}PtCl_2(NH_3)_2]$, plus limited NH_3, or boiled with NCO^- (which releases NH_3 slowly by hydrolysis), yield $[PtCl(NH_3)_3]^+$. Heating any of these with excess NH_3 produces colorless $[Pt(NH_3)_4]^{2+}$, which can be crystallized as the soluble salt $[Pt(NH_3)_4]Cl_2 \cdot H_2O$. Evaporating this with excess 6-M HCl yields yellow $[\textit{trans}\text{-}PtCl_2(NH_3)_2]$. Both *cis* and *trans* Cl^- ions can be replaced by other anions, and the NH_3 by organic bases.

A sequence designed for "cisplatin" is:

$$[PtCl_4]^{2-} + 4\ I^- \rightarrow [PtI_4]^{2-} + 4\ Cl^-$$

$$[PtI_4]^{2-} + 2\ NH_3 \rightarrow [\textit{cis}\text{-}PtI_2(NH_3)_2] + 2\ I^-$$

$$[\textit{cis}\text{-}PtI_2(NH_3)_2] + 2\ Ag^+ + 2\ H_2O \rightarrow [\textit{cis}\text{-}Pt(H_2O)_2(NH_3)_2] + 2\ AgI\downarrow$$

$$[\textit{cis}\text{-}Pt(H_2O)_2(NH_3)_2] + 2\ Cl^- \rightarrow [\textit{cis}\text{-}PtCl_2(NH_3)_2] + 2\ H_2O$$

The lower concentration of NH_3 from hot $NH_4CH_3CO_2$ is better:

$$[PtCl_4]^{2-} + 2\ NH_4^+ + 2\ CH_3CO_2^- \rightarrow$$

$$[cis\text{-}PtCl_2(NH_3)_2]\downarrow + 2\ Cl^- + 2\ CH_3CO_2H$$

The formula $[Pt^{IV}X_n(NH_3)_{6-n}]^{(4-n)+}$, with X often a halogen, summarizes a vast field of complexes, still excluding those with more than two different ligands. At least for X = Cl, we have every value $0 \le n \le 6$, and with all the stereoisomers. For example, treatment of $[PtCl_6]^{2-}$ with HPO_4^{2-} in 5-M NH_3 with refluxing and cooling yields a white product:

$$[PtCl_6]^{2-} + 6\ NH_3 + HPO_4^{2-} \rightarrow [PtCl(NH_3)_5]PO_4\downarrow + 5\ Cl^- + NH_4^+$$

Six-molar HCl can convert this to white $[PtCl(NH_3)_5]Cl_3$. As with many other complexes of various metals, the coordinated Cl^- can be replaced by (non-aqueous) $CF_3SO_3^-$ ("triflate"), which can then be replaced especially nicely by other ligands. Refluxing $[PtCl(NH_3)_5]Cl_3$ with OH^-, followed by HCl, yields $[Pt(NH_3)_5(H_2O)]Cl_4$.

Aqueous NH_4^+ with $[PtCl_6]^{2-}$ gives a yellow, crystalline precipitate of $(NH_4)_2[PtCl_6]$, insoluble in ethanol, slightly soluble in H_2O, soluble in an excess of the alkalis and reprecipitated by HCl.

Mellor lists hundreds of Pt ammines and related complexes [3].

Metallic Pd is precipitated from solutions by $N_2H_5^+$.

The reaction of $H_2[PtCl_6]$ and $N_2H_6Cl_2$ gives red $[PtCl_4]^{2-}$:

$$[PtCl_6]^{2-} + 2\ N_2H_5^+ + 2\ H_2O \rightarrow$$

$$[PtCl_4]^{2-} + N_2\uparrow + 2\ NH_4^+ + 2\ H_3O^+ + 2\ Cl^-$$

Platinum and gold may be separated from most other metals by precipitation with excess $N_2H_5^+$ in dilute HCl. The precipitate is almost entirely Pt, Au, Hg, and some Cu. In alkaline or acetic-acid solution, N_2H_4 (but not NH_2OH) reduces platinum species to Pt. Stoichiometric amounts of $N_2H_5^+$ with warm, acidic solutions of $[PtCl_6]^{2-}$ or $[PtBr_6]^{2-}$ give Pt^{II}, from which K^+ precipitates brown $K_2[PtBr_4]$:

$$2\ [PtBr_6]^{2-} + N_2H_5^+ + 5\ H_2O \rightarrow 2\ [PtBr_4]^{2-} + N_2\uparrow + 5\ H_3O^+ + 4\ Br^-$$

Large cations, Cat^+, and N_3^- precipitate non-explosive $Cat_2[Pd(N_3)_4]$, $[AsPh_4]_2[\{Pd(N_3)_2\}_2(\mu\text{-}1,1\text{-}N_3)_2]$ and so on.

Aqueous N_3^- and $[PtCl_4]^{2-}$ form $[Pt_2(N_3)_6]^{2-}$ or, with much excess N_3^-, $[Pt(N_3)_4]^{2-}$. From $[trans\text{-}PtCl_2(NH_3)_4]_2^{2+}$ it yields $[trans\text{-}PtCl(N_3)(NH_3)_4]^{2+}$.

Oxidized nitrogen. Expected complexes of Pt^{II} and NO tend to oxidize the NO to NO_2, but $[Pt(NH_3)_4]Cl_2$ and NO can form $[PtCl(NO)(NH_3)_4]$; nitric acid and $[Pt(NO_2)_4]^{2-}$ give $[Pt(NO_3)(NO_2)_4(NO)]^{2-}$.

Cooling $[Pd(NH_3)_4]^{2+}$ to 10–15°C with nitrite and formic acid precipitates a light-yellow product:

$$[Pd(NH_3)_4]^{2+} + 2\ NO_2^- + 2\ HCHO_2 \rightarrow$$

$$[\textit{trans}\text{-}Pd(NO_2)_2(NH_3)_2]\downarrow + 2\ NH_4^+ + 2\ CHO_2^-$$

Adding I^- forms yellow $[\textit{trans}\text{-}PdI_2(NH_3)_2]$ immediately. Also:

$$[PtCl_4]^{2-} + 4\ NO_2^- \rightarrow [Pt(NO_2\text{-}\kappa N)_4]^{2-} + 4\ Cl^-$$

$$[PtCl_6]^{2-} + 6\ NO_2^-\ (\text{hot}) \rightarrow [Pt(NO_2\text{–}\kappa N)_4]^{2-} + 2\ NO_2\uparrow + 6\ Cl^-$$

The $[Pt(NO_2)_4]^{2-}$ is inert even to H_3O^+, OH^- and H_2S.

Cold $[\textit{cis}\text{-}Pt(NH_3)_2(H_2O)_2]^{2+}$ and NO_2^- form $[\textit{cis}\text{-}Pt(NH_3)_2(NO_2\text{-}\kappa N)_2]$.

Aqueous $[Pt(NH_3)_5(H_2O)]Cl_4$ and HNO_2 ($NaNO_2$ plus HCl) at 0 °C form $[Pt(NH_3)_5(NO_2\text{-}\kappa O)]Cl_3$, rearranging to $[Pt(NH_3)_5(NO_2\text{-}\kappa N)]Cl_3$.

Hot HNO_3 or cold, concentrated HNO_3 dissolves Pd and yields $[Pd(NO_3)_2(H_2O)_2]$, soluble in dilute HNO_3, but dilution, evaporation or standing precipitates a basic nitrate. Palladium dissolves more easily in hot HNO_3/HCl, and excess AlkCl yields $Alk_2[PdCl_6]$:

$$Pd + 2\ NO_3^- + 4\ Cl^- + 4\ H_3O^+ \rightarrow [PdCl_4]^{2-} + 2\ NO_2\uparrow + 6\ H_2O$$

Nitric acid has no effect on Pt, but hot HNO_3/HCl dissolves platinum (slowly for the coarse metal), yielding mainly $[PtCl_6]^{2-}$, with variable amounts of $[PtCl_4(NO)_2]$, NO, NO_2 etc., ruling out any single equation.

Melting KNO_3 and KOH together and heating with Pt give $K_2PtO_3\cdot aq$.

Fluorine species. Palladium(2+) and HF form violet or brown PdF_2.

10.2.3 Reagents Derived from the 3rd-to-5th-Period Non-Metals, Silicon through Xenon

Phosphorus species. Dissolved Pd and HPH_2O_2 precipitate Pd^0.

Phosphinate, $PH_2O_2^-$, reduces $PtCl_4$ or $[PtCl_6]^{2-}$ to red $[PtCl_4]^{2-}$ but not to Pt; $[Pt(CN)_4]^{2-}$ is also not reduced.

Phosphonic acid, H_2PHO_3, forms a yellow-green, air-stable, dinuclear diphosphonato complex from $[PtCl_4]^{2-}$ on a steam bath for 3 h. It decomposes in water over 24 h, but is more stable at low pH:

$$2\ [PtCl_4]^{2-} + 8\ H_2PHO_3 + 4\ K^+ \rightarrow$$

$$K_4[Pt^{II}_2(P^{III}_2H_2O_5)_4]\cdot 2H_2O\downarrow + 8\ HCl\uparrow + 2\ H_2O\uparrow$$

Phosphates give a brown precipitate with Pd^{II} but do not generally precipitate platinum salts.

Arsenic species. Arsane gas (AsH_3) and $PtCl_4$ give a black precipitate.

Arsenites and arsenates give precipitates with Pt^{IV}, soluble in HNO_3.

Reduced chalcogens. Sulfane, H_2S, does not notably tarnish Pd or Pt.

From Pd^{II}, H_2S or S^{2-} precipitates dark brown to black $PdS \cdot 2H_2O$, insoluble in "$(NH_4)_2S$", but soluble in HCl and aqua regia.

From Pt^{II} solutions, H_2S precipitates brownish to black PtS, insoluble in acids, slightly soluble in alkali sulfides. "Ammonium sulfide" in excess with Pt^{IV} forms $[PtS_3]^{2-}$. With H_2S, Pt^{IV} salts form PtS_2, black, slowly soluble in alkali sulfides, insoluble in acids except aqua regia, readily converted by air to $PtOS \cdot aq$.

Note on separating Pt and Au from Sn, As and Sb: see gold sulfides in **11.3.3 Reduced chalcogens**.

Adding H_2PtCl_6 dropwise to much $(NH_4)_2S_x$, mainly $(NH_4)_2S_5$, forms reddish, chirally resolvable $[Pt^{IV}(\eta^2\text{-}S_5)_3]^{2-}$ and $(NH_4)_2[Pt(S_5)_3] \cdot 2H_2O$, with turbidity in a few hours and some reduction to $[Pt^{II}(\eta^2\text{-}S_5)_2]^{2-}$, red-orange, faster in base, S^{2-} and especially CN^-, which also takes it farther to colorless $[Pt(CN)_4]^{2-}$, see equation; $(NH_4)_2S_x$ reoxidizes the $[Pt(S_5)_2]^{2-}$:

$$[Pt(S_5)_3]^{2-} + 17\ CN^- \rightarrow [Pt(CN)_4]^{2-} + 13\ NCS^- + 2\ S^{2-}$$

Air but not bright light is acceptable. Aqueous $[N(C_3H_7)_4]^+$ precipitates $[NPr_4]_2[Pt(S_5)_2]$ in good yield, interrupting the reduction by CN^-.

Thiocyanate ion, even in the presence of SO_2, does not precipitate Pd (distinction from Cu) but with excess reagent forms $[Pd(SCN\text{-}\kappa S)_4]^{2-}$, precipitable with $[NEt_4]^+$ and so on, or, without excess, red $Pd(SCN)_2$.

From $[Pt(\eta^2\text{-}C_2O_4)(NO_2)_2]^{2-}$ and SCN^-, $[cis\text{-}Pt(\eta^1\text{-}C_2O_4)(NO_2)_2(SCN)]^{3-}$, $[trans\text{-}Pt(\eta^1\text{-}C_2O_4)(NO_2)(SCN)_2]^{3-}$ and finally $[Pt(SCN)_4]^{2-}$ are made. Likewise $[Pt(\eta^2\text{-}C_2O_4)_2]^{2-}$ and SCN^- give $[Pt(\eta^1\text{-}C_2O_4)_2(SCN)_2]^{4-}$. Other Pt^{II} also form $[Pt(SCN)_4]^{2-}$.

With Pt^{IV} and SCN^- we get $[Pt(SCN)_6]^{2-}$ and many mixed complexes.

Thiourea, "Tu", $SC(NH_2)_2$, distinguishes between the *cis* and *trans* isomers of $PtCl_2(NH_3)_2]$ in hot solution, forming on cooling, a yellow $[PtTu_4]Cl_2$ or white $[trans\text{-}Pt(NH_3)_2Tu_2]Cl_2$, respectively. Various other treatments with $[cis\text{-}PtCl_2(NH_3)_2]$, however, can also yield $[cis\text{-}PtCl_2(NH_3)Tu]$ and $[PtClTu_3]^+$, all with $SC(NH_2)_2\text{-}\kappa S$.

The soft Pt^{II} in $[PtCl_4]^{2-}$ reacts with 1,2-dithiooxalate through the S, as expected, to form stable $[Pt(C_2O_2S_2\text{-}\kappa^2 S)_2]^{2-}$ with five-membered rings. Dithioacetate and others also give chelates and bridged compounds.

Oxidized chalcogens. The results of mixing $S_2O_3^{2-}$ with $[PdCl_4]^{2-}$ depend on the ratio of the two; 1:1 precipitates all the Pd as PdS and PdS_2O_3; 2:1 gives a likely chelate, soluble $[Pd(S_2O_3\text{-}\kappa S,\kappa O)_2]^{2-}$, brown, and a likely polymer, insoluble $K_2Pd(\mu\text{-}S_2O_3)_2$, yellow-brown.

Thiosulfate and $[PtCl_4]^{2-}$ can form either $[PtCl_2(S_2O_3\text{-}\kappa S,\kappa O)]^{2-}$ or $[Pt(S_2O_3\text{-}\kappa S,\kappa O)_2]^{2-}$, but in warm acidic solution it precipitates PtS.

Cooling warm $K_2[PtCl_4]$ and $KHSO_3$ forms white $K_6[Pt(SO_3)_4]$. Sulfur dioxide precipitates Pd^0 from the nitrate or sulfate, but not the chloride. Saturated with SO_2, $PdCl_2$ plus NaOH give $Na_6[Pd(SO_3\text{-}\kappa S)_4]\cdot 2H_2O$.

Platinum(II) oxide and hydroxide are soluble in aqueous SO_2, but not in the other oxoacids unless freshly precipitated. Sulfur dioxide or HSO_3^- reduces $PtCl_4$ or $[PtCl_6]^{2-}$, not to Pt, but to stable, colorless Pt^{II} complexes which do not respond to the usual reagents for Pt and require long boiling with HCl to remove the SO_2, although this is still a good route to the red $[PtCl_4]^{2-}$.

Water, $PdCl_2$ and Ag_2SO_3 yield $[Pd(SO_3\text{-}\kappa S)(H_2O)_3]$. Adding the right amount of concentrated NH_3 to this in solution precipitates white $[Pd(SO_3\text{-}\kappa S)(NH_3)_3]$ or a yellow-orange $[Pd(SO_3\text{-}\kappa S)(H_2O)_{3-n}(NH_3)_{n<3}]$. Dehydration in order to get $\eta^2\text{-}SO_3^{2-}$, a la SO_4^{2-}, decomposes them instead. On silica gel they detect CO in gasses by replacing an H_2O with the CO and then quickly forming visible Pd.

Water, $PdCl_2$, K_2SO_3 and $K_2S_2O_5$ yield $K_2Pd(\mu\text{-}SO_3)_2\cdot H_2O$.

Aqueous SO_3^{2-} and $[Pt(NH_3)_5(H_2O)]^{4+}$ produce $[Pt(SO_3\text{-}\kappa O)(NH_3)_5]^{2+}$, which isomerizes then to $[Pt(SO_3\text{-}\kappa S)(NH_3)_5]^{2+}$.

Palladium is slowly dissolved by boiling H_2SO_4. Dilute H_2SO_4 has no effect on Pt. Hot concentrated H_2SO_4 slowly forms $Pt(SO_4)_2$. Platinum (II) sulfate dissolves in dilute H_2SO_4.

Sulfuric acid dissolves PtO_2 or $Pt(OH)_4\cdot aq$, possibly giving $Pt(SO_4)_2$.

From $PdCl_2$, $[Te(OH)_6]$, NaClO and NaOH, one can prepare $Na_5H_3[Pd^{IV}(TeO_6)_2]\cdot 4H_2O$.

Aqueous $K_2[PdCl_6]$, $K_2[S_2O_8]$ and KCN form colorless $K_2[Pd(CN)_6]$ in low yield. This precipitates Cs, Mn^{II}, Fe^{II}, Co^{II}, Ni, Zn, Cd, Ag etc. salts. Ion exchange and vacuum evaporation at 25 °C give $(H_3O)_2[Pd(CN)_6]$.

Aqueous $S_2O_8^{2-}$ and $[Pt(NH_3)_4]^{2+}$ form mainly $[Pt(OH)(SO_4)(NH_3)_4]^+$, $[Pt(SO_4)_2(NH_3)_4]$ and $[Pt(OH)_2(NH_3)_4]^{2+}$, all probably *trans*, hydrolyzing in base to $[trans\text{-}Pt(OH)_2(NH_3)_4]^{2+}$; Br^- gives $[trans\text{-}PtBr_2(NH_3)_4]^{2+}$. The insoluble $[Pt(SO_4)_2(NH_3)_4]$ and Ba^{2+} quickly form $[Pt(OH)(SO_4)(NH_3)_4]^+$ with a highly inert SO_4 ligand, plus $BaSO_4$ and H_3O^+.

Reduced halogens. Palladium is slowly dissolved by boiling HCl.

The aqueous acids HX have no effect on Pt.

Substituting Cl^- or Br^- for H_2O in $[Pd(H_2O)_4]^{2+}$ and in $[PdCl_3(H_2O)]^-$ or $[PdBr_3(H_2O)]^-$ in turn, and the reverse, are much faster than for Pt^{II}. The $[(Pd,Pt)(H_2O)_4]^{2+}$ and X^- ions form $[MX(H_2O)_3]^+$, $[trans\text{-}MX_2(H_2O)_2]$ etc. faster for X^- as $Cl^- < Br^- < SCN^- < I^-$ via an associative mechanism.

Platinum oxides and hydroxides are soluble in HCl:

$$Pt(OH)_2 + 4\,Cl^- + 2\,H_3O^+ \rightarrow [PtCl_4]^{2-} + 4\,H_2O$$

$$Pt(OH)_4\cdot aq + 6\,Cl^- + 4\,H_3O^+ \rightarrow [PtCl_6]^{2-} + 8\,H_2O$$

Palladium(II) chloride, bromide and iodide form complex ions, $[PdX_4]^{2-}$, more or less readily. Many of the ordinary complexes are more soluble in water than are the simple salts.

Palladium dioxide is readily soluble in HCl, and then:

$$[PdCl_6]^{2-} \leftrightarrows [PdCl_4]^{2-} + Cl_2\uparrow$$

Aqueous I^- precipitates Pd^{2+} as PdI_2, black, visible even at 20 μM in solution. It is insoluble also in ethanol, but soluble in excess I^-.

Platinum(II) chloride dissolves in HCl as $[PtCl_4]^{2-}$, but also forms some $[PtCl_6]^{2-}$ and Pt.

Dilute HCl dissolves $PtCl_2$ with difficulty, forming red $[PtCl_4]^{2-}$, but the bromide and iodide are practically insoluble in HBr and HI, respectively. Even 10-M HCl forms no higher species than $[PtCl_4]^{2-}$.

The ammine $[Pt(NH_3)_4]^{2+}$ reacts with limited HCl to give [*trans*-$PtCl_2(NH_3)_2$], showing the trans effect, by which the first Cl^- ligand, more than the NH_3, promotes trans replacement.

The Cl^- ion and $[Pt(\eta^2\text{-}C_2O_4)_2]^{2-}$ form $[Pt(\eta^1\text{-}C_2O_4)_2Cl_2]^{4-}$. Warm and excess Cl^- produce $[PtCl_4]^{2-}$ and intermediates.

Aqueous Br^- and $[Pt(NH_3)_4]^{2+}$ form some [*trans*-$PtBr_2(NH_3)_4]^{2+}$ in acidified H_2O_2 and in neutral or acidified $S_2O_8{}^{2-}$ solutions. Also Br^- and $[PtI(NH_3)_5]^{3+}$ exchange halides, catalyzed by $[Pt(NH_3)_4]^{2+}$ but quenched by Ce^{IV}. Many other halide exchanges occur.

Halogen-bridged anions have been found in $(Et_4N)_2[(PtX_2)_2(\mu\text{-}X)_2]$, for example, with X = Br or I.

The reaction of I^- with [*trans*-$PtCl_2(NH_3)_2$] to form [*trans*-$PtClI(NH_3)_2$] exemplifies many reactions omitted here. The reaction of I^- with $[PtCl_4]^{2-}$ depends on concentrations and exposure to air:

$$[PtCl_4]^{2-} + 2\ I^- \rightarrow PtI_2\downarrow + 4\ Cl^-$$

$$2\ PtI_2\downarrow + 2\ I^- \leftrightarrows [Pt_2I_6]^{2-}$$

$$[Pt_2I_6]^{2-} + 2\ I^- \leftrightarrows 2\ [PtI_4]^{2-}$$

$$[PtI_4]^{2-} + {}^1/_2\ O_2 + 2\ I^- + 2\ CO_2 + H_2O \rightarrow [PtI_6]^{2-} + 2\ HCO_3{}^-$$

In general, halides (beyond F^-) form complex ions with both Pt^{II} and Pt^{IV}, namely $[PtX_4]^{2-}$ and $[PtX_6]^{2-}$, where X is Cl, Br, or I. The chlorides of K^+ and $NH_4{}^+$ form the yellow $K_2[PtCl_6]$ and $(NH_4)_2[PtCl_6]$, slightly soluble in H_2O, insoluble in ethanol. The $[PtBr_6]^{2-}$ ion is reddish. The softer, larger halide ions tend to substitute for the smaller ones, and the larger ones make the best bridges. Various substitutions of Y for X on Pt^{IV} go via reduction to Pt^{II} or Pt^{III} and then reoxidation of the Pt.

Acidified $[PdCl_6]^{2-}$ solutions and much excess CsI give the elusive $Cs_2[PdI_6]$, stable in humid air.

Iodide colors a solution of $PtCl_4$ red to brown (sensitive to 0.3 mmol Pt) (Fe^{III}, Cu^{II}, and other oxidants interfere) and may precipitate black PtI_4. Excess of KI forms $K_2[PtI_6]$, brown, slightly soluble, and unstable enough that platinum may be determined volumetrically by treating $[PtCl_6]^{2-}$ with excess I^- and titrating the liberated iodine with thiosulfate, which shifts the Pt^{II}-Pt^{IV} equilibrium completely toward reduction (as the O_2 and acidic CO_2 three paragraphs above shift it toward oxidation):

$$[PtCl_6]^{2-} + 2\ I^- + 2\ S_2O_3^{2-} \rightarrow PtI_2\downarrow + [S_4O_6]^{2-} + 6\ Cl^-$$

However, one may also prepare $K_4[PtI_4][PtI_6]$.

Elemental and oxidized halogens. Chlorine and HCl, or Br_2 and HBr, react with Pt or Pt^{II} to form $[PtCl_6]^{2-}$, $[PtBr_6]^{2-}$ or mixtures, and also yield $\sim(H_3O)_2[PtCl_6]\cdot(2,4)H_2O$ or $PtBr_4$. The potassium salts are insoluble.

Heating (aqueous) $K_2[Pt(CN)_4]$ with excess Cl_2, Br_2 or I_2 forms, after cooling, $K_2[$*trans*-$Pt(CN)_4X_2]$, pale-yellow, bright-yellow and brown, respectively. Ammonia and $K_2[Pt(CN)_4X_2]$ precipitate $[Pt(CN)_4(NH_3)_2]$, and aqueous KOH then ionizes one NH_3 to $K[Pt(CN)_4(NH_2)(NH_3)]$, but $AgNO_3$ leads to $Ag[Pt(CN)_4(NH_3)_2]NO_3$.

The oxidation by bromine is a step toward making an electrically conducting ionic solid, first a rapid reaction, then a slow one:

$$[Pt(CN)_4]^{2-} + Br_2 + + 2\ K^+ + 2\ H_2O \rightarrow K_2[\textit{trans}\text{-}PtBr_2(CN)_4]\cdot 2H_2O\downarrow$$

This is then mixed with five times as much of the starting material in water and made ice-cold, forming the desired lustrous, copper-colored linear polymer, formula $\sim K_2PtBr_{1/3}(CN)_4\cdot 3H_2O$, which may be otherwise written $K_2PtBr_{0.3}(CN)_4\cdot 3H_2O$, with some valence electrons free to roam, all Pt atoms equivalent and in non-integral oxidation states, and with rather unstable hydration. The final reaction is approximately thus:

$$5\ K_2[Pt(CN)_4]\cdot 3H_2O + K_2[PtBr_2(CN)_4]\cdot 2H_2O + H_2O \rightarrow$$

$$6\ K_2PtBr_{1/3}(CN)_4\cdot 3H_2O\downarrow$$

Dissolution restores the $[Pt(CN)_4]^{2-}$ and $[PtBr_2(CN)_4]^{2-}$. With various cations, HF_2^-, N_3^-, Cl^- etc. may replace the Br^-. Two $C_2O_4^{2-}$ may replace four CN^- in cation-deficient salts such as $(K_{2n},Mg_n,\mathbf{3d}_n)[Pt(C_2O_4)_2]\cdot mH_2O$, with n a little less than 1. See **Other reactions** below about "Pt^{III}".

Mixing Pd^{II}, X^- and X_2, with X = Cl or Br, forms $[PdX_6]^{2-}$, but PdX_4 and $H_2[PdX_6]$ cannot be isolated. A typical chloride solution of Rh, Ir, Pd and Pt from ores can be treated with HCl, evaporated, the Ir and Pt precipitated by NH_4Cl as

$(NH_4)_2[MCl_6]$, and the $[PdCl_4]^{2-}$ removed from the solution as $(NH_4)_2[PdCl_6]$ after adding Cl_2:

$$[PdCl_4]^{2-} + 2\ NH_4^+ + Cl_2 \rightarrow (NH_4)_2[PdCl_6]\downarrow$$

Chlorine and aqueous $[PdCl_2(NH_3)_2]$ give $[PdCl_4(NH_3)_2]$.

Chlorine and $[Pt(NH_3)_4]^{2+}$ form [*trans*-$PtCl_2(NH_3)_4]^{2+}$. Chlorine and [*cis-/trans*-$PtCl_2(NH_3)_2$] produce [*cis-/trans*-$PtCl_4(NH_3)_2$], respectively, each lemon-yellow. For the *cis*, Cl_2 is introduced slowly for 3 h at 75–80 °C, to avoid forming $[PtCl_6]^{2-}$ at higher *T*. For the *trans*, it is for 1 h at 100 °C. Each is nearly insoluble in cold water and not attacked even by concentrated H_2SO_4, but long boiling with Ag^+ releases all the Cl^-. The *trans* form dissolves in OH^- without releasing NH_3. Also, Cl_2, Br_2 or I_2 oxidizes [*trans*-$Pt(CN)_2(NH_3)_2$] to [*trans-trans*-$Pt(CN)_2X_2(NH_3)_2$].

At 20 °C, Cl_2 and $[PtBr(NH_3)_5]^{3+}$ form [*trans*-$PtBr(NCl_2)(NH_3)_4]^{2+}$, which then, at 100 °C, goes to [*trans*-$PtBrCl(NH_3)_4]^{2+}$.

Bromine and $[Pt(NH_3)_4]^{2+}$ very quickly form $[PtBr(OH)(NH_3)_4]^{2+}$.

Aqueous $HClO_3$ plus HCl dissolve platinum (slowly for massive Pt), and oxidize Pt^{II} to $[PtCl_6]^{2-}$.

Aqueous $[Pd(OH)_6]^{2-}$, $H_2IO_6^{3-}$ and KOH give $K_7[Pd(IO_6)_2]OH$.

10.2.4 Reagents Derived from the Metals Lithium through Uranium, plus Electrons and Photons

Oxidation. In, e.g., Cl^- media, Pt^{II} is oxidized to Pt^{IV}, sometimes via Pt^{III}, by Ce^{IV}, VO_2^+, $[Cr_2O_7]^{2-}$, MnO_4^-, Fe^{III}, $[IrCl_6]^{2-}$, $[AuCl_4]^-$ and others. Also:

$$[Pt(NH_3)_4]^{2+} + 2\ [Fe(CN)_6]^{3-} \rightarrow [Pt^{IV}(NH_3)_4\{-NCFe^{II}(CN)_5\}_2]^{4-}$$

Treating $[Pt^{II}(NH_3)_4](NO_3)_2$ with $Na_3[Mo^V(CN)_8]$ soon gives an intense red, trinuclear $[\{Mo^{IV}(CN)_7{-}CN{-}\}_2\{$*trans*-$\mu$-$Pt^{IV}(NH_3)_4\}]^{4-}$.

A Pt anode in KCN solution forms $[Pt(CN)_4]^{2-}$. Partial further oxidation by electrolysis of $[Pt(CN)_4]^{2-}$ can produce relatively large needles of metallic-appearing, polymeric $\sim K_{7/4}Pt(CN)_4 \cdot {}^3/_2H_2O$ with a fractional oxidation state. A somewhat similar treatment of $[Pt(C_2O_4)_2]^{2-}$ gives $\sim K_{5/3}Pt(C_2O_4)_2 \cdot 2H_2O$; cf. **Elemental halogens** above.

Anodic electrolysis of Pd^{2+} gives PdO_2 or $Pd(OH)_4$, losing O_2 easily.

Anodic oxidation of $[Pt(OH)_6]^{2-}$ gives PtO_4^{2-}, but $PtO_2 \cdot aq$ and KOH at 0 °C form PtO_3, which loses O_2 extremely easily.

Reduction. Palladium(0) is precipitated from solutions by Mg, Mn, Fe, FeS, Zn, Cd, Hg, Cu, Cu^+, Al, Sn, Sn^{II}, Pb, PbS, Sb, SbH_3, Bi, etc.

Without O_2, Fe^{2+} and Pd^{2+} precipitate black Pd^0 immediately. Oxygen delays this until it is all gone. Oxygen may oxidize a Pd^I intermediate faster than Fe^{2+} reduces it.

The V^{2+} and Cr^{2+} ions reduce $[PtCl_6]^{2-}$ and/or $[PtCl(NH_3)_5]^{3+}$ etc. to Pt^{II}, apparently with one-equivalent outer-sphere (V^{2+}) and two-equivalent inner-sphere (Cr^{2+}) steps. The V^{2+} via Pt^{III} is quite fast.

Boiling Fe^{2+} with Pt^{IV} precipitates metallic Pt, the reduction being hindered by acids, but helped, rather unexpectedly, by the oxidants $HgCl_2$ or NO_3^-. Iron(2+) may thus be used to separate both Au and Pt from Sn, As and Sb. Platinum salts are reduced to Pt also by metallic Fe, Co, Ni, Cu, Ag, Zn, Cd, Hg, Al, Sn, Pb and Bi, although many of these are very slow with Pt (and Ru, Rh, Os and Ir), but not the more labile Pd, complexes. Stibane (SbH_3) and $PtCl_2$ precipitate $Sb_2O_3 \cdot aq$ and Pt.

Copper(1+), CuCl and $SnCl_2$ reduce Pt^{IV} to Pt^{II}; also see **Other reactions** next for $SnCl_2$.

Photons may reduce $[PtCl_6]^{2-}$ to $[PtCl_5]^{2-}$, $t_{1/2} \sim \mu s$, labile toward $*Cl^-$ ($^{36}Cl)^-$, catalyzing isotopic exchange with quantum yields of hundreds:

$$[PtCl_6]^{2-} + [Pt(*Cl)_5]^- \leftrightarrows [Pt(*Cl)_5Cl]^{2-} + [PtCl_5]^{2-}$$

Other reactions. Aqueous K^+ with $[PtCl_6]^{2-}$ precipitates $K_2[PtCl_6]$, very similar to $(NH_4)_2[PtCl_6]$ (see NH_4^+ under **Reduced nitrogen** above), usable to determine these alkalis quantitatively. The bromo- and iodo-complexes are less satisfactory. The salt $Na_2[PtCl_6]$ is very soluble and is decomposed by light in alkaline solution, forming PtO_2.

Mixing $[PdCl_4]^{2-}$ or $[PtCl_4]^{2-}$ with $[MoS_4]^{2-}$ or $[WS_4]^{2-}$, better in aqueous CH_3CN, yields $[(Pd,Pt)^{II}\{\eta^2\text{-}(Mo,W)^{VI}S_4\}_2]^{2-}$.

Neither $[Fe(CN)_6]^{3-}$ nor $[Fe(CN)_6]^{4-}$ affects Pt^{II}. With $PtCl_4$ the potassium salts both precipitate $K_2[PtCl_6]$. Excess of $[Fe(CN)_6]^{4-}$ gives first a green precipitate, then, with still more reagent, a yellow solution.

Aqueous $K_2[Pt(CN)_4]$ and $K_2[PtCl_4]$ react to precipitate $Pt(CN)_2 \cdot aq$.

Colorless $[Pt(NH_3)_4]^{2+}$ and red $[PtCl_4]^{2-}$ precipitate "Magnus' Green Salt", $[Pt(NH_3)_4][PtCl_4]$, whose metal–metal bonds affect the color, one of many, e.g., $[Cu(NH_3)_4][PtCl_4]$. Partial oxidation of the Green Salt yields a photochromic $[Pt(NH_3)_4][Pt(NH_3)_4Cl_2](HSO_4)_4$ among others.

Aqueous $[Pt(NH_3)_4]^{2+}$ catalyzes, via a bridged activated complex:

$$[PtCl(NH_3)_5]^{3+} + Cl^- + H_3O^+ \rightarrow [\mathit{trans}\text{-}PtCl_2(NH_3)_4]^{2+} + NH_4^+ + H_2O$$

The old formula $PtBr_3(NH_3)_2$ exemplifies those suggesting Pt^{III}, but really having bromo Pt^{II}—Pt^{IV} bridges; if one mixes $[\mathit{trans}\text{-}PtBr_2(NH_3)_2]$ with $[\mathit{trans}\text{-}PtBr_4(NH_3)_2]$, one finds as the result the linear polymer $[\mathit{trans}\text{-}Pt^{II}Br_2(NH_3)_2](\mu\text{-}Br)[\mathit{trans}\text{-}Pt^{IV}Br_2(NH_3)_2](\mu\text{-}Br)$. A rather similar bridging occurs in $K_4[PtI_4][PtI_6]$ (which is not $K_2[PtI_5]$).

The complexes $[PdX_4]^{2-}$ with X = Cl, Br or I, and Ag^+ give AgX and $[PdX_{4-n}(H_2O)_n]^{(2-n)-}$, which are acidic.

The complex [*cis*-$PtCl_2(NH_3)_2$] and Ag_2O yield [*cis*-$Pt(NH_3)_2(OH)_2$] and $[\{Pt(NH_3)_2\}_2(\mu\text{-}OH)_2]^{2+}$. Silver(1+) and [*trans*-$PtClI(NH_3)_2$] produce [*trans*-$PtCl(NH_3)_2(H_2O)]^+$ (and AgI), yet another example of many.

Concentrated HCl, $GeHCl_3$ and $PtCl_4{}^{2-}$ with Ge:Pt::5:1 form a red solution; $[NMe_4]^+$ precipitates cream-colored $[NMe_4]_2[Pt^{IV}H(GeCl_3)_5]$, but Ge:Pt::2:1 give a red solution, then yellow $[NMe_4]_2[Pt^{II}Cl_2(GeCl_3)_2]$.

Tin dichloride colors aqueous Pt^{II} deep red (distinction from Ir, Pd and Au). Adding much $[SnCl_3]^-$ in 3-M HCl to $[PtCl_4]^{2-}$ produces, e.g., trigonal bipyramidal $[Pt^{II}(Sn^{II}Cl_3\text{-}\kappa Sn)_5]^{3-}$ in very complex solutions. Small amounts of $SnCl_2$ with dilute Pt give a golden-yellow color.

Anodic treatment of $K_2[Pt(CN)_4]$ in (aqueous) HF and KHF_2 forms the mixed-valence $K_2[Pt(CN)_4](HF_2)_{0.3}\cdot aq$.

Red light isomerizes $[Pt(NH_3)_4Cl(NO\text{-}\kappa N)]Cl_2$ to NO-*κO* [4].

References

1. Meyer F, Kozlowski H, in McCleverty JA, Meyer TJ (eds) (2004) Comprehensive coordination chemistry II. Elsevier, Amsterdam, vol 6, p 463
2. Femoni C et al (2005) Chem Comm 46:5769
3. Mellor JW (1937) Inorganic and theoretical chemistry, vol. XVI, Longmans, London, p 350
4. Schaniel Detal (2007) Phys Chem Chem Phys 9:5149

Bibliography

See the general references in the Introduction, specifically [116], [121] and [313], and some more-specialized books [4–10]. Some articles in journals discuss: Pt^{III} or Pt^{II}/Pt^{IV} complexes [11]; mixed-valence complexes of Pt etc. [12]; isomerization mechanisms of square-planar complexes [13]; and the cis and trans effects [14].

4. Coombes JS (1992) Platinum 1992. Johnson Matthey, London
5. Hartley FR (ed) (1991) Chemistry of the platinum group metals. Elsevier, Amsterdam
6. Robson GG (ed) Platinum 1985. Johnson Matthey.
7. Belluco U (1974) Organometallic and coordination chemistry of platinum. Academic, New York
8. Hartley FR (1973) The chemistry of platinum and palladium. Wiley, New York
9. Lewis CL, Ott WL (1970) Analytical chemistry of nickel. Pergamon, Oxford
10. Gibalo IM (1967) Schmorak J (trans) (1968) Analytical chemistry of nickel. Ann Arbor-Humphrey, Ann Arbor
11. Woolins DJ, Kelly PT (1985) Coord Chem Rev 65:115
12. Clark RJH (1984) Chem Soc Rev 13:219
13. Anderson GK, Cross RJ (1980) Chem Soc Rev 9:185
14. Hartley FR (1973) Chem Soc Rev 2:163

11 Copper through Roentgenium

11.1 Copper, $_{29}Cu$

Oxidation numbers in classical compounds: (I), (II) and (III), as in Cu_2O, "cuprous" oxide, CuO, "cupric" oxide, and $Na_9[Cu^{III}(TeO_6)_2]\cdot 16H_2O$.

11.1.1 Reagents Derived from Hydrogen and Oxygen

Dihydrogen. Acidified solutions of Cu^{2+} oxidize H_2 to H_3O^+.

Copper(II) catalyzes the reductions of Cr^{VI}, Fe^{III}, Tl^{III}, IO_3^- etc., by (relatively slowly) forming CuH^+, which is then rapidly oxidized:

$$Cu^{2+} + H_2 + H_2O \leftrightarrows CuH^+ + H_3O^+$$

$$\underline{CuH^+ + H_2O \rightarrow Cu^{2+} + 2\ e^- + H_3O^+}$$

$$H_2 + 2\ H_2O \rightarrow 2\ e^- + 2\ H_3O^+$$

Water. For Cu^{II} the sulfate is efflorescent in dry air; the crystallized chloride and chlorate are deliquescent; the acetate is efflorescent. Copper(II) borate, basic carbonate, cyanide, oxalate, phosphate, arsenite, sulfide, and the hexacyanoferrates (II and III) are insoluble.

Hydrated Cu^{2+} is often square pyramidal $[Cu(H_2O)_4(H_2O)]^{2+}$, but also distorted octahedral in $[Cu(H_2O)_6](ClO_4)_2$ and $(NH_4)_2[Cu(H_2O)_6](SO_4)_2$, or square planar in $[Cu(H_2O)_4]SO_4\cdot H_2O$.

Seawater and some freshwater contain Cu complexes as $CuOH^+$, $Cu(OH)_2$, $CuCO_3$, $CuHCO_3^+$, $Cu(CO_3)_2^{2-}$, $CuSO_4$, CuCl, $[CuCl_2]^-$, and $CuCl^+$. Natural brines may contain $[CuCl_3]^{2-}$. Hot natural waters may contain $[CuCl_4]^{2-}$. Some other natural waters may contain $Cu(NH_3)_n^{2+}$ or $H_mCu^{II}S_n^{(2n-m-2)-}$ and polysulfido and thiosulfato complexes.

Oxonium. Copper does not readily dissolve and release H_2 from H_3O^+.

Oxonium ion, H_3O^+, from, e.g., H_2SO_4, or even HNO_3 when cold and very dilute, converts Cu_2O to Cu and Cu^{2+}.

Hydroxide. Hydroxide ion, OH^-, precipitates yellow CuOH from Cu^I, insoluble in excess reagent. Copper(I) oxide, Cu_2O, is insoluble in H_2O, soluble in NH_3, scarcely soluble in OH^-. Aqueous Cu^+ or CuOH dismutates in water at all pH values.

Limited amounts of OH^-, with Cu^{II}, precipitate basic salts of a lighter blue than the hydroxide, with such compositions as $Cu_4SO_4(OH)_6·aq$, depending on conditions. From Cu^{II}, including $Cu(NH_3)_4^{2+}$, sufficient OH^- precipitates blue $Cu(OH)_2$, changed by boiling to black CuO but soluble in acids, NH_3, CN^- or hot NH_4^+, slightly soluble in rather concentrated OH^-, completely so if tartrate, citrate, glycerol or other chelators are present (Fehling's solution). Boiled alone, this solution is fairly stable, but reductants such as glucose, N_2H_4 or arsenite precipitate yellow Cu_2O. The solubility in tartrate (without excess OH^-) is a separation from Zn and Cd; in OH^- and glycerol, a separation from Cd.

Di- and trioxygen. In moist air containing CO_2, Cu becomes coated with a film of "verdigris", a basic Cu^{II} carbonate, which protects it from further action by air or water.

Cold CH_3CO_2H slowly dissolves Cu in the air.

Aqueous H_2S has virtually no action on finely divided Cu at ordinary temperatures, but air with it causes a vigorous oxidation.

Cold HCl and HBr attack Cu appreciably only in the presence of air.

Moist air readily oxidizes Cu^I salts; CuCl and HCl give $CuCl_n^{(n-2)-}$.

Air and NH_3 partly oxidize CuCN to $[Cu^{II}(NH_3)_4][Cu^I(CN)_2]_2$.

Ozone does not oxidize Cu^{II} even if alkaline and hot.

11.1.2 Reagents Derived from the Other 2nd-Period Non-Metals, Boron through Fluorine

Boron species. Aqueous $[CuCl_3]^{2-}$ and $[BH_4]^-$ form a somewhat stable intermediate, possibly $[CuHCl_n]^{n-}$, and H_2.

Carbon oxide species. Solutions of CuCl both in NH_3 and in concentrated HCl absorb CO, and $CuCl·CO·2H_2O$ can be isolated.

Copper(1+) with CO_3^{2-} precipitates yellow Cu_2CO_3.

Carbonate, CO_3^{2-}, with Cu^{II}, precipitates greenish-blue basic salts, the composition variable, depending on the temperature and concentration. Adding $NaHCO_3$ can yield $Na_2Cu(CO_3)_2·3H_2O$, or $Na_2Cu(CO_3)_2$ by boiling with saturated $NaHCO_3$ and Na_2CO_3 for 24 h. Boiling with only Na_2CO_3 ultimately gives the oxide, CuO. The $AeCO_3$ do not precipitate $CuCO_3$ in the cold; basic carbonates are precipitated on boiling.

Although the composition of many basic salts is indefinite, a definite complex (and μ_4-OH!) copper anion can be isolated from a solution of Cu^{II} in excess K_2CO_3 and $KHCO_3$ by precipitation with $[Co(NH_3)_6]^{3+}$ at 25°C as a green product, $[Co(NH_3)_6]_3[Cu_4(\mu_4\text{-}OH)(CO_3)_8]·2H_2O$, which is stabilized actually by 40 H-bonds [1].

Cyanide species. Aqueous CN^- precipitates white CuCN from Cu^I solutions not too strongly acidic. This and other Cu^I salts are readily soluble in excess

CN^-, forming especially $[Cu(CN)_4]^{3-}$ and other complexes, from which sulfides or OH^- give no precipitate. Some examples of solids are: $KCu(CN)_2$, $Na_2[Cu(CN)_3]\cdot 3H_2O$, $K_3[Cu(CN)_4]$, $K[Cu_2(CN)_3]\cdot H_2O$ and $(Rb,Cs)_2[Cu_3(CN)_5]$.

With Cu^{II} salts in cool non-acidic media, CN^- gives a transient violet $[Cu(CN)_4]^{2-}$ and precipitates green $\sim Cu^{II}[Cu^{I}(CN)_2]_2\cdot aq$ or yellowish $Cu(CN)_2$, readily soluble in excess with decomposition to $[Cu^{I}(CN)_n]^{(n-1)-}$ and $(CN)_2$. Heating the precipitates gives white CuCN and $(CN)_2$.

In ammonia the overall reaction may be simplified and written as:

$$2\,[Cu(NH_3)_4]^{2+} + 9\,CN^- + H_2O \rightarrow$$

$$2\,[Cu(CN)_4]^{3-} + CNO^- + 2\,NH_4^+ + 6\,NH_3$$

The following (non-elementary) steps are given:

$$4\,CN^- + [Cu(NH_3)_4]^{2+} \rightarrow [Cu(CN)_4]^{2-} + 4\,NH_3$$

$$[Cu(CN)_4]^{2-} + CN^- \rightarrow [Cu(CN)_4]^{3-} + {}^1/_2\,(CN)_2$$

$$(CN)_2 + H_2O + 2\,NH_3 \rightarrow CNO^- + CN^- + 2\,NH_4^+$$

Cyanide also dissolves CuO, $Cu(OH)_2$, the carbonate, sulfides, etc., which change rapidly to $[Cu(CN)_4]^{3-}$. In these solutions the $c(Cu^+)$ is too low to precipitate Cu_2S with H_2S (separation from Cd).

With NCS^- and NH_3, the borderline hard-or-soft Cu^{2+} forms both $[Cu(NH_3)_2(NCS\text{-}\kappa N)_2]$ and $[Cu(NH_3)_4](SCN\text{-}\kappa S)_2$ if we may thus show the latter "semi-coordinated" axial SCN^- with a long Cu–S bond.

Some "simple" organic species. Copper(I) solutions absorb alkenes. Ethene, Cu and $[Cu(H_2O)_6](ClO_4)_2$ form an explosive product:

$$Cu + [Cu(H_2O)_4]^{2+} + 2\,C_2H_4 + 2\,ClO_4^-$$

$$\rightarrow 2\,[Cu(\eta^2\text{-}C_2H_4)(H_2O)_2]ClO_4\downarrow$$

Formate and Cu^{2+} produce $Cu(CHO_2)_2\cdot 4H_2O$, whose structure exposes a dilemma in formulating various substances, whether to write it, with its two "semi-coordinate" H_2O (long Cu–O bonds) and two lattice H_2O, as $[Cu(\eta^2\text{-}CHO_2)_2(H_2O)_2]\cdot 2H_2O$ or perhaps as $[Cu(\eta^2\text{-}CHO_2)_2](H_2O)_2\cdot 2H_2O$.

Acetate and Cu^{II} yield dark-green $[\{-Cu(H_2O)\}_2(\mu\text{-}CH_3CO_2)_4]$ or $Cu_2(CH_3CO_2)_4\cdot 2H_2O$; excess can give, e.g., $Ca[Cu(CH_3CO_2)_4]\cdot 6H_2O$.

Oxalate, $C_2O_4^{2-}$, precipitates white Cu^I oxalate from Cu^I solutions not too strongly acidic. Oxalate (or equivalently, $H_2C_2O_4$ buffered with the weakly basic $CH_3CO_2^-$) precipitates from Cu^{II} salts, light-blue $CuC_2O_4\cdot{}^1/_2H_2O$, insoluble in acetic acid (distinction from Cd). Excess $Na_2C_2O_4$ gives $Na_2[Cu(C_2O_4)_2]\cdot 2H_2O$, and with NH_3 added we can have $[trans\text{-}Cu(NH_3)_2(\mu\text{-}\eta^2\text{-}C_2O_4)]\cdot 2H_2O$.

Copper(I) oxide, red, is precipitated by reducing alkaline Cu^{II}, e.g., in Fehling's tartrate solution, heated with, say, glucose.

Reduced nitrogen. All ordinary salts of copper, except CuS, CuSe and CuTe, are soluble in NH_3, often giving salts of square-planar $[Cu(NH_3)_4]^{2+}$ with weakly, axially coordinated $[BF_4]^-$, NO_2^-, NO_3^-, ClO_4^-, I_3^- etc. Many others are known, for example $[Cu(\eta^2\text{-}CO_3)(NH_3)_2]$, $[Cu(\eta^2\text{-}C_2O_4)(NH_3)_2]\cdot 2H_2O$, $K[Cu(NH_3)_5][PF_6]_3$, $Cu(NH_3)_2(\mu\text{-}NCS)_2$, and $[CuBr_2(NH_3)_2]$, plus $[Cu(NH_3)_4][PtCl_4]$, $[Cu(NH_3)_4][Cu_4(CN)_6]$ and $[Cu(NH_3)_2(NCS)_3Ag]$. The solution of CuCN in NH_3 may form $[Cu(NH_3)_2]^+$ and $[Cu(CN)_2]^-$. Copper(II) oxide is insoluble in NH_3 in the absence of NH_4^+.

Ammonia and "$(NH_4)_2CO_3$", with Cu^I, precipitate and redissolve CuOH, forming a colorless solution that turns blue on exposure to air; OH^- precipitates CuOH from the unoxidized solution.

Ammonia, added in small amount to Cu^{II}, precipitates pale blue basic salts; in equivalent amount, it precipitates the deep blue hydroxide (in both cases acting like OH^-). The precipitate is soluble in excess of the reagent, forming $[Cu(NH_3)_4]^{2+}$, deep blue (separation from Bi). No precipitate of $Cu(OH)_2$ occurs with a moderate concentration of NH_4^+. The blue color found with NH_3 is a good test for Cu^{II} in a solution freed from other **d**- or **p**-block metals (sensitivity, 0.7 mM, less in the presence of Fe). "Ammonium carbonate" solution acts like the NH_3 that it contains.

Sulfate, Cu^{2+} and NH_3 can yield $[Cu(NH_3)_4(H_2O)]SO_4$, square-pyramidal (no longer unusual), with H_2O at the pyramid's "top". Unlike H_2O, only four NH_3 ligands occur with aqueous Cu^{2+}, e.g., with $NH_4[Cu(NH_3)_4](ClO_4)_3\cdot NH_3$.

Both $N_2H_5^+$ (from $N_2H_6SO_4$ or $N_2H_6Cl_2$ in water) and NH_3OH^+ reduce $CuCl_2$ to white CuCl, which, when moist, darkens in the air. Copper(I) oxide, red, is precipitated on reduction of Cu^{II} by alkaline NH_2OH.

Oxidized nitrogen. Copper(II) nitrite is not easily obtained; air oxidizes it to the NO_3^-. However, NO_2^-, Cu^{2+} and K^+ form the unusual $K_3[Cu(NO_2)_5]$, i.e., $K_6[Cu(NO_2)_3(O_2N)_2][Cu(NO_2)_2(ONO)_2(O_2N)]$, or we may write $K_6[Cu(NO_2\text{-}\kappa N)_3(NO_2\text{-}\kappa^2 O)_2][Cu(NO_2\text{-}\kappa N)_2(NO_2\text{-}\kappa O)_2(NO_2\text{-}\kappa^2 O)]$, having ligancies of 7 and 6, plus some "semi-coordinated" long Cu–O bonds.

Adding excess KNO_2 to equivalent amounts of $Cu(NO_3)_2$ and $Pb(CH_3CO_2)_2$ in CH_3CO_2H gives dark-green $K_2Pb[Cu(NO_2\text{-}\kappa N)_6]$.

Dilute nitric acid is the most practical solvent for copper, although it is more readily dissolved by HNO_2. The major reaction is:

$$3\ Cu + 8\ H_3O^+ + 2\ NO_3^- \rightarrow 3\ Cu^{2+} + 2\ NO\uparrow + 12\ H_2O$$

Copper(I) oxide is oxidized and dissolved vigorously by HNO_3, unless cold and very dilute, when it yields both Cu and Cu^{2+}. Otherwise nitric acid rapidly oxidizes Cu^I to Cu^{II} or $[Cu(NO_3)_2(H_2O)_2]\cdot nH_2O$.

Nitric acid oxidizes and dissolves CuCN and the sulfides as Cu^{II}.

Fluorine species. Cold HF attacks Cu appreciably only in air.

Aqueous HF and Cu_2O give Cu and $CuF_2\cdot H_2O$.

11.1.3 Reagents Derived from the 3rd-to-5th-Period Non-Metals, Silicon through Xenon

Phosphorus species. Copper dichloride is reduced to CuCl by PH_3.

Ions of Cu^+ and Cu^{2+} are reduced to Cu by P_4.

Copper(2+), slightly acidified with HCl, precipitates CuCl when treated with $PH_2O_2^-$ or H_2PHO_3; boiling the Cu^{2+} with excess $PH_2O_2^-$ precipitates Cu. At 70 °C, Cu^{2+} reacts with $PH_2O_2^-$ to precipitate largely CuH, which decomposes rapidly.

Phosphonic acid, H_2PHO_3, and Cu^{II} acetate precipitate an unstable phosphonate, $CuPHO_3 \cdot 2H_2O$.

Phosphoric acid, H_3PO_4, with Cu_2O gives Cu and Cu^{II} phosphates.

Aqueous HPO_4^{2-} precipitates a bluish-white Cu^{II} phosphate. The $[P_2O_7]^{4-}$ ion precipitates $Cu_2[P_2O_7] \cdot 2H_2O$, soluble in NH_3, inorganic acids, and excess reagent; it is not precipitated in the presence of tartrate or $S_2O_3^{2-}$ (separation from Cd).

Arsenic species. Copper(I) oxide, Cu_2O, red, arises from alkaline Cu^{II}, e.g., Fehling's solution (containing tartrate), heated with AsO_3^{3-}.

Arsenites precipitate from nearly neutral solutions of Cu^{II} salts, other than the acetate, green copper(II) arsenite, composition variable. It is known as "Scheele's Green" or "Paris Green", and is readily soluble in acids and NH_3, and decomposed by concentrated OH^-. From Cu^{II} acetate, arsenites precipitate, on boiling, "Schweinfurt" green or "Imperial" green. This is a mixture of Cu^{II} arsenite and acetate, readily soluble in NH_3 and acids, decomposed by OH^-. Soluble arsenates precipitate $Cu_3(AsO_4)_2$, blue green, readily soluble in acids and NH_3.

Reduced chalcogens. Copper(I) salts are precipitated or transposed by H_2S or S^{2-}, forming Cu_2S, black, possessing the same solubilities as CuS.

With Cu^{II} salts, H_2S or S^{2-} gives black CuS (accompanied by small amounts of Cu_2S and S), produced alike in acidic solution (distinction from Mn, Fe, Co, Ni), and in alkaline solution (distinction from Sn, As, Sb). The solubility of CuS in water is 3 μM at 18 °C. The precipitate is soluble in CN^- (distinction from Cd, Hg, Pb and Bi); easily soluble in 2-M HNO_3, especially if a small amount of nitrite is present (distinction from Hg); insoluble in S^{2-} and only slightly soluble in S_x^{2-} (distinction from Sn, As, Sb); insoluble in hot dilute H_2SO_4 (distinction from Cd); and dissolved with difficulty by concentrated HCl (distinction from Sb).

When Cu is precipitated as CuS, it carries down soluble sulfides, especially of Zn, depending on the acidity and relative concentrations. The precipitate tends to be colloidal from a cold solution of low acidity.

Copper(I) salts, when boiled with sulfur, go partly to Cu_2S:

$$4\ CuCl + {}^1/_8\ S_8 \rightarrow Cu_2S\downarrow + 2\ Cu^{2+} + 4\ Cl^-$$

Thiocyanate precipitates white CuSCN from Cu^I solutions not too strongly acidic. Its solubility is 2 μM at 18 °C, and it is formed by SO_2 for the reduction and complete precipitation of copper even from Cu^{II}.

Copper may be determined then by titrating the CuSCN, e.g.:

$$4\ CuSCN + 7\ IO_3^- + 14\ H_3O^+ + 7\ Cl^- \rightarrow$$

$$4\ Cu^{2+} + 4\ SO_4^{2-} + 4\ HCN + 7\ ICl + 19\ H_2O$$

$$5\ CuSCN + 7\ MnO_4^- + 21\ H_3O^+ \rightarrow$$

$$5\ Cu^{2+} + 5\ SO_4^{2-} + 5\ HCN + 7\ Mn^{2+} + 29\ H_2O$$

Thiocyanate dissolves CuCN and forms, e.g., $K_3[Cu(CN)_3(SCN)]$.

From Cu^{II}, SCN^- precipitates black $Cu(SCN)_2$, unstable, gradually changing to white CuSCN (H_2SO_4 hastens the change), soluble in NH_3. With reducing agents, e.g., SO_2, CuSCN is precipitated at once (separation from Zn and Cd):

$$2\ Cu^{2+} + SO_2 + 2\ SCN^- + 6\ H_2O \rightarrow 2\ CuSCN\downarrow + SO_4^{2-} + 4\ H_3O^+$$

Oxidized chalcogens. Boiling with $S_2O_3^{2-}$ rapidly converts Cu into Cu_2S. Thiosulfate in neutral or acidified Cu^{2+} gives a reddish-brown precipitate of Cu_2S and S, which gradually becomes black (a distinction from Cd if the solution is fairly acidic).

Copper dichloride is reduced to CuCl by $S_2O_4^{2-}$.

Sulfur dioxide affects Cu only slightly, but it reduces Cu^{2+} to Cu^I. Both SO_2 and sulfites reduce $CuCl_2$ to CuCl. Copper(I) oxide dissolves in SO_2 solutions to give Cu_2SO_3.

A good preparation of white CuCN, without the $(CN)_2$ generated in the absence of a reductant, is from Cu^{II}, HSO_3^- and CN^- at 60 °C.

Dilute sulfuric acid has only a slight effect on Cu; hot and concentrated, it dissolves copper while releasing SO_2:

$$Cu + 4\ H_2SO_4 \rightarrow CuSO_4 + SO_2\uparrow + 2\ H_3O^+ + 2\ HSO_4^-$$

The Cu turns black during the process, however, apparently due to the formation also of Cu_2S and/or CuS, with reactions such as ($n = 1$ or 2):

$$(3 + n)\ Cu + 8\ H_2SO_4 \rightarrow Cu_nS\downarrow + 3\ CuSO_4 + 4\ H_3O^+ + 4\ HSO_4^-$$

$$Cu_2S + 12\ H_2SO_4 \rightarrow 2\ CuSO_4 + 5\ SO_2\uparrow + 6\ H_3O^+ + 6\ HSO_4^-$$

$$CuS + 8\ H_2SO_4 \rightarrow CuSO_4 + 4\ SO_2\uparrow + 4\ H_3O^+ + 4\ HSO_4^-$$

Sulfate and Cu^{2+} produce, for example, the common $CuSO_4{\cdot}5H_2O$, with "semi-coordinate" SO_4 (long Cu–O bonds) and one lattice H_2O, i.e., $[Cu(H_2O)_4](SO_4){\cdot}H_2O$, or we may write $[\{Cu(H_2O)_4(\mu\text{-}SO_4)\}_n]{\cdot}nH_2O$, $n \rightarrow \infty$. We also have $(NH_4)_2[Cu(H_2O)_6](SO_4)_2$, for example.

Copper(II), TeO_6^{6-} and NaOH form $Na_9[Cu^{III}(TeO_6)_2]{\cdot}16H_2O$.

Reduced halogens. Both HBr and HI, and hot 5-M HCl, dissolve Cu, giving CuX and $[CuX_2]^-$. Impurities greatly affect the solubility of Cu in these acids. In 5-M Cl^-, Cu^{II} rivals Fe^{III} as an oxidant.

Copper(I) oxide, Cu_2O, is soluble in HCl and HBr. Aqueous HI forms CuI. Copper(I) chloride and bromide are soluble in Cl^-.

Concentrated HCl dissolves CuCN, reprecipitated by water.

Trigonal bipyramidal Cu occurs in $Cs_3[CuCl_5]$.

If a small amount of Cu^{2+} is added to concentrated HBr, or to a mixture of Br^- and either H_2SO_4 or H_3PO_4, an intense purplish-red color, especially from $[CuBr_4]^{2-}$, is obtained, said to be more sensitive than the $[Fe(CN)_6]^{4-}$ or S^{2-} test, detecting 0.03 μmol of Cu in a drop of the bromide solution. Of the common metals, only iron interferes.

Boiling $CuBr_2$ with KBr forms $KCuBr_2$.

Aqueous HI precipitates, from concentrated copper salts, even from CuCN, white CuI, colored yellow to brown by some of the iodine liberated from Cu^{II}, and soluble in CN^-, NH_3, $S_2O_3^{2-}$ and I^-:

$$Cu^{2+} + 2\,I^- \rightarrow CuI\downarrow + {}^1/_2\,I_2$$

This reaction underlies a determination of Cu, the liberated iodine being titrated with a standardized reductant such as $S_2O_3^{2-}$:

$$Cu^{2+} + I^- + S_2O_3^{2-} \rightarrow CuI\downarrow + {}^1/_2\,[S_4O_6]^{2-}$$

$$2\,Cu^{2+} + SO_2 + 2\,I^- + 6\,H_2O \rightarrow 2\,CuI\downarrow + 4\,H_3O^+ + SO_4^{2-}$$

$$Cu^{2+} + Fe^{2+} + I^- + Cl^- \rightarrow CuI\downarrow + FeCl^{2+}$$

Halides dissolve CuCN, apparently forming separate, not mixed, bromo/iodo and cyano complexes.

Elemental and oxidized halogens. Chlorine, bromine and iodine all attack Cu to an extent increasing in the order given.

Copper, only slightly attacked by H_2SO_4, is readily dissolved if ClO_3^- is added, reducing it practically quantitatively to Cl^-.

Copper(II) iodate, $Cu(IO_3)_2$, pale blue, is obtained by adding IO_3^- to concentrated Cu^{2+}. Its solubility is 3.3 mM at 25 °C.

Copper(II), IO_6^{5-} and NaOH may form $Na_7[Cu^{III}(\eta^2\text{-}IO_6)_2]\cdot 20H_2O$.

11.1.4 Reagents Derived from the Metals Lithium through Uranium, plus Electrons and Photons

Oxidation. Copper and Cu^I are oxidized to Cu^{II} by Fe^{III}, which is reduced to Fe^{2+}, and by Pt^{IV}, Ag^+, Au^{III}, Hg_2^{2+} and Hg^{2+} solutions, these ions being reduced to the metals. Also $[Co(NH_3)_5Cl]^{2+}$ oxidizes Cu^I.

Light (274 nm), $[CuCl_3]^{2-}$ and 1-M H_3O^+ yield H_2 and $[CuCl_3]^-$.

Reduction. Ions of Cu^+ and Cu^{2+} are reduced to Cu by metallic Mg, Fe, Co, Ni, Zn, Cd, Al, Sn, Pb and Bi. A bright strip of Fe, in Cu^{II} acidified with HCl, receives a bright copper coating, recognizable in 0.1_3-mM Cu. A Zn-Pt couple precipitates the Cu on the Pt, confirmed with 18-M H_2SO_4, Br^- and air, see **Reduced halogens** above. A novel but useful way of showing such a reaction might be (cf. **7.2.2 Fluorine species**):

$$\begin{array}{cccccccccc}
Cu^{2+} & Zn^{2+} & & Cu^{2+} & Zn^{2+} & & & Cu^{2+} & Zn^{2+} \\
| & | & \rightarrow & / \backslash & / \backslash & & \rightarrow & | & | \\
H_2O & 2\,e^-\ H_2O & & H_2O\ \ 2\,e^- & H_2O & & & H_2O\ \ 2\,e^- & H_2O
\end{array}$$

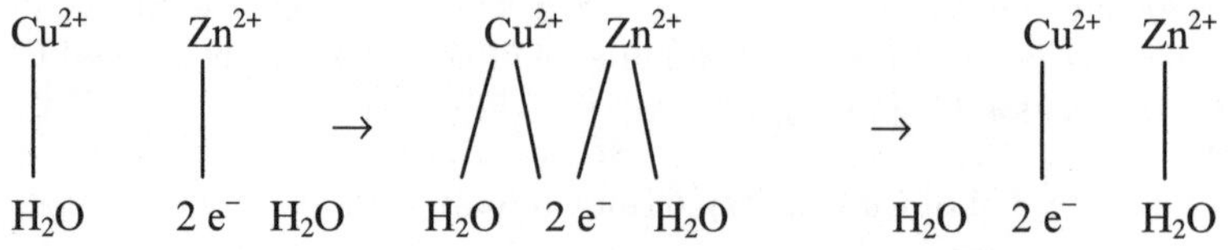

Copper(2+) and $Ti_2(SO_4)_3$ precipitate Cu (sensitivity, 20 μM).
Aqueous V^{2+}, V^{III} or Cr^{2+}, no excess, and Cu^{II} yield Cu^I.
Copper(2+) is reduced to CuCl by $[SnCl_3]^-$.
A Cu^{II} salt, treated with $[Sn(OH)_3]^-$, gives at first a greenish precipitate of Cu^{II} hydroxide (from the excess base), which rapidly changes to CuOH, yellowish brown, and may then be reduced to Cu.

Other reactions. Metal surfaces catalyze the dismutation of Cu^+.
Dichromate ion, $[Cr_2O_7]^{2-}$, does not precipitate Cu^{2+}; CrO_4^{2-} forms a brownish-red precipitate, soluble in NH_3 or dilute acids.
Copper(I) solutions, not too strongly acidic, precipitate $[Fe(CN)_6]^{4-}$ and $[Fe(CN)_6]^{3-}$ as their salts, brownish and white, respectively.
The presence of Cu^{II} in a solution free from other **d**- or **p**-block ions may be confirmed by using $[Fe(CN)_6]^{4-}$ to precipitate reddish brown $Cu_2[Fe(CN)_6]$, after acidifying with CH_3CO_2H. This is insoluble in dilute acids; decomposed by OH^-; soluble in NH_3, and is a very sensitive test for copper. Also precipitated, however, depending on the conditions, may be brown $K_2Cu_3[Fe(CN)_6]_2$ and yellow $K_2Cu[Fe(CN)_6]$. In acidic solution 15-μM Cu can be detected, and in neutral solution 10-μM or better. The sensitivity of the test is increased and the interference of dissolved Fe decreased with F^-. In dilute solution no precipitate appears, the solution becoming pink to red. Aqueous $[Fe(CN)_6]^{3-}$ precipitates $Cu_3[Fe(CN)_6]_2$, greenish yellow, insoluble in HCl.
Copper(0) and the appropriate Cu^{II} compounds yield such slightly soluble or dissociated Cu^I species as CuCl, CuBr and $[Cu(NH_3)_2]^+$ (with both oxidation and reduction of the Cu species). The hydrated Cu^+ ion, however, dismutates to Cu and Cu^{2+}.
Copper(II) in Cl^- dissolves chalcopyrite as various chloro complexes:

$$3\ Cu^{II} + CuFeS_2 \rightarrow 4\ Cu^I + Fe^{II} + 2\ S\downarrow$$

Oxygen reoxidizes the Cu^I quickly; thus Cu^{II} catalyzes the dissolution of $CuFeS_2$ by O_2 and Cl^- as aqueous Cu^{II} and Fe^{II}.

Freshly precipitated Cu_2S and CuS transpose $AgNO_3$, forming Ag_2S and $Cu(NO_3)_2$, plus Ag in the former case:

$$CuS + 2\,Ag^+ \rightarrow Ag_2S\downarrow + Cu^{2+}$$

$$Cu_2S + 4\,Ag^+ \rightarrow Ag_2S\downarrow + 2\,Ag\downarrow + 2\,Cu^{2+}$$

Freshly precipitated (mostly mono-) sulfides of Fe, Co, Zn, Cd, Sn and Pb (Bi^{III} is similar), written as M here, when boiled with CuCl in the presence of Cl^- give Cu_2S and the chloride of the metal, e.g.:

$$MS + 2\,CuCl \rightarrow Cu_2S\downarrow + MCl_2$$

With $CuCl_2$, CuS and a chloride of the metal are formed, except that SnS gives Cu_2S, CuCl and Sn^{IV}:

$$MS + Cu^{2+} \rightarrow CuS\downarrow + M^{2+}$$

$$SnS + 2\,Cu^{2+} + 4\,Cl^- \rightarrow Cu_2S\downarrow + SnCl_4 \text{ and perhaps}$$

$$2\,SnS + 4\,Cu^{2+} + 8\,Cl^- \rightarrow 4\,CuCl\downarrow + SnS_2\downarrow + SnCl_4$$

The tetraaqua or hexaaqua ion, $[Cu(H_2O)_4]^{2+}$, or $[Cu(H_2O)_4(H_2O)_2]^{2+}$, in crystals or in solution, is green or blue.

11.2 Silver, $_{47}$Ag

Oxidation numbers: (I), (II) and (III), as in Ag_2O, "argentous" oxide, unstable Ag^{2+}, and Ag_2O_3. For "AgO", $Ag^IAg^{III}O_2$.

11.2.1 Reagents Derived from Hydrogen and Oxygen

Dihydrogen. Hydrogen very slowly precipitates metallic silver from solution, but another reversible reaction may also occur:

$$Ag^+ + {}^1/_2\,H_2 + H_2O \rightarrow Ag\downarrow + H_3O^+$$

$$Ag^+ + H_2 + H_2O \leftrightarrows AgH\downarrow + H_3O^+$$

Water. At ordinary temperatures, silver is not affected by moisture. Silver oxide, Ag_2O, dissolves in water to about 0.1 mM as AgOH. The nitrite, $AgNO_2$, is slightly soluble in H_2O, but $AgNO_3$ is very soluble, along with AgF, $AgClO_3$ and (hygroscopic) $AgClO_4$. Silver forms a greater number of insoluble compounds than perhaps any other metal, although it is approached by Hg and Pb. In solubility, some common silver compounds may be arranged as follows: AgCl > AgCN > AgSCN > AgBr > AgI > Ag_2S, each a possible test for silver.

Aqueous Ag^+ is $[Ag(H_2O)_4]^+$, but most solids are anhydrous.

Some natural waters may contain HAgS or AgS^-.

Oxonium. Silver(I) oxide and carbonate react with nearly all acids (not aqueous CO_2), forming the corresponding salts. Black $Ag^IAg^{III}O_2$ acts similarly, sometimes with reduction; HF yields yellow AgF (soluble) and O_2, but $HClO_4$ gives unstable $[Ag(H_2O)_4]^{2+}$.

Hydroxide. The metal is not acted on by OH^-, hence the use of Ag crucibles for caustic fusions. Aqueous OH^- precipitates, from solutions of Ag^+, a grayish brown Ag_2O, slightly soluble in concentrated OH^- as $[Ag(OH)_2]^-$. Most silver compounds except AgI are transposed to Ag_2O on boiling with OH^-. However, this Ag_2O, which is strongly basic yet not very soluble, may be decomposed by light and heat.

Cold OH^- decomposes $Ag_4[Fe(CN)_6]$ to Ag and $Ag_3[Fe(CN)_6]$.

Peroxide. Hydrogen peroxide precipitates Ag from Ag^+.

Alkaline HO_2^- reduces $[Ag(OH)_4]^-$ to AgOH via Ag^{II}.

Di- and trioxygen. At ordinary temperatures, Ag is not affected by O_2.

When finely divided, silver is dissolved by NH_3 in the presence of oxygen. Silver nitrite, $AgNO_2$, is readily oxidized by O_2 to $AgNO_3$.

Ozone (O_3) and Ag^+ in low acidity yield black $Ag^IAg^{III}O_2$.

11.2.2 Reagents Derived from the Other 2nd-Period Non-Metals, Boron through Fluorine

Carbon oxide species. Bases with Ag^I and CO form CO_2 and Ag.

Aqueous CO_2, "carbonic acid", does not attack Ag or its oxides, but suspensions of Ag_2O are basic enough to absorb CO_2; also CO_3^{2-} or HCO_3^-, with Ag^+, precipitate white or yellowish Ag_2CO_3, and some Ag_2O.

Cyanide species. Aqueous CN^- dissolves all ordinary Ag compounds except Ag_2S. From neutral or slightly acidic solutions CN^- or HCN precipitates white AgCN, readily soluble in excess, forming especially $[Ag(CN)_2]^-$, but also $[Ag(CN)_3]^{2-}$ and $[Ag(CN)_4]^{3-}$. The AgCN is slowly soluble in hot, dilute H_3O^+. The ready solubility of nearly all silver compounds in CN^- allows us to separate silver from many minerals. However, the action on metallic Ag is much slower than on Au in the cyanide process for obtaining these metals from minerals. Some known salts are: $K[Ag(CN)_2]$, $K[Ag_2(CN)_3]\cdot H_2O$ and $K_3[Ag(CN)_4]$.

Fulminate and Ag^+ form the explosively sensitive AgCNO. A non-explosive complex is $[NEt_4][Ag(CNO)_2]$.

Some "simple" organic reagents. Hot formic acid reduces Ag^+:

$$2\ Ag^+ + HCO_2H + 2\ H_2O \rightarrow 2\ Ag\downarrow + CO_2\uparrow + 2\ H_3O^+$$

Non-reducing acids, however, react with fresh Ag_2O to give, e.g., $AgCH_3CO_2$, but metallic Ag is not affected by acetic acid.

Oxalic acid and $C_2O_4^{2-}$ precipitate Ag^+ as white $Ag_2C_2O_4$, somewhat soluble in HNO_3, difficultly soluble in H_2SO_4, readily soluble in NH_3, forming $[Ag(NH_3)_2]^+$. Heated under water, $Ag_2C_2O_4$ does not decompose; heated dry, or catalyzed by Cu^{2+}, it decomposes explosively to Ag and CO_2; exposed to sunlight, it partially decomposes.

In the gradual reduction of silver by organic reagents (aldehydes, tartrates, etc.) the metal may be obtained as a bright silver coating, or mirror, on the inner surface of a test tube or beaker if the glass surface is quite clean. An aqueous solution of Ag^+, treated with various organic compounds such as dextrin, sugar or starch, gives, on addition of OH^-, a brown suspension of colloidal silver. In 50 mL of solution, 0.4 mM Ag^+ can be detected. Ammonia interferes. However, treatment of $[Ag(NH_3)_2]^+$ with concentrated OH^- and a quite small amount of glycerol precipitates grayish silver in a very sensitive test.

Reduced nitrogen. Ammonia dissolves all insoluble silver compounds except Ag_2S and AgI, but AgBr only a little. The easy dissolution of AgCl in NH_3 separates it from $PbCl_2$ and $[Hg_2Cl_2]$.

Ammonia in neutral solutions of Ag^+ first precipitates Ag_2O, readily soluble in excess of the reagent, finally giving (and no higher ammine):

$$Ag^+ + 2\ NH_3 \rightarrow [Ag(NH_3)_2]^+$$

Very acidic solutions give no precipitate due to the NH_4^+ formed, which lowers the $[OH^-]$. Some salts are $[Ag(NH_3)_2]NO_3$ and $[Ag(NH_3)_2]_2(SO_4)_2$.

Silver(I) oxide and cyanide also dissolve in NH_4^+. Caution! Do not dissolve Ag_2O in NH_3 without NH_4^+, to avoid the highly explosive "fulminating silver" (written as Ag_3N_4, Ag_3N or $AgNH_2$), formed with OH^- above pH 12.9. Ammonium acetate and "$(NH_4)_2(CO_3)$" prevent it.

A superior way to prepare pure AgCN begins with an ammoniacal solution of equivalent amounts of Ag^+ and CN^-, perhaps $[Ag(NH_3)_2]^+$ and $[Ag(CN)_2]^-$, followed by removing the NH_3 in a current of air.

Aqueous Ag^+ is complexed by N_2H_4 and NH_2OH, then reduced to Ag.

The N_3^- ion and Ag^+ precipitate explosive AgN_3 and form $[Ag(N_3)_2]^-$.

Alkaline N_3^- reduces $[Ag(OH)_4]^-$ via $[Ag(OH)_3N_3]^-$ forming N_2 and, depending on conditions, white AgN_3, colorless $[Ag(N_3)_n]^{(n-1)-}$, dark Ag_2O or AgOH, completely soluble in dilute NH_3, or Ag_2O_2.

Oxidized nitrogen. Solutions of $AgNO_2$ with KNO_2 or $Ba(NO_2)_2$ yield $K[Ag(NO_2)_2]\cdot{}^1/_2H_2O$ or $Ba[Ag(NO_2)_2]_2\cdot H_2O$.

Aqueous NO_2^- does not reduce alkaline $[Ag(OH)_4]^-$.

The best solvent for Ag is HNO_3 (2:1 or about 11 M), containing a little nitrite, in the absence of which the reaction is very sluggish:

$$Ag + 2\ H_3O^+ + NO_3^- \rightarrow Ag^+ + 3\ H_2O + NO_2\uparrow$$

Silver nitrate, $AgNO_3$, is only slightly soluble in 16-M HNO_3. Cold, dilute HNO_3 dissolves all common silver compounds except AgCl, AgBr, $AgBrO_3$, AgI, $AgIO_3$, AgCN and AgSCN.

Aqua regia changes AgI to the more tractable AgCl plus ICl.

Aqueous HNO_3 oxidizes $Ag_4[Fe(CN)_6]$ to $Ag_3[Fe(CN)_6]$.

Fluorine species. Aqueous HF and F^- do not precipitate AgF.

11.2.3 Reagents Derived from the 3rd-to-5th-Period Non-Metals, Silicon through Xenon

Silicon species. Silane reduces Ag^+ to Ag and blackens $AgNO_3$ paper.

Phosphorus species. Phosphane (PH_3) and P_4 reduce Ag^+ to the metal.

Alkaline $PH_2O_2^-$ and $[Ag(OH)_4]^-$ form PHO_3^{2-} and Ag^I.

Aqueous PHO_3^{2-} (much more slowly than the isoelectronic SO_3^{2-}) and $[Ag(OH)_4]^-$ form Ag^I, and H_2PHO_3 reduces Ag^+ to Ag.

Monohydrogen phosphate, HPO_4^{2-}, gives a yellow precipitate with Ag^+, darkening on exposure to light. The composition of the precipitate approaches Ag_3PO_4, but probably is a mixture of that compound with Ag_2HPO_4, the relative quantity of each depending on conditions:

$$HPO_4^{2-} + 2\ Ag^+ \rightarrow Ag_2HPO_4\downarrow$$

$$2\ HPO_4^{2-} + 3\ Ag^+ \rightarrow Ag_3PO_4\downarrow + H_2PO_4^-$$

The precipitate is soluble in acids, NH_3 and "$(NH_4)_2CO_3$". Diphosphate precipitates white $Ag_4[P_2O_7]$ with the same solubilities as Ag_3PO_4, except that it is insoluble in acetic acid. The $[P_2O_7]^{4-}$ and $[Ag(OH)_4]^-$ form a complex that gives Ag_2O_2 and O_2 at $pH > 9$, but is stable if $6 < pH < 8$.

Arsenic species. Silver(1+) is reduced to Ag by AsH_3, As and As^{III}:

$$AsH_3 + 6\ Ag^+ + 9\ H_2O \rightarrow 6\ Ag\downarrow + H_3AsO_3 + 6\ H_3O^+$$

Arsenite precipitates Ag_3AsO_3, bright yellow, insoluble in H_2O, readily dissolved or transposed by both acids and bases. Arsenate precipitates Ag_3AsO_4, brown, insoluble in H_2O, soluble in H_3O^+ and NH_3.

Alkaline AsO_3^{3-} reduces $[Ag(OH)_4]^-$ to AgOH.

Arsenite, AsO_3^{3-}, reduces AgCl, but not AgBr or AgI, to Ag.

Reduced chalcogens. Silver is tarnished to Ag_2S by H_2S (unless pure and dry), S^{2-}, and many organic compounds containing sulfur. Alkali sulfides convert silver halides to Ag_2S.

Sulfane (H_2S) and alkali sulfides precipitate Ag_2S, black, from Ag^+. It is soluble in HNO_3 (over 8 dM) (distinction from Hg), soluble in CN^-:

$$Ag_2S + 4\ CN^- \rightarrow 2\ [Ag(CN)_2]^- + S^{2-}$$

insoluble in NH_3 and S^{2-} (distinction from As, Sb and Sn); converted to AgCl by boiling with Cl^-:

$$2\ AgCl + S^{2-} \leftrightarrows Ag_2S + 2\ Cl^-$$

In concentrated $AgNO_3$ either H_2S or S_8 gives $[Ag_3S]NO_3$, light sensitive. Carbon disulfide gives the same, albeit less expected, result when shaken with $AgNO_3$ in 2-M HNO_3 for 24 h in the dark:

$$6\ Ag^+ + CS_2 + 2\ NO_3^- + 6\ H_2O \rightarrow 2\ [Ag_3S]NO_3\downarrow + CO_2\uparrow + 4\ H_3O^+$$

The Ag^+ ion precipitates Se^{2-} and Te^{2-} as Ag_2Se and Ag_2Te, which dissolve in concentrated Ag^+ as colorless or pale-yellow $[Ag_8Te]^{6+}$ etc.

The Se_x^{2-} or Te_x^{2-} ions with Ag^+ produce a great variety of structures of complexes well beyond our scope, especially when other metal ions are included, depending on conditions and on the large organic cations which are sometimes used for isolation:

$$n\ Ag^+ + m\ Y_x^{2-} \rightarrow [Ag_nY_p(Y_q)_r]^{s-} \text{ with } Y = Se \text{ or } Te$$

Thiocyanate precipitates AgSCN, white, insoluble in water, HNO_3, or slight excess of reagent, but readily soluble in the concentrated reagent. It is reprecipitated on dilution and is readily soluble in NH_3, especially when warmed. Cooling then yields glistening, colorless crystals. The AgSCN is decomposed by Cl_2, Br_2 or I_2. It is distinctly soluble in concentrated Ag^+. Hot concentrated H_2SO_4 dissolves AgSCN even in the presence of excess Ag^+ (separation from AgCl).

Thiocyanate dissolves AgBr moderately.

Solids such as $NH_4Ag(SCN)_2$ contain no $[Ag(SCN)_2]^-$, but only AgSCN molecules, each surrounded by three more-distant SCN^- ions.

The less-stable $SeCN^-$ gives more-stable complexes and salts, including $K[Ag(SeCN)_2]$ and $K_2[Ag(SeCN)_3]$.

Dissolving AgSCN or AgSeCN in concentrated aqueous and/or ethanolic $AgNO_3$ (sometimes with $AgCF_3CO_2$) or $AgClO_4$ yields $Ag_2SCN(NO_3,ClO_4)$, $Ag_2SeCN(ClO_4)$, $Ag_3SCN(NO_3)_2$, or $Ag_3SeCN(NO_3)_2$, with the last one having a μ_6–1,2,3,4κSe:5,6κN corrugated layer.

Thiourea, $CS(NH_2)_2$, Tu, forms $AgTu_2SCN$, $AgTu_2Cl$, $AgTu_3ClO_4$ etc. Adding $AgNO_3$ to $CS(NH_2)_2$ in HNO_3 gives $Ag_2(Tu\text{-}\kappa S)_3(NO_3)_2$.

Elemental and oxidized chalcogens. Tellurium reduces Ag^+ to Ag.

Thiosulfate gives a white precipitate of $Ag_2S_2O_3$ (usually gray due to a little Ag_2S), readily soluble in excess, forming $[Ag(S_2O_3\text{-}\kappa S)_n]^{(2n-1)-}$ or with multiple Ag. The product is readily decomposed by warm H_2O:

$$Ag_2S_2O_3 + 2\ H_2O \rightarrow Ag_2S\downarrow + H_3O^+ + HSO_4^-$$

The thiosulfate ion, $S_2O_3^{2-}$, dissolves all common silver compounds (requiring an excess for Ag_2S), giving for example:

$$AgI + 2\ S_2O_3^{2-} \rightarrow I^- + [Ag(S_2O_3\text{-}\kappa S)_2]^{3-}$$

Some solids are $NaAgS_2O_3 \cdot H_2O$ and $(NH_4)_7[Ag(S_2O_3\text{-}\kappa S)_4] \cdot 2NH_4Cl$.

Alkaline $S_2O_3^{2-}$ reduces $[Ag(OH)_4]^-$ to Ag^I.

Dithionite reduces the ammine to silver:

$$2\ Ag(NH_3)_2^+ + S_2O_4^{2-} + 2\ H_2O \rightarrow 2Ag\downarrow + 2\ SO_3^{2-} + 4\ NH_4^+$$

Aqueous SO_2 and SO_3^{2-} precipitate silver sulfite, Ag_2SO_3, white, from Ag^+. It resembles precipitated AgCl, rapidly darkening on exposure to light. It is reduced to Ag by excess of SO_2 and is soluble in excess of SO_3^{2-}. Its solubility in water is less than 0.2 mM. Boiling water tends to decompose Ag_2SO_3 to Ag and $Ag_2[S_2O_6]$:

$$2\ Ag_2SO_3 \rightarrow Ag_2[S_2O_6] + 2\ Ag\downarrow$$

Silver sulfite, Ag_2SO_3, is soluble in NH_3, but then Ag precipitates fairly readily. Treatment of Ag_2SO_3 with a strong acid liberates SO_2.

Aqueous SO_3^{2-} and $[Ag(OH)_4]^-$ form SO_4^{2-} and AgOH.

At pH 10.5 and 70–90 °C, Ag^+ and SO_2–O_2^{2-} (concurrently derived from SO_2, $^1/_2\ O_2$ and 2 OH^-, or SO_3^{2-} and $^1/_2\ O_2$) yield $Ag^IAg^{III}O_2$, ≤ 98 %.

Silver is soluble in 14-M H_2SO_4, especially when hot and aerated:

$$2\ Ag + 3\ H_3O^+ + HSO_4^- \rightarrow 2\ Ag^+ + SO_2\uparrow + 5\ H_2O$$

$$2\ Ag + {}^1/_2\ O_2 + 2\ H_3O^+ \rightarrow 2\ Ag^+ + 3\ H_2O$$

The ions HSO_4^- and SO_4^{2-} precipitate white Ag_2SO_4 from concentrated solutions of Ag^+. The product is slightly soluble in H_2O, more soluble in HNO_3; the solubility is increased by H_2SO_4 and decreased by SO_4^{2-}. It reacts with Fe^{2+} to give Ag:

$$Ag_2SO_4 + 2\ Fe^{2+} + 2\ H_2O \rightarrow 2\ Ag\downarrow + \text{e.g. } FeOH^{2+} + FeSO_4^+ + H_3O^+$$

Cold, dilute H_2SO_4 does not affect AgCl; the hot 18-M acid converts AgCl to Ag_2SO_4 with release of HCl.

Aqueous H_2SO_4 slowly decomposes $Ag_4[Fe(CN)_6]$.

Peroxodisulfate and Ag^+ first give $Ag_2[S_2O_8]$, which, in the presence of H_2O, becomes black $Ag^IAg^{III}O_2$, with some Ag_3O_4, and HSO_4^-:

$$^1/_2\ Ag_2[S_2O_8] + 2\ H_2O \rightarrow {}^1/_2\ Ag^IAg^{III}O_2\downarrow + H_3O^+ + HSO_4^-$$

In base also, at 80–90 °C, it gives:

$$2\ AgOH + [S_2O_8]^{2-} + 2\ OH^- \rightarrow Ag^IAg^{III}O_2\downarrow + 2\ SO_4^{2-} + 2\ H_2O$$

This compound is a very active oxidant. In acid, we find:

$$Ag^+ + {}^1/_2\ [S_2O_8]^{2-} + H_3O^+ \rightarrow Ag^{2+} + HSO_4^- + H_2O$$

An interesting test for Ag uses its catalysis (via Ag^{2+}) of the oxidation of Mn^{2+} to MnO_4^- by $[S_2O_8]^{2-}$: To a slightly acidic portion of the original solution add $K_2[S_2O_8]$. Boil to oxidize other reductants; then add a drop of a dilute $MnSO_4$ solution, a little more $K_2[S_2O_8]$, and boil again. The reddish-purple color of MnO_4^- indicates Ag.

A rather more stable form of the rare Ag^{II} can be made by slowly adding excess aqueous Ag^+ and pyridine to cold (aqueous) $[S_2O_8]^{2-}$:

$$Ag^+ + 4\ C_5H_5N + {}^3/_2\ [S_2O_8]^{2-} \rightarrow [Ag(C_5H_5N)_4][S_2O_8]\downarrow + SO_4^{2-}$$

This is nearly insoluble in water, but HNO_3 dissolves it as Ag^{II}. It oxidizes H_2O_2 to O_2, NH_3 to N_2, I^- to I_2, and various organic species.

Reduced halogens. Concentrated (12 M) HCl is without action on Ag; concentrated HI releases H_2 and forms $[AgI_n]^{(n-1)-}$.

Some natural brines may contain $[AgCl_n]^{(n-1)-}$, with $0 \le n \le 4$.

The precipitation of Ag^+ by HCl from acidic or ammoniacal solution as curdy, white AgCl is a sensitive test for Ag^+, excess Hg^{2+} being absent, 3mM being recognizable in good lighting. The precipitate is made more compact and easy to separate by vigorously shaking the mixture:

$$[Ag(NH_3)_2]^+ + Cl^- + 2\ H_3O^+ \rightarrow AgCl\downarrow + 2\ NH_4^+ + 2\ H_2O$$

It turns violet to brown on exposure to light. It is fusible without decomposition, slightly more soluble in HNO_3 than in water, insoluble in low concentrations of Cl^-, but slightly soluble in higher concentrations, forming, e.g., $[AgCl_3]^{2-}$ and $[AgCl_4]^{3-}$. For analysis, AgCl dissolves least in ~ 2.5-mM Cl^-.

Silver chloride dissolves in NH_3 or "$(NH_4)_2CO_3$" as $[Ag(NH_3)]_2^+$. When mixed with much $[Hg_2Cl_2]$ (from precipitation of a possibly unknown mixture by Cl^-), little if any AgCl dissolves, because the $[Hg_2Cl_2]$ and NH_3 produce Hg, which displaces the Ag^I as insoluble Ag.

Silver chloride is soluble as complexes in CN^- or $S_2O_3^{2-}$. It is fairly soluble in Hg^{2+} because $HgCl^+$ is so slightly ionized.

Concentrated (16 M) HNO_3 has slight effect on AgCl. Concentrated (18 M) H_2SO_4 completely transposes even the fused chloride on long boiling (due to removal of Cl as gaseous HCl at the high *T*).

Aqueous Br^- hardly converts AgCl to AgBr, but I^- readily forms AgI.

Bromide precipitates very light-yellow AgBr, slightly soluble in excess Br^-, much less soluble than AgCl in NH_3, soluble in CN^- and $S_2O_3^{2-}$, and moderately soluble in SCN^-. Iron(2+) in sunlight does not act on AgBr; boiling HNO_3 has no effect; hot H_2SO_4 decomposes it:

$$2\ AgBr + 2\ H_3O^+ + 2\ HSO_4^- \rightarrow Ag_2SO_4\downarrow + SO_2\uparrow + Br_2\uparrow + 4\ H_2O$$

Aqueous I^- precipitates pale-yellow AgI, even from $[Ag(NH_3)_2]^+$:

$$[Ag(NH_3)_2]^+ + I^- \rightarrow AgI\downarrow + 2\ NH_3$$

(distinction from AgCl), soluble in excess I^- as $[AgI_2]^-$ etc., but the simple salt is reprecipitated on dilution with H_2O.

Silver iodide, AgI, is decomposed by 16-M HNO_3 (distinction from AgCl and AgBr). It is moderately soluble in CN^-, insoluble in "$(NH_4)_2CO_3$" (separation from AgCl), but is slightly soluble in $S_2O_3^{2-}$, less soluble in SO_3^{2-} or SCN^-, soluble in concentrated Ag^+ as colorless or pale-yellow $[Ag_3I]^{2+}$ etc. Even in the light, AgI is not affected by Fe^{2+}.

The tendency of silver halides to form complex anions increases in the order: AgCl < AgBr < AgI, giving, e.g., Cs_2AgCl_3 and K_2AgI_3.

Elemental and oxidized halogens. Chlorine, bromine and iodine, when added to a solution of Ag^+, form the corresponding halide and halate, the hypohalite being an intermediate product:

$$6\ Ag^+ + 3\ Br_2 + 9\ H_2O \rightarrow 5\ AgBr\downarrow + AgBrO_3\downarrow + 6\ H_3O^+$$

Suspensions of Ag_2O or Ag_2CO_3, treated with Cl_2, have also been used to produce $AgClO_3$.

At 0 °C the reaction of silver sulfate with bromine is:

$$Ag_2SO_4 + 2\ Br_2 + 3\ H_2O \rightarrow 2\ AgBr\downarrow + HSO_4^- + H_3O^+ + 2\ HBrO$$

When heated, the instability of HBrO yields the net result:

$$5\ Ag_2SO_4 + 6\ Br_2 + 13\ H_2O \rightarrow$$

$$10\ AgBr\downarrow + 5\ HSO_4^- + 2\ BrO_3^- + 7\ H_3O^+$$

The acids $HClO_3$, $HBrO_3$ and HIO_3 all act on Ag:

$$6\ Ag + 6\ H_3O^+ + ClO_3^- \rightarrow AgCl\downarrow + 5\ Ag^+ + 9\ H_2O$$

Again, however, Ag (or Ag^+, Ag_2O or Ag_2CO_3) and $HClO_3$ have also been used to arrive at white $AgClO_3$. Silver(+) from $AgNO_3$ is very effective. The product, a powerful oxidant, darkens slightly in light.

Bromate and Ag^+ form white $AgBrO_3$, slightly soluble in water, soluble in NH_3. Iodate precipitates $AgIO_3$, white, insoluble in water, slightly soluble in HNO_3, readily soluble in 3-M NH_3.

Xenon species. Xenon difluoride oxidizes Ag^I:

$$XeF_2 + 2\ AgOH + 2\ OH^- \rightarrow Xe\uparrow + Ag^IAg^{III}O_2\downarrow + 2\ F^- + 2\ H_2O$$

11.2.4 Reagents Derived from the Metals Lithium through Uranium, plus Electrons and Photons

Oxidation. Metallic silver precipitates gold and platinum from their solutions, thus being oxidized to the corresponding halides and so on, and reduces some Cu^{II} to Cu^I, Hg^{2+} to Hg_2^{2+}, and MnO_4^- to $MnO_2 \cdot aq$.

Anodic Ag and 1.2-M OH^- form yellow $[Ag(OH)_4]^-$, square planar, stable for ~ 2 h at 25 °C but with $t_{1/2} < 30$ min at pH 13, becoming black $Ag^IAg^{III}O_2$ and O_2. At pH 11, the $t_{1/2}$ of Ag^{III} is seconds, or minutes if with 1-dM to 1-M $[B(OH)_4]^-$, CO_3^{2-}, HPO_4^{2-} (stablest), $[P_2O_7]^{4-}$ or $HAsO_4^{2-}$, giving Ag^I {but $[P_2O_7]^{4-}$ causes reduction to yellow-brown $[Ag(P_2O_7)_2]^{6-}$, later Ag_2O_2} and much longer even near pH 7 if with H_6TeO_6 or $H_4IO_6^-$.

A Pt anode gives either: $Ag_7O_8NO_3$ (in $AgNO_3$); black Ag_3O_4, i.e., $Ag^{II}Ag^{III}{}_2O_4$ (in 2-dM AgF and 1.8-M NaF at 0 °C); or black, shiny Ag_2O_3 (in 1-dM $Ag[BF_4]$, $Ag[PF_6]$ or $AgClO_4$), less stable than Ag_3O_4, slowly releasing O_2, quickly releasing it in H_3O^+.

Reduction. Metallic silver (only slightly soluble in Hg) is precipitated from solution by: Mg, Mn, Fe, FeS, Cu, Cu^+, Zn, Cd, Hg, Al, Sn, Sn^{II}, Pb, PbS, Sb, SbH_3, Bi, etc. Various metals such as Mg, Fe, Zn, Cd, and Al also reduce (undissolved) silver halides to Ag, especially in the presence of acid. Iron(2+) incompletely reduces Ag^+ in the cold; on boiling, the Fe^{III} initially formed is reduced and the Ag dissolved again.

Alkaline $[Mo(CN)_8]^{4-}$, MnO_4^{2-} or $[Fe(CN)_6]^{4-}$ reduces $[Ag(OH)_4]^-$ to AgOH via Ag^{II}.

Boiling $Ag_4[Fe(CN)_6]$ alone will yield Ag and $Ag_3[Fe(CN)_6]$.

In base, silver species are also reduced by Mn^{II}, Hg^I, Sb^{III} and Bi^{III}.

The reactions of Ag_2S with certain metals are illustrated here:

$$Ag_2S + Fe + 2\ H_3O^+ \rightarrow 2\ Ag\downarrow + Fe^{2+} + H_2S\uparrow + 2\ H_2O$$

$$Ag_2S + Hg \rightarrow 2\ Ag\downarrow + HgS\downarrow$$

If the Hg is in excess, an amalgam is formed.

An amalgam of tin and mercury reduces insoluble compounds of silver in the wet way: the silver becomes Ag_{Hg}, and the tin becomes Sn^{IV}.

A solution of $[Ag(NH_3)_2]^+$ in a great excess of NH_3 gives a very sensitive test for tin as $[Sn(OH)_4]^{2-}$ in the presence of OH^-, precipitating Ag. Antimony does not interfere under these conditions.

Stibane reduces Ag^+, with Ag_3Sb as an apparent intermediate:

$$2\ SbH_3 + 12\ Ag^+ + 15\ H_2O \rightarrow 12\ Ag\downarrow + Sb_2O_3 \cdot aq\downarrow + 12\ H_3O^+$$

The well-known sensitivity of AgX to light is the greatest for AgBr.

Other reactions. Both oxidation and reduction of silver(I, II) are shown by the otherwise extremely slow reduction of Ce^{IV} by Tl^+ when catalyzed by Ag^+; see **13.5.4 Oxidation.**

Chromate added to an excess of Ag^+ gives brownish-red Ag_2CrO_4. The product is insoluble in H_2O, readily soluble in HNO_3, H_2SO_4, or NH_3. The solubility in acetic acid depends on its concentration. One obtains $Ag_2[Cr_2O_7]$, bright red, when $[Cr_2O_7]^{2-}$ is added to acidified Ag^+. The compound becomes Ag_2CrO_4 on boiling with H_2O:

$$Ag_2[Cr_2O_7] + 2\ H_2O \rightarrow Ag_2CrO_4\downarrow + H_3O^+ + HCrO_4^-$$

Silver dichromate is insoluble in H_2O, soluble in HNO_3 and NH_3.

Aqueous $[Fe(CN)_6]^{4-}$ precipitates $Ag_4[Fe(CN)_6]$, white, from Ag^+ or AgCl, but not from AgBr or AgI. It is soluble in NH_3 but not in NH_4^+. Boiling with NH_3 decomposes it completely:

$$Ag_4[Fe(CN)_6] + 2\ NH_3 + 2\ H_2O \rightarrow$$

$$Fe(OH)_2\downarrow + 2\ AgCN\downarrow + 2\ [Ag(CN)_2]^- + 2\ NH_4^+$$

$$2\ Ag_4[Fe(CN)_6] + 6\ NH_3 + 3\ H_2O \rightarrow$$

$$Fe_2O_3 \cdot aq\downarrow + 6\ [Ag(CN)_2]^- + 2\ Ag\downarrow + 6\ NH_4^+$$

Aqueous $[Fe(CN)_6]^{3-}$ precipitates $Ag_3[Fe(CN)_6]$, reddish brown, becoming more yellowish and compact on heating. The precipitate is transposed by OH^- to Ag_2O and $[Fe(CN)_6]^{3-}$, and readily soluble in NH_3.

Certain insoluble sulfides, when boiled with Ag^+, give Ag_2S:

$$CuS + 2\ Ag^+ \rightarrow Ag_2S\downarrow + Cu^{2+}$$

Silver halides dissolve in Ag^+, but less than in X^-, forming Ag_2X^+ and Ag_3X^{2+}. Aqueous Ag^+ and AgCN form $(AgCNAg)^+$ etc. Concentrated, hot $AgNO_3$ dissolves AgCN, leading to $[Ag_3(CN)_2]NO_3$ crystals.

Dissolving AgOH in $[Ag(CN)_2]^-$ and OH^- yields $[Ag(CN)(OH)]^-$.

11.3 Gold, $_{79}$Au and Roentgenium, $_{111}$Rg

Oxidation numbers: (I) and (III), as in AuCl, "aurous" chloride, and Au_2O_3, "auric" oxide. Most "Au^{II}" species, e.g., "AuO" or "AuSe", are mixtures: $Au^IAu^{III}O_2$ or $Au^IAu^{III}Se_2$, etc. Liquid NH_3 can yield Au^{-I} in salt-like CsAu, stabilized by relativity. Relativity also makes metallic Au yellow. (Non-aqueous) F_2 can produce $[(Au^VF_4)_2(\mu\text{-}F)_2]$.

The stability of Au^I in $[AuL_2]^+$, $[AuL_2]^-$ or $[AuL_2]^{3-}$ is in the order for L as CN^- >> $(S_2O_3\text{-}\kappa S)^{2-}$ > $[CSe(NH_2)_2\text{-}\kappa Se]$ > $[CS(NH_2)_2\text{-}\kappa S]$ > $NH_3 \approx I^-$ > $(SCN\text{-}\kappa S)^-$ > Br^- > Cl^- >> H_2O. The similar order CN^- >> NH_3 > I^- > $(SCN\text{-}\kappa S)^-$ > Br^- > Cl^- >> H_2O applies to $[Au^{III}L_4]^{3+}$ or $[Au^{III}L_4]^-$.

The oxidation states calculated relativistically for $_{111}$Rg to be stable in water: (–I), (III) and (V). Thus, in addition to predicting some well-known properties of Cu, Ag and Au, relativistic quantum mechanics predicts stability (chemical, not nuclear!) in RgH and $[RgF_6]^-$.

11.3.1 Reagents Derived from Hydrogen and Oxygen

Water. Aqueous Au^+ is unstable to dismutation, especially in alkalis. It is much more acidic (hydrolyzed) than Ag^+ in spite of the former's greater radius, due to strong relativistic effects. No $[Au(H_2O)_4]^{3+}$ is found, but $[AuCl_4]^-$ hydrolyzes to mixed $H_2O\text{-}OH^-\text{-}Cl^-$ species.

The cyanide AuCN is only slightly soluble in water.

Water forms a colloidal solution of Au_2S, and slowly decomposes Au_2S_3 to Au_2S and S.

The gold(I) ions $[AuX_2]^-$, with X = Cl, Br or SCN, but not CN, dismutate in water to Au and Au^{III}, but can be stabilized in excess X^-:

$$3\,[AuX_2]^- \leftrightharpoons [AuX_4]^- + 2\,Au\downarrow + 2\,X^-$$

Water hydrolyzes AuF_3 and $[AuF_4]^-$ (from non-aqueous sources).

The chloride $[Au_2Cl_6]$ is deliquescent, and $[Au_2Br_6]$ dissolves readily; they are $[(AuX_2)_2(\mu\text{-}X)_2]$ with *quadro*-Au^{III}. The triiodide and $[AuI_4]^-$ are decomposed by H_2O, without excess I^-, forming AuI. The complex chlorides, bromides, iodides and cyanides are mostly soluble in H_2O.

Seawater appears to contain gold as $[AuCl(OH)]^-$, $[AuCl_2]^-$, $[AuBrCl]^-$ etc. Other natural waters, with varying uncertainties, may contain $[Au(CN)_2]^-$, $[Au(SCN)_4]^-$, $[Au(S_2O_3)_2]^{3-}$, $[AuCl_2]^-$, $[AuCl_4]^-$, $[AuClBr]^-$, $[AuBr_4]^-$ or $[AuI_2]^-$, also $H_mAu^IS_n^{(2n-m-1)-}$ in hot waters.

Oxonium. The metal, and the oxides and hydroxides of gold, are insoluble in H_2O and in dilute oxoacids.

Hydroxide. A mixture of Au and Au^{III}, purple, is obtained by treating a solution of $[AuBr_2]^-$ with a slight excess of OH^- and boiling. Gold(III) hydrous oxide, $Au_2O_3·aq$, brown (variable), is formed on treating $[Au_2Cl_6]$ with just enough OH^-, but is hard to purify. Dried over $CaCl_2$ at 100 °C, it loses water, finally forming Au_2O_3, easily decomposed to Au on further heating. The Au^{III} oxide dissolves in OH^- as $[Au(OH)_4]^-$.

Peroxide. Solutions of HO_2^-, and Na_2O_2, reduce gold compounds to the metal (distinction from Pt and Ir), but H_2O_2 with HCl dissolves Au.

Di- and trioxygen. When finely divided, gold dissolves in excess HI plus O_2, although O_3 actually reduces neutral $AuCl_3$ to Au:

$$2\ Au + 6\ H_3O^+ + 8\ I^- + {}^3/_2\ O_2 \rightarrow 2\ [AuI_4]^- + 9\ H_2O.$$

11.3.2 Reagents Derived from the Other 2nd-Period Non-Metals, Boron through Fluorine

Boron species. Borohydride reduces gold species, e.g.:

$$4\ Au_2O_3 + 3\ [BH_4]^- \rightarrow 8\ Au\downarrow + 3\ [B(OH)_4]^-$$

Carbon oxide species. Aqueous $[Au(N_3)_4]^-$ and CO yield $[Au^I(NCO)_2]^-$.

Carbonate ion, like OH^-, precipitates hydrous gold oxides.

Cyanide species. One method of separating gold from ore is the cyanide process, in which the ore is leached with a dilute (e.g., less than 20 mM), aerated solution of CN^-:

$$2\ Au + 4\ CN^- + {}^1/_2\ O_2 + H_2O \rightarrow 2\ [Au(CN)_2]^- + 2\ OH^-$$

Then zinc can precipitate the gold, leaving $[Zn(CN)_3]^-$ or $[Zn(CN)_4]^{2-}$:

$$2\ [Au(CN)_2]^- + Zn \rightarrow 2\ Au\downarrow + [Zn(CN)_3]^- + CN^-$$

Gold(I) oxide and the frequently obtained mixtures of Au^0 with Au^{III} (hydrous) oxides, "Au_2O", react readily with CN^-, giving $[Au(CN)_2]^-$, the only important Au^I cyano complex; H_2S has no action on it, but hot, concentrated H_2SO_4 decomposes it; Tl^+ precipitates yellow $Tl[Au(CN)_2]$. Ion exchange and evaporation give the acid $H[Au(CN)_2]$. This acid may be the only thermodynamically stable cyano-complex acid with respect to the dissociative loss of HCN. Gold(I) cyanide, AuCN (from heating $H[Au(CN)_2]$ at 110 °C), is a stable, pale-yellow powder, slightly soluble in H_2O, but soluble in OH^-, CN^-, NH_3, "$(NH_4)_2S$" and $S_2O_3^{2-}$.

Likewise $Au_2O_3·aq$ and CN^- readily form $[Au(CN)_4]^-$ or $K[Au(CN)_4]·H_2O$, again the most important Au^{III} cyano complex. Another source is $[AuCl_4]^-$ plus CN^-. And again, ion exchange and evaporation yield the stable, strong acid,

$[H(H_2O)_2][Au(CN)_4]$, which precipitates a Ag^+ salt; warming $[Au(CN)_4]^-$ gives $[Au(CN)_2]^-$ and $(CN)_2$.

Fulminate (CNO^-) reduces $AuCl_4^-$ to $[Au(CNO)_2]^-$.

Some "simple" organic reagents. Ethyne (C_2H_2), aldehydes (e.g., CH_2O), $H_2C_2O_4$ and its anions, sugar etc. precipitate gold often from either acidic or alkaline solutions and especially with heating (separation from Pd, Pt, Cu, Hg, Sn, Pb, As, Sb and other metals that mostly form acid-insoluble sulfides). More specifically, many reductants plus $AuCl_4^-$ form colloidal gold, but hydroquinone yields a powder, plus quinone. Also, $H_2C_2O_4$, added to, e.g., $[AuCl_3OH]^-$, free from HNO_3 and excess of HCl, slowly (faster with heat) but completely reduces the gold to flakes or a mirror on a clean container wall:

$$[AuCl_3OH]^- + {}^3/_2\, H_2C_2O_4 + H_2O \rightarrow Au\downarrow + 3\, CO_2\uparrow + 3\, Cl^- + 2\, H_3O^+$$

Dissolving Au_2O_3 in neat CH_3CO_2H forms Au^{III} acetate, but this rapidly decomposes to metallic gold in water.

Reduced nitrogen. Ammonia, or "$(NH_4)_2CO_3$", added to either $Au_2O_3 \cdot aq$, $[AuCl_4]^-$ or $[Au_2Cl_6]$, especially if followed by hot water, gives a grayish or dirty-yellow precipitate of "fulminating gold", a very explosive mixture including, for example, $Au_2O_3 \cdot xNH_3$ and perhaps $AuCl(NH_2)_2$. If prepared in the presence of OH^-, its sensitivity is markedly increased.

Treating $[AuCl_4]^-$ with gaseous NH_3, buffered with NH_4NO_3, produces $[Au(NH_3)_4](NO_3)_3$. The acidic $[Au(NH_3)_4]^{3+}$ has a pK_a of 7.5, and $H[Au(NH_3)_3(OH)]^{3+}$ is a strong acid. From Br^- and $[Au(NH_3)_4]^{3+}$ one can isolate relatively inert [*trans*-$AuBr_2(NH_3)_2$]Br, and then $[AuBr_4]^-$ salts.

Gold compounds are reduced to the metal completely by $N_2H_5^+$, NH_3OH^+, N_2H_4 and NH_2OH in acidic, neutral, or alkaline solutions.

Gold(I) and gold(III), plus N_3^-, form $[Au(N_3)_2]^-$ and $[Au(N_3)_4]^-$.

Oxidized nitrogen. Aqueous NO_2^- precipitates the metal even from dilute solutions of gold compounds. The precipitate may be colloidal.

Nitric acid alone does not attack gold, but with HCl (in aqua regia) dissolves it as $[AuCl_4]^-$. Similarly, HBr and HNO_3 yield $[AuBr_4]^-$.

11.3.3 Reagents Derived from the 3rd-to-5th-Period Non-Metals, Silicon through Xenon

Phosphorus species. Phosphane gives Au_3P and other products from $Au_2O_3 \cdot aq$. Elemental P_4 and black P reduce gold compounds to Au.

Phosphane quickly reduces Au_2Cl_6 to Au^I, but excess PH_3 precipitates black Au_3P, stable in warm H_2O, and leaves ortho- and meta-phosphates in solution. Drying leads to oxidation. Excess Au_2Cl_6, however, yields some Au. A small amount forms a blood-red colloid. The Au_3P with concentrated HNO_3 gradually forms NO and Au; with concentrated H_2SO_4, SO_2; HCl may restore PH_3 and form $[AuCl_4]^-$.

Aqueous PO_4^{3-} does not produce a precipitate with $[Au_2Cl_6]$, but $[P_2O_7]^{4-}$ seems to form a double salt. The acids HPH_2O_2 and H_2PHO_3, as well as their salts, precipitate Au.

Arsenic species. Arsenic and As^{III} reduce gold compounds to the metal.

Reduced chalcogens. Gold is not tarnished or affected at all by H_2S.

Acidified $[Au(CN)_2]^-$ and H_2S yield Au_2S. It is insoluble in dilute acids, but soluble in CN^-, S_x^{2-} and aqua regia.

Sulfane precipitates, from cold neutral or acidified Au^{III}, black Au_2S_3:

$$2\,[AuCl_3OH]^- + 3\,H_2S + 2\,H_2O \rightarrow Au_2S_3\downarrow + 4\,H_3O^+ + 6\,Cl^-$$

From a hot solution the metal is obtained. The trisulfide is insoluble in dilute HNO_3 or HCl, but soluble in CN^-, S^{2-} and aqua regia.

Gold and As mixtures may be precipitated by H_2S, dissolved in S_2^{2-}, and reprecipitated by HCl; then HNO_3 dissolves only the As_2S_5. Or, if Au and Pt in a mixture are in solution with S_2^{2-}, they may be separated from Sn, As and Sb by dissolving the reprecipitated (by HCl) sulfides in HCl plus $KClO_3$ (or HNO_3), evaporating the Cl_2, and, after adding excess KOH (because NaOH would precipitate stibates), boiling with $CCl_3CH(OH)_2$ (chloral hydrate), which precipitates only the Pt and Au.

Concentrated Na_2S dissolves Au_2S_n and $[AuCl_4]^-$ salts; then, if not too dilute, $[AsPh_4]Cl$ completely precipitates $[AsPh_4]_4[Au_{12}S_8]$, cubic with S at the corners and Au on the edge centers. Gold(I) cyanide, Na_2Se and $[NEt_4]Cl$ in methanol form a cryptand, precipitated by ether as $[NEt_4]_3[NaAu_{12}Se_8]$, with Na centered in the cubes, larger than $[Au_{12}S_8]$.

Thiocyanate and $[AuX_n]^-$ quickly form $[Au(SCN\text{-}\kappa S)_n]^-$; X = Cl or Br and $n = 2$ or 4. Also, $[Au(SCN)_2]^-$ and $(-SCN)_2$ yield $[Au(SCN)_4]^-$ which, however, slowly restores $[Au(SCN)_2]^-$, in effect catalyzing the decomposition of some SCN^- or $(-SCN)_2$, e.g., overall:

$$3\,[AuX_4]^- + 7\,SCN^- + 10\,H_2O \rightarrow$$

$$3\,[Au(SCN)_2]^- + HSO_4^- + HCN + 12\,X^- + 6\,H_3O^+$$

(Much cyanide inhibits it some.) This can go farther to metallic Au.

Organic sulfides, R_2S, can reduce and complex gold(III):

$$[AuCl_4]^- + 3\,R_2S + 3\,H_2O \rightarrow AuCl(SR_2)_2\downarrow + R_2SO + 2\,H_3O^+ + 3\,Cl^-$$

Thiourea, $CS(NH_2)_2$, leaches gold from its ores (faster than CN^- does), with Fe^{III} as oxidant, as $[Au^I\{CS(NH_2)_2\}_2]^+$, and reduces Au^{III} to Au^I.

Dithioacetic acid, CH_3CS_2H, and $AuCl_4^-$, form what may be written as $[Au^I_4(\mu\text{-}CH_3CS_2)_4]$ with S_4 symmetry.

Gold(I) and $R_2NCS_2^-$, *N,N*-dialkyldithiocarbamates, with R = CH_3 etc., yield the dinuclear $[Au^I_2(R_2NCS_2)_2]$; Au^{III} gives $[Au^{III}(R_2NCS_2)_2]^+$.

Elemental and oxidized chalcogens. Elemental Se and Te reduce gold compounds to the metal.

Thiosulfate reduces and complexes Au^{III} to linear $[Au^I(S_2O_3\text{-}\kappa S)_2]^{3-}$, useful in medicine, and to $Na_3[Au(S_2O_3)_2]\cdot 2H_2O$; in great dilution a purple color is first formed.

Alkaline SO_3^{2-} complexes Au^I as ~$[Au(SO_3\text{-}\kappa S)_2]^{3-}$ for electroplating, but also in some conditions, like SO_2, reduces gold compounds to Au:

$$2\,[AuCl_3OH]^- + 3\,SO_2 + 11\,H_2O \rightarrow$$

$$2\,Au\downarrow + 3\,HSO_4^- + 7\,H_3O^+ + 6\,Cl^-$$

Aqueous $[Au(OH)_4]^-$ and Na_2SO_3 give $Na_5[Au(SO_3\text{-}\kappa S)_4]\cdot 5H_2O$; excess SO_3^{2-} reduces it to $[Au^I(S_2O_3\text{-}\kappa S)_2]^{3-}$.

A rare sulfate, $AuSO_4$, long thought to be $Au^IAu^{III}(SO_4)_2$, contains two five-membered rings in $[(Au^{II}\text{–}Au^{II})(\mu\text{-}SO_4)_2]$.

Gold dissolves in concentrated H_2SeO_4 but not H_2SO_4.

Reduced halogens. Gold and HCl alone do not react, but Au and HCl or HBr, plus concentrated HNO_3, Cl_2, Br_2 or an anode, form yellow $H_3O[AuCl_4]\cdot nH_2O$ or red-brown $H_3O[AuBr_4]\cdot nH_2O$.

Gold(I) also readily forms complexes with Cl^-, Br^-, and I^-. The $[AuCl_2]^-$ and $[AuBr_2]^-$ are colorless, $[AuI_2]^-$ yellow. The first two, but not $[AuI_2]^-$, dismutate in water but not excess X^-, to Au and Au^{III}.

The gold oxides, hydrous or anhydrous, dissolve in HX, e.g.:

$$Au_2O_3.aq + 8\,Cl^- + 6\,H_3O^+ \rightarrow 2\,[AuCl_4]^- + 9\,H_2O$$

The softer Br^- and Au^{III} push the following to the right, $\log K \approx 7$, and intermediate steps show the stronger trans effect of Br^- over Cl^-:

$$[AuCl_4]^- + 4\,Br^- \leftrightarrows [AuBr_4]^- + 4\,Cl^-$$

Iodide, added in small portions to Au^{III} (with H_3O^+ from hydrolysis) precipitates yellow AuI when equivalent quantities are combined. The precipitate is insoluble in H_2O but soluble in excess reagent:

$$[AuBr_3OH]^- + 3\,I^- + H_3O^+ \rightarrow AuI\downarrow + 3\,Br^- + 2\,H_2O + I_2$$

Gradually adding $[AuCl_3OH]^-$ to I^- forms, first a dark-green solution of $[AuI_4]^-$, then a dark-green precipitate of AuI_3, very unstable, decomposed by H_2O and changed in the air to AuI and I_2 vapor, but various concentrations of $[AuCl_4]^-$ and I^- form $[AuCl_2]^-$, $[AuI_2]^-$, ICl, I_2, $[ICl_2]^-$, $[I_2Cl]^-$ and $[I_3]^-$. The $[AuBr_4]^-$ ion acts somewhat similarly.

Mixed salts are known, many formerly believed to be of "Au^{II}", e.g.: $Cs_2[AuCl_2][AuCl_4]$ "$CsAuCl_3$", $Rb_2[AuBr_2][AuBr_4]$ and $K_2[AuI_2][AuI_4]$, variously prepared by the thermolysis of Au^{III} or in water, although very high pressures can give true Au^{II}. Also "$AuCl_2$" is $[(AuCl_4)_2(\mu\text{-}Au)_2]$, in a chair-like ring with square-planar $AuCl_4$, linear $AuCl_2$ and bent Au_2Cl.

Elemental and oxidized halogens. Chlorine, as a gas or in aqueous solution, converts gold to $[Au_2Cl_6]$; bromine water forms $[Au_2Br_6]$, both bridged $[(AuX_2)_2(\mu\text{-}X)_2]$. Gold with Cl_2 and Cl^- or with Br_2 and Br^- forms $[AuX_4]^-$; evaporation at 40 °C with K^+ gives reddish-purple $K[AuBr_4]$, not light sensitive; in air this goes to $K[AuBr_4]\cdot 2H_2O$.

Treating $[Au(CN)_2]^-$ with X_2 (X = Cl, Br, I) gives [*trans*-$Au(CN)_2X_2]^-$, only partly for X = I, colored pale yellow, yellow and black, respectively; I_3^- is much faster than I_2. Cyanide can then convert these (for X = Cl or Br) to $[Au(CN)_4]^-$. The dichloro complex, treated with KN_3 or KSCN, forms a yellow, explosive $K[Au(CN)_2(N_3)_2]$ or orange or dark-red (SCN linkage isomers?) $K[Au(CN)_2(SCN)_2]$.

From $[AuCl_3OH]^-$, IO_3^- precipitates yellow $Au(IO_3)_3$, slightly soluble.

11.3.4 Reagents Derived from the Metals Lithium through Uranium, plus Electrons and Photons

Oxidation. Aerated $[Fe(CN)_6]^{4-}$ dissolves finely divided Au, perhaps as:

$$6\ Au + 2\ [Fe(CN)_6]^{4-} + 2\ O_2 + H_2O \rightarrow$$

$$6\ Au(CN)_2^- + Fe_2O_3\cdot aq\downarrow + 2\ OH^-$$

likewise aerated HCl with $CuCl_2$ as a catalyst. Warm, concentrated H_2SO_4 and oxidants such as $MnO_2\cdot aq$, $KMnO_4$ or PbO_2 dissolve Au.

Reduction. Relativity causes Au and anhydrous Cs to yield halide-like Cs^+Au^- instead of a metallic alloy CsAu; Rb acts similarly.

Many reagents have been suggested for the detection of gold, often involving its reduction to the colloidal metal, which imparts various colors to the system, depending on the particle size.

Gold compounds are reduced to the metallic state by numerous reagents, including: the elements Mg, Fe, Co, Ni, Pd, Pt, Cu, Ag, Zn, Cd, Hg, Al, Sn, Pb, Sb and Bi; and the ions Ti^{3+}, $V^{2+/3+}$, VO^{2+}, $Cr^{2+/3+}$, Mn^{2+}, Fe^{2+}, Cu^+, Hg_2^{2+}, Sn^{II} and Sb^{III}, as well as light. Iron(2+), for example, may give a brown or black precipitate (separation from $[PdCl_4]^{2-}$ and $[PtCl_6]^{2-}$, e.g., from aqua regia). Aqueous $[Fe(CN)_6]^{4-}$ reduces Au^{III} to $[Au(CN)_2]^-$. The reduction of $[AuCl_4]^-$ to $[AuCl_2]^-$, initiated by Fe^{2+} or light, first yields a transient, labile species, perhaps $Au^{II}Cl_3^-$, as shown by our finding of the rapid exchange of isotopic Cl^-. Reduction by $[PtCl_4]^{2-}$, however, is by a single two-electron step, transferring Cl.

Tin(II) chloride, added to neutral or acidified gold chloride, gives the "Purple of Cassius", a mixture of hydrated SnO_2 and Au.

Cathodic e^- deposit Au from solutions in CN^-, S^{2-} or SCN^-.

Photons appear to reduce $[AuCl_4]^-$ to $[AuCl_3]^-$, labile toward $^*Cl^-$, thus catalyzing isotopic exchange with quantum yields of hundreds:

$$[AuCl_4]^- + [Au(^*Cl)_3]^- \leftrightarrows [AuCl_3]^- + [Au(^*Cl)_3Cl]^-$$

Other reactions. Mixing $[AuCl_4]^-$ and $[Au(CN)_4]^-$ in a 1:3 ratio gives $K[AuCl(CN)_3]$. This plus saturated KBr yield $K[AuBr(CN)_3]$.

Quickly mixing $Na_3[Au(S_2O_3)_2]\cdot 2H_2O$ and $[PPh_4]Br$ with a little excess of fresh $(NH_4)_2[WS_4]$ and treating further yields dark-red $[PPh_4]_2[Au_2(WS_4)_2]$ with 8-membered rings –Au–S–W–S–Au–S–W–S– having nearly linear Au^IS_2 and nearly tetrahedral $W^{VI}S_4$.

Reference

1. Abrahams BF, Haywood MG, Robson R (2004) Chem Comm 2004:938

Bibliography

See the general references in the Introduction, and some more-specialized books [2–7].

2. Schmidbaur H (ed) (1999) Gold: chemistry, biochemistry and technology. Wiley, New York
3. Fritz JJ, Koenigsberger E (1996) Copper(I) halides and pseudohalides. IUPAC, Blackwell, London
4. Miyamoto H, Woolley EM, Salomon M (1990) Copper and silver halates. IUPAC, Blackwell, London
5. Karlin KD, Zubieta J (eds) (1986) Biological and inorganic copper chemistry. Adenine Press, Guilderland NY
6. Puddephatt RJ (1978) The chemistry of gold. Elsevier, Amsterdam
7. Dozinel CM, Man SL (trans) (1963) Modern methods of analysis of copper and its alloys, 2nd ed. Elsevier, Amsterdam

12 Zinc through Mercury

12.1 Zinc, $_{30}Zn$

Oxidation number in classical compounds: (II), as in Zn^{2+}.

12.1.1 Reagents Derived from Hydrogen and Oxygen

Water. At least with dilute anions Zn^{2+} is $[Zn(H_2O)_6]^{2+}$, although the ligancy falls from six to four with much $HClO_4$; cf. Br^- below in **12.1.3**.

Zinc nitrate (6 H_2O), halides (fluoride excepted), and chlorate are deliquescent; the sulfate (7 H_2O) is efflorescent.

Zinc basic carbonate, cyanide, oxalate, phosphate, arsenate, sulfide, periodate, hexacyanoferrate(II and III), and hexacyanocobaltate(III) are insoluble in water; the sulfite is sparingly soluble.

Pure water (free of air) does not oxidize zinc.

Zinc(2+) is hydrolyzed to $Zn_2(\mu\text{-}OH)^{3+}$, $Zn_4(OH)_4{}^{4+}$ etc.

Seawater and some freshwater contain traces of Zn^{II} complexes as $ZnOH^+$, $Zn(OH)_2$, $ZnCO_3$, $ZnHCO_3{}^+$, $ZnSO_4$ and $ZnCl_n{}^{(n-2)-}$. Some other natural waters may contain $H_mZnS_n{}^{(2n-m-2)-}$, or $[ZnF_4]^{2-}$ (in hot waters).

Oxonium. Pure Zn dissolves very slowly in acids or alkalis. Impurities, or contact with Au, Pt, etc., accelerate the reactions, hence the ready solution of commercial Zn.

Hydroxide. Zinc dissolves in alkaline solutions, with release of H_2:

$$Zn + 2\ OH^- + 2\ H_2O \rightarrow [Zn(OH)_4]^{2-} + H_2\uparrow$$

Aqueous OH^- precipitates Zn^{2+} as $Zn(OH)_2$, white, soluble in excess, at first "quasi-colloidally", then forming a mixture, especially of a tetrahedral zincate, $[Zn(OH)_n(H_2O)_{4-n}]^{(n-2)-}$, depending on the $c(OH^-)$, and more at ambient T than when heated. A $c(OH^-)$ of 1 dM dissolves almost none. All common Zn salts, except ZnS, are soluble in OH^-.

Dioxygen. Aqueous O_2 oxidizes Zn in contact with Fe.

12.1.2 Reagents Derived from the Other 2nd-Period Non-Metals, Boron through Fluorine

Carbon oxide species. Alkali carbonates precipitate, from solutions of Zn^{2+}, basic carbonates such as $Zn_5(CO_3)_2(OH)_6{\cdot}H_2O$, white, soluble in "$(NH_4)_2CO_3$", readily in OH^- or NH_3. Alkaline-earth carbonates have no action at ambient *T*, but upon boiling, precipitate all of the Zn^{2+}.

Cyanide species. Alkali cyanides precipitate zinc cyanide, $Zn(CN)_2$, white, soluble in excess of the reagent, forming $[Zn(CN)_4]^{2-}$. One can isolate, e.g., $K_2[Zn(CN)_4]$, rather like $K_2[(Cd,Hg)(CN)_4]$, by evaporation.

Fulminate ion complexes Zn^{2+} up to $[Zn(CNO)_4]^{2-}$.

In some salts $[Zn(SCN)_4]^{2-}$ is more stable than $[Zn(NCS)_4]^{2-}$, but water favors complexes with NCS^- up to $[Zn(NCS\text{-}\kappa N)_4]^{2-}$.

Some "simple" organic reagents. Zinc dissolves in dilute CH_3CO_2H:

$$Zn + 2\ CH_3CO_2H \rightarrow Zn(CH_3CO_2)_2 + H_2\uparrow$$

Basic carboxylates, *tetrahedro*-$[Zn_4(\mu_4\text{-}O)(\mu\text{-}RCO_2)_6]$, form a bit like those of Be, but also with much $[Zn_3O(RCO_2)_3]^+$ and $[Zn_4O(RCO_2)_4]^{2+}$.

Solutions of $C_2O_4^{2-}$ precipitate Zn^{2+} as a white zinc oxalate, $ZnC_2O_4{\cdot}2H_2O$, soluble in acids and alkalis.

Reduced nitrogen. Except for ZnS and $Zn_2[Fe(CN)_6]$, all common Zn salts are soluble in NH_3. Ammonia precipitates Zn^{2+} partly as $Zn(OH)_2$ if NH_4^+ is absent. Excess NH_3 dissolves the $Zn(OH)_2$ as $[Zn(NH_3)_4]^{2+}$ and can yield, e.g., $[Zn(NH_3)_4]I_2$.

Zinc(2+), N_2H_4 and N_3^- or NO_3^- form explosive $[Zn(N_2H_4)_2(N_3)_2]$ or $[Zn(N_2H_4)_3(NO_3)_2]$. The N_3^- alone gives $[Zn(N_3)_4]^{2-}$.

Hydroxylamine and Zn^{II} produce either $[Zn(Cl,Br)_2(NH_2OH\text{-}\kappa N)_2]$ or $[Zn(Cl,Br)_2(NH_2OH\text{-}\kappa O)_2]$ depending on the procedure; cf. Cd, **12.2.2**.

Oxidized nitrogen. Nitrite complexes Zn^{2+}, unlike Cd^{2+}, only up to $Zn(NO_2)_2$. Nitrate, however, goes up to $[Zn(\eta^2\text{-}NO_3)_4]^{2-}$.

Zinc with Cu (Zn-Cu couple) reduces NO_3^- and NO_2^- to NH_3.

Zinc dissolves in very dilute HNO_3 without releasing gas, but in moderately dilute, cold HNO_3 releasing chiefly N_2O, and in more concentrated HNO_3 releasing NO. Concentrated HNO_3 dissolves little Zn, the nitrate being very sparingly soluble in that medium:

$$4\ Zn + 10\ H_3O^+ + NO_3^- \rightarrow 4\ Zn^{2+} + NH_4^+ + 13\ H_2O$$

$$4\ Zn + 10\ H_3O^+ + 2\ NO_3^- \rightarrow 4\ Zn^{2+} + N_2O\uparrow + 15\ H_2O$$

$$3\ Zn + 8\ H_3O^+ + 2\ NO_3^- \rightarrow 3\ Zn^{2+} + 2\ NO\uparrow + 12\ H_2O$$

12.1.3 Reagents Derived from the 3rd-to-5th-Period Non-Metals, Silicon through Xenon

Phosphorus species. Slowly adding NH_3 to Zn^{2+} and H_3PO_4 precipitates white $ZnNH_4PO_4$ quantitatively. Other conditions give $Zn_3(PO_4)_2 \cdot 4H_2O$. They are soluble in alkalis and nearly all acids. Gravimetrically, the $ZnNH_4PO_4$ may be ignited to $Zn_2[P_2O_7]$ and weighed as such.

A sample of various other phosphates may include $Zn_5[P_3O_{10}]_2 \cdot 17H_2O$ and $Ba_2Zn[cyclo\text{-}P_3O_9]_2 \cdot 10H_2O$.

Arsenic species. Aqueous Zn^{2+} is precipitated by arsenite or arsenate ion, forming the corresponding white, gelatinous salts, readily soluble in alkalis and acids, including arsenic acids.

Reduced chalcogens. Sulfane, H_2S, precipitates some of the Zn^{2+} as ZnS, white, from solutions not too acidic; with enough $CH_3CO_2^-$ to consume the H_3O^+, precipitation is complete. Alkaline solutions of S^{2-} precipitate ZnS completely except for a small solubility in excess sulfide, although a large excess of NH_3, OH^- or Cl^- etc. tends to inhibit the reaction.

Zinc may be precipitated from mixtures with Mn, Co and Ni from an ammoniacal solution by HS^-. Digestion of the precipitate with cold, dilute HCl dissolves the MnS and ZnS. Then the solution is boiled to eliminate H_2S, and the Zn changed to $[Zn(OH)_4]^{2-}$ by an excess of OH^- plus some Br_2, H_2O_2, or ClO^-, which precipitate the Mn^{2+} as $MnO_2 \cdot aq$. The excess oxidant is destroyed and the resulting solution tested for Zn^{2+}, perhaps by adding H_2S to give ZnS.

Gravimetrically, zinc may be precipitated as the sulfide from dilute H_2SO_4 or formic-acid solution, converted to the oxide or sulfate, and weighed as such.

Oxidized chalcogens. Concentrated SO_3^{2-} precipitates Zn^{2+} as a basic zinc sulfite; if the solution is too dilute for immediate precipitation, boiling will precipitate a bulky white basic sulfite.

An easily crystallized salt is $K_2[Zn(H_2O)_6](SO_4)_2$.

Hot, concentrated H_2SO_4 dissolves Zn:

$$Zn + 4\ H_2SO_4 \rightarrow ZnSO_4 + SO_2\uparrow + 2\ H_3O^+ + 2\ HSO_4^-$$

Reduced halogens. The hexacyanoferrate(II) is insoluble in HCl.

Many salts such as $Alk_2[Zn(Cl,Br)_4]$ can be crystallized. At least with Br^- we appear to have $[ZnBr_n(H_2O)_{6-n}]^{(2-n)+}$ and $[ZnBr_n(H_2O)_{4-n}]^{(n-2)-}$ in water, with $n \leq 1$ and $n \geq 2$, respectively. Note the changing ligancy (c.n.) also with Cd^{2+} and Br^- or I^- below. Large cations stabilize $[ZnI_4]^{2-}$.

Some batteries contain "$ZnCl_3^-$" and $[ZnCl_4]^{2-}$. Many "$ZnCl_3^-$", "$ZnBr_3^-$", or "ZnI_3^-" ions are actually $[(ZnX_2)_2(\mu\text{-}X)_2]^{2-}$.

Oxidized halogens. Zinc with Cu (Zn-Cu couple), reduces ClO_3^-, BrO_3^- and IO_3^- to Cl^-, Br^- and I^-.

Perchlorate and Zn^{2+} give rise to $[Zn(H_2O)_6](ClO_4)_2$ crystals.

Periodate forms a white precipitate with Zn^{2+}. In the cold, NH_4^+ and NH_3 prevent precipitation, but boiling overcomes their interference.

12.1.4 Reagents Derived from the Metals Lithium through Uranium, plus Electrons and Photons

Oxidation. Zinc forms Zn^{II} while reducing aqueous Fe, Ru, Os, Co, Rh, Ir, Ni, Pd, Pt, Cu, Ag, Au, Cd, Hg, In, Tl, Sn, Pb, Sb, Bi or Te to M^0. With an acid it likewise reduces $[Cr_2O_7]^{2-}$ to Cr^{III}, Mn(>II) to Mn^{2+}, and iron(III) to Fe^{2+}. With Cu (Zn-Cu couple) it reduces $Fe(CN)_6^{3-}$ to $Fe(CN)_6^{4-}$. In Leclanché cells with NH_4Cl and MnO_2 (in ordinary flashlight batteries) it is oxidized to $Zn(OH)_2$, $Zn(NH_3)_2Cl_2$ etc. In alkaline batteries it becomes $[Zn(OH)_4]^{2-}$ and ZnO.

Reduction. Magnesium precipitates Zn from an acetic-acid solution.

The complete electrolytic deposition of Zn is difficult to attain.

Other reactions. Chromate ion, but not dichromate, precipitates with Zn^{2+} a yellow chromate, readily soluble in acids and alkalis.

Aqueous $[Fe(CN)_6]^{4-}$ yields such salts as white $K_2Zn[Fe(CN)_6]$ and $Zn_2[Fe(CN)_6]$. The $[Fe(CN)_6]^{3-}$ ion precipitates $Zn_3[Fe(CN)_6]_2$, yellowish, or $CsZn[Fe(CN)_6]$, for example.

The Zn^{2+} ion may be titrated with $[Fe(CN)_6]^{4-}$, determining the endpoint potentiometrically or with uranyl acetate as external indicator.

Triple salts, $(NH_4)_2(Zn,Cd,Hg)CoCl_6 \cdot 2H_2O$, also of Ni^{II}, are known.

An interesting solid double salt, $2Hg(CN)_2 \cdot Zn(NO_3)_2 \cdot 7H_2O$, contains $[(H_2O)_4Zn(NCHgCN)_2]^{2+}$ with two *trans*-N on the octahedral Zn.

12.2 Cadmium, $_{48}$Cd

Oxidation number: (II), as in CdO.

12.2.1 Reagents Derived from Hydrogen and Oxygen

Water. Aqueous Cd^{2+} is very acidic $[Cd(H_2O)_6]^{2+}$, forming Cd_2OH^{3+} and $[CdOH(H_2O)_5]^+$. The carbonate, cyanide, oxalate, phosphate, sulfide and hexacyanoferrates(II and III) are insoluble, the fluoride slightly soluble. The chloride and bromide are deliquescent; the iodide is not, but all three dissolve in water or ethanol. Some natural waters may contain polysulfido and thiosulfato complexes as well as $H_mCdS_n^{(2n-m-2)-}$.

Hydroxide. Hydroxide ion, in the absence of citrate and similar chelators, precipitates white $Cd(OH)_2$ from Cd^{II}. This dissolves only in *concentrated* OH^- (distinction from Sn and Zn) as $[Cd(OH)_4]^{2-}$. It absorbs CO_2 from the air, but readily loses H_2O on heating, forming a mixture of the oxide and hydroxide. The sulfate can give $Cd_2(OH)_2SO_4$.

The oxide and hydroxide are soluble in NH_3 and in acids; soluble in cold OH^- plus tartrate, reprecipitated as CdO on boiling (distinction from Cu^{II}). Fresh $Cd(OH)_2$ is distinctly soluble in Cl^-, Br^-, I^- and SCN^-.

12.2.2 Reagents Derived from the Other 2nd-Period Non-Metals, Boron through Fluorine

Carbon oxide species. Aqueous CO_3^{2-} precipitates $CdCO_3$, white, insoluble in excess of the reagent. "Ammonium carbonate" forms the same precipitate, which, however, dissolves in excess. Barium carbonate in the cold completely precipitates Cd as $CdCO_3$ from its dissolved salts.

Cyanide species. Aqueous CN^- precipitates $Cd(CN)_2$, white, soluble in excess reagent as $[Cd(CN)_4]^{2-}$. Evaporation isolates $K_2[Cd(CN)_4]$.

Aqueous NCS^- does not precipitate Cd^{II} (distinction from copper), but it forms aqueous complexes up to $[Cd(NCS)_n(SCN)_{4-n}]^{2-}$.

Some "simple" organic species. The many observed formates include $AlkCd(CHO_2)_3$, $K_3Cd(CHO_2)_5$ and $BaCd(CHO_2)_4 \cdot 2H_2O$.

Oxalic acid and $C_2O_4^{2-}$ precipitate cadmium oxalate from Cd^{2+}, white, soluble in inorganic acids and NH_3.

Reduced nitrogen. The carbonate, cyanide, oxalate, phosphate, and hexacyanoferrates(II and III) dissolve in NH_3. Ammonia and Cd^{2+} precipitate $Cd(OH)_2$, soluble in excess NH_3. Cooling a concentrated Cd^{II} salt in excess NH_3 (first dissolved by warming) crystallizes it with halides as $Cd(NH_3)_n^{2+}$, $n = 2$ to 4, or 6 with much NH_3.

Cadmium(2+), N_2H_4 and NO_3^- form explosive $[Cd(N_2H_4)_2(NO_3)_2]$ or $[Cd(N_2H_4)_3(NO_3)_2]$. With N_3^- we have $Cd(N_3)_n^{(n-2)-}$; $1 \le n \le 5$.

Hydroxylamine and Cd^{II} form $[Cd(Cl,Br)_2(NH_2OH\text{-}\kappa N)_2]$ but not $(NH_2OH\text{-}\kappa O)$; cf. Zn in **12.1.2.**

Oxidized nitrogen. Nitrite complexes Cd^{2+} up to $[Cd(NO_2)_4]^{2-}$.

Cadmium dissolves readily in HNO_3, releasing nitrogen oxides. The acid also dissolves all well-known compounds of Cd as at least $CdNO_3^+$.

Cadmium dissolves in NH_4NO_3 quietly, producing no gas, possibly as:

$$4\,Cd + NO_3^- + 9\,NH_4^+ + 3\,H_2O \rightarrow$$

$$2\,[Cd(NH_3)_2(H_2O)_2]^{2+} + 2\,[Cd(NH_3)_3(H_2O)]^{2+}$$

12.2.3 Reagents Derived from the 3rd-to-5th-Period Non-Metals, Silicon through Xenon

Phosphorus species. Phosphate ions precipitate cadmium phosphate, white, readily soluble in acids, complexed by at least one phosphate group. Diphosphate ion, $[P_2O_7]^{4-}$, precipitates the diphosphate; this is soluble in excess reagent and in inorganic acids, but not in dilute CH_3CO_2H. The reaction is not hindered by the presence of either tartrate or $S_2O_3^{2-}$ (separation from copper). Other phosphates include $NH_4Cd(HPO_4)OH$, Cd[*catena*-PO_3]$_2$ and Cd_3[*cyclo*-P_3O_9]$_2 \cdot 14H_2O$.

Arsenic species. Arsenite and arsenate ions precipitate the corresponding cadmium salts, readily soluble in acids and NH_3.

Reduced chalcogens. Sulfane (H_2S) and S^{2-} precipitate, from slightly acidic or alkaline solutions, CdS, yellow, insoluble in excess reagent, in NH_3 or in CN^- (distinction from copper); soluble in 2-M HNO_3, hot 3 to 4-M H_2SO_4, and in saturated NaCl (distinction from copper).

A solution of copper and cadmium salts, very dilute, when applied to a filter paper or porous porcelain plate, gives a ring of the Cd^{II} beyond that of the Cu^{II}, both easily detected by H_2S.

Thiourea, $CS(NH_2)_2$, Tu, forms $[Cd(\eta^{3/2}\text{-}CH_3CO_2)_2(Tu\text{-}\kappa S)_2]$ and both $[Cd(Tu\text{-}\kappa S)_4](ClO_4)_2$ and $[Cd(Tu\text{-}\kappa S)_6](ClO_4)_2$, where we try the symbol $\eta^{3/2}\text{-}CH_3CO_2$ to show that one Cd–O bond is weaker than the other.

Oxidized chalcogens. Aqueous $S_2O_3^{2-}$ does not precipitate Cd^{II} (distinction from Cu^{II}), but gives, e.g., $(Rb,NH_4)_2[Cd(S_2O_3\text{-}\kappa S\text{-}\kappa O)_2]$.

Cadmium dissolves slowly in hot, rather dilute H_2SO_4, releasing H_2 and forming at least $CdSO_4$ and $Cd(SO_4)_2^{2-}$ complexes.

Slow evaporation of $Cd(CH_3CO_2)_2$ and H_2SeO_3 at 22 °C yields $4CdSeO_3 \cdot 3H_2O$; at 40 °C, $CdSeO_3$. Excess H_2SeO_3 can give $CdSe_2O_5$.

Reduced halogens. Cadmium dissolves slowly in hot, rather dilute HCl, producing H_2. The carbonate, cyanide, oxalate, phosphate, sulfide and hexacyanoferrates(II and III) dissolve in HCl, forming $[CdCl_2(H_2O)_4]$, $[CdCl_4]^{2-}$, $CdCl_2 \cdot {}^5/_2H_2O$ etc. Species in $CdBr_2$ or CdI_2 are $[Cd(H_2O)_6]^{2+}$, $[CdX(H_2O)_5]^+$, $[CdX_2(H_2O)_4]$, $[CdX_3]^-$ and $[CdX_4]^{2-}$ in equilibrium. Note the changing ligancy (c.n.) also with Zn^{2+} and Br^- above.

All Cd compounds are soluble in excess of I^-, especially forming $[CdI_4]^{2-}$, with no precipitate (distinction from copper).

Oxidized halogens. Perchlorate and Cd^{2+} give rise to $[Cd(H_2O)_6](ClO_4)_2$ crystals. Adding ClO_4^- to an ammoniacal solution of Cd^{2+} completely precipitates the cadmium as $[Cd(NH_3)_4](ClO_4)_2$.

12.2.4 Reagents Derived from the Metals Lithium through Uranium, plus Electrons and Photons

Oxidation. Metallic cadmium precipitates the corresponding metal from solutions of Co, Pt, Cu, Ag, Au, Hg, Sn, Pb and Bi.

In the discharge of rechargeable Ni-Cd batteries, Cd becomes $Cd(OH)_2$; recharging, essentially by definition, reverses this.

Reduction. Metallic Mg, Zn and Al precipitate Cd from Cd salts.

Other reactions. Chromate ion, CrO_4^{2-}, precipitates yellow cadmium chromate only from concentrated solutions of Cd^{2+}.

With Cd^{2+} the $[Fe(CN)_6]^{4-}$ ion yields such salts as white $K_2Cd[Fe(CN)_6]$ and $Cd_2[Fe(CN)_6]$. The $[Fe(CN)_6]^{3-}$ ion, however, precipitates a yellow $Cd_3[Fe(CN)_6]_2$ or $KCd[Fe(CN)_6]$ etc. These all dissolve in HCl or NH_3.

12.3 Mercury, $_{80}Hg$ (and Ununbium, $_{112}Uub$)

Oxidation numbers for Hg: (I) and (II), as in: Hg_2O, "mercurous" oxide, HgO, "mercuric" oxide, and $Hg^{II}O^{-I}_2$ [mercury(II) peroxide]. The "mercurous" ion is dimercury(2+) or dimercury(I), Hg_2^{2+}. Relativity greatly strengthens the Hg–Hg bond but may also explain the metal's liquidity. Relativity shrinks and stabilizes mercury's $6s^2$ orbital, making Hg rather like a noble gas but favoring linear sp hybrids with more (1/2 s) low-energy s nature than the tetrahedral sp^3 (1/4 s). The many (linear) organo-mercury compounds are excluded here.

One example of the continuing difficulties of interpreting reports is opposing statements in the same article [1]: We read, "Since many ligands bind the Hg^{2+} ion very strongly, the number of coordination compounds of Hg_2^{2+} is limited," because of course lowering the $c(Hg^{2+})$ shifts the equilibrium $Hg_2^{2+} \rightleftarrows Hg + Hg^{2+}$ to the right, but the next page of the same review states, "The myth that the dimercury(I) species Hg_2^{2+} forms few coordination compounds has been exploded."

In addition to predicting some well-known properties of Zn, Cd and Hg, relativistic quantum mechanics predicts very low polarizability and van der Waals forces, and a low boiling point for $_{112}Uub$, temporarily called ununbium, the next member of this Group, recently synthesized. The oxidation states calculated to be stable in water: (0), (II) and (IV).

12.3.1 Reagents Derived from Hydrogen and Oxygen

Dihydrogen. Acidic Hg^{2+} and Hg_2^{2+}, plus H_2, form Hg and H_3O^+.

Water. Elemental Hg dissolves up to about 10^{-7} M.

Salts such as $Hg_2(NO_3)_2{\cdot}2H_2O$ contain $[\{-Hg(H_2O)\}_2]^{2+}$. Acidified solutions of Hg^{2+}, e.g., the perchlorate, contain $[Hg(H_2O)_6]^{2+}$.

Dimercury(I) oxide is insoluble in water; HgO is soluble up to 0.2 mM at 25 °C. Aqueous Hg^{2+} is much more acidic (hydrolyzed) than even Cd^{2+} in spite of the former's greater radius, due to strong relativistic effects. This converts many Hg^{2+} salts in water, if not acidified, to basic salts, and forms polynuclear species such as perhaps $[\{Hg(H_2O)\}_2(\mu\text{-}OH)]^{3+}$. Aqueous Hg_2^{2+} is also very acidic. Many Hg^{I} and Hg^{II} compounds are either insoluble or require free acid to prevent hydrolysis (more so with "hard" anions), which precipitates a basic compound. Even $[HgCl_2]$ can give $[(HgCl)_2O]$ or $[(HgCl)_3(\mu_3\text{-}O)]Cl$.

The insoluble Hg^{I} compounds (i.e., almost all) are usually prepared for analysis by dissolving with oxidants, yielding only Hg^{II}.

The basic carbonate and the oxalate of Hg^{II} are insoluble in water. Dimercury(I) acetate, $Hg_2(CH_3CO_2)_2$, dissolves to about 20 mM. Mercury(II) cyanide and acetate are very soluble; $[Hg(CN)_2]$, not hydrated, is the only soluble binary **d**- or **p**-block cyanide.

Dimercury(I) nitrate, $Hg_2(NO_3)_2$, is soluble but see HNO_3 below.

Mercury(II) nitrate, $Hg(NO_3)_2$, is deliquescent and soluble in a small amount of water; dilution results in the precipitation of a basic compound readily soluble in HNO_3.

Water dismutates Hg_2F_2 to Hg, HgO and HF; HgF_2 in much water is entirely hydrolyzed to HgO and HF.

Mercury(II) phosphate, arsenite, arsenate and sulfide are insoluble.

Dimercury(I) sulfate, Hg_2SO_4, is slightly soluble in water (1 mM), but soon decomposes with precipitation of a basic compound; it is soluble in HNO_3 and dilute H_2SO_4. Likewise $HgSO_4$ reacts with water, a precipitate of the basic sulfate being formed; this is prevented if free H_2SO_4 is present.

The solubilities of the non-fluoride mercury(II) halides in water at 25 °C are: $[HgCl_2]$ 2.5 dM, $[HgBr_2]$ 16 mM and $[HgI_2]$ 0.13 mM.

Mercury(II) chromate is hydrolyzed, and the hexacyanoferrates(II and III) are insoluble.

Seawater and some freshwater contain traces of Hg^{II} complexes as $HgOH^+$, $Hg(OH)_2$ and $HgCl_n^{(n-2)-}$. Some other natural waters may contain $H_mHg^{II}S_n^{(2n-m-2)-}$, and some natural brines may contain $[HgI_4]^{2-}$.

Oxonium. Both Hg_2^{2+} and Hg^{2+} require an excess of free acid to prevent hydrolysis and hold them in solution. For $Hg_2(NO_3)_2$, see HNO_3 below.

The oxide HgO is soluble in acids except H_3PO_4 and H_3AsO_4.

Hydroxide. Mercury is unaffected by treatment with alkalis.

Aqueous OH^- precipitates, from solutions of dimercury(I) salts, Hg_2O, black, insoluble in excess alkali, readily transposed by acids.

From dissolved mercury(II), OH^- precipitates reddish-brown basic compounds when added in less than equivalent amounts, but the yellow oxide, HgO, when in excess. It is somewhat more soluble in OH^- than in water. If the original solution of mercury(II) is strongly acidic the precipitation of HgO may be incomplete due to forming a stable complex ion, e.g., $[HgCl_4]^{2-}$, from the acid's anion.

Peroxide. Mercury peroxide, $Hg^{II}O_2$, reddish brown, has been prepared by treating $Hg(NO_3)_2$ with an excess of H_2O_2 at 0 °C or in ethanol. It is fairly stable in air but slowly decomposed by water.

Di- and trioxygen. Insoluble Hg^I compounds are rather inert in air, but O_3 and $[Hg_2Cl_2]$ or $[Hg_2Br_2]$ give some Hg_2OX_2. Air very slowly, or O_3 more quickly, oxidizes aqueous Hg_2^{2+}:

$$Hg_2^{2+} + O_3 \rightarrow HgO\downarrow + Hg^{2+} + O_2\uparrow$$

12.3.2 Reagents Derived from the Other 2nd-Period Non-Metals, Boron through Fluorine

Carbon oxide species. Mercury(II) oxidizes CO in water:

$$Hg^{2+} + CO + 2\ H_2O \rightarrow [HgCO_2H]^+ + H_3O^+$$

$$[HgCO_2H]^+ + H_2O \rightarrow Hg + CO_2\uparrow + H_3O^+$$

$$Hg^{2+} + Hg \rightarrow Hg_2^{2+}$$

$$\overline{\hspace{2em}}$$

$$2\ Hg^{2+} + CO + 3\ H_2O \rightarrow Hg_2^{2+} + CO_2\uparrow + 2\ H_3O^+$$

Aqueous CO_3^{2-}, HCO_3^- and $AeCO_3$ precipitate from Hg_2^{2+} yellow impure Hg_2CO_3, which readily decomposes into Hg and HgO when heated and which darkens in light.

Mixing Hg^{2+} or $[HgCl_2]$ with CO_3^{2-} or HCO_3^- in various ways, temperatures and so on may yield reddish-brown $HgCO_3$, a brownish basic carbonate such as $Hg_3O_2CO_3$, a basic chloride, and/or complexes, $HgHCO_3^+$ or $HgOHCO_3^-$. The Ae carbonates precipitate basic compounds from $Hg(NO_3)_2$ or $HgSO_4$ but not from $[HgCl_2]$.

Cyanide species. Aqueous HCN and CN^- decompose $(Hg^I)_2$ into metallic Hg and $[Hg(CN)_2]$. Mercury(II) forms the same readily soluble, un-ionized, white cyanide. Aqueous HCN and HgO provide a good route to this. The firmly bound, relatively non-toxic $[Hg(CN)_2]$ releases very little of the highly poisonous Hg^{2+} and CN^-.

No precipitate is obtained from $[Hg(CN)_2]$ with OH^-, CO_3^{2-}, NH_3 or Ag^+, but it does show a markedly alkaline reaction in water, and H_2S yields HgS. Excess CN^- forms $[Hg(CN)_3]^-$ or $[Hg(CN)_4]^{2-}$, and $(Alk,Tl)_2[Hg(CN)_4]$. The first two steps in forming $[Hg(CN)_4]^{2-}$ are relatively slow, the others fast. With X = N_3, NCO, SCN, Cl, Br or I, $[Hg(CN)_2]$ and AlkX form $Alk[Hg(CN)_2X]\cdot nH_2O$, but $[Hg(CN)_2]$ and

$Ae(NCS)_2$ give, e.g., $Ae[Hg(CN)_2(SCN)_2(H_2O)_2]\cdot 2H_2O$. However, KCN and $[Hg(Cl,Br)_2]$ can yield $K_2[HgX_4][Hg(CN)_2]\cdot 2H_2O$.

Mixing $[Hg(CH_3CO_2)_2]$ with (K,Rb,Cs)NCO forms double salts, $2Alk[Hg(NCO)_3]\cdot[Hg(NCO)_2]$.

The mutual diffusion of initially separate $HgCl_2$ and H_2CN_2 solutions yields crystalline $[Hg^{II}_3(NCN)_2]Cl_2$ and $[Hg^{II}_3(NCN)_2][HgCl_4]$.

Thiocyanate and Hg_2^{2+} give $Hg_2(SCN)_2$, then Hg and $Hg(SCN)_2$.

Some "simple" organic species. Methanol, Hg and HNO_3 yield the explosive fulminate, $[Hg(CNO)_2]$; adding (K,Rb,Cs)CNO forms the very explosive $Alk_2[Hg(CNO)_4]$. The large $[AsPh_4]^+$ ion gives more stability.

Cold HCO_2H reduces Hg^{2+} only to Hg_2^{2+}.

Dimercury(I) oxide is completely soluble in neat, "glacial" CH_3CO_2H. Cooling a hot solution of HgO in 9-M CH_3CO_2H yields $Hg(CH_3CO_2)_2$. Mercury(I and II) acetates both darken in the light.

Oxalic acid and $C_2O_4^{2-}$ precipitate, from dimercury(I) salts, white $Hg_2C_2O_4\cdot H_2O$, insoluble in dilute HNO_3 or H_2SO_4. This becomes dirty yellow after long contact with cold water, darkens in hot water, and is slightly more soluble in $H_2C_2O_4$ than in water. Aqueous $C_2O_4^{2-}$ and Hg_2^{2+} form stable complexes. The Hg^{2+} ion, but not $[HgCl_2]$, precipitates $C_2O_4^{2-}$ as HgC_2O_4, white, explosive, readily soluble in HCl, insoluble in cold water or $H_2C_2O_4$, difficultly soluble in HNO_3.

Reduced nitrogen. Ammonia and HgO yield "Millon's base", $Hg_2N(OH)\cdot 2H_2O$. Ammonia and "$(NH_4)_2CO_3$" form from soluble and insoluble Hg_2^{II} salts, mixtures of mercury(0) and mercury(II) amides. For example, $[Hg_2Cl_2]$ gives a black mixture of $HgNH_2Cl$ and Hg.

Mercury(II)-ammonia compounds have been divided into three groups: (1) additive compounds; (2) ammonolyzed compounds in which NH_2, NH or N takes the place of the acid anion in a mercury(II) compound; and (3) both hydrolyzed and ammonolyzed compounds.

If $[HgCl_2]$ is slowly added to a hot mixture of NH_3 and NH_4^+ the "fusible white precipitate", $[Hg(NH_3)_2]Cl_2$, is formed, an example of (1) above. However, if NH_3 is added to $[HgCl_2]$ the "infusible white precipitate", $HgNH_2Cl$, is obtained, an example of (2) above. The addition of NH_3 to $[HgI_2]$ or, more readily, the reaction between NH_3 and "Nessler's reagent", $[HgI_4]^{2-}$ and OH^-, precipitates the reddish-brown iodide of "Millon's base" [an example of (3) above], i.e., Hg_2NI. This is a test for NH_3 and is sometimes inconveniently over-sensitive. Rather similar bromides are $[Hg(NH_3)_2]Br_2$, $HgNH_2Br$ and Hg_2NBr.

Aqueous $Hg(NO_3)_2$ and NH_3 may produce ~$NH_4Hg_3(NH)_2(NO_3)_2$. Above 270 °C this goes to $[Hg_2N]NO_3$, N_2O, NH_3, N_2 and H_2O, then further above 380 °C to HgO and N_2O.

Bubbling NH_3 into $Hg(ClO_4)_2$ and cooling forms $[Hg(NH_3)_4](ClO_4)_2$.

Dissolving HgO or $HgCl_2$ in $AlkSO_3NH_2$ gives $AlkHg(SO_3N\text{-}\kappa N)$.

In the presence of an alkali, NH_2OH and N_2H_4 reduce most Hg^{II} to Hg. Strong acids make the reductions incomplete.

The Hg_2^{2+} ion and N_3^- form explosive $Hg_2(N_3)_2$; Hg^{2+} and N_3^-, also HgO and HN_3, give $[Hg(N_3)_2]$, in either a less-explosive "α" or a more-explosive "β" form, plus $[Hg(N_3)_3]^-$ and $[Hg(N_3)_4]^{2-}$ from the former.

Oxidized nitrogen. Aqueous KNO_2 and $Hg(NO_3)_2$ give a ligancy (c. n.) of eight in $K_3[Hg(NO_2\text{-}\kappa^2O)_4]NO_3$.

Nitric acid with a little HNO_2 is the most effective solvent for Hg. Iron(III) decelerates and Mn^{2+} accelerates solution. Dilute acid, hot or cold, is effective; the concentrated acid becomes hot and possibly violently reactive. At ambient *T*, excess HNO_3 forms Hg^{2+} and $Hg(NO_3)_2 \cdot nH_2O$ with NO and some NO_2; excess Hg gives Hg_2^{2+}:

$$3\ Hg + 8\ H_3O^+ + 2\ NO_3^- \rightarrow 3\ Hg^{2+} + 2\ NO\uparrow + 12\ H_2O$$

Nitric acid and Hg_2O form Hg_2^{2+}, oxidized by excess acid to Hg^{2+}.

Free HNO_3 is necessary in solutions of $Hg_2(NO_3)_2$ to prevent the precipitation of a basic salt. On standing, this acid gradually oxidizes the Hg_2^{2+} to Hg^{2+}; this is prevented by the presence of Hg, but then the excess HNO_3 is gradually consumed and the basic salt precipitates.

Boiling HNO_3 slowly oxidizes and dissolves $[Hg_2Cl_2]$:

$$^3/_2\ [Hg_2Cl_2] + 4\ H_3O^+ + NO_3^- \rightarrow 3\ [HgCl]^+ + NO\uparrow + 6\ H_2O$$

Mercury(II) bromide is decomposed by warm HNO_3.

Mercury compounds are more soluble in concentrated HNO_3 than in water or the dilute acid. Nitric acid dissolves all common insoluble compounds of mercury except HgS, which, however, may be converted to the more soluble complex $Hg_3S_2(NO_3)_2$, white, by boiling with concentrated acid. All dimercury(I) salts are oxidized to mercury(II) compounds by excess of HNO_3.

Fluorine species. Neither gaseous nor aqueous HF attacks Hg.

Yellow HgO dissolves in HF, not too dilute, and forms $HgF_2 \cdot 2H_2O$.

12.3.3 Reagents Derived from the 3rd-to-5th-Period Non-Metals, Silicon through Xenon

Silicon species. Silane does not blacken $[HgCl_2]$ (cf. $AgNO_3$) paper.

Phosphorus species. Phosphinic and phosphonic acids reduce Hg^{II} compounds to Hg^I and Hg; with the latter reagent Hg is obtained only at higher *T*. From solutions of $[HgCl_2]$ or in the presence of Cl^-, first the white $[Hg_2Cl_2]$, then gray Hg is formed. Heating promotes this.

Mercury(II) compounds may be determined by precipitation as $[Hg_2Cl_2]$ after reduction with H_2PHO_3 or HPH_2O_2 (with H_2O_2 to prevent further reduction to Hg).

Phosphoric acid and its anions precipitate from dimercury(I) salts, white Hg_3PO_4, that is, $(Hg_2)_3(PO_4)_2$, if the reagent is in excess, but a basic nitrate phosphate, somewhat yellowish, if $Hg_2(NO_3)_2$ is in excess. The $(Hg_2)_3(PO_4)_2$ is soluble in HNO_3, insoluble in H_3PO_4.

From Hg^{2+}, even with excess acid, white $Hg_3(PO_4)_2$ is precipitated by HPO_4^{2-}; somewhat soluble in hot water; soluble in Cl^-, less readily in HNO_3; insoluble in H_3PO_4. Phosphoric acid does not produce a precipitate from solutions of $[HgCl_2]$, nor does HPO_4^{2-} give the normal phosphate, but on standing for some time with the latter, a portion of the mercury separates as a dark-brown precipitate.

Aqueous $[P_2O_7]^{4-}$ gives, with Hg_2^{2+}, $(Hg_2)_2[P_2O_7]\cdot H_2O$, white, darkening on heating, soluble in excess and in HNO_3. Stable complexes of Hg_2^{2+} arise with both $[P_2O_7]^{4-}$ and $[P_3O_{10}]^{5-}$.

Mercury(2+) and $[P_2O_7]^{4-}$ precipitate white $Hg_2[P_2O_7]$, turning yellow, soluble in excess especially as $Hg(OH)[P_2O_7]^{3-}$, and in acids and Cl^-.

Arsenic species. Arsenous acid or salts, plus Hg_2^{2+}, precipitate a yellowish-white $(Hg_2)_3(AsO_3)_2$, soluble in HNO_3, intensely yellow when air dried. Treating $[HgCl_2]$ with AsO_3^{3-} precipitates a white, slightly soluble $Hg_3(AsO_3)_2$. It turns yellow on standing, perhaps due to oxidation of some of the AsO_3^{3-} to AsO_4^{3-}.

Arsenic acid or $HAsO_4^{2-}$ precipitates, with Hg_2^{2+}, a yellow to orange product. The Hg^{2+} ion yields yellow, slightly soluble $Hg_3(AsO_4)_2$, readily soluble in HCl.

Reduced chalcogens. Dry H_2S at ordinary temperatures does not react with Hg; oxygen yields HgS, but water vapor retards the reaction.

Hydrogen sulfide and S^{2-} precipitate from dimercury(I) salts at room temperature HgS, black, and metallic mercury. Dimercury(I) sulfide, Hg_2S, is said to be stable only below 0 °C.

Aqueous $[HgCl_2]$ and sulfides first precipitate white Hg_2Cl_2S, rapidly changing on addition of more reagent through higher ratios of sulfide over chloride to yellow, red, brown and finally black HgS. This succession of colors is characteristic of mercury. Although black HgS is the final form in analysis, it is meta-stable, vermillion (cinnabar) being the stable one; thus the black is more reactive and may be converted into the red by grinding in a mortar or by sublimation.

Mercury(II) cyanide reacts with H_2S, but not with other common acids, to release HCN.

Mercury(II) sulfide is insoluble in dilute HNO_3, very slightly soluble in "$(NH_4)_2S$" or the polysulfide, less so in the latter than in the former, fairly soluble in OH^- or S^{2-}, readily soluble in a mixture of the two (separation from Cu, Ag, Pb and Bi), insoluble in cold HS^-, soluble in CS_3^{2-} (separation from Cu, Pb and Bi) from which it is reprecipitated as HgS by H_3O^+, soluble in aqua regia or in HCl plus ClO_3^-, distinctly soluble in concentrated HCl with liberation of H_2S, much more easily soluble in HBr or HI.

A separation of Hg^{2+} from the other acid-insoluble sulfides is based on the insolubility of HgS in 2-M HNO_3 (separation from Cu, Cd, Pb and Bi) and in "$(NH_4)_2S$" (separation from Sn, As, Sb etc.). After the sulfide is dissolved in aqua regia or HCl plus ClO_3^-, and the excess oxidant is decomposed, mercury is confirmed by adding $SnCl_2$, which gives $[Hg_2Cl_2]$ (white) or Hg (black) or a mixture (gray).

The hydride H_2Se and Se^{2-}, or H_2Te and Te^{2-}, react with Hg^{2+} to produce dark violet HgSe or white HgTe.

Thiocyanate gives a gray precipitate with Hg_2^{2+}; with a moderate $c(Hg^{2+})$, it precipitates $Hg(SCN)_2$. This burns to a large, spongy ash called "Pharaoh's serpents", dissolves in hot water, and with excess NCS^- gives (K,Rb,Cs)$Hg(SCN)_3$ or (Ae,Co,Cu)$[Hg(SCN)_4]\cdot nH_2O$. Also see $Hg[Co(NCS)_4]$ in **9.1.4 Other reactions**.

Mixing (aqueous) $[Hg(Cl,Br,I)_2]$ with $Hg(SCN)_2$ yields HgX(SCN).

Thiocyanate and $[Hg(CN)_2]$ give $[Hg(CN)_2(SCN)_2]^{2-}$.

Thiourea, $SC(NH_2)_2$-κS, Tu, forms complexes with $[HgCl_2]$ and thus yields [HgTuCl]Cl and $[HgTu_n]Cl_2$, with $2 \leq n \leq 4$. Their water solubilities are small and decreasing as n is $4 > 3 > 1 > 2$.

Aqueous $SeCN^-$ and $Hg(CH_3CO_2)_2$ yield $Hg(SeCN)_2$, $[Hg(SeCN)_3]^-$ and $[Hg(SeCN)_4]^{2-}$, which are more stable than the thiocyanates. Some salts are $AlkHg(SeCN)_3$ and $M[Hg(SeCN)_4]$; M = Co, Cu, Zn, Cd or Pb.

Elemental and oxidized chalcogens. Sulfur attacks mercury, forming HgS. Among all the metals, possibly excepting Pd, Hg may have the greatest affinity for sulfur.

Thiosulfate and Hg yield HgS.

Dimercury(I) nitrate forms with $S_2O_3^{2-}$ a grayish-black precipitate, part of the mercury remaining in solution; $[Hg_2Cl_2]$ forms a soluble complex and metallic mercury.

Thiosulfate complexes Hg^{2+} as $(S_2O_3\text{-}\kappa S)^{2-}$ but then decomposes, depending on the ratio of reagent to Hg^{II} and the total concentration. Thus $[HgCl_2]$:$[S_2O_3^{2-}]$ in the ratio 3:2 gives white $Hg_3Cl_2S_2$:

$$3\ [HgCl_2] + 2\ S_2O_3^{2-} + 6\ H_2O \rightarrow$$

$$Hg_3Cl_2S_2\downarrow + 4\ Cl^- + 4\ H_3O^+ + 2\ SO_4^{2-}$$

and $[HgCl_2]$:$[S_2O_3^{2-}]$ at 1:1 gives black HgS (sometimes red):

$$[HgCl_2] + S_2O_3^{2-} + 3\ H_2O \rightarrow HgS\downarrow + 2\ Cl^- + 2\ H_3O^+ + SO_4^{2-}$$

but $[HgCl_2]$:$[S_2O_3^{2-}]$ at 1:4 also splits the added $S_2O_3^{2-}$ ions:

$$[HgCl_2] + 4\ S_2O_3^{2-} \rightarrow HgS\downarrow + 4\ S\downarrow + 2\ Cl^- + 3\ SO_4^{2-}$$

Mercury(II) iodide is soluble in $S_2O_3^{2-}$.

Mercury is not attacked by cold solutions of SO_2, either alone or in the presence of HCl or H_2SO_4.

Sulfur dioxide and SO_3^{2-} form in Hg_2^{2+} solutions a gray-black precipitate, mostly $Hg_2SO_3 \cdot {}^1/_2H_2O$. At ambient T this slowly becomes Hg_2SO_4, Hg and SO_2; warm water quickly gives Hg and H_2SO_4.

Sulfur dioxide precipitates from solutions of Hg^{2+}, gray Hg; from [$HgCl_2$] or in the presence of Cl^-, first the white [Hg_2Cl_2], then gray Hg. Mercury(II) oxide reacts with SO_2 in water to give Hg and H_2SO_4.

Mercury(2+) gives, with SO_3^{2-}, a voluminous white precipitate containing dimercury(I); [$HgCl_2$] gives no precipitate in the cold but is reduced by boiling with SO_2 to [Hg_2Cl_2] and then to Hg.

Selenite, however, precipitates $HgSeO_3$ from Hg^{2+}.

Suspensions of HgO and H_2SeO_3, or of $HgSO_4$ and SeO_2, at 57 to 97 °C for one to a few weeks, form $HgSeO_3 \cdot {}^1/_2H_2O$ or $HgSeO_3$.

A suspension of HgO and H_2SeO_3, plus excess NH_3 to a pH of 8 to 9, in a few weeks at 37 to 57 °C, yield $Hg(NH_3)SeO_3$.

Selenium disulfide can detect one part (by mass) of Hg in four million parts of air in a four-minute exposure, by forming black HgS.

Dilute or concentrated H_2SO_4 at 25 °C has no effect on Hg. The hot concentrated acid forms Hg_2SO_4 with excess Hg, otherwise $HgSO_4$:

$$2\ Hg + 4\ H_2SO_4 \rightarrow Hg_2SO_4 + SO_2\uparrow + 2\ H_3O^+ + 2\ HSO_4^-$$

$$Hg + 4\ H_2SO_4 \rightarrow HgSO_4 + SO_2\uparrow + 2\ H_3O^+ + 2\ HSO_4^-$$

Sulfuric acid and Hg_2O form Hg_2SO_4, which on boiling with excess acid becomes $HgSO_4$. Aqueous SO_4^{2-} precipitates from sufficiently concentrated dimercury(I) solutions, Hg_2SO_4, white, sparingly soluble (1 mM) in cold water, but soon decomposing with precipitation of a basic compound. It is decomposed by boiling water into dirty-yellow HgO and Hg. It is darkened by light, soluble in dilute HNO_3, and more soluble in dilute H_2SO_4 than in H_2O. Sulfate does not precipitate Hg^{II}.

Dimercury(I) chloride dissolves in concentrated $(NH_4)_2SO_4$.

Mercury dibromide, unlike the dichloride, undergoes some decomposition with warm concentrated H_2SO_4, giving some Br_2.

Slightly acidified $Hg_2(NO_3)_2$ and excess H_2SeO_4 precipitate dark-brown Hg_2SeO_4.

Reduced halogens. Cold HCl does not attack Hg.

Aqueous Cl^- and Hg_2^{2+}, or HCl and Hg_2O, precipitate white [Hg_2Cl_2], "calomel", soluble in concentrated Cl^-. This separates Hg_2^{2+} sharply from Hg^{II}. Mercury(II) oxide etc. are soluble in Cl^-, Br^- and I^-, forming numerous homoleptic and mixed complexes up to $[HgX_4]^{2-}$, and many salts; some stabilities are $[HgCl_4]^{2-}$ < $[HgBr_4]^{2-} < [HgI_4]^{2-}$.

Aqueous Cl^-, Br^- or I^- makes $[Hg(CN)_2X_n]^{n-}$ from $[Hg(CN)_2]$, and among the solids at least $K[Hg(CN)_2I]$ has been well studied.

Aqueous HBr reacts with Hg slowly in the cold, more rapidly when hot, but HI reacts quickly, yielding $[Hg_2I_2]$ or $[HgI_2]$, (or $[HgI_4]^{2-}$ with excess HI) depending on conditions.

Bromide and I^- ions both displace Cl^- quantitatively from $[Hg_2Cl_2]$.

Aqueous Br^- and Hg_2^{2+} precipitate $[Hg_2Br_2]$, white, insoluble in dilute HNO_3; ammonia converts it to $Hg_2Br(NH_2)(OH)_2 \cdot H_2O$, possibly thus:

$$2\,[Hg_2Br_2] + 4\,NH_3 + 3\,H_2O \rightarrow$$

$$Hg_2Br(NH_2)(OH)_2 \cdot H_2O\downarrow + 2\,Hg\downarrow + 3\,Br^- + 3\,NH_4^+ \text{ or}$$

$$2\,[Hg_2Br_2] + NH_3 + 4\,H_2O \rightarrow$$

$$Hg_2Br(NH_2)(OH)_2 \cdot H_2O\downarrow + [HgBr_2]\downarrow + HgBrOH\downarrow + 2\,H_2\uparrow$$

Not too dilute Hg^{ii} and Br^- form white $[HgBr_2]$, soluble in excess Hg^{ii} or Br^-.

Aqueous I^- precipitates from Hg_2^{2+}, greenish yellow $[Hg_2I_2]$, insoluble in water, dilute HNO_3 or ethanol (distinction from $[HgI_2]$), soluble in Hg_2^{2+} and Hg^{2+}, decomposed by excess I^- into Hg and $[HgI_4]^{2-}$. Dimercury(I) chloride is transposed to $[Hg_2I_2]$ by I^-, excess of I^- acting as above. Dimercury(I) nitrate forms $[Hg_2I_2]$ and Hg^{II}.

Mercury diiodide is obtained on adding I^- to a mercury(II) solution. The unstable pale yellow $[HgI_2]$ first formed rapidly changes to the stable red modification, insoluble in cold water. The $[HgI_2]$ dissolves in excess of I^-, giving the complexes $[HgI_3]^-$ and $[HgI_4]^{2-}$; such solutions do not give the normal reactions of Hg^{2+}. The Na or K compound, as Thoullet's reagent, may be used to determine the refractive index in mineralogy. Mercury diiodide is soluble in NH_4^+; moderately soluble in Cl^-, Hg^{2+}, $[HgCl_2]$, $[Hg(CH_3CO_2)_2]$, etc.; readily soluble in Br^-.

Excess OH^- and $[HgI_4]^{2-}$ are called Nessler's reagent, used to detect NH_3, giving a yellow to brown precipitate, the iodide of Millon's base, Hg_2NI. Boiling $[HgI_2]$ with OH^- forms HgO and $[HgI_4]^{2-}$.

Mercury diiodide is transposed by CN^-, giving $[Hg(CN)_2]$ and I^-.

Ammonia converts $[Hg_2I_2]$ into $Hg(NH_3)_2I_2$ plus Hg.

Mercury diiodide is soluble in concentrated HNO_3, forming $Hg(IO_3)_2$; with more dilute acid, $HgI(NO_3)$, white, separates on cooling.

Mercury diiodide is soluble in SO_3^{2-} or $S_2O_3^{2-}$ (the latter solution on heating gives red HgS). It is not attacked by H_2SO_4 in the cold, but when heated, the compound is decomposed.

Mercury diiodide is fairly soluble in concentrated HCl (without decomposition), one of the best agents for recrystallizing $[HgI_2]$.

Mercury diiodide is reduced by $SnCl_2$, finally forming Hg and Sn^{IV}:

$$[HgI_2] + SnCl_2 \rightarrow Hg\downarrow + \text{e.g. } [SnCl_2I_2]$$

Elemental and oxidized halogens. Elementary Cl_2, Br_2 and I_2 all attack Hg forming Hg_2X_2 with excess Hg, but HgX_2 with excess halogen.

Aqua regia or Cl_2 dissolves $[Hg_2Cl_2]$ forming $[HgCl_4]^{2-}$ or $[HgCl_2]$.

Chlorine dismutates on passing over cold, fresh HgO; alternately CCl_4 is a useful solvent for the Cl_2 and product Cl_2O; H_2O, however, gives mainly HClO. The brown Hg precipitate is $HgCl_2 \cdot 2HgO$ or $Hg_3O_2Cl_2$:

$$2\ Cl_2 + 3\ HgO \rightarrow Cl_2O\uparrow + Hg_3O_2Cl_2\downarrow$$

$$2\ Cl_2 + 3\ HgO + H_2O \rightarrow 2\ HClO + Hg_3O_2Cl_2\downarrow$$

Metallic Hg and Hg_2^{2+} are oxidized to Hg^{II} by $HClO_3$.

Dissolving HgO in either $HClO_3$, $HBrO_3$ or $HClO_4$ produces $Hg(ClO_3)_2 \cdot 2H_2O$, $Hg(BrO_3)_2 \cdot 2H_2O$ or $Hg(ClO_4)_2 \cdot 6H_2O$.

Bromate and Hg_2^{2+} precipitate $Hg_2(BrO_3)_2$, white (changing to a yellow basic compound on heating in water), soluble in HCl, difficultly soluble in HNO_3. Mercury(2+) yields, when treated with BrO_3^-, $Hg(BrO_3)_2$, yellowish; soluble up to 3 mM in cold and 30 mM in boiling water; slightly soluble in HNO_3; easily decomposed by HCl. Adding bromate to $[HgCl_2]$ gives no precipitate, due to the low $c(Hg^{2+})$.

Iodate precipitates, from Hg_2^{2+} solution, white $Hg_2(IO_3)_2$, difficultly soluble in water; not affected by boiling water or cold HNO_3. Mercury(II) iodate, precipitated from IO_3^- and Hg^{2+}, is white; it is completely converted to $[HgI_2]$ and O_2 on heating; soluble in HI or HBr with release of I_2 or Br_2; slightly soluble in CN^-, $S_2O_3^{2-}$, Cl^-, Br^-, or I^-; insoluble in OH^-, $B_4O_7^{2-}$, CO_3^{2-}, CH_3CO_2H, NH_3, HF, HPO_4^{2-}, SO_3^{2-}, ClO_3^-, BrO_3^-, IO_3^- or $[HgCl_2]$. Mercury(II) chloride does not precipitate IO_3^-.

12.3.4 Reagents Derived from the Metals Lithium through Uranium, plus Electrons and Photons

Oxidation. Manganate(VII) in the cold oxidizes the metal to Hg_2O, when hot to HgO. Free Hg precipitates Ag, Au and Pt from their solutions, thus being oxidized, and it reduces Hg^{2+} to Hg_2^{2+}.

Reduction. Reducing agents, such as Mg, Fe, Co, Cu, Cu^+, Zn, Cd, Al, Sn, Sn^{II}, Pb and Bi precipitate, from Hg_2^{2+} and Hg^{2+}, gray Hg; from $[HgCl_2]$ or in the presence of Cl^-, first the white $[Hg_2Cl_2]$, then gray Hg. Heating promotes this. Atomic-absorption spectroscopy uses the reduction of small amounts to Hg_{aq} by, e.g., Cr^{2+}, for environmental analysis.

Anhydrous $[HgCl_2]$, moistened with ethanol, is reduced by Fe, a bright strip of which is corroded soon after immersion in the sample to be tested (distinction from $[Hg_2Cl_2]$).

A clean strip of Cu, placed in a slightly acidic solution of Hg_2^{2+} or Hg^{2+} becomes coated with gray Hg, and when gently rubbed with cloth or paper presents the tin-white luster of the metal:

$$Hg_2^{2+} + Cu \rightarrow 2\ Hg\downarrow + Cu^{2+}$$

The coating is driven off by heat. This is a good test for Hg, but it will not differentiate the two oxidation states.

A solution of $Hg_2(NO_3)_2$ exposed to air may be kept free from Hg^{2+} for a short time by placing some metallic Hg in the bottle to reduce it back down to Hg^{I}. After standing some weeks a basic dimercury(I) nitrate crystallizes out, which may be dissolved by adding HNO_3.

The very common reduction by $SnCl_2$ is shown here:

$$2\ [HgCl_2] + SnCl_2 \rightarrow [Hg_2Cl_2]\downarrow + SnCl_4$$

$$[Hg_2Cl_2] + SnCl_2 \rightarrow 2\ Hg\downarrow + SnCl_4$$

Other reactions. Chromate or $[Cr_2O_7]^{2-}$ gives with Hg_2^{2+} a precipitate of Hg_2CrO_4, yellow, brown or reddish depending on conditions; less soluble in CrO_4^{2-} than in water; soluble in HNO_3. Mercury(2+) is precipitated by CrO_4^{2-} as a light yellow precipitate, rapidly darkening in color; readily soluble in acids and $[HgCl_2]$. Mercury dichloride forms a precipitate with CrO_4^{2-}, but not with $[Cr_2O_7]^{2-}$.

Equivalent amounts of HgO and H_2WO_4, boiled a few minutes until the orange HgO is gone, precipitate pale-yellow $HgWO_4$.

Aqueous $[Fe(CN)_6]^{4-}$ gives with aqueous dimercury(I) a gelatinous, pale-yellow precipitate that soon becomes bluish green; with mercury(II) a gelatinous white precipitate, slowly becoming blue on standing. Aqueous $[Fe(CN)_6]^{3-}$ gives with Hg_2^{2+} yellowish to green $(Hg_2)_3[Fe(CN)_6]_2$; with $[HgCl_2]$ no precipitate; with Hg^{2+} reddish-brown, gelatinous $Hg_3[Fe(CN)_6]_2$, turning yellow on standing.

Aqueous $[Hg(CN)_2]$ forms stable complexes with $[Mo(CN)_8]^{4-}$, $[Fe(CN)_6]^{4-}$, $[Fe(CN)_6]^{3-}$, $[Ru(CN)_6]^{4-}$, $[Ni(CN)_4]^{2-}$ and others.

In microscopic analysis, $K_2[Hg(SCN)_4]$ may be used as a "Group reagent", particularly for the detection of Fe^{III}, Co^{2+}, Cu^{2+}, Zn^{2+} and Cd^{2+}.

A solid double salt, $2Hg(CN)_2 \cdot Zn(NO_3)_2 \cdot 7H_2O$, has the interesting structure [*trans*-$Zn(NCHgCN)_2(H_2O)_4](NO_3)_2 \cdot 3H_2O$.

The (symmetrical) Hg_2^{2+} ion may be viewed formally as a complex:

$$\mathrm{p}\boldsymbol{K} = 2.2: \quad (Hg^{II}Hg^{0})^{2+} \leftrightarrows Hg^{2+} + Hg_{liq}$$

$$\mathrm{p}\boldsymbol{K} \approx 6.9: \quad Hg_{liq} \leftrightarrows Hg_{aq}$$

$$\mathrm{p}\boldsymbol{K} \approx 9.1: \quad (Hg^{II}Hg^{0})^{2+} \leftrightarrows Hg^{2+} + Hg_{aq}$$

with a stability near that of $HgBr^+$:

$$pK = 9.05: \qquad HgBr^+ \leftrightarrows Hg^{2+} + Br^-$$

Aqueous OH^-, CN^-, S^{2-} etc. also tie up Hg^{2+} more strongly than Hg_2^{2+} (with a lower charge density), pushing the first equilibrium to the right.

Mercury(2+) dissolves $[Hg_2Cl_2]$, forming Hg_2^{2+} and the stable $[HgCl_2]$.

Aqueous $[Hg(CN)_2]$ dissolves HgO, crystallizing as $[(HgCN)_2(\mu\text{-}O)]$. In another con-mutation or comproportionation, $[Hg(CN)(CH_3CO_2)]$ is easily obtained as crystals from $[Hg(CN)_2]$ and $[Hg(CH_3CO_2)_2]$.

At 60 °C, $HgCl_2$ and NH_4Cl dissolve some $HgNH_2Cl$. Cooling yields:

$$2\ HgNH_2Cl + HgCl_2 + 2\ NH_4^+ + 2\ Cl^- \rightarrow$$

$$[Hg(NH_3)_2][HgCl_3]_2\!\downarrow + 2\ NH_3$$

Reference

1. Brodersen K, Hummel HU in Wilkinson G, Gillard RD, McCleverty JA (eds) (1987) Comprehensive coordination chemistry, vol 5. Pergamon, Oxford, p 1048.

Bibliography

See the general references in the Introduction, and a few more-specialized books [2–5].

2. McAuliffe CA (ed) (1977) The chemistry of mercury. Macmillan, London
3. Aylett BJ (1975) The chemistry of zinc, cadmium and mercury. Pergamon, New York
4. Farnsworth M, Kline CH (1973) Zinc chemicals. Zinc Institute, New York
5. Makarova LG, Nesmeyanov AN (eds) (1967) The organic compounds of mercury, methods of elemento-organic chemistry, vol 4. North Holland, Amsterdam

13 Boron through Thallium, the Triels

13.1 Boron, $_5B$

Oxidation number in classical compounds: (III), as in both $[BH_4]^-$ (due to the low electronegativity of B) and $[B(OH)_4]^-$.

Non-classical [*closo*-$CB_{11}H_{12}]^-$ etc., from non-aqueous sources, are some of the most inert and weakly coordinating anions available.

13.1.1 Reagents Derived from Hydrogen and Oxygen

Water. Boron is practically insoluble in water. However, and only with amorphous boron, H_2O (containing either NH_3 or H_2S) yields H_3BO_3 and H_2. Similarly, H_2O and exposure to light slowly yield the same.

Water hydrolyzes boron hydrides to H_3BO_3, H_2 etc., quickly for B_2H_6 and B_5H_{11}, slowly for B_4H_{10} and $B_{10}H_{14}$, only on heating B_5H_9 and B_6H_{10}.

Boiling water oxidizes $[BH_4]^-$ to a mixture containing various polyborate anions, simplified here, while releasing H_2 gas:

$$[BH_4]^- + 4\ H_2O \rightarrow \sim[B(OH)_4]^- + 4\ H_2\uparrow$$

The oxide B_2O_3 reacts slowly with water to form, first a "metaboric acid", $(HBO_2)_n$, then "orthoboric acid", H_3BO_3. Both are white and nicely crystalline. The very weak acid H_3BO_3 dissolves in H_2O at 21 °C up to about 8 dM. The solubility is lowered by the presence of many other acids, such as HNO_3, H_2SO_4 or HCl. Although H_3BO_3 is readily soluble at temperatures nearer the boiling point of water, there is an appreciable loss of the acid at 80 °C and above due to its volatility.

The borates of the Group-1 metals are soluble, and of the Group-2 metals, somewhat soluble in water, but those of most others are not.

"Borax", i.e., $Na_2[B_4O_5(OH)_4]\cdot 8H_2O$ or "$Na_2B_4O_7\cdot 10H_2O$", in water becomes a mixture, $[B_iO_j(OH)_k]^{(2j+k-3i)-}$, formally condensed from equal amounts of H_3BO_3 and $[B(OH)_4]^-$; the average value of the negative charge, $2j + k - 3i$, on all the condensed ions would be the average of $i/2$, absent further ionization.

However, we have only $[B(OH)_4]^-$ if dilute or at pH > 11, but with pH 5–11 one finds, e.g., $[B_3O_3(OH)_4]^-$, $[B_4O_5(OH)_4]^{2-}$ and $[B_5O_6(OH)_4]^-$.

Oxonium. Magnesium boride and 4-M HCl at 60 °C give small yields of boron hydrides, otherwise prepared in non-aqueous reactions.

Aqueous H_3BO_3 is displaced from its salts by nearly all acids, often including even aqueous CO_2. It may be made from borax and, e.g., hot H_2SO_4 or HCl. Cooling gives fairly pure crystals of H_3BO_3:

$$B_iO_j(OH)_k{}^{(2j+k-3i)-} + (2j + k - 3i)\ H_3O^+ \leftrightarrows$$

$$i\ H_3BO_3\downarrow + (3j + 2k - 6i)\ H_2O$$

Hydroxide. Amorphous boron and OH^- quickly yield borate ions and H_2, but crystalline boron is not attacked even by hot, concentrated OH^-.

The BH_3OH^- ion survives several hours at a $pH \geq 12.5$.

Peroxide. Crystalline B (slowly) or amorphous B (quickly) and H_2O_2 yield $HB(OH)_2(O_2)$, i.e., $B(OH)_2(O_2H)$; Na_2O_2 forms the slightly soluble sodium "perborate", $NaBO_3 \cdot 4H_2O$, i.e., $Na_2[B_2(OH)_4(O_2)_2] \cdot 6H_2O$ or $Na_2[\{B(OH)_2\}_2(\mu\text{-}O_2)_2] \cdot 6H_2O$, relatively stable but a powerful oxidant and common bleaching agent. It reacts much like H_2O_2.

Dioxygen. Air spontaneously inflames B_2H_6, B_5H_9 and B_5H_{11}, but not pure B_4H_{10}, B_6H_{10}, or $B_{10}H_{14}$ etc., although O_2 also quickly oxidizes solutions of the hydrides to H_3BO_3.

13.1.2 Reagents Derived from the Other 2nd-Period Non-Metals, Boron through Fluorine

Carbon oxide species. Borax can be made by treating colemanite, the slightly soluble mineral $Ca_2B_6O_{11} \cdot 5H_2O$, with Na_2CO_3, represented only formally by the following, without the complexity of the actual $B_iO_j(OH)_k{}^{(2j+k-3I)-}$:

$$Ca_2B_6O_{11} \cdot 5H_2O + 2\ CO_3{}^{2-} + 6\ H_2O \rightarrow$$

$$2\ CaCO_3\downarrow + 2\ H_3BO_3 + 4\ [B(OH)_4]^-$$

and then evaporating the filtrate in air to crystallize the borax:

$$4\ H_3BO_3 + 8\ [B(OH)_4]^- + 6\ Na^+ + CO_2 + 8\ H_2O \rightarrow$$

$$3\ Na_2B_4O_5(OH)_4 \cdot 8H_2O\downarrow + CO_3{}^{2-}$$

Cyanide species. Amorphous boron and HCN or $(CN)_2$ yield H_3BO_3.

Some "simple" organic reagents. Borates interfere in some analyses but may be removed by repeated evaporation nearly to dryness with concentrated HCl plus CH_3OH, forming the volatile $(CH_3)_3BO_3$.

Borates and 1,2-$C_2H_4(OH)_2$, 1,2,3-$C_3H_5(OH)_3$, 1,2-$C_6H_4(OH)_2$ etc. easily form chelates, especially but not only with 5-membered rings, yielding $[(-CH_2O)_2B(OH)_2]^-$, $[\{(-CH_2O)_2\}_2B]^-$ and so on. Oxalic and hydroxycarboxylic acids also give rise to many chelates.

Reduced nitrogen. Some alkaline-earth borates, although somewhat soluble in H_2O, are insoluble in NH_3, perhaps due to the formation of different borates.

Oxidized nitrogen. Amorphous boron and HNO_3 quickly yield H_3BO_3.

Fluorine species. Crystalline boron is not attacked by aqueous HF.

Boron(III) oxide is soluble in dilute HF.

Adding boric acid slowly to cold concentrated HF (not in glass!) and letting it stand a few hours at room temperature yields:

$$H_3BO_3 + 4\ HF \rightarrow [BF_4]^- + H_3O^+ + 2\ H_2O$$

Adding KOH and cooling gives the easily isolated and preserved $K[BF_4]$, although water slowly hydrolyzes it.

Alternately, NH_4HF_2 may be used instead of HF to prepare $[BF_4]^-$ and $NH_4[BF_4]$, useful for the (non-aqueous) preparation of BF_3:

$$H_3BO_3 + 2\ NH_4HF_2 \rightarrow [BF_4]^- + NH_3 + NH_4^+ + 3\ H_2O:$$

The $[BF_4]^-$ ion coordinates M^{n+} even more weakly than does ClO_4^-.

The interference of borates in some analyses may be prevented by evaporation with HF or with F^- plus H_2SO_4, releasing the volatile BF_3.

13.1.3 Reagents Derived from the 3rd-to-5th-Period Non-Metals, Silicon through Xenon

Phosphorus species. Treating $K[BH_4]$ with 15-M (85 %) H_3PO_4 yields the hazardous B_2H_6:

$$[BH_4]^- + H_3PO_4 \rightarrow {}^1/_2\ B_2H_6\uparrow + H_2\uparrow + H_2PO_4^-$$

Oxidized chalcogens. Hot H_2SO_4 attacks crystalline boron; $[S_2O_8]^{2-}$ oxidizes it very slowly, amorphous boron more rapidly, to H_3BO_3.

Boron(III) oxide is soluble in warm, concentrated H_2SO_4.

Hot, concentrated H_2SO_4 can be used to make BF_3:

$$6\ [BF_4]^- + B_2O_3 + 9\ H_2SO_4 \rightarrow 8\ BF_3\uparrow + 9\ HSO_4^- + 3\ H_3O^+$$

Reduced halogens. Crystalline boron is not attacked by boiling HCl.

Oxidized halogens. (Only amorphous) boron and $HClO_3$ easily yield H_3BO_3 and $HClO_2$. Similarly, HIO_3 or IO_4^- easily yields H_3BO_3 and I_2.

13.1.4 Reagents Derived from the Metals Lithium through Uranium, plus Electrons and Photons

Oxidation. Boron hydrides, $B_4H_{10} > B_2H_6 > B_5H_9 > B_{10}H_{14}$, are quickly oxidized to H_3BO_3 by MnO_4^-, Ni^{2+}, Cu^{2+}, Ag^+, etc.

Boron, only if amorphous, not crystalline, is quickly oxidized to H_3BO_3 or borates by MnO_4^-, by $FeCl_3$ (going to Fe^{2+}), by aqueous $PdCl_n$ and $PtCl_n$ (to Pd and Pt), by Ag^+ (to Ag and H_3O^+), and by Au_2Cl_6 to Au or AuCl, plus Cl^- where indicated.

Anodes and $[BH_4]^-$ in various conditions produce borates by way of BH_3OH^- on the electrode and with some release of H_2 [from BH_3OH^-, $BH_2(OH)_2^-$ and $BH(OH)_3^-$ as possible sources].

Other reactions. Borax precipitates most of the common metallic cations, not Alk^+, from neutral solutions, and not from ammines of Co, Ni, Cu, Ag, Zn or Cd in aqueous NH_3. Rather concentrated solutions of Ae^{2+} form bulky white precipitates of AeB_2O_4, some hydrated. Many other salts give basic salts of the original anion, e.g., sulfate, or of borates. Thorium(4+) and hot $B_4O_7^{2-}$ form a borate. The Cr^{3+}, Fe^{3+} and Al^{3+} ions give their hydroxides. Borax and $Hg_2(NO_3)_2$ yield a basic nitrate, and Hg^{2+} forms its oxide. Tin(II) yields its hydroxide or dismutates into Sn and Sn^{IV}, but Bi^{III} gives a borate.

However, H_3BO_3 dissolves calcined MgO slowly, hydrated but not anhydrous $MgCO_3$ quickly, $Ca(OH)_2$ with boiling, but not $CaCO_3$ or $BaCO_3$. Thorium(4+) and H_3BO_3 yield a flocculent borate. Boric acid dissolves the hydroxides of Mn^{2+}, Fe^{2+}, Zn^{2+}, Al^{3+} etc. Red-brown Fe^{III} acetate solution loses its color with H_3BO_3, and Tl_2CO_3 gives $Tl_2B_4O_7 \cdot 2H_2O$.

Boric acid forms various complexes with V^V, Mo^{VI}, W^{VI}, P^V and As^{III}.

13.2 Aluminum, $_{13}Al$

Oxidation number: (III) as in Al^{3+}.

13.2.1 Reagents Derived from Hydrogen and Oxygen

Water. Powdered aluminum, when boiled with water, releases hydrogen, forming the hydroxide.

The oxide, Al_2O_3, and hydrous oxide, $Al_2O_3 \cdot aq$, are insoluble in water.

Aqueous Al^{III} is octahedral in $[Al(H_2O)_6]^{3+}$, which can be crystallized as hydrated nitrates or perchlorate among others, and, e.g., $[AlSO_4(H_2O)_5]^+$. Inorganic salt solutions are acidic due to hydrolysis. Aging and high concentrations take it through

$[\{Al(H_2O)_4\}_2(\mu\text{-}OH)_2]^{4+}$ and $[\{Al(H_2O)_4\}_2(OH)_2](SO_4)_2$ to higher polymers, but smaller ions arise with H_3O^+. A 2.5/1 ratio of $[OH^-]/[Al^{III}]$ yields the persistent $[Al_{13}O_4(OH)_{25}(H_2O)_{11}]^{6+}$–whose sulfate and chloride dissolve only slightly–and $[Al_{13}O_4(OH)_{24}(H_2O)_{12}]^{7+}$, whose Raman spectrum is like that of solids: $Na[Al_{13}O_4(OH)_{24}(H_2O)_{12}](SO_4,SeO_4)_4$.

The normal acetate is soluble, the basic acetate, $Al(OH)_2CH_3CO_2$, insoluble. Aluminum phosphate is insoluble in water.

The sulfide of aluminum cannot be prepared in the wet way; the Al_2S_3 prepared in the dry way is completely hydrolyzed by water.

The anhydrous sulfate is insoluble. The "alums" (double sulfates with Alk^+ or NH_4^+), $M^IAl(SO_4)_2\cdot 12H_2O$, i.e., $[M^I(H_2O)_6][Al(H_2O)_6](SO_4)_2$, are less soluble for high-*Z* Alk–see Table 1.2–but melt readily.

The chloride is deliquescent.

Oxonium. The oxide, Al_2O_3, if not too strongly ignited, and $Al_2O_3\cdot aq$ dissolve readily in dilute H_3O^+, but corundum, crystallized Al_2O_3, is insoluble.

Hydroxide. Aqueous OH^- and either Al (releasing H_2) or Al_2O_3 (if not too strongly ignited) readily yield aluminates; one example would be:

$$Al + 2\ OH^- + 4\ H_2O \rightarrow [Al(OH)_5(H_2O)]^{2-} + {}^3/_2\ H_2\uparrow$$

Aqueous OH^- with Al^{3+} precipitates $Al_2O_3\cdot aq$, colorless to grayish-white, gelatinous, insoluble in water, soluble in low or high $c(OH^-)$ to form ions such as $[\{Al(OH)_3\}_2(\mu\text{-}O)]^{2-}$ and $[Al_2(OH)_8]^{2-}$ up to $[Al(OH)_6]^{3-}$ respectively, or, at $< 10\text{-}\mu M\ Al^{III}$, $[Al(OH)_4(H_2O)_2]^-$ etc.

Dioxygen. Pure aluminum is scarcely oxidized in either dry or moist air; the powder, however, is gradually oxidized.

13.2.2 Reagents Derived from the Other 2nd-Period Non-Metals, Boron through Fluorine

Boron species. Solutions of alkali borates precipitate Al^{III} as $Al_2O_3\cdot aq$, due to hydrolysis of the reagent.

Carbon oxide species. Aluminum is attacked by aqueous CO_3^{2-}.

Aqueous CO_3^{2-} and Al^{3+} precipitate gelatinous $Al_2O_3\cdot aq$:

$$2\ Al^{3+} + 6\ CO_3^{2-} + 3\ H_2O \rightarrow Al_2O_3\cdot aq\downarrow + 6\ HCO_3^-$$

sparingly soluble in CO_3^{2-}, but much less so in HCO_3^-. However, passing CO_2 into (alkaline) $[Al(OH)_4]^-$ at 80 °C yields crystalline $Al(OH)_3$.

$BaCO_3$ on digestion in the cold for some time, completely precipitates Al^{III} as the hydrous oxide, mixed with a little basic salt:

$$4\ Al^{3+} + 6\ BaCO_3 \rightarrow 2\ Al_2O_3.aq\downarrow + 6\ Ba^{2+} + 6\ CO_2\uparrow$$

Solutions of $Al(OH)_4^-$ yield a precipitate of $Al_2O_3 \cdot aq$ by careful neutralization with acids, including $CO_2 \cdot aq$.

Cyanide species. Cyanide and Al^{3+} precipitate $Al_2O_3 \cdot aq$.

Some "simple" organic reagents. Fresh $Al_2O_3 \cdot aq$ after dissolution in HCO_2H yields crystals of $Al(OH)(HCO_2)_2 \cdot H_2O$ or, with excess acid, $Al(HCO_2)_3 \cdot 3H_2O$.

Metallic Al is not appreciably attacked by cold, dilute CH_3CO_2H.

Aluminum acetate is decomposed upon boiling, forming the insoluble basic acetate (separation of Fe and Al from others):

$$Al(CH_3CO_2)_3 + 2\ H_2O \rightarrow Al(OH)_2(CH_3CO_2)\downarrow + 2\ CH_3CO_2H$$

The basic acetate is best formed thus: to the solution of Al^{III} add CO_3^{2-} or "$(NH_4)_2CO_3$" sufficient to neutralize any excess acid, but not enough to form a precipitate. Next add an excess of acetate, dilute and boil for some time. Separate while hot, for cooling reverses the reaction.

Oxalates do not precipitate Al^{III}. Precipitated aluminum hydroxide reacts well on boiling with aqueous $HC_2O_4^-$ to give $[Al(C_2O_4)_3]^{3-}$ and, say, $K_3[Al(C_2O_4)_3] \cdot 3H_2O$ on evaporation. This anion, however, unlike those of **d**-block M^{III}, is too labile to be resolved chirally.

Chelators, e.g., oxalic, lactic, tartaric or citric acid or their salts greatly hinder or prevent the precipitation of $Al_2O_3 \cdot aq$ from acidic, neutral or basic solutions. Citrate can be mono-, di- or tri-dentate.

Monocarboxylate ions give $Al(RCO_2)^{2+}$ and $Al(RCO_2)_2^+$ complexes, albeit with less stability.

Reduced nitrogen. Treating aluminum salts with NH_3 in the cold precipitates gelatinous $Al_2O_3 \cdot aq$; when hot, less hydrated, ~AlO(OH). The $Al_2O_3 \cdot aq$ is sparingly soluble in NH_3, much less in NH_4^+. Excess NH_4^+ and aluminate also yield a hydroxide, more compact and washed more readily than that obtained by neutralizing an acidic solution:

$$2\ Al(OH)_4^- + 2\ NH_4^+ \rightarrow Al_2O_3 \cdot aq\downarrow + 2\ NH_3 + 5\ H_2O$$

Aluminum is precipitated from unknown mixtures with Cr^{III} and Fe^{III} as the hydrous oxide by NH_3 in the presence of NH_4^+. It is separated from $Fe_2O_3 \cdot aq$ by warming with OH^-; from $Cr_2O_3 \cdot aq$ by boiling, first with OH^- and H_2O_2 to oxidize $Cr(OH)_4^-$ to CrO_4^{2-}, then with excess NH_4^+, which precipitates $Al_2O_3 \cdot aq$. Many of the confirmatory tests applied to this precipitate use organic compounds.

Oxidized nitrogen. Nitric acid produces passivity with Al, but in the presence of small amounts of some other ions, e.g., Hg^{2+}, it dissolves rapidly, forming NO in concentrated HNO_3, and NH_4NO_3 in the dilute.

Fluorine species. The slightly soluble AlF_3 and other Al^{III} dissolve in aqueous HF or F^- forming $[AlF_m(OH)_n(H_2O)_{6-m-n}]^{(m+n-3)-}$. Concentrated solutions in HF on standing deposit $AlF_3 \cdot 3H_2O$.

13.2.3 Reagents Derived from the 3rd-to-5th-Period Non-Metals, Silicon through Xenon

Phosphorus species. Alkali phosphates, e.g., HPO_4^{2-}, precipitate aluminum phosphate, $AlPO_4$, white, insoluble in H_2O and acetic acid, soluble in H_3O^+ and in OH^- (distinction from $FePO_4$).

Aqueous H_3PO_4 strongly complexes Al^{3+}, especially as $[Al(HPO_4)_3]^{3-}$, but also with polymers and more hydronated ("protonated") species.

Aluminum(III) may be separated from PO_4^{3-} by dissolving in HCl, adding tartaric acid, NH_3 and "magnesia mixture", and digesting some time to precipitate $MgNH_4PO_4$. Phosphate may also be precipitated by Sn^{IV}, and the excess Sn is removed more easily than is tartaric acid.

Arsenic species. Metallic Al with As and H_3O^+ yields AsH_3; in alkalis, As^{III} apparently becomes first As and then AsH_3; As^V is unaffected.

Aluminum(III) is precipitated by alkali arsenites and arsenates, but not by the corresponding acids.

Reduced chalcogens. Sulfane, H_2S, does not precipitate Al^{III} from acidic or neutral solution; from alkaline $Al(OH)_4^-$, H_2S precipitates $Al_2O_3 \cdot aq$ if enough is used to neutralize the alkali (distinction from Zn, which is rapidly precipitated as ZnS from solutions not too alkaline).

$$2\ Al(OH)_4^- + 2\ H_2S \rightarrow Al_2O_3 \cdot aq\downarrow + 2\ HS^- + 5\ H_2O$$

Similar precipitations of $Al_2O_3 \cdot aq$ occur with HS^- and NH_4^+ (forming S^{2-} and NH_3), also from Al^{3+} with HS^- and NH_3 (forming H_2S and NH_4^+), all from the "$(NH_4)_2S$" mixture.

Oxidized chalcogens. From neutral solutions of Al^{III}, $S_2O_3^{2-}$ forms $Al_2O_3 \cdot aq$, S and aqueous SO_2, e.g.:

$$2\ AlSO_4^+ + 3\ S_2O_3^{2-} \rightarrow$$

$$Al_2O_3 \cdot aq\downarrow + 3\ S\downarrow + 3\ SO_2 + 2\ SO_4^{2-}$$

Sulfite also precipitates $Al_2O_3 \cdot aq$ with liberation of SO_2:

$$2\ Al^{3+} + 3\ SO_3^{2-} \rightarrow Al_2O_3 \cdot aq\downarrow + 3\ SO_2$$

Neither of these precipitates Fe, thus separating Al (and Cr) from Fe.

Dilute H_2SO_4 attacks Al slowly, releasing hydrogen; the hot, concentrated acid dissolves it readily, with release of SO_2.

Aluminum, chromium and iron(III) sulfates form double salts, called alums, with the alkali sulfates. Perhaps the best known of this group is the so-called common alum, $KAl(SO_4)_2 \cdot 12H_2O$ (not to be confused with commercial alum, i.e., crude aluminum sulfate). The alums are usually less soluble than their constituent sulfates and may be crystallized on adding a saturated solution of alkali sulfate, especially of NH_4^+ or larger M^+, to a concentrated solution of Al^{III}, Cr^{III} or Fe^{III} sulfate.

Selenate, but no tellurate alums, are quite like the sulfate alums, e.g., $[K(H_2O)_6][Al(H_2O)_6](SeO_4)_2$.

Reduced halogens. Dilute or concentrated HCl, HBr or HI dissolves aluminum readily, with release of H_2, yielding, e.g., $[Al(H_2O)_6]Cl_3$.

Elemental halogens. The metal is attacked by the halogens, forming the corresponding halide.

13.2.4 Reagents Derived from the Metals Lithium through Uranium, plus Electrons and Photons

Oxidation. Metallic Al is oxidized to Al^{III} and it precipitates, as metals from their solutions, Be (only in alkalis), Co, Ni, Pd, Pt, Cu, Ag, Au, Cd, Hg, Tl, Sn, Pb, Bi (incompletely), Se and Te; Fe^{III} is reduced to Fe^{2+}, Sb in acids, e.g., HCl, becomes SbH_3; in alkalis Sb^{III} or Sb^V becomes Sb.

Other reactions. Alkali chromates precipitate Al^{III} as $Al_2O_3 \cdot aq$.

Aqueous $Fe(CN)_6^{4-}$ and Al^{III} produce slowly in the cold, more rapidly upon heating, a white precipitate that gradually turns green.

13.3 Gallium, $_{31}Ga$

Oxidation numbers: (I), (II) and (III), as in Ga^+, $[(-GaCl_3)_2]^{2-}$ and Ga^{3+}.

13.3.1 Reagents Derived from Hydrogen and Oxygen

Water. Metallic gallium is only slightly affected by water at room temperature, but action is vigorous at the boiling point.

Water and $[(-GaX_3)_2]^{2-}$ slowly form H_2 and hydroxo Ga^{III} complexes.

Hydrated Ga^{III} is $[Ga(H_2O)_6]^{3+}$ in oxoanion salts and acidic media.

Treating Ga_2Cl_4 with cold water under, e.g., N_2 precipitates Ga and $Ga_2O_3 \cdot aq$, releases H_2, and provides a few-centimolar Ga^I in solution, which lasts a few hours at 0 °C before decomposing further. This Ga^I easily reduces I_3^-, $HCrO_4^-$, $FeSCN^{2+}$, $[IrCl_6]^{2-}$, etc., but not $[CoCl(NH_3)_5]^{2+}$, $[CoBr(NH_3)_5]^{2+}$ etc., with Ga^{II} as the intermediate in some cases.

Anhydrous GaF_3 is slightly soluble. The nitrate, sulfate, chloride, chlorate, perchlorate, bromide, bromate and iodide are all soluble in water but hydrolyze and polymerize readily. On boiling, basic salts separate. Dilute acids readily dissolve the slightly soluble $Ga(IO_3)_3 \cdot 2H_2O$. The chloride and perchlorate are deliquescent.

Some hot natural waters may contain $[GaCl_4]^-$ etc.

Oxonium. Gallium(III) oxide, Ga_2O_3, is almost insoluble in acids, which precipitate the hydrous trioxide from gallates. On standing this goes to GaO(OH), also formed as crystals at 110 °C from $Ga_2O_3 \cdot aq$.

Acids dissolve Ga_2S_3.

Hydroxide. Gallium dissolves in dilute OH^-, releasing H_2.

Gallium(III) oxide, Ga_2O_3, is difficultly soluble in alkalis. Some products of fusion with alkalis and other procedures are more amenable to further treatment, and they have various structures with formulas such as Li_5GaO_4, $Na_8Ga_2O_7$, KNa_2GaO_3 and $MgGa_2O_4$. The amphoteric hydrous trioxide, much more acidic than $Al_2O_3 \cdot aq$, is readily soluble in OH^-, forming a gallate, $Ga(OH)_4^-$, which polymerizes, but less than Al^{III}.

Gallium is separated from Rth, U, Ti, Fe, In and Tl by using the greater solubility of $Ga_2O_3 \cdot aq$ in OH^-.

Concentrated OH^- dissolves Ga_2S_3 as gallates and thiogallates.

Peroxide, di- and trioxygen. These generally oxidize $Ga^{<III}$ to Ga^{III}.

13.3.2 Reagents Derived from the Other 2nd-Period Non-Metals, Boron through Fluorine

Carbon oxide species. Boiling Ga^{III} with $BaCO_3$ or soluble CO_3^{2-} precipitates GaO(OH).

Cyanide species. Gallium(III) and NCS^- form cations as well as $Ga(NCS)_3 \cdot 3H_2O$ and $Ga(NCS\text{-}\kappa N)_n^{(n-3)-}$, with $n \leq 6$.

Some "simple" organic reagents. Boiling with $CH_3CO_2^-$ precipitates Ga^{III} as GaO(OH).

Hot solutions of $Ga(NO_3)_3$ and $H_2C_2O_4$ yield $Ga_2(C_2O_4)_3 \cdot 4H_2O$. Cationic and anionic complexes up to $[Ga(C_2O_4)_3]^{3-}$, as well as basic ones, are also known.

Lactate, tartrate and citrate form chelates as well, although OH^- displaces the organic anions, giving $Ga(OH)_4^-$.

Reduced nitrogen. A white, gelatinous, gallium hydrous oxide, $Ga_2O_3\cdot aq$, is obtained when a solution of Ga^{III} is treated with NH_3 (tartrates etc. interfere). The product is soluble in excess of the reagent or "$(NH_4)_2CO_3$", but separates again upon boiling.

Oxidized nitrogen. Cold, dilute HNO_3 has little effect on gallium. Gallium with concentrated HNO_3 does not dissolve much in the cold, but at 40–50 °C does so, releasing NO_x, although the surface is sometimes passivated, perhaps surprisingly for a liquid (at those temperatures) metal. It dissolves slowly in aqua regia.

Fluorine species. Gallium is quite slow to dissolve even in 30-M HF, but it then yields H_2 and $GaF_3\cdot 3H_2O$ after evaporation. The solutions may contain $[GaF_n(H_2O)_{6-n}]^{(n-3)-}$ with n up to 4, at least.

13.3.3 Reagents Derived from the 3rd-to-5th-Period Non-Metals, Silicon through Xenon

Phosphorus species. Gallium(III) and $H_2PO_4^-$ form the very insoluble $GaPO_4\cdot 2H_2O$. Basic phosphates arise above pH 7. Acidic media give at least $Ga(H_nPO_4)^{n+}$ with n up to 3. Diphosphate yields $Ga_4[P_2O_7]_3$ and complex anions.

Arsenic species. With As^V, neutralizing excess acid gives $GaAsO_4\cdot 2H_2O$.

Reduced chalcogens. Gallium(III) alone is not precipitated by H_2S, but in a weakly acidic or ammoniacal solution containing Mn, Ag, Zn or As^{III}, Ga_2S_3 coprecipitates completely. Gallium is separated (non-precipitating) from Cu, Hg, Pb, As^V etc. by H_2S in 3-dM HCl solution.

Gallium(III) and $SeCN^-$ give rise to $[Ga(SeCN)_4]^-$ among others.

Oxidized chalcogens. In neutral or slightly acidic solution, Ga^{III} and HSO_3^- give a white precipitate. Indium(III) and much Zn^{II} interfere.

Gallium dissolves very slowly in either dilute or concentrated H_2SO_4.

Adding H_2SO_4 prevents volatilization of the otherwise volatile $GaCl_3$.

Reduced halogens. Gallium dissolves slowly in either 1-M or 12-M HCl, producing $[Ga_2Cl_6]^{2-}$, i.e., $[(-GaCl_3)_2]^{2-}$, and H_2. Rather similar are HBr and HI. The fresh $[Ga_2X_6]^{2-}$ ions are quite strong reductants.

Gallium(III) plus HX or other X^- form $[GaX_4]^-$, but dilution favors hydrated cations, up to $[Ga(H_2O)_6]^{3+}$ or hydroxo anions.

In some analyses Ga^{III} and Fe^{III} are separated from other species by extracting the former two from a 6-M HCl solution with ether. The Fe^{III} is reduced to Fe^{2+}, e.g., by means of Hg, and the Ga^{III} is extracted again.

Elemental and oxidized halogens. Aqueous halogens easily oxidize Ga^{I} and $[(-GaX_3)_2]^{2-}$, e.g., to $[GaX_4]^-$.

Gallium is recovered from zinc flue dust by dissolving the dust in much HCl, adding ClO_3^-, distilling out the $GeCl_4$, and leaving $[GaCl_4]^-$.

Hot, concentrated $HClO_4$ (caution!) quickly dissolves Ga, and the very deliquescent salt crystallizes on cooling, apparently arising two ways:

$$7\ Ga + 24\ ClO_4^- + 24\ H_3O^+ + 6\ H_2O \rightarrow$$

$$7\ [Ga(H_2O)_6](ClO_4)_3\downarrow + {}^3/_2\ Cl_2\uparrow$$

$$8\ Ga + 27\ ClO_4^- + 24\ H_3O^+ + 12\ H_2O \rightarrow$$

$$8\ [Ga(H_2O)_6](ClO_4)_3\downarrow + 3\ Cl^-$$

13.3.4 Reagents Derived from the Metals Lithium through Uranium, plus Electrons and Photons

Oxidation. Aqueous Ag^+ and Hg^{2+} easily oxidize $[(-GaX_3)_2]^{2-}$ to Ga^{III}.

Reduction. Fractional cathodic electrolysis of a slightly acidic solution of the sulfates of Zn, Ga and In separates the In almost completely from the Ga, and the latter metal entirely from the Zn.

Metallic gallium is probably best prepared by electrolyzing an alkaline gallate solution.

Other reactions. Aqueous $[Fe(CN)_6]^{4-}$ precipitates, from a 4-M HCl solution of Ga^{III}, white $Ga_4[Fe(CN)_6]_3$. This characteristic reaction can detect < 2 µmol of gallium, while the chloro-complexes of other metals, possibly in a higher c(HCl), prevent their precipitation.

13.4 Indium, $_{49}$In

Oxidation numbers: (I), (II) and (III), as in In^+, $[(-InCl_3)_2]^{2-}$ and In_2O_3.

13.4.1 Reagents Derived from Hydrogen and Oxygen

Water. Indium does not decompose water, even at the boiling point.

Hydrated In^{III} is $[In(H_2O)_6]^{3+}$ in oxoanion salts and acidic media.

Anhydrous InF_3 is insoluble, but $InF_3 \cdot 3H_2O$ and In^{III} nitrate, sulfate, alums and heavier halides are soluble. They hydrolyze readily above pH 3, forming slightly soluble basic salts. The sulfate is very hygroscopic, the chloride deliquescent.

Oxonium. Indium dissolves in H_3O^+. The black monoxide, InO, is slowly soluble in acids, but light yellow In_2O_3 is readily soluble.

Hydroxide. Indium does not dissolve in OH^-.

Treating In^{III} with OH^- precipitates gelatinous, white $In_2O_3 \cdot aq$, less acidic than $Ga_2O_3 \cdot aq$, slightly soluble, especially if fresh, in high $c(OH^-)$, apparently up to $[In(OH)_6]^{3-}$, reprecipitated on boiling.

13.4.2 Reagents Derived from the Other 2nd-Period Non-Metals, Boron through Fluorine

Carbon oxide species. Barium carbonate or soluble CO_3^{2-} precipitates In^{III} as $In_2(CO_3)_3$, insoluble in CO_3^{2-}, soluble in "$(NH_4)_2CO_3$", but reprecipitated if the solution is boiled.

Cyanide species. Aqueous CN^- with In^{3+} forms a white precipitate of $In(CN)_3$, soluble in excess of the reagent. If the resulting solution is diluted and boiled, $In_2O_3 \cdot aq$ separates.

Some "simple" organic reagents. The $In_2O_3 \cdot aq$, but not the metal, dissolves in cold CH_3CO_2H. When boiled with $CH_3CO_2^-$, In^{III} forms a basic indium(III) acetate only slightly soluble in water.

Indium dissolves somewhat in $H_2C_2O_4$, but this and $C_2O_4^{2-}$ precipitate In^{III}, not too dilute, as $In_2(C_2O_4)_3 \cdot 6H_2O$, insoluble in NH_3.

Reduced nitrogen. Treating In^{III} with NH_3 forms a gelatinous white precipitate of $In_2O_3 \cdot aq$, insoluble in NH_3.

Oxidized nitrogen. Indium dissolves in HNO_3 as In^{3+}.

Fluorine species. Fluoride and In^{3+} give rise to $[InF_n(H_2O)_{6-n}]^{(n-3)-}$.

13.4.3 Reagents Derived from the Heavier Non-Metals, Silicon through Xenon

Phosphorus species. Aqueous HPO_4^{2-} gives a voluminous white, practically insoluble, orthophosphate with In^{III}, but salts of $[In(PO_4)_2]^{3-}$, and a complex with $H_2PO_4^-$, are known.

Reduced chalcogens. Sulfane, H_2S, added to a neutral or acetic-acid solution of In^{III}, precipitates a yellow In_2S_3, soluble in H_3O^+ or S^{2-}, partly soluble in hot $(NH_4)_2S_x$, forming a white residue. Cooling gives a voluminous white precipitate. Sulfane passed into an alkaline solution, or "$(NH_4)_2S$" added to a neutral solution of In^{III}, forms white In_2S_3.

With In^{III} the SCN^- ion has no visible effect.

Oxidized chalcogens. Thiosulfate ion, added to neutral In^{III}, precipitates indium sulfite; acids give the sulfide. Neither reaction is complete.

An important source of In is zinc flue dust. The formation of a basic sulfite is an interesting step in one way to separate In from Fe etc.

One first treats a sample with not quite enough HCl to dissolve all of the metal. After some time, a spongy deposit separates, containing Fe, Cu, Cd, In, Pb etc. It is dissolved in HNO_3 and evaporated with H_2SO_4 to remove any Pb. The solute is heated to boiling and then made slightly alkaline with NH_3, which precipitates $Fe_2O_3 \cdot aq$ and $In_2O_3 \cdot aq$. After separation and washing with NH_4NO_3 the residue is dissolved in the minimum amount of HCl. This solution is barely neutralized with NH_3 and, after adding excess HSO_3^-, boiled 15 to 20 minutes in order to precipitate a white $\sim In_4(OH)_6(SO_3)_3 \cdot 5H_2O$. This may then be dissolved in H_2SO_4 and reprecipitated as $In_2O_3 \cdot aq$ with NH_3, repeating the process several times to achieve 99.5 % purity.

Indium dissolves in H_2SO_4 as In^{III}. Light-yellow In_2O_3 is readily soluble, especially in hot, dilute H_2SO_4. Sulfato complexes are also found, and some solid compounds are $NH_4In(SO_4)_2 \cdot 4H_2O$ and $H_3OIn(SO_4)_2 \cdot 4H_2O$, with both containing $[trans\text{-}In(\eta^2\text{-}SO_4)_2(H_2O)_2]^-$.

Selenate and In^{3+} give crystals of $In_2(SeO_4)_3 \cdot 8H_2O$, and we have $NH_4[trans\text{-}In(\eta^2\text{-}SeO_4)_2(H_2O)_2] \cdot 2H_2O$ rather like the sulfate; tellurite and tellurate also form double or complex salts.

Reduced halogens. Indium dissolves in HX as In^{III}, including $[InCl_5(H_2O)]^{2-}$, $[cis\text{-}In(Br,Cl)_4(H_2O)_2]^-$ and $[InI_4]^-$ (I^- being the largest).

Elemental halogens. Indium dissolves in aqueous X_2 as In^{III}.

13.4.4 Reagents Derived from the Metals Lithium through Uranium, plus Electrons and Photons

Oxidation. Aqueous In^I of a few-decimolar concentration may be prepared electrolytically or from In_{Hg} (amalgam) and Ag(1+) in CH_3CN and is stable for hours except in strong acids or air.

Indium(I) reduces $[Co(NH_3)_5X]^{2+}$, where X = C_2O_4H, N_3, NCS, Cl, Br or I etc., going through the metastable In^{II}.

Reduction. Metallic In is readily obtained by electrolyzing a 3-M acidic sulfate solution containing 2-M In^{III} and 1-M citric acid. A Pt anode, In or Fe cathode, and a current density of 2 A/dm^2 yield a thick, compact deposit of the metal.

Other reactions. Some vanadates precipitate In^{III}.

Neutral In^{III} plus CrO_4^{2-} form a yellow precipitate, but $[Cr_2O_7]^{2-}$ does not. Aqueous MoO_4^{2-} precipitates $In_2(MoO_4)_3 \cdot 2H_2O$.

Aqueous $[Fe(CN)_6]^{4-}$ precipitates In^{III} as white ~$In_4[Fe(CN)_6]_3$, soluble (much more than the Ga^{III}) in HCl, but $[Fe(CN)_6]^{3-}$ has no visible effect.

13.5 Thallium, $_{81}$Tl (and Ununtrium, $_{113}$Uut)

Oxidation numbers for Tl: (I), (II) and (III), as in TlOH, "thallous" hydroxide, $[(-TlCl_3)_2]^{2-}$ and Tl_2O_3, "thallic" oxide.

Relativistic quantum mechanics predicts a higher electronegativity for Uut (temporarily named ununtrium for the next element, recently synthesized, in this group) than for Tl, In, Ga or even Al, but also a surprising stability for $UutF_6^-$ with oxidation number (V), but not $UutF_5$ or perhaps Uut^{3+}. In addition, Uut^+ may be more like Ag^+ than like Tl^+.

13.5.1 Reagents Derived from Hydrogen and Oxygen

Dihydrogen. Copper(2+) catalyzes the reduction of Tl^{3+}; see **1.0.4**:

$$Tl^{3+} + H_2 + 2\ H_2O \rightarrow Tl^+ + 2\ H_3O^+$$

Water. Thallium(I) oxide reacts with H_2O to form yellow TlOH, which dissolves up to 1.5 M at 20 °C. Thallium(III) oxide is insoluble in H_2O and is only slightly affected when boiled with it.

Thallium(I) resembles the higher-Z Alk^+, in that (omitting hydration) TlOH, TlF, Tl_2CO_3, $TlCH_3CO_2$, $TlNO_2$, $TlNO_3$, Tl_2SO_4 and $Tl_3[Fe(CN)_6]$ are soluble, but $Tl_3[Co(NO_2)_6]$ and Tl_2PtCl_6 are insoluble. It also resembles Pb^{II} in that (Tl_3PO_4 and Tl_3AsO_3 slightly), Tl_2S, Tl_2CS_3, Tl_2CrO_4 and $Tl_4[Fe(CN)_6]$ are not soluble, TlSCN is slightly soluble, the salts of Cl^-, Br^- and I^- are rather insoluble in cold water but more soluble in hot water, and $TlClO_3$ and $TlClO_4$ are soluble. Additionally, Tl_2SO_3 and $Tl_2C_2O_4$ are moderately soluble.

The thallium(III) halides, nitrate and sulfate are soluble in water but hydrolyze much more readily than the corresponding thallium(I) salts, giving rise at least to $TlOH^{2+}$ and $Tl(OH)_2^+$. The acidity of $[Tl(H_2O)_6]^{3+}$, $pK_a = 1.2$, is greater than that for the hydrated B^{III}, Al^{III}, Ga^{III} or In^{III}, due to relativity, even though Tl^{3+} is the largest. Strangely, many writers say that such a number (1.2) is near "unity" but never that 2 is "duality".

Hot H_2O reduces Tl^{III} to Tl^+.

Oxonium. The dissolution of Tl in HCl is slow. Thallium(III) oxide is readily soluble in the common acids when freshly precipitated, but after drying, it reacts with HCl to release Cl_2, and with H_2SO_4 to give O_2.

Hydroxide. A dilute solution of Tl^+ does not precipitate OH^-.

Air-free Tl_2SO_4 and $Ba(OH)_2$ provide aqueous TlOH (and $BaSO_4\downarrow$).

Treating Tl^{III} with a base precipitates a feebly basic, brown $Tl_2O_3\cdot aq$ that is very insoluble in water and in excess reagent.

Peroxide. Thallium(1+) is oxidized to Tl^{III} by Na_2O_2. Thallium(I) hydroxide is readily oxidized by H_2O_2.

Di- and trioxygen. Thallium is oxidized by O_2 in water, using ethanol to help separate the yellow, crystalline TlOH. Heating TlOH above 100 °C forms black Tl_2O. Oxygen easily changes TlOH to Tl_2O_3.

Paper soaked in TlOH solution has been suggested to detect ozone, because a very small amount of O_3 will turn the paper brown (Tl_2O_3).

13.5.2 Reagents Derived from the Other 2nd-Period Non-Metals, Boron through Fluorine

Carbon oxide species. Thallium(I) hydroxide is about as basic as NaOH. It rapidly absorbs CO_2 from the air. A dilute solution of Tl^+ gives no precipitate with CO_3^{2-}, which, however, in air, precipitates brown $Tl_2O_3\cdot aq$, insoluble in excess.

Cyanide species. Aqueous CN^- does not dissolve TlI; it yields TlCN but no complex anions from the acetate. The compound $Tl[Tl(CN)_4]$, from Tl^+, Tl^{3+} and CN^-, is one of various examples of mixed oxidation states that may appear misleadingly [if written as $Tl(CN)_2$ in this case], to be derived from Tl^{II}. (Other examples are an oxalate, sulfate and selenate.)

Thallium(+) forms both TlOCN and TlNCS, and very weak anionic complexes from the latter.

Some "simple" organic reagents. Acetic acid and TlOH or Tl_2CO_3 give $TlCH_3CO_2$. With Tl^{III} we get Tl^{III} acetate. The mixed compound $Tl[Tl(CH_3CO_2)_4]$, is derived from Tl^+, Tl^{3+} and $CH_3CO_2^-$.

Reduced nitrogen. Ammonia neither precipitates Tl^+ nor dissolves TlI. Thallium(III) forms brown $Tl_2O_3\cdot aq$, insoluble in excess.

Hydroxylamine reduces Tl^{III} to Tl^I.

Thallium(+) and N_3^- give a yellow, not extremely explosive, TlN_3.

Oxidized nitrogen. Aqueous Tl_2SO_4 and $Ba(NO_2)_2$ yield $TlNO_2$ (and $BaSO_4\downarrow$). Nitrite appears to complex Tl^{III} as $Tl(NO_2\text{-}\kappa N)_n^{(3-n)+}$ with $n < 4$ but also to reduce it thus:

$$Tl(NO_2)_2^+ + 2\ H_2O \rightarrow Tl^+ + NO_3^- + HNO_2 + H_3O^+$$

The best solvent for Tl is HNO_3, which forms mainly Tl^+ with perhaps a little Tl^{3+}. Nitric acid also dissolves TlOH and Tl_2CO_3 as Tl^+. Concentrated HNO_3 and Tl_2O_3 give $Tl(NO_3)_3{\cdot}3H_2O$, a strong oxidant that easily decomposes to $Tl_2O_3{\cdot}2H_2O$.

Fluorine species. Aqueous F^- and Tl^+ give $TlF_n^{(n-1)-}$ with $n \le 4$.

13.5.3 Reagents Derived from the 3rd-to-5th-Period Non-Metals, Silicon through Xenon

Phosphorus species. Aqueous $H_2PO_4^-$ precipitates Tl^+ only partly:

$$3\ Tl^+ + 3\ H_2PO_4^- \leftrightarrows Tl_3PO_4\downarrow + 2\ H_3PO_4$$

but HPO_4^{2-} or PO_4^{3-} quickly gives white, crystalline Tl_3PO_4.

Thallium(III) oxide reacts with H_3PO_4 to precipitate $TlPO_4{\cdot}2H_2O$.

Arsenic species. Arsenite reduces Tl^{III} to Tl^+.

Reduced chalcogens. Hydrogen sulfide does not precipitate thallium from strongly acidic solution, but separations from other ions based on this fact have little value because the thallium is carried down with the sulfide formed. In acetic-acid, neutral, or alkaline solution Tl^+ may be completely precipitated as Tl_2S. The precipitate is rapidly oxidized to Tl_2SO_4 and/or $Tl_2S_2O_3$, Tl_2O or Tl_2O_3 on exposure to air. It is practically insoluble in OH^-, CO_3^{2-}, CN^-, NH_3 and "$(NH_4)_2S$"; slightly soluble in HCl and acetic acid; readily soluble in HNO_3.

Thallium(III) sulfide is not obtained when Tl^{III}, in either acidic or alkaline solution, is treated with a sulfide, but Tl^+ and sulfur are formed.

Thallium(1+) reacts with SCN^- to give a white crystalline precipitate of TlSCN (≡ TlNCS), soluble in excess, and TlSeCN from $SeCN^-$.

Oxidized chalcogens. When Tl^+ is treated with $S_2O_3^{2-}$, a white precipitate is formed. It is not affected by OH^-, but addition of an acid is said to produce Tl_2S. Thiosulfate dissolves TlI.

For Tl with aqueous SO_2, action is slow and Tl_2SO_3 is formed; this arises also from Tl_2CO_3 plus SO_2.

Sulfur dioxide reduces Tl^{III} to Tl^+.

Dilute H_2SO_4 and Tl slowly produce Tl_2SO_4; the concentrated acid readily effects solution. Thallium(I) sulfate forms a series of alums similar to those of the alkali sulfates.

Thallium(III) oxide dissolves slowly in dilute H_2SO_4, finally giving $Tl_2(SO_4)_3 \cdot 7H_2O$; OH^- then forms $Tl(OH)SO_4 \cdot 2H_2O$. The warm acid causes decomposition to Tl^+ and O_2.

Reduced halogens. Thallium(1+), when treated with Cl^-, forms TlCl, white, somewhat curdy, becoming compact on standing. The product, like $PbCl_2$, is soluble in hot water, from which cooling gives crystals.

Cold HCl readily forms Tl^{III} chloride and complexes from Tl_2O_3. Heating $TlCl_3$ gives TlCl and Cl_2. The reaction is reversible, which makes possible a separation from Ag^+.

The chemical reactions of the bromides and iodides resemble those of the chloride. Light darkens them all. The periods 3–5 anionic complexes of Tl^+ in water are less stable in the series $Cl^- > Br^- > I^-$. The similarities of Tl to Ag, and some resemblances of In to Cu, Sn to Zn, Pb to Cd, Sb to Ga, and Bi to In, albeit more in physical than in chemical properties, are reminiscent of the knight's move in chess [1].

Thallium tribromide apparently decomposes thus:

$$2\ TlBr_3 \rightarrow \text{yellow } Tl[TlBr_4] + Br_2 \text{ by itself}$$

$$4\ TlBr_3 \rightarrow \text{red } Tl_3[TlBr_6]\downarrow + 3\ Br_2 \text{ in water}$$

Thallium(I) iodide, when freshly precipitated, is yellow, but on standing becomes distinctly green; it is also only slightly soluble in $S_2O_3^{2-}$ (distinction from PbI_2). The product of Tl^{3+} and I^-, black TlI_3, is actually $Tl^+[I_3]^-$ in the solid, not $Tl^{3+}(I^-)_3$.

Elemental and oxidized halogens. Thallium and Tl^+, when treated with Cl_2 as the gas, or from nitric-hydrochloric acid (aqua regia), yield Tl^{III}. Treating TlCl with Cl_2 or TlBr with Br_2 results in $TlX_3 \cdot 4H_2O$.

Bromine and $Tl(CH_3CO_2)$ form stable $TlBr_2(CH_3CO_2)$. From Tl_2SO_4 we get stable $Tl^I[Tl^{III}Br_2SO_4]$.

Aqueous ClO^- oxidizes Tl^+ to Tl^{III} in the cold.

Dithallium trioxide dissolves in $HClO_3$, $HBrO_3$, HIO_3 or $HClO_4$, forming the corresponding Tl^{III} salt.

Aqueous Tl_2SO_4 and either $Ba(ClO_3)_2$ or $Ba(BrO_3)_2$ are a good source of $TlClO_3$ or $TlBrO_3$ respectively (and $BaSO_4\downarrow$).

Iodate gives, with Tl^+, a white precipitate of $TlIO_3$.

Perchloric acid reacts with TlOH or Tl_2CO_3 and forms $TlClO_4$.

13.5.4 Reagents Derived from the Metals Lithium through Uranium, plus Electrons and Photons

Oxidation. Silver(I) catalyzes the very slow oxidation of Tl^+ by Ce^{IV}:

$$Ce^{IV} + Ag^{I} \leftrightarrows Ce^{III}\ Ag^{II}$$

$$Tl^{I} + Ag^{II} \rightarrow Tl^{II} + Ag^{I}$$

$$\underline{Tl^{II} + Ce^{IV} \rightarrow Tl^{III} + Ce^{III}\ (\text{fast})}$$

$$2\ Ce^{IV} + Tl^{I} \rightarrow 2\ Ce^{III} + Tl^{III}$$

Thallium(1+) is oxidized to Tl^{III} by MnO_4^- and PbO_2, and by anodic treatment, which can give, e.g., thallium(III) perchlorate.

Anodes form: Tl_2O_3, dark brown to black, from Tl^+; $Tl(O_2)$, a superoxide, from Tl_2SO_4; or Tl^{3+} in $HClO_4$ via Tl_2^{4+}. The $Tl(O_2)$ is insoluble in H_2O, OH^- and dilute H_3O^+, but dilute HCl releases O_2.

Light (254 nm), Tl^I, Cl^- and O_2 give Tl^{III} and H_2O_2, destroying the fluorescence of the unoxidized solution:

$$TlCl + O_2 + 2\ H_3O^+ + 3\ Cl^- + \gamma \rightarrow [TlCl_4]^- + H_2O_2 + 2\ H_2O$$

Reduction. Thallium(III) is reduced to Tl^+ by Fe^{2+}, $SnCl_2$, and so on. Metallic Mg, Zn, or Al will reduce various thallium species to Tl.

The two-electron oxidant Tl^{III} forms Tl^I and CrO^{2+} from Cr^{2+}.

Light and Tl^{3+} yield Tl^+ and O_2.

Other reactions. Thallium(1+) gives, with CrO_4^{2-}, a yellow precipitate of Tl_2CrO_4. It yields $Tl_4[Fe(CN)_6]\cdot 2H_2O$ from $[Fe(CN)_6]^{4-}$. Soluble Prussian Blue, non-toxic $KFe[Fe(CN)_6]$, precipitates toxic Tl^+ as $TlFe[Fe(CN)_6]$ and prevents absorption from the digestive tract. Thallium(1+), unlike **d**-block M^+ and M^{2+}, is not precipitated by $[Co(CN)_6]^{3-}$, but it, rather like K^+, precipitates $[Co(NO_2)_6]^{3-}$ as a light-red $Tl_3[Co(NO_2)_6]$. It also, again compare K^+, precipitates $[PtCl_6]^{2-}$ as a pale-orange $Tl_2[PtCl_6]$.

The reaction of $TlCl_3$ with $AgClO_4$ provides $Tl(ClO_4)_3$ (and $AgCl\downarrow$).

Depending on conditions, TlCl and $TlCl_3$ form $Tl[TlCl_4]$ or $Tl_3[TlCl_6]$. Also known are $Tl[TlBr_4]$ and $Tl_3[TlBr_6]$.

Reference

1. Laing M (1999) Educ Chem 36:160

Bibliography

See the general references in the Introduction, and some more-specialized books [2–21]. Some articles in journals discuss: new thallium chemistry [22]; aqueous aluminates, silicates and aluminosilicates [23]; indium's lower oxidation states [24]; boron electrochemistry [25]; and indium complexes [26].

2. Davidson MG, Hughes AK, Marder TB, Wade K (2000) Contemporary boron chemistry. Royal Society of Chemistry, Cambridge
3. King RB (ed) (1999) Boron chemistry at the millennium. Elsevier, Amsterdam
4. Downs AJ (ed) (1993) Chemistry of aluminium, gallium, indium and thallium. Blackie, London
5. Robinson GH (ed) (1993) Coordination chemistry of aluminium. VCH, New York
6. Hart DL (ed) (1990) Aluminum chemicals science and technology handbook. American Ceramic Society, Washington
7. Housecroft CE (1990) Boranes and metalloboranes: structure, bonding, and reactivity. Horwood, Chichester
8. Liebman JF, Greenberg A, Williams RE (eds) (1988) Advances in boron and the boranes: a volume in honor of Anton B. Burg. VCH, New York
9. Muetterties EL (ed) (1975) Boron hydride chemistry. Academic, San Diego
10. Greenwood NN, Thomas BS (1973) The chemistry of boron. Pergamon, Oxford
11. Lee AG (1971) The chemistry of thallium. Elsevier, Amsterdam
12. Dymov AM, Savostin AP (1970) Schmorak J (trans) (1970) Analytical chemistry of gallium. Ann Arbor Science, Ann Arbor
13. Muetterties EL (ed) (1967) The chemistry of boron and its compounds. Wiley, New York
14. Sheka IA, Chaus IS, Mityureva TT (1966) The chemistry of gallium. Elsevier, Amsterdam
15. Adams RM (ed) (1964) Boron, metallo-boron compounds and boranes. Wiley, New York
16. Nemodruk AA, Karalova ZK (1964) Kondor R (trans) (1969) Analytical chemistry of boron. Ann Arbor-Humphrey, Ann Arbor
17. Lipscomb WN (1963) Boron hydrides. Benjamin Cummings, San Francisco
18. Busev AI (1962) Greaves JT (trans) (1962) The analytical chemistry of indium. Macmillan, New York
19. Korenman IM (1960) Lerman Z (trans) (1963) Analytical chemistry of thallium. Ann Arbor-Humphrey, Ann Arbor
20. Ludwick MT (1959) Indium. Indium Corp of America, New York
21. Kemp PH (1956) The chemistry of borates. Borax Ltd, Valencia, CA
22. Glaser J (1995) Adv Inorg Chem 43:1
23. Swaddle TW, Salerno J, Tregloan PA (1994) Chem Soc Rev 23:319
24. Tuck DG (1993) Chem Soc Rev 22:269
25. Morris JH, Gysling HJ, Reed D (1985) Chem Rev 85:51
26. Tuck DG (1975) in Lippard SJ (ed) Prog Inorg Chem 19:243

14 Carbon through Lead, the Tetrels

14.1 Carbon, $_6C$

Oxidation numbers in the simplest compounds: (–IV), (–II), (0), (II), (IV), as in CH_4, CH_3OH, CH_2O, CO and CO_2. Most of these, however, are well described with the older concept of valence, usually four. Still, oxidation number provides one needed criterion for sequencing here.

14.1.1 Reagents Derived from Hydrogen and Oxygen

Water. Carbon monoxide is slightly soluble in H_2O, 1.3 mM at 8 °C.

Hydrogen cyanide is completely miscible with water. The cyanides of Alk^+, most Ae^{2+}, Au^{III} and Hg^{II} are soluble; $Ba(CN)_2$ is sparingly soluble. The solutions react alkaline.

Aqueous HCN (or CN^-) hydrolyzes slowly, forming ammonium formate, more readily in the light:

$$HCN + 2\,H_2O \rightarrow HCO_2^- + NH_4^+$$

Warming with a dilute acid or alkali yields, as expected, HCO_2H and NH_4^+, or HCO_2^- and NH_3, respectively.

Acetic acid, CH_3CO_2H, is completely miscible with water. Its salts are all readily soluble except that the silver and dimercury(I) salts are sparingly soluble. Certain basic salts, such as those of Fe^{III} or Al^{III}, are insoluble. Many acetates are soluble in ethanol.

Anhydrous $H_2C_2O_4$ is hygroscopic; it has been recommended as a drying agent for certain work, but its solubility in H_2O is only 1.06 M at 20 °C. Oxalates of the alkalis plus Be^{2+} and Fe^{3+} are soluble in H_2O. Nearly all others are insoluble to slightly soluble.

Water slowly hydrolyzes cyanogen, $(CN)_2$, to $[-(CO)(NH_2)]_2$, known as "oxamide", and then to $(NH_4)_2C_2O_4$.

Water dissolves CO_2 up to 3 cM at 25 °C, but only ~ 0.05 mM of this is H_2CO_3, which is weakly acidic to litmus, and the CO_2 to H_2CO_3 equilibrium is slow. Surprisingly many of the best compilations show significant internal discrepancies on the following pK values. Just one example is 2.8 for $[H_2CO_3]/[CO_2]$; 3.88 for $[H_3O^+][HCO_3^-]/[H_2CO_3]$; and 6.35 for $[H_3O^+][HCO_3^-]/[CO_2]$, where $2.8 + 3.88 \neq 6.35$. Worse, many others confuse $[H_3O^+][HCO_3^-]/[H_2CO_3]$ with $[H_3O^+][HCO_3^-]/$

$[CO_2]$, even recently. Finally anyway, the p*K* for $[H_3O^+][CO_3^{2-}]/[HCO_3^-]$ is 10.33. We state the first-ionization equilibria thus (omitting $[H_2O]$ from the p*K*):

$$CO_2 + 2\ H_2O \leftrightarrows H_2CO_3 + H_2O \leftrightarrows H_3O^+ + HCO_3^-$$

The alkali carbonates are soluble in H_2O, the hydrogencarbonates less so than the normal salts; other carbonates are insoluble to slightly soluble. The presence of some other salts, especially of NH_4^+, prevents the precipitation of some carbonates, notably $MgCO_3$, by forming HCO_3^-. Many of the carbonates are soluble in H_2O saturated with CO_2, again forming HCO_3^-, as in the dissolution of limestone to form caves and then stalagmites and stalactites by evaporation:

$$CaCO_3 + H_2O + CO_2 \leftrightarrows Ca^{2+} + 2\ HCO_3^-$$

Boiling also removes the excess CO_2, thus reprecipitating the carbonate, as in tea kettles and water heaters.

Water hydrolyzes cyanate, and boiling hastens it:

$$2\ NCO^- + 3\ H_2O \rightarrow 2\ NH_3\uparrow + CO_2\uparrow + CO_3^{2-}$$

Hydrogen isocyanate in acidic solutions goes faster:

$$HNCO + H_3O^+ \rightarrow NH_4^+ + CO_2\uparrow$$

The cyanates of the alkali metals are soluble in water; most of the others are insoluble to slightly soluble. When freshly prepared, solutions of the alkali cyanates react neutral to phenolphthalein. They gradually decompose on standing:

$$NCO^- + 2\ H_2O \rightarrow NH_3 + HCO_3^-$$

$$NCO^- + NH_4^+ \leftrightarrows CO(NH_2)_2 \text{ (urea)}$$

In 1-dM aqueous solution NH_4NCO is 92 % converted to $CO(NH_2)_2$.

Calcium cyanate in water at 80 °C forms $CaCO_3$, NH_4^+ and some urea:

$$NCO^- + Ca^{2+} + 2\ H_2O \rightarrow NH_4^+ + CaCO_3\downarrow$$

Silver cyanate is a white solid, soluble in water to the extent of 0.5 mM at 22 °C, readily soluble in NH_3.

Cyanamide can be made from the commercial calcium salt by hydrolysis, followed with H_2SO_4:

$$2\ CaCN_2 + 2\ H_2O \rightarrow 2\ HCN_2^- + Ca^{2+} + Ca(OH)_2\downarrow$$

$$HCN_2^- + Ca^{2+} + HSO_4^- \rightarrow H_2CN_2 + CaSO_4\downarrow$$

It is toxic to the skin. Acid or much alkali hydrolyzes it further to urea, $CO(NH_2)_2$, but a mild alkali gives a dimer, "cyanoguanidine", $H_2NC(=NH)NHCN$, neutral and slightly soluble. Cyanamide with H_2S yields thiourea, $CS(NH_2)_2$, but HCl forms a dihydrochloride.

Oxonium. The hydrides of **d**-block metals with simple carbon ligands have aqueous acidities (although not of H-C bonds) listed together here for easy comparisons as pK_a, mostly at 25 °C, where $K_a = [H_3O^+][X^-]/[HX]$:

Gr. 5: $[HV(CO)_6]$ strong
Gr. 7: $[HMn(CO)_5]$ 7.1; $[HRe(CO)_5]$ very weak
Gr. 8: $[H_2Fe(CO)_4]$ 4.0; $[HFe(CO)_4]^-$ ~ 12.7 (at 20 °C)
Gr. 9: $[HCo(CO)_4]$ strong; $[HCo(CN)_5]^{3-}$ ~ 20

Strong acids transpose acetates, forming CH_3CO_2H.

The simple cyanides are transposed by H_3O^+ and fairly strong acids, more or less readily, liberating HCN, which escapes from a concentrated or hot solution. For the alkali and alkaline-earth metals even aqueous CO_2 is effective. Free CN^- is detected in the presence of $[Fe(CN)_6]^{n-}$ by slightly acidifying the sample, warming, and passing a relatively inert gas (H_2, N_2, CO_2 or Ar) through the solution. The hexacyanoferrates(3− or 4−) do not release HCN under 80 °C. Any HCN is collected in water or an alkali for possible examination.

Aqueous HCN is a weak acid, scarcely reddening litmus. The odor is characteristic, somewhat resembling that of bitter almonds. Although HCN and the cyanides are indeed very poisonous, their toxicity is popularly exaggerated, in comparison with that of, e.g., H_2S, although the stronger odor of the latter gives more warning.

Oxalates are readily transposed by an excess of the strong acids:

$$CaC_2O_4.H_2O + 2\,H_3O^+ \leftrightarrows Ca^{2+} + H_2C_2O_4 + 3\,H_2O$$

An acid and a carbonate readily give CO_2:

$$CaCO_3 + 2\,H_3O^+ \rightarrow Ca^{2+} + CO_2\uparrow + H_2O$$

In fact, carbonic acid, forming CO_2, is completely displaced from all carbonates by all stronger acids, e.g., $H_2C_2O_4$, HNO_3, H_3PO_4, H_2SO_4, HCl, $HClO_3$, and even by H_2S for metals forming insoluble sulfides.

The decomposition of carbonates by acids is usually attended by marked effervescence of gaseous CO_2. With normal carbonates in the cold, adding a small amount of acid (up to neutralizing half the base) does not cause effervescence, because a hydrogencarbonate is formed:

$$CO_3^{2-} + H_3O^+ \rightarrow HCO_3^- + H_2O$$

When there is much free alkali present (as in testing caustic alkalis for slight admixtures of carbonate) perhaps no effervescence will be obtained, for by the time all of the alkali is neutralized there is enough water present to dissolve the small amount of gas liberated. If, however, the alkali solution is added to the acid dropwise, so that the latter is constantly in excess, a fairly small amount of carbonate will give a perceptible effervescence. The effervescence of CO_2 is distinguished from that of H_2S or SO_2 by the lack of odor, and from that of H_2 by the latter's flammability. We note that CO_2 is also released on adding concentrated H_2SO_4 to oxalates (along with CO) or cyanates.

Cyanates, when treated with H_3O^+, yield mainly isocyanic acid, HNCO, with little cyanic acid, HOCN (not isolated). The ionic salts are identical; $NCO^- \leftrightarrow OCN^-$, and these are all called cyanates. Most **d**-block metal ions prefer attachment to the N in complexes, but these salts too are still called cyanates. The acids are somewhat stable only at low temperatures, the stability of the solute varying inversely with the concentration. The solution effervesces with the escape of CO_2 (distinction from cyanides), the pungent odor of the acid also being perceptible:

$$HNCO + H_3O^+ \rightarrow NH_4^+ + CO_2\uparrow$$

Isocyanic acid may be determined by decomposing it with, e.g., H_2SO_4 as the source of H_3O^+, and titrating the excess acid. Acid hydrolyzes SCN^- too to NH_4^+ and COS, but SCN^-, $SeCN^-$ or $TeCN^-$ also go to H_2Q and elemental Q_n ($Q = S$, Se or Te). depending on conditions.

Hydroxide. Carbon monoxide reacts with OH^- to form formates:

$$CO + OH^- \rightarrow CHO_2^-$$

The OH^- ion, in boiling solution, strongly alkaline, gradually decomposes CN^- with production of NH_3 and formate:

$$CN^- + 2\,H_2O \rightarrow CHO_2^- + NH_3$$

The hexacyanoferrates(II and III) finally yield the same products.

Sawdust heated with OH^- yields $C_2O_4^{2-}$, which may be converted to $H_2C_2O_4$ by precipitating $CaC_2O_4 \cdot H_2O$ and removing the Ca with H_2SO_4.

Carbon dioxide is rapidly absorbed by hydroxides of the alkalis and of the alkaline earths, forming normal or hydrogencarbonates:

$$OH^- + CO_2 \rightarrow HCO_3^-$$

$$Sr(OH)_2 + CO_2 \rightarrow SrCO_3\downarrow + H_2O$$

Cyanogen chloride reacts to form cyanate:

$$NCCl + 2\,OH^- \rightarrow NCO^- + Cl^- + H_2O$$

Peroxide. Hydrogen peroxide with HCN forms "oxamide":

$$2\,HCN + H_2O_2 \rightarrow [-CO(NH_2)]_2$$

Dissolving K, Rb or Cs carbonates in 10-M H_2O_2 and crystallizing gives $Alk_2CO_3 \cdot 3H_2O_2$. However, passing CO_2 into OH^- in H_2O_2, i.e., HO_2^-, at –5 to –20 °C, yields $Na_2[(CO_2)_2(\mu\text{-}O_2)] \cdot aq$ or $(K,Rb,Cs)_2[(CO_2)_2(\mu\text{-}O_2)]$, i.e., $Alk_2C_2O_6$, the peroxodicarbonates; also $(Na,K,Rb)HO_2$ plus CO_2 form $NaHCO_4 \cdot H_2O$ or $(K,Rb)HCO_4$, all unstable, and $Li_2CO_4 \cdot H_2O$ arises from LiOH, H_2O_2 and CO_2.

Dioxygen. Dissolved CN^-, exposed to the air, takes up some O_2 to form the cyanate, and commercial KCN usually contains KNCO:

$$CN^- + {}^1/_2\, O_2 \rightarrow NCO^-$$

14.1.2 Reagents Derived from the Other 2nd-Period Non-Metals, Boron through Fluorine

Carbon oxide species. Ionic cyanides are partially transposed to carbonates by aqueous CO_2.

An excess of CO_3^{2-} partially transposes the alkaline-earth oxalates, and vice versa, in accord with the Law of Mass Action:

$$AeC_2O_4\downarrow + CO_3^{2-} \leftrightarrows AeCO_3\downarrow + C_2O_4^{2-}$$

Cyanide species. The acid HCN does not act on $H_2C_2O_4$.

Some "simple" organic species. Acetic acid is less acidic even than $HC_2O_4^-$ and does not dissolve oxalates. Certain of the salts do dissolve appreciably in $H_2C_2O_4$, and form hydrogenoxalates.

If CH_3CO_2H or a salt of it is warmed with H_2SO_4 and a small amount of an alcohol, the characteristic pungent and fragrant odor of the corresponding ester is obtained. Pentanol is often used because of the well-known banana-like odor of the ester, pentyl acetate ("amyl" acetate). Both the alcohol and the ester, however, are toxic.

Reduced nitrogen. Ammonia, treated with CO_2, produces an "ammonium carbonate" solution, not really mostly $(NH_4)_2CO_3$, but mainly a mixture of NH_4^+, NH_3, HCO_3^- and carbamate, $CO_2NH_2^-$. That is, the first equilibrium here lies rather to the right:

$$CO_3^{2-} + NH_4^+ \leftrightarrows HCO_3^- + NH_3 \leftrightarrows CO_2NH_2^- + H_2O$$

Solutions of CS_2 in various polar organic solvents react with gaseous or aqueous NH_3 to form the dithiocarbamate, $NH_4CS_2NH_2$, although H_2O greatly reduces its purity. It is stable for several days if dry at 0 °C, and in cold water for several weeks. One may determine it by weighing white $Zn(CS_2NH_2)_2$, and Cu^{2+} gives a yellow, flocculent precipitate.

Oxidized nitrogen. Nitrous acid seems not to act on $H_2C_2O_4$.

Aqueous HNO_3, C_2H_5OH and Ag or Hg, or their salts, form the highly explosive fulminates, AgCNO or $Hg(CNO)_2$. Then ice-cold water and Na_{Hg} convert the $Hg(CNO)_2$ to CNO^-, which complexes **d**-block cations. This and other methods have produced fulminato complexes of at least Fe^{II}, Ru^{II}, Co^{III}, Rh^{III}, Ir^{III}, Ni^{II}, Pd^{II}, Pt^{II}, Cu^{I}, Ag^{I}, Au^{I}, Zn^{II}, Cd^{II} and Hg^{II}. Such complexes, if precipitated by large cations ($[NR_4]^+$, $[AsPh_4]^+$) are not explosive. The salts of Alk^+ or Ae^{2+}, however, are very explosive. The stabilities of $Hg(CNO)_2$ and $[Hg(CNO)_4]^{2-}$ toward dissociation are a little less than those of the corresponding cyanides.

Concentrated HNO_3 and sawdust (especially softer woods), starch or sugar yield $H_2C_2O_4$. The continued action of the HNO_3, after, say, the sugar is all oxidized to $H_2C_2O_4$, converts the latter to CO_2:

$$3\ H_2C_2O_4 + 2\ NO_3^- + 2\ H_3O^+ \rightarrow 6\ CO_2\uparrow + 2\ NO\uparrow + 6\ H_2O$$

14.1.3 Reagents Derived from the 3rd-to-5th-Period Non-Metals, Silicon through Xenon

Phosphorus species. The acids HPH_2O_2, H_2PHO_3 and H_3PO_4 have no action upon $H_2C_2O_4$.

Arsenic species. Heating (dry) a suspected acetate with arsenic(III) oxide forms the very repulsive and poisonous vapor of “cacodyl oxide”. This highly sensitive test should be made under a good hood with great caution, using small quantities:

$$4\ CH_3CO_2^- + As_2O_3 \rightarrow [As(CH_3)_2]_2O\uparrow + 2\ CO_3^{2-} + 2\ CO_2\uparrow$$

Oxalic acid reduces As^V to As^{III} and becomes CO_2.

Reduced chalcogens. Heating a cyanide with sulfur or $(NH_4)_2S_x$ produces SCN^- (writing the most electronegative atom at the end in both NCO^- and SCN^-), reducing its toxicity:

$$CN^- + S_2^{2-} + NH_4^+ \rightarrow SCN^- + HS^- + NH_3$$

The SCN^- test for CN^- is more sensitive than that involving “Prussian Blue”. To the sample in an evaporating dish add 1–2 drops of yellow $(NH_4)_2S_x$. Digest on the water-bath until the mixture is colorless and free from sulfide. Slightly acidify with HCl (which should not liberate any H_2S) and add a drop of $FeCl_3$. Blood-red $FeSCN^{2+}$ will appear if cyanide was present in the original material to the extent of 0.04 mM.

Aqueous HSCN does not act on $H_2C_2O_4$.

Oxidized chalcogens. Heating a formate, oxalate, or $[Fe(CN)_6]^{4-}$ with concentrated H_2SO_4 yields CO:

$$HCHO_2 + H_2SO_4 \rightarrow CO\uparrow + H_3O^+ + HSO_4^-$$

$$H_2C_2O_4 + H_2SO_4 \rightarrow CO\uparrow + CO_2\uparrow + H_3O^+ + HSO_4^-$$

$$H_4[Fe(CN)_6] + 2\ H_2SO_4 + 6\ H_3O^+ \rightarrow$$

$$\text{e.g. } Fe(HSO_4)_2 + 6\ CO\uparrow + 6\ NH_4^+$$

Concentrated H_2SO_4 (18 M, 96 %) decomposes all cyanides:

$$HCN + H_3O^+ \rightarrow CO\uparrow + NH_4^+$$

possibly beginning with H_2SO_4 as a catalyst:

$$HCN + H_2SO_4 \leftrightarrows HCNH^+ + HSO_4^-$$

Anhydrous acetates, with concentrated H_2SO_4, give pure CH_3CO_2H, but if the H_2SO_4 is in excess and heat is applied, the mixture will blacken with separation of carbon.

Warming a suspected oxalate with concentrated H_2SO_4 after decomposing any carbonates with dilute H_2SO_4 releases CO_2 and CO:

$$H_2C_2O_4 + H_2SO_4 \rightarrow CO_2\uparrow + CO\uparrow + HSO_4^- + H_3O^+$$

The CO_2 is detectable by $Ba(OH)_2$, and the CO by its combustibility.

Heating peroxosulfates with $CH_3CO_2^-$ gives CO_2, CH_4 and other C_mH_n, apparently in part from a catalyzed breakdown of the acetate.

Oxalic acid, treated with a peroxodisulfate, dilute H_2SO_4 and a small amount of Ag^+ catalyst, is quantitatively converted to CO_2:

$$H_2C_2O_4 + [S_2O_8]^{2-} + 2\ H_2O \rightarrow 2\ CO_2\uparrow + 2\ SO_4^{2-} + 2\ H_3O^+$$

Reduced halogens. Instead of (non-)reactions between CN^- and halides here we may compare CN^- and X^- to justify calling CN^- a pseudohalide:

1. Weak oxidants such as Cu^{2+} oxidize CN^- and I^- to $(CN)_2$ and I_2.
2. Base hydrolyzes $(CN)_2$ or X_2 to CN^- and CNO^- or to X^- and XO^-.
3. Each complexes soft cations, e.g., as $[Cu(CN)_2]^-$ and $[CuCl_2]^-$.
4. Each precipitates Ag^+ and Hg_2^{2+}, and some Ag salts dissolve in NH_3.
5. Each forms (pseudo)interhalogens such as ICN and ICl.

Other pseudohalides, at least partly, are NCS^-, N_3^- and $S_2O_3^{2-}$.

Elemental and oxidized halogens. In sunlight CO reacts with Cl_2 or Br_2:

$$CO + Cl_2 + 3\,H_2O \rightarrow CO_2\uparrow + 2\,H_3O^+ + 2\,Cl^-$$

Solid I_2O_5 can be used to detect and determine CO. One part of CO in 30,000 of air is readily revealed this way by the reactions of I_2, q.v.:

$$I_2O_5 + 5\,CO \rightarrow 5\,CO_2\uparrow + I_2\uparrow$$

Chlorine and HCN form cyanogen chloride, CNCl, NCCl or ClCN, where each formula, like others, fails to show either the sequence of atoms (NCCl or ClCN), consistency with other formulas or the regular order of electronegativities (C < Cl < N). (The different conventional orders in H_2O and OH^-, however, already reveal our low priority for consistency in any case.) Iodine acts similarly, but less markedly:

$$HCN + Cl_2 + H_2O \rightarrow NCCl + H_3O^+ + Cl^-$$

Toxic CN^- can be destroyed by ClO^- etc., then OH^- (pH > 11):

$$CN^- + ClO^- + H_2O \rightarrow NCCl + 2\,OH^-$$

$$NCCl + 2\,OH^- \rightarrow NCO^- + Cl^- + H_2O$$

$$CN^- + H_2O_2 \rightarrow NCO^- + H_2O$$

$$NCO^- + H_3O^+ + H_2O \rightarrow HCO_3^- + NH_4^+ \; (pH < 7)$$

Chlorine and $H_2C_2O_4$ yield HCl and CO_2, or (more readily) Cl^- and HCO_3^- or CO_3^{2-} at pH ≥ 7. The reaction with Br_2 is similar.

Aqueous NCO^- and Br_2 water give CO_2, NH_4^+, N_2 and Br^-.

Warmed with I_2 in water, AgNCO forms AgI, $AgIO_3$, CO_2 and urea.

Oxalic acid and HClO give Cl_2 and CO_2. If the $H_2C_2O_4$ is in excess, HCl is formed. The action is faster at pH ≥ ~7, forming Cl^- and CO_3^{2-}. Chloric acid, $HClO_3$, forms CO_2 and varying amounts of Cl_2 and HCl. A high temperature and excess of $H_2C_2O_4$ favor the production of HCl. With $HBrO_3$, Br_2 and CO_2 result; excess warm $H_2C_2O_4$ gives HBr.

Aqueous NCO^- treated with BrO^- yields CO_3^{2-}, CHO_2^-, N_2 and Br^-.

Iodic acid, HIO_3, reacts with $H_2C_2O_4$ to give CO_2 and I_2. With mixtures of ClO_3^-, BrO_3^- and IO_3^-, the ClO_3^- is the first decomposed, then the BrO_3^- and finally the IO_3^-, even though BrO_3^- has nearly the same reduction potential as ClO_3^-.

14.1.4 Reagents Derived from the Metals Lithium through Uranium, plus Electrons and Photons

Oxidation. Warm $[Cr_2O_7]^{2-}$ and H_2SO_4 oxidize CO to CO_2.

Oxalic acid reduces $[Cr_2O_7]^{2-}$ to Cr^{III}, catalyzed by Mn^{2+}.

Oxalic acid, with H_3O^+, reduces $MnO_2\cdot aq$:

$$MnO_2\cdot aq + H_2C_2O_4 + 2\ H_3O^+ \rightarrow Mn^{2+} + 2\ CO_2\uparrow + 4\ H_2O$$

Acetic acid is not oxidized by MnO_4^-, even on boiling. It does not reduce alkaline copper, and reduces Au^{III} only in alkaline solution.

Acidic solutions of MnO_4^-, $PdCl_4^{2-}$ and Hg^{2+}, and basic mixtures with Ag^I, oxidize CO to CO_2, e.g.:

$$2\ MnO_4^- + 3\ CO + 2\ H_3O^+ \rightarrow 2\ MnO_2\cdot aq\downarrow + 3\ CO_2\uparrow + 3\ H_2O$$

with the uncatalyzed mechanism apparently starting with:

$$MnO_4^- + CO \rightarrow [:(C{=}O){-}O{-}MnO_3]^-$$

$$[:(C{=}O){-}O{-}MnO_3]^- + 3\ H_2O \rightarrow MnO_4^{3-} + CO_2\uparrow + 2\ H_3O^+$$

Strong catalysts for this are Ag^+ and Hg^{2+}, which may first form, e.g., $[Ag{-}(CO){-}O{-}MnO_3]$, which then gives CO_2, MnO_3^- and Ag^+ again. The unstable MnO_4^{3-} or MnO_3^- would quickly go to $MnO_2\cdot aq$ and MnO_4^-.

A strong oxidant, MnO_4^-, and CN^- form NCO^- and, e.g., $MnO_2\cdot aq$. The weaker Cu^{II} in alkaline solution precipitates yellowish $Cu(CN)_2$, which soon goes to white CuCN and gaseous $(CN)_2$.

Aqueous MnO_4^- in cold alkaline solution has no action on $C_2O_4^{2-}$; in hot acidic solution rapid oxidation occurs:

$$2\ MnO_4^- + 5\ H_2C_2O_4 + 6\ H_3O^+ \rightarrow 2\ Mn^{2+} + 10\ CO_2\uparrow + 14\ H_2O$$

Aqueous oxalate may be detected by precipitation in a neutral, alkaline or acetic-acid solution as CaC_2O_4. The precipitate is washed and dissolved in hot, dilute H_2SO_4. If the resulting solution is treated with dilute MnO_4^-, the first few drops will be reduced very slowly to the colorless Mn^{2+}, after which the purple color will disappear rapidly until all of the $C_2O_4^{2-}$ has been oxidized to CO_2. In fact, oxalates may be titrated with MnO_4^-

Oxalic acid and $FeCl_3$ give Fe^{2+} and CO_2 in sunlight or actinometers.

Oxalic acid reduces $Co_2O_3\cdot aq$ and NiO_2 to M^{II}, forming CO_2.

Carbon monoxide may be detected by its reduction of $PdCl_2$ to Pd (also forming CO_2 and HCl), which may be detected with MoO_4^{2-}.

A white, crystalline precipitate of $Ag_2C_2O_4$ is explosive when dry:

$$Ag_2C_2O_4 \rightarrow 2\ Ag\downarrow + 2\ CO_2\uparrow$$

Oxalic acid reduces Au_2Cl_6, slowly in the dark, rapidly in sunlight:

$${}^1/_2\ Au_2Cl_6 + {}^3/_2\ H_2C_2O_4 + 3\ H_2O \rightarrow Au\downarrow + 3\ CO_2\uparrow + 3\ H_3O^+ + 3\ Cl^-$$

Mercury(II) oxidizes CO in water:

$$Hg^{2+} + CO + 2\ H_2O \rightarrow [HgCO_2H]^+ + H_3O^+$$

$$[HgCO_2H]^+ + H_2O \rightarrow Hg + CO_2\uparrow + H_3O^+$$

$$Hg^{2+} + Hg \rightarrow Hg_2^{2+}$$

$$2\ Hg^{2+} + CO + 3\ H_2O \rightarrow Hg_2^{2+} + CO_2\uparrow + 2\ H_3O^+$$

Lead dioxide oxidizes HCN, forming $Pb(CN)_2$ and $(CN)_2$:

$$PbO_2 + 4\ HCN \rightarrow Pb(CN)_2\downarrow + (CN)_2\uparrow + 2\ H_2O$$

Oxalic acid or $C_2O_4^{2-}$ boiled with $[HgCl_2]$ in the sunlight gives $[Hg_2Cl_2]$ and CO_2. In the absence of light a trace of $[S_2O_8]^{2-}$, $MnO_2 \cdot aq$, MnO_4^- or HNO_2, or of HNO_3 plus Mn^{2+}, promotes the reaction:

$$H_2C_2O_4 + 2\ [HgCl_2] + 2\ H_2O \rightarrow$$

$$[Hg_2Cl_2]\downarrow + 2\ CO_2\uparrow + 2\ H_3O^+ + 2\ Cl^-$$

Lead dioxide forms PbC_2O_4 and CO_2 with $H_2C_2O_4$:

$$PbO_2 + 2\ H_2C_2O_4 \rightarrow PbC_2O_4\downarrow + 2\ CO_2\uparrow + 2\ H_2O$$

Bismuth(V) forms bismuth(III) oxalate and CO_2.

Reduction. Aqueous CN^- can be reduced to CH_3NH_2 by strong reductants such as $[Co(CN)_5]^{3-}$.

Most reducing agents have no effect on oxalic acid at ordinary temperatures. It is reduced, however, by Na_{Hg} and H_2O, by Mg and H_2O, or by Zn and H_3O^+, first to $(CHO)–CO_2^-$, then to $CH_2(OH)–CO_2^-$.

Other reactions of carbon monoxide and carbonate species. Carbon monoxide may be determined by absorption in CuCl and measuring the loss of volume. Aqueous $[CuCl_2]^-$ reversibly gives compounds such as $CuCl \cdot CO \cdot H_2O$.

Alkali hydroxides, AlkOH, can be made by treating aqueous Alk_2CO_3 with the oxide or hydroxide of Ca, Sr or Ba:

$$CO_3^{2-} + Ca(OH)_2 \rightarrow 2\ OH^- + CaCO_3\downarrow$$

Carbonate may be determined by weighing as, e.g., $CaCO_3$.

To detect CO_2, a gas is sometimes passed into a solution of $Ca(OH)_2$, $Ba(OH)_2$, or ammoniacal Ca^{2+}, Ba^{2+}, or lead acetate, whereupon a white precipitate or turbidity of the corresponding carbonate is formed:

$$CO_2 + Ca(OH)_2 \rightarrow CaCO_3\downarrow + H_2O$$

$$CO_2 + Ba^{2+} + 2\ NH_3 \rightarrow BaCO_3\downarrow + 2\ NH_4^+ + H_2O$$

$$CO_2 + Pb(CH_3CO_2)_2 + 2\ NH_3 + H_2O \rightarrow$$

$$PbCO_3\downarrow + 2\ NH_4^+ + 2\ CH_3CO_2^-$$

The solutions of $Ca(OH)_2$ and $Ba(OH)_2$ furnish more sensitive tests for CO_2 than the ammoniacal solutions of Ca^{2+} and Ba^{2+} but less so than the basic lead acetate. The latter is so quickly affected by atmospheric CO_2 that it cannot be preserved in bottles partly full and frequently opened, nor can it be diluted except with recently boiled water.

Carbonates of the alkaline-earth metals are readily converted to soluble hydrogencarbonates by excess CO_2; hence the disappearance of the $CaCO_3$ or $BaCO_3$ precipitate in the tests described above may be evidence of an excess of CO_2, not of its absence. Solutions thus obtained will effervesce on heating, with escape of CO_2 and reprecipitation of the normal carbonate:

$$Ae^{2+} + 2\ HCO_3^- \leftrightarrows AeCO_3\downarrow + CO_2\uparrow + H_2O$$

This equilibrium also produces stalagmites and stalactites in caves, as well as the deposit from boiling hard water in a tea kettle.

Hydrogencarbonate in a carbonate may be detected by precipitating the CO_3^{2-} with Ca^{2+} and adding NH_3 to the filtrate. A precipitate of $CaCO_3$ is produced even if < 2-cM HCO_3^- is present, although the separation is slow with the lower concentrations. (Conversely, a carbonate may be detected in a hydrogencarbonate solution by phenolphthalein, which turns pink if CO_3^{2-} is present.)

Dissolved species similar to Cr^{III}, Fe^{III}, Al^{III} and Sn^{IV} are precipitated as hydroxides by $BaCO_3$, while Mn^{2+}, Co^{2+} and Ni^{2+}, for example, are not.

All non-alkali-metal M^{n+} precipitate or hydrolyze CO_3^{2-}. The Ae^{2+}, Mn^{2+}, Fe^{2+}, Ag^+, Cd^{2+} or Hg_2^{2+} (or Pb^{2+} in the cold) ions give normal carbonates, but the Ag_2CO_3 or Hg_2CO_3 (or Cu^{II} basic carbonate) decompose, quickly on heating, to the oxide or to HgO and Hg. Other Cr^{III}, Fe^{III}, Co^{3+}, HgX_2 (halide), Al, Sn or Sb species yield hydroxides, oxides or basic non-carbonate salts; still other metal species form basic carbonates, e.g., $Pb_3(OH)_2(CO_3)_2$ with heat.

Stable complexes $[M(CS_3)_2]^{2-}$ arise from CS_3^{2-} and Co^{2+}, Ni^{2+}, Pd^{2+} or Pt^{2+}. At least the Ni complex is oxidized by S_8, and even I_2, to $[Ni(CS_4)_2]^{2-}$, probably with 5-membered rings of –Ni–S–(CS)–S–S–. We note that the S, but not the C or Ni, is oxidized (or sulfurized?).

Other reactions of cyano species. Cyanates of the Alk, Ae, Mn, Co, Ni and Zn cations arise from passing $(CN)_2$ gas into the hydroxides, e.g.:

$$2\ Mg(OH)_2 + 2\ (CN)_2 \rightarrow Mg(NCO)_2\downarrow + Mg^{2+} + 2\ CN^- + 2\ H_2O$$

Treating CN^- with even slightly acidic cations at 100 °C forms HCN:

$$2\ CN^- + Mg^{2+} + 2\ H_2O \rightarrow Mg(OH)_2\downarrow + 2\ HCN\uparrow$$

Certain metals (Cd, Hg, Sn, Pb, As, Sb and Bi) are dissolved by CN^- or become oxides or hydroxides while absorbing oxygen (but not oxidizing the cyanide unit); others (Mg, Fe, Co, Ni, Cu, Zn and Al) react rather similarly but produce hydrogen:

$$Cd + 4\ CN^- + {}^1/_2\ O_2 + H_2O \rightarrow [Cd(CN)_4]^{2-} + 2\ OH^-$$

$$2\ As + 2\ CN^- + {}^3/_2\ O_2 + 3\ H_2O \rightarrow 2\ H_2AsO_3^- + 2\ HCN$$

$$Cu + 4\ CN^- + H_2O \rightarrow [Cu(CN)_4]^{3-} + OH^- + {}^1/_2\ H_2\uparrow$$

$$Al + CN^- + 4\ H_2O \rightarrow Al(OH)_4^- + HCN\uparrow + {}^3/_2\ H_2\uparrow$$

Many of the **d**-block metal cyanides, insoluble in water, readily dissolve in an excess of CN^- with formation of a complex ion. We can generally distinguish two classes of cyanocomplexes formed by adding excess CN^- to the initial precipitate:

Class I. Cyanocomplexes that are not affected by OH^- but are decomposed by dilute H_3O^+:

$$[Ni(CN)_4]^{2-} + 2\ H_3O^+ \rightarrow Ni(CN)_2\downarrow + 2\ HCN + 2\ H_2O$$

These resemble the thiocomplexes. The principal ions of this class are: $[Ni(CN)_4]^{2-}$, $[Cu(CN)_4]^{3-}$, $[Ag(CN)_2]^-$, $[Au(CN)_4]^-$, $[Zn(CN)_4]^{2-}$, $[Cd(CN)_4]^{2-}$ and $[Hg(CN)_4]^{2-}$.

Class II. Cyanocomplexes that, as precipitates, are transposed (but not quickly decomposed) by dilute OH^-, and are converted to acids without immediate decomposition by dilute H_3O^+:

$$Cu_2[Fe(CN)_6] + 4\ OH^- \rightarrow 2\ Cu(OH)_2\downarrow + [Fe(CN)_6]^{4-}$$

$$[Fe(CN)_6]^{4-} + 2\ H_3O^+ \leftrightarrows H_2[Fe(CN)_6]^{2-} + 2\ H_2O$$

These more inert cyanocomplexes correspond on the latter point to $[PtCl_6]^{2-}$. The most common cyanocomplexes in this class are: $[Cr(CN)_6]^{3-}$, $[Mn(CN)_6]^{3-}$, $[Fe(CN)_6]^{3-}$, $[Fe(CN)_6]^{4-}$ and $[Co(CN)_6]^{3-}$.

The thermodynamic stabilities of some cyanocomplexes are poorly known but may be: $Fe^{II} > Cr^{II} > V^{II} > Mn^{II}$ and $Co^{III} > Fe^{III} > Mn^{III} \sim Cr^{III}$.

Aqueous Fe^{2+}, added to saturation, precipitates from CN^- (not HCN) $Fe_2[Fe(CN)_6]$, white if free from iron(III), otherwise yellowish red due to $Fe_2O_3 \cdot aq$. This is sometimes written as $FeFe[Fe(CN)_6]$ or $Fe[FeFe(CN)_6]$ to suggest structural differences not described here, somewhat as we write HPH_2O_2 for distinct H atoms in phosphinic acid. The precipitate is soluble in excess CN^- with formation of $[Fe(CN)_6]^{4-}$. Solutions of Fe^{III} yield, with CN^-, a precipitate of $Fe_2O_3 \cdot aq$ and HCN. A small amount of the $Fe_2O_3 \cdot aq$ will dissolve in excess CN^-, forming $[Fe(CN)_6]^{3-}$.

The production of a "Prussian Blue", e.g., $KFe^{III}[Fe^{II}(CN)_6] \cdot aq$, is a fairly sensitive test for cyanides. A small amount of the sample is treated with Fe^{2+} and a few drops of an alkali. After shaking, a drop or two of $FeCl_3$ is added and the whole slightly acidified with H_2SO_4 (to dissolve the hydroxides) whereupon the blue precipitate will appear if CN^- was in the original sample. The test can detect 0.8-mM cyanide.

Silver(+) precipitates CN^-, not from $[Hg(CN)_2]$, as $Ag[Ag(CN)_2]$, white, insoluble in dilute HNO_3, soluble in NH_3, hot "$(NH_4)_2CO_3$", excess CN^-, and $S_2O_3^{2-}$, as in the case of AgCl. Cold, concentrated HCl decomposes $Ag[Ag(CN)_2]$ with liberation of HCN (distinction from AgCl). When well washed, then gently ignited, $Ag[Ag(CN)_2]$ yields Ag, soluble in HNO_3 (distinction and separation from AgCl). The complex $[Ag(CN)_2]^-$ is slowly decomposed by acetic acid, readily by HNO_3 or H_2SO_4, while precipitating $Ag[Ag(CN)_2]$ and liberating HCN.

Cyanide may be determined by titration in an ammoniacal iodide solution with standardized $AgNO_3$, using the opalescence of AgI as the endpoint. The NH_3 prevents the temporary, local precipitation of $Ag[Ag(CN)_2]$. The main reactions are:

$$[Ag(NH_3)_2]^+ + 2\ CN^- \rightarrow [Ag(CN)_2]^- + 2\ NH_3$$

$$[Ag(NH_3)_2]^+ + I^- \rightarrow AgI\downarrow + 2\ NH_3$$

We may determine HNCO by adding Ag^+ and titrating the excess.

Aqueous $[Ag(NH_3)_2]^+$ precipitates a yellow silver salt of cyanamide.

Aqueous Hg_2^{2+} and CN^- (unlike halides) give Hg and $[Hg(CN)_2]$.

Lead cyanate may be prepared by treating NCO^- with Ba^{2+} to remove any CO_3^{2-}, then adding Pb^{2+} to the filtrate. The lead cyanate precipitated is comparatively stable and serves well for the preparation of HNCO.

Other reactions of "simple" organic species. Acetates may be detected by developing a blue color when treated with $La(NO_3)_3$, iodine and a little NH_3, and gradually heated to boiling. Homologs, sulfates and anions precipitating La^{III} interfere.

Oxalic acid and $C_2O_4^{2-}$ precipitate numerous ions, but many of the precipitates are soluble in excess $C_2O_4^{2-}$, forming complex or double salts, e.g., $AgNH_4C_2O_4$. Well-defined exceptions are Ca, Sr and Ba; their precipitates are normal oxalates.

Various metals, when finely divided, are attacked by $H_2C_2O_4$ (without reducing the carbon), releasing H_2. Aqueous $H_2C_2O_4$ releases H_2S from MnS and FeS but not from CoS, NiS or ZnS etc. It appears inactive with $H_4[Fe(CN)_6]$ or $H_3[Fe(CN)_6]$.

Oxalic acid yields oxalates from the oxides, hydroxides or carbonates of Na, K, Mg, Ca, Sr, Ba, Cr^{III}, Mn, Fe^{II}, Fe^{III}, Co, Ni, Cu^{II}, Ag, Zn, Cd, Hg^{I}, Hg^{II}, Al, Sn^{II}, Pb, Bi and many others. Adding $H_2C_2O_4$ to many soluble salts of the above metals also forms oxalates, except those of the alkalis, Mg, Cr^{III}, Fe^{III}, Al and Sn^{IV}, which are not precipitated. Antimony(III) becomes a basic salt. An excess of $H_2C_2O_4$ may form a hydrogenoxalate with those listed.

14.2 Silicon, ${}_{14}Si$

Oxidation number: (IV), as in both SiH_4 and SiO_2. For Si, however, the older concept of valence, normally four, may serve well.

14.2.1 Reagents Derived from Hydrogen and Oxygen

Water. The catena silanes, Si_nH_{2n+2} ($n > 1$), are all decomposed by H_2O, especially if alkaline, forming $SiO_2\cdot aq$ (silicic acid) and H_2.

The alkali-metal silicates are soluble, but the other silicates are not, or only slightly so. However, natural waters contain H_4SiO_4, $H_3SiO_4^-$, $H_2SiO_4^{2-}$, $(Mg,Ca)H_3SiO_4^+$, $(Mg,Ca)H_2SiO_4$, $FeH_3SiO_4^{2+}$, polymers etc.

Oxonium. Silane, SiH_4, is stable toward weak acids, but it reduces strong acids, releasing H_2.

Acids, including aqueous CO_2, plus the soluble (alkali-metal) silicates such as "water glass", readily precipitate gels of orthosilicic acid, $H_4SiO_4\cdot aq$, slightly soluble, or polysilicic acids, $H_{2m}Si_nO_{m+2n}$, from the various ortho- or polysilicates. The simpler acids, however, undergo condensation and polymerization, except in very dilute solutions. The complexities of these acids and their salts resemble somewhat those of the boric, molybdic and tungstic acids. The set time for the gels depends greatly on many factors, falling to a minimum of a few seconds as the pH falls to nearly 7 but then rising rapidly with further acidification until it falls again.

Although anhydrous SiO_2 is insoluble in inorganic acids, they can peptize fresh H_4SiO_4 as a silica sol. This can then be stabilized with a little alkali, or allowed to precipitate again as a gel. Silicic acids are dehydrated by evaporation to dryness, to the much less soluble SiO_2.

Hydroxide. The alkali in glass can destabilize SiH_4.

Commercial sodium silicate, "water glass", $\sim Na_4SiO_4$, may be made by dissolving SiO_2 in NaOH.

Dioxygen. The catena silanes, Si_nH_{2n+2}, are all spontaneously flammable and tend to explode, in air, to form SiO_2.

14.2.2 Reagents Derived from the Other 2nd-Period Non-Metals, Boron through Fluorine

Reduced nitrogen. Silicates and NH_4^+ precipitate silicic acids, e.g.:

$$SiO_4^{4-} + 4\ NH_4^+ \rightarrow H_4SiO_4\cdot aq\downarrow + 4\ NH_3$$

Fluorine species. Silicon and silica, SiO_2, are practically insoluble in water or acids, except HF, which forms SiF_4, non-corrosive but readily hydrolyzed (see below):

$$Si + 4\ HF \rightarrow SiF_4\uparrow + 2\ H_2\uparrow$$

$$SiO_2 + 4\ HF \rightarrow SiF_4\uparrow + 2\ H_2O$$

Silicic acids and F^- also form gaseous SiF_4, $[SiF_6]^{2-}$ and silicates.

Concentrated H_2SO_4 can both help provide HF and be a sink for the H_2O, to drive the above more strongly to the right, with the net result:

$$SiO_2 + 2\ CaF_2 + 4\ H_2SO_4 \rightarrow SiF_4\uparrow + 2\ CaSO_4\downarrow + 2\ H_3O^+ + 2\ HSO_4^-$$

This reaction provides a test applicable to silicic acid, silica or a silicate. One treats a sample with HF (or CaF_2 as above, warmed). Some of the released SiF_4 is absorbed in a drop of H_2O suspended over the reaction mixture. Hydrolysis precipitates silicic acid, visible in the drop, e.g.:

$$3\ SiF_4 + 8\ H_2O \rightarrow 2\ [SiF_6]^{2-} + H_4SiO_4\downarrow + 4\ H_3O^+$$

Most hexafluorosilicates are soluble, but those of Na, K (translucent and gelatinous) and Ba dissolve only slightly, less so in dilute ethanol.

Aqueous or gaseous HF also etches glass, e.g.:

$$Na_2CaSi_6O_{14} + 28\ HF \rightarrow 2\ NaF + CaF_2\downarrow + 6\ SiF_4\uparrow + 14\ H_2O$$

14.2.3 Reagents Derived from the 3rd-to-5th-Period Non-Metals, Silicon through Xenon

Phosphorus species. Phosphoric acid is a useful solvent for some substances that resist other reagents. Heated at 230 °C for about three hours, Si dissolves as Si^{IV}, and SiC is entirely decomposed in the same time; dilution of the solutions gives no precipitate.

14.2.4 Reagents Derived from the Metals Lithium through Uranium, plus Electrons and Photons

Reduction. Silane, SiH_4, blackens $AgNO_3$ paper but not $[HgCl_2]$ paper. This action of SiH_4 on Ag^+ is rather similar to that of AsH_3, but the low electronegativity of Si implies that both the Si^{IV} and the Ag^I are reduced by the change of H^{-I} to H^I:

$$Si^{IV}H^{-I}_4 + 4\,Ag^+ + 4\,H_2O \rightarrow Si\downarrow + 4\,Ag\downarrow + 4\,H^I_3O^+$$

Other reactions. Solutions of the alkali silicates precipitate the simple cations of all other metals as insoluble silicates.

The precipitate from a silicate solution, NaOH and $CuSO_4$, if heated for 10 d at 230 °C, yields a microporous zeolite, $Na_2Cu_2Si_4O_{11}\cdot 2H_2O$, that can lose H_2O reversibly while keeping the framework [1].

Silicic acid may be detected by a non-specific method beginning by forming a stable heteropoly acid, yellow $H_4SiMo_{12}O_{40}$. The unknown is treated with a neutral solution of ammonium molybdate in a test tube, and the mixture is slightly acidified, forming the complex; then a few drops of $SnCl_2$ are added, producing a deep-blue "Molybdenum Blue" if more than 0.01-mM silicic acid is present.

14.3 Germanium, $_{32}$Ge

Oxidation numbers: (II) and (IV), as in GeO and GeO_2.

14.3.1 Reagents Derived from Hydrogen and Oxygen

Water. Germanium is insoluble in H_2O, but GeO_2 is slightly soluble.

The sulfides, GeS, red, and GeS_2, white, are slightly soluble in H_2O.

The halides of Ge^{IV} are decomposed by H_2O.

Oxonium. Perchloric acid dissolves $Ge(OH)_2$ as Ge^{2+}.

Germanium dioxide, GeO_2, is slightly soluble in H_3O^+.

Germanates acidified to a pH of 5 precipitate $GeO_2\cdot aq$.

Hydroxide. Germanium is insoluble in OH^-, but GeO dissolves readily.

Treating Ge^{2+} with OH^- precipitates $Ge(OH)_2$. It dissolves in excess OH^- and is yellow when first formed but turns red on heating; it is slightly less acidic than acetic acid, CH_3CO_2H.

Germanium dioxide dissolves much more readily in hot concentrated OH^- than in acids, forming the germanates, $[GeO(OH)_3]^-$, $[GeO_2(OH)_2]^{2-}$ and even $[\{Ge(OH)_4\}_8(OH)_3]^{3-}$ in solution, and $[Ge(OH)_6]^{2-}$ in solids,

Solutions of Ge^{IV} may give no precipitate with OH^- because of the ready conversion to $[Ge(OH)_6]^{2-}$.

The sulfides, GeS and GeS_2, are soluble in OH^-.

14.3.2 Reagents Derived from the Other 2nd-Period Non-Metals, Boron through Fluorine

Boron species. Pouring germanate and excess $[BH_4]^-$ under N_2 or Ar into 14 to 17-M CH_3CO_2H yields GeH_4, also with some Ge_2H_6 etc.:

$$[Ge(OH)_6]^{2-} + [BH_4]^- + 3\ CH_3CO_2H \rightarrow$$

$$GeH_4\uparrow + H_3BO_3 + 3\ CH_3CO_2^- + 3\ H_2O$$

Some "simple" organic species. Equivalent amounts of the following reactants to make a 2-dM solution of the complex are refluxed to dissolve them; then cooling and ethanol bring down a white product:

$$GeO_2 + K_2C_2O_4{\cdot}H_2O + 2\ H_2C_2O_4{\cdot}2H_2O \rightarrow$$

$$K_2[Ge(C_2O_4)_3]{\cdot}H_2O\downarrow + 6\ H_2O$$

Reduced nitrogen. If air is kept out, NH_3 precipitates cream-colored germanium(II) hydroxide from $GeCl_2$.

Oxidized nitrogen. Nitric oxide changes GeS_2 to GeO_2.

Germanium reacts with HNO_3 to form GeO_2, and dissolves in aqua regia. The sulfides, GeS and GeS_2, are soluble in aqua regia with separation of sulfur.

Fluorine species. Hydrogen fluoride dissolves GeO_2 as a complex which is then precipitable as $Ba[GeF_6]$:

$$GeO_2 + 6\ HF \rightarrow [GeF_6]^{2-} + 2\ H_3O^+$$

14.3.3 Reagents Derived from the 3rd-to-5th-Period Non-Metals, Silicon through Xenon

Phosphorus species. Excess HPH_2O_2 and GeO_2 form $Ge^{II}PHO_3$, soluble in hydrogen halides.

Heating $GeCl_4$ in 3 to 6-M HCl, or GeS_2 in 5-M HCl, with excess phosphinic acid at 100 °C for a few hours reduces it to $[GeCl_n]^{(n-2)-}$, stable as 2–4 dM Ge^{II} for some weeks:

$$[GeCl_6]^{2-} + HPH_2O_2 + 3\ H_2O \rightarrow$$

$$[GeCl_4]^{2-} + H_2PHO_3 + 2\ Cl^- + 2\ H_3O^+$$

$$GeS_2 + HPH_2O_2 + 4\ Cl^- + 2\ H_3O^+ \rightarrow$$

$$[GeCl_4]^{2-} + H_2PHO_3 + 2\ H_2S\uparrow + H_2O$$

Then NH_3 gives $Ge(OH)_2$, only slightly soluble in 6-M $HClO_4$, eventually forming $GeO_2 \cdot aq$, but easily redissolved by 6-M HCl. Reducing GeS_2 with HPH_2O_2 alone yields dark-gray GeS.

The red crystals of GeI_4 plus HPH_2O_2 form yellow plates of GeI_2:

$$GeI_4 + HPH_2O_2 + 3\ H_2O \rightarrow GeI_2\downarrow + H_2PHO_3 + 2\ I^- + 2\ H_3O^+$$

Reduced chalcogens. Sulfide (S^{2-}) and Ge^{2+} precipitate a brown GeS. Sulfane (H_2S) and $Ge(OH)_2$ yield reddish-brown GeS, air-stable when dry, also formed in alkalis or weak acids, soluble in strong acids. Thus, stable in air, it is a good source for other Ge^{II} species. In contrast, Ge^{IV}, with H_2S, gives white GeS_2 only from at least 6-M H_3O^+.

Sulfane, H_2S, precipitates white GeS_2, readily soluble in "$(NH_4)_2S$".

These sulfides, GeS and GeS_2, are soluble in alkali sulfides.

Selane, H_2Se, passed into aqueous $GeCl_2$, gives dark-brown GeSe.

Oxidized chalcogens. Germanium is insoluble in dilute H_2SO_4, but oxidized to GeO_2 by the concentrated acid.

The sulfides, GeS and GeS_2, are insoluble in H_2SO_4.

Reduced halogens. Germanium is insoluble in HCl, but GeO dissolves readily in it. Germanium(II) in 6-M HCl is stable for weeks, but decomposes in hours on great dilution.

Aqueous HCl precipitates the hydroxide from germanate and then redissolves it as $GeCl_4$:

$$[Ge(OH)_6]^{2-} + 6\ H_3O^+ + 4\ Cl^- \rightarrow GeCl_4 + 12\ H_2O$$

Boiling is avoided due to the volatility of $GeCl_4$, bp 83 °C. Six-molar HCl prevents hydrolysis. Evaporating a solution of GeO_2 or GeS_2 in excess HCl to dryness vaporizes it completely. The solubility of $GeCl_4$ in cold 12-M HCl is less than 1 cM, but is much greater in dilute HCl.

The monosulfide, GeS, dissolves in hot, concentrated HCl, but the disulfide, GeS_2, contrarily, requires much HCl even to be precipitated.

Hot, concentrated HI partly dissolves GeS and, on cooling, forms orange or russet GeI_2, stable if dry, but slowly hydrolyzed by moisture, much more soluble in hot than in cold HI:

$$GeS + 2\,I^- + 2\,H_3O^+ \leftrightarrows GeI_2\downarrow + H_2S\uparrow + 2\,H_2O$$

Boiling concentrated HI with GeO_2 for a few minutes under, say, CO_2 (to prevent aerial oxidation of the HI), gives reddish-orange GeI_4, insoluble in concentrated HI. Water dissolves and hydrolyzes this to a colorless, clear, acidic solution.

Excess Ge^{II} reduces I_3^-, forming Ge^{IV}, catalyzed by H_3O^+, with a rate nearly independent of *c*(oxidant) during most of the reaction, implying initiation via unimolecular conversion of Ge^{II} to an activated cation.

14.3.4 Reagents Derived from the Metals Lithium through Uranium, plus Electrons and Photons

Oxidation. Iron(III) oxidizes Ge^{II} slowly in HCl, unless catalyzed by Cu^{II}, $[IrCl_6]^{2-}$, $[PtCl_6]^{2-}$ etc., via Ge^{III} as intermediate in some cases. Not appreciably reactive are UO_2^{2+}, $[Co(Cl,Br)(NH_3)_5]^{2+}$ and so on.

Reduction. Zinc and Ge^{IV} in acid precipitate Ge as a dark brown slime.

Other reactions. Ammonia and Mg^{2+} precipitate Mg_2GeO_4 from acidic Ge^{IV} in one method, after drying, to determine Ge by weight.

Solutions of Ge^{2+} give a white precipitate with $[Fe(CN)_6]^{4-}$.

14.4 Tin, $_{50}$Sn

Oxidation numbers: (II) and (IV), as in SnO, "stannous" oxide and "stannites", and SnO_2, "stannic" oxide and "stannates".

Tin(IV) oxide dihydroxide, $SnO(OH)_2\cdot aq$, also treated as an acid, $H_2SnO_3\cdot aq$, has been said to occur in at least two modifications, i.e., Frémy's stannic and metastannic acids, and Berzelius' α- and β-stannic acids, with the first of each being more reactive, hydrated, bulky, soluble gels than the second, more polymeric species, which might be written as $(SnO_2)_\infty\cdot aq$, or simply as $SnO_2\cdot aq$. Previous formulas, e.g., $(H_2SnO_3)_5$, were normally too precise.

A spectrum actually exists, also for many other polyvalent metals like iron, but the distinction, although formerly made too sharply for tin, seems under-used in modern writing, as if we could never say, e.g., "more dilute" or "warmer", even absent great precision, although repeating, e.g., "the fresher, more hydrated, less polymeric form" would be verbose. The distinction remains useful in the laboratory, certainly for tin, and we apply it here. The absence of simple, approved terms is understandable for such a spectrum, but still problematic.

The possible Greek letters and prefixes already have various meanings, but so do most words. Let us use "ortho-" and "meta-" for the simpler and the more polymeric forms, in turn. These may be usefully reminiscent of the "ortho" in orthophosphoric acid, H_3PO_4, and the "meta" in the polymeric metaphosphoric acid, $(HPO_3)_\infty$, even though such terms can be applied more precisely in these cases. For the hydrous oxides, then, we will sometimes have approximately the (ortho) $H_2SnO_3 \cdot aq$, $SnO(OH)_2 \cdot aq$ or $Sn(OH)_4 \cdot aq$, and the (meta) $(SnO_2)_n \cdot aq$, or $SnO_2 \cdot aq$, with some ambiguity still being inevitable. For many other metals, formulas like $M_2O_3 \cdot aq$ merely mean any indefinite aquation.

14.4.1 Reagents Derived from Hydrogen and Oxygen

Water. Tin(II and IV) oxides, and ortho- and meta-tin(IV) hydrous oxides are insoluble in H_2O, as are the tin phosphates and sulfides. The salts $Sn(SO_4)_2 \cdot 2H_2O$, $Sn(ClO_4)_2 \cdot 3H_2O$, and $SnBr_4$ are deliquescent or very hygroscopic. Likewise $SnSO_4$, $SnCl_4$, $SnBr_2$, $SnBr_4$ and SnI_4 dissolve in small or moderate amounts of water with little or no precipitation of basic salts. A basic Sn^{II} sulfate resulting from dilution and hydrolysis at rather low pH is $Sn_3O(OH)_2SO_4$. Tin dichloride is soluble in less than two parts of H_2O by weight, but more water yields a basic precipitate unless excess acid is present.

A little H_2O added to neat, liquid $SnCl_4$ combines exothermically to form crystals of $SnCl_4 \cdot 5H_2O$, readily soluble in excess H_2O.

Some natural waters may contain $[SnS_3]^{2-}$, or (in hot waters) $[Sn(OH)_6]^{2-}$, $SnCO_3(OH)_3^-$ or $[SnF_6]^{2-}$.

Oxonium. Highly acidified, hydrated Sn^{II} ions appear to be $[Sn(H_2O)_3]^{2+}$ and/or $[SnOH(H_2O)_2]^+$, or $[Sn_3(OH)_4]^{2+}$ in lower acidity.

Acidifying ortho- or meta-stannates(IV) yields the related hydroxides.

Tin(II) oxide and oxalate are soluble in acids. Freshly precipitated ortho-tin(IV) hydroxide is soluble or peptized in many acids, even giving $[Sn(OH)_n(H_2O)_{6-n}]^{(4-n)+}$ in $HClO_4$. Tin dioxide and meta-tin(IV) oxide/hydroxide are insoluble in most acids except concentrated H_2SO_4.

Hydroxide. Tin dissolves in OH^- very slowly in the air or, in hot alkali releasing H_2, apparently forming the powerful reductant, $[Sn(OH)_3]^-$, as an intermediate, but also with $Na_2[\{Sn(OH)_2\}_2(\mu\text{-}O)]$ etc.:

$$Sn + OH^- + {}^1/_2\, O_2 + H_2O \rightarrow [Sn(OH)_3]^-$$

$$Sn + 2\, OH^- + 4\, H_2O \rightarrow [Sn(OH)_6]^{2-} + 2\, H_2\uparrow$$

Tin(II) starts precipitating at a pH ~ 2. The precipitate is gelatinous and slowly absorbs oxygen from the air, forming white $H_2SnO_3 \cdot aq$.

If mixed quite slowly, OH^- and Sn^{II} precipitate a white $[Sn_6O_4(OH)_4]$, i.e., [*octahedro*-Sn_6-*tetrahedro*-$(\mu_3\text{-}O)_4$-*tetrahedro*-$(\mu_3\text{-}OH)_4$]:

$$3\,[SnCl_3]^- + 6\,OH^- \rightarrow {}^1/_2\,[Sn_6O_4(OH)_4]\downarrow + 2\,H_2O + 9\,Cl^-$$

$${}^1/_2\,[Sn_6O_4(OH)_4] + 3\,OH^- + 2\,H_2O \rightarrow 3\,[Sn(OH)_3]^-$$

These Sn(II) hydroxides and oxide all dissolve in excess OH^- (and H_3O^+) but not in NH_3, thus being amphoteric, but the low oxidation state confers more basicity than acidity, as opposed to Sn^{IV}.

Heating a solution of stannate(II) gives black crystalline SnO, but boiling concentrated OH^- and stannate(II) causes dismutation:

$$2\,[Sn(OH)_3]^- \rightarrow Sn\downarrow + [Sn(OH)_6]^{2-}$$

Tin dioxide is difficultly soluble in OH^-.

Aqueous OH^- precipitates Sn^{IV} as ortho-$SnO(OH)_2\cdot aq$ or $H_2SnO_3\cdot aq$ (see **14.4 Tin** above), soluble in excess OH^-, but not in NH_3 or CO_3^{2-}:

$$[SnCl_6]^{2-} + 4\,OH^- \rightarrow SnO(OH)_2\cdot aq\downarrow + 6\,Cl^- + 2\,H_2O$$

$$SnO(OH)_2\cdot aq + 2\,OH^- + H_2O \rightarrow [Sn(OH)_6]^{2-}$$

The white product, when dried, looks like gelatin. It is amphoteric, but turns wet blue litmus red, acting like a dibasic $H_2SnO_3\cdot aq$, i.e., revealing more acidity than basicity; hence many stannates(IV), such as $M^I{}_2[Sn(OH)_6]\cdot aq$, are known, some with variable compositions.

Heated, the ortho hydrous oxides change rapidly to a meta form.

Meta-Sn^{IV} salts with alkalis precipitate meta-$SnO_2\cdot aq$; less acidic than the ortho form; not readily soluble in NaOH; insoluble in NH_3 and CO_3^{2-}; soluble in not-too-concentrated KOH; excess KOH precipitates potassium meta-stannate(IV), soluble in water. Tartrate prevents the precipitation of Sn^{II} hydrous oxide by OH^- and CO_3^{2-}; similarly (ortho) $SnCl_4$ gives no precipitate, but the action of meta-Sn^{IV} chlorides is unaffected—see **14.4.3 Reduced halogens** below about some of these.

Tin(II) sulfide, SnS, is decomposed by concentrated OH^-:

$$6\,SnS + 6\,OH^- \rightarrow 3\,Sn\downarrow + 2\,[SnS_3]^{2-} + [Sn(OH)_6]^{2-}$$

Di- and trioxygen. Stannate(II) absorbs oxygen on standing, more rapidly in the presence of a little tartrate as complexant:

$$2\,[Sn(OH)_3]^- + O_2 \rightarrow [Sn(OH)_6]^{2-} + SnO_2\cdot aq\downarrow$$

Ozone quickly oxidizes $[SnCl_3]^-$ and HCl to $[SnCl_6]^{2-}$.

14.4.2 Reagents Derived from the Other 2nd-Period Non-Metals, Boron through Fluorine

Boron species. Pouring $[Sn(OH)_3]^-$ and $[BH_4]^-$ under N_2 or Ar into excess 6-M HCl gives a low yield of SnH_4, also with some Sn_2H_6 etc.:

$$4\ [Sn(OH)_3]^- + 3\ [BH_4]^- + 7\ H_3O^+ \rightarrow 4\ SnH_4\uparrow + 3\ H_3BO_3 + 7\ H_2O$$

Carbon oxide species. Aqueous CO_3^{2-} precipitates from Sn^{II}, white Sn^{II} hydrous oxide, $[Sn_6O_4(OH)_4]$, insoluble in excess CO_3^{2-}, in a limited distinction from Sb. (It is also precipitated by $BaCO_3$ in the cold):

$$3\ [SnCl_3]^- + 6\ CO_3^{2-} + 4\ H_2O \rightarrow$$

$$\tfrac{1}{2}\ [Sn_6O_4(OH)_4]\downarrow + 9\ Cl^- + 6\ HCO_3^-$$

Tin(IV) is precipitated by CO_3^{2-} as white $SnO(OH)_2 \cdot aq$, somewhat soluble in excess CO_3^{2-}:

$$[SnCl_6]^{2-} + 4\ CO_3^{2-} + 3\ H_2O \rightarrow SnO(OH)_2 \cdot aq\downarrow + 6\ Cl^- + 4\ HCO_3^-$$

$$SnO(OH)_2 \cdot aq + 2\ CO_3^{2-} + 3\ H_2O \leftrightarrows [Sn(OH)_6]^{2-} + 2\ HCO_3^-$$

or, with less carbonate:

$$[SnCl_6]^{2-} + 2\ CO_3^{2-} + H_2O \rightarrow SnO(OH)_2 \cdot aq\downarrow + 6\ Cl^- + 2\ CO_2\uparrow$$

Cyanide species. Aqueous CN^- precipitates hydrous oxides, but not true cyanides, from tin solutions, by hydrolysis.

Some "simple" organic reagents. Oxalic acid forms a white, crystalline precipitate of SnC_2O_4 with a nearly neutral solution of $SnCl_2$, soluble in HCl; excess Cl^- prevents the precipitation. If a nearly neutral solution of $SnCl_2$ is added dropwise to a solution of $C_2O_4^{2-}$, the white precipitate formed at once dissolves in the excess of reagent, but $SnCl_4$ gives no precipitate with oxalates.

Reduced nitrogen. All the common tin compounds are insoluble in NH_3 or hydrolyzed and precipitated by it. Thus NH_3 gives approximately $[Sn_3O(OH)_2]SO_4$ with $SnSO_4$.

Heated in 2-M NH_3 to ~ 65 °C, $[Sn_6O_4(OH)_4]$ goes to black SnO.

Oxidized nitrogen. Nitrous acid and Sn^{II} give Sn^{IV}.

Concentrated HNO_3 rapidly converts tin into meta-tin(IV) hydrous oxide, insoluble in acids:

$$3\ Sn + 4\ NO_3^- + 4\ H_3O^+ \rightarrow 3\ SnO_2 \cdot aq\downarrow + 4\ NO\uparrow + 6\ H_2O$$

Dilute HNO_3 dissolves tin without release of gas, becoming NH_4^+:

$$4\,Sn + 10\,H_3O^+ + NO_3^- \rightarrow 4\,Sn^{2+} + 13\,H_2O + NH_4^+$$

The resulting aqueous Sn^{II} nitrate, also made from SnO or $[Sn_6O_4(OH)_4]$ and dilute HNO_3, is fairly stable but may explode on evaporation.

The solid nitrates of tin are not stable. Tin(II) nitrate is deliquescent and soon decomposes on standing exposed to the air. A basic nitrate resulting from hydrolysis at rather low pH is $[Sn_6(OH)_8](NO_3)_4$.

Aqua regia dissolves tin easily:

$$3\,Sn + 16\,H_3O^+ + 4\,NO_3^- + 18\,Cl^- \rightarrow 3\,[SnCl_6]^{2-} + 4\,NO\uparrow + 24\,H_2O$$

Tin dichloride warmed with HCl and HNO_3 forms Sn^{IV} and NH_4^+:

$$4\,[SnCl_3]^- + NO_3^- + 10\,H_3O^+ + 12\,Cl^- \rightarrow$$

$$4\,[SnCl_6]^{2-} + NH_4^+ + 13\,H_2O$$

With HCl absent the reaction is closer to:

$$6\,SnCl_2 + 4\,H_3O^+ + 4\,NO_3^- \rightarrow$$

$$3\,SnCl_4 + 3\,SnO_2\cdot aq\downarrow + 4\,NO\uparrow + 6\,H_2O$$

Tin(II) sulfide is oxidized by HNO_3 to meta-tin(IV) hydrous oxide. Other Sn^{II} salts and SnO, as well as freshly precipitated (ortho) $SnO(OH)_2\cdot aq$, when heated with HNO_3 likewise yield the much less soluble $SnO_2\cdot aq$. Nitrous acid gives similar results.

14.4.3 Reagents Derived from the 3rd-to-5th-Period Non-Metals, Silicon through Xenon

Phosphorus species. Phosphinic acid, HPH_2O_2, does not precipitate Sn^{II} or Sn^{IV}, nor are these ions reduced when boiled with the acid. Phosphinate ion, $PH_2O_2^-$, produces a white precipitate when added to $SnCl_2$, soluble in excess of HCl; no precipitate is formed with $SnCl_4$.

Dissolving Sn in 15-M (85 %) H_3PO_4 at 170 °C, cooling to 130 °C and adding boiling water yield white, very insoluble $SnHPO_4\cdot{}^1/_2H_2O$.

Phosphoric acid and its anions precipitate from Sn^{II}, not too strongly acidic, white Sn^{II} phosphate, composition variable, soluble in various acids and OH^-, insoluble in H_2O. A white gelatinous precipitate is formed with $SnCl_4$, soluble in HCl and OH^-, insoluble in HNO_3 and CH_3CO_2H. If any $SnCl_4$ is dissolved in excess OH^- before adding HPO_4^{2-} and the mixture is then acidified with HNO_3, any tin is completely precipitated as Sn^{IV} phosphate (separation from all but a little Sb). The hydrated dioxide, $SnO_2\cdot aq$, forms a tin(IV)-phosphate gel with H_3PO_4.

Tin(II and IV) metaphosphates and diphosphates have been prepared. Tin(II) forms strong complexes with HPO_4^{2-} and $[P_2O_7]^{4-}$.

Arsenic species. Arsenite plus $SnCl_2$ yield white Sn^{II} arsenite, sometimes with some AsH_3 and Sn^{IV}:

$$3\ SnCl_2 + 2\ AsO_3^{3-} + 2\ H_2O \rightarrow Sn_3(AsO_3)_2.2H_2O\downarrow + 6\ Cl^-$$

Heating this in either acid or base slowly gives As and Sn^{IV}.

Tin(II) is oxidized to Sn^{IV} by As^V in excess HCl.

Adding $SnCl_2$ to $H_2AsO_4^-$ in aqueous CH_3CO_2H precipitates the voluminous and flocculent tin(II) hydrogenarsenate, $SnHAsO_4\cdot{}^1/_2H_2O$. It decomposes on heating, partly to As, As_2O_3 and $SnO_2\cdot aq$.

Added to $SnCl_4$, arsenite ions precipitate Sn^{IV} orthoarsenite, white, likewise decomposing on heating.

A gelatinous, white $SnOHAsO_4\cdot{}^9/_2H_2O$, is precipitated on adding HNO_3 to a mixture of $[Sn(OH)_6]^{2-}$ and AsO_4^{3-}.

Reduced chalcogens. Sulfane (H_2S) or S^{2-} precipitates, from neutral or acidic Sn^{II}, a dark-brown SnS, unless prevented by $H_2C_2O_4$ or much Cl^-:

$$SnCl_3^- + H_2S + 2\ H_2O \leftrightarrows SnS\cdot aq\downarrow + 2\ H_3O^+ + 3\ Cl^-$$

This is insoluble in dilute, soluble in moderately concentrated, HCl. It is readily oxidized and dissolved by S_x^{2-}, forming trithiostannate(IV):

$$SnS + S_2^{2-} \rightarrow [SnS_3]^{2-}$$

The normal (not poly-) alkali sulfides dissolve scarcely any SnS at room temperature, but concentrated S^{2-} decomposes it:

$$2\ SnS + S^{2-} \rightarrow [SnS_3]^{2-} + Sn\downarrow$$

Hydroxide dissolves it, but the common explanation:

$$2\ SnS + 3\ OH^- \rightarrow [Sn(OH)_3]^- + SnS_2^{2-}$$

implies, contrary to the just-noted insolubility (without decomposition) of SnS in S^{2-}, that SnS_2^{2-} is inert after all. Conceivably we get e.g.:

$$2\ SnS + 4\ OH^- \rightarrow 2\ [SnS(OH)_2]^{2-}$$

In any case, acids reprecipitate the SnS. Aqueous CO_3^{2-} and NH_3 do not dissolve it (distinction from As and, with respect only to CO_3^{2-}, from Sb). Aqua regia dissolves it:

$$3\ SnS + 16\ H_3O^+ + 18\ Cl^- + 4\ NO_3^- \rightarrow$$

$$3\ [SnCl_6]^{2-} + 3\ S\downarrow + 4\ NO\uparrow + 24\ H_2O$$

Nitric acid converts it to the insoluble meta-tin(IV) hydrous oxide:

$$3\ SnS + 4\ H_3O^+ + 4\ NO_3^- \rightarrow$$

$$3\ SnO_2\cdot aq\downarrow + 4\ NO\uparrow + 3\ S\downarrow + 6\ H_2O$$

(distinction from As, but any such separation is poor because the precipitate adsorbs the otherwise soluble H_3AsO_4, also iron, etc.). Hydrogen peroxide in alkaline solution oxidizes and dissolves it:

$$3\ SnS + 3\ OH^- + 3\ HO_2^- + 3\ H_2O \rightarrow [SnS_3]^{2-} + 2\ [Sn(OH)_6]^{2-}$$

Sulfane, H_2S, precipitates tin(IV) as SnS_2, yellow, having generally the same solubilities as SnS except that SnS_2 is moderately soluble in S^{2-}. The following equations give some important reactions:

$$[SnCl_6]^{2-} + 2\ H_2S\uparrow + 4\ H_2O \leftrightarrows SnS_2\downarrow + 4\ H_3O^+ + 6\ Cl^-$$

$$SnS_2 + HS^- + NH_3 \rightarrow [SnS_3]^{2-} + NH_4^+$$

$$[SnS_3]^{2-} + 2\ H_3O^+ \rightarrow SnS_2\downarrow + H_2S\uparrow + 2\ H_2O$$

$$SnS_2 + S_2^{2-} \rightarrow [SnS_3]^{2-} + S\downarrow$$

$$SnS_2 + 2\ S_2^{2-} \rightarrow [SnS_3]^{2-} + S_3^{2-}$$

$$3\ SnS_2 + 6\ OH^- \rightarrow [Sn(OH)_6]^{2-} + 2\ [SnS_3]^{2-}$$

$$[Sn(OH)_6]^{2-} + 2\ [SnS_3]^{2-} + 6\ H_3O^+ \rightarrow 3\ SnS_2\downarrow + 12\ H_2O$$

$$3\ SnS_2 + 16\ H_3O^+ + 18\ Cl^- + 4\ NO_3^- \rightarrow$$

$$3\ [SnCl_6]^{2-} + 4\ NO\uparrow + 3\ S\downarrow + 24\ H_2O$$

$$3\ SnS_2 + 4\ H_3O^+ + 4\ NO_3^- \rightarrow$$

$$3\ SnO_2\cdot aq\downarrow + 4\ NO\uparrow + 6\ S\downarrow + 6\ H_2O$$

Boiling SnS_2 1 h with Cs_2S leads to colorless $Cs_8[Sn_{10}O_4S_{20}]\cdot 13H_2O$.

Sulfane does not precipitate SnS_2 in the presence of excess OH^-, $H_2C_2O_4$ (distinction from As and Sb), excess H_3PO_4 (distinction from Sn^{II} and Sb^{III}), HF (distinction from Sn^{II} and Sb^{III}), or excessive HCl.

Oxidized chalcogens. Aqueous $S_2O_3^{2-}$ forms no precipitate with Sn^{II}, but acid produces SO_2, which oxidizes the Sn^{II} to Sn^{IV}, e.g.:

$$SO_2 + 3\,[SnCl_3]^- + 6\,H_3O^+ + 9\,Cl^- \rightarrow 3\,[SnCl_6]^{2-} + H_2S\uparrow + 8\,H_2O$$

The H_2S may or may not precipitate SnS_2 as above, depending on conditions. Excess $S_2O_3^{2-}$ will react with the $SnCl_6^{2-}$, producing a white precipitate of Sn^{IV} sulfide and hydrous oxide. Sulfur dioxide and sulfites react with Sn^{II} as already indicated.

Tin dissolves slowly in H_2SO_4, over 6 M, with liberation of H_2:

$$Sn + H_3O^+ + HSO_4^- \rightarrow SnSO_4 + H_2\uparrow + H_2O$$

Hot concentrated H_2SO_4 dissolves tin rapidly, liberating SO_2 and S:

$$Sn + 4\,H_2SO_4 \rightarrow SnSO_4 + SO_2\uparrow + 2\,HSO_4^- + 2\,H_3O^+$$

$$2\,SnSO_4 + SO_2 + 4\,H_2SO_4 \rightarrow$$

$$2\,Sn(SO_4)_2 + S\downarrow + 2\,HSO_4^- + 2\,H_3O^+$$

The sulfates of tin are formed by dissolving the freshly precipitated hydrous oxides in H_2SO_4 and evaporating at a gentle heat; at 130–200 °C $SnSO_4$ is oxidized to $Sn(SO_4)_2$:

$$SnSO_4 + 2\,H_2SO_4 \rightarrow Sn(SO_4)_2 + 2\,H_2O\uparrow + SO_2\uparrow$$

Cold, concentrated H_2SO_4 does not dissolve $SnO_2\cdot aq$.

Tin(II) is oxidized to Sn^{IV} also by $[S_2O_8]^{2-}$.

Reduced halogens. Tin dissolves in HCl slowly when cold and dilute, but rapidly when hot and concentrated. Hot HBr and HI also dissolve it:

$$Sn + 2\,H_3O^+ + 3\,Cl^- \rightarrow [SnCl_3]^- + H_2\uparrow + 2\,H_2O$$

Aqueous HCl is found to change meta-SnO_2.aq to a soluble product written as "$Sn_5O_5Cl_2(OH)_8$", which upon further addition of acid becomes insoluble "$Sn_5O_5Cl_4(OH)_6$". Conflicting older reports and some shortage of modern data do not justify much space here, but the complexity is clear. These various polymeric meta forms, in HCl, change definitely but gradually to $[SnCl_6]^{2-}$ or $H_2[SnCl_6]\cdot 6H_2O$.

Tin monosulfide is soluble in not-too-dilute HCl, releasing H_2S.

Concentrated HCl dissolves SnS_2 as $[SnCl_6]^{2-}$ (separation from As).

Iodide, added to concentrated $SnCl_2$, first forms a yellow precipitate, soluble in excess $SnCl_2$. More KI precipitates yellow (soon turning to dark orange) needle-like or rosette crystals. Adding a drop of $SnCl_2$ to excess KI, gives the yellow precipitate, which remains unless more $SnCl_2$ is added, giving the orange variety. Each form is soluble in HCl, OH^-, or ethanol, and sparingly soluble in H_2O with some decomposition.

Concentrated I^- precipitates SnI_4, yellow, from very concentrated $SnCl_4$, readily soluble in H_2O and HCl to a colorless solution. Aqueous HI does not release I_2 with Sn^{IV} (distinction from As^V and Sb^V).

Elemental and oxidized halogens. Tin(II) is oxidized to Sn^{IV} by Cl_2, Br_2 and I_2. They react more vigorously in alkaline than in acidic solution, but then react at least partly as hypohalites, XO^-.

Tin is attacked by ClO^- and dissolved by $HClO_3$.

Aqueous $SnCl_2$ is oxidized to Sn^{IV} by HClO, $HClO_2$, $HClO_3$, $HBrO_3$ or HIO_3. Chlorate rapidly oxidizes $SnCl_2$ to Sn^{IV}. Bromate or iodate, plus $SnCl_2$, form yellowish to white salts that quickly decompose, liberating Br_2 or I_2. However, fresh tin(II) hydroxide dissolves in $HClO_3$, forming Sn^{II} chlorate which, as a solid, soon decomposes explosively.

The three halates all form precipitates with $SnCl_4$, soluble in HCl without liberating halogen.

14.4.4 Reagents Derived from the Metals Lithium through Uranium, plus Electrons and Photons

Oxidation. Metallic tin is oxidized by ions of Pt, Cu, Ag, Au, Hg and Bi, which are reduced to the metallic state, while Sn goes at least to Sn^{II}.

General note: Tin(II) chloride is one of the most convenient and efficient ordinary discriminating reductants for operations in the wet way. Because the products of its oxidation, e.g., $SnCl_4$, are soluble in the solvents of $SnCl_2$, no tin compound is precipitated as a result of its reducing action in chloride solutions, but many other metals do yield precipitates and are thus identified in analysis, e.g., Ag, Au (forms Cassius Purple), Hg, As, and Bi.

Tin(II) is oxidized to Sn^{IV} by the ordinary (especially oxo) species of Cr^{VI}, Mo^{VI}, $Mn^{>II}$, Fe^{III} including $[Fe(CN)_6]^{3-}$, Co^{III}, Ni^{III}, Pd, Pt, Cu, Pb^{IV}, Sb^V and Bi^V, all in both acid and base, and by Bi^{III}, only in base.

Tin(II) and $(NH_4)_2MoO_4$ give a blue partly reduced molybdenum. The color varies with the concentration and acidity (Sb, if present, gives a green color). A freshly prepared reagent detects 6-μM Sn. Or, adding SCN^- to the molybdate in HCl, and then the Sn^{II}, produces a red color.

Tin(II) does not reduce $[CoCl(NH_3)_5]^{2+}$, $[CoBr(NH_3)_5]^{2+}$ etc.

Aqueous $[Ag(NH_3)_2]^+$ oxidizes $[Sn(OH)_3]^-$ and gives metallic Ag (a sensitive test for stannite). Adding excess OH^- to an unknown removes most **d**- or **p**-block metals that are not amphoteric (leaving Cr, Zn, Al, Sn, Pb, As and Sb); of these only tin(II) precipitates Ag from a cold, strongly ammoniacal solution. Compounds of As^{III} and Sb^{III} give the black precipitate of metallic silver if the solution is boiled.

A solution of $[HgCl_2]$ reacts with $SnCl_2$ forming $SnCl_4$ and a precipitate of $[Hg_2Cl_2]$, white, or Hg, gray, or a mixture of the two, depending on the temperature and relative amounts of reagents.

Light (254 nm), very dilute $[SnCl_3]^-$, Cl^- and O_2 give $[SnCl_6]^{2-}$ and some H_2O_2, destroying the fluorescence of the unoxidized solution.

Reduction. Tin(IV) is reduced by Fe, Ni, Cu, Al, Sn, Pb and Sb to Sn^{II}, which may be detected by means of $[HgCl_2]$, $(NH_4)_2MoO_4$, I^-, Bi^{3+} etc.

Tin(II and IV) salts are reduced to tin by Mg, Fe, Zn, Cd and Al.

Other reactions. If $SnCl_2$ is carefully added to CrO_4^{2-} in excess, an abundant yellow precipitate is obtained without much apparent reduction of the chromium. Aqueous CrO_4^{2-} added to $SnCl_4$ gives a bright yellow precipitate, soluble in excess of $SnCl_4$, insoluble in H_2O, difficultly soluble in HCl. Dichromate also gives precipitates with $SnCl_2$ and $SnCl_4$. Note **6.2.5** on heteropolymolybdates and tungstates.

Aqueous $[Fe(CN)_6]^{4-}$ precipitates, from $SnCl_2$ solutions, $Sn_2[Fe(CN)_6]$, white, soluble in hot, concentrated HCl. Tin tetrachloride gives a greenish white, gelatinous precipitate soluble in hot, concentrated HCl, but reprecipitated on cooling (distinction from Sb).

Aqueous $[Fe(CN)_6]^{3-}$ with $SnCl_2$ precipitates white $Sn_3[Fe(CN)_6]_2$, readily soluble in HCl, subject to some internal oxidation and reduction on warming. No precipitate is formed by $[Fe(CN)_6]^{3-}$ with $SnCl_4$.

14.5 Lead, $_{82}$Pb (and Ununquadium, $_{114}$Uuq)

Oxidation numbers of lead: (II) and (IV), as in PbO, "plumbous" oxide, PbO_2, "plumbic" oxide, and Pb_3O_4, i.e., $Pb^{II}_2Pb^{IV}O_4$, "red lead", not quite stiochiometric. Relativity gives us the inert pair in Pb^{II}.

Relativistic quantum mechanics, applied to Uuq, recently synthesized, predicts chemical stability for Uuq^{2+} but not for $UuqO_2$, $UuqF_4$ or $UuqCl_4$. It also predicts that Uuq will be only about as reactive as Hg.

14.5.1 Reagents Derived from Hydrogen and Oxygen

Water. The hydrated Pb^{II} ion is $[Pb(H_2O)_6]^{2+}$.

Among the classical Pb^{II} compounds, the borate, carbonate, cyanide, oxalate, phosphate, sulfide, sulfite, sulfate, iodate, chromate, and hexacyanoferrate(II), plus Pb_3O_4 and PbO_2, are practically insoluble.

Lead monoxide, yellowish (in massicot) or reddish (in litharge), is slightly soluble in water. The halides are also slightly soluble in cold water, more so in hot. The hexacyanoferrate(III) is slightly soluble.

Seawater and some freshwater contain traces of Pb^{2+} complexes as $PbOH^+$, $PbCO_3$, $PbHCO_3^+$, $PbSO_4$, $Pb(SO_4)_2^{2-}$ and $PbCl_n^{(n-2)-}$.

Other natural waters may contain $Pb(CO_3)_2^{2-}$, $Pb(HCO_3)_2$, $Pb(HCO_3)_3^-$ $Pb_2(CO_3)_2Cl^-$, $H_mPb^{II}S_n^{(2n-m-2)-}$ and polysulfido and thiosulfato complexes. Some natural brines may contain $[PbCl_3]^-$, $[PbCl_4]^{2-}$, $PbBr^+$, $PbBr_2$, $[PbBrCl_2(OH)]^{2-}$, $[PbBr_2Cl(OH)]^{2-}$, $Pb_4Cl_4(OH)_4$ and so on.

Oxonium. Hydrochloric acid attacks lead slowly, releasing hydrogen.

Alloys of lead are best dissolved by first treating with HNO_3, or HCl with HNO_3 or ClO_3^- in some cases. If a white residue is left, it is washed with water and, if not dissolved, is treated with HCl in which it will usually be soluble.

The chemical properties of lead present some strange contrasts. The metal resists the action of H_2SO_4 or HCl much better than do iron, zinc or tin; yet it is readily attacked by weak organic acids, it dissolves slowly even in water, and it is quickly corroded by moist air.

Lead monoxide, PbO, and the hydroxides, $Pb_2O(OH)_2$ and $Pb_3O_2(OH)_2$, actually $[Pb_6O_4(OH)_4]$ structurally, react readily with acids, forming the corresponding compounds, soluble or insoluble:

$$PbO + 2\ H_3O^+ \rightarrow Pb^{2+} + 3\ H_2O$$

With reducing agents, e.g., H_2O_2, CH_2O, C_2H_5OH, $H_2C_2O_4$, Cl^-, Br^-, I^-, free metal, etc., acids easily dissolve Pb_3O_4 or PbO_2 as Pb^{II}.

Strong non-reducing acids, e.g., HNO_3, H_2SO_4 and $HClO_4$, separate the oxidation states in Pb_3O_4, precipitating Pb^{II} salts where expected:

$$Pb_3O_4 + 4\ H_3O^+ \rightarrow 2\ Pb^{2+} + PbO_2\downarrow + 6\ H_2O$$

Hydroxide. Nearly all compounds of lead are soluble in OH^-; PbS, PbSe and PbTe are notable exceptions.

Aqueous OH^- and lead(II) form white $Pb_3O_2(OH)_2$ [there is no simple $Pb(OH)_2$], slightly soluble, the resulting solution reacting alkaline, also soluble in excess OH^- and in certain ligands, e.g., chloride:

$$3\ Pb^{2+} + 6\ OH^- \rightarrow \sim Pb_3O_2(OH)_2\downarrow + 2\ H_2O$$

The exact composition depends on the temperature and concentrations of the reactants. Concentrated OH^- yields $[Pb(OH)_4]^{2-}$.

The dioxide is slowly soluble in OH^- as $[Pb(OH)_6]^{2-}$:

$$PbO_2 + 2\ OH^- + 2\ H_2O \rightarrow [Pb(OH)_6]^{2-}$$

Peroxide. Alkalis, Pb^{II} and O_2^{2-} give PbO_2, a very strong oxidant, or $[Pb(OH)_6]^{2-}$. Digesting the PbO_2 with NH_3 yields some NO_3^-; triturated with a little sulfur or sugar, PbO_2 starts a fire; with phosphorus, it detonates.

Di- and trioxygen. Water, air and Pb slowly form the Pb^{II} oxide hydroxide. Slowly if neutral, quickly if basic, Pb^{II} and O_3 precipitate dark-brown PbO_2, and dark-brown PbS becomes white $PbSO_4$.

14.5.2 Reagents Derived from the Other 2nd-Period Non-Metals, Boron through Fluorine

Carbon oxide species. Lead(2+) with CO_3^{2-} or HCO_3^- precipitates $PbCO_3$ in the cold, basic lead carbonate when hot, which varies with conditions but is chiefly $Pb_3(CO_3)_2(OH)_2$. Boiling many lead salts with freshly precipitated $BaCO_3$ completely precipitates them.

Cyanide species. Lead(2+) and CN^- precipitate $Pb(CN)_2$ (separation from Tl)—lead acetate gives $Pb_3O_2(CN)_2$—white, sparingly soluble in a large excess of the reagent, reprecipitated on boiling. In the presence of dilute H_2SO_4, HCN reduces Pb^{IV} to $PbSO_4$.

Some "simple" organic species. Lead(II) oxide-hydroxide is soluble in solutions of such anions as acetate, citrate, tartrate, Edta etc. Lead sulfate dissolves in concentrated solutions of acetate or tartrate ions.

Because $Pb(C_2H_3O_2)_2$ is very slightly ionized, many insoluble compounds of lead dissolve in $CH_3CO_2^-$ (but not CH_3CO_2H), and $CH_3CO_2^-$ can even leach Pb^{II} from some ores.

Aqueous $Pb(C_2H_3O_2)_2$ precipitates many (and the "subacetate", basic acetate, more) organic acids, color substances, resins, gums, etc. The "subacetate" exceptions include $H(CH_2)_nCO_2H$ ($n < 5$) and lactic acid.

The oxides Pb_3O_4 or PbO_2 react with certain acids in the cold forming compounds of lead(IV); e.g., concentrated acetic acid forms $[Pb(CH_3CO_2)_4]$. These compounds are very unstable, decomposing to give the lead(II) compound when warmed.

Oxalic acid and $C_2O_4^{2-}$ precipitate PbC_2O_4, white, from Pb^{2+}; soluble in HNO_3, $C_2O_4^{2-}$ and hot Cl^-; insoluble in CH_3CO_2H. Oxalic acid and very many organic compounds, with dilute H_2SO_4, reduce Pb^{IV} to $PbSO_4$.

Reduced nitrogen. Ammonia precipitates Pb^{2+} as white basic compounds, insoluble in water and in excess reagent. Examples include $Pb_2OCl(OH)$ and $Pb_3O_2(NO_3)(OH)$. Excess of NH_3 (free from CO_3^{2-}) gives no precipitate with the acetate at ordinary concentrations, due to the low concentrations of both Pb^{2+} and OH^-.

Freshly made PbO_2 oxidizes NH_3 to NO_2^- and NO_3^- in a few hours.

Oxidized nitrogen. Nitrous acid, in the presence of dilute H_2SO_4, reduces Pb^{IV} to $PbSO_4$.

Eight-M HNO_3 is the best of the common solvents for Pb:

$$3\ Pb + 8\ H_3O^+ + 2\ NO_3^- \rightarrow 3\ Pb^{2+} + 2\ NO\ (\text{and } NO_2)\uparrow + 12\ H_2O$$

The presence of nitrite hastens the action. Concentrated (16-M) HNO_3 is less effective because lead nitrate is insoluble in this acid and forms a protective coating on the metal, preventing further action.

Lead nitrate is readily soluble in water; its solubility is greatly increased by the presence of moderate amounts of added NO_3^-, a complex ion being formed. The nitrate reacts with PbO to form a basic nitrate. This is slightly soluble, but is also precipitated on adding NO_3^- to a solution of basic lead acetate.

Fluorine species. Hydrofluoric acid and F^- precipitate, from solutions of lead(II) salts, PbF_2, white, sparingly soluble in water or HNO_3; practically insoluble in HF; more soluble in HCl; slowly soluble in OH^-. Its solubility in alkali halides increases with the size and polarizability of the halide ion. The properties of PbF_2 are closer to those of PbO than to those of the other halides. It is decomposed by H_2SO_4, forms an oxy-fluoride in the presence of NH_3, and is little affected by Cl_2, Br_2 or I_2.

14.5.3 Reagents Derived from the 3rd-to-5th-Period Non-Metals, Silicon through Xenon

Phosphorus species. Lead(IV) and dilute H_2SO_4, together with either P_4, HPH_2O_2 or H_2PHO_3, form $PbSO_4$.

The higher oxides of lead are reduced by HPH_2O_2, and H_2PHO_3, forming $Pb_3(PO_4)_2$. Lead(II) oxide gives $Pb(PH_2O_2)_2$.

The basic, normal, and acid phosphonites of lead are all white and soluble in dilute H_3O^+ (but not CH_3CO_2H).

The ions HPO_4^{2-} and PO_4^{3-} precipitate from lead acetate or nitrate, normal lead phosphate, $Pb_3(PO_4)_2$, white, slightly soluble in CH_3CO_2H, soluble in HNO_3 or OH^-, converted to PbI_2 by I^-:

$$3\ Pb(CH_3CO_2)_2 + 2\ HPO_4^{2-} \rightarrow$$

$$Pb_3(PO_4)_2\downarrow + 4\ CH_3CO_2^- + 2\ CH_3CO_2H$$

Lead(II) salts and $[P_2O_7]^{4-}$ yield white, amorphous $Pb_2[P_2O_7]$, soluble in excess reagent, dilute acids or OH^-; insoluble in NH_3, CH_3CO_2H or aqueous SO_2 (probably transposed by the latter). Insoluble white, crystalline lead metaphosphates (several forms), readily decomposed by acids, arise from $(PO_3^-)_n$ and Pb^{2+}.

Arsenic species. Arsane or As^{III}, in the presence of dilute H_2SO_4, reduces Pb^{IV} to $PbSO_4$.

Arsenites precipitate from Pb^{2+}, bulky, white $Pb_3(AsO_3)_2 \cdot aq$, difficultly soluble in water, readily soluble in dilute acids and OH^-. Arsenate precipitates white lead arsenate from neutral or slightly alkaline Pb^{II}, soluble in OH^- and HNO_3; insoluble in CH_3CO_2H. It may be a mixture of $Pb_3(AsO_4)_2$ and $PbHAsO_4$, depending on conditions.

Reduced chalcogens. Sulfane (H_2S) and S^{2-} precipitate from neutral, slightly acidic, or alkaline solutions of lead compounds, PbS, brownish black; insoluble in dilute H_3O^+, OH^-, CO_3^{2-} or sulfides. Sulfane is sometimes used as a test for Pb. The test is not characteristic, for many of the metals form black precipitates with this reagent, but it is very sensitive, a brown color appearing in solutions as dilute as 5 μM, and S^{2-} will detect 2-μM Pb^{2+}. Lead sulfide may be obtained in good condition for later separation with enough care in regulating the amount of reagent used, but too much H_2S makes the precipitate colloidal.

Freshly precipitated metallic sulfides such as MnS, FeS, CoS, NiS and CdS will also precipitate PbS. Sulfane and S^{2-} transpose other freshly precipitated lead compounds to PbS, thus making them amenable to treatment by HNO_3, for example.

Moderately dilute HNO_3 ($2\text{–}2^1/_2$ M) dissolves PbS, separating sulfur:

$$3\ PbS + 8\ H_3O^+ + 2\ NO_3^- \rightarrow 3\ Pb^{2+} + 3\ S\downarrow + 2\ NO\uparrow + 12\ H_2O$$

Some of the sulfur, especially if the acid is 16 M, is oxidized to sulfate which will precipitate a portion of the lead unless enough HNO_3 is present to hold the $PbSO_4$ in solution. Some sulfur is always oxidized when HNO_3 acts on sulfides, and to a degree dependent on the $c(H_3O^+)$, temperature, and duration of contact.

$$3\ PbS + 8\ H_3O^+ + 8\ NO_3^- \rightarrow 3\ PbSO_4\downarrow + 8\ NO\uparrow + 12\ H_2O$$

If a solution is too strongly acidic, especially with HCl (at least 1.4 M), either no precipitation of PbS with H_2S takes place, or a red double compound, Pb_2Cl_2S, is formed incompletely. Chloride salts lower very distinctly the concentration of HCl necessary to prevent the precipitation of PbS from dilute lead chloride solution.

Iron trichloride oxidizes PbS, forming $PbCl_2$, Fe^{2+} and sulfur. The reaction takes place in the cold, more rapidly when warm. Iodine reacts readily with PbS even in a dry mixture.

Sulfane (H_2S), in the presence of dilute H_2SO_4, reduces Pb^{IV} to $PbSO_4$.

Gray PbSe, and white PbTe are precipitated by H_2Se and H_2Te from Pb^{2+} in low acidity.

Lead(2+) yields, with SCN^-, white $Pb(SCN)_2$, soluble in excess of the reagent and in HNO_3. Aqueous HSCN, in the presence of dilute H_2SO_4, reduces Pb^{IV} to $PbSO_4$.

Oxidized chalcogens. Lead thiosulfate, PbS_2O_3, white, is precipitated by adding $S_2O_3^{2-}$ to a lead(II) solution. The precipitate is readily soluble in excess of reagent, forming $[Pb(S_2O_3)_3]^{4-}$. On boiling, especially in the presence of Cl^-, the lead is quantitatively precipitated as the sulfide. The salt PbI_2 dissolves in $S_2O_3^{2-}$.

Aqueous SO_3^{2-} precipitates $PbSO_3$, white, less soluble in water than the sulfate; slightly soluble in aqueous SO_2; decomposed by H_2SO_4, HNO_3, HCl, H_2S, and S^{2-}; not by cold H_3PO_4 nor by acetic acid. Sulfur dioxide, in the presence of dilute H_2SO_4, reduces Pb^{IV} to $PbSO_4$.

Dilute H_2SO_4 has slight action on Pb; the concentrated acid is almost without effect in the cold, but when hot it slowly changes the metal to the sulfate with release of SO_2. A portion of the compound dissolves in the acid but is precipitated on the addition of water.

Aqueous SO_4^{2-} with Pb^{2+} in neutral or acid solution precipitates $PbSO_4$, white, not readily changed or permanently dissolved by acids, except H_2S, slightly soluble in strong acids, soluble in OH^-, moderately soluble in concentrated solutions of complexing anions such as acetate, tartrate, citrate, and Edta, soluble in warm $S_2O_3^{2-}$ solution (which decomposes on stronger heating with precipitation of PbS, insoluble in $S_2O_3^{2-}$), distinction and separation from $BaSO_4$, which does not dissolve in $S_2O_3^{2-}$ or acetate.

The test for lead using SO_4^{2-} is about one 25th as sensitive as that employing H_2S or SO_3^{2-}; yet lead is quantitatively separated as the sulfate by precipitation with SO_4^{2-} in moderate excess. The $PbSO_4$ is unusually compact and forms slowly in dilute solution. When heated with CrO_4^{2-}, $PbSO_4$ becomes yellow $PbCrO_4$. Excess of I^- also transposes $PbSO_4$ (to PbI_2), a distinction from barium. Repeatedly washing $PbSO_4$ with aqueous Cl^- likewise completely transposes the lead (to $PbCl_2$).

Water dissolves $PbSO_4$ up to 0.14 mM at ambient T, more in the presence of HNO_3 or HCl; it is almost completely transposed to the nitrate by standing several days in cold, concentrated HNO_3; insoluble in HF or in aqueous ethanol even when dilute; slightly soluble in concentrated H_2SO_4, depending markedly on the concentration of acid; less soluble in dilute H_2SO_4 than in water; more soluble in HCl than in HNO_3; transposed and dissolved by excess of HCl, HBr or HI; solutions in acetate, tartrate, citrate, nitrate, or chloride are not readily precipitated in most cases by NH_3 or SO_4^{2-}; soluble in OH^-, especially on warming.

Hot, concentrated H_2SO_4 and PbO_2 form $PbSO_4$ and oxygen.

Aqueous $[Pb(OH)_4]^{2-}$ and $[S_2O_8]^{2-}$ yield PbO_3^{2-}.

Much NH_3 with $[S_2O_8]^{2-}$ and $Pb(CH_3CO_2)_2$ precipitate PbO_2.

Reduced halogens. Aqueous Cl^- precipitates, from Pb^{2+} solutions not too dilute, $PbCl_2$, white. Its solubility is greater when warm, but even the cold supernate gives good tests for Pb^{2+} with H_2S, SO_4^{2-}, CrO_4^{2-}, etc. (Failure of such tests may be due to excess Cl^- or H_3O^+.) Lead dichloride is precipitated slowly when the solution is rather dilute.

The solubility of $PbCl_2$ is affected by other chlorides, falling to a minimum (the common-ion effect) and then rising again with $c(Cl^-)$ (formation of complexes

$[PbCl_4]^{2-}$ or even $[PbCl_6]^{4-}$). Minimal solubilities are found a little above 1-M Cl^-, with some dependence on the cation, and solubilities around 4 mM. The chloride is more soluble in HNO_3 than in water. The chloride, bromide and iodide are insoluble in ethanol. The iodide is moderately soluble in solutions of I^-.

Lead sulfate is soluble in cold, saturated NaCl, depositing crystals of $PbCl_2$ after some time, with complete transposition.

Galena, PbS, dissolves in HCl, up to 5 dM if hot, as chloro complexes.

Lead(IV) is somewhat stable in 11-M HCl as $[PbCl_6]^{2-}$, but the acids HCl, HBr and HI, with dilute H_2SO_4, reduce Pb^{IV} to $PbSO_4$. One may crystallize mixed Pb^{II} and Pb^{IV} as solid $[Co(NH_3)_6]_2[Pb^{II}Cl_6][Pb^{IV}Cl_6]$.

Bromide precipitates $PbBr_2$, white, somewhat less soluble in cold water than the chloride; soluble in excess of concentrated Br^- as a complex ion, which is decomposed with precipitation of $PbBr_2$ by dilution with water; soluble in OH^-.

Iodide precipitates PbI_2, much less soluble in water than the chloride or bromide, soluble in hot, moderately concentrated HNO_3 and in OH^-, forming complexes. These ions are decomposed by adding water, precipitating the PbI_2. This is not precipitated in the presence of excess $CH_3CO_2^-$ (not CH_3CO_2H). It dissolves easily in $S_2O_3^{2-}$. Freshly precipitated PbO_2 oxidizes I^- to I_2. Lead diiodide is appreciably soluble in a warm or hot solution; so precipitation may not take place immediately on adding the I^-. In such a case cooling will cause the formation of beautiful golden-yellow crystals of PbI_2. If the original precipitate is flocculent it may be converted to the crystalline form by dissolving in hot water and cooling slowly (characteristic of lead).

Lead dioxide is decomposed by HCl, HBr or HI, liberating the halogen and forming the corresponding lead halide:

$$PbO_2 + 4\,H_3O^+ + 4\,X^- \rightarrow PbX_2\downarrow + X_2 + 6\,H_2O$$

Elemental and oxidized halogens. Elemental Cl_2 and Br_2 attack Pb slowly but I_2 does not dissolve Pb in water.

Halogens, ClO^-, etc. convert alkaline Pb^{II} to PbO_2 or PbO_3^{2-}.

Bromate and Pb^{2+} precipitate $Pb(BrO_3)_2{\cdot}H_2O$. Lead acetate, however, may give the dangerously explosive $PbBrO_3(CH_3CO_2)$.

Iodate precipitates, from solutions of Pb^{2+}, white $Pb(IO_3)_2$, insoluble in CH_3CO_2H, difficultly soluble in HNO_3.

14.5.4 Reagents Derived from the Metals Lithium through Uranium, plus Electrons and Photons

Oxidation. Lead dioxide is formed by treating $[Pb(OH)_4]^{2-}$ with alkaline $[Fe(CN)_6]^{3-}$ (catalyzed by OsO_4), MnO_4^- etc. Metallic lead becomes Pb^{II} and precipitates the metals from solutions of Pt, Cu, Ag, Au, Hg or Bi. Lead dioxide is also formed at anodes with Pb^{2+}.

Reduction. Magnesium, Cr, Mn, Fe, Cu, Zn and Cd, but neither Co nor Ni, precipitate Pb from aqueous $Pb(CH_3CO_2)_2$.

Lead(IV), including Pb_3O_4, with dilute H_2SO_4, is reduced to $PbSO_4$ or Pb by Mg, Fe, Zn, Cd, Al, Sn etc., and to $PbSO_4$ by Cr^{III}, $Mn^{<VII}$, Fe^{2+}, $H_2[Fe(CN)_6]^{2-}$, Cu^{I}, Hg_2^{2+}, Sn^{II} and Sb^{III}, e.g.:

$$Pb_3O_4 + 3\ HSO_4^- + 2\ H_2[Fe(CN)_6]^{2-} + H_3O^+ \rightarrow$$

$$3\ PbSO_4\downarrow + 2\ [Fe(CN)_6]^{3-} + 5\ H_2O$$

Iron, Ni, Cu, Sn, Sb and Bi scarcely reduce Pb^{2+}. Cobalt or Al reduces $Pb(NO_3)_2$ a little; Zn or Cd reduces it slowly but completely.

Aluminum with HCl, better if including $[HgCl_2]$, reduces $Pb^{II,\,IV}$ to Pb.

Lead compounds and hot $[Sn(OH)_4]^{2-}$ precipitate black Pb.

Other reactions. Chromate and $[Cr_2O_7]^{2-}$ readily precipitate the bulky, yellow $PbCrO_4$, whether the solution is hot or cold, soluble in OH^- (distinction from Ba and Bi), insoluble in excess of chromic acid (distinction from Ba), insoluble in NH_3 (distinction from Ag), insoluble in CH_3CO_2H (distinction from Bi), soluble in 3-M HNO_3, and decomposed by moderately concentrated HCl. The chromate test is the most dependable of the classical chemical tests for Pb^{2+}.

Molybdate offers a means of separating lead from Mn, Co, Ni, Cu and Zn. The hot, slightly acidic (HCl or HNO_3) solution is first treated with an excess of MoO_4^{2-}; then $CH_3CO_2^-$ is added to reduce the acidity and thus insure complete precipitation of the $PbMoO_4$.

Aqueous $[Fe(CN)_6]^{4-}$ precipitates $Pb_2[Fe(CN)_6]$, white, insoluble in water or dilute acids. A freshly prepared solution of $[Fe(CN)_6]^{3-}$ gives no visible action with $[Pb(CH_3CO_2)_2]$. Subsequent addition of NH_3 causes the formation of a reddish-brown precipitate. On reaction with $Pb(NO_3)_2$, however, $[Fe(CN)_6]^{3-}$ gives $Pb_2[Fe(CN)_6](NO_3)(H_2O)_5$.

The hydrated oxide of lead(II) is very soluble in $[Pb(CH_3CO_2)_2]$ in the absence of CO_2, forming basic lead acetate.

Electrolysis both oxidizes and reduces Pb^{II} in storage batteries:

$$2\ PbSO_4\downarrow + 4\ H_2O \leftrightharpoons Pb\downarrow + PbO_2\downarrow + 2\ HSO_4^- + 2\ H_3O^+$$

where charging is to the right and discharging to the left.

A type of flow battery being developed [2] uses methanesulfonic acid because of the solubility of $Pb(CH_3SO_3)_2$, overcoming the problems of precipitated $PbSO_4$, or of $[BF_4]^-$, $[SiF_6]^{2-}$ and ClO_4^- salts in other potential flow batteries. The high $c(Pb^{2+})$ shifts the equilibrium to the right, lowering the E from > 2 V to ~ 1.5 V:

$$2\ Pb^{2+} + 6\ H_2O \leftrightharpoons Pb\downarrow + PbO_2\downarrow + 4\ H_3O^+$$

References

1. Brandão P, Almeida Paz FA, Rocha J (2005) Chem Comm 2005:171
2. Hazza A, Pletcher D, Wills R (2004) Phys Chem Chem Phys 6:1773.

Bibliography

See the references in the Introduction, and some more-specialized books [3–13], although [3] is mainly outside our scope. Some articles in journals discuss Si–Pb cluster compounds [14], valence and related concepts [15], the homogeneous hydrogenation of CO_2 [16], and lead poisoning and the fall of Rome [17].

3. Smith PJ (ed) (1998) Chemistry of tin, 2nd edn. Blackie, Glasgow
4. Kumar Das VGK, Weng NS, Gielen M (eds) (1992) Chemistry and technology of silicon and tin. Oxford University Press, Oxford.
5. Harrison PG (ed) (1989) Chemistry of tin. Blackie, Glasgow
6. Urry G (1989) Elementary equilibrium chemistry of carbon. Wiley, New York
7. Iler RK (1979) The chemistry of silica. Wiley, New York
8. Patai S (ed) (1977) The chemistry of cyanates and their thio derivatives. Wiley, New York
9. Newman AA (1975) Chemistry and biochemistry of thiocyanic acid and its derivatives. Academic, London
10. Glockling F (1969) The chemistry of germanium. Academic, San Diego
11. Davydov VI (1966) Germanium. Gordon & Breach, New York
12. Voinovitch IA, Debras-Guedon J, Louvrier J (1962) Kondor R (trans) Seijffers E (ed) (1966) The analysis of silicates. Israel Program for Scientific Translations, Jerusalem
13. Williams HE (1948) Cyanogen compounds, their chemistry, detection and estimation. Edward Arnold, London
14. Schnepf A (2007) Chem Soc Rev 36: 745
15. Smith DW (2005) J Chem Educ 82:102
16. Jessop PG, Ikariya T, Noyori R (1995) Chem Rev 95:259
17. Gilfillan SC (1965) J Occupat Med 7(2):53

15 Nitrogen through Bismuth, the Pentels

15.1 Nitrogen, $_7N$

Oxidation numbers: (−III), (−II), (−I), (I), (II), (III), (IV), (V) and other, as in NH_3, N_2H_4, NH_2OH, $H_2N_2O_2$, NO, HNO_2, NO_2, NO_3^- and N_3^-.

15.1.1 Reagents Derived from Hydrogen and Oxygen

Dihydrogen. Nascent hydrogen with compounds of N usually forms NH_3, always the ultimate product in an alkaline solution.

Water. The radius of NH_4^+ is close to that of K^+, but NH_4^+ salt solubilities resemble more those of Rb^+ and Cs^+. We cannot isolate neutral NH_4, however, as we can metallic Alk.

Diazane (hydrazine), N_2H_4, is a colorless, hygroscopic liquid. Like sulfuric acid, N_2H_4 reacts vigorously with H_2O when diluted. It is decomposed by heating into N_2 and NH_3. The monohydrate and solutions are much more stable than the anhydrous hydride.

Hydroxylamine, NH_2OH, the thermodynamically strongest nitrogen reductant, going to N_2, is unstable, decomposing slowly at ambient *T*. An aqueous solution reacts alkaline and soon decomposes:

$$3\ NH_2OH \rightarrow N_2\uparrow + NH_3 + 3\ H_2O$$

Therefore NH_2OH is generally used as the salts NH_3OHCl, $(NH_3OH)_2SO_4$, or NH_3OHNO_3, which are more stable although less soluble in water than the base itself.

The triazadienides (trinitrides or "azides") of the alkali metals (NaN_3 etc.) are readily soluble in H_2O; those of the alkaline earths are also soluble, while those of the other metals are slightly soluble to insoluble or, as with Th, Zr, Al and Sn^{IV}, hydrolyzed to hydroxides.

The two oxides N_2O and NO, although somewhat soluble, are neutral in water; the higher oxides, N_2O_3, NO_2, N_2O_4 and N_2O_5, are acid anhydrides forming, with water, HNO_2 and/or HNO_3. Although N_2O may be obtained by dehydration from $H_2N_2O_2$, "hyponitrous acid", the thermodynamically strongest nitrogen oxidant (going to N_2), it does not combine with H_2O or hydroxides to form this acid or its salts; so this goes only one way:

$$H_2N_2O_2 \rightarrow N_2O\uparrow + H_2O$$

Nitrogen monoxide likewise does not combine directly with H_2O to form an acid. It cannot, however, be kept over water because of the action of dissolved oxygen and of the H_3O^+, which on long contact produce HNO_2, along with $H_2N_2O_2$ and ultimately N_2.

Nitrous acid, a moderately strong acid, has not been isolated. Its aqueous solution, freshly prepared by adding N_2O_3 to cold water, is blue, but the color soon fades and brown fumes (from NO + O_2 forming NO_2) are released:

$$3\ HNO_2 \rightarrow H_3O^+ + NO_3^- + 2\ NO$$

Nitrites hydrolyze in water sufficiently to give a slightly alkaline reaction which increases with the age of the solution.

Silver nitrite is only slightly soluble; the other normal nitrites are soluble, but many basic nitrites as well as some complex compounds, e.g., $K_3Co(NO_2)_6$, are insoluble.

Dinitrogen tetraoxide dissolves in water, forming a blue-green solution containing HNO_3 and HNO_2. The latter reacts further as above.

Most normal nitrates are soluble, but a few are decomposed by H_2O:

$$Bi(NO_3)_3 + 3\ H_2O \rightarrow BiONO_3\downarrow + 2\ H_3O^+ + 2\ NO_3^-$$

Oxonium. The N_3^-, triazadienide, trinitride or "azide" ion, reacts with strong acids to give the very explosive triazadiene, hydrogen trinitride, hydrogen "azide" or "hydrazoic acid", HN_3. The pure acid is a colorless, mobile liquid with a penetrating odor. It is very irritating to the skin. It readily explodes with marked violence. It is soluble in water and ethanol. Aqueous solutions of less than 1 M are relatively safe, but when boiled with H_3O^+ they slowly decompose to form N_2 and NH_4^+:

$$3\ HN_3 + H_3O^+ \rightarrow 4\ N_2\uparrow + NH_4^+ + H_2O$$

This acid shows marked activity, dissolving a number of metals with release of hydrogen.

Adding solid $NaNO_2$ to 2-M H_2SO_4 generates NO:

$$3\ NO_2^- + 2\ H_3O^+ \rightarrow 2\ NO\uparrow + NO_3^- + 3\ H_2O$$

Constant-boiling, ordinary concentrated, HNO_3 is about 16 M. The so-called fuming acid is a solution of a variable amount of NO_2 in HNO_3; it should be kept in a cool, dark place to avoid decomposition.

Hydroxide. Aqueous OH^- releases N_2H_4, NH_2OH or NH_3 from the salts:

$$N_2H_6SO_4 + 2\ OH^- \rightarrow N_2H_4 + SO_4^{2-} + 2\ H_2O$$

On long contact with OH^-, NO will form NO_2^- and N_2O; above 125 °C the reaction is faster, but the products are NO_2^- and N_2.

The reaction of NO_2 with OH^- produces nitrate and nitrite:

$$2\ NO_2 + 2\ OH^- \rightarrow NO_2^- + NO_3^- + H_2O$$

with CaO much of the reaction goes farther:

$$2\ CaO + 5\ NO_2 \rightarrow 2\ Ca(NO_3)_2 + {}^1/_2\ N_2$$

Peroxide. Diazanium with H_2O_2 gives N_3^- and a small or an equivalent amount of NH_4^+, depending on conditions:

$$2\ N_2H_5^+ + 2\ H_2O_2 \rightarrow HN_3 + NH_4^+ + H_3O^+ + 3\ H_2O$$

Hydrogen peroxide plus NO yield HNO_2 and HNO_3.

Di- and trioxygen. Neat N_2H_4 and O_2 slowly form N_2 and H_2O. Ozone unexpectedly oxidizes $N_2H_5^+$ to similar amounts of NH_4^+, N_2 and NO_3^-.

Nitrogen oxide, N_2O, does not react at 25 °C with O_2 or O_3.

When exposed to the air NO becomes NO_2 or N_2O_4:

$$2\ NO + O_2 \rightarrow 2\ NO_2$$

15.1.2 Reagents Derived from the Other 2nd-Period Non-Metals, Boron through Fluorine

Some "simple" organic species. Propanone is used to protect diazane during and after its synthesis, with later recovery by hydrolysis:

$$2\ Me_2CO + H_2N\text{–}NH_2 \leftrightarrows Me_2C\text{=}N\text{–}N\text{=}CMe_2 + 2\ H_2O$$

Nitrous acid is sometimes removed by using urea or formaldehyde, likewise nitric acid with formaldehyde. (The volume of N_2 can be measured for determination, using alkali to remove the CO_2). Thus:

$$2\ HNO_2 + CO(NH_2)_2 \rightarrow 2\ N_2\uparrow + CO_2\uparrow + 3\ H_2O$$

$$4\ HNO_2 + 3\ CH_2O \rightarrow 2\ N_2\uparrow + 3\ CO_2\uparrow + 5\ H_2O$$

$$4\ NO_3^- + 5\ CH_2O + 4\ H_3O^+ \rightarrow 2\ N_2\uparrow + 5\ CO_2\uparrow + 11\ H_2O$$

Ammonium nitrate heated in glycerol, $(CH_2OH)_2CHOH$, gives an almost quantitative yield of N_2.

Ammonia vapor, often from the reduction of other nitrogen compounds, may be detected with wet red litmus paper, turning blue.

Reduced nitrogen. Nitrous acid and $N_2H_5^+$, diazanium ion, generate HN_3, triazadiene, but also NH_4^+, N_2 and N_2O:

$$N_2H_5^+ + HNO_2 \rightarrow HN_3 + H_2O + H_3O^+$$

The reaction of NH_2Cl with N_2H_4 may not be suprising:

$$N_2H_4 + 2\ NH_2Cl \rightarrow N_2\uparrow + 2\ NH_4^+ + 2\ Cl^-$$

Nitrous acid oxidizes HN_3:

$$HN_3 + HNO_2 \rightarrow N_2\uparrow + N_2O\uparrow + H_2O$$

Oxidized nitrogen. Dinitrogen oxide, N_2O, an oxidant, is distinguished from O_2 by its faint odor, taste and inertness toward NO.

Nitrous acid oxidizes both NH_4^+ and $CO(NH_2)_2$ (see above) to N_2:

$$HNO_2 + NH_4^+ \rightarrow N_2\uparrow + H_3O^+ + H_2O$$

These two reactions are often used to remove NO_2^- from solution.

With high acidity and an apparently complicated dependence on conditions, $N_2H_5^+$ and HNO_2 form NH_4N_3, sometimes with little NH_4^+.

Nitrous acid, via NO^+, and N_3^- form $N^-{=}N^+{=}N{-}N{=}O$, which breaks down to N_2 and N_2O.

Nitrous acid and NH_2OH give *trans*-$(=N{-}OH)_2$, "(bis)hyponitrous acid", as one might expect from a "simple" dehydration: $HON(H_2 + O)NOH$. This acid and excess HNO_2 react:

$$H_2N_2O_2 + HNO_2 \rightarrow N_2\uparrow + H_3O^+ + NO_3^-$$

Nitric acid and NH_3OH^+ go to N_2O, and HN_3 gives N_2, N_2O and NO.

Nitric acid decomposes all nitrites, forming nitrates. The liberated HNO_2 quickly decomposes reversibly, although lower $c(HNO_3)$ and $c(HNO_2)$ form NO_2, also reversibly:

$$3\ HNO_2 \leftrightarrows H_3O^+ + NO_3^- + 2\ NO\uparrow$$

Nearly all nitrates are less soluble in HNO_3 than in H_2O. The barium salt dissolves only slightly in the concentrated acid.

15.1.3 Reagents Derived from the 3rd-to-5th-Period Non-Metals, Silicon through Xenon

Phosphorus species. The N_2H_4 and $N_2H_5^+$ species go to N_2, reducing the peroxodiphosphates, $H_n(PO_3)_2O_2^{(4-n)-}$, and the peroxomonophosphates to PO_4^{3-} and the various hydrogenphosphates, catalyzed by Cl^-, Br^- and I^-.

Reduced chalcogens. Sulfane (H_2S) with HN_3 forms N_2 and NH_3. Sulfane and HNO_2 give NO, S and $S_2O_3^{2-}$.

Aqueous N_3^-, refluxed with CS_2 for two days, forms the azidodithiocarbonate $CS_2N_3^-$, whose **d**- or **p**-block metal salts are very explosive. Treatment with cold, concentrated HCl precipitates the somewhat explosive acid HCS_2N_3, rather stable for a day or two if cold and dark. Aqueous I_3^- oxidizes the anion to the highly explosive pseudohalogen $[N_3(CS)S-]_2$.

Oxidized chalcogens. Aqueous SO_2 and NO form N_2O.

Passing NO into alkaline SO_3^{2-} yields a "(bis)hyponitrite" adduct, [*cis*-$O_2N_2SO_3]^{2-}$, i.e., $[^-O-N=N^+(-SO_3^-)-O^-]$.

Nitrous acid and SO_2 first form $[HSO_3 \cdot NO]$, then $[H_2N_2O_2]$, NH_3OH^+ (well synthesized by stopping here) and finally NH_4^+, along with SO_4^{2-}.

At 0 °C, NO_2^-, HSO_3^- and K^+ yield $K_2[N(OH)(SO_3)_2] \cdot 2H_2O$, but heating NO_2^- with HSO_3^- (perhaps from OH^- plus excess SO_2) yields the nitridotrisulfonate ion, although NO_2^- and SO_3^{2-} do not react:

$$4\,HSO_3^- + NO_2^- \rightarrow [N(SO_3)_3]^{3-} + SO_3^{2-} + 2\,H_2O$$

The potassium salt is very slightly soluble in cold water, and is stable in base, hydrolyzed quite quickly in acid, but slowly at pH 7, and may be kept if pure in a dessicator for a month or so. It precipitates with lead acetate but not with Mn^{2+}, Fe^{2+}, Co^{2+}, Cu^{2+}, Ag^+, Cd^{2+} or Hg^{2+}.

Nitrous acid may be prepared by adding H_2SO_4 to a solution of $Ba(NO_2)_2$ and removing the $BaSO_4$ precipitate, but excess H_2SO_4, concentrated or dilute, decomposes nitrites and HNO_2, producing HNO_3 and NO_2 or NO, depending on the concentration and temperature, e.g:

$$3\,NO_2^- + 4\,H_2SO_4 \rightarrow HNO_3 + 2\,NO\uparrow + 4\,HSO_4^- + H_3O^+$$

One can distill HNO_3 from $NaNO_3$ plus concentrated H_2SO_4 by heating to 130 °C, leaving "nitre cake", $NaHSO_4$.

Peroxo(mono)sulfate oxidizes $N_2H_5^+$ to N_2, but the disulfate, in high acidity and with an apparently complicated dependence on conditions, gives, sometimes with little NH_4^+:

$$N_2H_5^+ + 2\,HSO_3(O_2)^- \rightarrow N_2\uparrow + 2\,HSO_4^- + H_3O^+ + H_2O$$

$$2\,N_2H_5^+ + 2\,[S_2O_8]^{2-} + 5\,H_2O \rightarrow NH_4^+ + HN_3 + 4\,SO_4^{2-} + 5\,H_3O^+$$

The peroxo ion $HSO_3(O_2)^-$, more effectively than $SO_3(O_2)^{2-}$, also oxidizes HN_3 to N_2O, and, more rapidly, N_3^- to N_2.

Reduced halogens. Aqueous HCl reacts with HNO_3 in aqua regia:

$$NO_3^- + 3\ Cl^- + 4\ H_3O^+ \rightarrow NO\uparrow + 2\ H_2O + {}^3/_2\ Cl_2\uparrow \text{ or}$$

$$NO_3^- + 3\ Cl^- + 4\ H_3O^+ \rightarrow NOCl + 2\ H_2O + Cl_2\uparrow$$

Aqueous HI reduces NO to NH_4^+, but a simple way to prepare NO is to add 9-M H_2SO_4 dropwise to a concentrated solution of about 4 mol KNO_2 (a large excess) to 1 mol KI:

$$HNO_2 + I^- + H_3O^+ \rightarrow NO\uparrow + {}^1/_2\ I_2 + 2\ H_2O$$

A mixture of a nitrite and I^- liberates I_2 on addition of acetic acid. Using starch to detect the iodine, this test will reveal 2-μM NO_2^-. Many other ions interfere. The oxides NO_2 and N_2O_4 also oxidize I^- to I_2.

Elemental and oxidized halogens. Also see **17.2.2 Reduced nitrogen.**
Diazane, N_2H_4, with Cl_2, Br_2 or I_2 (X_2 here) yields N_2, e.g.:

$$N_2H_5^+ + 2\ X_2 + 5\ CH_3CO_2^- \rightarrow N_2\uparrow + 4\ X^- + 5\ CH_3CO_2H$$

Dinitrogen oxide, N_2O, does not react at 25 °C with the halogens.
Bromine and HBrO oxidize $N_2H_5^+$ to N_2, becoming Br^- themselves.
Bromine oxidizes NO_2^- to NO_3^-.
Iodine oxidizes $N_2H_5^+$ to N_2, but slowly at low pH.
In CO_3^{2-} solution NH_2OH may be oxidized quantitatively by I_2.
Iodine and $(=NOH)_2$ produce nitrate and nitrite, approximately:

$$H_2N_2O_2 + 3\ I_2 + 10\ H_2O \rightarrow NO_3^- + HNO_2 + 6\ I^- + 7\ H_3O^+$$

Mixing equivalent amounts of NH_3 and ClO^- at 0 °C yields chloroamine, NH_2Cl, about as soluble in ether as in water:

$$NH_3 + ClO^- \rightarrow NH_2Cl + OH^-$$

Then a large excess of hot ammonia can be added to give diazane efficiently in the presence of a small amount of glue, gelatin, etc. to complex the trace **d**- or **p**-block metal ions that catalyze decomposition:

$$2\ NH_3 + NH_2Cl \rightarrow NH_2NH_2 + Cl^- + NH_4^+$$

but there is also some of the destructive reaction:

$$N_2H_4 + 2\ NH_2Cl \rightarrow N_2\uparrow + 2\ NH_4^+ + 2\ Cl^-$$

Then the solution may be boiled down and finally N_2H_4 distilled out, followed by H_2SO_4 at 0 °C to produce crystals of the much more stable and slightly soluble diazanediium or hydrazinium(2+) sulfate, $N_2H_6SO_4$.

The oxidations of N_2H_4 by ClO^- and of $N_2H_5^+$ by $HClO_2$ are fast and slow, respectively, and ClO_3^- reacts with $N_2H_5^+$ only at high T and acidity or with a catalyst:

$$N_2H_4 + 2\ ClO^- \rightarrow N_2\uparrow + 2\ Cl^- + 2\ H_2O$$

$$N_2H_5^+ + HClO_2 \rightarrow N_2\uparrow + Cl^- + 2\ H_3O^+$$

Aqueous BrO_3^- seems unique in giving fair amounts of NH_4^+ and HN_3, along with N_2, from $N_2H_5^+$ at 25 °C. Also:

$$3\ N_2H_5^+ + 2\ IO_3^- \rightarrow 3\ N_2\uparrow + 2\ I^- + 3\ H_3O^+ + 3\ H_2O$$

$$N_2H_5^+ + \text{excess } 2\ IO_4^- \rightarrow N_2\uparrow + 2\ IO_3^- + H_3O^+ + H_2O$$

Aqueous HClO oxidizes NO to HNO_3.

Aqueous NO_2^- in HCO_3^- reacts with ClO^- to form Cl^- and NO_3^-.

Solutions of ClO_3^-, BrO_3^- or IO_3^- may be boiled with N_2H_4 without reaction, although iodates react with $N_2H_5^+$ in strongly acidic solution. A little CuO, however, produces immediate oxidation even in the cold:

$$3\ N_2H_5^+ + 2\ ClO_3^- \rightarrow 3\ N_2\uparrow + 2\ Cl^- + 3\ H_3O^+ + 3\ H_2O$$

With high acidity and an apparently complicated dependence on conditions, ClO_3^- and BrO_3^- also give, sometimes without much NH_4^+:

$$6\ N_2H_5^+ + 2\ ClO_3^- \rightarrow 3\ NH_4^+ + 3\ HN_3 + 2\ Cl^- + 3\ H_3O^+ + 3\ H_2O$$

Nitrous acid reduces BrO_3^-, ClO_3^- and IO_3^- to the free halogens, decreasing in activity surprisingly in the order mentioned, but nitrite also reduces $AgBrO_3$ to AgBr, for example.

Diazane (N_2H_4) and IO_3^- yield N_2 with little of either N_3^- or NH_4^+.

15.1.4 Reagents Derived from the Metals Lithium through Uranium, plus Electrons and Photons

Oxidation. The $AlkN_3$ and $Ae(N_3)_2$ salts of N_3^- are not explosive; most others are, with oxidation of N_3^- to N_2.

Many reports are incomplete and contradictory, but from Ce^{4+}, Mn^{3+}, Fe^{3+}, Co^{3+} or Cu^{2+} we have half reactions, as in electrolysis, below:

$$N_2H_5^+ + H_2O \rightarrow {}^1/_2\ N_2\uparrow + NH_4^+ + H_3O^+ + e^-$$

Somewhat different results come from VO_4^{3-}, CrO_4^{2-}, MoO_4^{2-}, $Fe(CN)_6^{3-}$, Ag^I, Hg^{II} or Tl^{3+}, reacting with N_2H_4 or $N_2H_5^+$, causing half reactions such as these:

$$N_2H_5^+ + 5\ CH_3CO_2^- \rightarrow N_2\uparrow + 5\ CH_3CO_2H + 4e^-$$

$$N_2H_4 + 4\ OH^- \rightarrow N_2\uparrow + 4\ H_2O + 4e^-$$

Even with excess reductant, $N_2H_6^{2+}$ generally reduces V^V only to V^{IV}:

$$2\ VO_2^+ + {}^1/_2\ N_2H_6^{2+} + H_3O^+ \rightarrow 2\ VO^{2+} + {}^1/_2\ N_2\uparrow + 3\ H_2O$$

Interestingly, the non-reduction of Ag^+ in such solutions argues against the intermediate (reductant) V^{III}, although some conditions do yield V^{III} and NH_3. Two V^V are found to cooperate in attacking one N_2H_4. Other effective oxidants are Ce^{IV}, UO_2^{2+} (in light), $Cr^{3+}(O_2^-)$, Cr^{VI}, Mo^{VI}, Mn^{III}, e.g., $[Mn(P_2O_7)_3]^{9-}$, Mn^{VII}, Fe^{III}, Fe^{VI}, $[Co(CO_3)_3]^{3-}$, $[Co(C_2O_4)_3]^{3-}$, $[IrCl_6]^{2-}$, $[IrBr_6]^{2-}$, Ni^{III}, $[PtCl_6]^{2-}$, $[Cu(NH_3)_4]^{2+}$, $[AuCl_4]^-$ etc., thus:

$$4\ HCrO_4^- + 3\ N_2H_5^+ + 13\ H_3O^+ \rightarrow 4\ Cr^{3+} + 3\ N_2\uparrow + 29\ H_2O$$

with a complicated formation of NH_4^+ catalyzed by Mn^{2+}, and:

$$FeO_4^{2-} + N_2H_4 \rightarrow Fe(OH)_2\downarrow + N_2\uparrow + 2\ OH^-$$

without Fe^{III} as an intermediate.

The $[Cr_2O_7]^{2-}$ ion, $MnO_2{\cdot}aq$, MnO_4^-, Ag_2O or PbO_2, and NO give NO_3^-.

Nitrous acid reduces all ordinary oxidants, which form NO_3^-; it reduces $[Cr_2O_7]^{2-}$ to Cr^{III}, and MnO_4^- to Mn^{2+}:

$$2\ MnO_4^- + 6\ HNO_2 \rightarrow 2\ Mn^{2+} + 5\ NO_3^- + NO_2^- + 3\ H_2O$$

At pH 2.4 to 4.4 we have, and $[W(CN)_8]^{3-}$ is similar:

$$4\ [Mo(CN)_8]^{3-} + N_2H_5^+ + 5\ H_2O \rightarrow 4\ [Mo(CN)_8]^{4-} + N_2\uparrow + 5\ H_3O^+$$

To supplement and restate, N_2H_4 plus $MnO_2{\cdot}aq$, MnO_4^- or $Fe_2O_3{\cdot}aq$ yield N_2, NH_4^+ and a little N_3^-. Iron(III) with $N_2H_5^+$ forms Fe^{2+}. Copper(2+) with $N_2H_5^+$ in acids is changed to Cu^+; in alkaline solution, e.g., Fehling's solution, it becomes Cu (separation from Zn and Sn). Diazane precipitates metallic Ag, Au and Hg from solution. With HgO or $[HgCl_2]$ it gives N_2 and only a little of both NH_4^+ and N_3^-.

Aqueous MnO_4^- oxidizes HN_3 to N_2.

The reaction of $[Fe(CN)_6]^{3-}$ and N_2H_4 to give $[Fe(CN)_6]^{4-}$ and N_2 is faster in alkalis than in acids, and complicated with other products by O_2, by Cu^{2+} catalysis, and perhaps by potentiometric-electrode catalysis. The simultaneous hydrogenation of olefins shows even the quantitative generation of unstable N_2H_2.

Iron(III) in acidic solution oxidizes NH_3OH^+; if boiled, the reaction may be made quantitative, e.g.:

$$4\ FeCl^{2+} + 2\ NH_3OH^+ + 5\ H_2O \rightarrow 4\ Fe^{2+} + N_2O\uparrow + 6\ H_3O^+ + 4\ Cl^-$$

Hydroxylamine may be determined by titrating the Fe^{2+} with MnO_4^-.

Hydroxylamine reduces Pt^{IV}, Ag^+, Au^{III} and Hg^{2+} to the metals.

Fehling's solution detects NH_2OH, sensitive to about 0.3 mM:

$$4\ Cu^{2+}(cpx) + 2\ NH_2OH + 8\ OH^- \rightarrow 2\ Cu_2O\downarrow + N_2O\uparrow + 7\ H_2O$$

but Cu, Ag and Hg ions also yield $[trans\text{-}(O\text{–}N{=})_2]^{2-}$, a "hyponitrite".

Thallium(III) reacts with $N_2H_5^+$ at least as follows; acetate and Cl^- slow the reaction by forming less reactive Tl^{III} complexes:

$$2\ TlOH^{2+} + N_2H_5^+ + H_2O \rightarrow 2\ Tl^+ + N_2\uparrow + 3\ H_3O^+$$

The electrolytic oxidation of diazane gives several products, but often involving N_2H_4 in the mechanism, therefore faster at moderately high pH. Generally, however, HN_3 or N_3^- is an important product only at high T and low pH:

$$N_2H_5^+ + H_2O \rightarrow {}^1/_2\ N_2\uparrow + NH_4^+ + H_3O^+ + e^-$$

$$N_2H_4 + OH^- \rightarrow {}^1/_2\ N_2\uparrow + NH_3 + H_2O + e^-$$

$$2\ N_2H_5^+ + 5\ H_2O \rightarrow NH_4^+ + HN_3 + 5\ H_3O^+ + 4\ e^-$$

$$2\ N_2H_4 + 5\ OH^- \rightarrow NH_3 + N_3^- + 5\ H_2O + 4\ e^-$$

$$N_2H_5^+ + 5\ H_2O \rightarrow N_2\uparrow + 5\ H_3O^+ + 4\ e^-$$

$$N_2H_4 + 4\ OH^- \rightarrow N_2\uparrow + 4\ H_2O + 4\ e^-$$

Light promotes the oxidation of $N_2H_5^+$ by UO_2^{2+}:

$$2\ N_2H_5^+ + UO_2^{2+} + \gamma + 2\ H_3O^+ \rightarrow N_2\uparrow + U^{4+} + 2\ NH_4^+ + 4\ H_2O$$

Reduction of nitrogen(<III). We do not yet have the long-sought cheap industrial fixation or reduction of N_2, but it is reduced to N_2H_4 (often with the strongest reductants) or NH_3 (separately, not after N_2H_4) in water or similar solvents, which also always release H_2, by reductants usually combined with catalysts such as: Na_{Hg}, Mg^{2+} and Mo^{III}; Na_{Hg} and Ti^{II}; $Ti_2O_3\cdot aq$, $Mg(OH)_2$ and Mo^{III}; $V(OH)_2$ and $Mg(OH)_2$; $Cr(OH)_2$ and Mo^{III}; cathodic e^-, $Ti_2O_3\cdot aq$ and Mo^{III}. The yields are often low.

In an alkaline solution $Fe(OH)_2$ reduces NH_2OH:

$$2\ Fe(OH)_2 + NH_2OH \rightarrow Fe_2O_3.aq\downarrow + NH_3 + 2\ H_2O$$

The acid HN_3 is readily reduced by Mn, Fe, Cu, Zn etc., plus H_3O^+, but without producing H_2:

$$M + 3\ HN_3 + H_3O^+ \rightarrow M(N_3)_2\downarrow + N_2\uparrow + NH_4^+ + H_2O$$

Aqueous N_3^- is reduced by $Fe(OH)_2$ to NH_3 and N_2H_4.

Tin(II) and HCl, both concentrated, convert HN_3 to N_2 plus NH_4^+.

Dinitrogen oxide, N_2O, does not react at 25 °C with alkali metals.

Aqueous Cr^{2+} reduces NO to NH_3OH^+.

Gaseous NO is reduced by $LiAlH_4$ to "(bis)hyponitrite", $[(=NO^-)_2]$.

Tin(II) in acidic solution changes NO to NH_3OH^+ and NH_4^+; in alkaline solution the product is *trans*-$N_2O_2^{2-}$.

Reduction of Nitrite. Nitrite ion, NO_2^-, plus Na_{Hg} (amalgam) at 0 °C give [*trans*-$(=N{-}O^-)]^{2-}$, a "(bis)hyponitrite". Then Ag^+ precipitates yellow Ag_2[*trans*-N_2O_2]. Adding HCl in ether gives [*trans*-$(=N{-}OH)_2$], the acid. Alkalis yield a very reactive $M^I[HN_2O_2]$ and a hard-to-reduce $M^I_2[N_2O_2]$; note their stability during formation using Na_{Hg}.

Aqueous Ti^{II} reduces the nitrosodisulfonate anion.

Iron(II) in 1-M H_2SO_4 reduces nitrites to nearly pure NO, e.g.:

$$HNO_2 + Fe^{2+} + 2\,HSO_4^- \rightarrow NO\uparrow + Fe(SO_4)_2^- + H_3O^+$$

In neutral or alkaline solution $Fe(OH)_2$ will quantitatively reduce NO_2^- and NO_3^- to NH_3.

Metallic zinc and CH_3CO_2H slowly reduce NO_2^- to NH_4^+; the action (forming NH_3) is rapid with Al and OH^-. When this is to be used as a test, any NH_4^+ must first be removed by boiling with OH^- before adding the Al. Nothing else interferes with this test. The result is delayed, however, by strong oxidants such as ClO_3^- or $[Cr_2O_7]^{2-}$.

Reduction of Nitrate. One way to make NH_2OH is to electrolyze a cold solution of 7-M H_2SO_4 to which 10-M HNO_3 is slowly added. A mercury or amalgamated-lead cathode yields NH_2OH in an 80 % yield.

Many metals are passive in concentrated HNO_3 but dissolve readily in a dilute solution. Ions like MnO_4^- or ClO_3^- are catalytic.

If, with concentrated H_2SO_4, a crystal of iron(II) sulfate or a small piece of Cu is added to a concentrated solution or residue of a nitrate, the mixture will give off abundant brown vapors of NO_2:

$$NO_3^- + 3\,FeSO_4 + 6\,H_2SO_4 \rightarrow NO\uparrow + 3\,FeSO_4^+ + 6\,HSO_4^- + 2\,H_3O^+$$

$$2\,NO_3^- + 3\,Cu + 12\,H_2SO_4 \rightarrow 2\,NO\uparrow + 3\,Cu^{2+} + 12\,HSO_4^- + 4\,H_3O^+$$

$$2\,NO + O_2 \rightarrow 2\,NO_2\uparrow$$

However, if the nitrogen monoxide is liberated in a cold solution containing excess $FeSO_4$ and H_2SO_4, instead of being released the NO combines with the Fe^{2+} to form brown $FeNO^{2+}$ and, e.g., $FeSO_4^+$:

$$NO_3^- + 4\,FeSO_4 + 7\,H_2SO_4 \rightarrow$$

$$FeNO^{2+} + 3\,FeSO_4^+ + 2\,H_3O^+ + 8\,HSO_4^-$$

This "brown-ring" test is sensitive to about 2-mM NO_3^-. Some ions interfere, e.g., ClO_3^-, Br^-, I^-, $[Cr_2O_7]^{2-}$; also the solution to be tested must not be deeply colored. The brown complex is formed by Fe^{2+} more easily from nitrite, NO_2^-, even when acidified only with dilute CH_3CO_2H.

When reacting with HNO_3, the metals Cu, Ag, Hg and Bi differ from others in that the main nitrogen species is NO:

$$3\ Cu + 2\ NO_3^- + 8\ H_3O^+ \rightarrow 3\ Cu^{2+} + 2\ NO\uparrow + 12\ H_2O$$

With Zn and other metals, the main products vary materially, depending on conditions and the ratio between acid and metal.

A simple way to prepare NO is to drop 6.0–6.5 M HNO_3 onto a mixture of Cu and Pt. Another is to treat Hg with a mixture of HNO_3 and concentrated H_2SO_4.

In a test for nitrate by way of nitrite, the solution is treated with Zn and acetic acid. At short intervals a sample of the liquid is taken and a drop of I^- is added, followed by a drop of CCl_4, to detect any iodine liberated. Other oxidants like As^V, ClO_3^-, BrO_3^-, IO_3^-, etc., are reduced before the NO_3^-. The test will detect 0.2-mM nitrate:

$$NO_3^- + Zn + 2\ CH_3CO_2H \rightarrow NO_2^- + Zn^{2+} + 2\ CH_3CO_2^- + H_2O$$

$$HNO_2 + I^- + CH_3CO_2H \rightarrow NO\uparrow + {}^1/_2\ I_2 + CH_3CO_2^- + H_2O$$

One sometimes prefers other ways of making the nascent hydrogen, say with Na_{Hg}, Mg plus H_3PO_4, Cd plus CH_3CO_2H or Al plus OH^-. We note that in any case the NO_2^- is only an intermediate in the reduction of the NO_3^-. If the reaction period is too short, insufficient NO_2^- will have been formed; if too long, the nitrogen will have been reduced beyond NO_2^-. Hence the need to take samples at short intervals.

The oxide N_2O may be prepared by boiling a mixture of about 10 mL of 16-M HNO_3, 20 mL of 12-M HCl and 60 mmol of $SnCl_2$:

$$2\ NO_3^- + 4\ SnCl_3^- + 12\ Cl^- + 10\ H_3O^+ \rightarrow$$

$$N_2O\uparrow + 4\ SnCl_6^{2-} + 15\ H_2O$$

Nitrates may be made by dissolving the metal in HNO_3, often forming NO, except for Cr, Pt, Au or Al, which are scarcely or not attacked by this acid; also, As gives H_3AsO_4, Sb forms Sb_2O_5, and with an excess of hot HNO_3 Sn gives metastannic acid.

Upon long boiling the chlorides of most ordinary metals are completely decomposed by HNO_3, forming nitrates and, e.g., NO, with no chlorine remaining. However, the chlorides of Pt, Ag, Au and Hg are not attacked, and the chlorides of Sn and Sb become oxides.

Other reactions. The only metal that reacts with (very dilute) HNO_3 to give H_2 is magnesium. (Reducing the H_3O^+ need not include reducing the NO_3^-.) The known nitrates can be made by adding HNO_3 to metallic oxides, hydroxides or carbonates.

Aqueous N_3^- precipitates Rth^{3+} as basic salts, and thorium, uranium, zirconium and aluminum ions as hydroxides.

Ammonia may be detected by means of $Mn^{2+} + H_2O_2$:

$$2\ NH_3 + Mn^{2+} + H_2O_2 \rightarrow 2\ NH_4^+ + MnO_2{\cdot}aq\downarrow$$

Treating a mixture of NH_2OH, $(NH_4)_2S_x$ and NH_3 with a little Mn^{2+} develops an evanescent purple color, a sensitive test for NH_2OH.

A red color is a sensitive test for NO_2^- when CH_3CO_2H, $C_2O_4^{2-}$, Mn^{2+} and H_2O_2 are added, in that order, to a suspected NO_2^- solution.

With Fe^{III} and N_3^- a red solution of, e.g., $[Fe(N_3)_3]$ is obtained.

One-M $FeSO_4$ readily dissolves NO, forming a brown $[Fe(H_2O)_5NO]^{2+}$, decomposed at 100 °C. Aqueous Mn^{2+}, Co^{2+} or Ni^{2+} also absorbs NO, but without a color change.

Adding NO_2^- to an almost colorless solution of $Fe(CN)_6^{3-}$ acidified with CH_3CO_2H, gives a greenish-yellow color (distinction from NO_3^-). The test is sensitive to 0.04 mM.

Mixing KNO_2, $\mathbf{3d}^{2+}$ and Ca^{2+}, Ba^{2+} or Pb^{2+} (M^{2+}), sometimes with ethanol and a pH buffer, yields $K_2M[\mathbf{3d}(NO_2)_6]$; **3d** = Fe, Co, Ni or Cu.

The hope for non-biological nitrogen fixation under mild conditions created interest in the following reaction; see **8.2.2** for more:

$$[Ru(NH_3)_5(H_2O)]^{2+} + N_2 \rightarrow [Ru(NH_3)_5(N_2)]^{2+} + H_2O$$

Ammonia (included here as reactant, not reagent) and Cu^{II} (catalyst) can leach Co, Ni, Cu or Ag (taken together here for brevity) from, e.g., S^{2-} ores. The Cu^{II} oxidizes the ores, and O_2 reoxidizes the Cu^I quickly.

Copper(2+) and N_3^- form a red-brown precipitate of $Cu(N_3)_2$.

White silver trinitride or "azide", AgN_3, is somewhat similar in properties to the silver halides, especially AgCl and AgBr.

Adding NO_2^- to aqueous Ag^+ precipitates white $AgNO_2$.

Nessler's reagent, $[HgI_4]^{2-}$ plus OH^-, detects traces of NH_3:

$$NH_3 + 2\ [HgI_4]^{2-} + 3\ OH^- \rightarrow \text{brownish } Hg_2NI{\cdot}H_2O\downarrow + 7\ I^- + 2\ H_2O$$

15.2 Phosphorus, $_{15}P$

Oxidation numbers: (−III), (I), (III), (IV) and (V), etc., as in PH_3, HPH_2O_2, H_2PHO_3, $H_4[P_2O_6]$ and H_3PO_4.

The P in phosphane, PH_3, is often considered to be P(−III), partly in analogy with NH_3. The electronegativities of hydrogen and phosphorus on various scales, however, are about the same, so that we could regard them as being P^0 and H^0 in

PH_3. In HPH_2O_2 or H_2PHO_3, then, H^0 in P–H leads to P^{III} or P^{IV} although their losses of four or two electrons, or eight from PH_3, in forming H_3PO_4 make the most common assignments convenient. This exemplifies the problems with the electronegativity concept [Taube H (~1953) personal comment].

Phosphorus comes mainly as the white, molecular P_4 (very reactive) and various allotropic lattices in red P (of medium reactivity), black P (the least reactive) and others. Only white P_4 is poisonous.

15.2.1 Reagents Derived from Hydrogen and Oxygen

Hydrogen. Nascent hydrogen (Zn + H_2SO_4) and H_2PHO_3 yield PH_3.

Water. Phosphane (phosphine), PH_3, a colorless, poisonous, odoriferous gas, may be made by adding water or dilute H_3O^+ to Ca_3P_2, Zn_3P_2 or AlP:

$$Ca_3P_2 + 6\,H_2O \rightarrow 2\,PH_3\uparrow + 3\,Ca(OH)_2\downarrow$$

Various other phosphides, also from non-aqueous sources, such as K_2P_5, MnP_2, Hg_3P_4, Sn_3P and PbP_5, are usually brittle solids, decomposing in water or dilute acids to form various phosphanes.

Water dissolves a trace of P_4, 0.1 mM calculated as P, but organic solvents and especially CS_2 dissolve far more. Red phosphorus is insoluble in these solvents. In most reactions of phosphorus, however, the white and red varieties finally react similarly, the latter with much less intensity, and frequently requiring the aid of heat.

Phosphinic ("hypophosphorous") acid, HPH_2O_2, a colorless, syrupy liquid, is completely miscible with water. The salts are all soluble in H_2O, and a number of them will dissolve in ethanol.

Phosphonic acid, "phosphorous acid", H_2PHO_3, a crystalline, highly deliquescent solid, is made by hydrolyzing PCl_3:

$$PCl_3 + 6\,H_2O \rightarrow H_2PHO_3 + 3\,H_3O^+ + 3\,Cl^-$$

This is almost completely miscible with H_2O. Its alkali salts are soluble; most of the others are not (distinction from phosphinates). True phosphorous acid would be the isomeric H_3PO_3, i.e., $P(OH)_3$, of which we do have organic derivatives.

Diphosphonic acid, $(HPHO_2)_2O$, $H_2P_2H_2O_5$, is obtained by the interaction of PCl_3 and limited water:

$$2\,PCl_3 + 11\,H_2O \rightarrow H_2P_2H_2O_5 + 6\,H_3O^+ + 6\,Cl^-$$

Phosphorus(V) oxide, P_4O_{10}, is snow-white and a very efficient drying agent. It will remove H_2O even from concentrated H_2SO_4. On contact with H_2O it dissolves with a hissing sound to form varieties of phosphoric acid: ortho-, H_3PO_4; di-,

$H_4[P_2O_7]$; and meta-, HPO_3. The latter is polymerized to $(HPO_3)_n$, but the simpler formula is sometimes used at least for its salts, e.g., $AgPO_3$.

Phosphoric acid may be prepared from P_4O_{10} and excess H_2O. Evaporating aqueous H_3PO_4 under low pressure gives $H_3PO_4 \cdot {}^1/_2H_2O$. Pure H_3PO_4 is a translucent, crystallizable and very deliquescent solid.

Diphosphoric acid is a soft, glass-like solid or opaque crystalline mass. The crystals separate from the syrupy solution at −10 °C. They are very soluble in, but practically unhydrolyzed by, water at 25 °C. The acid is broken down by boiling water to H_3PO_4. Diphosphoric acid yields four classes of salts, i.e., (of univalent cations) $M^I{}_nH_{4-n}[P_2O_7]$.

Metaphosphoric acid may be obtained by the spontaneous hydration of P_4O_{10} by the air, or by adding the calculated amount of H_2O to P_4O_{10}.

Solutions of metaphosphates may be heated, but the presence of a strong acid causes hydrolysis to the ortho form. At room temperature the metaphosphoric acids revert to H_3PO_4 in a few days:

$$(HPO_3)_n + n\ H_2O \rightarrow n\ H_3PO_4$$

All of the phosphoric acids are readily soluble in water. Alkali dihydrogenphosphates (primary phosphates) in solution react acidic; the hydrogenphosphates (secondary phosphates) and normal (tertiary) phosphates are alkaline. The latter are easily converted, even by aqueous CO_2, to the hydrogenphosphates.

A number of non-alkali dihydrogenphosphates (primary phosphates) are soluble in H_2O, e.g., $Ca(H_2PO_4)_2$. The normal (tertiary) and (mono) hydrogenphosphates are insoluble except those of the alkalis.

The non-alkali di- and metaphosphates are insoluble in water.

The solubilities of PFO_3^{2-} salts resemble those of SO_4^{2-}. Those of $PF_2O_2^-$ salts resemble those of the ClO_4^- salts, but are a little higher for the slightly soluble ones. Aqueous PFO_3^{2-} hydrolyzes slowly, faster in acid, and $PF_2O_2^-$ is slowly hydrolyzed to PFO_3^{2-}, especially when warm.

The phosphorus chlorides, bromides and iodides react with H_2O, forming the corresponding hydrogen halide and a phosphorus acid.

Deficient warm water reacts interestingly with white phosphorus and P_2I_4 (prepared in situ from I_2 and a little excess of P_4 in CS_2), partly thus, with PH_4I subliming:

$$^{13}/_4\ P_4 + {}^5/_2\ P_2I_4 + 32\ H_2O \rightarrow 10\ PH_4I\uparrow + 8\ H_3PO_4$$

or with simpler stoichiometry:

$$^1/_4\ P_4 + {}^1/_2\ P_2I_4 + 4\ H_2O \rightarrow PH_4I\uparrow + H_3PO_4 + HI\uparrow$$

Oxonium. The solubilities of the phosphates of Mg, Ca, Sr, Ba, Ni, Zn, Pb and others generally rise rapidly with increase in acidity.

Hydroxide. Boiling white (yellow) P_4 with OH^- releases PH_3 and leaves $PH_2O_2^-$ and some PO_4^{3-}; red phosphorus is not affected:

$$P_4 + 3\ OH^- + 3H_2O \rightarrow PH_3\uparrow + 3\ PH_2O_2^-$$

On boiling $PH_2O_2^-$ with excess base, first PHO_3^{2-}, then PO_4^{3-}, is formed with release of hydrogen:

$$PH_2O_2^- + OH^- \rightarrow PHO_3^{2-} + H_2\uparrow$$

$$PHO_3^{2-} + OH^- \rightarrow PO_4^{3-} + H_2\uparrow$$

Phosphinic acid, HPH_2O_2, although containing three hydrogen atoms, is monobasic and forms only one series of salts, e.g., $NaPH_2O_2$, $Ba(PH_2O_2)_2$, etc.; it is, however, a stronger acid than H_3PO_4:

$$2\ HPH_2O_2 + Mg(OH)_2 \rightarrow Mg^{2+} + 2\ PH_2O_2^-$$

The acid may be prepared by first warming P_4 with $Ba(OH)_2$ until PH_3 ceases to be released. Any excess Ba^{2+} is removed by precipitation with CO_2. The filtrate is evaporated to crystallize the barium salt. This is weighed, dissolved in H_2O and treated with the calculated amount of H_2SO_4. The filtrate may be concentrated and the solid HPH_2O_2 isolated by evaporating the H_2O below 110 °C; otherwise:

$$2\ HPH_2O_2 \rightarrow PH_3\uparrow + H_3PO_4$$

Phosphonic acid, H_2PHO_3, is a dibasic acid, stronger than H_3PO_4, reacting with bases to form salts of the types $NaHPHO_3$ and Na_2PHO_3.

Hypophosphoric acid, $H_4[P_2O_6]$, forms four series of salts, all four hydrogen ions being removable by OH^-.

One H^+ of H_3PO_4 may be titrated with OH^-, using methyl orange as an indicator, forming dihydrogenphosphates (primary salts), $M^IH_2PO_4$. The second H^+ may be titrated using phenolphthalein as the indicator, forming (mono) hydrogenphosphates (secondary salts), $M^I_2HPO_4$. The third H^+ may also be removed by OH^-, forming normal phosphates (tertiary salts), $M^I_3PO_4$.

Metaphosphoric acid, HPO_3, occurs only in cyclic polymers, but is a monobasic acid, neutralizing only one OH^- per P atom.

Precipitating $Na_4[P_4O_{12}]\cdot 4H_2O$ from water with ethanol purifies it. Then adding minimal water at < 40 °C plus 3 NaOH (an excess) for each *cyclo*-tetraphosphate ion slowly yields the *catena*-tetraphosphate:

$$[P_4O_{12}]^{4-} + 2\ OH^- \rightarrow [P_4O_{13}]^{6-} + H_2O$$

Ethanol (an equal volume) also separates this sodium salt, plus some NaOH, in the lower, sirupy layer of two liquid layers.

Phosphorus trichloride sulfide, PCl_3S, reacts with hot aqueous OH^-:

$$PCl_3S + 6\ OH^- \rightarrow PO_3S^{3-} + 3\ Cl^- + 3\ H_2O$$

After crystallization by cooling and dissolving the sodium salts in water at 40–45 °C, adding excess methanol yields a white thiophosphate $Na_3PO_3S \cdot 12H_2O$. This can be titrated with highly acidified iodine:

$$2\ PO_3S^{3-} + I_3^- + 2\ H_3O^+ \rightarrow [(-SPO_3H)_2]^{2-} + 3\ I^- + 2\ H_2O$$

Peroxide. Hydrogen peroxide does not oxidize $H_4[P_2O_6]$, which may be elucidated as $[(-PO_3H_2)_2]$. This and other hypophosphates are much more stable than phosphinates or phosphonates toward oxidants.

Peroxophosphoric acid, H_3PO_5, may be prepared by treating P_4O_{10} with cold 10-M (30 %) H_2O_2.

Dioxygen. Ordinary white P_4, when freshly prepared, is a transparent solid but becomes coated with a thin white film when placed in water containing air. At low temperatures P_4 is oxidized slowly in the air, with a characteristic odor, due in part to some ozone formed. One of the products of the slow oxidation of P_4 in moist air is hypophosphoric (hexaoxodiphosphoric) acid, $H_4[P_2O_6]$, forming small, hygroscopic, colorless crystals, decomposing at ~ 70 °C into H_2PHO_3 and $(HPO_3)_n$.

In a finely divided state (as obtained from the evaporation of a CS_2 solution) P_4 ignites spontaneously at temperatures at which the compact phosphorus may be kept for days. It must be kept under water. Ordinary P_4, along with at least a little O_2, is luminous in the dark, hence the name phosphorus. The presence of CS_2, H_2S, SO_2, Cl_2 or Br_2 prevents the glowing.

The oxide P_4O_6, a snow-white solid obtained by burning phosphorus in a limited amount of air, smells somewhat like P_4. Air oxidizes it to P_4O_{10}. It is slowly soluble in cold water to form phosphonic acid; in hot water the action is more complex:

$$^1/_2\ P_4O_6 + 3\ H_2O \rightarrow 2\ H_2PHO_3$$

$$P_4O_6 + 6\ H_2O \rightarrow PH_3\uparrow + 3\ H_3PO_4$$

Phosphonic acid, H_2PHO_3, is a strong reducing agent, changing to H_3PO_4 even on exposure to the air.

15.2.2 Reagents Derived from the Other 2nd-Period Non-Metals, Boron through Fluorine

Carbon oxide species. Carbonate ion, when boiled with the alkaline-earth phosphates, converts each to the corresponding carbonate, MCO_3, sufficiently completely for qualitative purposes.

Some "simple" organic species. Acetic acid transposes and dissolves most common phosphates except those of Fe^{III}, Al and Pb.

The Ae diphosphates are difficultly soluble in CH_3CO_2H.

Reduced nitrogen. Adding P_4O_{10} in small portions to concentrated NH_3 at about 5–10 °C gives mostly the tetrametaphosphate, $(NH_4)_4[P_4O_{12}]$, precipitated by methanol. The dropwise addition of $POCl_3$ to ice-cold 5-M NH_3 yields $HPO_3NH_2^-$, plus some $PO_2(NH_2)_2^-$ and $PO(NH_2)_3$:

$$POCl_3 + 5\ NH_3 + 2\ H_2O \rightarrow HPO_3NH_2^- + 4\ NH_4^+ + 3\ Cl^-$$

Further manipulation produces the solid ammonium and other salts and, from $HClO_4$, the acid.

Oxidized nitrogen. Phosphorus(<V) is readily oxidized to H_3PO_4 by HNO_2, HNO_3, aqua regia, or other moderately strong oxidants. For determination this may be followed by precipitation and weighing as $(NH_4)_3[PMo_{12}O_{40}]$ or $Mg_2[P_2O_7]$.

Nitric acid transposes or dissolves all phosphates but that of Sn^{IV}.

Fluorine species. Concentrated H_3PO_4 reacts reversibly with HF:

$$H_3PO_4 + HF \leftrightarrows H_2PO_3F + H_2O$$

15.2.3 Reagents Derived from the 3rd-to-5th-Period Non-Metals, Silicon through Xenon

Phosphorus species. All phosphates are soluble in H_3PO_4 except those of Hg, Sn, Pb and Bi. Incidentally, excess H_3PO_4 may be used to completely remove all NO_3^-, SO_4^{2-} or Cl^- by volatilization as the acid (SO_4^{2-} as SO_3) upon evaporation and heating on a sand bath.

Nearly all diphosphates ("pyrophosphates"), but not $Ag_4[P_2O_7]$, dissolve in excess $[P_2O_7]^{4-}$ (distinction from orthophosphates), and then fail to show many of the ordinary reactions of their metal ions.

Arsenic species. Phosphinic acid, HPH_2O_2, reduces arsenites and arsenates to As in HCl solution.

Reduced chalcogens. Phosphoric acid, being a very weak oxidant, is not reduced by any of the reducing acids. The phosphates of metals forming acid-insoluble sulfides are transposed by H_2S; alkali sulfides transpose these and many other phosphates. The metal sulfide remains as a precipitate except with Cr or Al, which form hydroxides. Phosphoric acid or a phosphate will be found in the solution.

Treating $Pb_2[P_2O_7]$ with H_2S has been one route to $H_4[P_2O_7]$:

$$Pb_2[P_2O_7] + 2\ H_2S \rightarrow H_4[P_2O_7] + 2\ PbS\downarrow$$

Oxidized chalcogens. Aqueous SO_2 transposes the phosphates of Mg, Ca, Sr, Ba, Mn, Ag and Pb to sulfites.

Phosphinic acid (HPH_2O_2) and SO_2 or H_2SO_4, also phosphorus with concentrated H_2SO_4, form H_2PHO_3, H_3PO_4, S and H_2S or SO_2, depending on conditions.

Phosphonic acid and hot concentrated H_2SO_4 yield H_3PO_4 and SO_2.

Sulfuric acid transposes all phosphates to sulfates and dissolves most.

Heating pulverized phosphate rock with H_2SO_4 gives impure H_3PO_4:

$$Ca_3(PO_4)_2 + 3\ H_2SO_4 \rightarrow 2\ H_3PO_4 + 3\ CaSO_4\downarrow$$

Similarly, "superphosphate", $Ca(H_2PO_4)_2$ plus gypsum, $CaSO_4.2H_2O$, used as a fertilizer, may be made by treating ground phosphate rock with 7-M (50 %) sulfuric acid. Substituting H_3PO_4 for H_2SO_4 produces the "double" or "triple superphosphate":

$$Ca_3(PO_4)_2 + 2\ H_3O^+ + 2\ HSO_4^- + 2\ H_2O \rightarrow$$

$$Ca(H_2PO_4)_2\downarrow + 2\ CaSO_4{\cdot}2H_2O\downarrow$$

Treating $Ba_2[P_2O_7]$ with H_2SO_4 has been one route to $H_4[P_2O_7]$:

$$Ba_2[P_2O_7] + 2\ H_3O^+ + 2\ HSO_4^- \rightarrow H_4[P_2O_7] + 2\ BaSO_4\downarrow + 2\ H_2O$$

Reduced halogens. Phosphane forms phosphonium salts with HBr or HI, but the chloride is obtained only under high pressure:

$$PH_3 + HI \rightarrow PH_4I$$

The acids HCl, HBr and HI transpose all phosphates to halides, and dissolve most of them.

Elemental and oxidized halogens. Aqueous Cl_2, Br_2 or I_2 oxidize P_4 or HPH_2O_2:

$$^1/_4\ P_4 + {}^5/_2\ Cl_2 + 9\ H_2O \rightarrow H_3PO_4 + 5\ H_3O^+ + 5\ Cl^-$$

Hypophosphoric acid, $H_4[P_2O_6]$, is not oxidized by dilute ClO^-.

At pH 5.3, $NaClO_2$ with yellow or red P yields a hypophosphate:

$$P + ClO_2^- + Na^+ + CH_3CO_2^- + 4\ H_2O \rightarrow$$

$$^1/_2\ Na_2H_2[P_2O_6]{\cdot}6H_2O\downarrow + Cl^- + CH_3CO_2H$$

Its low solubility in cold water allows a separation from $HPHO_3^-$ and $H_2PO_4^-$, or as $Na_4[P_2O_6]{\cdot}10H_2O$ at pH 10.

Aqueous HPH_2O_2 and the Cl, Br and I oxoanions, except ClO_4^-, yield phosphonate, phosphate and Cl^-, Br^- or I^- in either acidic or alkaline mixture, but hot, concentrated $HClO_4$ oxidizes HPH_2O_2 etc. to H_3PO_4.

Aqueous ClO_3^-, BrO_3^- or IO_3^- oxidize P_4 to H_3PO_4, just as with the halogens themselves.

15.2.4 Reagents Derived from the Metals Lithium through Uranium, plus Electrons and Photons

Oxidation. Phosphinic acid, HPH_2O_2, is a powerful reductant, readily oxidized to H_2PHO_3 and H_3PO_4: Cr^{VI}, $Mn^{>II}$, Fe^{III}, Co^{III} and Ni^{IV} are reduced in acidic and sometimes basic solution to Cr^{III}, Mn^{2+}, Fe^{2+} or Fe, Co^{2+} or Co, and Ni^{2+} or Ni respectively. Copper(2+) on boiling forms CuH (separation from Cd); Ag^+ forms Ag, and Hg^{2+} gives first Hg_2^{2+} and then Hg in either acid or base; Pb^{IV} becomes Pb^{2+}, and Bi^{III} becomes Bi in the presence of either alkalis or CH_3CO_2H.

Hypophosphoric acid, $H_4[P_2O_6]$, i.e., $[-PO(OH)_2]_2$, and its salts are much more stable than phosphinates or phosphonates toward oxidants, and $[Cr_2O_7]^{2-}$ does not oxidize the acid.

Highly alkaline solutions of MnO_4^- and P_4 yield $[P_2O_6]^{4-}$. The acid is oxidized to H_3PO_4 slowly by MnO_4^- in the cold, rapidly when heated.

Phosphinic acid may be distinguished from phosphonic acid with MnO_4^-, which oxidizes HPH_2O_2 immediately, H_2PHO_3 only later. They are also distinguished by adding Cu^{2+} to the free acid and warming to 55 °C. With HPH_2O_2 a dark red precipitate is first formed which, at 100 °C, decomposes, deposits Cu and releases H_2, but H_2PHO_3 gives no intermediate compound.

White P_4 with a solution of Pt, Cu, Ag or Au salts finally yields a precipitate of the corresponding metal, or CuH.

Warm phosphinate or phosphonate species reduce $CuCl_2$ to CuCl, releasing no H_2, but an excess of $PH_2O_2^-$ yields CuH.

Phosphane (e.g., from P_4 + OH^-) may be detected by its reaction with a test paper impregnated with, e.g., 6-dM $CuSO_4$, 1-dM $AgNO_3$. or 4-dM $[HgCl_2]$. The silver reagent will reveal 1 ppm PH_3 in a gas. Another way to use the $AgNO_3$ paper is to hang it in a closed flask above the sample to be tested, then place the flask in wam water (30–40 °C). The paper will soon turn black if even a small amount of white phosphorus was present, the Ag^+ being reduced to Ag by the vapor.

Phosphonic acid reduces Cu^{2+} or Hg^{2+} to the lower oxidation state and then to the metal, and Ag^+, Au^I or Au^{III}, to the metal. This way, and with many higher metallic oxides, the acid becomes a phosphate.

Phosphinic acid may be determined by heating with $[HgCl_2]$ below 60 °C, drying and weighing the precipitate of $[Hg_2Cl_2]$ as such:

$$H_2PO_2^- + 4\,[HgCl_2] + 5\,H_2O \rightarrow 2\,[Hg_2Cl_2]\downarrow + H_3PO_4 + 4\,Cl^- + 3\,H_3O^+$$

Anodes, HPO_4^{2-} and a little F^- and $[Cr_2O_7]^{2-}$ form (peroxo) H_3PO_5 salts with a high current density, $H_4[P_2O_8]$ salts with a low one.

Reduction. Treating HPH_2O_2 with Zn and H_3O^+ releases PH_3. Zinc and H_3O^+ do not reduce $H_4[P_2O_6]$ (distinction from HPH_2O_2 and H_2PHO_3).

Other reactions of phosphorus(<V). Phosphane reacts with some metal ions to produce phosphides, with some of uncertain composition.

Phosphane may be determined approximately by passing the gas into a solution of $[HgCl_2]$, filtering off the yellow precipitate and titrating the filtrate with 1-dM OH^-, using methyl orange as the indicator. The composition of the precipitate is variable, but three moles of HCl are always liberated per mole of phosphane, e.g.:

$$PH_3 + 3\ [HgCl_2] + 3\ H_2O \rightarrow \sim[P(HgCl)_3] + 3\ H_3O^+ + 3\ Cl^-$$

Phosphonate, PHO_3^{2-}, precipitates Ca^{2+}, Sr^{2+} and Ba^{2+}, while phosphinate, $PH_2O_2^-$, does not, providing a useful distinction.

Silver(+) and $H_4[P_2O_6]$ form a white precipitate that does not blacken in the light (distinction from HPH_2O_2 and H_2PHO_3).

Other reactions of monophosphates. Phosphoric acid (trihydrogen phosphate) reacts with the oxides and hydroxides of the alkalis and alkaline earths and with other freshly precipitated oxides and hydroxides except perhaps Sb_2O_3. It also decomposes all carbonates with release of CO_2. Phosphates are formed in these reactions, but their compositions depend on the conditions.

Phosphoric acid dissolves some metals, such as Mg, Fe and Zn, with release of hydrogen.

Phosphoric acid is a useful solvent for some substances that resist other reagents. Heated at 230 °C, Zr dissolves in only a few minutes. Dilution of the solution gives no precipitate.

Phosphoric acid gives no precipitate with ordinary salts of the metals whose sulfides are soluble in 0.3-M HCl (in the case of Fe^{III}, a distinction from di- and metaphosphoric acids) but does form precipitates with Ag^+ and Pb^{2+}. Concentrated H_3PO_4, however, does not precipitate Ag^+.

The PO_4^{3-} and HPO_4^{2-} ions form precipitates from the hydrated cations and some other species of all but the alkali metals.

Phosphate precipitates Mg^{2+}, Ca^{2+}, Sr^{2+}, Ba^{2+}, Mn^{2+} and Zn^{2+} as normal (tertiary) salts, $M_3(PO_4)_2$, from decidedly alkaline, perhaps ammoniacal, media (better without air for Mn^{2+}), otherwise as the hydrogen-phosphates (secondary phosphates), $MHPO_4$. Aqueous Ba^{2+} with HPO_4^{2-} precipitates largely $BaHPO_4$ and perhaps a little $Ba_3(PO_4)_2$. Aqueous $H_2PO_4^-$ does not precipitate these cations. The ions Mg^{2+}, Mn^{2+} and Zn^{2+} may be quantitatively precipitated from ammoniacal solution as crystalline, hydrated $MNH_4PO_4 \cdot aq$, which can be ignited to $M_2[P_2O_7]$. Phosphates may be determined by weighing as $Mg_2[P_2O_7]$.

Aqueous PO_4^{3-} gives, with Cr^{III}, Fe^{III} and Al^{III}, mostly the normal phosphates, $CrPO_4$, $FePO_4$, and $AlPO_4$. The Fe^{III} salt is only slightly soluble in CH_3CO_2H. It can therefore be used to separate PO_4^{3-} from ions whose precipitation by it interferes in analysis.

Ammonium molybdate will precipitate, preferably from HNO_3 solution, yellow $(NH_4)_3[PMo_{12}O_{40}]$. This is a sensitive test for phosphate. The system must be acidic because the precipitate is soluble in OH^-. On the other hand, a large excess of HNO_3 decreases the sensitivity of the test. The reagent reacts only slowly with meta- or diphosphates, after they have been converted to the ortho form by the HNO_3.

Tartrates and similar ligands should be destroyed with concentrated HNO_3 before adding the molybdate. Arsenates should not interfere if precipitation is effected in the cold. It is better, however, to remove any As^V with H_2S because a temperature of 30–40 °C promotes precipitation of the $(NH_4)_3[PMo_{12}O_{40}]$. (If the system becomes too hot, MoO_3 will precipitate.) Halides should be removed with Ag^+ before testing for phosphate. Aqueous $[Fe(CN)_6]^{4-}$ forms a red-brown precipitate with the acidified molybdate, hence may cause some uncertainty.

Phosphates may be determined by weighing the $(NH_4)_3[PMo_{12}O_{40}]$. This precipitate may also be dried and titrated with OH^-, although not with the maximum accuracy, using phenolphthalein as the indicator:

$$(NH_4)_3[PMo_{12}O_{40}] + 23\ OH^- \rightarrow$$

$$HPO_4^{2-} + 12\ MoO_4^{2-} + 3\ NH_4^+ + 11\ H_2O$$

and, at that endpoint, along with a little:

$$HPO_4^{2-} + NH_4^+ \leftrightarrows H_2PO_4^- + NH_3$$

Molybdate catalyzes the hydrolysis of $[P_2O_7]^{4-}$.

Aqueous Ag^+ with the several orthophosphate ions forms Ag_3PO_4 (completely in the presence of $CH_3CO_2^-$), yellow, darkening on exposure to light, insoluble in H_2O, soluble in HNO_3 and NH_3. This color distinguishes orthophosphate ions from di- and metaphosphates, which yield white $Ag_4[P_2O_7]$ and $AgPO_3$.

If HPO_4^{2-} is added to a neutral solution of Ag^+, the pH falls, as shown by litmus, but with PO_4^{3-} the solution remains neutral (distinction between HPO_4^{2-} and PO_4^{3-}):

$$2\ HPO_4^{2-} + 3\ Ag^+ \rightarrow Ag_3PO_4\downarrow + H_2PO_4^-$$

$$PO_4^{3-} + 3\ Ag^+ \rightarrow Ag_3PO_4\downarrow$$

If Sn or fresh Sn^{IV} is added to phosphate acidified with HNO_3, the phosphate is precipitated as a stannic hydroxide phosphate. This serves to remove phosphate before precipitating other metals.

Aqueous HPO_4^{2-} precipitates with Pb^{2+}, white $PbHPO_4$, $Pb_3(PO_4)_2$, or a mixture, slightly soluble in CH_3CO_2H, soluble in OH^- or H_3O^+.

Bismuth salts with orthophosphates form $BiPO_4$, insoluble in water or dilute H_3O^+. The monohydrogen and dihydrogen salts apparently are not known. Most other metallic species precipitate a normal phosphate (at high pH), hydrogenphosphate or basic phosphate, except that with, e.g., Hg and Sb chlorides it is an oxide or oxide chloride.

An excess of PO_3S^{3-}, with $CH_3CO_2^-$ in case of low pH, colors Co^{II} deep blue and can thus identify it even with Ni^{II} present.

Aqueous $[PF_6]^-$ coordinates $M^{(n+)}$ even more weakly than does ClO_4^-.

Other reactions of poly- and metaphosphates. Diphosphates soluble in excess $[P_2O_7]^{4-}$ include those of $U^{VI}O_2$, Mn, Fe, Co, Ni, Cu, Zn, Hg^I and Al, but not those of Cr^{III} or Hg^{II}.

The acid $H_4[P_2O_7]$ forms a precipitate with Fe^{III}, Ag^+ and Pb^{2+}, but not with Mg^{2+}, Ca^{2+}, Sr^{2+}, Ba^{2+} or Fe^{2+}.

Aqueous $[Co(NH_3)_6]^{3+}$ gives an orange precipitate with $[P_2O_7]^{4-}$ (not too dilute), but not from an acetic-acid solution.

The diphosphate ion, treated with acetic acid and Cu^{2+} or Cd^{2+}, gives a blue or white precipitate respectively. Zinc acetate with $H_4[P_2O_7]$, and Zn^{2+} with $[P_2O_7]^{4-}$ or $H_2[P_2O_7]^{2-}$, precipitate white $Zn_2[P_2O_7]$. Some of the above reactions distinguish di- from ortho- and meta-phosphates.

Aqueous Ag^+ with diphosphate ion forms $Ag_4[P_2O_7]$, white, insoluble in H_2O, soluble in HNO_3 and NH_3.

The ions Mg^{2+}, Ba^{2+}, Ca^{2+}, Fe^{2+} and Al^{III} are not precipitated by $(HPO_3)_n$ but Fe^{III} is precipitated (distinction from H_3PO_4). However, after adding an excess of NH_3, Mg^{2+} gives a white precipitate, soluble in NH_4^+.

Aqueous $[Co(NH_3)_6]^{3+}$ forms a brownish-yellow precipitate with $(PO_3^-)_n$ in CH_3CO_2H solution (distinction from the ortho- and di- ions).

Some, but not all, of the metaphosphoric acids precipitate Ag^+ and Pb^{2+}. The white precipitates are soluble in moderately dilute HNO_3. Lead salts also form white precipitates with $[P_2O_7]^{4-}$; the diphosphate, $Pb_2[P_2O_7]$, dissolves in excess $[P_2O_7]^{4-}$. The $Pb(PO_3)_2$ is insoluble in excess $(PO_3^-)_n$. The $AgPO_3$ is soluble in NH_3, and seems soluble or insoluble in excesses of various $(PO_3^-)_n$ ions.

Metaphosphate ions do not form a precipitate even in a fairly concentrated cold solution of Zn^{2+}.

Other reactions using high temperatures. Heating H_3PO_4 to 215 °C yields diphosphoric ("pyrophosphoric") acid, $H_4[P_2O_7]$. (A better way is to heat until a sample, cooled and dissolved in H_2O, gives no yellow precipitate with Ag^+.) The anion may also be obtained by heating $Na_2HPO_4 \cdot 7H_2O$. It melts in its water of crystallization and then, on removal of the H_2O, solidifies as $Na_4[P_2O_7]$, which may be converted to $H_4[P_2O_7]$. In general, hydrogenphosphates (secondary phosphates), $M^I_2HPO_4$, are changed to normal diphosphates, $M^I_4[P_2O_7]$, on ignition. The same is true of normal (tertiary) orthophosphates in which one hydron has been replaced by a cation forming a volatile product, e.g., $Mg(H_2O)_6NH_4PO_4$:

$$2\ Mg(H_2O)_6NH_4PO_4 \rightarrow Mg_2[P_2O_7]\downarrow + 2\ NH_3\uparrow + 13\ H_2O\uparrow$$

Diphosphoric acid, when heated to a dull red, gives a meta acid, $(HPO_3)_n$. Heating H_3PO_4 to dense white fumes also produces a transparent, very deliquescent, glassy mass of "glacial phosphoric acid", chiefly $(HPO_3)_n$. Dihydrogenphosphates (primary phosphates), $M^IH_2PO_4$, also become metaphosphates upon ignition; the same is true of monohydrogen or normal (tertiary) phosphates that have only one hydron displaced by a metallic ion whose oxide is non-volatile:

$$NaNH_4HPO_4 \rightarrow NaPO_3 + NH_3\uparrow + H_2O\uparrow$$

Dihydrogendiphosphates, $M^I_2H_2[P_2O_7]$, upon ignition also form the metaphosphates, M^IPO_3. When di- or metaphosphates are fused with an excess of a non-volatile oxide, hydroxide or carbonate, the normal (tertiary) phosphate is formed:

$$K_3H[P_2O_7] + 3\ KHCO_3 \rightarrow 2\ K_3PO_4 + 3\ CO_2\uparrow + 2\ H_2O\uparrow$$

15.3 Arsenic, $_{33}$As

Oxidation numbers: (III), as in As_2O_3, "arsenous" oxide, and arsenite salts, and (V), as in "arsenic" oxide, As_2O_5, and arsenate salts. Arsane, AsH_3, is often considered to show As(−III), partly in analogy with NH_3. The Allred-Rochow and other electronegativities of H and As, however, are the same or nearly so, so that we may regard AsH_3 as $As^0H^0_3$.

15.3.1 Reagents Derived from Hydrogen and Oxygen

Water. Arsenic is insoluble in water.

The solubility of As_2O_3 is only about 1 dM, forming 2-dM H_3AsO_3.

Arsenic(V) oxide, As_2O_5, is a white, amorphous mass and is slowly deliquescent, combining with H_2O to form H_3AsO_4 in a concentrated solution. Crystallization gives $H_3AsO_4 \cdot {}^1/_2H_2O$ at room temperature, or $HAsO_3 \cdot {}^1/_3H_2O$ at 100 °C.

Alkali arsenites are quite soluble in water; the alkaline-earth ones are slightly soluble, and most **d**-, **p**- or **f**-block metal salts are insoluble.

Alkali arsenates, and alkaline-earth hydrogen arsenates, are soluble.

Precipitated As_2S_3 may become colloidal on treatment with pure water. It is reprecipitated by solutions of most inorganic salts, and these, or acids, prevent its peptization in water. Boiling water slowly decomposes the sulfide, forming H_3AsO_3 and H_2S.

The trihalides $AsCl_3$, $AsBr_3$ and AsI_3 are decomposed by small amounts of water into the corresponding oxyhalide, e.g., AsOCl. A further addition of H_2O decomposes these compounds into As_2O_3 and the halogen acid.

Some natural waters contain $HAsS_2$ or AsS_2^-, or, if hot, AsF_5^{2-} etc.

Oxonium. Arsenites are all easily dissolved in acids. Arsenates are all soluble in inorganic acids, including H_3AsO_4.

Hydroxide. Hot OH^- dissolves As, forming AsH_3 and likely $H_2AsO_3^-$.

Aqueous OH^- has no effect on arsane, AsH_3.

No acids (hydroxides) of As_2O_3 have been isolated; with bases it forms salts, arsenites, as if derived from hypothetical ortho-, di-, or metaarsenous acids, H_3AsO_3, $H_4As_2O_5$ or $(HAsO_2)_n$. The alkali arsenites are usually meta-compounds, M^IAsO_2; the Ae and **d**- or **p**-block arsenites are usually ortho-compounds, such as $M^{II}_3(AsO_3)_2$ or Ag_3AsO_3.

Aqueous OH^- reacts with As_2O_5 forming arsenates.

Peroxide. Arsenic(0) is oxidized to As^{III} by H_2O_2, and by excess to As^V.

Dioxygen. Arsenic is slowly oxidized in moist air at ordinary temperatures to As_2O_3, "white arsenic".

15.3.2 Reagents Derived from the Other 2nd-Period Non-Metals, Boron through Fluorine

Boron species. Arsenites plus $[BH_4]^-$, added under N_2 or Ar to an excess of 2-M H_2SO_4, release AsH_3, also with some As, As_2H_4 etc.:

$$4\ H_2AsO_3^- + 3\ [BH_4]^- + 7\ H_3O^+ \rightarrow 4\ AsH_3\uparrow + 3\ H_3BO_3 + 10\ H_2O$$

Carbon species. Aqueous CO_3^{2-} reacts with As^{III} and As^V oxides, forming soluble arsenites and arsenates.

Reduced nitrogen. Ammonia does not act on arsenic. Concentrated NH_3 decomposes AsH_3 incompletely with the separation of arsenic.

Oxidized nitrogen. Nitrites in acid readily decompose arsane:

$$AsH_3 + 3\ HNO_2 \rightarrow As\downarrow + 3\ NO\uparrow + 3\ H_2O$$

but AsH_3 is also oxidized to As^V by HNO_2 and HNO_3.

Nitric acid oxidizes arsane to arsenic acid:

$$3\ AsH_3 + 8\ H_3O^+ + 8\ NO_3^- \rightarrow 3\ H_3AsO_4 + 8\ NO\uparrow + 12\ H_2O$$

and may be used instead of $AgNO_3$ to separate As and Sb in the Marsh test. The nitric acid solution is evaporated to dryness and the residue thoroughly washed with water. The solution thus obtained is tested for arsenic. The residue may be dissolved in HCl or aqua regia and tested for antimony with H_2S.

Nitric acid readily oxidizes As to H_3AsO_3, and with excess to H_3AsO_4.

A suspension of As_2O_3 in concentrated HNO_3 yields H_3AsO_4:

$$3\ As_2O_3 + 4\ H_3O^+ + 4\ NO_3^- + 3\ H_2O \rightarrow 6\ H_3AsO_4 + 4\ NO\uparrow$$

Nitric acid easily oxidizes As_2S_3 and other $As^{<V}$ to H_3AsO_4:

$$3\ As_2S_3 + 10\ H_3O^+ + 10\ NO_3^- \rightarrow$$

$$6\ H_3AsO_4 + 9\ S\downarrow + 10\ NO\uparrow + 6\ H_2O$$

Fluorine species. When As^{III} and As^{V} are treated with HF plus H_2SO_4, only the As^{III} is volatilized.

15.3.3 Reagents Derived from the 3rd-to-5th-Period Non-Metals, Silicon through Xenon

Phosphorus species. Phosphinic acid, HPH_2O_2, in the presence of hot concentrated HCl reduces all oxidized compounds of arsenic to the element. Boiling a few mL of a solution containing 0.1 mg of arsenic with 10 mL of concentrated HCl and 2 dg of calcium phosphinate ("hypophosphite") will give a good test:

$$4\ AsCl_3 + 3\ HPH_2O_2 + 18\ H_2O \rightarrow$$

$$4\ As\downarrow + 3\ H_3PO_4 + 12\ H_3O^+ + 12\ Cl^-$$

If a small amount .of arsenic is present, only a yellowish-brown color will develop even upon prolonged heating. In milder conditions HPH_2O_2 reduces As^V to As^{III}.

Arsenic species. Arsane reacts with As^{III}, and both become As.

Reduced chalcogens. Sulfane (H_2S) precipitates lemon-yellow As_2S_3 from acidic, but not alkaline, solutions of arsenites, even in concentrated HCl. Chelators such as citric acid hinder this, but do not wholly prevent it if much HCl is present. In water the sulfide tends to become colloidal, breaking down when the solution is boiled:

$$As_2S_3 + 6\ H_2O \leftrightarrows 2\ H_3AsO_3 + 3\ H_2S\uparrow$$

Similarly, H_2S, H_3AsO_4 and concentrated HCl form As_2S_5, insoluble in and coagulated by the HCl, best at 0 °C.

Alkali sulfides dissolve the sulfides, e.g.:

$$As_2S_3 + HS^- + NH_3 \rightarrow 2\ AsS_2^- + NH_4^+$$

Arsenic(III) sulfide also dissolves in OH^- and CO_3^{2-}, forming arsenites and thioarsenites. The thioarsenites react with acids, forming As_2S_3:

$$2\ AsS_2^- + 2\ H_3O^+ \rightarrow As_2S_3\downarrow + H_2S\uparrow + 2\ H_2O \text{ or}$$

$$3\ AsS_2^- + AsO_3^{3-} + 6\ H_3O^+ \rightarrow 2\ As_2S_3\downarrow + 9\ H_2O$$

The solubility of the As sulfides in yellow $(NH_4)_2S_2$ separates As with Sn and Sb from the other (acid-insoluble) sulfides, oxidizing the As_2S_3:

$$As_2S_3 + 4\ S_2^{2-} \rightarrow 2\ [AsS_4]^{3-} + S_3^{2-}$$

$$As_2S_5 + 6\ S_2^{2-} \rightarrow 2\ [AsS_4]^{3-} + 3\ S_3^{2-}$$

It may also be separated from Sn and Sb by boiling with concentrated HCl, leaving solid As_2S_3 but dissolving the Sn and Sb sulfides.

A neutral or alkaline solution of arsenate treated with "$(NH_4)_2S$" forms $[AsS_4]^{3-}$. Adding acid yields As_2S_5 at once, faster than with H_2S, and even more on warming. The acid coagulates it.

The bases OH^- or CO_3^{2-} dissolve As_2S_3 readily with the formation of arsenite and AsS_2^-; As_2S_3 is also soluble in S^{2-}, HS^- and S_x^{2-}, forming AsS_2^- and, e.g., $[AsS_4]^{3-}$. Freshly precipitated As_2S_3 is soluble in HSO_3^- (separation from Sn and Sb):

$$2\ As_2S_3 + 3\ HSO_3^- + 3\ H_2O \rightarrow 3\ H_2AsO_3^- + H_3AsO_3 + 9\ S\downarrow$$

with a variable further reaction of some of the sulfur with HSO_3^- to form $S_2O_3^{2-}$ and SO_2. Arsenic(III) sulfide also dissolves readily in NH_3 plus H_2O_2, giving $HAsO_4^{2-}$ and SO_4^{2-}.

Arsenic(V) sulfide is insoluble in water; soluble in HNO_3 or in Cl_2 water, yielding H_3AsO_4; soluble in OH^-, CO_3^{2-}, S^{2-} and "$(NH_4)_2S$", forming arsenates and tetrathioarsenate (also with a little $[AsO_3S]^{3-}$ etc.):

$$4\ As_2S_5 + 24\ OH^- \rightarrow 5\ [AsS_4]^{3-} + 3\ AsO_4^{3-} + 12\ H_2O$$

$$As_2S_5 + 3\ HS^- + 3\ NH_3 \rightarrow 2\ [AsS_4]^{3-} + 3\ NH_4^+$$

At low pH As^V is reduced to As^{III} by HSCN.

Oxidized chalcogens. Aqueous H_3AsO_3 or H_3AsO_4, boiled with $S_2O_3^{2-}$, forms As_2S_3 or As_2S_5. Intermediates, including SO_3^{2-}, $[S(SO_3)_2]^{2-}$, i.e., $[O_3S\text{–}S\text{–}SO_3]^{2-}$, and [*catena*-$S_3(SO_3)_2]^{2-}$, i.e., $[O_3S\text{–}S\text{–}S\text{–}S\text{–}SO_3]^{2-}$, may arise when excess $S_2O_3^{2-}$ precipitates As^V from aqueous HCl.

In solution As^V is reduced to As^{III} by SO_2 and $S_2O_3^{2-}$.

Boiling with BaS_2O_3 removes As^{III} and As^{V} impurities from H_2SO_4:

$$2\ H_3AsO_3 + 3\ BaS_2O_3 \rightarrow As_2S_3\downarrow + 3\ BaSO_4\downarrow + 3\ H_2O$$

Arsenic(III and V) are reduced to elemental arsenic by $S_2O_4^{2-}$.

Warm aqueous SO_2 oxidizes arsane, AsH_3, to As and As_2S_3, but it readily reduces arsenic acid to arsenous acid:

$$4\ AsH_3 + 3\ SO_2 \rightarrow 2\ As\downarrow + As_2S_3\downarrow + 6\ H_2O$$

$$H_3AsO_4 + SO_2 + 3\ H_2O \rightarrow H_3AsO_3 + SO_4^{2-} + 2\ H_3O^+$$

Concentrated H_2SO_4 at room T decomposes AsH_3 to brown flakes of As, soluble on heating. The hot acid oxidizes it directly to H_3AsO_3.

Sulfuric acid, dilute and cold, does not react with As; heat and concentrated acid form As_2O_3 and SO_2, more easily from As_2S_3.

Reduced halogens. Arsenic is not attacked by concentrated hydrochloric acid at ordinary temperatures, but is slowly attacked by the hot acid in the presence of air, forming $AsCl_3$.

Aqueous HCl and AsH_3 form H_3AsO_3 and H_2, but removing AsH_3 from H_2S by passing it through hot, dilute HCl tends to be incomplete.

Hot concentrated HCl decomposes As_2S_3 very slightly.

If As^{III} and As^{V} are treated with concentrated HCl and then distilled in a current of gaseous HCl, the As^{III} goes into the distillate as $AsCl_3$, the As^{V} being slowly reduced. Usually $N_2H_5^+$, CuCl, or other reductant is added to facilitate this quantitative separation of As from all other metals except Ge and from other nonvolatile organic and inorganic material. The $AsCl_3$ is distilled below 108 °C and absorbed in water. This solution may be tested for As^{III} by the usual methods. The As, if need be, can readily be separated from the HCl by adding a little excess ClO^-, forming Cl_2, H_2O and H_3AsO_4, and evaporating the Cl_2.

Concentrated, but not dilute, HBr reduces H_3AsO_4 to H_3AsO_3:

$$H_3AsO_4 + 3\ Br^- + 2\ H_3O^+ \rightarrow H_3AsO_3 + Br_3^- + 3\ H_2O$$

The hydrolysis of $AsBr_3$ is usefully reversible. Heating As_2O_3 with concentrated HBr at 120 °C (below its bp) up to 20 min and cooling yields the white or colorless $AsBr_3$.

Even dilute HI reduces H_3AsO_4 to H_3AsO_3, liberating I_2. This detects as little as 100 ppm of arsenate in the presence of As_2O_3:

$$H_3AsO_4 + 2\ I^- + 2\ H_3O^+ \rightarrow H_3AsO_3 + I_2 + 3\ H_2O$$

Arsenic triiodide can be precipitated with much I^- from, say, half-M As^{III} in rather concentrated HCl. The AsI_3 in air slowly liberates I_2.

Elemental and oxidized halogens. Chlorine or bromine in water oxidizes arsenic or arsane (AsH_3) first to arsenous, then to arsenic acid:

$$As + {}^{5}/_{2}\ Cl_2 + 9\ H_2O \rightarrow H_3AsO_4 + 5\ H_3O^+ + 5\ Cl^-$$

Chlorine water or hypochlorite decomposes As_2S_3 readily:

$$As_2S_3 + 5\ Cl_2 + 18\ H_2O \rightarrow 2\ H_3AsO_4 + 3\ S\downarrow + 10\ H_3O^+ + 10\ Cl^-$$

$$As_2S_3 + 14\ Cl_2 + 51\ H_2O \rightarrow 2\ H_3AsO_4 + 3\ HSO_4^- + 31\ H_3O^+ + 28\ Cl^-$$

The XO^- ions oxidize AsH_3 to As^V, As to As^{III} and As^{III} to As^V, e.g.:

$$AsH_3 + 4\ ClO^- + 2\ OH^- \rightarrow HAsO_4^{2-} + 4\ Cl^- + 2\ H_2O$$

A halate, ClO_3^- or BrO_3^-, with a little $AgNO_3$ as a catalyst, oxidizes AsH_3 to As^V. Arsenic also is oxidized to As^{III} by ClO_3^-, BrO_3^- or IO_3^-, and As^{III} by excess to As^V. Then too, the acids $HClO_3$ and $HBrO_3$ oxidize As^{III} to As^V, forming some halogen from a side reaction, and HIO_3 oxidizes AsH_3 to As^{III}, then to As^V, leaving iodine:

$$^{3}/_{2}\ As_4O_6 + 2\ BrO_3^- + 9\ H_2O \rightarrow 6\ H_3AsO_4 + 2\ Br^-$$

$$5\ H_3AsO_3 + 2\ IO_3^- + 2\ H_3O^+ \rightarrow 5\ H_3AsO_4 + I_2 + 3\ H_2O$$

15.3.4 Reagents Derived from the Metals Lithium through Uranium, plus Electrons and Photons

Oxidation. Cerium(IV), catalyzed by OsO_4, oxidizes H_3AsO_3 to H_3AsO_4. Also, As^{III} is oxidized to As^V with H_3O^+ by $[Cr_2O_7]^{2-}$ and PbO_2; by oxo-compounds of $Mn^{>II}$, $Co^{>II}$ and $Ni^{>II}$; and in alkaline mixture by CrO_4^{2-}, $[Fe(CN)_6]^{3-}$, CuO, Hg_2O, HgO and PbO_2.

Boiling CrO_4^{2-} with arsenites and HCO_3^- gives a Cr^{III} arsenate.

Aqueous MnO_4^- oxidizes AsH_3 and As to As^V.

A Cu^{II}-tartrate complex, Fehling's solution, tests for arsenite:

$$AsO_3^{3-} + 2\ [Cu(C_4H_3O_6)_2]^{4-} + 2\ H_2O \rightarrow AsO_4^{3-} + Cu_2O\downarrow + 4\ C_4H_4O_6^{2-}$$

Aqueous Ag^+ oxidizes both arsenic and arsane to As^{III}:

$$6\ Ag^+ + AsH_3 + 9\ H_2O \rightarrow 6\ Ag\downarrow + 6\ H_3O^+ + H_3AsO_3$$

Aqueous AsO_3^{3-} is oxidized by AgCl but not AgBr or AgI.

Reduction. "Nascent hydrogen", generated by any metal and acid that readily generate H_2, i.e., Mg, Fe, Zn, Sn, etc., with H_2SO_4 or HCl, reduces As^{III} or As^V to AsH_3. Some free metals alone, such as Mg, Cu, Zn, Cd etc., also reduce As^V and

As^{III} to metallic arsenic, more or less completely. In alkaline solution, As^{III} is reduced to AsH_3 by Na_{Hg}, Zn plus OH^-, Al plus OH^-, etc. (separation from Sb).

Copper(II) with arsane yields copper arsenide:

$$3\ Cu^{2+} + 2\ AsH_3 + 6\ H_2O \rightarrow Cu_3As_2\downarrow + 6\ H_3O^+$$

The oxidation states of zero in AsH_3 make this a reduction of As to the –III oxidation state [with H going from (0) to (I)].

Aqueous $[SnCl_3]^-$ in hot, concentrated HCl reduces As^{III} (As^V slowly) to the flocculent, dark brown element in Bettendorff's test:

$$2\ AsCl_3 + 3\ [SnCl_3]^- + 3\ Cl^- \rightarrow 2\ As\downarrow + 3\ [SnCl_6]^{2-}$$

Copper, Cd, Sn, Pb, and Sb compounds do not interfere. The $[SnCl_3]^-$ should be fresh, with HCl ≥ 8 M for rapid and complete precipitation. Mercury interferes, giving a gray precipitate capable of being gathered into globules. Organic matter lowers the sensitivity of the test somewhat.

Other special tests have been developed, chiefly to detect small amounts of arsenic. Preliminary concentration, if needed, may be achieved by distilling it from concentrated HCl.

The Marsh-Gutzeit Test: Arsenic, from all of its soluble compounds, is reduced (by "nascent hydrogen") in acidic solution to arsane, AsH_3, a colorless gas smelling like garlic:

$$H_3AsO_4 + 4\ Zn + 8\ H_3O^+ \rightarrow AsH_3\uparrow + 4\ Zn^{2+} + 12\ H_2O$$

The hydrogen with any arsane is passed into a test tube containing 1-dM $AgNO_3$. The gas, before the actual test, is allowed to bubble through the $AgNO_3$ solution for several minutes, and should produce no appreciable black precipitate or suspension, proving the reagents free from arsenic. Then the solution to be tested is added to the gas-generating flask in small amounts at a time through a funnel tube. If much arsenic is present the solution is blackened almost immediately, precipitating Ag and forming H_3AsO_3 as shown under **Oxidation** above.

The arsenous acid in the $AgNO_3$ solution can be confirmed by the usual tests after removing the excess Ag^+ with dilute HCl. (In a non-aqueous alternative or confirmation, the gas is passed through a hot glass tube where any arsane decomposes and deposits an arsenic "mirror" even with small amounts.)

To generate arsane (with the hydrogen), magnesium or iron may be used instead of zinc, with either HCl or H_2SO_4. Arsane will not be formed in the presence of oxidants such as HNO_3, the halogens, ClO^-, or ClO_3^-. It can also be produced from As^{III} compounds by alkaline reductants. Sodium amalgam (1:8 by weight or nearly NaHg), Zn and NaOH, or Al and NaOH may be used.

The most important interference in the Marsh Test for As is Sb, which gives stibane, SbH_3, similar in many ways to AsH_3. Although Ge and Se also interfere, they are so uncommon that the danger is small. Mercury, especially $[HgCl_2]$, also fluorides and sulfites, should be absent. (The latter form H_2S, which combines with AsH_3 in the heated tube, if used, to give As_2S_3 but no mirror.) The use of $AgNO_3$ clearly requires the absence of anything other than As which gives a black precipitate, e.g., sulfides, phosphorus compounds that would form PH_3, or Sb and Bi compounds.

Reinsch's Test: If As^{III} is boiled with HCl and a thin strip of bright Cu foil, the As is deposited on the copper as a gray film, not As, but apparently a copper arsenide. (Note that As^{V} is reduced far less readily.) With much As, the coating separates as scales. The HCl must be at least 3-M overall. This determines very well the presence or absence of As in HCl (diluted to 3 M). A trace of arsenic (4 ppm), if present, will soon coat the foil.

Other reactions. The oxides of arsenic do not act as bases with oxo acids, but with metallic oxides they form arsenites and arsenates.

Magnesium salts with NH_4^+ and NH_3 precipitate arsenates slowly, but finally completely, as a white, crystalline salt, easily soluble in acids:

$$H_2AsO_4^- + Mg^{2+} + 2\ NH_3 + 6\ H_2O \rightarrow Mg(H_2O)_6NH_4AsO_4\downarrow + NH_4^+$$

The reagents should first be mixed, and the clear "magnesia mixture" be used to make sure of enough NH_4^+ to avoid precipitating $Mg(OH)_2$. [Compare with the similar $Mg(H_2O)_6NH_4PO_4$.] Magnesium arsenite is insoluble in water, soluble in NH_4^+ (distinction from arsenate).

Ammonium molybdate, $(NH_4)_2MoO_4$, when warmed to 60–70 °C with an arsenate (but not As^{III}) in HNO_3, precipitates yellow $(NH_4)_3[AsMo_{12}O_{40}]$, ammonium 12-molybdoarsenate. In appearance and properties it resembles the phosphate compound, except that the latter is precipitated in the cold. The limit of sensitivity is claimed to be one in 100 million.

Arsenites precipitate Fe^{III} salts, and fresh $Fe_2O_3\cdot aq$ forms with As_2O_3, variable basic Fe^{III} arsenites or adsorption products, scarcely soluble in CH_3CO_2H, but soluble in HCl. Water slowly and sparingly dissolves some of the As^{III}, but a large excess of $Fe_2O_3\cdot aq$ retains nearly all of it. The basic Fe^{III} arsenates(III) change partly to basic Fe^{II} arsenates(V). Arsenic acid, with $CH_3CO_2^-$ and Fe^{III}, precipitates yellowish-white $FeAsO_4$.

Aqueous Cu^{2+} precipitates from neutral solutions of arsenites, green copper arsenite, $\sim Cu_3(AsO_3)_2\cdot aq$ or $CuHAsO_3$ ("Scheele's Green"), soluble in NH_3 and dilute acids. Copper acetate, in boiling solution, precipitates a green copper acetate-arsenite, approximately $Cu_2(C_2H_3O_2)(AsO_2)_3$ ("Paris Green"), soluble in NH_3 and acids.

Copper arsenate, $Cu_3(AsO_4)_2$, greenish blue, is precipitated by Cu^{2+} from solutions of arsenates, the solubility and conditions of precipitation being the same as for arsenites.

Aqueous Ag^+ precipitates from neutral solutions of arsenites, silver arsenite, Ag_3AsO_3, bright yellow, readily soluble in dilute acids or NH_3.

Arsenates are precipitated in neutral solution as silver arsenate, Ag_3AsO_4, reddish brown, having the same solubilities as the arsenite.

Most of the **d**- or **p**-block species oxidize arsane, but sometimes only the H atoms, not the As (to the extent to which we regard separate formal charges within molecules as real). See the introduction of section **15.3**. Silver(1+) as shown under **Oxidation**, oxidizes both. Gold salts and [$HgCl_2$], however, form As from AsH_3. The oxidation states of 0 for both As and H in AsH_3, then, make this really neither oxidation nor reduction of the As; we may well say that only the H is oxidized:

$$AsH_3 + AuCl_4^- + 3\ H_2O \rightarrow As\downarrow + Au\downarrow + 4\ Cl^- + 3\ H_3O^+$$

15.4 Antimony, $_{51}$Sb

Oxidation numbers: (III) and (V), as in Sb_2O_3, "antimonious" or "antimonous" oxide and stibite salts, and even in SbH_3 [i.e., not as antimony(−III)], and (V) as in Sb_2O_5, "antimonic" oxide and stibates.

15.4.1 Reagents Derived from Hydrogen and Oxygen

Water. The trioxide, Sb_2O_3, dissolves in water at 15 °C to the extent of only 30 μM. The trichloride, $SbCl_3$, is very deliquescent, decomposed by water, forming approximately SbOCl.

The pentoxide is insoluble, but its hydrate, $Sb_2O_5\cdot aq$, is sparingly soluble as "antimonic acid", H[$Sb(OH)_6$], $pK_a = 2.55$.

The least soluble Alk[$Sb(OH)_6$] is that of Na, 3.3 mM at 25 °C.

Antimony(III) tartrate and potassium Sb^{III} tartrate (tartar emetic), $K_2[Sb_2(C_4H_2O_6)_2]\cdot{\sim}3H_2O$, are soluble.

Solutions of $[SbCl_4]^-$ form, when diluted with H_2O, a series of oxychlorides from SbOCl to $Sb_4Cl_2O_5$, depending on the amount of water. The precipitates dissolve in tartaric acid (distinction from BiOCl) and, with enough Cl^- and H_3O^+ (or OH^-), are formed partly or not at all:

$$[SbCl_4]^- + 3\ H_2O \leftrightarrows SbOCl\downarrow + 2\ H_3O^+ + 3\ Cl^-$$

More water, by the mass-action effect and dilution, causes nearly complete precipitation. If the precipitate is washed with water, the acid is gradually displaced, finally leaving Sb_2O_3.

The halides $SbBr_3$ and SbI_3 are deliquescent and require moderately concentrated acid to keep them in solution.

The pentachloride, $SbCl_5$, is a liquid, very readily combining with a small amount of water to form crystals containing one or four molecules of H_2O. Adding more H_2O decomposes this, forming a basic salt, possibly $SbOCl_3$; if, however, a little HCl is added first, forming $[SbCl_6]^-$, any amount of water, added at one time, may be introduced without immediately precipitating the basic salt.

In general, H_2O decomposes inorganic species of Sb, except fluorides and chalcogenides, and organic chelates such as tartrates and citrates, precipitating basic salts. The inorganic salts require some free acid (not acetic) to keep them in solution.

Some natural waters contain $Sb^{III}S_n^{(2n-3)-}$, $HSb_2S_4^-$ etc., and some hot waters may contain $Sb^{III}(OH)_n^{(3-n)+}$.

Oxonium. Carefully neutralizing alkaline Sb^{III} with an acid (not hydrofluoric, tartaric or citric) precipitates the hydrous oxide, which is at once dissolved by more acid in many cases:

$$2\ [Sb(OH)_4]^- + 2\ H_3O^+ \rightarrow Sb_2O_3{\cdot}aq\downarrow + 7\ H_2O$$

Hydroxide. Stibane ($Sb^{III}H^{-I}_3$ or nearly $Sb^0H^0_3$) passed into 6-M OH^- forms H_2 and metallic Sb rather rapidly; and with air, a little $[Sb(OH)_4]^-$.

Aqueous OH^- precipitates from acidulated Sb^{III} salts, unless prevented by, say, organic hydroxoacids, Sb^{III} hydrous oxide, $Sb_2O_3{\cdot}aq$:

$$2\ [SbCl_4]^- + 6\ OH^- \rightarrow Sb_2O_3{\cdot}aq\downarrow + 8\ Cl^- + 3\ H_2O$$

white, bulky, readily becoming crystalline on boiling, sparingly soluble in water, readily soluble in excess of OH^-:

$$Sb_2O_3{\cdot}aq + 2\ OH^- + 3\ H_2O \rightarrow 2\ [Sb(OH)_4]^- \text{ or}$$

$$[SbCl_4]^- + 4\ OH^- \rightarrow [Sb(OH)_4]^- + 4\ Cl^-$$

A sodium salt is the most stable and least soluble such salt in water; the potassium salt is readily soluble in dilute KOH solution, but decomposed by pure water. Part of the Sb is precipitated from the alkaline solution on long standing (24 hours).

Antimony(V) oxide, Sb_2O_5, is slowly soluble in concentrated KOH.

Antimony(V) salts are precipitated by OH^- under the same conditions as the Sb^{III} salts. The $Sb_2O_5{\cdot}aq$ in various degrees of hydration has been reported as ortho-, pyro-, or meta-antimonic acid, but there are no discrete pure acids such as an ortho-acid, H_3SbO_4, analogous to H_3PO_4 and H_3AsO_4. The highest oxidation state of Sb, like those of its neighbors in the **5p**-row, has a ligancy (c. n.) of six, as in $[Sb(OH)_6]^-$.

The $Sb_2O_5{\cdot}aq$ precipitate is insoluble in NH_3 or "$(NH_4)_2CO_3$"; easily soluble in KOH or an excess of K_2CO_3, but insoluble in NaOH due to the insolubility of the sodium salt, $Na[Sb(OH)_6]$. Other antimonate(VI) or hexahydroxoantimonate(VI) salts are many and include both $[Mg(H_2O)_6][Sb(OH)_6]_2$ and $Ag[Sb(OH)_6]$.

A "meta-acid" however, approximately $(HSbO_3)_n$, is reported to be easily soluble in all the fixed alkalis (i.e., including NaOH but not NH_3).

15.4.2 Reagents Derived from the Other 2nd-Period Non-Metals, Boron through Fluorine

Boron species. Aqueous $[BH_4]^-$ reduces Sb^{III} completely to stibane, SbH_3, perhaps as:

$$4\ [Sb(OH)_4]^- + 3\ [BH_4]^- \rightarrow 4\ SbH_3\uparrow + 3\ [B(OH)_4]^- + 4\ OH^-$$

The Sb^{III} tartrate complex plus $[BH_4]^-$, added under N_2 or Ar to an excess of 2-M H_2SO_4, release SbH_3, also giving some Sb etc.:

$$2\ [Sb_2(C_4H_2O_6)_2]^{2-} + 3\ [BH_4]^- + 7\ H_3O^+ + 2\ H_2O \rightarrow$$

$$4\ SbH_3\uparrow + 3\ H_3BO_3 + 4\ H_2C_4H_4O_6$$

Carbon oxide species. Aqueous CO_3^{2-} (but not NH_3) decomposes SbH_3 to release H_2 and precipitate dark brown Sb.

Antimony(III) hydrous oxide, $Sb_2O_3{\cdot}aq$, is formed by treating a solution of $[SbCl_4]^-$ with CO_3^{2-}:

$$2\ [SbCl_4]^- + 3\ CO_3^{2-} \rightarrow Sb_2O_3.aq\downarrow + 8\ Cl^- + 3\ CO_2\uparrow$$

Excess cold CO_3^{2-} dissolves a small amount of the oxide, but adding HCO_3^- (cold) to $[Sb(OH)_4]^-$ almost completely precipitates it:

$$Sb_2O_3{\cdot}aq + 2\ CO_3^{2-} + 5\ H_2O \leftrightarrows 2\ [Sb(OH)_4]^- + 2\ HCO_3^-$$

Boiling with an excess of CO_3^{2-} drives out the acidic CO_2 and makes the compound fairly soluble (distinction from Sn):

$$Sb_2O_3{\cdot}aq + CO_3^{2-} + 4\ H_2O \rightarrow 2\ [Sb(OH)_4]^- + CO_2\uparrow$$

Cyanide species. Aqueous CN^- gives a white precipitate with Sb^{III} salts, $Sb_2O_3{\cdot}aq$, soluble in excess CN^-.

Some "simple" organic reagents. Acetic acid produces a precipitate from solutions of stibites or stibates if tartaric acid etc. are absent.

Freshly precipitated Sb^{III} oxide is soluble in oxalic acid, but the antimony soon slowly, but completely, separates as a white crystalline precipitate unless $C_2O_4^{2-}$ is present as a ligand. The precipitate of antimony oxalate dissolves in HCl.

The hydrated trioxide, $Sb_2O_3{\cdot}aq$, is fairly soluble in glycerol. The pentoxide, Sb_2O_5, dissolves in an alkaline solution of glycerol.

The trioxide, Sb_2O_3, is soluble in tartaric acid. The long-known tartar emetic contains the tartrate chelate $[Sb_2(C_4H_2O_6)_2]^{2-}$, but is still often called the "antimonyl" complex salt $KSbOC_4H_4O_6$. We may note the equivalence with one empirical (disputed) formula of the hydrate:

$$KSbOC_4H_4O_6 \equiv KSbC_4H_4O_7 \equiv {}^1/_2\, K_2[Sb_2(C_4H_2O_6)_2]\cdot 2H_2O$$

The oxide chloride SbOCl is soluble in oxalic, tartaric or citric acids, but not in the non-chelating acetic acid.

Aqueous $[SbS_2]^-$ plus CH_3CO_2H give reddish orange Sb_2S_3.

Fresh Sb^V oxide, Sb_2O_5, dissolves readily in, e.g., oxalic acid or tartaric acid without reduction, and does not separate on standing.

Ion-exchange resins charged with H_3O^+ convert solutions of $K[Sb(OH)_6]$ to the quite acidic $H[Sb(OH)_6]$ or $[Sb(OH)_5(H_2O)]$, see **Water** above, but concentrated solutions also may contain condensed polymers, sometimes called stibates or pyro- or meta-stibates.

Reduced nitrogen. Antimony(III) oxide, Sb_2O_3, is formed by treating a solution of $[SbCl_4]^-$ with NH_3. It is insoluble in NH_3 or "$(NH_4)_2CO_3$".

Oxidized nitrogen. The best solvent for antimony may be hydrochloric acid mixed with only a small amount of HNO_3. The complex $[SbCl_4]^-$ is first formed, but if sufficient HNO_3 is present, this is rapidly changed to $[SbCl_6]^-$. However, if too much HNO_3 is present, the corresponding oxides are precipitated. Antimony(III) compounds in general are oxidized to Sb^V by HNO_3.

A warm, dilute solution of HNO_3 and HF is an excellent solvent for Sb and its alloys. Nitric acid combined with tartaric acid also easily dissolves antimony.

Antimony(III) oxide, Sb_2O_3, is formed by the action of dilute nitric acid upon Sb. Antimony is attacked but not dissolved by hot HNO_3 forming Sb_2O_3 or Sb_2O_5 depending on the concentration of the acid:

$$2\ Sb + 2\ H_3O^+ + 2\ NO_3^- \rightarrow Sb_2O_3\downarrow + 2\ NO\uparrow + 3\ H_2O$$

$$6\ Sb + 10\ H_3O^+ + 10\ NO_3^- \rightarrow 3\ Sb_2O_5\downarrow + 10\ NO\uparrow + 15\ H_2O$$

The dry, ignited oxide, Sb_2O_3, is only slightly soluble in HNO_3; the moist, freshly precipitated oxide, however, dissolves readily in the dilute or concentrated acid, warm or cold. Under certain conditions of concentration a portion of the antimony separates upon standing as a white crystalline precipitate.

The pentoxide, Sb_2O_5, is formed by treating Sb, Sb_2O_3, or Sb_2O_4 with concentrated nitric acid. It is a citron-yellow powder, reddening moist blue-litmus paper.

The oxides or hydrated oxides (acids) are precipitated from solutions of the stibites or stibates upon neutralization with HNO_3 or other inorganic acids, the freshly formed precipitate readily dissolving in an excess of the acid, giving

$Sb(OH)_2^+$ or Sb^{3+} (in concentrated acid) if complexing ligands are absent. Antimony(III) nitrate is quite unstable and Sb^V nitrate apparently is not known.

If antimony and arsenic compounds occurring together are treated with 16-M HNO_3, the antimony oxide precipitate may contain arsenic.

15.4.3 Reagents Derived from the 3rd-to-5th-Period Non-Metals, Silicon through Xenon

Phosphorus species. Antimony(V) is reduced to Sb^{III} by $PH_2O_2^-$.

Antimony phosphates seem unknown. Aqueous HPO_4^{2-} does not precipitate Sb salts as such (separation from Sn).

Reduced chalcogens. Antimony dissolves on heating with S_x^{2-}.

Sulfane precipitates from not too acidic solutions of Sb^{III}, Sb_2S_3:

$$2\ SbCl_3 + 3\ H_2S + 6\ H_2O \rightarrow Sb_2S_3\downarrow + 6\ H_3O^+ + 6\ Cl^-$$

orange-red; boiling this sulfide or passing H_2S into the hot solution for a long time gives a black variety. In neutral solution (where, e.g., tartrate must be present) the precipitation is incomplete. In OH^- solution only soluble thio complexes are formed.

Sulfane (H_2S) precipitates Sb^V from dilute HCl as orange Sb_2S_5.

The trisulfide is slowly decomposed by boiling water:

$$Sb_2S_3 + 3\ H_2O \rightarrow Sb_2O_3\downarrow + 3\ H_2S\uparrow$$

and is soluble in boiling OH^- as $[Sb(OH)_4]^-$ and thio complexes. These are oxidized on standing, by O_2, rapidly by S. From the alkaline solutions HCl precipitates Sb_2S_3, Sb_2S_5, or a mixture:

$$3\ SbS_2^- + [Sb(OH)_4]^- + 4\ H_3O^+ \rightarrow 2\ Sb_2S_3\downarrow + 8\ H_2O$$

$$2\ [SbS_4]^{3-} + 6\ H_3O^+ \rightarrow Sb_2S_5\downarrow + 3\ H_2S\uparrow + 6\ H_2O$$

The trisulfide is insoluble in "$(NH_4)_2CO_3$" (distinction from As); insoluble in CO_3^{2-} in the cold, but completely dissolved upon warming (distinction from Sn):

$$2\ Sb_2S_3 + 2\ CO_3^{2-} + 2\ H_2O \rightarrow 3\ SbS_2^- + [Sb(OH)_4]^- + 2\ CO_2\uparrow$$

Antimony(III) sulfide is soluble in hot tartaric acid, slowly soluble in oxalic acid, soluble in citric acid, and easily soluble in the last two if NO_3^-, NO_2^- or ClO_3^- is present.

Antimony trisulfide is sparingly soluble in hot NH_3.

Dilute HNO_3 gives Sb_2O_3:

$$Sb_2S_3 + 2\ H_3O^+ + 2\ NO_3^- \rightarrow Sb_2O_3\downarrow + 3\ S\downarrow + 2\ NO\uparrow + 3\ H_2O$$

Depending on concentrations, antimony trisulfide dissolves sparingly in "$(NH_4)_2S$", readily in S^{2-} and S_2^{2-}:

$$Sb_2S_3 + S^{2-} \rightarrow 2\ [SbS_2]^-$$

$$Sb_2S_3 + 4\ S_2^{2-} \rightarrow 2\ [SbS_4]^{3-} + S_3^{2-} \text{ or}$$

$$Sb_2S_3 + 3\ S_2^{2-} \rightarrow 2\ [SbS_4]^{3-} + S$$

Dissolution of Sb_2S_3 or Sb_2S_5 with S_2^{2-} as $[SbS_4]^{3-}$ separates it from any concomitant Cu, Cd, Hg (mostly), Pb and Bi. The filtrate may be acidified and digested with hot concentrated HCl; thus the soluble Sb and Sn complexes are separated from As_2S_5.

Dilute H_2SO_4 is almost without action on Sb_2S_3.

The trisulfide is soluble in cold concentrated, and in hot 6-M HCl (distinction from As):

$$Sb_2S_3 + 6\ H_3O^+ + 8\ Cl^- \rightarrow 2\ [SbCl_4]^- + 3\ H_2S\uparrow + 6\ H_2O$$

and is slowly decomposed on boiling with rather concentrated NH_4Cl:

$$Sb_2S_3 + 6\ NH_4^+ + 8\ Cl^- \rightarrow 2\ [SbCl_4]^- + 6\ NH_3\uparrow + 3\ H_2S\uparrow$$

The pentoxide, Sb_2O_5, is slowly soluble in "ammonium sulfide".

Sulfane (H_2S) and S^{2-} under conditions similar to those mentioned above, precipitate Sb^V sulfide, Sb_2S_5, orange, from solutions of Sb^V salts. This is mostly like the trisulfide in solubilities, including in "$(NH_4)_2S$":

$$Sb_2S_5 + 3\ HS^- + 3\ NH_3 \rightarrow 2\ [SbS_4]^{3-} + 3\ NH_4^+$$

$$Sb_2S_5 + 3\ S^{2-} \rightarrow 2\ [SbS_4]^{3-}$$

$$4\ Sb_2S_5 + 18\ OH^- \rightarrow 5\ [SbS_4]^{3-} + 3\ [Sb(OH)_6]^-$$

resulting also in mixed oxo- or hydroxothiostibates.

It is insoluble in "$(NH_4)_2CO_3$" and sparingly soluble in cold NH_3 (distinction from Sb_2S_3) more readily when warmed, but boiling causes precipitation of Sb_2S_3 and S. On boiling with water, Sb_2S_5 slowly decomposes into Sb_2O_3, H_2S and S; on warming with HCl it dissolves and is also reduced:

$$Sb_2S_5 + 6\ H_3O^+ + 8\ Cl^- \rightarrow 2\ [SbCl_4]^- + 3\ H_2S\uparrow + 2\ S\downarrow + 6\ H_2O$$

Elemental and oxidized chalcogens. Stibane (the H, not the Sb) is oxidized by sulfur to Sb_2S_3 in the sunlight, slowly at room temperature and rapidly at 100 °C:

$$2\ Sb^{III}H^{-I}_3 + {}^3/_4\ S_8 + \gamma \rightarrow Sb_2S_3\downarrow + 3\ H_2S\uparrow$$

Hot $S_2O_3^{2-}$ and OH^- dissolve Sb.

The metal is also slowly dissolved by hot, concentrated H_2SO_4:

$$2\ Sb + 12\ H_2SO_4 \rightarrow Sb_2(SO_4)_3 + 3\ SO_2\uparrow + 6\ HSO_4^- + 6\ H_3O^+$$

The trioxide, Sb_2O_3, is soluble in H_2SO_4.
Antimony tetroxide, Sb_2O_4, is slightly soluble in H_2SO_4.
The sulfate, $Sb_2(SO_4)_3$, dissolves in moderately concentrated H_2SO_4.
All salts of antimony when boiled with $S_2O_3^{2-}$ precipitate the sulfide:

$$2\ SbCl_3 + 3\ S_2O_3^{2-} + 9\ H_2O \rightarrow Sb_2S_3\downarrow + 3\ SO_4^{2-} + 6\ H_3O^+ + 6\ Cl^-$$

Some conditions yield red "antimony vermillion", Sb_2S_3, with Sb_2O_3.
Sulfur dioxide reduces Sb^V to Sb^{III}. Sulfates of Sb are not prepared by precipitation, but by boiling the oxides with concentrated H_2SO_4. They are readily hydrolyzed by H_2O.

Reduced halogens. Antimony is insoluble in air-free HCl, but air and HCl oxidize the metal slowly to $[SbCl_4]^-$.

Aqueous HCl, or any other inorganic acid except HF, carefully added to a solution of Sb salts in OH^- precipitates the corresponding oxide, hydrated oxide, oxychloride (SbOCl) etc., often soluble in more acid or concentrated Cl^- and so on. The trioxide, Sb_2O_3, dissolves in HCl:

$$Sb_2O_3 + 6\ H_3O^+ + 8\ Cl^- \rightarrow 2\ [SbCl_4]^- + 9\ H_2O$$

Potassium Sb^{III} tartrate (tartar emetic), precipitates SbOCl when treated with dilute HCl.

The trichloride $SbCl_3$ boils out from HCl at over 108 °C.

Hot concentrated HCl slowly dissolves Sb_2O_4 as $[SbCl_4]^-$, $[SbCl_6]^-$ etc. We may crystallize mixtures of Sb^{III} and Sb^V as $Cs_4[Sb^{III}Cl_6][Sb^VCl_6]$ for example, or $Rb_{16}[Sb^{III}Cl_6]_5[Sb^VCl_6]$. Likewise HBr and NH_4Br yield black, diamagnetic (hence not Sb^{IV}) $(NH_4)_4[Sb^{III}Br_6][Sb^VBr_6]$.

The pentoxide, Sb_2O_5, is soluble in HCl as $[SbCl_6]^-$.

Iodide ion, added to $[SbCl_4]^-$, precipitates yellow SbI_3 (in the absence of oxalic or tartaric acids or excessive halide), soluble in, e.g., HCl.

Aqueous HI (or I^- in acid) dissolves and reduces Sb_2O_5, and reduces $[SbCl_6]^-$ etc., liberating I_3^- (with excess I^-) or I_2 (detected by shaking with CCl_4) (distinction from Sn^{IV}), for example:

$$Sb_2O_5 + 14\ I^- + 10\ H_3O^+ \rightarrow 2\ [SbI_4]^- + 2\ I_3^- + 15\ H_2O$$

$$[SbCl_6]^- + 2\ I^- \rightarrow [SbCl_4]^- + 2\ Cl^- + I_2$$

Alkaline solution reverses this with what is then IO^-:

$$[Sb(OH)_4]^- + IO^- + H_2O \rightarrow [Sb(OH)_6]^- + I^-$$

Elemental and oxidized halogens. Chlorine or Br_2 converts Sb to Sb^{III} or Sb^{V} (and Sb^{III} to Sb^{V}), depending on the amount of reagent and *T*; alkalis may give Sb^{V}. Aqueous HCl containing Br_2 dissolves Sb well.

Stibane with excess I_2 in water yields other Sb^{III}; excess SbH_3 forms Sb.

15.4.4 Reagents Derived from the Metals Lithium through Uranium, plus Electrons and Photons

Oxidation. Antimony(III) in acidic, neutral or alkaline solution, rapidly reduces CrO_4^{2-} or $[Cr_2O_7]^{2-}$ to Cr^{III}. Acidic solutions of Sb^{III} reduce MnO_4^{2-} and MnO_4^- to Mn^{2+}; in alkaline solution the product is $MnO_2 \cdot aq$. The Sb^{III} becomes Sb^{V}. Without other reductants these can be quantitative.

Antimony(III) compounds are oxidized to Sb^{V} also by Ag_2O in the presence of OH^-. An Sb^{III} compound when evaporated on a water bath with $Ag(NH_3)_2^+$ gives a black precipitate of Ag and Sb^{V}. A solution of $Sb(OH)_4^-$ when treated with Ag^+ gives a heavy black precipitate of Ag, insoluble in NH_3, and thus separated from any Ag_2O precipitated. If, instead of Ag^+, a solution of $[Ag(NH_3)_2]^+$ containing a large excess of 1-M NH_3 is added, no precipitation occurs in the cold (distinction from $[Sn(OH)_3]^-$), nor upon heating until the excess NH_3 has been driven off.

Antimony(III) is oxidized to Sb^{V} by $[AuCl_4]^-$ in HCl solution, gold being deposited as a yellow precipitate. The Sb is precipitated as Sb_2O_5 unless enough acid is present to dissolve the oxide:

$$4\,[AuCl_4]^- + 3\,Sb_2O_3 + 18\,H_2O \rightarrow$$

$$4\,Au\downarrow + 3\,Sb_2O_5\downarrow + 12\,H_3O^+ + 16\,Cl^-$$

Reduction. Antimony(III and V) compounds are reduced to the metal by Na_{Hg}, Mg, Fe, Cu, Zn, Cd, Sn, Pb and Bi; but in the presence of dilute acids and metals that release hydrogen the product is also, in part, SbH_3.

More specifically, an aqueous antimony salt, when heated with iron wire in the presence of HCl, or with metallic Mg, Zn or Sn, gives a black precipitate of Sb (distinction from Sn). Copper in concentrated HCl (Reinsch's test) becomes coated with a violet deposit if Sb is present. If a drop of solution containing antimony is placed on a silver coin, or other silver, and the coin touched through the drop with a piece of tin or zinc, an identifying black spot of Sb will form. Sulfides interfere.

Stibane, SbH_3, the secondary product (relative to Sb) in the moderately rapid generation of H_2 by Zn and H_2SO_4 or HCl, contains reduced H:

$$Sb_2O_3 + 6\,Zn + 12\,H_3O^+ \rightarrow 6\,Zn^{2+} + 2\,SbH_3\uparrow + 15\,H_2O$$

Stibane is a colorless, odorless gas even more poisonous than arsane. It explodes at 200 °C unless diluted with 75 % or more of H_2. The Sb in stibane is deposited as the metal when the gas is passed into a concentrated solution of OH^- or through a U tube filled with solid KOH or soda lime (distinction and separation from As). Aqueous CO_3^{2-} and Group-2 hydroxides also decompose SbH_3, to Sb and H_2.

When SbH_3 is passed into a solution of Ag^+, the silver is reduced and the Sb converted to the oxide (or acid), which is only slightly soluble in H_2O (distinction from As). The precipitate should be washed free from excess Ag^+ and H_3AsO_3 if present, then treated with dilute HCl (or tartaric acid), which dissolves the antimony and leaves the silver. The filtrate may then be tested for Sb. (Because AgCl is somewhat soluble in HCl, enough may be present to interfere with the test for Sb. Therefore it is best to add first a drop or two of I^- and remove any AgI formed.)

Neutral or alkaline Sb^{III} or Sb^{V} with either Zn, Al or Na_{Hg} give Sb, not SbH_3 (distinction from As^{III}).

Tin dichloride reduces Sb^{V} to Sb^{III} but not Sb (distinction from As).

Other Reactions. Slightly acidified Fe^{III} chloride, Cu^{2+} or $[HgCl_2]$ completely transpose Sb_2S_3 to chlorides or basic compounds.

Stibane, SbH_3, and $PtCl_2$ precipitate Pt and $Sb_2O_3 \cdot aq$.

Aqueous Ag^+ and SbH_3 produce $Sb_2O_3 \cdot aq$ and Ag.

15.5 Bismuth, $_{83}Bi$ (and Ununpentium, $_{115}Uup$)

Oxidation numbers: (III) and (V), as in Bi_2O_3 and the unstable Bi_2O_5. Relativity gives us the inert pair in Bi^{III}. The next member of the Group, Uup, recently synthesized, is predicted by relativistic quantum mechanics to occur in water as Uup^+ and Uup^{3+}.

15.5.1 Reagents Derived from Hydrogen and Oxygen

Water. The solubility of $Bi(OH)_3$ in H_2O at 20 °C is about 6 μM. On long contact with H_2O, Bi_2O_5 decomposes to $Bi(OH)_3$ and O_2.

From non-aqueous sources, BiF_5, the only stable halide of Bi^{V}, reacts violently with H_2O, forming O_3, OF_2 and a brown solid.

Many of the bismuth salts are insoluble in water. Most of those that are soluble hydrolyze to form an insoluble oxy-salt. In such cases, the presence of the corresponding free acid will prevent hydrolysis. E.g., $BiCl_3$, $BiBr_3$, $Bi(NO_3)_3$, $Bi_2(SO_4)_3$ will dissolve readily in a dilute solution of HCl, HBr, HNO_3 or H_2SO_4, respectively. The halides require more free acid than does the nitrate. Bismuth chloride, bromide and sulfate are deliquescent; the nitrate less so.

Basic nitrates are obtained when bismuth nitrate is treated with hot or cold water until no more HNO_3 is extracted. Bismuth sulfate, $Bi_2(SO_4)_3$, is hydrolyzed slowly by cold water, rapidly by hot water, forming dibismuthyl sulfate, $(BiO)_2SO_4$, and perhaps some $Bi(OH)SO_4$.

Oxonium. Bismuth does not dissolve in H_3O^+ alone.

Hydroxide. Aqueous OH^- precipitates from solutions of bismuth salts, bismuth hydroxide, $Bi(OH)_3$, or such basic salts as $Bi_2O_2(OH)(NO_3)$, white, becoming yellow on boiling, with the formation of ~BiO(OH); only slightly soluble in excess OH^- (distinction from Sb and Sn); slightly soluble in Cl^-. Before precipitation, basic cations, especially $[Bi_6(\mu_3\text{-}O)_4(\mu_3\text{-}OH)_4]^{6+}$, are formed. These are found also in such salts as $[Bi_6O_4(OH)_4](NO_3)_6 \cdot H_2O$ and $[Bi_6O_4(OH)_4](ClO_4)_6 \cdot 7H_2O$.

The trioxide and $Bi(OH)_3$ are soluble in HCl, HNO_3 and H_2SO_4; insoluble in CO_3^{2-}, as well as in NH_3. All precipitates of $Bi(OH)_3$ retain some of the corresponding oxide salt.

The pentoxide, Bi_2O_5, is weakly acidic, somewhat soluble in concentrated OH^-, forming $[Bi(OH)_6]^-$ (acidification reprecipitates it); is insoluble in dilute HNO_3 and H_2SO_4; is decomposed with formation of $Bi(NO_3)_3$ when heated in concentrated HNO_3. In concentrated H_2SO_4 or fuming HNO_3 it is soluble at 0 °C, slightly soluble in dilute HF, readily soluble in the concentrated acid, soluble in HCl, HBr or HI with release of Cl_2, Br_2 or I_2 and the formation of BiX_3 or $[BiX_4]^-$, attacked by H_2S to form Bi_2S_3, and by SO_2 to give $Bi_2(SO_4)_3$.

Peroxide. Bismuth dissolves in acids, especially CH_3CO_2H, but far more extensively with H_2O_2. At high pHs, HO_2^- oxidizes Bi^{III} to Bi^V.

Di- and trioxygen. Bismuth dissolves in HCl with O_2:

$$2\ Bi + {}^3/_2\ O_2 + 8\ Cl^- + 6\ H_3O^+ \rightarrow 2\ [BiCl_4]^- + 9\ H_2O$$

Ozone oxidizes alkaline, not acidic, Bi^{III} to Bi^V.

15.5.2 Reagents Derived from the Other 2nd-Period Non-Metals, Boron through Fluorine

Carbon oxide species. Alkali carbonates precipitate white bismuthyl carbonate, $(BiO)_2CO_3$, containing more or less hydroxide and other impurities, insoluble in excess reagent. Alkaline-earth carbonates react similarly.

Cyanide species. Aqueous CN^- forms bismuth cyanide, which is rapidly hydrolyzed to $Bi(OH)_3$, insoluble in excess reagent, soluble in acids.

Some "simple" organic species. The hydroxide $Bi(OH)_3$ is reduced to the metal by alkaline methanal or on warming with glucose, e.g.:

$$2\ Bi(OH)_3 + 3\ CH_2O + 3\ OH^- \rightarrow 2\ Bi\downarrow + 3\ HCO_2^- + 6\ H_2O$$

Oxalic acid and $C_2O_4^{2-}$ precipitate white $Bi_2(C_2O_4)_3$, soluble in strong acids if not too dilute. Bismuth acetate is of minor importance.

The presence of acetic, citric, and other organic acids prevents the precipitation of Bi^{III} by an excess of water, and $Bi(OH)_3$ is soluble in alkaline solutions with, e.g., glycerol, tartrate, or citrate.

Reduced nitrogen. Ammonia precipitates either $Bi(OH)_3$ or white $Bi_2O_2(OH)NO_3$ etc., insoluble in excess (distinction from Cu and Cd).

Diazane (N_2H_4) and hydroxylamine do not reduce Bi^{III} in alkaline solution (distinction from Cu, Ag and Hg).

Oxidized nitrogen. Nitric acid (initiated by some HNO_2) or aqua regia dissolves Bi well. At 65 °C the reaction with HNO_3 is essentially:

$$Bi + 4\,H_3O^+ + NO_3^- \rightarrow Bi^{3+} + 6\,H_2O + NO\uparrow$$

Allowing bismuth nitrate to crystallize from hot HNO_3 yields $Bi(NO_3)_3 \cdot 5H_2O$. If air dried, ~$Bi(OH)_2NO_3$ is obtained. Hot water converts this into bismuthyl hydroxide nitrate, ~$(BiO)_2(OH)NO_3$.

Fluorine species. The fluoride BiF_3 is the most stable of the halides, insoluble in and not hydrolyzed by H_2O, soluble in hot HCl, HNO_3 or H_2SO_4, soluble in excess HF and F^-.

15.5.3 Reagents Derived from the 3rd-to-5th-Period Non-Metals, Silicon through Xenon

Phosphorus species. Phosphinic acid, HPH_2O_2, gives a white precipitate of bismuth phosphinate, $Bi(PH_2O_2)_3$, which slowly turns gray, hastened by heating, forming metallic Bi, something like this:

$$4\,Bi(PH_2O_2)_3 \rightarrow 2\,Bi\downarrow + Bi_2(PHO_3)_3\downarrow + 3\,H_2PHO_3 + {}^3/_2\,P_4\downarrow + 6\,H_2O$$

Phosphonic acid, H_2PHO_3, precipitates white bismuth phosphonate, $Bi_2(PHO_3)_3 \cdot 3H_2O$, practically insoluble in water, not affected by OH^-, slowly decomposed by I^-, and acted upon at once by H_2S.

Orthophosphates quantitatively precipitate Bi^{III} from a nitrate solution as $BiPO_4$, white; insoluble in dilute HNO_3; readily soluble in concentrated HCl; partly decomposed by boiling with NH_3; completely decomposed by OH^-. Chloride solutions and H_3PO_4 give no precipitate. Soluble phosphates, however, do form a precipitate (soluble in HCl).

Bismuth diphosphate, $Bi_4[P_2O_7]_3$, white, is obtained when $Bi(NO_3)_3$ is treated with $[P_2O_7]^{4-}$. It is soluble in excess reagent, decomposed, on warming, to $BiPO_4$, insoluble in CH_3CO_2H, soluble in hot HCl or HNO_3, decomposed by hot H_2SO_4. Metaphosphoric acid and metaphosphates give a white precipitate of bismuth metaphosphate, $Bi(PO_3)_3$, if, after mixing, the solution is made alkaline with NH_3. The precipitate is readily converted to $BiPO_4$ by boiling.

Arsenic species. Bismuth arsenite, $BiAsO_3$, is obtained by treating a bismuth nitrate-mannite solution with H_3AsO_3 or AsO_3^{3-}. The precipitate is readily soluble in HNO_3, not affected by OH^- or CO_3^{2-}. Bismuth arsenate is obtained by treating a bismuth nitrate-mannite solution with H_3AsO_4 or AsO_4^{3-}. The precipitate is readily soluble in HCl, less so in HNO_3; slowly decomposed by OH^-.

Reduced chalcogens. Sulfane does not affect Bi^0, but it precipitates from slightly acidified Bi^{III} a dark-brown Bi_2S_3, insoluble in cold, dilute H_3O^+ and alkali sulfides, but soluble in hot 2-M HNO_3 (separation from Hg), in dilute H_2SO_4 (separation from Pb), and in hot concentrated HCl. Its insolubility in $(NH_4)_2S_x$ separates it from Mo, Sn, As and Sb.

The precipitate in base with S^{2-}, however, is soluble in excess reagent, varying with the $c(OH^-)$. The solubility in water is 0.35 μM at 18 °C. The trisulfide is insoluble in CN^-. It reacts with $FeCl_3$, although the Fe^{III} reacts with only the S^{2-}, and only the Cl^- directly with the Bi^{3+}, thus:

$$Bi_2S_3 + 6\ FeCl_3 \rightarrow 2\ [BiCl_4]^- + 3\ S\downarrow + 6\ Fe^{2+} + 10\ Cl^-$$

Concentrated SCN^- gives deep reddish-brown complexes (best obtained with a solid reagent).

A solution of dithiocarbamate, $CS_2NH_2^-$, prepared by shaking CS_2 with concentrated NH_3, gives, with Bi^{III} in neutral or slightly alkaline solution, a yellow-orange precipitate, which becomes rust-colored upon adding acetic acid, while the solution becomes yellow.

Oxidized chalcogens. Thiosulfate, $S_2O_3^{2-}$, when heated with a solution containing Bi^{III}, precipitates Bi_2S_3. Dithionite, $S_2O_4^{2-}$, in excess, precipitates from slightly acidic solutions, Bi_2S_3 contaminated with S and Bi. A large excess of reagent or acid precipitates Bi at ambient T.

Neither sulfur dioxide nor H_2SO_4 gives a precipitate with Bi^{III}. Any precipitate formed by SO_3^{2-} or SO_4^{2-} is due to hydrolysis and the separation of a basic salt of Bi. Metallic bismuth is not affected by cold or dilute H_2SO_4, but is soluble in hot, concentrated H_2SO_4:

$$2\ Bi + 6\ H_2SO_4 \rightarrow Bi_2(SO_4)_3 + 3\ SO_2\uparrow + 6\ H_2O$$

Reduced halogens. Bismuth is insoluble in HCl alone.

Chloride and Bi^{3+} precipitate bismuth oxychloride, BiOCl, white, if not too much free acid is present. A solution of $BiCl_3$ in water, acidified with HCl, likewise precipitates BiOCl upon dilution:

$$[BiCl_4]^- + 3\ H_2O \leftrightarrows BiOCl\downarrow + 2\ H_3O^+ + 3\ Cl^-$$

This dissolves readily in more acid (partial distinction from Ag, Pb and Hg^{I}), forming chloro-complexes. It is white, insoluble in tartaric acid (distinction from Sb), completely converted by H_2S into Bi_2S_3.

Both BiOCl and BiOBr, but not BiOI, are transposed by OH^-:

$$BiOCl + OH^- \rightarrow BiO(OH)\downarrow + Cl^-$$

The bromide is insoluble in aqueous I^-. The iodide is decomposed by concentrated H_2SO_4 or by HNO_3, giving I_2, but is not affected by CH_3CO_2H, alkalis, or H_2S.

Aqueous Br^- does not make a precipitate with $BiCl_3$, but from Bi^{3+} it precipitates white bismuth oxybromide, BiOBr, which, by long treatment with water, is converted to Bi_2O_3. The oxybromide is soluble in HCl, HBr and HNO_3; difficultly soluble in concentrated H_2SO_4. The presence of Br^- prevents the precipitation of BiOCl by H_2O, and also dissolves any oxychloride already precipitated. A moderately high c of Br^- (or Cl^-) is necessary to prevent precipitation on dilution.

Aqueous Bi^{III} (unless strongly acidic) and a little I^- precipitate brown BiI_3. Any excess reagent dissolves this as a deep-yellow complex:

$$BiCl_3 + 3\ I^- \rightarrow BiI_3\downarrow + 3\ Cl^-$$

$$BiI_3 + I^- \rightarrow [BiI_4]^-$$

This excellent test is sensitive to 5 μM. The BiI_3 is re-precipitated and slowly hydrolyzed by adding H_2O. Prolonged boiling gives red BiOI, which is soluble in HCl, decomposed by HNO_3 or H_2SO_4, converted to the oxide by OH^-, not affected by CH_3CO_2H or Cl^-. The triiodide is decomposed by HNO_3, giving I_2; is not affected by dry H_2S but is quickly converted to Bi_2S_3 by alkali sulfides.

A red salt, Tl_2BiI_5, soluble to 40 μM in H_2O, is formed by adding I^- to an acidic solution containing Tl^+ and Bi^{III}. (This reaction is good, except for toxicity, to detect Bi.)

Elemental and oxidized halogens. Chlorine, with much OH^-, oxidizes Bi^{III} to Bi^V–see **Oxidation** just below. Chlorate ion, added to warm Bi^{III}, precipitates white $BiOClO_3$ on cooling. The BrO_3^- and IO_3^- ions, with Bi^{III}, also both give white precipitates, of uncertain compositions. The bromate is readily soluble, the iodate slightly soluble, in HNO_3.

15.5.4 Reagents Derived from the Metals Lithium through Uranium, plus Electrons and Photons

Oxidation. Metallic bismuth goes to Bi^{III} while reducing salts of Pt, Ag, Au and Hg to the metals.

A "tetroxide", ~Bi_2O_4, yellow to brown, a Bi^{III}-Bi^V mixture, can arise from treating the trioxide anodically or with various oxidants in alkaline solution. This is unaffected by NH_3, attacked only slowly by cold, dilute inorganic acids, but decomposed energetically by concentrated acids. It reacts very slowly with (aqueous) SO_2 giving $Bi_2(SO_4)_3$, and is not decomposed by cold dilute H_2SO_4. It is decomposed even at -15 °C by concentrated HCl, releasing Cl_2 and forming $BiCl_3$ or $[BiCl_4^-]$ but is insoluble in concentrated KOH.

Oxidizing Bi_2O_3 in concentrated NaOH, conveniently by electrolysis, produces an often poorly characterized pentoxide or "sodium bismuthate", $NaBiO_3$ or $Na[Bi(OH)_6]$, scarlet red, brown or black.

Reduction. Bismuth metal, dark gray, is precipitated quickly and completely from solution by Mg, Fe, Cu, Zn, Cd, Al, Sn, and Pb.

Stannite reduces all Bi compounds to black Bi (a very sensitive test):

$$2\ Bi(OH)_3 + 3\ [Sn(OH)_3]^- + 3\ OH^- \rightarrow 2\ Bi\downarrow + 3\ [Sn(OH)_6]^{2-}$$

(The stannite is prepared as needed by adding to some $SnCl_2$ enough OH^- to redissolve the white precipitate formed at first. With too much alkali or heat, metallic Sn or black Sn^{II} oxide may be precipitated.)

Other reactions. Chromate ion, CrO_4^{2-}, or $[Cr_2O_7]^{2-}$ added to a not too acidic solution of Bi^{III} precipitates yellow to orange $(BiO)_2[Cr_2O_7]$, going to completion with an acetate buffer; soluble in inorganic acids; difficultly soluble in acetic acid; slightly soluble in OH^- unless hot (distinction from $PbCrO_4$). The sensitivity of this test is ~1 mM.

Aqueous $[Fe(CN)_6]^{4-}$ gives a yellowish white precipitate; $[Fe(CN)_6]^{3-}$ yields a yellow to brown precipitate. Both are soluble in HCl.

Bibliography

See the general references in the Introduction, and more-specialized books [1–19]. Some articles in journals include: the prediction of formation constants of metal–ammonia complexes in water using density-functional theory [20]; an introduction to a thematic issue on NO chemistry [21]; the oxidation of diazane, N_2H_4, in water [22]; dinitrogen complexes [23]; open-chain polyphosphorus hydrides [24]; peroxynitrites [25]; nitrogen fixation [26 and 27]; NO on **d**-block metals [28]; mechanisms of nitrogen-compound reactions [29]; N_2 fixation [30]; and "common" Bi^+ [31].

1. Norman NC (ed) (1998) Chemistry of arsenic, antimony, and bismuth. Blackie, London
2. Durif A, Averbuch-Pouchot MT (1996) Topics in phosphate chemistry. World Scientific, Singapore
3. Corbridge DEC (1995) Phosphorus, 5th ed. Elsevier, Amsterdam
4. Schmidt EW (1984) Hydrazine and its derivatives. Wiley, New York
5. Goldwhite H (1981) Introduction to phosphorus chemistry. Cambridge University, Cambridge

6. Gibson AH, Newton WE (eds) (1980) Current perspectives in nitrogen fixation, proc 4th int symp. Elsevier, Amsterdam
7. Heal HG (1980) The inorganic heterocyclic chemistry of sulfur, nitrogen and phosphorus. Academic, London
8. Emsley J, Hall D (1976) The chemistry of phosphorus. Harper & Row, London
9. Greenfield S, Clift M (1975) Analytical chemistry of the condensed phosphates. Pergamon, Oxford
10. Colburn CB (ed) (vol 1 1966; vol 2 1973) Developments in inorganic nitrogen chemistry. Elsevier, Amsterdam
11. Smith JD (1973) The chemistry of arsenic, antimony and bismuth. Pergamon, Oxford
12. Allcock HR (1972) Phosphorus-nitrogen compounds. Academic, New York
13. Halmann M (ed) (1972) Analytical chemistry of phosphorus compounds. Wiley, New York
14. Streuli CA, Averell PR (eds) (1970) The analytical chemistry of nitrogen and its compounds, part 1. Wiley, New York
15. Corbridge DEC (1966–1969) Topics in phosphorus chemistry, vols 3–6. Wiley, New York
16. Jolly WL (1964) The inorganic chemistry of nitrogen. Benjamin Cummings, San Francisco
17. Van Wazer JR (1958) and (1961) Phosphorus and its compounds, vols 1 and 2. Interscience, New York
18. Audrieth LF, Ogg BA (1951) The chemistry of hydrazine. Wiley, New York
19. Franklin EC (1935) The nitrogen system of compounds. Reinhold, New York
20. Hancock RD, Bartolotti LJ (2004) Chem Commun 2004:534
21. Richter-Addo GB, Legzdins P, Burstyn J (2002) Chem Rev 102:857
22. Stanbury DM (1998) in Karlin KD (ed) Prog Inorg Chem 47:511
23. Hidai M, Mizobe Y (1995) Chem Rev 95:1115
24. Baudler M, Glinka K (1994) Chem Rev 94:1273
25. Edwards JO, Plumb RC (1994) in Karlin KD (ed) Prog Inorg Chem 41:599
26. Henderson RA, Leigh GJ, Pickett CJ (1983) Adv Inorg Chem Radiochem 27:198
27. Bossard GE, George TA (1981) Inorg Chim Acta 54:L239
28. McCleverty JA (1979) Chem Rev 79:53
29. Stedman G (1979) Adv Inorg Chem Radiochem 22:113
30. Chatt J, Dilworth JR, Richards RL (1978) Chem Rev 78:589
31. Smith DW (1975) J Chem Educ 52:576 and Smith GP, Davis HL (1973) Inorg Nucl Chem Lett 9:991

16 Oxygen through Polonium, the Chalcogens

16.1 Oxygen, $_8O$

Oxidation numbers: (−II), (−I) and (II), as in H_2O, H_2O_2 and OF_2, plus fractional values in, say, KO_2.

16.1.1 Reagents Derived from Hydrogen and Oxygen

Water. Hydrogen peroxide is miscible with water in all proportions; Na_2O_2 is readily soluble with the generation of much heat.

Ozone, trioxygen, does not form H_2O_2 with H_2O.

The reaction of the very red ozonides (from O_3 + solid AlkOH, where Alk = Na, K, Rb or Cs) with H_2O is quite violent:

$$2\,O_3^- + H_2O \rightarrow {}^5/_2\,O_2\uparrow + 2\,OH^-$$

Oxonium. The binding energies of oxygenated ligands are in the order: $H_3O^+ > Na^+ > K^+ \approx NH_4^+$. Also see Table 1.1 for acidities.

Hydroxide. Adding 10-M H_2O_2 to a little excess of NaOH at 15 °C, followed by ethanol at 15 °C, gives $Na_2O_2\cdot 8H_2O$. This reacts readily with CO_2 and melts at 30 °C, yielding O_2.

Ozone decomposes, catalyzed by OH^-:

$$2\,O_3 \rightarrow 3\,O_2\uparrow$$

Peroxide. Hydrogen peroxide (as both reagent and reactant here) is volatile, explosive when concentrated, irritating to the skin and weakly acidic. It easily liberates oxygen by dismutating, either on warming or with a catalyst such as OH^-:

$$2\,H_2O_2 \rightarrow O_2\uparrow + 2\,H_2O$$

Fused Na_2O_2 will give a steady stream of O_2 if H_2O is added dropwise. A modification is to warm a mixture of Na_2O_2 and $Na_2SO_4\cdot 10H_2O$.

Ozone and H_2O_2 react quite slowly, but catalyzed by Mn^{2+} or OH^-:

$$O_3 + H_2O_2 \rightarrow 2\ O_2\uparrow + H_2O$$

Both Na_2O_2 and BaO_2 decompose, releasing O_2 and OH^-.

16.1.2 Reagents Derived from the Other 2nd-Period Non-Metals, Boron through Fluorine

Some "simple" organic species. Charcoal catalyzes the decomposition of H_2O_2 into H_2O and O_2.

"Hyperol" is a stable, easily handled compound of H_2O_2 and urea, $CO(NH_2)_2$, containing a little citric acid as a stabilizer.

Oxalic acid, $H_2C_2O_4$, and H_2O_2 become CO_2 and H_2O.

Oxidized nitrogen. Nitrous acid (as NO^+) oxidizes H_2O_2 to O_2.

Fluorine species. Passing F_2 into 5-dM OH^- gives ≤ 80 % OF_2, some O_2:

$$2\ F_2 + 2\ OH^- \rightarrow OF_2\uparrow + 2\ F^- + H_2O$$

16.1.3 Reagents Derived from the 3rd-to-5th-Period Non-Metals, Silicon through Xenon

Phosphorus species. With H_2O_2, HPH_2O_2 forms H_3PO_4.

Arsenic species. Hydrogen peroxide oxidizes $As^{<V}$ to As^V.

Reduced chalcogens. Sulfides and H_2O_2 form sulfates, particularly in alkaline solution.

Depending on conditions, O_3 oxidizes H_2S and S^{2-} to S, SO_3^{2-} and SO_4^{2-}, but see also $S_2O_3^{2-}$ below:

$$O_3 + S^{2-} \rightarrow SO_3^{2-}$$

Oxidized chalcogens. With H_2O_2, thiosulfates at first form disulfanedisulfonates, but then various other products arise:

$$2\ S_2O_3^{2-} + H_2O_2 \rightarrow [S_4O_6]^{2-} + 2\ OH^-$$

Neutral $S_2O_3^{2-}$ mostly reduces all three atoms in O_3, mainly according to the first reaction here, but with some of the second too:

$$S_2O_3^{2-} + O_3 \rightarrow SO_2 + SO_4^{2-}$$

$$S_2O_3^{2-} + O_3 \rightarrow [S_2O_6]^{2-}$$

The same products, plus O_2, arise in base, likewise with S^{2-} or SO_3^{2-}.
With H_2O_2, sulfites become sulfates, especially in alkaline solution.
Ozone is reduced by SO_2, at least partly releasing no O_2:

$$^1/_3\ O_3 + SO_2 + 2\ H_2O \rightarrow HSO_4^- + H_3O^+$$

Ten-M H_2O_2 can be made by adding Na_2O_2 (formed by heating Na in air or oxygen) to ice-cold 2.3-M H_2SO_4 and distilling. Hydrogen peroxide may also be prepared by treating BaO_2 (from heating BaO with O_2) with H_2SO_4 and separating the precipitated $BaSO_4$.

With H_2O_2, concentrated H_2SO_4 forms H_2SO_5, that is, $H_2SO_3(O_2)$ or peroxosulfuric acid; H_2O_2 completely decomposes $(NH_4)_2[S_2O_8]$, a peroxodisulfate, containing $[(SO_3)_2(\mu\text{-}O_2)]^{2-}$, and liberates both N_2 and O_2.

Concentrated H_2SO_4 and O_3 do not generate H_2SO_5.

Reduced halogens. Ozone oxidizes most non-metals to their highest oxidation states, but HCl, HBr and HI go to the free halogen, and the following may be used to detect O_3 (with some H_2O_2 also produced):

$$O_3 + 2\ I^- + 2\ H_3O^+ \rightarrow O_2\uparrow + I_2 + 3\ H_2O$$

$$O_3 + 2\ I^- + H_2O \rightarrow O_2\uparrow + I_2 + 2\ OH^-$$

The I_2 and OH^- then go on, as is usual, to IO_3^- and I^-.

One may determine H_2O_2 by reaction with I^- and H_2SO_4, and titration of the liberated iodine with $S_2O_3^{2-}$.

Elemental and oxidized halogens. Reactions of H_2O_2 with halogen compounds depend on conditions. The halides may be oxidized to the halogens and, conversely, the halogens may be reduced to the halides.

Treating $Ca(OH)_2$ with I_2 and H_2O_2 forms the iodide and O_2:

$$Ca(OH)_2 + I_2 + H_2O_2 \rightarrow$$

$$Ca^{2+} + 2\ I^- + O_2\uparrow + 2\ H_2O$$

Periodic acid is reduced by H_2O_2 to iodic acid; in a dilute solution of periodic acid the reduction is to iodine; the H_2O_2 becomes O_2.

Alkaline but not neutral or acidified iodate, with O_3, forms periodate.

16.1.4 Reagents Derived from the Metals Lithium through Uranium, plus Electrons and Photons

Oxidation. In acid, H_2O_2 changes MnO_4^- to Mn^{2+}. Thus, peroxide may be oxidized and determined by titration with MnO_4^-:

$$2\ MnO_4^- + 5\ H_2O_2 + 6\ H_3O^+ \rightarrow 2\ Mn^{2+} + 5\ O_2\uparrow + 14\ H_2O$$

An example of the oxidation, in base, of H_2O_2 is:

$$2\ [Fe(CN)_6]^{3-} + OH^- + HO_2^- \rightarrow 2\ [Fe(CN)_6]^{4-} + H_2O + O_2\uparrow$$

Also, alkaline peroxide changes Cu^{II} complexes (tartrates in Fehling's solution) to Cu_2O, becoming O_2.

Likewise H_2O_2 goes to O_2 when reducing Ag_2O, Au_2O_3 and HgO to the metals. The separation of a gray precipitate, Ag, when an alkaline unknown is treated with $[Ag(NH_3)_2]^+$ is a good test for peroxide:

$$[Ag(NH_3)_2]^+ + {}^1/_2\ H_2O_2 \rightarrow Ag\downarrow + {}^1/_2\ O_2\uparrow + NH_4^+ + NH_3$$

Ozone may be prepared by the electrolytic oxidation of H_2O, e.g., in 1.7-M H_2SO_4, using platinum electrodes (although platinum black or sponge, oxides of Fe, Cu, Ag, Au etc. catalyze its decomposition):

$$3\ H_2O \rightarrow {}^1/_3\ O_3\uparrow + 2\ H_3O^+ + 2\ e^-$$

The electrolytic oxidation of OH^- gives rather pure oxygen:

$$2\ OH^- \rightarrow {}^1/_2\ O_2\uparrow + H_2O + 2\ e^-$$

Saturated K_2CO_3 at -20 °C with an anode forms $K_2[(CO_2)_2(\mu\text{-}O^{-I}{}_2)]$.

Beta (e^-), γ and α rays produce $OH^\bullet$, $HO_2^\bullet$ and H_2O_2 in H_2O.

Reduction. Moist O_3 is reduced to oxides by nearly all metals except Ru, Rh, Ir, Pd, Pt and Au, some requiring heating.

Oxygen may be determined by absorbing it in $CrCl_2$ or CuCl and measuring the contraction in volume of the gas sample, e.g.:

$${}^1/_2\ O_2 + 2\ Cr^{2+} + 2\ Cl^- + 2\ H_3O^+ \rightarrow 2\ CrCl^{2+} + 3\ H_2O$$

It is also reduced rapidly by V^{2+}.

Ozone oxidizes Cr^{III} to CrO_4^{2-} in high pH, or to $[Cr_2O_7]^{2-}$.

Treating Cr^{III} and Mn^{2+} with Na_2O_2 gives CrO_4^{2-} and $MnO_2 \cdot aq$.

Neutral Mn^{2+} and O_3 precipitate brown $MnO_2 \cdot aq$ quantitatively, with part of the O_3 being reduced to O^{-II}, perhaps thus:

$$O_3 + Mn^{2+} + 3\ H_2O \rightarrow O_2\uparrow + MnO_2 \cdot aq\downarrow + 2\ H_3O^+$$

but slightly acidic solutions also give some MnO_4^-, or only this above ~ 1-M H_3O^+, then Mn^{III} sulfato or chloro complexes above ~ 4-M H_3O^+. The required concentration of acid varies somewhat, however, with that of the Mn^{2+}. The practically colorless Mn^{2+} can serve as invisible ink, to be made brown with O_3.

The reduction of the oxidant H_2O_2 is greatly catalyzed by Fe^{2+}. Various metals such as Mo and Tl are oxidized by H_2O_2 to the highest state, but Pt and Au are not attacked. Iron becomes Fe^{III}, and the following, in acid, contrasts with the basic case, in **Oxidation** above:

$$H_2[Fe(CN)_6]^{2-} + {}^1/_2\, H_2O_2 \rightarrow [Fe(CN)_6]^{3-} + H_3O^+$$

Neutral and acidified Fe^{II} reduces O_3 in two ways:

$$O_3 + 2\, Fe^{2+} + 2\, H_3O^+ \rightarrow O_2\uparrow + 2\, Fe^{3+} + 3\, H_2O$$

$$O_3 + 6\, Fe^{2+} + 6\, H_3O^+ \rightarrow 6\, Fe^{3+} + 9\, H_2O$$

but basic solutions produce FeO_4^{2-} (see section **8.1.1**).

Ozone oxidizes $[Fe(CN)_6]^{4-}$ to $[Fe(CN)_6]^{3-}$ but does not attack the $[Fe(CN)_5CO]^{3-}$ ion. Ozone does not oxidize Ni^{2+} but converts Co^{2+} to a dark-brown product. The sulfides go to the higher oxides plus SO_4^{2-}. Palladium(II) becomes PdO_2, or Pd^{IV} complexes, in base.

Metallic Cu, Ag, Hg, Pb and Bi reduce H_2O_2 as it dissolves them in "glacial" (neat) CH_3CO_2H.

In water O_3 is one of the strongest oxidants known, being reduced while changing nearly all metals except Pt and Au, or their compounds except sulfides, to their highest oxidation states, but limited to Cu^{II} and Ag^{II} (or Ag^I/Ag^{III}) and to Fe^{III} in acid, itself becoming O_2:

$$O_3 + Hg_2^{2+} \rightarrow O_2\uparrow + HgO\downarrow + Hg^{2+}$$

$$O_3 + Tl^+ + OH^- \rightarrow O_2\uparrow + \sim TlO(OH)\downarrow$$

$$4\, O_3 + PbS \rightarrow PbSO_4\downarrow + 4\, O_2\uparrow$$

$$O_3 + Pb^{2+} + 3H_2O \rightarrow O_2\uparrow + PbO_2\downarrow + 2\, H_3O^+$$

The last reaction is slow in neutral, fast in basic solutions. The dark-brown PbO_2, on a basic-lead-acetate paper, is a good test for O_3. Many other insoluble sulfides—e.g., of Cu, Zn, Cd and Sb—behave like PbS.

The unique role of oxygen in a book on reactions in its simplest dihydride calls for tabulating its electrode potentials, as in Table 16.1, even though actual reactions do not depend solely on thermodynamics.

Other reactions. Aqueous OH^- precipitates M^{n+}, except Alk^+ and Tl^+, as oxides or hydroxides. Of these, the $Ae(OH)_2$ except $Mg(OH)_2$ are moderately soluble. Excess KOH easily dissolves the most amphoteric Be, Cr, Zn, Al, Sn, Pb and Sb precipitates.

Concentrated OH^- ion (treated for brevity as the reactant, not reagent, here) can leach Zr, Hf, Nb, Ta, Mo, W, Al, Sn and Pb from some ores.

Table 16.1 Standard reduction potentials for oxygen at 25 °C

One-electron reactions, acidic and alkaline	$E°/V$
$O_2 + e^- + H_3O^+ \leftrightarrows HO_2 + H_2O$	– 0.125
$HO_2 + e^- + H_3O^+ \leftrightarrows H_2O_2 + H_2O$	1.515
$H_2O_2 + e^- + H_3O^+ \leftrightarrows HO + 2\ H_2O$	0.68
$HO + e^- + H_3O^+ \leftrightarrows H_2O + H_2O$	2.85
$O_2 + e^- \leftrightarrows O_2^-$	0.33
$O_2^- + e^- + H_2O \leftrightarrows HO_2^- + OH^-$	0.20
$HO_2^- + e^- + H_2O \leftrightarrows HO + 2\ OH^-$	–0.29
$HO + e^- \leftrightarrows OH^-$	2.02
Two-electron reactions, acidic and alkaline	
$O_2 + 2\ e^- + 2\ H_3O^+ \leftrightarrows H_2O_2 + 2\ H_2O$	0.695
$H_2O_2 + 2\ e^- + 2\ H_3O^+ \leftrightarrows 2\ H_2O + 2\ H_2O$	1.763
$O_2 + 2\ e^- + H_2O \leftrightarrows HO_2^- + OH^-$	–0.065
$HO_2^- + 2\ e^- + H_2O \leftrightarrows 3\ OH^-$	0.867
Four-electron reactions, acidic and alkaline	
$O_2 + 4\ e^- + 4\ H_3O^+ \leftrightarrows 6\ H_2O$	1.229
$O_2 + 4\ e^- + 2\ H_2O \leftrightarrows 4\ OH^-$	0.401

Some substances, e.g., $MnO_2 \cdot aq$, Pt, Ag and Au, catalyze the (redox) dismutation of H_2O_2 into H_2O and O_2. It is surprising then that the $Co(NH_3)_5{}^{3+}$ in $[CoO_2(NH_3)_5]^{2+}$ stabilizes the O_2^- against dismutation.

Many tests for a peroxide have been suggested. Some good ones are the formation of the blue peroxochromium complex, as in **6.1.1**, and the yellow peroxotitanium complex, as in **4.1.1**.

16.2 Sulfur, $_{16}$S

Oxidation numbers: (–II), (–I), (II), (IV) and (VI), etc., as in H_2S, S_2^{2-}, $S_2O_3^{2-}$, SO_2 and SO_4^{2-}, plus fractional values in S_n^{2-}. Another important series of compounds has sulfur in fractional oxidation states in the anions $[S_n(SO_3)_2]^{2-}$, usually written $[S_{n+2}O_6]^{2-}$, containing chains of sulfur atoms with $n \geq 0$, capped at each end and having the traditional names dithionate (for $n = 0$), trithionate, etc. For $n \geq 1$, we now have the systematic names sulfanedisulfonate, disulfanedisulfonate, and so on. Some chemical differences plus tradition, however, have led to keeping the name dithionate for $[S_2O_6]^{2-}$ itself, even though "thio" has its own different meaning. Also the names thionates and thionic acids are convenient for the whole group and so are still used collectively.

16.2.1 Reagents Derived from Hydrogen and Oxygen

Hydrogen. Nascent hydrogen, e.g., from Zn and H_3O^+, reduces SO_2:

$$SO_2 + 6\,H \rightarrow H_2S\uparrow + 2\,H_2O$$

Water. At 20 °C, H_2S is soluble up to about 1.3 dM, is feebly acidic to litmus, and is readily boiled out of its solutions.

Sulfides of the alkalis are soluble in water; all others are insoluble or hydrolyzed, generally forming H_2S and the metal hydroxide.

Long continued boiling with water more or less completely decomposes the sulfides of Mn, Fe, Co, Ni, Ag, Sn, As and Sb. Sulfides are unaffected for: Mo, Pt, Cu, Au, Zn, Cd and Hg.

The thiocyanates of the alkalis, the alkaline earths and most of the other metals are soluble. Some small solubilities, however, are found with: CuSCN, 0.04 mM; AgSCN, 1.5 μM; $Hg(SCN)_2$, 2 mM; TlSCN, 1.2 mM; $Pb(SCN)_2$, 1.4 mM, and with Sn^{II} thiocyanate, all at about 20 °C. All of these thiocyanates are soluble in excess of the reagent, forming complexes such as $[Ag(SCN)_3]^{2-}$ or $[Hg(SCN)_4]^{2-}$.

The various forms of free sulfur are insoluble in water.

The $[S_2O_4]^{2-}$ ion is somewhat unstable:

$$2\,[S_2O_4]^{2-} + H_2O \rightarrow S_2O_3^{2-} + 2\,HSO_3^-$$

The larger number of thiosulfates are soluble in water; those of Ba, Ag and Pb are only slightly soluble.

When boiled, a solution of $S_2O_3^{2-}$ forms SO_4^{2-} and H_2S:

$$S_2O_3^{2-} + H_2O \rightarrow SO_4^{2-} + H_2S\uparrow$$

All dithionates are soluble, most more than $K_2[S_2O_6]$, which dissolves to nearly 3 dM. Salts (only normal salts are known) of Alk^+ and Ae^{2+} are the most stable, and they remain so even in boiling water.

The salts of $H_2[S_3O_6]$ are more stable than the acid, but in water they gradually decompose. Many of the salts are readily soluble, but the Ag^+, Hg_2^{2+} and Hg^{2+} compounds are only slightly so.

Water and $[S_3O_6]^{2-}$ slowly form sulfate and thiosulfate.

Aqueous $H_2[S_4O_6]$ is the most stable "thionic acid" but has not been isolated. Its dilute solution can be boiled without decomposition. A concentrated solution decomposes into S, SO_2 and H_2SO_4. Its acidity is as great as that of $H_2[S_2O_6]$. The salts of $H_2[S_4O_6]$ are generally soluble. In the solid state they may be kept for a month or more, but they readily decompose in solution, especially when warmed. The alkali salts are more stable than those of Ba^{2+}, Cu^{2+}, etc.

The salts of $H_2[S_5O_6]$ are all readily soluble and all unstable. Both the solid and solution readily decompose to give S. A little H_2SO_4 or HCl retards this but does not completely arrest it:

$$[S_5O_6]^{2-} \rightarrow SO_4^{2-} + 3\ S\downarrow + SO_2\uparrow$$

The salt $K_2[S_6O_6]$ is stable if dry but quickly decomposed in water.

Sulfur dioxide is readily soluble in water with release of heat, up to 1.7 M at 20 °C; it forms $SO_2 \cdot H_2O$ (or $H_2O \cdot SO_2$) but almost no H_2SO_3 [i.e., $SO(OH)_2$]. It is volatile with a strong odor, is toxic although the anions are much less so, and is quickly expelled by boiling. The HSO_3^- or SO_3^{2-} residue in treated foods is negligible except for allergic people.

Aqueous solutions of SO_2 decompose quite slowly at room temperature, finally forming HSO_4^-, H_3O^+ and S. Decomposition is inhibited by a strong acid, such as HCl, but accelerated by Co^{2+}. Solutions of HSO_3^- decompose faster than do those of SO_3^{2-}.

The p*K* for $[H_3O^+][HSO_3^-]/[SO_2 \cdot H_2O]$ is 1.8:

$$SO_2 \cdot H_2O + H_2O \leftrightarrows H_3O^+ + HSO_3^-$$

Both the hydrogen sulfite, $SO_2(OH)^-$, and sulfonate, $H{-}SO_3^-$, tautomers of HSO_3^-, occur in water, but solid $RbHSO_3$ and $CsHSO_3$ are sulfonates.

The sulfites of the alkali metals, Alk_2SO_3, are readily soluble in water; the normal sulfites of all other metals are insoluble to slightly soluble.

Neat H_2SO_4 mixes with water in all proportions, releasing much heat; it absorbs water from the air (hence its use in desiccators), and quickly removes the elements of water from many organic compounds, leaving carbon, with characteristic charring:

$$H_2SO_4 + H_2O \leftrightarrows H_3O^+ + HSO_4^-$$

The sulfates of Sr, Ba, Hg^I and Pb are insoluble in water, those of Ca and Hg^{II} slightly soluble.

Peroxodisulfuric acid, i.e., $[(SO_3H)_2(\mu\text{-}O_2]$, and the anhydride, $[S_2O_7]$, are hydrolyzed in steps, first giving H_2SO_5, i.e., $H_2SO_3(O_2)$. We note that H_2SO_4 in the solvent inhibits the decomposition:

$$H[S_2O_8]^- + 2\,H_2O \leftrightarrows HSO_4^- + HSO_5^- + H_3O^+$$

$$HSO_5^- + H_2O \leftrightarrows HSO_4^- + H_2O_2$$

Oxonium. For laboratory purposes H_2S is often made by treating a sulfide such as FeS with a dilute strong acid (not HNO_3):

$$FeS + 2\,H_3O^+ \rightarrow Fe^{2+} + H_2S\uparrow + 2\,H_2O$$

When SCN^- is boiled with dilute H_3O^+ some isothiocyanic acid, HNCS, distils over, while some of it decomposes. As with the oxygen analog—see **14.1.1 Oxonium**—the isomeric thiocyanic acid, HSCN, has not been isolated, but the anions are the same: $NCS^- \leftrightarrow {}^-NCS \equiv SCN^-$. In cold water HNCS is fairly stable toward acids if not too concentrated. It is a little more strongly acidic than HNO_3. At room temperature HNCS is a gas that is stable for about a day. If cooled, it condenses to a white pungent smelling mass of crystals stable for only a short time. The solid melts at 5°C, forming a liquid that readily polymerizes to a yellow solid. Concentrated HCl or H_2SO_4, added in excess to aqueous SCN^-, gradually produces HCN and a yellow precipitate of "perthiocyanic acid", $C_2N_2S_3H_2$, slightly soluble in hot H_2O, from which it crystallizes in needles, soluble in ethanol and ether.

Acidification of $S_2O_3^{2-}$ gives many products, formally from $H_2S_2O_3$, i.e., the unstable orange-yellow *cyclo*-S_6 favored by low T and extraction by toluene (yields of 15 % at best), with S_8, amorphous S and others:

$$S_2O_3^{2-} + 2\,H_3O^+ \rightarrow H_2S_2O_3 + 2\,H_2O \rightarrow {}^1/_6\,S_6 + SO_2 + 3\,H_2O$$

$$3\,H_3O^+ + 2\,S_2O_3^{2-} \rightarrow H_2S\uparrow + H[S_3O_6]^- + 3\,H_2O$$

Such reactions are promoted by weak acids (CO_2 or H_3BO_3) as well as strong, although the latter act more quickly. The release of both sulfur and SO_2 by dilute acid is characteristic of $S_2O_3^{2-}$. Caution: Oxidants can be activated by the acid and will also precipitate sulfur even from only S^{2-}. This test for $S_2O_3^{2-}$ takes at least five minutes when using a cold 6–7 mM solution of $S_2O_3^{2-}$.

The thionic acids, $H_2[S_{n+2}O_6]$ or $[S_n(SO_3H)_2]$, $1 \le n \le 4$, are readily formed by the action of H_3O^+ on $S_2O_3^{2-}$; also see **16.2.3 Reduced chalcogens** below about "Wackenroder's solution". Higher acidities yield more of the higher-n species. The pure acids cannot be isolated at ambient T. Evaporation, even at room temperature, decomposes them, mostly as follows, but also with some H_2S:

$$H_2[S_{n+2}O_6] + H_2O \rightarrow HSO_4^- + SO_2\uparrow + nS\downarrow + H_3O^+$$

In a cold solution $H_2[S_5O_6]$ may be kept for a few months, but even under the best conditions it slowly decomposes to give S and other sulfur acids, including $H_2[S_3O_6]$ and $H_2[S_4O_6]$. If such a solution is heated, it decomposes rapidly as above.

Some corresponding acids may be obtained by adding, e.g., $HClO_4$ or $H_2[SiF_6]$ to the anionic solutions. The Ae salts are stable and isolable.

A convenient laboratory method for the preparation of SO_2 in a Kipp generator has used the decomposition with dilute H_2SO_4 of cubes made of three parts calcium sulfite and one part calcium sulfate.

Hot, concentrated HCl decomposes dithionate:

$$[S_2O_6]^{2-} + H_3O^+ \rightarrow HSO_4^- + SO_2\uparrow + H_2O$$

Two points may be of interest concerning the acid strength of H_2SO_4. One is that $H_2[S_2O_7]$, which may be present in "fuming" H_2SO_4 and "oleum" containing "free" SO_3, make it equivalent to 25-M aqueous H_2SO_4 or more. The other is that because of its high boiling point, it displaces all of the volatile inorganic acids; on the other hand it is displaced, when heated above its own boiling point, by the weaker (in water) but less volatile boric, silicic and phosphoric acids.

Hydroxide. Sulfur dissolves readily in hot OH^-, e.g., with n equal to 5, or a smaller integer by using enough base:

$$3\ Ca(OH)_2 + (2n + 2)\ S \rightarrow 3\ Ca^{2+} + 2\ S_n^{2-} + S_2O_3^{2-} + 3\ H_2O$$

Ethanol can separate these, by dissolving the polysulfides. Acids readily decompose the products, with the separation of sulfur.

With a high $c(OH^-)$, $[S_3O_6]^{2-}$ decomposes to SO_3^{2-} and $S_2O_3^{2}$:

$$2\ [S_3O_6]^{2-} + 6\ OH^- \rightarrow S_2O_3^{2-} + 4\ SO_3^{2-} + 3\ H_2O$$

If $[S_4O_6]^{2-}$ is treated with an alkali, $S_2O_3^{2-}$ and $[S_3O_6]^{2-}$ are formed:

$$4\ [S_4O_6]^{2-} + 6\ OH^- \rightarrow 5\ S_2O_3^{2-} + 2\ [S_3O_6]^{2-} + 3\ H_2O$$

If the system is hot some S^{2-} is obtained.

Bases easily decompose $[S_5O_6]^{2-}$; the separation of S on adding an alkali is a rather sensitive test for it.

Sulfur dioxide reacts with OH^- to form SO_3^{2-} and then, with more SO_2, HSO_3^- in solution or $[S_2O_5]^{2-}$ in solids. (The HSO_3^- does not persist in solids, nor $[S_2O_5]^{2-}$ in solution.) Soluble sulfites must be quite dry or air-free to be preserved. Even air-free, wet HSO_3^- slowly decomposes:

$$3\ HSO_3^- \rightarrow SO_4^{2-} + HSO_4^- + S\downarrow + H_2O$$

In base, $[S_2O_6]^{2-}$ is unstable.

Peroxide. Hydrogen peroxide slowly acts upon H_2S to form S. Alkaline sulfide solutions readily undergo oxidation to SO_4^{2-}:

$$HS^- + 4\ HO_2^- \rightarrow SO_4^{2-} + H_2O + 3\ OH^-$$

Mercury(II) sulfide is readily soluble in dilute HCl containing H_2O_2.

Thiocyanate, SCN^-, plus H_2O_2 in acidic or neutral solution form H_2SO_4 and, under some conditions, HCN, NH_4^+, S, CO_2 or $(SCN)_x$.

With H_2O_2, sulfides and sulfites become sulfates, particularly in alkaline solution. However, dithionate also arises in low yield, with SO_4^{2-}, from SO_2 plus H_2O_2 in acid, whereas thiosulfate can give $[S_3O_6]^{2-}$:

$$2\ SO_2 + H_2O_2 + 2\ H_2O \rightarrow [S_2O_6]^{2-} + 2\ H_3O^+$$

$$2\ S_2O_3^{2-} + 4\ H_2O_2 \rightarrow [S_3O_6]^{2-} + SO_4^{2-} + 4\ H_2O$$

Hydrogen peroxide in aqueous CH_3CO_2H readily oxidizes $S_2O_3^{2-}$:

$$2\ S_2O_3^{2-} + H_2O_2 + 2\ CH_3CO_2H \rightarrow [S_4O_6]^{2-} + 2\ CH_3CO_2^- + 2\ H_2O$$

In alkaline solution, however, $S_2O_3^{2-}$ and HO_2^- form SO_4^{2-}.

Ice-cold $S_2O_3^{2-}$ and H_2O_2 form the sulfanedisulfonate:

$$3\ S_2O_3^{2-} + 4\ H_2O_2 \rightarrow 2\ [S_3O_6]^{2-} + 3\ H_2O + 2\ OH^-$$

Cold aqueous $[S_2O_6]^{2-}$ is oxidized to SO_4^{2-} by Na_2O_2.

Hydrogen peroxide plus concentrated H_2SO_4 yield H_2SO_5, i.e., $H_2SO_3(O_2)$, peroxosulfuric acid; H_2O_2 decomposes $(NH_4)_2[S_2O_8]$, the peroxodisulfate, and liberates both N_2 and O_2.

Di- and trioxygen. Dry H_2S is stable, but an aqueous solution exposed to the air slowly deposits sulfur, hastened by finely divided Ni.

Air slowly oxidizes $SO_2{\cdot}H_2O$ and all sulfite salts mainly to HSO_4^- and SO_4^{2-}. See **16.2.4 Oxidation of sulfites** for more. However, dithionate can be made in low yield from SO_2 and O_2:

$$2\ SO_2 + {}^1/_2\ O_2 + 3\ H_2O \rightarrow [S_2O_6]^{2-} + 2\ H_3O^+$$

although air also oxidizes $H_2[S_2O_6]$ to H_2SO_4.

Ozone, not in excess, converts H_2S and S^{2-} to SO_4^{2-} and the various intermediates, $S_2O_3^{2-}$, $[S_2O_4]^{2-}$, SO_3^{2-} and $[S_n(SO_3)_2]^{2-}$ ($n \geq 0$).

Ozone and CuS, ZnS, CdS, PbS, Sb_2S_3 etc. form sulfates.

16.2.2 Reagents Derived from the Other 2nd-Period Non-Metals, Boron through Fluorine

Carbon oxide species. Insoluble sulfates are decomposable for analysis by long boiling with concentrated Alk_2CO_3 or, more readily, by fusion with Alk_2CO_3. Each method produces alkali sulfates soluble in water, and metal carbonates soluble in HCl after removing the sulfate:

$$BaSO_4 + CO_3^{2-} \rightarrow SO_4^{2-} + BaCO_3\downarrow$$

Cyanide species. Cyanide and sulfur combine directly; polysulfides are merely diminished if cyanide is not in excess, e.g.:

$$CN^- + {}^1/_8\, S_8 \rightarrow CNS^-$$

$$2\ CN^- + S_4^{2-} \rightarrow 2\ CNS^- + S_2^{2-}$$

A solution of $S_2O_3^{2-}$ also forms the corresponding thiocyanate:

$$S_2O_3^{2-} + CN^- \rightarrow SO_3^{2-} + CNS^-$$

Some "simple" organic species. Precipitated sulfur, as obtained during analysis when HCl is added to a polysulfide extract of metallic sulfides, is soluble in benzene or low-boiling petroleum ether. This has value when looking for traces of As or Sb sulfides.

Mixtures of sulfur and various organic compounds, e.g., paraffin, when warmed, give off H_2S. This method may be very convenient because regulating the temperature easily controls the release of gas.

The water-soluble metallic sulfates are insoluble in ethanol, which precipitates them from moderately concentrated aqueous solutions. Ethanol, added to aqueous hydrogensulfates, e.g., $KHSO_4$, precipitates normal salts, e.g., K_2SO_4, and leaves H_2SO_4 in solution.

Mixing HSO_3^- and CH_2O forms $HOCH_2SO_3^-$ and releases heat. Then adding NH_3 gives $H_2NCH_2SO_3^-$ with more heat. Finally, acidification and cooling crystallize white $H_3N^+{-}CH_2SO_3^-$, useful as a primary-standard weak acid for titrating carbonate-free alkalis, a bit slowly, with phenolphthalein; boiling releases SO_2 and other products. The anion precipitates Ba^{2+} and to some extent Cu^{2+}.

Reduced nitrogen. Shaking CS_2 with concentrated NH_3 readily yields thiocyanate or dithiocarbamate, depending on conditions:

$$CS_2 + 3\ NH_3 \rightarrow SCN^- + 2\ NH_4^+ + HS^-$$

$$CS_2 + 2\ NH_3 \rightarrow CS_2NH_2^- + NH_4^+$$

Sulfur, boiled with NH_3, forms HS^- and $S_2O_3^{2-}$, also some SO_4^{2-}.

If catalyzed by Ag^+, NH_3 is partly oxidized to N_2 by $[S_2O_8]^{2-}$:

$$4\ NH_3 + {}^3/_2\ [S_2O_8]^{2-} \rightarrow {}^1/_2\ N_2\uparrow + 3\ SO_4^{2-} + 3\ NH_4^+$$

but NH_4^+ is also attacked by excess oxidant in the presence of Ag^+:

$$NH_4^+ + 4\ [S_2O_8]^{2-} + 13\ H_2O \rightarrow 10\ H_3O^+ + 8\ SO_4^{2-} + NO_3^-$$

Hydroxylammonium ion in acid partly oxidizes $H_2[S_4O_6]$:

$$H[S_4O_6]^- + NH_3OH^+ + 2\ H_2O \rightarrow 2\ HSO_4^- + 2\ S\downarrow + NH_4^+ + H_3O^+$$

In alkaline solution, however, the ion is slightly reduced:

$$[S_4O_6]^{2-} + 2\ NH_2OH + 3\ OH^- \rightarrow 2\ S_2O_3^{2-} + NO_2^- + NH_3 + 3\ H_2O$$

Aqueous N_3^- and iodine, if catalyzed by either a soluble or an insoluble sulfide, release N_2:

$$2\ N_3^- + I_2 \rightarrow 3\ N_2\uparrow + 2\ I^-$$

This is a very sensitive test for a sulfide, thiocyanate or thiosulfate (containing $S^{<0}$), but not for S or $S^{>0}$; pure I_2 and N_3^- alone do nothing. One may place one drop of the unknown onto a spot plate, add one drop of starch iodine and then one drop of 3-dM NaN_3. If the liquid contains more than 1 μM $S_2O_3^{2-}$ the blue color quickly disappears.

Oxidized nitrogen. Aqueous SO_3^{2-} absorbs NO and forms $[SO_3N_2O_2]^{2-}$, i.e., *N*-nitrosohydroxylamine-*N*-sulfonate, a "hyponitrite" derivative:

$$SO_3^{2-} + 2\ NO \rightarrow [ON\text{–}N(\text{–}O)\text{–}SO_3]^{2-}$$

The Na, K, NH_4, etc. salts are white, and the disodium salt is much more soluble than the other two. They decompose to SO_4^{2-} and N_2O, slowly in cold water, rapidly in hot or in acids.

Sulfane (H_2S) and HNO_2 give NO, S and $S_2O_3^{2-}$.

Aqueous $K_2S_2O_3$, KNO_2 and HCl form $K_2[S_6O_6]$.

Heating nitrite with HSO_3^- (perhaps from KOH plus excess SO_2) yields nitridotrisulfonate:

$$4\ HSO_3^- + NO_2^- \rightarrow [N(SO_3)_3]^{3-} + SO_3^{2-} + 2\ H_2O$$

For more, see p. 367.

Nitrous acid and sulfamate, $SO_3NH_2^-$, sometimes used to destroy HNO_2, give HSO_4^- and N_2.

Sulfane and HNO_3 yield NO and S. Hot and concentrated HNO_3 results in H_2SO_4.

Mercury(II) sulfide is practically insoluble in hot, dilute (2 M) HNO_3. A more concentrated acid may form the slightly soluble, white, $\sim Hg_3S_2(NO_3)_2$. Mercury(II) sulfide is readily soluble in aqua regia. Practically all other sulfides are soluble in HNO_3.

Thiocyanate, SCN^-, plus HNO_3 or HNO_2 form H_2SO_4 and NO and, under some conditions, CO_2, HCN, NH_4^+ or S. Nitric acid containing nitrogen oxides, acting in hot, concentrated solutions of thiocyanates, precipitates a "perthiocyanogen" or "pararhodane", $(SCN)_x$, with a yellow to red color, sometimes even blue. It may be formed in the test for iodine, and mistaken for that element, with starch or CS_2. If boiled with a solution of OH^- it re-forms thiocyanate.

Nitric acid reacts with S to form NO and H_2SO_4. Complete oxidation requires concentrated acid and long boiling. Crystallized S_8 is attacked more slowly than the amorphous or sublimed types.

Thiosulfate reduces HNO_3 to NO with the liberation of S.

Sulfur dioxide plus HNO_3 or HNO_2 form NO and H_2SO_4, the latter reacting more readily than the former.

Dithionates are stable at 25 °C toward HNO_3, but boiling gives H_2SO_4.

Insoluble **d**- or **p**-block sulfates all dissolve in HNO_3.

16.2.3 Reagents Derived from the 3rd-to-5th-Period Non-Metals, Silicon through Xenon

Phosphorus species. The acids of phosphorus are not reduced by H_2S.

Sulfur dioxide plus PH_3, HPH_2O_2 or H_2PHO_3 produce H_3PO_4.

In acid, HPH_2O_2 reduces $[S_2O_8]^{2-}$ to SO_4^{2-} or HSO_4^-.

The PO_4^{3-} ion, acting like the OH^- from its hydrolysis, when boiled with sulfur forms S_x^{2-} and $S_2O_3^{2-}$, plus HPO_4^{2-}.

The oxide P_4O_{10} removes H_2O even from the strong desiccant H_2SO_4:

$$^1/_2\ P_4O_{10} + H_2SO_4 \rightarrow {}^2/_n\ (HPO_3)_n + SO_3$$

Arsenic species. Cold $S_2O_3^{2-}$ with arsenite and HCl yield $H[S_5O_6]^-$ apparently free from $H[S_3O_6]^-$ and $H[S_4O_6]^-$.

Reduced chalcogens. Aqueous HNCS may be obtained by treating $Pb(SCN)_2$ with H_2S and filtering out the PbS.

At 25 °C, S_8 and S^{2-} or HS^- first form mainly S_4^{2-} and S_5^{2-}, changing from pale yellow to deep red at high *n*, reverting in acid to S and H_2S:

$$NH_3 + HS^- + {}^n/_8\ S_8 \rightarrow NH_4^+ + S_{n+1}^{2-}$$

The actual reactions (among S^{-II}, S^0, S^{IV} and intermediates) below and in the next subsection are related but depend on conditions too complicated or sometimes unclear to elaborate.

Amorphous S (not S_8), one product of acidifying $S_2O_3^{2-}$ or of mixing SO_2 and H_2S, insoluble in CS_2, is solubilized later with H_2S water:

$$2\ H_2S + SO_2 \rightarrow 3\ S\downarrow + 2\ H_2O$$

Under controlled conditions, HS^- and HSO_3^- give very pure $S_2O_3^{2-}$:

$$2\ HS^- + 4\ HSO_3^- \rightarrow 3\ S_2O_3^{2-} + 3\ H_2O$$

One route to $[S_3O_6]^{2-}$ is by:

$$H_2S + 6\ HSO_3^- + 2\ SO_2 \rightarrow 3\ [S_3O_6]^{2-} + 4\ H_2O$$

One way to prepare $H_2[S_5O_6]$ is to pass H_2S slowly into aqueous SO_2 at 0 °C. The filtrate is "Wackenroder's solution", which, with KCH_3CO_2, yields $K_2S_5O_6$. Beside $H_2[S_5O_6]$ it contains some $H_2[S_3O_6]$ and $H_2[S_4O_6]$, all unstable, as noted above. (Excess H_2S produces only S):

$$5\ H_2S + 5\ SO_2 \rightarrow H[S_5O_6]^- + 5\ S\downarrow + H_3O^+ + 3\ H_2O$$

Sulfane decomposes all of ammoniacal $[S_5O_6]^{2-}$ but not $[S_4O_6]^{2-}$:

$$[S_5O_6]^{2-} + 5\ HS^- + 7\ NH_4^+ \rightarrow 10\ S\downarrow + 7\ NH_3 + 6\ H_2O$$

Elemental and oxidized chalcogens. Thiosulfates are prepared by boiling sulfur in a solution of sulfite:

$$SO_3^{2-} + {}^1/_8\ S_8 \rightarrow S_2O_3^{2-}$$

Boiling also decomposes $S_2O_3^{2-}$ in slightly more alkaline solution:

$$2\ S_2O_3^{2-} \rightarrow S^{2-} + [S_3O_6]^{2-}$$

Aqueous SO_2 and sulfur react differently at room temperature, e.g.:

$$2\ SO_2 + {}^1/_4\ S_8 + 4\ H_2O \rightarrow [S_3O_6]^{2-} + H_2S\uparrow + 2\ H_3O^+$$

$$2\ SO_2 + {}^1/_8\ S_8 + {}^1/_2\ O_2 + 3\ H_2O \rightarrow [S_3O_6]^{2-} + 2\ H_3O^+$$

Strong oxidants readily convert the $S_nO_6^{2-}$ anions to SO_4^{2-}.

Treating $S_2O_3^{2-}$ with SO_2 produces $[S_4O_6]^{2-}$, with small amounts of $[S_3O_6]^{2-}$ and $[S_5O_6]^{2-}$. Traces of arsenite favor forming the $[S_4O_6]^{2-}$ (but cf. the different result under **Arsenic species** above). However, treating $K_2S_2O_3$ with sulfur dioxide yields $K_2[S_3O_6]$:

$$2\ K_2S_2O_3 + 3\ SO_2 \rightarrow 2\ K_2[S_3O_6] + S\downarrow$$

A good one-electron reductant, SO_2^-, is in equilibrium with $S_2O_4^{2-}$.

The sulfites of Ae^{2+}, and some others, are soluble in aqueous SO_2 but are reprecipitated on boiling. The alkalis form hydrogensulfites ("bisulfites"), which can be precipitated from excess SO_2. Evaporation then yields disulfites, $M_2[S_2O_5]$. The sulfites are decomposed by all ordinary acids except H_3BO_3, (aqueous) CO_2, or, in some cases, H_2S.

If a solution of HSO_3^- is allowed to stand for any length of time, $[S_3O_6]^{2-}$, and presumably SO_4^{2-}, are formed.

Thiosulfates can be oxidized to $[S_4O_6]^{2-}$ by SeO_2.

Dilute H_2SO_4 has no action on H_2S. The hot, concentrated acid gives:

$$3\ H_2SO_4 + H_2S \rightarrow S\downarrow + SO_2\uparrow + 2\ HSO_4^- + 2\ H_3O^+$$

Sulfides that dissolve in dilute H_2SO_4 release H_2S, e.g., MnS, FeS, ZnS, CdS, etc. If a sulfide requires concentrated H_2SO_4 for its solution, S_8 and SO_2 are formed (sometimes only SO_2), e.g., CuS, HgS and Bi_2S_3.

Cold concentrated H_2SO_4 has no action on FeS; heat yields SO_2:

$$2\ FeS + 20\ H_2SO_4 \rightarrow Fe_2(SO_4)_3 + 9\ SO_2\uparrow + 10\ H_3O^+ + 10\ HSO_4^-$$

In general, if concentrated H_2SO_4 is used on a sulfide that would be soluble in dilute acid, no H_2S is released:

$$ZnS + 8\ H_2SO_4 \rightarrow ZnSO_4 + 4\ SO_2\uparrow + 4\ H_3O^+ + 4\ HSO_4^-$$

With a small amount of H_2O present, this reaction becomes:

$$ZnS + HSO_4^- + 3\ H_3O^+ \rightarrow Zn^{2+} + S\downarrow + SO_2\uparrow + 5\ H_2O$$

Cold concentrated H_2SO_4 has no action on S; heat slowly yields SO_2:

$$S + 4\ H_2SO_4 \rightarrow 3\ SO_2\uparrow + 2\ H_3O^+ + 2\ HSO_4^-$$

Aqueous HNCS may be obtained by adding the calculated amount of H_2SO_4 to remove the barium from a solution of $Ba(NCS)_2$, but thiocyanates are decomposed by concentrated H_2SO_4.

Thiocyanate, SCN^-, plus $[S_2O_8]^{2-}$ in acidic or neutral solution form H_2SO_4 and, under some conditions, HCN, NH_4^+, S_8, CO_2 or $(SCN)_x$.

The $[S_2O_8]^{2-}$ ion with excess $S_2O_3^{2-}$ forms $[S(SO_3)_2]^{2-}$, but an excess of the former leads, perhaps counterintuitively, to $[S_2(SO_3)_2]^{2-}$.

Aqueous SO_2 and $HSO_3(O_2)^-$ quickly yield HSO_4^-.

Reduced halogens. The **d**- or **p**-block basic sulfates dissolve in HCl.

Aqueous HBr forms Br_2 and SO_2 if the H_2SO_4 is concentrated enough. A low concentration of HI reduces H_2SO_4 to SO_2. Much HI reduces it further to H_2S. Aqueous HI may also reduce S to H_2S.

The halides Cl^-, Br^- and I^- react with $[S_2O_8]^{2-}$ to form the elements, although I^- is partly oxidized to IO_3^-. (Iodide in excess, however, is suitable to determine $[S_2O_8]^{2-}$, when followed by back titration.) The ions Fe^{2+} and Cu^{2+} markedly accelerate the reactions, a mixture of the two apparently being more effective than either alone. With Ag^+ as a catalyst in the presence of HNO_3, the halides are oxidized not to the free elements but to halates; $I^- \rightarrow IO_3^-$ etc.

Elemental and oxidized halogens. Chlorine, with H_2S in excess, forms HCl and S; Cl_2 in excess yields HCl and H_2SO_4. Chlorine in the presence of H_2O reacts with S to form HCl and H_2SO_4. Bromine gives corresponding results, *mutatis mutandis*. Iodine reacts only thus:

$$H_2S + I_2 + 2\ H_2O \rightarrow 2\ I^- + S\downarrow + 2\ H_3O^+$$

Thiocyanate, SCN^-, plus Cl_2 or Br_2, in acidic or neutral solution form H_2SO_4 and, under some conditions, HCN, NH_4^+, S_8, CO_2 or $(SCN)_x$.

Nascent chlorine, acting (similarly to HNO_3 plus NO_2) in hot, concentrated SCN^-, precipitates "perthiocyanogen", $(SCN)_x$.

Thiocyanate (SCN^-) plus I_2 and HCO_3^- form SO_4^{2-}, I^- and others.

Sulfur(0) appears unreactive to iodine or iodine compounds.

Disulfanedisulfonates are readily prepared by adding iodine to thiosulfates, which are also titrated this way, in H_2O or CH_3CO_2H:

$$2\ S_2O_3^{2-} + I_2 \rightarrow [S_4O_6]^{2-} + 2\ I^-$$

In alkaline solution the main product may be SO_4^{2-}. Chlorine and bromine can react similarly to, but more vigorously than, iodine. An excess of either forms sulfuric acid from $S_2O_3^{2-}$:

$$S_2O_3^{2-} + 4\ Cl_2 + 13\ H_2O \rightarrow 2\ HSO_4^- + 8\ H_3O^+ + 8\ Cl^-$$

Low yields of dithionate, with SO_4^{2-}, arise from SO_2 plus Cl_2, Br_2 or I_2:

$$2\ SO_2 + Cl_2 + 6\ H_2O \rightarrow [S_2O_6]^{2-} + 2\ Cl^- + 4\ H_3O^+$$

Dithionates are stable at 25 °C against oxidation by Br_2, but boiling gives the sulfates. Cold aqueous $[S_2O_6]^{2-}$ is not affected by ClO^- or BrO^-.

Iodine apparently oxidizes $[S_3O_6]^{2-}$ to SO_4^{2-}.

Mercury(II) sulfide is readily soluble in ClO^- or $HClO_3$.

Toward thiosulfate, ClO^-, BrO^- and IO^- react like the halogens (above) although under some conditions, especially in a dilute solution, the reaction is more complicated:

$$3\ S_2O_3^{2-} + 5\ ClO^- \rightarrow 2\ SO_4^{2-} + [S_4O_6]^{2-} + 5\ Cl^-$$

With H_2S in excess, $HClO_3$ forms HCl and S_8; with $HClO_3$ in excess, HCl and H_2SO_4. In the presence of H_2O, $HClO_3$ reacts with S_8 to form HCl and H_2SO_4. Bromic acid gives corresponding results in each case.

Thiocyanate, SCN^-, plus ClO_3^- and HCl form H_2SO_4 and, under some conditions, HCN, NH_4^+, S_8, CO_2 or $(SCN)_x$. Thiocyanate plus BrO^- form SO_4^{2-}, Br^- and other products. Thiocyanate plus $HBrO_3$ form H_2SO_4, HBr etc. Thiocyanate plus HIO_3 yield H_2SO_4 and free iodine, which may react further to form ICN.

The chlorate ion, with HCl, completely oxidizes $[S_5O_6]^{2-}$ to SO_4^{2-}.

Chloric, bromic and iodic(V) acids, HXO_3, are first reduced by $S_2O_3^{2-}$ to the corresponding halogens, and then, with an excess of $S_2O_3^{2-}$, to the halides, always accompanied by the separation of S.

The salts of $[S_3O_6]^{2-}$ are not oxidized by ClO_3^- or IO_3^-, but an acidic solution rapidly forms SO_4^{2-} and some S.

The halogen oxoacids, except $HClO_4$, react with SO_2 to yield halides and sulfate. The titration of SO_3^{2-} with excess ClO^-, followed by titrating the excess hypochlorite with I_2, has been used to determine S^{IV}.

However, a mixture of iodic acid, HIO_3, and starch is turned to violet/blue (with I_2) by traces of SO_3^{2-}, for which this is a test. The color is destroyed by an excess of the SO_3^{2-}, forming I^-.

Chlorates are decomposed by concentrated H_2SO_4:

$$3\ ClO_3^- + 3\ H_2SO_4 \rightarrow 3\ HSO_4^- + ClO_4^- + 2\ ClO_2\uparrow + H_3O^+$$

Iodic acid and H_2S form HI and S:

$$3\ H_2S + IO_3^- \rightarrow I^- + 3\ S\downarrow + 3\ H_2O$$

but much excess IO_3^- plus much HCl oxidize H_2S to HSO_4^-.

In different conditions H_2S and H_5IO_6 quickly yield S or HSO_4^-.

16.2.4 Reagents Derived from the Metals Lithium through Uranium, plus Electrons and Photons

Oxidation of reduced sulfur. The higher oxidation states of Cr and Mn react with H_2S to give products that depend largely on the acidity and temperature. Aqueous CrO_4^{2-} forms $Cr_2O_3\cdot aq$, S_8, S^{2-} and $S_2O_3^{2-}$; with a dichromate and sufficient H_3O^+, the reaction gives Cr^{III} and S_8.

Dilute (e.g., < 0.1 M) MnO_4^- reacts with H_2S giving mixtures, e.g.:

$$8\ MnO_4^- + 12\ H_2S \rightarrow 7\ MnS\downarrow + Mn^{2+} + 5\ SO_4^{2-} + 12\ H_2O$$

$$8\ MnO_4^- + 17\ H_2S \rightarrow 7\ MnS\downarrow + Mn^{2+} + 5\ S_2O_3^{2-} + 17\ H_2O$$

$$2\ MnO_4^- + 5\ H_2S \rightarrow 2\ MnS\downarrow + S\downarrow + S_2O_3^{2-} + 5\ H_2O$$

Sulfane (H_2S) and acidified $[Fe(CN)_6]^{3-}$ form $H_2[Fe(CN)_6]^{2-}$ and S_8.

Aqueous SCN^- reduces $[Cr_2O_7]^{2-}$, Co_2O_3.aq, NiO_2 and H_3AsO_4 to their lower oxidation states, while forming HSO_4^- or SO_4^{2-}. The higher oxidation states of Mn in acidic solution form Mn^{2+}, HCN and HSO_4^-. In alkaline mixture NCO^- and SO_4^{2-} are formed.

An alkaline thiocyanate with PbO forms NCO^-, SO_4^{2-} and Pb:

$$SCN^- + 4\ PbO + 2\ OH^- \rightarrow NCO^- + SO_4^{2-} + 4\ Pb + H_2O$$

The lead(IV) in PbO_2 and Pb_3O_4, in acid only, oxidizes HNCS apparently to $(CN)_2$, leaving Pb^{2+} and $PbSO_4$, tentatively giving:

$$7\ PbO_2 + 2\ SCN^- + 12\ H_3O^+ \rightarrow$$

$$5\ Pb^{2+} + 2\ PbSO_4\downarrow + (CN)_2\uparrow + 18\ H_2O$$

Oxidation of thiosulfates and polythionates. Dichromates are reduced to Cr^{III} by $S_2O_3^{2-}$. Permanganates, in neutral solution, become $MnO_2\cdot aq$; in acid the products are Mn^{2+}, SO_4^{2-} and $[S_2O_6]^{2-}$.

Cold aqueous $[S_2O_6]^{2-}$ is not affected by MnO_4^-.

Solutions of Fe^{III} are reduced to Fe^{2+} by $S_2O_3^{2-}$, which is oxidized:

$$2\ FeCl_3 + 2\ S_2O_3^{2-} \rightarrow 2\ Fe^{2+} + 6\ Cl^- + [S_4O_6]^{2-}$$

This provides for a volumetric analysis of $S_2O_3^{2-}$, with CNS^- as indicator.

Thiosulfates can also be oxidized to $[S_4O_6]^{2-}$ by Cu^{2+} and PbO_2.

If $[S_3O_6]^{2-}$ is treated with Cu^{2+}, Ba^{2+} and HCl and boiled for one hour, all of the sulfur is precipitated as $BaSO_4$. All other polythionic compounds decompose on heating, depositing sulfur.

Oxidation of sulfites. Dithionate can be made in low yield, along with SO_4^{2-}, from neutral HSO_3^- plus CrO_4^{2-}:

$$10\ HSO_3^- + 2\ CrO_4^{2-} \rightarrow Cr_2O_3\cdot aq\downarrow + 3\ [S_2O_6]^{2-} + 4\ SO_3^{2-} + 5\ H_2O$$

With SO_2, MnO_4^- ordinarily yields Mn^{2+} and sulfate:

$$2\ MnO_4^- + 5\ H_2O\cdot SO_2 + H_2O \rightarrow 5\ SO_4^{2-} + 2\ Mn^{2+} + 4\ H_3O^+$$

In ice-cold water, SO_2 and $MnO_2\cdot aq$ give a good yield of dithionate:

$$2\ H_2O\cdot SO_2 + MnO_2\cdot aq \rightarrow [S_2O_6]^{2-} + Mn^{2+} + 2\ H_2O$$

Then one can crystallize, e.g., $Ca[S_2O_6]$ after warming with $Ca(OH)_2$ [or $Ba(OH)_2$ or Na_2CO_3], followed by removing the solid $Mn(OH)_2$, then removing the excess $Ca(OH)_2$ with CO_2, evaporating, cooling, and adding ethanol. Similar dithionates are obtained with NiO_2 or Fe_2O_3.

The acid $H_2[S_2O_6]$ results from treating $Mn[S_2O_6]$ (just above) with $Ba(OH)_2$ and the filtrate from this with the calculated amount of H_2SO_4; this then gives a colorless solution which may be evaporated in a vacuum to a density of 1.347 g/mL. It decomposes on heating:

$$H_2[S_2O_6] \rightarrow H_2SO_4 + SO_2\uparrow$$

Aqueous SO_2 cannot be titrated directly with MnO_4^-, which gives low results. Adding excess MnO_4^-, followed by back titration, is satisfactory.

Dithionate can also be made in low yield, along with SO_4^{2-}, from neutral HSO_3^- plus MnO_4^-. Dithionates are stable at 25 °C against oxidation by MnO_4^-, but boiling gives the sulfates.

The sulfites of Fe^{III}, Cu, Ag and Hg are unstable, the SO_3^{2-} becoming SO_4^{2-} while reducing the cation.

Sulfur dioxide, $[Fe(CN)_6]^{3-}$ and H_3O^+ form $H_2[Fe(CN)_6]^{2-}$ and HSO_4^-.

Cobalt(3+) and nickel(>II) become Co^{2+} and Ni^{2+}, oxidizing SO_2 to $S^{>IV}$, especially to SO_4^{2-}. Compounds of Pt or Au precipitate Pt or Au.

With SO_2, PbO_2 forms $PbSO_4$; AsO_4^{3-} forms AsO_3^{3-} and SO_4^{2-}; and Sb^V becomes Sb^{III}. However, dithionates appear to be preparable in good yield by heating, e.g., $Na_2SO_3 \cdot 7H_2O$ with PbO_2 and H_2O. After adding CO_2 (to remove Pb^{II} from the alkaline solution), filtering, and adding lactic acid (whose salts remain dissolved after replacing the CO_3^{2-}), one concentrates the resulting solution to crystallize the desired salt.

Light (inefficiently) and O_2 convert $H_2O \cdot SO_2$, via $HSO_3(O_2)^-$, to HSO_4^-, making acid rain. A low pH, however, requires catalysts (impurities in dust?) such as TiO_2, Fe_2O_3, ZnO, CdS or complexes of Mn, Fe or Ni. Thus at a pH ~ 2, HSO_3^- and $[FeOH(H_2O)_5]^{2+}$ form transitory complexes. Photo-activated $H_2O \cdot SO_2$ also activates Fe^{2+} and H_3O^+ to form Fe^{III} and H_2.

Anodic treatment of SO_3^{2-} yields $[S_2O_4]^{2-}$.

Reduction. Sodium (amalgam) or Zn reduces SO_2 to $[S_2O_4]^{2-}$.

The alkali polysulfides slowly attack many metals, reducing the sulfur: Fe yields FeS; Ni forms NiS; Cu becomes CuS and Cu_2S; Ag gives Ag_2S; and Sn forms $[SnS_3]^{2-}$.

Sublimed S ("flowers"), when boiled with Hg_2^{2+}, form Hg and HgS; $SnCl_2$ gives SnS and $SnCl_4$. However, neither sulfates of Mn^{2+}, Fe^{2+}, Ni^{2+}, Zn^{2+} or Cd^{2+}, nor AsO_3^{3-}, AsO_4^{3-}, Sb^{III} or Bi^{III} species precipitate MS_n thus.

Sulfur in the element and low oxidation states, e.g., in foods or inorganic substances, may be detected by the tarnishing of a piece of silver, forming dark-colored Ag_2S, or by the formation of purple $[Fe(CN)_5NOS]^{2-}$ from $[Fe(CN)_5NO]^{2-}$, both after boiling with OH^-.

Treated with Zn and an acid, thiocyanates yield H_2S and sometimes CS_2, NH_3, CH_3NH_2 and other products. Distilling the SCN^- with Al and HCl quantitatively decomposes it into NH_3, C and H_2S:

$$3\ SCN^- + 4\ Al + 18\ H_3O^+ \rightarrow$$

$$3\ NH_4^+ + 3\ C\downarrow + 3\ H_2S\uparrow + 4\ Al^{3+} + 18\ H_2O$$

Zinc and HCl reduce the $[S_nO_6]^{2-}$ anions to H_2S.

Metallic Al in alkaline solution quickly reduces $S_2O_3^{2-}$ to S^{2-}. Zinc and HCl act similarly, but release H_2S due to the acid, of course.

Exposing $Zn[Fe(CN)_5NO]$ to NH_3 vapors, then to SO_2, makes it rose-red, possibly the well-known $[Fe(CN)_5(NOS)]^{4-}$, thus detecting SO_2.

The weakly acidic aqueous SO_2, and Fe, Cu, Zn or Sn, form dithionite, $S_2O_4^{2-}$, detectable with methylene blue.

Strong acid and Zn give H_2S, perhaps via nascent H:

$$3\ Zn + 6\ H_3O^+ + SO_2 \rightarrow 3\ Zn^{2+} + H_2S\uparrow + 8\ H_2O$$

Sodium amalgam or zinc in acid reduce dithionate also:

$$[S_2O_6]^{2-} + (Zn, 2\ Na_{Hg}) + 4\ H_3O^+ \rightarrow 2\ SO_2 + (Zn^{2+}, 2\ Na^+) + 6\ H_2O$$

Tin(II) reacts with SO_2 to form first SnS, then SnS_2. If the concentration of H_3O^+ is high, the Sn^{II} is oxidized directly to Sn^{IV}:

$$3\ SnCl_3^- + SO_2 + 6\ H_3O^+ + 9\ Cl^- \rightarrow 3\ SnCl_6^{2-} + H_2S\uparrow + 8\ H_2O$$

Tin(II) sulfite decomposes with reduction of S and oxidation of Sn.

Dilute sulfuric acid has no action on Cu, Ag, Hg, Pb (except very slowly, as in a storage battery) or Bi. It also is not reduced by and does not oxidize any of the lower metallic oxides.

Neither the dilute nor the concentrated acid attacks Rh, Ir, Pt or Au, although fuming H_2SO_4 readily attacks Pt.

Both dilute and concentrated H_2SO_4 dissolve the dangerously reactive alkali and alkaline-earth metals, and they dissolve or attack Th, Mn, Fe, Co, Ni, Zn, Cd, Al and Sn, although concentrated H_2SO_4 does not attack Fe. The dilute or cold, concentrated acid gives H_2 with the above. The hot, concentrated acid attacks Cu, Ag, Hg and the other metals, releasing SO_2. Varying acidity and T produce the two gasses in various ratios. Copper, e.g., from 130 to 170 °C, reacts approximately thus:

$$6\ Cu + 12\ H_2SO_4 \rightarrow 4\ CuSO_4 + Cu_2S\downarrow + SO_2\uparrow + 6\ H_3O^+ + 6\ HSO_4^-$$

Secondary reactions often occur, yielding sulfides, free sulfur, thionates, etc. Reactions releasing only H_2 do not belong with the reduction of sulfur, but are mentioned here for convenience and completeness.

The hot, concentrated acid is also reduced to SO_2 by some $M^{>0}$: Fe^{2+} is oxidized to Fe^{III}; Hg_2O forms $HgSO_4$; and $SnCl_2$ forms first SO_2, then H_2S, which may react with the Sn^{IV} formed to precipitate SnS_2.

Cathodic e^- reduce HSO_3^- to $S_2O_4^{2-}$.

Other reactions of sulfides. The common impurities in H_2S are H_2 and AsH_3. The former may come from Fe in the FeS used (with acid), the latter from arsenic in the FeS or the acid. Better H_2S is made by saturating $Mg(OH)_2$ suspended in H_2O with H_2S. Heating the mixture releases pure H_2S. Good H_2S also arises from dropping H_2O onto Al_2S_3:

$$Mg(OH)_2\downarrow + 2\ H_2S\uparrow \leftrightarrows Mg^{2+} + 2\ HS^- + 2\ H_2O$$

$$Al_2S_3 + 3\ H_2O \rightarrow Al_2O_3.aq\downarrow + 3\ H_2S\uparrow$$

Sulfane, H_2S, precipitates alkaline-earth sulfides under certain conditions, but they are readily hydrolyzed.

Classical qualitative analysis has the HCl-insoluble sulfide Group: Mo, W, Re, the Pt metals, Cu, Au, Cd, Hg, Ge, Sn, Pb, As, Sb, Bi, Se and Te.

Sulfane passed into oxometalates of at least V, Mo, W and Re, gives successively $[MO_{4-m}S_m]^{n-}$. The Alk^+ and NH_4^+ salts are soluble in water decreasingly from Na^+ through Cs^+. The NH_4^+ salts, even at ambient T, are rather unstable. The others are somewhat unstable too, all the more at low pH, becoming sulfides and oxometalates, or undergoing redox.

Metallic Cu and colorless "$(NH_4)_2S$" (i.e., no S_x^{2-}), without O_2, slowly give some Cu_2S and H_2, but Ag seems inert:

$$2\ Cu + HS^- + NH_4^+ \rightarrow Cu_2S\downarrow + H_2\uparrow + NH_3$$

Some metals however, e.g., Cu, Ag (forming tarnished silver) and Hg, are converted into their sulfides when exposed to moist sulfides and air:

$$2\ Ag + H_2S + {}^1/_2\ O_2 \rightarrow Ag_2S\downarrow + H_2O$$

Many insoluble metallic sulfides are broken down on treatment with Zn and dilute HCl (thus reducing the cation but not the sulfide ion):

$$CuS + Zn + 4\ Cl^- + 2\ H_3O^+ \rightarrow Cu\downarrow + H_2S\uparrow + ZnCl_4^{2-} + 2\ H_2O$$

Sulfane precipitates some M_mS_n, from $M^{>0}$ in either acidic, neutral or alkaline solutions. Arsenic sulfides, however, precipitate only when a free acid, or salt that is not alkaline to litmus, is present.

Aqueous Zn^{2+} precipitates ZnS incompletely from a solution of NH_3 saturated with H_2S. The same reagent diluted with an equal volume of NH_3 (to dissociate more of the H_2S and HS^-) gives a good separation.

Aqueous Cd^{II}, Sn^{II} and Pb^{II}, and H_2S, precipitate MS only partly from a solution much over 5 dM in HCl, due to lowering both the $c(M^{2+})$ from, e.g., $[MCl_4]^{2-}$ and the $c(S^{2-})$ from H_2S by the common-ion effect. With some redox in contrast, iron(III) becomes ~FeS with liberation of S; Hg_2^{2+} forms HgS and Hg; and AsO_4^{3-} may form a mixture of As_2S_5, As_2S_3 and S. In acidic solution some other M^{II} and M^{III} ions are not disturbed except by coprecipitation. E.g., if HgS is precipitated from an acidic solution containing Zn^{2+} or Cd^{2+} it always carries down some of these ions, and SnS precipitated in the presence of Co^{2+} almost invariably carries down some of that ion. Also, MnS alone, unlike ZnS, dissolves readily in CH_3CO_2H, but coprecipitated MnS and ZnS, on digestion with this acid, leave a residue of ZnS with up to 24 % MnS.

All of the oxides and hydroxides of the alkali and alkaline-earth metals), as well as the hydroxides (generally more readily than the oxides) and non-ignited oxides of Mn^{II}, Fe^{II}, Co^{II}, Ni^{II}, Cu^{I}, Cu^{II}, Ag^{I}, Zn^{II}, Cd^{II}, Hg^{II}, Sn^{II}, Sn^{IV}, Pb^{II}, Sb^{III}, Sb^{V}, Bi^{III} etc., react with moist H_2S at ambient T to form sulfides, with no oxidation or reduction.

Alkali sulfides transpose salts of the **d**- or **p**-block metals, e.g.:

$$PbBr_2 + HS^- + NH_3 \rightarrow PbS\downarrow + 2\ Br^- + NH_4^+$$

Most products are sulfides, but Cr^{III} and Al^{III} yield a hydroxide; dimercury(I) salts form HgS and Hg; Fe^{III} in acidic solution becomes Fe^{2+}, in an alkaline solution FeS_x. The fresh sulfide of any metal will precipitate any other metal ion if the latter sulfide is less soluble.

The precipitated sulfides have strongly marked colors: ZnS is white; MnS is pinkish when obtained by rapid precipitation (the more stable, green form differs in particle size and crystalline structure); CdS, SnS_2, As_2S_3 and As_2S_5, are yellow; Sb_2S_3, orange; SnS, brown; FeS, FeS_x, Cu_2S, CuS and PbS are black; and HgS forms complexes with the unaffected $[HgCl_2]$ of the sort xHgS·(1-x)$HgCl_2$ which vary in color from white through yellow, orange and brown to black as the value of x increases.

Aqueous $[Fe(CN)_5NO]^{2-}$ and a sulfide with alkali develop a transient purple $[Fe(CN)_5(NOS)]^{4-}$, a very sensitive and characteristic test.

One step in preparing a sulfide unknown for titration is to absorb the H_2S, after displacement by HCl, in ammoniacal $ZnSO_4$ or $CdCl_2$:

$$H_2S + [Zn(NH_3)_4]^{2+} \rightarrow ZnS\downarrow + 2\ NH_4^+ + 2\ NH_3$$

(After acidification the dissolved H_2S is titrated with I_2 or IO_3^-.)

Filter paper moistened with $Pb(CH_3CO_2)_2$ solution turns black with PbS when exposed to H_2S. A drop or two of NH_3 on the paper, or the use of $[Pb(OH)_4]^{2-}$ in place of $Pb(CH_3CO_2)_2$, increases greatly the sensitivity of the test. Even 0.05 nmol of H_2S has tested positive.

Other reactions of other reduced sulfur. The SCN^- ion, for detection and determination, may first be separated from $[Fe(CN)_6]^{3-}$ and $[Fe(CN)_6]^{4-}$ by precipitating them with Zn^{2+}. Any $[Fe(CN)_6]^{4-}$ may also be removed by adding an excess of Fe^{III}. Upon filtration, the red color of the thiocyanatoiron(III) appears in the filtrate or, without filtration, shaking with pentanol extracts the red complex. This color may also be used for detection (to ~ 2 μmol SCN^-). The ion may be determined by comparing the color with a standard.

Thiocyanate can be titrated with Cu^{2+} in the presence of SO_2, which can be seen as reducing the copper but not the SCN^- in this mixture:

$$2\ SCN^- + 2\ Cu^{2+} + SO_2 + 6\ H_2O \rightarrow 2\ CuSCN\downarrow + SO_4^{2-} + 4\ H_3O^+$$

Thiocyanates may be determined also gravimetrically by precipitating AgSCN, drying and weighing as such. Volumetrically, one may titrate the SCN^- with Ag^+ using an excess, which is determined by back titration with standardized SCN^- using Fe^{III} as the indicator.

Thiourea, $CS(NH_2)_2$ or Tu, binds through S strongly to soft ions (Cu^I, Ag^I, Au^I and Hg^{II}) but also even to Ln^{III} in $Ln(Tu\text{-}\kappa S)(ClO_4)_3 \cdot 10H_2O$. It reduces Cu^{II}, Au^{III}, Pt^{IV} and Te^{IV} to Tu-*κS* complexes of Cu^I, Au^I, Pt^{II} and Te^{II}, but boiling does precipitate simple sulfides.

Because the *trans*-effect decreases as Tu > Cl^- > NH_3, thiourea can distinguish [*cis*- and [*trans*-Pt(halide)$_2$(NH_3)$_2$]. Details are omitted here, but one can reason that thiourea must convert [*cis*-$PtCl_2(NH_3)_2$] finally to $[PtTu_4]^{2+}$, but the other isomer to [*trans*-$PtTu_2(NH_3)_2]^{2+}$.

A few more of the products already isolated are: [$MoTu_3Cl_3$]; some polymeric complexes $MTu_2(NCS)_2$, where M is Mn, Co, Ni or Cd; [*trans*-$NiTu_4Cl_2$]; [$PdTu_4$]Cl_2; [$ZnTu_2(NCS)_2$]; [$CdTu_2Cl_2$]; the polymeric complex $PbTu_2Cl_2$; and [$TeTu_4$]Cl_2.

Other reactions of sulfur and thiosulfates. Sulfur, boiled with the hydroxides of Mn, Co, Ni, Cu^I, Cu^{II}, Ag, Cd, Hg^I, Hg^{II}, Pb or Bi, but not Fe, Zn or Sn, dismutates to a metal sulfide and $S_2O_3^{2-}$, also some SO_4^{2-}.

Aqueous Ba^{2+} forms, with $S_2O_3^{2-}$, white barium thiosulfate, BaS_2O_3, slightly soluble in H_2O. The corresponding strontium salt is formed only from a fairly concentrated solution, and the calcium salt is readily soluble in H_2O (distinction from SO_3^{2-}). The Mn^{2+} and Zn^{2+} ions give no precipitate with $S_2O_3^{2-}$ (distinction from S^{2-}).

A solution of $RuCl_3$, made slightly alkaline with NH_3, turns deep red on boiling after adding $S_2O_3^{2-}$. The test will readily reveal 2-mM $S_2O_3^{2-}$.

Solutions of Cu^{2+} on long standing with $S_2O_3^{2-}$ form $Cu_2S_2O_3$ which is changed by boiling to Cu_2S and H_2SO_4 (separation from Cd):

$$2\ Cu^{2+} + 3\ S_2O_3^{2-} \rightarrow Cu_2S_2O_3\downarrow + [S_4O_6]^{2-}$$

$$Cu_2S_2O_3 + 3\ H_2O \rightarrow Cu_2S\downarrow + SO_4^{2-} + 2\ H_3O^+$$

If Ag^+ is added to $S_2O_3^{2-}$, a precipitate is obtained which is first white, then yellow, brown, and finally black (faster with heat). The white $Ag_2S_2O_3$ ultimately becomes Ag_2S, like the Cu_2S above, although one way of determining thiosulfate is to add excess Ag^+ to the $S_2O_3^{2-}$ and titrate the excess with CNS^-.

Excess $S_2O_3^{2-}$, however, first complexes and dissolves the thiosulfate precipitates of Ag and Pb; also those of $CaSO_4$, AgCl, AgBr, AgI, $[Hg_2Cl_2]$ (which precipitates HgS when warmed), $PbSO_4$, PbI_2 etc., e.g.:

$$Ag_2S_2O_3 + S_2O_3^{2-} \rightarrow 2\ Ag(S_2O_3\text{-}\kappa S)^-$$

$$AgCl + 2\ S_2O_3^{2-} \rightarrow [Ag(S_2O_3\text{-}\kappa S)_2]^{3-} + Cl^-$$

$$PbSO_4 + 3\ S_2O_3^{2-} \rightarrow [Pb(S_2O_3\text{-}\kappa S)_3]^{4-} + SO_4^{2-}$$

The precipitated white Hg^{II} salt decomposes almost instantly, but the Pb^{II} salt decomposes differently on standing and rapidly if warmed:

$$HgS_2O_3 + 3\ H_2O \rightarrow HgS\downarrow + SO_4^{2-} + 2\ H_3O^+$$

$$2\ PbS_2O_3 \rightarrow PbS\downarrow + Pb[S_3O_6]\downarrow$$

An interesting method for the preparation of sulfanedisulfonates involves heating a silver or dimercury(I) double thiosulfate:

$$2\ NaAgS_2O_3 \rightarrow Na_2[S_3O_6] + Ag_2S\downarrow$$

Likewise, when slightly alkaline $S_2O_3^{2-}$ is warmed with $[Pb(OH)_4]^{2-}$, the sulfur dismutates and gives $[S_3O_6]^{2-}$ and PbS.

Hot, acidic solutions of As or Sb precipitate As_2S_3 or Sb_2S_3 with $S_2O_3^{2-}$, a distinction from Sn, which is not precipitated.

Other reactions of polythionates. These do not precipitate Ba^{2+}.

A solution of $[S_3O_6]^{2-}$ treated with Cu^{2+} ultimately forms CuS; with Ag^+ a white precipitate is first formed, which gradually turns black.

The $[S_5O_6]^{2-}$ ion, but not $[S_4O_6]^{2-}$, gives a precipitate with $[Ag(NH_3)_2]^+$; this is used to detect $[S_5O_6]^{2-}$. At first a faint brown color appears, which darkens as the quantity of precipitate increases. Sulfites interfere.

Aqueous Hg_2^{2+} and $[S_3O_6]^{2-}$ give a black precipitate, but $[S_4O_6]^{2-}$ and $[S_5O_6]^{2-}$ both produce a yellow one.

An excess of $[HgCl_2]$ added to $[S_3O_6]^{2-}$ or $[S_4O_6]^{2-}$ gives the white $HgCl_2{\cdot}2HgS$. Tin(II) chloride also produces a white precipitate.

Ammoniacal mercury(II) cyanide forms HgS with $[S_5O_6]^{2-}$; a solution of Sn^{II} gives a dark brown precipitate.

Other reactions of sulfites. Sulfites of the common metals may be made from (1) SO_2 and the oxides, hydroxides or carbonates of the metals; (2) SO_2 and the M^{n+} of those metals whose sulfides (not sulfites) are insoluble in 3-dM HCl, except

Cu^{2+} and Cd^{2+}. Other sulfites are more acid soluble; and (3) SO_3^{2-} and the M^{n+} of the non-alkali metals except Cr^{III}. Any excess of SO_3^{2-} tends to form soluble complexes. Many of the compounds, often white, are soluble even in acetic acid. The solids are normal (i.e., containing SO_3^{2-}), except for Hg^{I}, which is acidic (i.e., HSO_3^-), and Cr, Al and Cu, which are basic (i.e., hydroxide sulfites).

Saturating cold Alk_2SO_3 solutions with SO_2 yields $AlkHSO_3$. Adding enough ethanol may isolate the solid. Disulfites also arise:

$$2\ M^{I}HSO_3 \rightarrow M^{I}_2[S_2O_5] + H_2O$$

The $CaSO_3$ from (1) or (3) above is less soluble than $CaSO_4$ in water. The $BaSO_3$ dissolves easily in dilute HCl (distinction from $BaSO_4$). Adding an oxidant to the solution precipitates $BaSO_4$ (a test for sulfite).

The sulfites of Ca, Sr, Ba, Hg and Pb are usually accompanied by sulfates because solutions of SO_3^{2-} nearly always contain SO_4^{2-}.

The Ag_2SO_3 turns brown when boiled. The $PbSO_3$, also made from $Pb(CH_3CO_2)_2$, does not blacken when boiled (unlike thiosulfate).

Sulfite forms various complexes such as $[Os(SO_3)_3(H_2O)_3]^{4-}$, $[Os(SO_3)_3(NH_3)_3]^{4-}$, $[OsO_2(SO_3)_4]^{6-}$, $[Co(SO_3)(NH_3)_5]^{+}$, $[Rh(SO_3)_3]^{3-}$, and [*trans*-$IrCl_4(SO_3)_2]^{5-}$, plus $[IrCl_2(SO_3)_4]^{7-}$, $[Ir(SO_3)_3(NH_3)_3]^{3-}$, $[Pd(SO_3)_4]^{6-}$, $[Pt(SO_3)_4]^{6-}$ and $[Hg(SO_3)_2]^{2-}$, generally with SO_3-κS, i.e., M–S bonding.

In the spectrochemical series (SO_3-κS) is similar to NH_3. The *trans*-effect is strong for SO_3^{2-}-κS, $S_2O_3^{2-}$-κS and $CS(NH_2)_2$-κS.

In the following we see both oxidation and reduction of S^{IV}.

Iron(II or III), also Cu^{I} or Cu^{II}, react with SO_2 with reduction in dilute acid, but oxidation in concentrated acid, perhaps thus:

$$2\ Cu^{2+} + SO_2 + 6\ H_2O \rightarrow 2\ Cu^{+} + SO_4^{2-} + 4\ H_3O^{+}$$

$$4\ Cu^{+} + SO_2 + 4\ H_3O^{+} \rightarrow 4\ Cu^{2+} + S\downarrow + 6\ H_2O$$

Rather similarly, Hg^{I}_2 and SO_2 in low $c(H_3O^+)$ precipitate gray Hg; with an acidity above 2 M, oxidation to Hg^{II} occurs:

$$2\ [Hg_2Cl_2] + SO_2 + 4\ H_3O^{+} + 12\ Cl^{-} \rightarrow 4\ [HgCl_4]^{2-} + S\downarrow + 6\ H_2O$$

A solution of $[HgCl_2]$ produces no change in the cold, but on boiling, the white $[Hg_2Cl_2]$ is precipitated with the formation of H_2SO_4. Further digestion produces Hg. These changes depend, however, on the acidity of the system as indicated above:

$$2\ [HgCl_2] + SO_2 + 5\ H_2O \rightarrow [Hg_2Cl_2]\downarrow + HSO_4^{-} + 2\ Cl^{-} + 3\ H_3O^{+}$$

Other reactions of sulfates. Sulfuric acid or SO_4^{2-}, with Ca^{2+} not too dilute, precipitate white calcium sulfate. Aqueous Ba^{2+} gives barium sulfate, white, insoluble in HCl or HNO_3, unlike all other salts except $BaSeO_4$ and $Ba[SiF_6]$. The precipitate formed in the cold is finely divided and hard to filter; better from a hot, acidic solution and allowed to digest for a short time. In a dilute solution, time should be allowed for complete precipitation. Aqueous Sr^{2+} has an intermediate reaction. Solutions of Pb^{2+} give a white precipitate of $PbSO_4$, not transposed by acids except H_2S, soluble in OH^-. The presence of ethanol or the absence of other acids makes the precipitation quantitative.

Hot, concentrated H_2SO_4 decomposes all ions of $Mn^{>II}$ to Mn^{II}, releasing O_2.

Sulfuric acid transposes the salts of nearly all other acids, forming sulfates. The displaced acids, or their decomposition products, are often more volatile and may be expelled by evaporation. The chlorides of Ag, Sn and Sb are transposed with difficulty, and the chlorides of Hg are not affected. The insoluble sulfates are best made by precipitation.

Some normal sulfates tend to form double salts or alums with the general formula $M^IM^{III}(SO_4)_2\cdot 12H_2O$, i.e., $[M^I(H_2O)_6][M^{III}(H_2O)_6](SO_4)_2$, as in $KAl(SO_4)_2\cdot 12H_2O$, $NaCr(SO_4)_2\cdot 12H_2O$ or $NH_4Fe(SO_4)_2\cdot 12H_2O$. Others give schönites or "Tutton salts", $(M^I)_2M^{II}(SO_4)_2\cdot 6H_2O$, with Alk^+ or NH_4^+, and Mg^{2+} or $\mathbf{3d}^{2+}$, but with the terms used inconsistently.

A test for insoluble sulfates involves boiling the unknown with 3-dM $Hg(NO_3)_2$ in 16-cM HNO_3. The formation of very small, yellow crystals of $Hg_3O_2SO_4$ or $HgSO_4\cdot 2HgO$, "turpeth mineral", indicates the presence of a sulfate. The action takes place in the cold with $CaSO_4$ or Hg_2SO_4, less readily with $SrSO_4$ or $PbSO_4$, and slowly with $BaSO_4$.

Peroxosulfuric acid, $H_2SO_3(O_2)$, is obtained by electrolyzing cold, concentrated H_2SO_4 (thus oxidizing the oxygen but not the sulfur).

Various peroxodisulfates, $M^I_2[S_2O_8]$, have been prepared, principally by electrolysis of a cold, concentrated solution of the sulfate or hydrogensulfate; e.g., $(NH_4)_2[S_2O_8]$ crystallizes out when a cold solution of $(NH_4)_2SO_4$ is used; or $Na_2[S_2O_8]$ is obtained from $NaHSO_4$. The salts are white, stable and moderately soluble in water; this includes $Ba[S_2O_8]$ (distinction from $BaSO_4$).

A solution of $[S_2O_8]^{2-}$ attacks many metals, forming either the ion (Mg, Fe, Co, Ni, Zn, Cd, etc.) or an oxide (Mn, Cu, Ag, Hg, Pb, Bi, etc.):

$$Zn + [S_2O_8]^{2-} \rightarrow Zn^{2+} + 2\ SO_4^{2-}$$

$$2\ Ag^+ + [S_2O_8]^{2-} + 4\ H_2O \rightarrow Ag^IAg^{III}O_2\downarrow + 2\ HSO_4^- + 2\ H_3O^+$$

Aqueous $[S_2O_8]^{2-}$ also, again with reduction of the O but not the S, oxidizes Ce^{III} to Ce^{IV}, Ti^{III} to Ti^{IV} (quantitatively), Cr^{III} to Cr^{VI} (Ag^+ promotes completion in acidic solution, as with other $[S_2O_8]^{2-}$ reactions), Mn^{2+} to $MnO_2\cdot aq$, Fe^{2+} to Fe^{III}, Co^{2+} to Co^{III}, Ni^{II} to Ni^{IV} (in OH^-), etc. Reductants suitable to determine $[S_2O_8]^{2-}$, with excess reductant, when followed by back titration, include Ti^{III} and Fe^{2+}.

The $[Fe(CN)_6]^{4-}$ and $[Fe(CN)_6]^{3-}$ ions react with $[S_2O_8]^{2-}$ in the presence of H_3O^+ to form HCN and a small amount of NH_4^+.

The SO_4^{2-} ion (treated for brevity as the reactant here) can leach Ln, U, Cr, Mn, Fe, Co, Ni, Cu, Zn and Cd from some ores.

16.3 Selenium, $_{34}$Se

Oxidation numbers: (–II), (IV) and (VI), as in H_2Se, H_2SeO_3 and SeO_4^{2-}, and other, as in Se_n^{2-}.

16.3.1 Reagents Derived from Hydrogen and Oxygen

Water. Water hydrolyzes Al_2Se_3 (from a non-aqueous source) to H_2Se, with some H_2, and $Al_2O_3\cdot aq$. The solubility of the gas at 25 °C is ~1 dM.

Selenium dioxide, SeO_2, is readily soluble in H_2O, forming H_2SeO_3. The solid, deliquescent acid (but also easily dehydrated in dry air) may be crystallized by evaporation at a low pressure over H_2SO_4. Solid selenic acid, H_2SeO_4, is also readily soluble and very hygroscopic.

The halides, e.g., $SeCl_4$, hydrolyze readily in H_2O.

Oxonium. Selenide ion, or, e.g., MgSe or FeSe, plus HCl or H_2SO_4 give H_2Se, a colorless gas smelling like H_2S but more penetrating, poisonous and soluble than H_2S, also acidic in water.

With 2.5-cM to 2.5-dM Se^{IV} at pH 3 to 7, the oxo form is mainly $HSeO_3^-$ with a little $H_nSe_2O_6^{(4-n)-}$, but is SeO_3^{2-} at pH > 8, and is H_2SeO_3, not SeO_2 like SO_2 or CO_2, in strong acids.

Hydroxide. Concentrated OH^- dissolves Se as Se^{2-}, brown Se_x^{2-} (slowly decolorized by O_2 and CO_2), SeO_3^{2-} and perhaps more.

Peroxide. Acidified H_2O_2 oxidizes SeO_2 or H_2SeO_3 to H_2SeO_4:

$$SeO_2 + H_2O_2 + H_2O \rightarrow HSeO_4^- + H_3O^+$$

Di- and trioxygen. Air quickly oxidizes moist H_2Se to red selenium flakes, and Se slowly to SeO_2. Ozone oxidizes H_2SeO_3 to H_2SeO_4.

16.3.2 Reagents Derived from the Other 2nd-Period Non-Metals, Boron through Fluorine

Cyanide species. Selenium dissolves in cyanide ion, forming $SeCN^-$. At pH ≤ 5 this reverts fully to red selenium.

Reduced nitrogen. Diazanium (hydrazinium, $N_2H_5^+$) reduces H_2SeO_3 to Se in warm H_2SO_4 solution. A red color or precipitate indicates the presence of Se, detectable at 0.07-mM.

The NH_3OH^+ ion also reduces H_2SeO_3 to red Se, changing to gray Se on warming; Se and Te can be separated by precipitating the Se from 5-M HCl solution with NH_3OH^+, leaving the Te dissolved.

Oxidized nitrogen. Selenium and selenides are dissolved as H_2SeO_3 not, like S_8, to oxidation state (VI), by hot, dilute or concentrated HNO_3.

The very toxic and malodorous gaseous H_2Se from some reactions can be precipitated as red Se_8 by bubbling it through dilute HNO_3.

16.3.3 Reagents Derived from the 3rd-to-5th-Period Non-Metals, Silicon through Xenon

Phosphorus species. Aqueous H_2SeO_3 and HPH_2O_2/H_2PHO_3 give red Se.

Reduced chalcogens. Aqueous H_2SeO_3 and H_2S precipitate ~SeS_2 (in eight-membered rings), lemon yellow, becoming bright red on heating, soluble in "$(NH_4)_2S$":

$$4\ H_2SeO_3 + 8\ H_2S \rightarrow Se_3S_5\downarrow + {}^1/_2\ Se_2S_6\downarrow \text{ etc.} + 12\ H_2O$$

Oxidized chalcogens. Slightly acidified $S_2O_3^{2-}$ and SeO_2 give $[SeS_4O_6]^{2-}$ at 0 °C, but $S_2O_3^{2-}$ catalyzes decomposition, requiring excess SeO_2:

$$SeO_2 + 4\ S_2O_3^{2-} + 4\ CH_3CO_2H \rightarrow$$

$$[SeS_4O_6]^{2-} + [S_4O_6]^{2-} + 4\ CH_3CO_2^- + 2\ H_2O$$

This breaks down to Se and $[S_4O_6]^{2-}$, faster with alkalis. Ethanol and ether crystallize a pale yellow-green $Na_2[SeS_4O_6]\cdot 3H_2O$.

Soluble sulfites and H_2Se precipitate selenium and sulfur together.

Sulfur dioxide reduces H_2SeO_3 to red Se, which becomes gray and brittle in a few hours. Selenium is precipitated even from concentrated HCl, separating it easily from Te, which reacts only in dilute acid.

To distinguish SeO_3^{2-} from SeO_4^{2-}, SO_2 may be used as the reductant. After the reduction of SeO_3^{2-}, any SeO_4^{2-} may be detected by boiling with HCl to reduce it also.

Selenium appears to be peptized in cold concentrated H_2SO_4 without oxidation, giving a green colored liquid (dilution with H_2O precipitates the Se). If the liquid is warmed, SO_2 is released and the color disappears, with the Se oxidized to SeO_2. Colorless crystals are obtained that sublime readily, have a disagreeable odor, and are hygroscopic.

Reduced halogens. Concentrated HCl dissolves SeO_2 or $SeCl_4$ as $[SeCl_6]^{2-}$, which forms $Alk_2[SeCl_6]$ and $(NH_4)_2[SeCl_6]$; the $Cs_2[SeCl_6]$ has the lowest solubility. The $Tl_2[SeCl_6]$ salt is similar. Corresponding reactions yield $[SeBr_6]^{2-}$ and $[SeI_6]^{2-}$.

Six to 12-M HCl reduces H_2SeO_4 reversibly to H_2SeO_3 on evaporating, with limited heating to avoid vaporizing $SeCl_4$:

$$HSeO_4^- + 2\ Cl^- + 3\ H_3O^+ \leftrightarrows H_2SeO_3 + Cl_2\uparrow + 4\ H_2O\uparrow$$

Barium selenate, $BaSeO_4$, dissolves therefore in HCl (distinction and separation from $BaSO_4$) because boiling reduces it to $BaSeO_3$.

Selenium may be separated from most elements as $SeBr_4$ (along with $GeBr_4$ and $AsBr_3$) by distillation from concentrated (9-M) HBr, containing a little Br_2 to prevent some interferences:

$$HSeO_4^- + 6\ Br^- + 7\ H_3O^+ \rightarrow SeBr_4\uparrow + Br_2\uparrow + 11\ H_2O$$

Iodide reduces both H_2SeO_3 and $HSeO_4^-$ to Se.

Elemental and oxidized halogens. Chlorine oxidizes H_2SeO_3 to H_2SeO_4.

Chlorate oxidizes H_2SeO_3 in acid in a convenient preparation of H_2SeO_4. It may be purified by adding Ba^{2+}, separating the precipitated $BaSeO_4$ and removing the Ba^{2+} with H_2SO_4 as the less soluble $BaSO_4$.

16.3.4 Reagents Derived from the Metals Lithium through Uranium, plus Electrons and Photons

Oxidation. Acidified permanganate oxidizes H_2SeO_3 to H_2SeO_4.

Anodic oxidation changes H_2SeO_3 to H_2SeO_4 in acid.

Reduction. Iron(2+) and tin(II) reduce H_2SeO_3 to red Se.

Other reactions. Selane, H_2Se, precipitates **d**- or **p**-block selenides from complexes of soft metallic ions, having almost the same solubilities as the corresponding sulfides. Many pure selenides of metals may be made by dropping salt solutions slowly into saturated aqueous H_2Se.

The selenites and selenates of the alkaline-earth metals are insoluble and may be precipitated by adding the metal ion to a neutral solution of an alkali selenite or selenate. Many of the selenites are soluble in excess of the corresponding acid. Selenates are less stable than selenites.

Aqueous $SeCN^-$ precipitates Cu^+, Ag^+, Hg_2^{2+}, Hg^{2+}, Tl^+ and Pb^{2+}. Brown $Cu(SeCN)_2$ comes from Cu^{2+} but quickly goes to black Cu_2Se and CuSe. Even in NH_3, AgSeCN is formed, but it darkens rapidly in sunlight.

16.4 Tellurium, $_{52}$Te and Polonium, $_{84}$Po (and Ununhexium, $_{116}$Uuh)

Oxidation numbers: (−II), fractional, (II), (IV) and (VI), as in H_2Te, H_2Po (also written well as $Po^{II}H_2$), Te_4^{2+}, Po^{2+}, H_2TeO_3, PoO_3^{2-} or Po^{4+}, H_6TeO_6 and PoO_3. The electronegativities, χ, of hydrogen and tellurium on various scales, however, are about the same, so that we could regard tellane as being $H^0{}_2Te^0$, but analogy with H_2Se makes $H^I{}_2Te^{-II}$ a convenient choice. In some ways Po resembles Pb more than it resembles Te. For Uuh, the next member of this Group, relativistic quantum mechanics predicts the existence of Uuh^{2+} and $UuhF_4$, and less stability for aqueous Uuh^{4+}.

16.4.1 Reagents Derived from Hydrogen and Oxygen

Water. Cold H_2O yields white H_2TeO_3, both from Te dissolved in HNO_3 and from $TeCl_4$; hot H_2O and warming H_2TeO_3 to 40 °C form TeO_2, dissolving very little as H_2TeO_3 in water. This is slightly amphoteric, least soluble at pH 4, otherwise giving, e.g., $TeO(OH)(NO_3,ClO_4)$ in concentrated acids (reprecipitating if diluted), or $HTeO_3^-$, TeO_3^{2-}, $Te_2O_5^{2-}$ etc. Tellurium trioxide is insoluble, but H_6TeO_6 dissolves easily.

Oxonium. Non-oxidizing acids with Te^{2-} give H_2Te, a colorless, very poisonous gas, smelling rather like H_2S, fairly soluble in H_2O.

Concentrated strong acids dissolve TeO_2 and H_2TeO_3.

Hydroxide. Concentrated OH^- and Te form a blood-red solution, which deposits Te on dilution. Fresh Po^{2+} and OH^- form dark-brown $Po(OH)_2$. The alkali hydroxides dissolve TeO_2 and H_2TeO_3. Hot concentrated OH^- dissolves TeO_3 and forms $H_4TeO_6^{2-}$ or $TeO_2(OH)_4^{2-}$, which rather easily condenses to $Te_2O_6(OH)_4^{4-}$. Dilute OH^- and Po^{IV} give voluminous, pale yellow, amphoteric $PoO_2 \cdot aq$, yielding K_2PoO_3, $Po(NO_3)_4$ or $Po(SO_4)_2$ with the corresponding reagents.

Peroxide. Tellurium dissolves slowly in H_2O_2 plus OH^- at 100 °C, forming a tellurate(VI); and H_2TeO_3 reacts with H_2O_2 giving H_6TeO_6:

$$H_2TeO_3 + H_2O_2 + H_2O \rightarrow H_6TeO_6$$

Di- and trioxygen etc. The air oxidizes H_2Te and Te^{2-} to gray metallic Te. It slowly oxidizes Te and Po to MO_2. Ozone and Te form H_6TeO_6.

Oxidants from Po's α particles and H_2O slowly change Po^{2+} to Po^{IV}.

16.4.2 Reagents Derived from the Other 2nd-Period Non-Metals, Boron through Fluorine

Cyanide species. Tellurium is soluble in warm concentrated cyanide, from which HCl precipitates Te.

Some "simple" organic species. Chelating aqueous H_6TeO_6 with an equal volume of glycerol, $C_3H_5(OH)_3$, enables titrating it as a monobasic acid with OH^-, using phenolphthalein as indicator.

Reduced nitrogen. Polonium(IV) and NH_3 give non-acidic $Po(OH)_4$.

In dilute acids NH_3OH^+ reduces Te compounds to Te. Selenium and tellurium can be separated, however, by precipitating the Se from 5-M HCl with NH_3OH^+, leaving the Te dissolved. Metallic Po is formed from $PoO_2 \cdot aq$ suspended in dilute OH^-, plus N_2H_4 or NH_2OH.

Oxidized nitrogen. At high *T*, HNO_3 and Te yield TeO_2 but not Te^{VI}. Nitric acid does not oxidize TeO_2, but HNO_3 or aqua regia oxidizes Te to a basic nitrate, H_2TeO_3 or H_6TeO_6. Aqueous HNO_3 dissolves the more metallic Po, first as rose Po^{2+}, unlike Te. Treating Po^{IV} hydroxide or chloride with 5-dM HNO_3 for 12 h and drying gives a basic nitrate; more HNO_3 yields $Po(NO_3)_4$.

Tellurium trioxide is insoluble in HNO_3; H_6TeO_6 is slightly soluble.

Fluorine species. The solubility of Po^{IV} hydroxide in HF rises with $c(HF)$, and there may be fluoro complexes.

16.4.3 Reagents Derived from the 3rd-to-5th-Period Non-Metals, Silicon through Xenon

Phosphorus species. Aqueous 2-M H_3PO_4 and $PoO_2 \cdot aq$, or a phosphate and $PoCl_4$, precipitate a white, gelatinous, basic Po^{IV} phosphate, with perhaps no Te analog, stable to NH_3 but not to OH^- or H_3O^+.

Reduced chalcogens. Dissolved Te and H_2S precipitate Te_xS_y or Te mixed with S. This resembles dark-brown SnS and dissolves readily in "$(NH_4)_2S$". Aqueous Po^{2+} and H_2S form black PoS, $pK_{sp} \sim 28.2$.

Oxidized chalcogens. Aqueous HCl or CH_3CO_2H plus $S_2O_3^{2-}$ dissolve TeO_2 as $[TeS_4O_6]^{2-}$ at 0 °C, but $S_2O_3^{2-}$ also catalyzes decomposition, requiring excess TeO_2:

$$TeO_2 + 4\ S_2O_3^{2-} + 4\ H_3O^+ \rightarrow [TeS_4O_6]^{2-} + [S_4O_6]^{2-} + 6\ H_2O$$

This breaks down to Te and $[S_4O_6]^{2-}$, faster with alkalis. Ethanol crystallizes yellowish $Na_2[TeS_4O_6]\cdot 2H_2O$. One can titrate with I_2:

$$[TeS_4O_6]^{2-} + 2\ I_2 + 6\ H_2O \rightarrow TeO_2\downarrow + [S_4O_6]^{2-} + 4\ I^- + 4\ H_3O^+$$

Metallic Po precipitates from $PoO_2\cdot aq$ in dilute OH^- plus $[S_2O_4]^{2-}$.

Reducing Te compounds to Te in slightly acidic solution by SO_2 easily separates Te from many other elements, especially Se, which is precipitated from concentrated acidic solution, Te only from dilute acids, or from more nearly neutral HSO_3^-.

Over some hours or days, H_2SO_4, 9 to 18 M, from 135 °C up to the bp, converts Te, TeO_2, TeO_3, H_6TeO_6 or $TeCl_4$ to TeO_3, soluble $TeO_3\cdot{}^3/_2H_2O$ or white $Te_2O_3SO_4$ (which, with H_2O, gives TeO_2 and H_2SO_4), releasing SO_2 when the Te is oxidized. However, 18-M H_2SO_4 and Te also form red Te_4^{2+} in a qualitative test.

Aqueous H_2SO_4 dissolves Po as Po^{2+}. Dilute H_2SO_4 or H_2SeO_4 form white, basic or normal Po^{IV} sulfates or selenates from $PoO_2\cdot aq$ or $PoCl_4$.

Reduced halogens. Aqueous HCl dissolves Po, not Te, as Po^{2+} and H_2.

Concentrated HCl dissolves TeO_2 as $[TeCl_6]^{2-}$. This forms $Alk_2[TeCl_6]$ (with K^+, Rb^+ or Cs^+) or yellow $(NH_4)_2[TeCl_6]$, which reverts to TeO_2 in H_2O or moist air. The $Cs_2[TeCl_6]$ is the least soluble; $Tl_2[TeCl_6]$ is similar. Solid TeO_3 is insoluble in cold HCl. Tellurates boiled with HCl release Cl_2 and are reduced to Te^{IV}, which precipitates as TeO_2 on adding H_2O (distinction from Se). Likewise concentrated HBr dissolves TeO_2 as $[TeBr_6]^{2-}$, which by adding, e.g., K^+ and evaporating can be crystallized as orange $K_2[TeBr_6]$. In water this reverts to TeO_2, but these TeX_6^{2-} salts are all stable in dilute acids. Tellurite ion, TeO_3^{2-}, treated with I^- in dilute acid, precipitates black TeI_4, which dissolves in excess reagent as red TeI_6^{2-} (distinction from Se).

Aqueous HCl dissolves PoO_2 as pink $PoCl_2$ and Cl_2, but adding Cl_2 can form yellow $PoCl_4$, and $[PoCl_6]^{2-}$ can be made. Dark-red $PoBr_4$ and NH_4Br yield $(NH_4)_2[PoBr_6]$ or, with heat, purple $PoBr_2$.

Elemental and oxidized halogens. Polonium and Cl_2 or Br_2 give PoX_2 or PoX_4, hydrolyzable, reducible by H_2S or SO_2 to Po^{2+}.

Tellurium, H_2TeO_3 and $TeCl_4$ react with ClO_3^- in acid to form H_6TeO_6. Following with concentrated HNO_3 precipitates the product.

16.4.4 Reagents Derived from the Metals Lithium through Uranium, plus Electrons and Photons

Oxidation. Tellurium and H_2TeO_3 react with $[Cr_2O_7]^{2-}$ in acid giving H_6TeO_6. Volumetrically, Te may be determined by oxidation with excess $[Cr_2O_7]^{2-}$ and titration of the excess with Fe^{2+}.

Boiling Te, TeO_2 or H_2TeO_3 with MnO_4^- and HNO_3 produces H_6TeO_6 rather insoluble in HNO_3, along with some $MnO_2 \cdot aq$ (dissolved later by H_2O_2); only the alkali and NH_4 salts are soluble, e.g., $Na_2[TeO_2(OH)_4]$ and polymeric $(NH_4)_2TeO_4$:

$$5\,TeO_2 + 2\,MnO_4^- + 6\,H_3O^+ + 16\,H_2O \rightarrow$$

$$5\,H_6TeO_6 \cdot 2H_2O\downarrow + 2\,Mn^{2+}$$

Reduction. A sensitive test for Te involves reduction with $TiCl_3$. The released H_2Te is decomposed to form a mirror as in the Marsh test for As. However, Po^0 is precipitated from acids by $TiCl_3$ or $[SnCl_3]^-$. Both Zn and $[SnCl_3]^-$ in acids precipitate black Te from its compounds.

Other reactions. Tellane, H_2Te, rather like H_2S, precipitates **d**- or **p**-block tellurides from complexes of their soft metallic cations. Bases and Zn^{2+}, Cd^{2+} and Hg^{2+} also form Alk_2Po, AePo and MPo from PoH_2.

Solutions of the alkali tellurites and tellurates form precipitates with ions of the metals, e.g.,: $BaTeO_4$, not isostructural with $BaSO_4$. In fact, telluric acid can be made by treating polymeric Na_2TeO_4 with Pb^{2+} or Ba^{2+} and transposing the salt with H_2SO_4. The H_6TeO_6 separates as colorless crystals on evaporating and then adding ethanol.

Bibliography

See the general references in the Introduction, and some more-specialized books [1–25]. Some articles in journals discuss: synthetic catalysts for the photo-oxidation of water to oxygen [26]; kinetics and mechanisms of metal-O_2 complexes [27]; **d**-block-catalyzed oxidation of sulfur(IV), especially in the atmosphere [28]; various metal-O_2 complexes [29]; mainly mono- and poly-selenide and telluride complexes with **d**-block elements [30]; heterogeneous redox catalysts for oxygen release [31]; the kinetics and equilibria of oxygen species and metal ions [32]; thio- and seleno-complexes of V, Nb, Ta, Mo, W and Re [33] and homonuclear sulfur species, not emphasizing aqueous reactions [34].

1. Steudel R (2003) Elemental sulfur and sulfur-rich compounds, 2 vols. Springer, Berlin Heidelberg New York
2. Jones CW (1999) Applications of hydrogen peroxide and derivatives. Royal Society of Chemistry, Cambridge
3. Foote CS, Valentine JS, Greenberg A, Liebmen JF (1995) Active oxygen in chemistry. Blackie, London
4. Sawyer DT (1991) Oxygen chemistry. Oxford University Press, Oxford
5. Martell AE, Sawyer DT (eds) (1988) Oxygen complexes and oxygen activation by transition metals. Plenum, New York
6. Horváth M, Bilitzky L, Hüttner J (1985) Ozone. Elsevier, Amsterdam
7. Müller A, Krebs B (eds) (1984) Sulfur: its significance for chemistry, for the geo-, bio-, and cosmosphere and technology. Elsevier, Amsterdam

8. Senning A (ed) (1971–1982) Sulfur in organic and inorganic chemistry, 4 vols. Dekker, New York
9. Rodgers MAJ, Powers EL (eds) (1981) Oxygen and oxy radicals. Academic, San Diego
10. Heal HG (1980) The inorganic heterocyclic chemistry of sulfur, nitrogen and phosphorus. Academic, London
11. Giguère PA (1975) Compléments au nouveau traité de chimie minéral: peroxyde d'hydrogène et polyoxydes d'hydrogène. Masson, Paris
12. Foss O (1975) IUPAC additional publication [on polythionic acids] (24th International Congress, Hamburg, 1973). Butterworth, London
13. Newman AA (ed) (1975) The chemistry and biochemistry of thiocyanic acid and its derivatives. Academic, London
14. Zingaro RA, Cooper WC (1974) Selenium. Van Nostrand Reinhold, New York
15. Karchmer JH (1970, 1972) The analytical chemistry of sulfur and its compounds, 2 parts. Wiley, New York
16. Cooper WC (ed) (1971) Tellurium. Van Nostrand Reinhold, New York
17. Nickless G (ed) (1968) Inorganic sulfur chemistry. Elsevier, Amsterdam
18. Bagnall KW (1966) The chemistry of selenium, tellurium and polonium. Elsevier, Amsterdam
19. Chizhikov DM, Shchastlivyi VP (1966) Elkin EM (trans) (1970) Tellurium and the tellurides. Collet's, London
20. Schroeter LC (1966) Sulfur dioxide. Pergamon, Oxford
21. Ardon M (1965) Oxygen: elementary forms and hydrogen peroxide. Benjamin Cummings, San Francisco
22. Chizhikov DM, Shchastlivyi VP (1965) Elkin EM (trans) (1968) Selenium and selenides. Collet's, London
23. Vol'nov, II (1964) Woroncow J (trans) Petrocelli AW (ed) (1966) Peroxides, superoxides and ozonides of alkali and alkaline earth metals. Plenum, New York
24. Edwards JO (ed) (1962) Peroxide reaction mechanisms. Wiley, New York
25. Machu W (1951) Das Wasserstoffperoxyd und die Perverbindungen. Springer, Berlin Heidelberg New York
26. Rüttinger W, Dismukes GC (1997) Chem Rev 97:1
27. Bakac A (1995) Prog Inorg Chem 43:267
28. Brandt C, Eldik Rv (1995) Chem Rev 95:119
29. Klotz IM, Kurtz DMJr (eds) (1994) Chem Rev 94:567
30. Roof LC, Kolis JW (1993) Chem Rev 93:1037
31. Mills A (1989) Chem Soc Rev 18:285
32. Taube H (1986) in Lippard SJ (ed) Prog Inorg Chem 34:607
33. Müller A et al (1981) Angew Chem Int Ed 20:934
34. Chivers T, Drummond I (1973) Chem Soc Rev 2:233

17 Fluorine through Astatine, the Halogens

17.1 Fluorine, $_9F$

Oxidation number: (–I) as in HF and even HFO, $H^IF^{-I}O^0$.

17.1.1 Reagents Derived from Hydrogen and Oxygen

Water. Hydrogen fluoride, HF, a colorless, intensely corrosive gas, even for glass, is readily soluble in water to form a weak acid (distinction from the other hydrogen halides), ionized in 1-dM solution to about 10 %. A constant boiling mixture at 112 °C is about 22 M, but the more common concentrated aqueous solution (~ 49 % HF) is about 29 M. Both the solution and its vapor act on the flesh with very little warning to produce burns that are painful and slow to heal. Unlike the other hydrogen halides, HF is not a reducing acid.

The alkali-metal fluorides mostly dissolve readily in water, react alkaline to litmus and are slightly corrosive to glass. The fluorides of the alkaline-earth metals (except BaF_2) are insoluble; of Li, Ba, Fe^{III}, Cu and Pb, slightly soluble; of Ag, Hg and Tl readily soluble.

Fluorine promptly decomposes water, liberating mainly oxygen.

Hydroxide. Passing F_2 into 5-dM OH^- yields a roughly 50–50 mixture of OF_2 and O_2:

$$2\ F_2 + 2\ OH^- \rightarrow OF_2\uparrow + 2\ F^- + H_2O$$

17.1.2 Reagents Derived from the Other 2nd-Period Non-Metals, Boron through Fluorine

Boron species. If boric acid, H_3BO_3, is treated with HF in excess, the product is $H[BF_4]$ (distinction from the other halogens).

Reduced nitrogen. Ammonia and $H_2[SiF_6]$ give NH_4F and silicic acids.

Fluorine species. Hydrogen fluoride and AlkF form many double salts, e.g., $K[HF_2]$ and $K[H_2F_3]$.

17.1.3 Reagents Derived from the 3rd-to-5th-Period Non-Metals, Silicon through Xenon

Silicon species. Fluorides may be detected by warming with concentrated H_2SO_4; the liberated HF etches an exposed (silicate) glass surface. Silica in the unknown interferes by changing HF to SiF_4, but this may be turned to advantage with the "hanging drop" test, which uses the reaction of SiF_4 with H_2O to precipitate silicic acids, e.g.:

$$3\ SiF_4 + 7\ H_2O \rightarrow H_2SiO_3 \cdot aq\downarrow + 2\ [SiF_6]^{2-} + 4\ H_3O^+$$

The unknown is treated as for the "etch" test except that silica is also added and the glass surface placed over the mixture carries a drop of water, which becomes turbid if SiF_4 is liberated. This is especially satisfactory to detect fluoride in silicates. Borates interfere in this as well as in the "etch" test. Fluorides may also be determined by heating with SiO_2 and H_2SO_4. The volatilized SiF_4 is collected in H_2O, and the $H_2[SiF_6]$ is either titrated with standard OH^- and phenolphthalein as an indicator or converted to K_2SiF_6 for gravimetry.

Oxidized chalcogens. Hydrogen fluoride is often prepared by treating fluorspar, CaF_2, with H_2SO_4 and distilling off the HF, which is absorbed in water. The operation is carried out in lead-lined apparatus.

Reduced halogens. Fluorine, no excess, will displace Cl_2 from Cl^-.

Elemental halogens. Fluorine and moist Cl_2 give HClO:

$$Cl_2 + F_2 + 2\ H_2O \rightarrow 2\ HClO + 2\ HF$$

17.1.4 Reagents Derived from the Metals Lithium through Uranium, plus Electrons and Photons

Non-redox reactions. Precipitation tests using Ca^{2+} or Sr^{2+} for the presence of F^- are not as satisfactory as their solubility figures suggest because of the colloidal nature of the precipitates; $La(CH_3CO_2)_3$ is better, and may also be used to determine fluoride. Calcium may still serve for gravimetry, albeit less well, by precipitating CaF_2.

Fluoride may be determined colorimetrically by its bleaching effect on peroxotitanic acid.

Aqueous F^- does not precipitate Ag^+, but, being basic, it hydrolyzes Hg_n^{2+} to oxides; note that HgF_2, unlike $[HgCl_2]$, is highly ionized.

Fluorides may be determined volumetrically by titration with Fe^{III}, using SCN^- as an indicator. The method is unsatisfactory for less than 1 mmol F^-, unless the endpoint is determined electrometrically. The methods by which the fluoride is precipitated as PbClF and weighed as such, or titrated by Fe^{III} with SCN^- as indicator, work well.

17.2 Chlorine, $_{17}$Cl

Oxidation numbers: (−I), (I), (III), (IV), (V) and (VII), as in Cl^-, ClO^-, ClO_2^-, ClO_2, ClO_3^- and ClO_4^-.

17.2.1 Reagents Derived from Hydrogen and Oxygen

Water. Hydrogen chloride, a colorless gas with an acrid, irritating odor, is readily soluble in water as the hydrochloric or "muriatic" acid of commerce. A common concentrated solution is about 36-% HCl and about 12 M. A constant-boiling or azeotropic solution is obtained at 110 °C; it contains 20.2-% HCl at 6.1 M.

All chlorides except AgCl, AuCl and $[Hg_2Cl_2]$ are soluble in water, CuCl, TlCl and $PbCl_2$ being only slightly soluble. The chlorides of Sn^{II}, Sb^{III} and Bi^{III} require the presence of some free acid to keep them in solution. Water decomposes PCl_3 and $AsCl_3$, and liberates HCl:

$$AsCl_3 + 6\,H_2O \rightarrow H_3AsO_3 + 3\,Cl^- + 3\,H_3O^+$$

The solubility of chlorine in water is 2.1 dM at 0 °C. Boiling completely removes the gas. Chlorine combines with ice-cold H_2O to form a clathrate, $4Cl_2 \cdot 29H_2O$ (which may plug the inlet tube!). The crystals, which decompose at nearly 10 °C, may soon be filtered off and dried between pieces of filter paper. Chlorine reacts slowly with H_2O to form HCl and HClO. In sunlight HClO quickly releases O_2.

Water, even in the presence of the dessicant anhydrous $CaCl_2$, which also attracts NH_3, forms Cl_2O from chloroazane (chloroamine):

$$NH_2Cl + H_2O \leftrightarrows NH_3 + HClO \leftrightarrows NH_4^+ + ClO^-$$

$$2\,HClO \leftrightarrows Cl_2O + H_2O \qquad [pK(0\,°C) = 2.45]$$

Dichlorine oxide, Cl_2O, is a very soluble, reddish-yellow gas. Warming can cause explosive decomposition into Cl_2 and O_2. Many organic solvents extract it effectively from water.

All hypochlorites (ClO^-) are soluble in water and are decomposed by boiling. The acid is easily distilled without much decomposition, however, from a sufficiently dilute solution. Salts of ClO^- and Alk^+ or Ae^{2+} arise on evaporating their alkaline solutions at low *T*.

The chlorites, salts of ClO_2^-, are fairly soluble in water, except for $AgClO_2$, 10 mM, and $Pb(ClO_2)_2$, 1.0 mM, respectively, at 0 °C; both are explosive on heating or impact.

All chlorates (ClO_3^-) are soluble in water, those of Hg, Sn and Bi requiring a little free acid. Iron(II) and dimercury(I) chlorates are very unstable. Potassium chlorate is the least soluble of the stable metallic chlorates, lithium chlorate one of the most soluble, at 66 g and 3.2 kg, in turn, per kg of water at 18 °C.

Perchloric acid and most of its salts are readily soluble, and the salts are deliquescent except NH_4ClO_4, $KClO_4$, $Hg_2(ClO_4)_2$ and $Pb(ClO_4)_2$. The solubilities of $KClO_4$ are only 5.5 cM at 0 °C but 1.4 M at 100°C, (distinction from $NaClO_4$, with 8.4 M and 11.0 M respectively). It is particularly insoluble in ethanol, especially if a small amount of $HClO_4$ is present. Certain perchlorates, e.g. $Mg(ClO_4)_2$ and $Ba(ClO_4)_2$, are very efficient drying agents.

Oxonium. This converts ClO_n^- to the acids. Thus $Ba(ClO_2)_2$ and H_2SO_4 yield aqueous $HClO_2$ (and solid $BaSO_4$).

The weak acid, HClO, has not been isolated. Its aqueous solution smells like Cl_2O. It decomposes more or less rapidly, depending on the concentration and exposure to light. In a 2.5-dM solution in darkness the acid lasts many months. At perhaps 5 M, even at 0 °C, decomposition is noted after a few hours. Sunlight or diffuse daylight promotes a composite reaction such as, approximately:

$$5\ HClO + 5\ H_2O \rightarrow ClO_3^- + 4\ Cl^- + O_2\uparrow + 5\ H_3O^+$$

A catalyst, such as $Co(OH)_2$, greatly hastens the loss of O_2. The oxidizing power of ClO^- is lowest in alkalis. It increases with increasing $c(H_3O^+)$, but below pH 7 the stability decreases rapidly.

Chlorous acid, $HClO_2$, is not isolable, is intensely yellow, and is unstable except when very dilute, decomposing even at 0 °C in ~10 min:

$$5\ HClO_2 \rightarrow 4\ ClO_2 + Cl^- + H_3O^+ + H_2O$$

The reddish-yellow color and pungent odor of ClO_2, boiling at 11 °C and soluble in water to ~ 1 M at ambient T and P, are easily recognized. Aqueous $HClO_2$ is a stronger oxidant than ClO_2^-. The chlorite ion, ClO_2^-, decomposes in two ways, yielding O_2 and Cl^- or ClO_3^- and Cl^-.

A cold solution of $HClO_3$ is colorless and odorless. It is a strong acid, much less stable than H_2SO_4, $HClO_4$ or its salts, but more stable than HNO_2. Dilute $HClO_3$, a strong oxidant, decomposes in three ways:

$$4\ ClO_3^- \rightarrow 3\ ClO_4^- + Cl^-$$

$$2\ ClO_3^- + 2\ H_3O^+ \rightarrow 2\ ClO_2 + {}^1\!/_2\ O_2\uparrow + 3\ H_2O$$

$$ClO_3^- \rightarrow Cl^- + {}^3\!/_2\ O_2\uparrow$$

Aqueous $HClO_3$ may be evaporated in a vacuum up to 6.1 M at 14 °C. This rather concentrated acid begins decomposing slowly. Further action rapidly gives mixed products, e.g. approximately as:

$$4\ ClO_3^- + 2\ H_3O^+ \rightarrow 2\ ClO_4^- + {}^3/_2\ O_2\uparrow + Cl_2\uparrow + 3\ H_2O$$

Concentrated $HClO_3$ oxidizes P_4 (phosphorescent) to H_3PO_4, S to H_2SO_4, and Cl^-, Br^- or I^- to X_2 and ClX, or, with excess $HClO_3$, to IO_3^-.

Hot, concentrated $HClO_4$ is explosive, but dilute $HClO_4$ does not oxidize HNO_2, SO_2, HCl (even if warm) or HI.

Hydroxide. Solutions containing more than 3-M hypochlorite, such as NaClO, are readily prepared by passing Cl_2 into cold OH^-:

$$Cl_2 + 2\ OH^- \rightarrow ClO^- + Cl^- + H_2O$$

Only a small amount of ClO_3^- is formed if the temperature is low enough and an excess of OH^- is maintained. Solutions containing about 7-dM NaClO are used as household bleaches. Chlorates, ClO_3^- salts, are easily prepared from Cl_2 and hot OH^-.

Chlorine is converted into bleaching powder, a mixture of $CaCl(ClO){\cdot}H_2O$, $Ca(OH)_2{\cdot}CaCl(ClO)$ etc. in varying ratio, by $Ca(OH)_2$.

Many salts of ClO_2^- have been isolated, often from ClO_2 and the metal hydroxide, but most are very unstable, some exploding when jarred:

$$2\ ClO_2 + 2\ OH^- \rightarrow ClO_2^- + ClO_3^- + H_2O$$

Peroxide. Chlorine and H_2O_2 form O_2 and HCl.

Aqueous HClO reacts with H_2O_2 to form HCl and O_2:

$$HClO + H_2O_2 \rightarrow H_3O^+ + Cl^- + O_2\uparrow$$

In base, ClO_2 dismutates to ClO_2^- and ClO_3^-, but H_2O_2 gives only:

$$2\ ClO_2 + HO_2^- + OH^- \rightarrow 2\ ClO_2^- + O_2\uparrow + H_2O$$

Aqueous H_2O_2, in the presence of a little HNO_3, completely reduces ClO_3^- to Cl^- (distinction from BrO_3^- and IO_3^-).

17.2.2 Reagents Derived from the Other 2nd-Period Non-Metals, Boron through Fluorine

Boron species. A practically pure solution of HClO may be made by distilling water, thirty parts by weight; bleaching powder, one part; and H_3BO_3 (weakly acidic), two to three parts (an excess).

Carbon oxide species. Aqueous HCl does not reduce CO_2.

Chlorine can form Cl_2O in CCl_4 with moist but solid Na_2CO_3:

$$2\ Cl_2 + 2\ Na_2CO_3 \cdot aq + H_2O \rightarrow Cl_2O + 2\ NaHCO_3 + 2\ NaCl$$

$$2\ Cl_2 + 2\ NaHCO_3 \rightarrow Cl_2O + 2\ CO_2\uparrow + H_2O + 2\ NaCl$$

From this, solutions in water generate HClO up to 5 M or more, nearly quantitatively under some conditions, but it decomposes to Cl_2, O_2 and some $HClO_3$, fastest near neutrality and slowly (as ClO^-) near pH 13.

All hypochlorites are unstable and decomposed by most acids:

$$Ca(ClO)_2 + CO_2 \rightarrow CaCO_3\downarrow + Cl_2\uparrow + {}^1/_2\ O_2\uparrow$$

Cyanide species. Aqueous HCl does not reduce HCN.

Chlorine and HCN form cyanogen chloride, NCCl and HCl.

Some "simple" organic species. Hot Cl_2 and $H_2C_2O_4$ or $C_2O_4^{2-}$ react:

$$H_2C_2O_4 + Cl_2 + 2\ H_2O \rightarrow 2\ CO_2\uparrow + 2\ Cl^- + 2\ H_3O^+$$

$$C_2O_4^{2-} + ClO^- + 2\ OH^- \rightarrow 2\ CO_3^{2-} + Cl^- + H_2O$$

Aqueous $HClO_3$ with $H_2C_2O_4$ or $C_2O_4^{2-}$ forms CO_2 and varying amounts of Cl_2 and Cl^-. Heat and excess reductant favor Cl^-. Free Cl_2 is recognized by its odor and, non-specifically, by bleaching litmus, etc.

Chloric Acid, $HClO_3$, first reddens, then bleaches, blue litmus paper.

Aqueous $HClO_4$ does not bleach but merely reddens blue litmus.

Reduced nitrogen. Chlorine plus NH_4^+ (preferably from the sulfate because Cl^- reverses this reaction) give NCl_3, extractable in CCl_4 and quite explosive, detonated even by finger grease on a vessel, but safer below about a 20-% concentration:

$$NH_4^+ + 3\ Cl_2 + 4\ H_2O \leftrightarrows NCl_3 + 3\ Cl^- + 4\ H_3O^+$$

Fairly pure N_2 may be prepared by the reaction:

$$8\ NH_3 + 3\ Cl_2 \rightarrow 6\ NH_4^+ + 6\ Cl^- + N_2\uparrow$$

but the ammonia must always be in excess to avoid the dangerous NCl_3.

Oxidized nitrogen. Nitrous acid and HCl form chiefly NO and Cl_2. Nitric acid yields NO_2Cl and Cl_2, or NOCl and Cl_2, or NO_2 and Cl_2, depending on conditions. With excess HCl the main reaction is:

$$NO_3^- + 3\ Cl^- + 4\ H_3O^+ \rightarrow NO\uparrow + {}^3/_2\ Cl_2\uparrow + 6\ H_2O$$

If dry HCl gas is led into a cold mixture of concentrated H_2SO_4 and HNO_3, the reaction becomes:

$$HNO_3 + HCl + H_2SO_4 \rightarrow NO_2 + {}^1/_2\,Cl_2\uparrow + H_3O^+ + HSO_4^-$$

Chlorine does not oxidize the aqueous oxides or acids of nitrogen.

Both HClO and $HClO_3$ react with HNO_2 to form HNO_3, which has no effect on ClO_3^-. However, concentrated $HClO_3$ and fuming HNO_3, mixed, are an especially strong oxidant, e.g. to destroy organic matter.

Fluorine species. Fluorine and moist Cl_2 give HClO:

$${}^1/_2\,Cl_2 + {}^1/_2\,F_2 + H_2O \rightarrow HClO + HF$$

17.2.3 Reagents Derived from the 3rd-to-5th-Period Non-Metals, Silicon through Xenon

Phosphorus species. Aqueous $HClO_3$ with PH_3 gives H_3PO_4.

With an excess of HCl, phosphinates, phosphonates, and phosphates are dissolved or transposed without reduction of the phosphorus.

Aqueous $P^{<V}$ and Cl_2 or ClO_n^- except ClO_4^-, yield phosphonate, phosphate and Cl^- in either acid or alkali, but hot, concentrated $HClO_4$ goes to Cl_2 or Cl^-. The reaction with $HClO_3$ may be explosive.

Arsenic species. Very concentrated HCl reduces As^V to $AsCl_3$.

Chlorine or HClO, and H_3AsO_3 or AsO_3^{3-}, readily form H_3AsO_4 or AsO_4^{3-} in acid or alkali. Volumetrically, they may be titrated with AsO_3^{3-} in base, the end-point being determined electrometrically.

Chlorous acid, $HClO_2$, in dilute acid, and chloric acid, $HClO_3$, slowly in strongly acidic solution, change H_3AsO_3 to H_3AsO_4, leaving Cl^-.

Reduced chalcogens. With H_2S, sulfur is first deposited, which an excess of Cl_2 changes to SO_4^{2-}. A sulfide in alkaline mixture is at once oxidized to SO_4^{2-} without apparent formation of the intermediate S.

Aqueous HClO or $HClO_2$, and H_2S form HSO_4^- and Cl^-.

Chlorine and HSCN yield NH_4^+, HSO_4^-, CO_2 and other products.

Hypochlorous acid reacts with SCN^- to give NCCl, SO_4^{2-} etc.

Aqueous $HClO_2$ or $HClO_3$, and SCN^-, form HSO_4^-, HCN and Cl^-:

$$2\,SCN^- + 3\,HClO_2 + 5\,H_2O \rightarrow 2\,HSO_4^- + 2\,HCN + 3\,Cl^- + 3\,H_3O^+$$

Oxidized chalcogens. Aqueous HCl does not reduce SO_2 or H_2SO_4. Thiosulfate merely forms $H_2S_2O_3$, which soon decomposes; see **16.2.1**.

Chlorine (in either acidic or alkaline solution), HClO, $HClO_2$ and $HClO_3$, but not $HClO_4$, oxidize $S^{<VI}$ to SO_4^{2-}. Some conditions lead to some free sulfur, but it

finally goes to SO_4^{2-}; also $S_2O_3^{2-}$ may yield various mixtures of SO_4^{2-} and $[S_4O_6]^{2-}$, e.g.:

$$5\ ClO^- + 3\ S_2O_3^{2-} \rightarrow 5\ Cl^- + 2\ SO_4^{2-} + [S_4O_6]^{2-}$$

To identify $ClO_{<4}^-$ it may be reduced to Cl^-, then identified e.g. by Ag^+. Chlorides, if originally present, should first be removed by Ag^+:

$$ClO_3^- + 3\ SO_2 + 6\ H_2O \rightarrow Cl^- + 3\ H_3O^+ + 3\ HSO_4^-$$

Concentrated H_2SO_4 generates HCl from NaCl etc. (with heating) or from concentrated HCl. Water absorbs this gas (e.g. from the NaCl source) to provide the common hydrochloric acid:

$$NaCl + H_2SO_4 \rightarrow NaHSO_4 + HCl\uparrow$$

$$NaCl + NaHSO_4 \rightarrow Na_2SO_4 + HCl\uparrow$$

Adding the calculated amount of H_2SO_4 to a solution of $Ba(ClO_3)_2$ and separating the $BaSO_4$ precipitate yields a solution of $HClO_3$.

One may use the oxidation of ClO_3^- by $[S_2O_8]^{2-}$ to prepare $HClO_4$.

Reduced halogens. The chlorides of Na, K, Ba, Th and NH_4^+ are nearly insoluble in concentrated HCl.

Aqueous $HClO_3$ with HBr slowly forms Br_2 and HCl:

$$ClO_3^- + 6\ Br^- + 6\ H_3O^+ \rightarrow Cl^- + 3\ Br_2 + 9\ H_2O$$

The reduction of $HClO_4$ by HBr is catalyzed by Ru salts.

Free Cl_2 may be recognized by its release of I_2 from I^-; starch-iodide paper turns blue, and this non-specific test reveals 1-μM aqueous Cl_2.

Volumetrically, Cl_2, HClO, $HClO_2$ or ClO_3^- may be treated with an excess of I^- in an acidic solution and the resulting I_2 titrated with $S_2O_3^{2-}$.

Analysis for HClO may use I^-, and any Cl_2 becomes Cl^- and consumes no H_3O^+, so one may simply measure the loss of acid:

$$HClO + 3\ I^- + H_3O^+ \rightarrow Cl^- + I_3^- + 2\ H_2O$$

Elemental and oxidized halogens. At high pH, Cl_2 oxidizes ClO^- and ClO_2^- to ClO_3^-:

$$ClO_2^- + Cl_2 + 2\ OH^- \rightarrow ClO_3^- + 2\ Cl^- + H_2O$$

Chlorine does not oxidize Br_2 in acidic solution, but Br^- becomes Br_2; in an alkaline medium BrO_3^- and Cl^- are formed from either Br^- or Br_2.

Chlorine oxidizes HI or neutral I^- to I_2:

$$2\ I^- + Cl_2 \rightarrow I_2\downarrow + 2\ Cl^-$$

Oxidation in acids goes on to IO_3^-; in alkalis, more directly to IO_4^-, $H_3IO_6^{2-}$, $H_2IO_6^{3-}$ and $H_2[I_2O_{10}]^{4-}$:

$$^1/_2\ I_2 + {}^5/_2\ Cl_2 + 9\ H_2O \rightarrow IO_3^- + 5\ Cl^- + 6\ H_3O^+$$

$$I^- + 4\ Cl_2 + 8\ OH^- \rightarrow IO_4^- + 8\ Cl^- + 4\ H_2O$$

Many decompositions involve internal oxidation and reduction but not of the same element (i.e. not dismutation). Thus, we can warm a solution of bleaching powder with a small amount of fresh cobalt(III) oxide as a catalyst for a smooth release of oxygen, and we can classify that here as a reaction of ClO^- with ClO^- :

$$2\ ClO^- \rightarrow 2\ Cl^- + O_2\uparrow$$

Aqueous $HClO_4$, unlike the lower oxoacids, does not oxidize HCl.

Aqueous $HClO_4$ oxidizes I_2 to H_5IO_6 with liberation of Cl_2:

$$^1/_2\ I_2 + ClO_4^- + H_3O^+ + H_2O \rightarrow H_5IO_6 + {}^1/_2\ Cl_2\uparrow$$

Small amounts of Cl_2 are made from dilute H_2SO_4 and bleaching powder, $CaCl(ClO)\cdot H_2O$ etc., with the same result from HClO and HCl:

$$ClO^- + Cl^- + 2\ H_3O^+ \rightarrow Cl_2\uparrow + 3\ H_2O$$

A solution containing both Cl^- and ClO^- can be made to yield Cl_2 in almost any desired concentration by properly adjusting the acidity. If a neutral chloride is used with HClO, chlorate is also formed, giving a variable ratio of oxidation states, e.g.:

$$6\ HClO + Cl^- \rightarrow ClO_3^- + 3\ Cl_2\uparrow + 3\ H_2O$$

Bromides and iodides react rather similarly.

Heating ClO^- also forms ClO_3^- and Cl^-. The rate is higher in a slightly acidic solution. Aqueous $HClO_2$ and HClO yield ClO_2 and Cl_2, but ClO^- forms ClO_3^- and Cl^-.

Aqueous $HClO_2$ and bromides in neutral solution do not react, but in acidic solution Br_2 is liberated; iodides react slowly in neutral solution to form I_2, although acid yields HIO_3.

One laboratory source of Cl_2 is HCl and $KClO_3$, with HCl in excess:

$$ClO_3^- + 5\ Cl^- + 6\ H_3O^+ \rightarrow 3\ Cl_2\uparrow + 9\ H_2O$$

With no excess of HCl, we find ClO_2, Cl_2O and Cl_2 in varying amounts.

Aqueous $HClO_3$ warmed with neutral Cl^- releases Cl_2 and leaves only the ClO_3^- salt, which may be better understood by writing old-fashioned neutral "molecules" to represent all of the spectator ions as:

$$6\ HClO_3 + 5\ KCl \rightarrow 5\ KClO_3 + 3\ Cl_2\uparrow + 3\ H_2O$$

instead of a simple net reaction showing no remaining ClO_3^-:

$$6\ H_3O^+ + ClO_3^- + 5\ Cl^- \rightarrow 3\ Cl_2\uparrow + 9\ H_2O$$

A neutral Br^- salt, warmed with $HClO_3$, releases Br_2 and possibly Cl_2, leaving only the extra ClO_3^- as a salt (as above). Any reaction between the bromine and the ClO_3^- is slow.

Iodine heated with ClO_3^- forms IO_3^- and ICl_3. Adding I^- to ClO_3^- in the presence of H_3O^+ produces I_2, but if the acid is acetic, no I_2 is obtained even after standing for some hours (distinction from IO_3^-).

Aqueous BrO_3^- is decomposed by boiling with HCl:

$$BrO_3^- + 5\ Cl^- + 6\ H_3O^+ \rightarrow {}^1/_2\ Br_2 + {}^5/_2\ Cl_2\uparrow + 9\ H_2O$$

Aqueous HIO_3 and HCl form ICl_3 and Cl_2, but not in dilute solution:

$$IO_3^- + 5\ Cl^- + 6\ H_3O^+ \rightarrow ICl_3 + Cl_2\uparrow + 9\ H_2O$$

Neat $HClO_4$ boils undecomposed at low pressure but decomposes or explodes when boiled at ordinary pressures. It produces very painful wounds that are slow to heal. At ordinary temperatures, especially when exposed to light, it gradually decomposes and ultimately explodes. The aqueous acid, however, is more stable and is not easily reduced.

17.2.4 Reagents Derived from the Metals Lithium through Uranium, plus Electrons and Photons

Oxidation. We find HCl oxidized to Cl_2 by: $[Cr_2O_7]^{2-} \rightarrow Cr^{III}$; $Mn^{>II} \rightarrow Mn^{2+}$; $FeO_4^{2-} \rightarrow Fe^{III}$; $Co_2O_3 \rightarrow Co^{2+}$, $NiO_2 \rightarrow Ni^{2+}$, $Pb_3O_4 \rightarrow [PbCl_4]^{2-}$, and $Bi^V \rightarrow [BiCl_4]^-$. The only metal oxides that oxidize HCl are those which can be formed by Cl_2 with OH^- but not an acid. Furthermore this reduction of the metal is not always to the original form; it cannot go below that oxidation state obtainable in the presence of an acid. Thus, iron(II) chloride with OH^- and Cl_2 produces $[FeO_4]^{2-}$, which with HCl therefore forms not Fe^{2+} but Fe^{III}, for the Fe^{2+} could be oxidized to Fe^{III} in the presence of an acid. This generalization is true for bromine and iodine as well as for chlorine.

In particular, one process for the laboratory production of chlorine uses the reaction of $MnO_2{\cdot}aq$ with HCl, or with Cl^- plus, say, H_2SO_4:

$$MnO_2{\cdot}aq + 2\ Cl^- + 4\ H_3O^+ \rightarrow Mn^{2+} + Cl_2\uparrow + 6\ H_2O$$

The Mn^{2+} may be converted back into $MnO_2{\cdot}aq$ and used again.

Dropping dilute HCl onto $KMnO_4$ also yields small amounts of Cl_2:

$$2\ MnO_4^- + 10\ Cl^- + 16\ H_3O^+ \rightarrow 5\ Cl_2\uparrow + 2\ Mn^{2+} + 24\ H_2O$$

Traces of aqueous HCl may be detected by boiling the solution with $MnO_2 \cdot aq$ and collecting the distillate in a mixture of I^- and starch to obtain the blue starch-iodine complex.

In dilute acid, $HClO_2$ and MnO_4^- become ClO_3^- and $MnO_2 \cdot aq$.

Aqueous $H_3[Fe(CN)_6]$ does not oxidize HCl.

Chlorine is produced from salt brine, NaCl:

$$Cl^- \leftrightarrows {}^1/_2\, Cl_2\uparrow + e^-$$

Electrolyzing a solution of Cl^- in a cell containing no diaphragm, i.e., one in which the Cl_2 released at the anode reacts with the OH^- formed at the cathode, gives ClO^-.

Chlorates are easily prepared by electrolyzing a *hot* chloride solution in a cell without a diaphragm. Repeated crystallization separates the chlorate from any contaminating chloride.

A good preparation of perchlorate is by further anodic oxidation:

$$ClO_3^- + 3\, H_2O \rightarrow ClO_4^- + 2\, e^- + 2\, H_3O^+$$

Due to its dangerous nature, anhydrous $HClO_4$ is prepared mainly for research purposes. A 70-% or nearly 12-M solution is readily obtained without great hazard and is commonly available. One starting material is the less soluble NH_4ClO_4 made from $NaClO_4$, from the electrolytic preparation of $NaClO_3$. A little hot aqua regia oxidizes the NH_4^+ from the NH_4ClO_4, leaving a dilute solution of $HClO_4$ containing a slight excess of the reagents. These are removed by boiling to a temperature of about 200 °C, after which the residual liquid is distilled at 60–95 °C under a pressure of 300–900 Pa, giving the concentrated solution.

Reduction of chlorine and hypochlorite. The various metals, as M^0, reduce moist Cl_2, mostly without heat, to chloride in low or high oxidation states, depending on the amount of Cl_2 used. Certain metals, e.g., Cu, Ag, Tl and Pb, form a protective coating of the chloride which retards further action. In some cases combination occurs with vivid incandescence. Without moisture many metals do not react.

A comparison of the electrode potentials of Cl_2 with those of Br_2 and I_2 shows reducibility falling as the atomic number Z rises. If all three have corresponding results, Cl_2 acts most rapidly. [In some cases, as with copper(I), Cl_2 oxidizes, although I_2 does not.]

Chlorine is reduced to Cl^- and takes metals to the same oxidation states as result from HCl acting on the oxide or hydroxide; thus, treating Co_2O_3 with HCl gives $CoCl_2$; hence Cl_2 and Co form $CoCl_2$, not $CoCl_3$, but Cu forms CuCl or $CuCl_2$. If any oxidation state rises in the presence of an acid due to Cl_2, Br_2 or I_2, the same increase, or more, will occur in the presence of OH^-. Thus, with Cl_2 and OH^-, but not with H_3O^+, we find $MnO_2 \cdot aq$, FeO_4^{2-}, $Co_2O_3 \cdot aq$, $NiO_2 \cdot aq$, PbO_2 and $[Bi(OH)_6]^-$.

Numerous metal$^{>0}$ species thus reduce Cl_2 while being oxidized as follows, in acidic or alkaline solution or both: $Cr^{III} \rightarrow CrO_4^{2-}$ (alkaline); $Mo^{<VI} \rightarrow Mo^{VI}$ (both); $Mn(OH)_2 \rightarrow MnO_2 \cdot aq$ (alkaline); $Fe^{2+} \rightarrow Fe^{III}$ (acidic) or FeO_4^{2-} (alkaline); $[Fe(CN)_6]^{4-} \rightarrow [Fe(CN)_6]^{3-}$ (both) but excess Cl_2 ultimately decomposes the $[Fe(CN)_6]^{3-}$; $Co(OH)_2 \rightarrow Co_2O_3 \cdot aq$ (alkaline); $Ni(OH)_2 \rightarrow NiO_2 \cdot aq$ (alkaline); $Cu^+ \rightarrow Cu^{II}$ (both); $Hg_2^{2+} \rightarrow Hg^{II}$ (both); $Sn^{II} \rightarrow Sn^{IV}$ (both); $Pb^{2+} \rightarrow PbO_2$ (alkaline); $Sb^{III} \rightarrow Sb^V$ (both); $Bi^{III} \rightarrow [Bi(OH)_6]^-$ (alkaline).

The reduction of HClO oxidizes MnO, FeO, CoO, NiO, SnO and PbO to more or less hydrated forms of $MnO_2 \cdot aq$, $Fe_2O_3 \cdot aq$, $Co_2O_3 \cdot aq$, NiO_2, $SnO_2 \cdot aq$ and PbO_2, in turn. Alkalis and Cr_2O_3 or Mn^{2+} form CrO_4^{2-} or $MnO_2 \cdot aq$, except in the presence of Ag^+ or Cu^{2+}, which promote oxidation to MnO_4^-; Sn^{II} changes to Sn^{IV}; and Sb^{III} becomes Sb^V.

Gravimetrically, hypochlorites may be reduced to chloride which is precipitated, dried and weighed as AgCl, but volumetry is prevalent.

Hypochlorite may be detected by shaking with Hg and reducing and precipitating it as yellowish-red Hg_2OCl_2. No related ions of Cl, i.e., Cl^-, ClO_2^-, ClO_3^- or ClO_4^-, act on Hg.

Reduction of chlorine(>I). In dilute acid, $HClO_2$ changes Mn^{2+} to $MnO_2 \cdot aq$, Fe^{2+} to Fe^{III}, etc.

Although Na_{Hg} surprisingly has no effect on ClO_3^-, Zn and dilute H_2SO_4, or Zn thinly coated with copper, completely reduce ClO_3^- to Cl^-, which is identified by Ag^+. (Chlorides, if originally present, should first be removed by Ag^+):

$$ClO_3^- + 3\ Zn + 6\ H_3O^+ \rightarrow Cl^- + 3\ Zn^{2+} + 9\ H_2O$$

Free aqueous $HClO_3$ is a powerful oxidant, and the hot, concentrated acid attacks all metals (M^0). When an excess of the reductant is used, the ClO_3^- becomes Cl^-. In general, metal ions of lower oxidation state, such as Ti^{III}, Fe^{2+}, Hg_2^{2+} and Sn^{II}, are oxidized to the higher state; Mn^{2+} is not oxidized by ClO_3^- alone (distinction from BrO_3^- and IO_3^-), but in the presence of HNO_3 rapid conversion to $MnO_2 \cdot aq$ occurs. Salts of Co, Ni and Pb appear not to be oxidized on boiling with ClO_3^- and HNO_3.

Titrating ClO_3^- to Cl^- with Ti^{III} determines well the ClO_3^-.

Aqueous $HClO_3$ with $H_4[Fe(CN)_6]$ first forms $H_3[Fe(CN)_6]$, but this decomposes if an excess of ClO_3^- is present.

Aqueous $HClO_4$ is reduced to Cl^- by Ti^{III}, V^{2+}, V^{III}, Ru^{II} or Sn^{II}, but not by Eu^{2+} or Cr^{2+} in spite of their greater reduction potentials. In the absence of other oxidants in fact, ClO_4^- may be reduced with an excess of $Ti_2(SO_4)_3$ and the excess titrated with MnO_4^-. The Cl^- from the reduction may also be determined as usual.

To detect ClO_4^- in the presence of other chlorine compounds, these others may be reduced by boiling with the Zn-Cu couple, to Cl^-, or with HCl, to Cl_2. The Cl^- may be removed as AgCl, the solute evaporated to dryness and the residue fused with Na_2CO_3, finally reducing any ClO_4^- to Cl^- (and O_2). An aqueous extract of the melt shows Cl^- well, e.g. as AgCl, if ClO_4^- was in the original sample.

Other reactions of chloride. Aqueous HCl acts on the following metals, forming chlorides and releasing H_2: the alkali and alkaline-earth metals (danger!); Cr, Mn, Fe, Co, Ni, Cu (very slowly), Zn, Cd, Al, Sn, Pb (slowly but completely). The "-ous", not "-ic" salts, are formed, if both exist. Metallic Pt, Ag, Au, Hg, As, Sb and Bi are insoluble in HCl.

Aqueous HCl dissolves or transposes all insoluble metal oxalates, phosphinates, phosphonates and phosphates.

The Cl^- ion can leach Ti, Zr, Hf, Nb, Ta, Cr, Mo, W, Fe, Ru, Os, Co, Rh, Ir, Ni, Pd, Pt, Cu, Ag, Au, Al, Sn, Pb, Sb, Se and Te from some ores. A high concentration of Cl^- dissolves many chlorides of **d**- or **p**-block metals, e.g., Pt, Cu^I, Ag, Au, Hg, Pb, Sb and Bi, by forming complexes such as $[PbCl_4]^{2-}$, as also with the often non-chloride ores above.

The sulfates of Ca, Sr, Hg^I and Pb are slowly but completely dissolved by hot concentrated HCl, but $BaSO_4$ is practically unaffected. The Ag_2SO_4 and Hg_2SO_4 salts are completely transposed by HCl; most other sulfates are more or less completely transposed (to chlorides).

In general, HCl is the best solvent for metallic oxides. It will dissolve all of the common oxides and hydroxides except (a) those that form insoluble chlorides, and (b) certain oxides which have been ignited: Cr_2O_3, Fe_2O_3, NiO_2, Al_2O_3 and SnO_2. In most of these cases, long-continued boiling will effect solution; Cr_2O_3, however, is quite inert.

Also with carbonates or sulfites, HCl yields chlorides free of other anions. Evaporation expels the excess acid, but any water of crystallization in **d**- or **p**-block chlorides may not be removed by heat alone; hydrolysis will usually drive off some HCl, leaving a basic salt.

The reaction of Cl^- with Cr^{VI} and concentrated H_2SO_4 is used to detect Cl^- in solids, especially in the presence of Br^-. Moisture must be absent:

$$4\ Cl^- + [Cr_2O_7]^{2-} + 9\ H_2SO_4 \rightarrow 2\ CrO_2Cl_2\uparrow + 9\ HSO_4^- + 3\ H_3O^+$$

The red-brown fumes of the chromyl chloride, CrO_2Cl_2, are easily seen. If in doubt, one collects the distillate in water and tests for $\sim CrO_4^{2-}$:

$$CrO_2Cl_2 + 5\ H_2O \rightarrow HCrO_4^- + 2\ Cl^- + 3\ H_3O^+$$

The chlorides of Hg do not respond to the test because they are not transposed by H_2SO_4. The chlorides of Ag, Sn, Pb and Sb are so slowly transposed that CrO_2Cl_2 may escape detection.

The sulfides of Mn, Fe^{II}, and Zn dissolve readily in HCl, releasing H_2S; those of Co, Ni, Cu, Ag, Cd, Sn, Pb, Sb and Bi dissolve if the acid is concentrated, but As_2S_3 and As_2S_5 are insoluble in cold, 12-M HCl, although very slowly soluble in the hot acid, while the escaping H_2S shifts the equilibrium.

The chlorides of Ag, Hg, Tl and Pb are best made by precipitation. To detect Cl^-, one may add Ag^+ to acidified Cl^-, precipitating curdy white AgCl, which darkens on exposure to light. Traces of Cl^- give only an opalescence. Interference comes from Br^-, I^- and S^{2-} salts.

The Cl^- in a green chloro-Cr^{III} complex, that is $[CrCl(H_2O)_5]^{2+}$, $[CrCl_2(H_2O)_4]^+$ or $[CrCl_3(H_2O)_3]$, is not promptly precipitated by Ag^+ or Hg_2^{2+}, and neither of the latter precipitates Cl^- from MoO_2Cl_2 in H_2SO_4.

Chloride in an insoluble chloride, e.g., AgCl, may be detected by digestion with Zn and H_2SO_4. Metallic Ag is precipitated and the Cl^- liberated, to be tested for further. Gravimetrically, Cl^-, perhaps from the reduction of ClO_n^-, is determined by precipitating, drying and weighing AgCl. Volumetrically, Cl^- may be precipitated with excess standard Ag^+. The AgCl is separated and the excess Ag^+ is titrated with SCN^- using Fe^{III} as the indicator. A shorter and slightly less accurate method is to titrate the Cl^- in a neutral solution directly with standard Ag^+, using a little CrO_4^{2-} as indicator.

Other reactions of chlorine and chlorine(>0). Chlorine dismutates on passing over cold, fresh, yellow HgO:

$$2\ Cl_2 + (n+1)\ HgO \rightarrow Cl_2O + HgCl_2 \cdot n HgO\downarrow$$

At 0 °C, the Cl_2O may be collected as a highly explosive liquid, bp 2°C. Alternately, CCl_4 is a useful solvent for the Cl_2 and Cl_2O. The Cl_2O is a very water-soluble, reddish-yellow, odoriferous gas. Water, however, gives mainly HClO, and it can yield at least a 5-M solution. The product can be distilled at low P from water after standing in contact for some hours, but even at 0 °C it soon becomes HCl and O_2.

Hypochlorite can be stabilized somewhat as $Ca(ClO)_2$, with a small percentage of $Ca(OH)_2$, $CaCl_2 \cdot aq$, $Ca(ClO_3)_2$ and $CaCO_3$, plus the H_2O:

$$\text{cold } 2\ HClO + Ca(OH)_2 \rightarrow Ca(ClO)_2 + 2\ H_2O$$

For commercial purposes, as a bleaching agent and as a disinfectant, an important hypochlorite is bleaching powder, made by the action of slaked lime, $Ca(OH)_2$, dismutating Cl_2:

$$Cl_2 + Ca(OH)_2 \rightarrow CaCl(ClO) \cdot H_2O \text{ etc.}$$

Although silver hypochlorite is soluble in H_2O, it dismutates very quickly, hence adding Ag^+ to ClO^- forms the white AgCl:

$$3\ ClO^- + 2\ Ag^+ \rightarrow 2\ AgCl\downarrow + ClO_3^-$$

A transient amethyst tint with slightly acidified $FeSO_4$ detects ClO_2^-.

The alkali metals reduce the H_3O^+, but not the ClO_4^-, in aqueous $HClO_4$, detected and weighed as $KClO_4$.

The acid $HClO_3$ attacks Mg, releasing H_2 and forming only a chlorate, not reducing the Cl; Fe, Cu, Zn and Sn give H_2, but also some chloride.

In some properties $HClO_4$ resembles H_2SO_4. Hot and concentrated, it is a very powerful oxidant. Cold and dilute, however, it is decidedly inert. Iron, Zn and some other metals reduce only the H_3O^+ (to H_2), leaving perchlorates. Some oxides (ignited Cr_2O_3) are easily dissolved, again without reducing the ClO_4^-. It is also not reduced by Fe^{2+}, nor by the Zn-Cu couple (distinction from chlorate), but is reduced by Ti^{III}.

17.3 Bromine, $_{35}$Br

Oxidation numbers: (−I), (I), (III), (IV), (V) and (VII), as in Br^-, BrO^-, BrO_2^-, BrO_2, BrO_3^- and BrO_4^-.

17.3.1 Reagents Derived from Hydrogen and Oxygen

Water. Aqueous HBr, hydrobromic acid, forms a constant-boiling mixture at 124 °C, which is the ordinary 48-% HBr, 8.7 M.

The bromides AgBr and $[Hg_2Br_2]$ are insoluble in water, TlBr moderately insoluble, and $PbBr_2$ slightly soluble; all other bromides are soluble. The bromides AgBr, $[Hg_2Br_2]$, TlBr and $PbBr_2$ are less soluble than the corresponding chlorides. An excess of Br^- appreciably increases their solubility. In ethanol the alkali bromides are slightly soluble, $CaBr_2$ and $[HgBr_2]$ soluble, and $[Hg_2Br_2]$ insoluble.

Bromine dissolves to a 2-dM solution at 20°C. It is much more soluble in HCl, HBr, KBr, $SrCl_2$, $BaCl_2$, in many other salt solutions, and in alcohols etc. than in water alone.

Water decomposes the phosphorus bromides to form HBr and the corresponding acids of phosphorus.

Evaporating water under vacuum from $HBrO_3$ solutions gives ~ 50 %, estimated to be around a 6-M solution, before decomposition is serious.

Barium bromate is soluble up to about 2 cM, silver bromate, 1 cM, thallium(I) bromate, 1 cM, and lead(II) bromate, 3 cM at 20 °C. Except for some basic bromates, all others are soluble.

Oxonium. Aqueous H_3O^+ yields bromic acid from metallic bromates. The gradual decomposition of the $HBrO_3$ may first yield HBr and O_2, but the HBr then immediately liberates the bromine of both acids:

$$5\ Br^- + BrO_3^- + 6\ H_3O^+ \rightarrow 3\ Br_2 + 9\ H_2O$$

Hydroxide. Bromine, rather like chlorine, reacts with cold alkali and alkaline-earth hydroxides to form a bromide and a hypobromite:

$$Br_2 + 2\ OH^- \rightarrow Br^- + BrO^- + H_2O$$

With heat, however, the BrO^- becomes Br^- and BrO_3^-; thus $AlkBrO_3$ and $Ae(BrO_3)_2$ may be obtained from their hydroxides with Br_2 at 100 °C:

$$3\ Br_2 + 6\ OH^- \rightarrow 5\ Br^- + BrO_3^- + 3\ H_2O$$

Peroxide. Aqueous H_2O_2 forms Br_2 from HBr at 100 °C (distinction from Cl^-), or with catalysis by peroxovanadium(V) species. Excess peroxide, however, can react violently with Br_2 in turn:

$$2\ Br^- + H_2O_2 + 2\ H_3O^+ \rightarrow Br_2 + 4\ H_2O$$

$$H_2O_2 + Br_2 + 2\ H_2O \rightarrow 2\ H_3O^+ + 2\ Br^- + O_2\uparrow$$

17.3.2 Reagents Derived from the Other 2nd-Period Non-Metals, Boron through Fluorine

Some "simple" organic species. Bromine is often detected by shaking its aqueous solution with CCl_4 or CS_2. The extract is reddish yellow to dark brown (with much Br_2). Other extractants include $CHCl_3$ and ether. Starch solution turns yellow with Br_2, but the reaction is less sensitive.

Urea reduces hypobromite:

$$3\ BrO^- + CO(NH_2)_2 \rightarrow 3\ Br^- + CO_2 + N_2 + 2\ H_2O$$

Air containing bromine vapor, 10 parts per million, can be detected readily if inhaled, by its irritating effect on the nose mucous membrane.

The acids $HBrO_3$ and dilute $H_2C_2O_4$ form CO_2 and Br_2, suitable for detection. An excess of hot $H_2C_2O_4$ yields Br^- instead of Br_2.

Oxidized nitrogen. Dilute HNO_2 has no action on Br^- (distinction from I^-), but HNO_3 forms NO and Br_2:

$$2\ NO_3^- + 6\ Br^- + 8\ H_3O^+ \rightarrow 2\ NO\uparrow + 3\ Br_2 + 12\ H_2O$$

Excess HNO_2 reduces $HBrO_3$ to HBr (excess $HBrO_3$ gives Br_2):

$$BrO_3^- + 3\ HNO_2 + 3\ H_2O \rightarrow Br^- + 3\ H_3O^+ + 3\ NO_3^-$$

Fluorine species. Perbromate, long thought unattainable, can be made by electrolysis, but more easily with F_2, in spite of the latter's hazard:

$$BrO_3^- + F_2 + 2\ OH^- \rightarrow BrO_4^- + 2\ F^- + H_2O$$

After several steps of purification, ion exchange can give the strong acid, which is stable even at 100 °C up to 6 M. The 12-M acid is unstable but not explosive, and is a vigorous oxidant even at 25 °C.

17.3.3 Reagents Derived from the 3rd-to-5th-Period Non-Metals, Silicon through Xenon

Phosphorus species. Phosphorus compounds, unlike hot H_2SO_4, do not oxidize HBr. Therefore, one way to prepare HBr is to add concentrated H_3PO_4 to concentrated aqueous NaBr. On heating the mixture, HBr is driven off and may be absorbed in H_2O.

Phosphorus(<V) and Br_2, BrO^- or BrO_3^- form Br^- and phosphonate and/or phosphate in either acid or alkali, depending on quantities etc. At least the alkaline BrO^- and $PH_2O_2^-$ reaction is catalyzed by Cu^{II}.

Arsenic species. Arsenic(V) and HBr form As^{III} and Br_2. The HBr must be concentrated and in excess, and the arsenic compound merely moistened with water. Much water reverses the reaction:

$$H_3AsO_4 + 2\ H_3O^+ + 2\ Br^- \leftrightarrows H_3AsO_3 + Br_2 + 3\ H_2O$$

(Much) Br_2 and AsH_3 form H_3AsO_3 first (then H_3AsO_4) and HBr.

Arsenous acid reduces BrO_3^- to Br^-, leaving H_3AsO_4.

Reduced chalcogens. Bromic acid and sulfides form S first, then SO_4^{2-}.

Aqueous SCN^- plus Br_2 or BrO_3^- form, among other products, SO_4^{2-}, and Br^- in either acidic or alkaline mixture, or Br_2 in acid..

Oxidized chalcogens. Sulfur(<VI) is oxidized to SO_4^{2-} by Br_2 (not I_2):

$$S_2O_3^{2-} + 4\ Br_2 + 13\ H_2O \rightarrow 2\ HSO_4^- + 8\ Br^- + 8\ H_3O^+$$

Bromic acid and $S^{<VI}$ become Br^- and SO_4^{2-}, although $S_2O_3^{2-}$ is changed to $[S_4O_6]^{2-}$, but depending partly on conditions:

$$BrO_3^- + 6\ HS_2O_3^- \rightarrow 3\ [S_4O_6]^{2-} + Br^- + 3\ H_2O$$

In one way to prepare HBr, sulfane (H_2S) or SO_2 is added to bromine in water until the yellow color disappears. The solution is then distilled, the first and last portions of the distillate being rejected due to contamination with H_2S or SO_2, and H_2SO_4 respectively:

$$SO_2 + Br_2 + 5\ H_2O \leftrightarrows HSO_4^- + 2\ Br^- + 3\ H_3O^+$$

This is partly reversed in attempts to distill gaseous HBr from concentrated H_2SO_4 plus e.g. concentrated HBr or NaBr, but constant-boiling HBr, ~8.7-M, can be distilled from KBr and 5 to 6-M H_2SO_4.

Bromic acid, $HBrO_3$, may be prepared by adding the calculated amount of dilute H_2SO_4 to $Ba(BrO_3)_2$ and removing the $BaSO_4$.

Aqueous $[S_2O_8]^{2-}$ and warm, acidic Br^- form Br_2 (separation from Cl^-).

Reduced halogens. Bromic acid ($HBrO_3$) and HCl release Cl_2 and Br_2, but excess $HBrO_3$ and HI yield HIO_3 and Br_2.

At 100 °C, 6-M $HBrO_4$ and Cl^- expel Cl_2 and Br_2. Dilute $HBrO_4$ at 25 °C oxidizes Br^- or I^- (not Cl^-) slowly; if 12-M with Cl^- it acts quickly.

One may determine Br_2, or Br^- after oxidation to Br_2 by passing the oxidant-free vapor into I^-, with excess I^- and titrating the I_2 with $S_2O_3^{2-}$.

Elemental and oxidized halogens. In seawater, Cl_2 releases Br_2:

$$2\ Br^- + Cl_2 \rightarrow Br_2 + 2\ Cl^-$$

Chlorine liberates bromine from all bromides, even fused AgBr.

Iodine or I^-, and Br_2, form IO_3^- and Br^- in an alkaline system. Aqueous I^- and Br_2 in acid become I_2 and Br^-.

Bromic acid is reduced to free bromine by I_2.

Aqueous HBrO is prepared similarly to, and resembles, HClO. It is fairly stable but decomposes slowly both of these ways:

$$5\ HBrO \rightarrow BrO_3^- + 2\ Br_2 + H_3O^+ + H_2O$$

$$2\ HBrO \rightarrow Br_2 + {}^1/_2\ O_2\uparrow + H_2O$$

Aqueous $HClO_3$ and HBr give Br_2 and HCl. If the chlorate is concentrated, other products may appear.

Aqueous HIO_3 and HBr form I_2 and Br_2, slowly at room temperature.

Under ordinary conditions Br_2 and ClO_3^- or ClO_4^- do not react.

Perchloric acid at ambient T does not liberate Br_2 from Br^-, but a little of some Ru salts makes the reaction rapid and smooth:

$$8\ Br^- + ClO_4^- + 8\ H_3O^+ \rightarrow 4\ Br_2 + Cl^- + 12\ H_2O$$

Xenon species. The following can be a source of BrO_4^-:

$$XeF_2 + BrO_3^- + H_2O \rightarrow Xe\uparrow + BrO_4^- + 2\ HF$$

17.3.4 Reagents Derived from the Metals Lithium through Uranium, plus Electrons and Photons

Oxidation. Chromium(VI) and HBr produce Cr^{III} and Br_2 (separation from Cl^- if dilute enough), depending largely on the $c(H_3O^+)$. Sulfates also promote this, perhaps by complexing and stabilizing the Cr^{III}.

A sensitive test for Br^- involves first reaction with $[Cr_2O_7]^{2-}$ in H_2SO_4 solution. The Br_2 is absorbed in chloroform, which is washed two or three times with water. Finally a little I^- solution is added. On shaking, the free I_2 liberated by the Br_2 colors the chloroform purple.

Although some oxides or hydroxides do not oxidize HBr, e.g.:

$$PbO + 2\ Br^- + 2\ H_3O^+ \rightarrow PbBr_2\downarrow + 3\ H_2O$$

other metal oxo or hydroxo species have no equivalent stable bromides and do oxidize it. Compounds of Mn, Co, Ni and Pb, with $M^{>II}$, become M^{II}; also FeO_4^{2-} and Bi^V form $[FeBr_4]^-$ and $[BiBr_4]^-$:

$$PbO_2 + 4\ Br^- + 4\ H_3O^+ \rightarrow PbBr_2\downarrow + Br_2 + 6\ H_2O$$

In particular, the following can be laboratory sources of Br_2:

$$6\ Br^- + [Cr_2O_7]^{2-} + 2\ HSO_4^- + 12\ H_3O^+ \rightarrow 3\ Br_2 + 2\ CrSO_4^+ + 19\ H_2O$$

$$2\ Br^- + MnO_2{\cdot}aq + 4\ H_3O^+ \rightarrow Br_2 + Mn^{2+} + 6\ H_2O$$

Aqueous MnO_4^- liberates all the Br_2 from Br^- in the presence of Cu^{II} (a separation of Br^- from Cl^-).

Excess concentrated HBr reduces moist solid $[Fe(CN)_6]^{3-}$ salts to $H_n[Fe(CN)_6]^{(4-n)-}$ and releases Br_2. This is reversed by much H_2O:

$$[Fe(CN)_6]^{3-} + Br^- + 2\ H_3O^+ \leftrightarrows H_2[Fe(CN)_6]^{2-} + {}^1/_2\ Br_2 + 2\ H_2O$$

At 0 °C, moist, fresh HgO gives Br_2O, extractable into CCl_4:

$$2\ Br_2 + (n+1)\ HgO \rightarrow Br_2O + HgBr_2{\cdot}n HgO\downarrow$$

Anodic electrolysis releases Br^- as Br_2, which is then blown out of solution by air and absorbed by OH^-, CO_3^{2-}, HCO_3^-, Fe, Cu, Zn, etc.

Reduction. Six-molar $HBrO_4$ at 100 °C becomes Br^- while oxidizing Ce^{3+} to Ce^{IV} nitrate complexes, Cr^{III} to $[Cr_2O_7]^{2-}$, and Mn^{2+} to $MnO_2{\cdot}aq$. The 3-M acid, usually sluggish, attacks stainless steel easily.

Bromine if moist, but often not if dry, unites with metals as bromides, but in one way to prepare HBr, Br_2 vapor is passed over (dry) heated aluminum to form Al_2Br_6, which is then converted to $AlBr_3{\cdot}6H_2O$. Heating this releases HBr and H_2O, leaving mostly $Al_2O_3{\cdot}aq$.

Bromine, less violently than Cl_2, is reduced by many ions (to Br^-). In either acids or alkalis we find: $Fe^{II} \rightarrow Fe^{III}$; $Cu^I \rightarrow Cu^{II}$; $Hg^I \rightarrow Hg^{II}$; $Sn^{II} \rightarrow Sn^{IV}$; $Sb^{III} \rightarrow Sb^V$. Only with alkalis do we find: $Cr_2O_3{\cdot}aq \rightarrow CrO_4^{2-}$; $Mn(OH)_2 \rightarrow MnO_2{\cdot}aq$; $Co(OH)_2 \rightarrow Co_2O_3{\cdot}aq$; $Ni(OH)_2 \rightarrow NiO_2{\cdot}aq$; $Pb(OH)_2 \rightarrow PbO_2$; and $Bi(OH)_3 \rightarrow Bi(OH)_6^-$.

The action of $HBrO_3$ on the free metals may differ from that of $HClO_3$. Although $HClO_3$ gives no free Cl_2, $HBrO_3$ yields Br_2 with many metals, nearly the full equivalent being released with Fe, Cu, Zn, Cd and Sn. Low oxidation states, for example Ti^{III}, Cr^{III}, Mn^{2+}, Fe^{2+}, Cu^+, Hg_2^{2+}, Sn^{II}, As^{III} and Sb^{III} are oxidized to Ti^{IV}, $[Cr_2O_7]^{2-}$, $MnO_2 \cdot aq$, Fe^{III}, Cu^{II}, Hg^{II}, Sn^{IV}, As^V and Sb^V respectively.

The acids $HBrO_3$ and $H_4[Fe(CN)_6]$ form $H_3[Fe(CN)_6]$ and HBr. An excess of $HBrO_3$ carries the oxidation (not of the Fe) further.

Other reactions. Aqueous K^+, Rb^+ or Cs^+ precipitates BrO_4^- as $AlkBrO_4$.

Dilute $HBrO_3$ and Na (take care!), Mg or Al give H_2 and bromates.

Aqueous HBr dissolves many metals, forming bromides and releasing H_2, e.g. with Fe, Co, Ni, Zn, Al, Sn and Pb.

Silver ion, Ag^+, precipitates Br^- as pale yellow silver bromide, AgBr, changing rapidly to a dark gray when exposed to light. The product is insoluble in, and not decomposed by, HNO_3, soluble in concentrated NH_3, nearly insoluble in concentrated "$(NH_4)_2CO_3$", slightly soluble in excess Br^-, and soluble in CN^- or $S_2O_3^{2-}$. It is slowly transposed by Cl_2.

Bromide may be determined by weighing it as AgBr, or by titrating excess Ag^+ with SCN^-, using Fe^{III} as the indicator.

Silver(+) dismutates Br_2 to AgBr and $AgBrO_3$ or (if cold) HBrO:

$$6\ Ag^+ + 3\ Br_2 + 9\ H_2O \rightarrow 5\ AgBr\downarrow + AgBrO_3\downarrow + 6\ H_3O^+$$

$$Ag^+ + Br_2 + 2\ H_2O \rightarrow AgBr\downarrow + HBrO + H_3O^+$$

Bromate, if more than ~6 mM, can be detected by precipitation as white $AgBrO_3$, soluble in NH_3, easily soluble in HNO_3. It is decomposed by HCl with release of Br_2 (distinction from AgBr). The ions Ba^{2+}, Hg_2^{2+} and Pb^{2+} also give precipitates, but require more BrO_3^-.

Aqueous Hg_2^{2+} yields pale-yellow $[Hg_2Br_2]$, soluble in excess Br^-.

Aqueous Pb^{2+} precipitates, from solutions not too dilute, white $PbBr_2$, appreciably soluble in excess Br^-.

17.4 Iodine, $_{53}I$ and Astatine, $_{85}At$ (and Ununseptium, $_{117}Uus$)

Oxidation numbers: (–I), (I), (III), (IV), (V) and (VII), as in I^-, At^-, IO^-, AtO^-, IO_2^-, IO_2, IO_3^-, AtO_3^-, H_5IO_6 and $AtO_4^- \cdot aq$. For Uus, to be the next member of this Group, relativistic quantum mechanics predicts that Uus, unsurprisingly in this case, can form Uus_2, with Uus^+ most stable, then Uus^{III}, Uus^V and, less stable, Uus^-.

17.4.1 Reagents Derived from Hydrogen and Oxygen

Dihydrogen. The reduction of iodate is catalyzed by Cu^{2+}:

$$IO_3^- + 3\ H_2 \rightarrow I^- + 3\ H_2O$$

Water. Hydrogen iodide is readily soluble in water, with which it forms three hydrates: $HI{\cdot}2H_2O$, $HI{\cdot}3H_2O$ and $HI{\cdot}4H_2O$, all melting between –50 °C and –30 °C. Water and HI form a widely available azeotropic mixture, boiling at 127 °C, containing 57- % HI at 7.6 M. Higher concentrations fume on exposure to air.

The iodides of Pd^{II}, Cu^{I}, Ag, Hg, Tl^{I} and Pb are insoluble in water. Iodides of the other ordinary metals are soluble, those of Sn, Sb and Bi requiring a little free acid. Lead iodide is slightly soluble in hot water, crystallizing out in golden yellow plates as the system cools.

The solubility of iodine in water at 25 °C is 1.2 mM. It differs from Cl_2 and Br_2 in forming no hydrate.

White I_2O_5 is very soluble as iodic acid, HIO_3, and a strong oxidant.

Iodic acid is white and crystalline. It forms various hydrogeniodates such as $M^{I}IO_3{\cdot}HIO_3$ and $M^{I}IO_3{\cdot}2HIO_3$. The very stable $KIO_3{\cdot}HIO_3$ arises on evaporating a solution containing equal amounts of KIO_3 and HIO_3. In analysis, KIO_3 and this salt are of interest because many reductants change them quantitatively to I_2 in dilute acids and to ICl in concentrated HCl. Most iodates are less soluble and more stable than the corresponding chlorates or bromates. In fact, IO_3^- is readily distinguished from ClO_3^- and BrO_3^- by the insolubility of $Ba(IO_3)_2$ and $Pb(IO_3)_2$. The solubilities/kg H_2O at about 20 °C are: $NaIO_3$, 90 g; KIO_3, 81 g; $Ca(IO_3)_2$, 3.7 g; $Ba(IO_3)_2$, 2.2 dg; $AgIO_3$, 6 cg; $TlIO_3$, 5.8 dg; and $Pb(IO_3)_2$ (25 °C), 3 cg.

Periodic acid, HIO_4, is quite soluble as H_5IO_6, which does not lose water at 100 °C, but most periodates are only slightly soluble. In highly acidic solutions, this "orthoperiodic acid" is a powerful oxidant.

The hydrolysis of PI_3 yields HI and H_2PHO_3.

Water forms IOF_5 and HF from IF_7.

Hydroxide. Iodates of the alkalis and alkaline earths are easily made by dismutation from iodine and the hydroxides, separating the also resulting iodides by fractional crystallization.

Iodates are soluble in OH^- if the relevant hydroxides are soluble.

Peroxide. Acidified I^- is oxidized by H_2O_2 to HIO, hence (with HI) to I_2, catalyzed by V^V peroxo complexes, Fe^{2+}, Cu^{2+} and other ions:

$$I^- + H_3O_2^+ \rightarrow HIO + H_2O$$

Precipitated iodine is a dark brown powder; large crystals have a metallic sheen. The vapor is violet and has a characteristic odor. In acidic solution excess H_2O_2 readily oxidizes I_2 further to IO_3^-.

Hydrogen peroxide reduces periodates to iodates, liberating O_2.

Di- and trioxygen. Oxygen and HI, but hardly neutral I^-, give I_2:

$$2\ I^- + {}^1/_2\ O_2 + 2\ H_3O^+ \rightarrow I_2\downarrow + 3\ H_2O$$

The reaction is promoted by traces of HNO_2.

Ozone, however, promptly liberates I_2 from any I^-.

17.4.2 Reagents Derived from the Other 2nd-Period Non-Metals, Boron through Fluorine

Carbon oxide species. Carbonate and $Ba(IO_3)_2$ give $BaCO_3$ and IO_3^-.

Some "simple" organic species. The iodides of Ca, Ba and Hg^{II} are soluble in ethanol; AgI and $[Hg_2I_2]$ are insoluble.

Iodine dissolves well in many organic solvents, and is generally brown in σ-donor chalcogen- or nitrogen-containing solvents, such as H_2O, ROH, R_2O, R_2CO, amines, RCN, R_2S, R_2Se, …; reddish brown or pink in π-donor alkenes and aromatic hydrocarbons such as C_6H_6, but violet like the vapor in weaker donors, such as CH_2Cl_2, CCl_4, CS_2, or aliphatic hydrocarbons. Iodine is readily detected by this color in, say, $CHCl_3$, or when warmed to release the vapor.

Iodic acid (or H_5IO_6) and $H_2C_2O_4$ form CO_2 and I_2, slowly when cold.

A very sensitive test for I_2 is based on the blue color formed with a cold solution of starch. This reaction will reveal 20-μM I_2 and is made more sensitive by the presence of SCN^-.

The large $[NBu_4]^+$ ion, for example, precipitates I_3^- and I_n^-.

Aqueous I_2 slowly bleaches litmus and stains the skin brown. Solid I_2 burns the skin, but aqueous or ethanolic solutions are good antiseptics.

Reduced nitrogen. Nitrogen triiodide ammoniate, $NI_3{\cdot}NH_3$, forms as crystals on dissolving iodine in aqueous NH_3, and is detonated by the slightest disturbance when dry, but is less explosive while wet.

Oxidized nitrogen. Aqueous HI and HNO_2 yield NO and I_2 (separation of I^- from Cl^- and Br^- in some conditions).

Iodic acid and HNO_2 produce HNO_3 with liberation of I_2.

Nitric acid and HI release I_2, which is oxidized slowly to HIO_3 with concentrated acid, a good way to make HIO_3, although HNO_2 is faster:

$$6\ I^- + 2\ NO_3^- + 8\ H_3O^+ \rightarrow 3\ I_2\downarrow + 2\ NO\uparrow + 12\ H_2O$$

$${}^3/_2\ I_2 + 5\ NO_3^- + 5\ H_3O^+ \rightarrow 3\ HIO_3 + 5\ NO\uparrow + 6\ H_2O$$

When small amounts of I^- are sought by testing for oxidation to I_2 (violet color in CCl_4), nitric acid is less liable than Cl_2 to cause errors, as more HNO_3 is required to oxidize the I_2 further (to colorless IO_3^-).

Treating I_2 with fuming HNO_3 gives very pure HIO_3 by evaporating the excess acid, but the process is slow and expensive, and the yield low.

Dilute HNO_3 converts $Na_3H_2IO_6$ into $NaIO_4$ on crystallization.

Most periodates are readily soluble in dilute HNO_3. The insolubility of barium nitrate in concentrated HNO_3, however, makes $Ba_3(H_2IO_6)_2$ well suited to prepare the acid, H_5IO_6.

17.4.3 Reagents Derived from the 3rd-to-5th-Period Non-Metals, Silicon through Xenon

Phosphorus species. Iodic acid and PH_3 form H_3PO_4 and I^- or I_2.

Iodine and P_4 suspended in H_2O yield HI and, with excess I_2, H_3PO_4:

$$^1/_4\, P_4 + {}^5/_2\, I_2 + 9\, H_2O \rightarrow H_3PO_4 + 5\, H_3O^+ + 5\, I^-$$

Distilling HI with red phosphorus, HPH_2O_2 or H_2PHO_3 purifies the HI nicely by reducing any I_3^- (from aerial oxidation) back to I^-:

$$I_3^- + HPH_2O_2 + 3\, H_2O \rightarrow 3\, I^- + H_2PHO_3 + 2\, H_3O^+$$

Phosphorus(<V) and IO_n^- form I^- and phosphonate and/or phosphate in either acid or alkali.

Heating an ionic iodide with concentrated H_3PO_4 and absorbing the gaseous HI in cold H_2O yields "hydriodic acid".

Arsenic species. Iodine suspended in H_2O and treated with AsO_3^{3-} gives I^- and AsO_4^{3-}, because H_3AsO_4 is more acidic than H_3AsO_3, tentatively:

$$I_2 + 3\, AsO_3^{3-} + H_2O \rightarrow 2\, I^- + AsO_4^{3-} + 2\, HAsO_3^{2-}$$

Aqueous I^- is inert to normal AsO_4^{3-}, but the hydrogenarsenates form As^{III} and I_2, so that changes of pH can reverse these two reactions:

$$2\, I^- + 3\, HAsO_4^{2-} \rightarrow I_2\downarrow + HAsO_3^{2-} + 2\, AsO_4^{3-} + H_2O$$

Iodic acid oxidizes As^{III} to As^V, and AsH_3 in excess to As, but an excess of oxidant yields H_3AsO_4.

Reduced chalcogens. One may prepare dilute HI by passing H_2S into finely divided iodine suspended in H_2O. The reaction is slow but good for removing H_2S from AsH_3 (some H_2S may also be oxidized to SO_4^{2-}):

$$I_2 + H_2S + 2\, H_2O \rightarrow 2\, I^- + S\downarrow + 2\, H_3O^+$$

If an insoluble but possible iodide is encountered, it may be treated with H_2S, the insoluble sulfide separated, the excess of H_2S boiled out, and the solution tested for I^-, e.g.:

$$2\ AgI + H_2S + 2\ H_2O \leftrightarrows Ag_2S\downarrow + 2\ H_3O^+ + 2\ I^-$$

Iodic acid and H_2S yield I_2 and S, and some aqueous HI and H_2SO_4:

$$2\ HIO_3 + 5\ H_2S \rightarrow I_2\downarrow + 5\ S\downarrow + 6\ H_2O$$

Thiocyanate and I^- or I_2 are oxidized slowly by O_2 in light, slowly by H_2O_2, and quickly by IO_3^-, showing SCN^- as a "pseudohalide":

$$I_2 + 4\ SCN^- + {}^1/_2\ O_2 + \gamma + 2\ H_3O^+$$

$$\rightarrow 2\ [I(SCN)_2]^- + 3\ H_2O$$

Without oxidant the same complex results from I^- and $(SCN)_2$, and I_2 and SCN^- appear to give $[I_2SCN]^-$. These complexes are rather stable < 10 °C, with $c(H_3O^+)$ at 1–2 M, and with a little excess SCN^-, but they decompose to HSO_4^- etc., as HSCN and HIO_3 (together) do also.

Oxidized chalcogens. Thiosulfate and sulfite species reduce HIO_3 and I_2 to I_2 and I^-. In many titrations $S_2O_3^{2-}$ produces disulfanedisulfonate, $(-S-SO_3^-)_2$, "tetrathionate". In one process to obtain I_2 from Chile saltpeter, IO_3^- is reduced by $NaHSO_3$ as SO_2 (no excess), e.g.:

$$2\ S_2O_3^{2-} + I_2 \rightarrow [S_4O_6]^{2-} + 2\ I^-$$

$$2\ IO_3^- + 5\ SO_2 + 7\ H_2O \rightarrow I_2\downarrow + 5\ HSO_4^- + 3\ H_3O^+$$

$$I_2 + \text{excess}\ SO_2 + 5\ H_2O \rightarrow 2\ I^- + HSO_4^- + 3\ H_3O^+$$

This (SO_2) is often used to detect IO_3^- because it acts rapidly in the cold, but traces of IO_3^- may escape detection, for a slight excess of sulfite reduces the I_2 at once to I^-, which does not color e.g. $CHCl_3$.

Some "clock" reactions use IO_3^- and $HSO_3^-/S_2O_3^{2-}$, and can give oscillations of pH and e.g. green-red-green with indicators [1].

Astatine (At_2) and SO_2 form At^-, coprecipitated with AgI or TlI.

Periodic acid oxidizes SO_2 to H_2SO_4. There is no separation of iodine if the reactants are present in equivalent amounts to form HIO_3.

Cold, dilute H_2SO_4 is inert to HI, but otherwise oxidizes it, requiring higher concentrations of H_2SO_4 at lower T, e.g. 25-% H_2SO_4 at 100 °C, 30 % at 60 °C, 35 % at 50 °C, 40 % at 45 °C and 50 % at 36 °C. Hot, concentrated H_2SO_4 decomposes all iodides, those of Ag, Hg and Pb slowly but completely, also yielding SO_2

and I_2. The I_2 is detected by the violet fumes, which condense on a cooler part of a test tube. Adding excess I^-, however, to boiling H_2SO_4 reduces the latter to H_2S:

$$2\,I^- + 6\,HSO_4^- \rightarrow I_2 + SO_2\uparrow + 5\,SO_4^{2-} + 2\,H_3O^+$$

$$2\,I^- + 5\,H_2SO_4 \rightarrow I_2\uparrow + SO_2\uparrow + 4\,HSO_4^- + 2\,H_3O^+$$

$$8\,I^- + HSO_4^- + 9\,H_3O^+ \rightarrow 4\,I_2\uparrow + H_2S\uparrow + 13\,H_2O\uparrow$$

In a common test for dilute I^- in acid, $[S_2O_8]^{2-}$ liberates I_2 at ordinary temperatures, more rapidly on boiling.

Boiling alkaline $[S_2O_8]^{2-}$ oxidizes iodate to periodate:

$$IO_3^- + [S_2O_8]^{2-} + 4\,OH^- \rightarrow H_2IO_6^{3-} + 2\,SO_4^{2-} + H_2O$$

Astatine and $[S_2O_8]^{2-}$ probably form AtO_3^-, coprecipitated with $AgIO_3$.

Reduced halogens. The interaction of HIO and HCl produces ICl. Thus ICl often results from either the reduction of IO_3^- or the oxidation of I^- in concentrated HCl. This acid and HIO_3 yield ICl and Cl_2 but no I_2.

Under different conditions, $Ba_5(IO_6)_2$, [from heating $Ba(IO_3)_2$] and concentrated HCl yield H_5IO_6, or H_5IO_6 and HCl form HIO_3 and Cl_2, then, conditionally, ICl_3:

$$H_5IO_6 + 2\,Cl^- + 2\,H_3O^+ \rightarrow HIO_3 + Cl_2\uparrow + 5\,H_2O$$

Hydrogen iodate, HIO_3, and HBr form Br_2 and I_2.

Iodine is much more soluble in I^- than in pure water, forming polyiodide salts with linear I_3^-, linear but bent I_5^-, i.e. $I(I_2)_2^-$ etc.

Iodic acid and HI form I_2 from both acids. However, an iodide alone, when acidified, will give a test for I_2 after a short time in the air. A good volumetric method for neutral IO_3^-, nevertheless, is to reduce it with I^- in CH_3CO_2H (weakly acidic) solution and titrate the I_2 with $S_2O_3^{2-}$.

Periodic acid and HI first form IO_3^-. Excess HI gives I_2 or I_3^-.

Elemental and oxidized halogens. Aqueous HI with Cl_2 in excess forms HCl and HIO_3; with excess of HI, HCl and I_2 are produced. Aqueous OH^- gives Cl^- and a periodate:

$$I^- + 4\,Cl_2 + 10\,OH^- \rightarrow H_2IO_6^{3-} + 8\,Cl^- + 4\,H_2O$$

When a solution is to be tested for I^-, a common reagent is chlorine water. The iodine is recognized by the violet color when shaken with e.g. $CHCl_3$. If only a small amount of I^- may be present, chlorine must be added very cautiously or the I^- will be oxidized to IO_3^- and no violet color obtained. (Nitric acid is better for this.)

As a source of I_2, seaweed ash filtrate is evaporated and the residue treated with H_2SO_4 and Cl_2. The liberated I_2 is collected and purified.

Chlorine or Br_2 reacts directly with iodine to form ICl or IBr. An excess of Cl_2 gives ICl_3. Water then yields I_2 and HIO_3. Treating I_2 in water with excess Cl_2 forms HIO_3 (periodate in alkali) directly. An alkaline system containing an excess of Br_2 forms iodate and bromide:

$$5\ ICl + 8\ H_2O \rightarrow 2\ I_2\downarrow + HIO_3 + 5\ Cl^- + 5\ H_3O^+$$

$${}^1/_2\ I_2 + {}^5/_2\ Cl_2 + 8\ H_2O \rightarrow HIO_3 + 5\ H_3O^+ + 5\ Cl^-$$

$$ {}^1/_2\ I_2 + {}^7/_2\ Cl_2 + 10\ OH^- \rightarrow 7\ Cl^- + H_2IO_6^{3-} + 4\ H_2O$$

$$ {}^1/_2\ I_2 + {}^5/_2\ Br_2 + 6\ OH^- \rightarrow IO_3^- + 5\ Br^- + 3\ H_2O$$

Aqueous HI or I^- plus Br_2 first form I_2 and HBr or Br^-.

Slowly adding Br_2 or Cl_2 to alkaline I^- at 80 °C provides a periodate:

$$I^- + 4\ Br_2 + 9\ OH^- \rightarrow H_3IO_6^{2-} + 8\ Br^- + 3\ H_2O$$

Bromine oxidizes astatine, perhaps to AtO^- or AtO_2^-.

Aqueous HClO oxidizes I^- to I_2, then to HIO_3 in acidic solution.

Astatine and ClO^- probably form AtO_3^-, coprecipitated with $AgIO_3$.

Hypoiodous acid, HIO, is very unstable. It quickly dismutates into (acidified) I^- and IO_3^-.

Treating alkaline IO_3^- with Cl_2 or (boiling) ClO^- is a very satisfactory way to prepare $H_3IO_6^{2-}$ or $H_2IO_6^{3-}$. A sodium salt, $Na_3H_2IO_6$, is fairly insoluble in the medium and separates as easily removed crystals:

$$IO_3^- + ClO^- + 2\ OH^- \rightarrow H_2IO_6^{3-} + Cl^-$$

Aqueous $HClO_3$ added to excess HI gives HCl and I_2. One convenient method to prepare HIO_3 is to oxidize I_2 with an excess of 25 % $HClO_3$, estimated to be around 3–4 M. Excess neutral ClO_3^-, with slight initial acidification by HNO_3, oxidizes I_2 also to HIO_3 at about 50 °C:

$$3\ I_2 + 5\ ClO_3^- + 3\ H_2O \rightarrow 6\ HIO_3 + 5\ Cl^-$$

Aqueous $HBrO_3$ reacts with a large amount of HI to produce I_2 and HBr; with excess $HBrO_3$ the iodine becomes HIO_3.

Iodic or periodic acid plus an iodide release iodine:

$$HIO_3 + 5\ I^- + 5\ H_3O^+ \rightarrow 3\ I_2\downarrow + 8\ H_2O$$

Periodic acid may be prepared by oxidizing I_2 with $HClO_4$:

$$ {}^1/_2\ I_2 + ClO_4^- + H_3O^+ + H_2O \rightarrow H_5IO_6 + {}^1/_2\ Cl_2\uparrow$$

17.4.4 Reagents Derived from the Metals Lithium through Uranium, plus Electrons and Photons

Oxidation. Aqueous HI and nearly all metallic oxides, hydroxides and carbonates (except ignited Cr_2O_3) form iodides, often, however, along with I_2 and a lower metallic oxidation state.

Iodide does not reduce CrO_4^{2-}, even if boiling and concentrated, but $[Cr_2O_7]^{2-}$ reacts with I^- giving I_2 as a solid in the cold (slowly), or as a vapor with heat (separation from Cl^-):

$$6\,I^- + 5\,[Cr_2O_7]^{2-} \rightarrow 3\,I_2\downarrow\uparrow + Cr_2O_3\cdot aq\downarrow + 8\,CrO_4^{2-}$$

Boiling AgI with $[Cr_2O_7]^{2-}$ and H_2SO_4 yields $AgIO_3$ and Cr^{III} but no I_2:

$$AgI + [Cr_2O_7]^{2-} + 2\,HSO_4^- + 6\,H_3O^+ \rightarrow$$

$$AgIO_3\downarrow + \text{e.g. } 2\,CrSO_4^+ + 10\,H_2O$$

Manganese(>II) and I^- become Mn^{2+} and I_2. Thus, seaweed-ash filtrate is evaporated and treated with H_2SO_4 and $MnO_2\cdot aq$.

Boiling I^- with excess MnO_4^- yields IO_3^- (distinction from Br^-):

$$2\,I^- + MnO_2\cdot aq + 4\,H_3O^+ \rightarrow I_2 + Mn^{2+} + 6\,H_2O$$

$$I^- + 2\,MnO_4^- + H_2O \rightarrow IO_3^- + 2\,MnO_2\cdot aq\downarrow + 2\,OH^-$$

Iron(III) yields Fe^{2+} and I_2 (distinction from Br^- and Cl^-). Aqueous HI and $H_3[Fe(CN)_6]$ give $H_4[Fe(CN)_6]$ and I_2, also partly in neutral solution:

$$I^- + [Fe(CN)_6]^{3-} + 2\,H_3O^+ \rightarrow H_2[Fe(CN)_6]^{2-} + {}^1/_2\,I_2 + 2\,H_2O$$

Iron(III) also oxidizes astatine.

Cobalt(III) and nickel(III or IV) oxides all yield M^{2+} and I_2 or I_3^- with acidified I^-. Similarly, Pb^{IV}, Sb^V and Bi^V become PbI_2. $[SbI_4^-]$ (or a basic salt) and $[BiI_4^-]$ in turn.

Electrolyzing I^- does not yield much HIO because of its instability.

Like the other halogenates, IO_3^- can be prepared by electrolyzing an iodide in alkaline solution, using a cell without a diaphragm:

$$I^- + 6\,OH^- \rightarrow IO_3^- + 3\,H_2O + 6\,e^-$$

Periodates may be prepared well by the further anodic electrolysis of an alkaline solution of iodate. A low current density, cold electrolyte, and a small amount of chromate ion favor the reaction.

Reduction. Warm, solid I_2 is reduced by and slowly oxidizes Ag and Pb (to iodides); more rapidly with Alk, Ae, Rth, An, Cr, Mn, Fe, Co, Ni, Cu, Zn, Cd, Hg, Al,

Sn, As, Sb and Bi. In either acidic or alkaline media, I_2 oxidizes e.g. Hg_2^{2+} to 2 HgI^+, and Sn^{2+} to Sn^{IV}. Only in an alkaline medium, Mn^{2+} goes to $MnO_2 \cdot aq$, Cr^{III} to Cr^{VI}, Fe^{2+} to Fe^{III}, Co^{2+} to Co^{III}, As^{III} to As^V, and Sb^{III} to Sb^V.

Excess H_5IO_6 readily oxidizes Mn^{2+} quantitatively to MnO_4^- in hot acidic solution, and it easily oxidizes Fe^{2+} or Cu^+ to Fe^{III} or Cu^{2+}, the former forming an insoluble periodate in nitric-acid solution.

Alkali iodides may be obtained by first preparing FeI_2, and then treating the product with the carbonate of the desired alkali:

$$Fe + I_2 \rightarrow FeI_2$$

$$FeI_2 + Rb_2CO_3 \rightarrow FeCO_3\downarrow + 2\ RbI, \text{ or really, of course:}$$

$$Fe^{2+} + CO_3^{2-} \rightarrow FeCO_3\downarrow$$

Iodine oxidizes $[Fe(CN)_6]^{4-}$ slowly and partially to $[Fe(CN)_6]^{3-}$.

Iodic acid and $H_4[Fe(CN)_6]$ yield $H_3[Fe(CN)_6]$ and I_2.

Some metal ions reduce HIO_3, e.g.: $Fe^{2+} \rightarrow Fe^{III}$ and I_2; and $Cu^+ \rightarrow CuI$ or Cu^{2+} and I_2. I.e., if the Cu^+ is in excess, the expected I^- gives CuI (less soluble than CuCl or CuBr from ClO_3^- or BrO_3^-). Also $Sn^{II} \rightarrow Sn^{IV}$ and I^- or I_2; and $Sb^{III} \rightarrow Sb^V$ and I^- or I_2; but $SbH_3 \rightarrow Sb$ and I^- or I_2. Any I^-, however, may complex e.g. Sn^{IV}.

Unlike the chlorates and perchlorates, both iodate and periodate are reduced or unaffected by the same reagents. Among the metals, Fe and Zn are readily attacked by periodic acid, Cu forms an iodate, but Hg, Sn and Pb are only slightly affected.

Traces of Pt from electrolytic oxidation apparently catalyze the slow reduction of H_5IO_6 by water to HIO_3.

Iodine suspended in H_2O and treated with Sn^{2+} yields HI.

Other reactions. Many polyhalogen salts can be made with large Cat^+; e.g. purple CsI_3 from CsI plus I_2; dark-orange $CsIBr_2$ from CsI plus Br_2; and with organic cations, e.g. yellow $[NMe_4]BrCl_2$ from $[NMe_4]Br$ plus Cl_2 in CH_3CO_2H; green $[NPr_4]I_7$ from $[NPr_4]I$ plus I_2 in EtOH; or yellow $[NBu_4][ICl_2]$ from $[NBu_4]Cl$ plus ICl in CH_3CO_2H. These show no isomerism, all having the higher-Z halogen atom at the center.

Aqueous Ba^{2+} precipitates IO_3^- as barium iodate, $Ba(IO_3)_2$, slightly soluble in cold, more soluble in hot water, insoluble in ethanol, soluble in hot, dilute HNO_3, readily soluble in cold dilute HCl; hence dilute solutions of HIO_3 should be neutralized before testing with Ba^{2+}. This iodate is readily separated from iodides by extraction of the latter with ethanol. When well washed, treated with a little SO_2 (no excess) and found to color $CHCl_3$ violet, its evidence for IO_3^- is conclusive. The precipitation of IO_3^- by Ba^{2+} separates it from ClO_3^- and BrO_3^-.

Remarkably, η^2-IO_6^{5-} greatly stabilizes often-unstable high oxidation states in $(Ce,Pd,Pt)^{IV}(IO_6)_2^{6-}$, $(Mn,Ni)^{IV}IO_6^-$, $[Mn^{IV}(IO_6)_3]^{11-}$, $(Fe,Co)^{III}IO_6^{2-}$, $(Fe,Co)^{III}{}_4(IO_6)_3^{3-}$, and $M^{III}(IO_6)_2^{7-}$ (M = Fe, Co, Cu, Ag or Au) etc.

Sufficiently concentrated Fe^{III} and IO_3^- precipitate pale-yellow $Fe(IO_3)_3$, slightly soluble in H_2O, readily soluble in excess iodate.

Palladium dichloride precipitates I^- as black PdI_2, insoluble in water or ethanol (distinction from Br^-), slightly soluble in excess I^-, but soluble in NH_3. This is a very good test for traces of I^-, visible to 6-μM PdI_2. Iodides may be determined by precipitation as PdI_2, which is dried and weighed, or ignited to Pd and weighed.

Adding Cu^{2+} to I^- yields white CuI mixed with iodine:

$$2\ Cu^{2+} + 4\ I^- \rightarrow 2\ CuI\downarrow + I_2\downarrow$$

If sufficient reductant, e.g., SO_2, is present to reduce the I_2 to I^-, or Cu^{II} to Cu^I, only the white CuI will be precipitated (distinction from Cl^- and Br^-) with no net redox action on the I^-. Iodide in the mineral waters of Java was said to be isolated as CuI.

Aqueous Ag^+ precipitates I^- as pale yellow silver iodide, AgI. The product blackens in the light without appreciable separation of iodine and becomes practically white when treated with NH_3. In the latter case washing with water restores the original color. Iodides may be determined by precipitation as AgI, and dried and weighed as such.

Volumetrically, iodides may be treated with an excess of Ag^+ and the excess titrated with SCN^-, using Fe^{III} as indicator; the AgI need not be removed. The Fe^{III} should be added after the I^- has been precipitated.

The action of Ag^+ on I_2 forms HIO_3 by dismutation:

$$5\ Ag^+ + 3\ I_2 + 8\ H_2O \rightarrow 5\ AgI\downarrow + 5\ H_3O^+ + HIO_3$$

If an insoluble but possible iodide is encountered, it may be treated with Zn and perhaps some H_2SO_4, and the solute tested for I^-:

$$2\ AgI + Zn \rightarrow 2\ Ag\downarrow + Zn^{2+} + 2\ I^-$$

A solution of Ag^+ precipitates IO_3^- as white silver iodate, $AgIO_3$, crystalline, soluble in NH_3, in excess of hot HNO_3 and in HIO_3. This may be dried and weighed for the determination of iodate.

Silver nitrate, added to a periodate, forms a precipitate the color and nature of which depend e.g. on the amount of extra HNO_3. Confusing reports find products varying from slate-colored $Ag_3H_2IO_6$ to orange $AgIO_4$. Boiling the silver periodates in water causes them to become dark red. The freshly prepared compounds are readily soluble in NH_3.

Solutions of Hg_2^{2+} precipitate I^- as $[Hg_2I_2]$, yellow to green. Aqueous Hg^{2+} reacts with I^- to form first a yellow HgI_2, quickly changing to a red form. The precipitate dissolves on stirring, forming HgI^+ with excess Hg^{2+}; or $[HgI_3]^-$ or $[HgI_4]^{2-}$ with little or more excess I^-, until nearly equivalent amounts (for HgI_2) are present, when the color deepens.

Dimercury(I), Hg_2^{2+}, gives a pale yellow precipitate with IO_3^-, insoluble in dilute HNO_3 but soluble in HIO_3. Mercury(2+) also gives a precipitate with IO_3^- (distinction from ClO_3^- and BrO_3^-).

Aqueous Hg_2^{2+} forms a yellow precipitate with periodates. Aqueous Hg^{II} yields a red-orange precipitate of $Hg_4I_2O_{11}$. This and sometimes the characteristic precipitates with Ag(+) are used to identify periodates.

Thallium(I) ions, Tl^+, added to I^-, precipitate yellow TlI.

Adding Pb^{2+} to cold, not too dilute, I^-, precipitates bright-yellow PbI_2, It is appreciably soluble in excess I^- and in hot water.

Lead(II) gives, with IO_3^-, a white precipitate of $Pb(IO_3)_2$. This may be dried and weighed to determine iodate.

Aqueous Pb^{2+} forms a white precipitate when added to a periodate solution slightly acidified with HNO_3. It turns yellow when heated.

Reference

1. Pfennig BW, Roberts RT (2006) J Chem Educ 83:1804

Bibliography

See the general references in the Introduction, and some more-specialized books [2–12]. Some articles in journals discuss mixed octahedral complexes [13]; heterogeneous redox catalysts for chlorine release [14]; and Cl_2O [15].

2. Banks RE (ed) (2000) Fluorine chemistry at the millennium. Elsevier, Amsterdam
3. Nakajima T, Zemva B, Tressaud A (eds) (2000) Advanced inorganic fluorides. Elsevier, Amsterdam
4. Howe-Grant M (ed) (1995) Fluorine chemistry: a comprehensive treatment. Wiley, New York
5. Miyamoto H, Salomon M (1987) Alkali metal halates, ammonium iodate and iodic acid. IUPAC, Blackwell, London
6. Schilt AA (1979) Perchloric acid and perchlorates. G. Frederick Smith Chemical Co, Columbus
7. Downs AJ, Adams CJ (1973) The chemistry of chlorine, bromine, iodine and astatine. Pergamon, Oxford
8. Emeléus HJ (1969) The chemistry of fluorine and its compounds. Academic, San Diego
9. Gutmann V (ed) (1967) Halogen chemistry. Academic, San Diego
10. Jolles ZE (1966) Bromine and its compounds. Academic, San Diego
11. Schumacher JC (1960) Perchlorates. Reinhold, New York
12. Simmons JH (1950) Fluorine chemistry. Academic, San Diego
13. Preetz W, Peters G, Bublitz D (1996) Chem Rev 96:977
14. Mills A (1989) Chem Soc Rev 18:285
15. Renard JJ, Bolker HI (1976) Chem Rev 76:487

18 Helium through Radon, the Aerogens

These elements are also known as the noble or inert gasses, although not all are inert, and "aerogens" is proposed as the Group name [1].

18.0 Helium, $_{2}$He through 18.2 Argon, $_{18}$Ar

Compounds of He, Ne and Ar in water are not known.

18.3 Krypton, $_{36}$Kr

Water. Water instantly decomposes KrF_2 as follows:

$$KrF_2 + H_2O \rightarrow Kr\uparrow + 2\ HF + {}^1/_2\ O_2\uparrow$$

18.4 Xenon, $_{54}$Xe and Radon, $_{86}$Rn (and Ununoctium, $_{118}$Uuo)

Oxidation numbers: (II), (IV), (VI) and (VIII), as in XeF_2 and Rn^{2+}, XeF_4, XeO_3 and $HXeO_6{}^{3-}$. Relativistic calculations suggest stability for Rn^{2+} and Rn^{IV}. Ununoctium is predicted to be most stable as Uuo^{IV}, then Uuo^{II} and Uuo^{-}, but with Uuo^{VI} less so.

For still higher *Z*, relativity and other strong effects make groupings and predictions based on ordinary periodic-chart trends unreliable.

18.4.1 Reagents Derived from Hydrogen and Oxygen

Water. Water completely decomposes the aerogen fluorides (colorless crystals and powerful oxidants from non-aqueous sources), very slowly for XeF_2, as follows (A = Xe or Rn):

$$AF_2 + H_2O \rightarrow A\uparrow + 2\ HF + {}^1/_2\ O_2\uparrow$$

The hydrolyses of XeF_4 (to Xe, XeO_3, HF and O_2) and XeF_6 (to $XeOF_4$, XeO_2F_2, XeO_3 and HF) are fast or violent.

Anhydrous XeO_3 slowly becomes Xe and O_2, is quite explosive, colorless, crystalline, deliquescent, soluble up to 11 M, feebly acidic (containing little H_2XeO_4) and stable in slightly acidified water.

Oxonium. To determine a base equivalent:

$$Na_4XeO_6{\cdot}aq + 4\ H_3O^+ \rightarrow XeO_3 + {}^1/_2\ O_2\uparrow + 4\ Na^+ + 6\ H_2O$$

The end point in back titrating excess acid in the latter reaction is found by potentiometry; XeO_3 destroys most indicators.

Hydroxide. Xenate slowly dismutates in OH^-, thereby giving $Na_4XeO_6{\cdot}nH_2O$, $K_4XeO_6{\cdot}9H_2O$ or Ba_2XeO_6:

$$2\ HXeO_4^- + 2\ OH^- \rightarrow XeO_6^{4-} + Xe\uparrow + O_2\uparrow + 2\ H_2O$$

Trioxygen. The stoichiometry is not certain for the following:

$$HXeO_4^- + O_3 + 3\ OH^- + 4\ Na^+ \rightarrow Na_4XeO_6{\cdot}aq\downarrow + O_2\uparrow + 2\ H_2O$$

The product is stable at ambient T if dry and airless, soluble to 25 mM, somewhat hydrolyzed, decomposing to Xe^{VI} and O_2 with a $t_{1/2}$ of several days, but reacting immediately below pH 7.

18.4.2 Reagents Derived from the Other 2nd-Period Non-Metals, Boron through Fluorine

Some "simple" organic species. Non-aqueous media adsorb and elute Rn^{2+} on cation-exchange resins, showing some metallicity for Rn.

18.4.3 Reagents Derived from the 3rd-to-5th-Period Non-Metals, Silicon through Xenon

Oxidized chalcogens. Concentrated H_2SO_4, slowly added to $Na_4XeO_4{\cdot}aq$ or Ba_2XeO_4, gives very explosive and strongly oxidizing XeO_4 gas.

Reduced halogens. Aqueous XeO_3 and Cl^-, Br^- or I^- yield Xe and X_2.

For analysis, one may titrate liberated I_3^- with $S_2O_3^{2-}$:

$$XeO_3 + 9\ I^- + 6\ H_3O^+ \rightarrow Xe\uparrow + 3\ I_3^- + 9\ H_2O$$

To determine the total oxidizing power:

$$Na_4XeO_6{\cdot}aq + 12\ I^- + 12\ H_3O^+ \rightarrow Xe\uparrow + 4\ I_3^- + 4\ Na^+ + 18\ H_2O$$

Oxidized halogens. The oxidant XeF_2 is stronger than BrO_4^-:

$$XeF_2 + BrO_3^- + H_2O \rightarrow Xe\uparrow + BrO_4^- + 2\ HF$$

18.4.4 Reagents Derived from the Metals Lithium through Uranium, plus Electrons and Photons

Reduction. Light generates Xe and O_2 slowly from aqueous XeO_3.

Reference

1. Noyes RM (1963) J Amer Chem Soc 85:2202

Bibliography

See the general references in the Introduction, and some more-specialized books [2–6]. Some articles in journals include: a renaissance in noble-gas chemistry [7]; new noble-gas chemistry [8], radon's non-nobility [9] and (older) new noble-gas chemistry [10].

2. Hawkins DT, Falconer WE, Bartlett N (1978) Noble gas compounds, a bibliography 1962-1976. Plenum, New York
3. Holloway JH (1968) Noble-gas chemistry. Methuen, London
4. Claassen HH (1966) The noble gases. Heath, Boston
5. Moody GJ, Thomas JDR (1964) Noble gases and their compounds. Macmillan, New York
6. Hyman HH (1963) Noble gas compounds. University of Chicago, Chicago
7. Christe KO (2001) Angew Chem Int Ed 40:1419
8. Holloway JH, Hope EG (1999) Adv Inorg Chem 46:51
9. Lee JD, Edmonds TE (1991) Educ Chem 27:152
10. Seppelt K, Lentz D (1982) in Lippard SJ (ed) Prog Inorg Chem 29:167

Appendix A: Periodic Charts

Hundreds of useful periodic charts have been published [Introduction ref. 1], and only a few newer ones will be shown here.

Because relativistic effects in the high-Z elements may still be generally under-recognized, we present a summary in a periodic chart as Fig. A.1, reprinted with permission from Pyykkö P., Chem. Rev. 88:586. Copyright (1988) the American Chemical Society.

Oxides and hydroxides, and their acidities and basicities, are crucial in aqueous inorganic chemistry; moreover, their periodic relationships are of great interest; we therefore present graphical comparisons next in Figs. A.2 through A.8 [1]. The solid and dashed curves represent the solubilities of the fresh precipitates (often more soluble) to aged products (less so) in turn. Supersaturated solutions are above the curves, and unsaturated ones below them. The element's aqua, hydroxo or oxo ion, calculated as mononuclear, is written as M'.

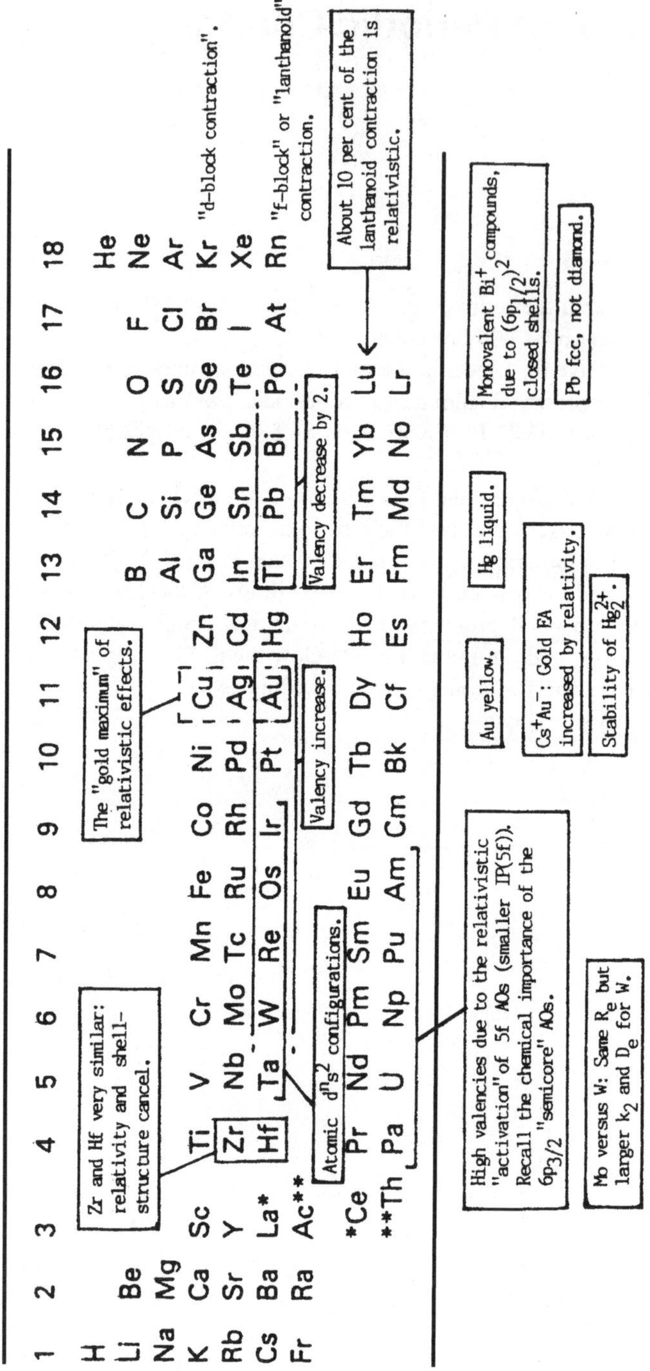

Fig. A.1. Some effects of relativity on properties of the elements

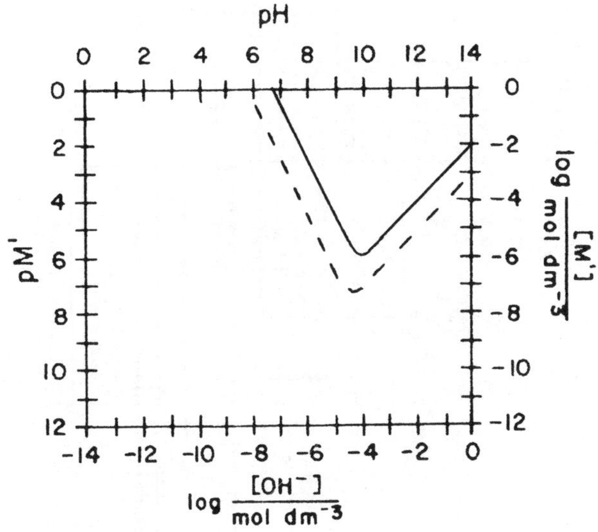

Fig. A.2. Key to Figs. A.3 through A.8

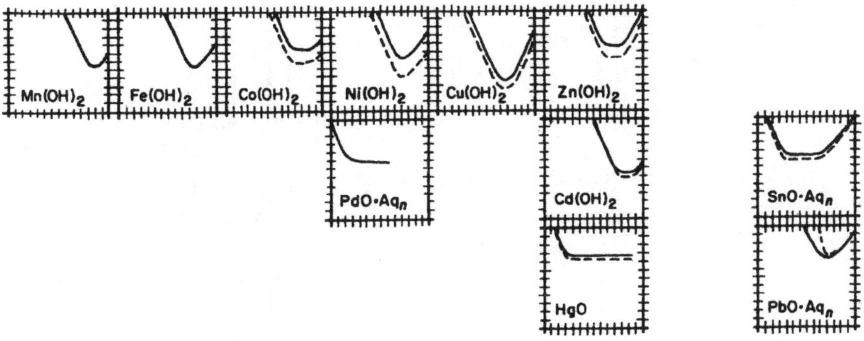

Fig. A.3. Solubility vs pH for oxides and hydrous oxides of $\mathbf{d}^{II}$ and $\mathbf{p}^{II}$

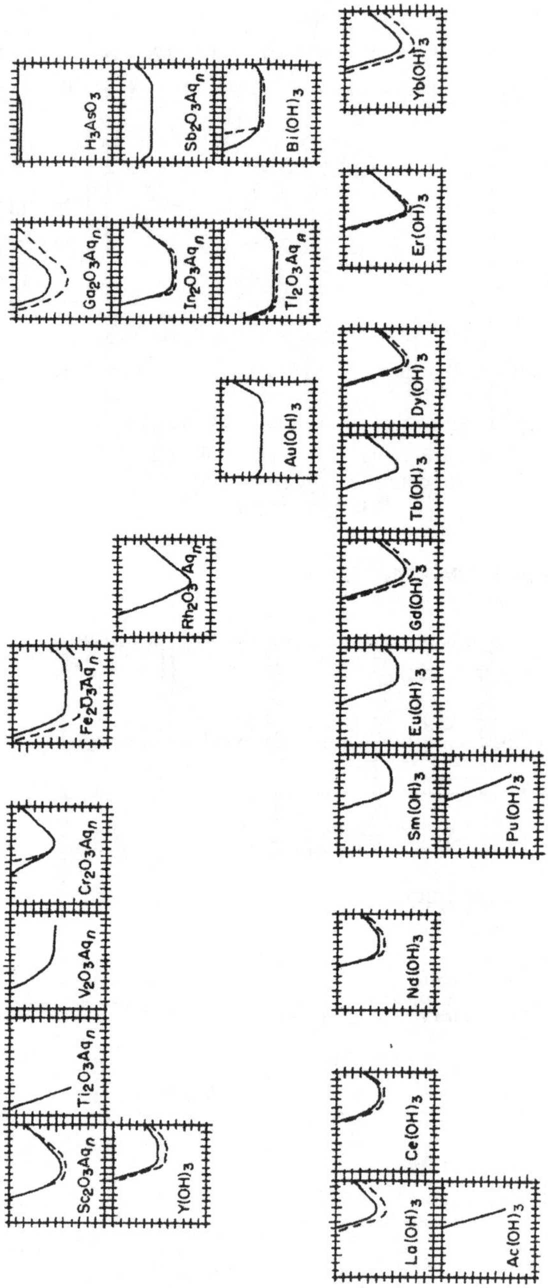

Fig. A.4. Solubility vs pH for oxides and hydrous oxides of $\mathbf{d}^{III}$, $\mathbf{p}^{III}$ and $\mathbf{f}^{III}$

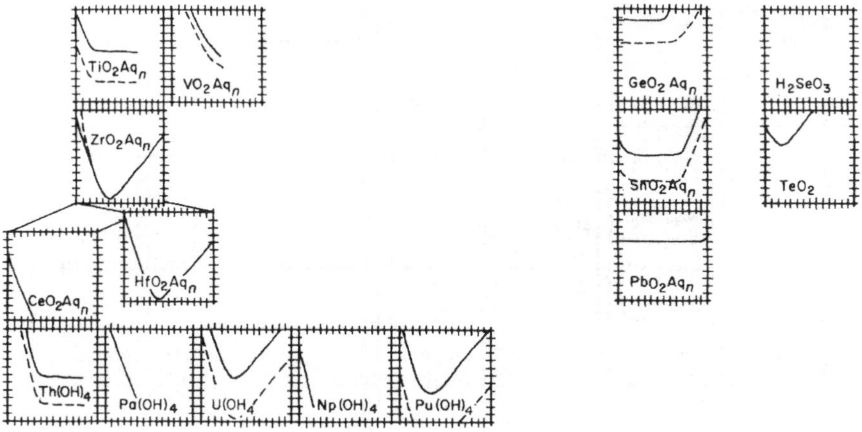

Fig. A.5. Solubility vs pH for oxides and hydrous oxides of $\mathbf{d}^{IV}$, $\mathbf{p}^{IV}$ and $\mathbf{f}^{IV}$

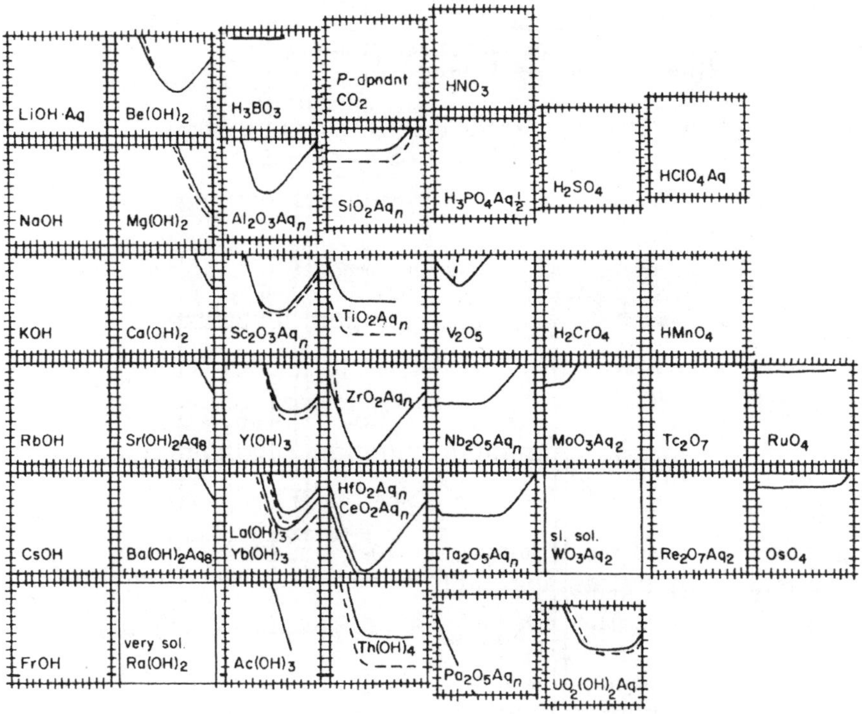

Fig. A.6. Solubility vs pH for hydroxide-oxides of noble-gas-type cations

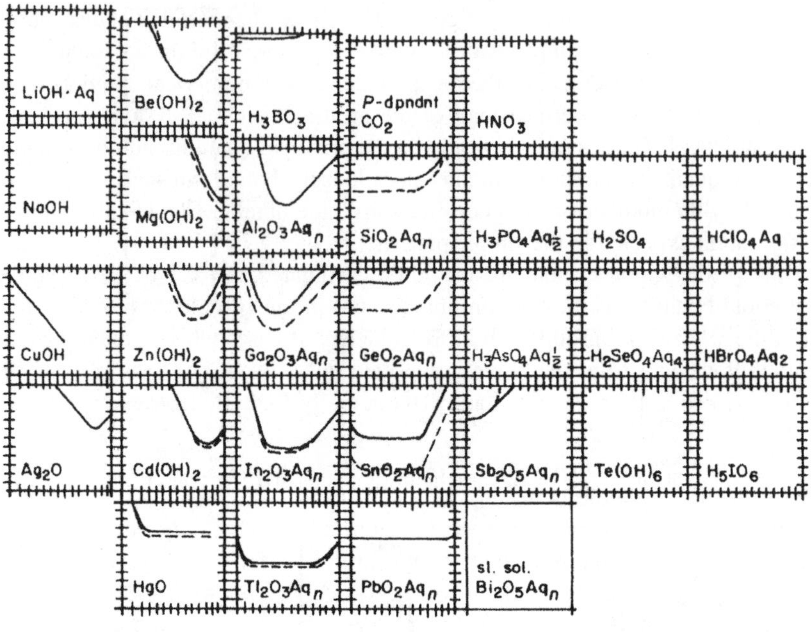

Fig. A.7. Solubility vs pH for hydroxide-oxides of $\mathbf{d}^{10}$ and related cations

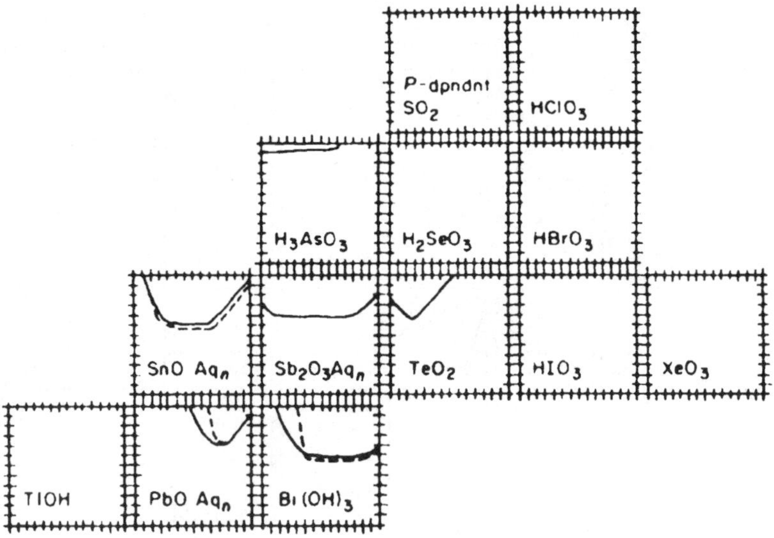

Fig. A.8. Solubility vs pH for hydroxide-oxides of $\mathbf{p}^2$ and $\mathbf{d}^{10}\mathbf{p}^2$ cations

The electronic configurations of isolated atoms in their ground states show an intriguing mixture of a simple pattern with exceptions. It might be convenient at times to be able to refer to the chemical elements with numerical symbols having digits based on that pattern instead of the arbitrary decimal system used in the atomic numbers. These symbols would be the principal quantum number n and the secondary quantum number l for the most loosely bound subshell of the atom, followed by the number of "valence electrons", the number in that subshell of the configuration expected in the simple pattern.

Thus H, the simplest element, would be designated as 1.0.1 or 101. The third digit could be taken as the column number, not in the 18-column periodic charts, but in the relevant subgroup even in the minority of cases where exceptional configurations deviate from the naively expected ones. In Ag for example, we have 4.2.9 or 429 ($n = 4$, $l = 2$) even though we actually find 10 electrons, not 9, in the **4d** subshell for an isolated atom of Ag in the ground state.

We use the hexadecimal number system with A through E for 10 through 14, in order to retain single digits for counting the subshell electrons. We may note that although the familiar atomic numbers have one, two or three digits, every element has three in this designation, which we put forth now for the first 120 elements in Table A.1, following Corbino's format [2] for the periodic chart with some extrapolation.

A very different periodic chart, Fig. A.9, of the medical uses of metallic species, may also be both interesting and useful.

Table A.1. A Periodic chart with numerical symbols based on idealized electronic configurations

01	02	11	12	13	14	15	16	21	22	23	24	25	26	27	28	29	2A	31	32	33	34	35	36	37	38	39	3A	3B	3C	3D	3E
101	102																														
H	He																														
201	202	211	212	213	214	215	216																								
Li	Be	B	C	N	O	F	Ne																								
301	302	311	312	313	314	315	316	321	322	323	324	325	326	327	328	329	32A														
Na	Mg	Al	Si	P	S	Cl	Ar	Sc	Ti	V	Cr	Mn	Fe	Co	Ni	Cu	Zn														
401	402	411	412	413	414	415	416	421	422	423	424	425	426	427	428	429	42A	431	432	433	434	435	436	437	438	439	43A	43B	43C	43D	43E
K	Ca	Ga	Ge	As	Se	Br	Kr	Y	Zr	Nb	Mo	Tc	Ru	Rh	Pd	Ag	Cd	La	Ce	Pr	Nd	Pm	Sm	Eu	Gd	Tb	Dy	Ho	Er	Tm	Yb
501	502	511	512	513	514	515	516	521	522	523	524	525	526	527	528	529	52A	531	532	533	534	535	536	537	538	539	53A	53B	53C	53D	53E
Rb	Sr	In	Sn	Sb	Te	I	Xe	Lu	Hf	Ta	W	Re	Os	Ir	Pt	Au	Hg	Ac	Th	Pa	U	Np	Pu	Am	Cm	Bk	Cf	Es	Fm	Md	No
601	602	611	612	613	614	615	616	621	622	623	624	625	626	627	628	629	62A														
Cs	Ba	Tl	Pb	Bi	Po	At	Rn	Lr	Rf	Db	Sg	Bh	Hs	Mt	Ds	Rg	Uub														
701	702	711	712	713	714	715	716																								
Fr	Ra	Uut	Uuq	Uup	Uuh	Uus	Uuo																								
801	802																														
Uue	Ubn																														

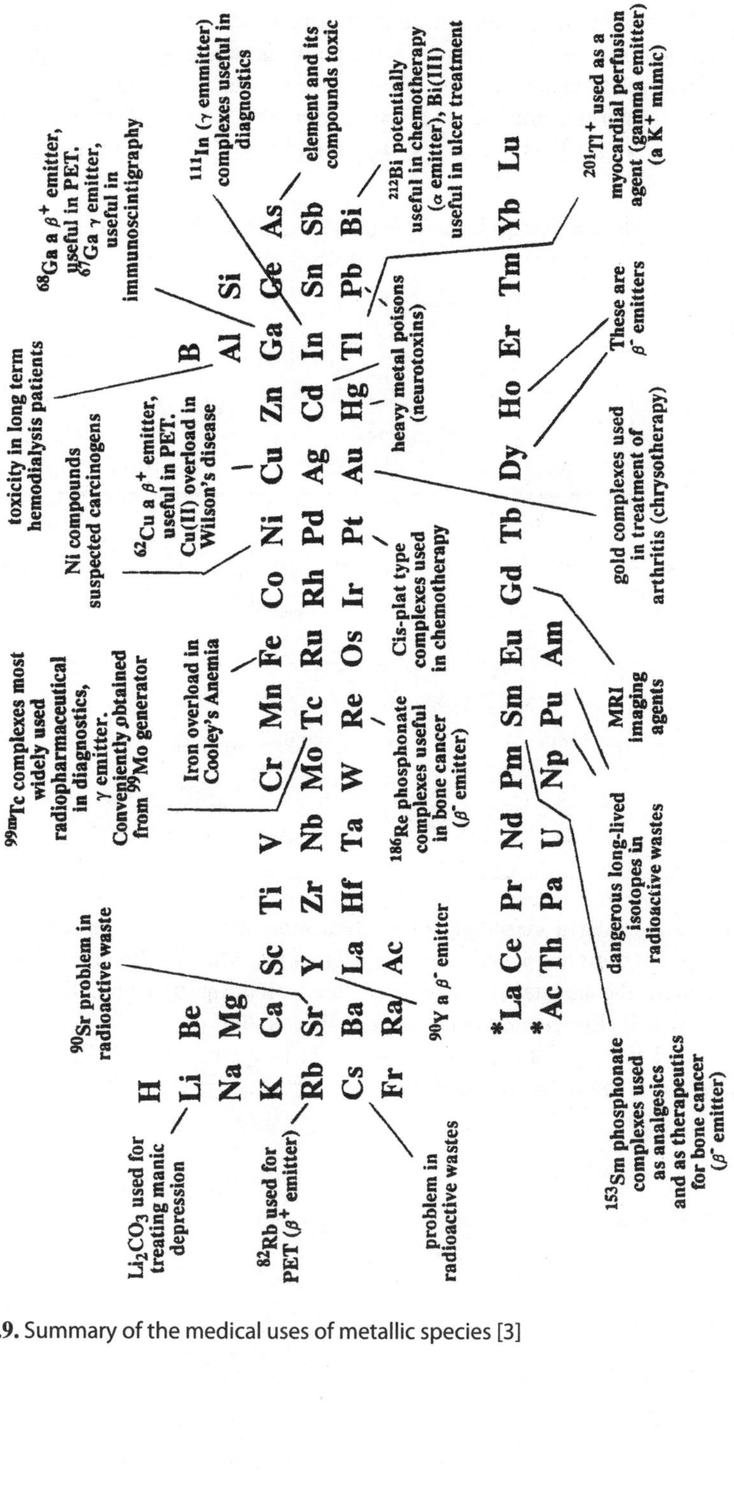

Fig. A.9. Summary of the medical uses of metallic species [3]

Table A.2, [4] modified, outlines the "Bromide" system of qualitative analysis for metals, whose analytical groups, with group numbers above each group, expose simpler relationships in the periodic chart than does the traditional H_2S scheme. The group reagents are: (1) HBr, a reductant, a large cation R^+, I^- and e.g. CH_2Cl_2; (2) $[Co(CN)_6]^{3-}$; (3) $C_6H_5(N_2O_2)^-$ or HPO_4^{2-}; (4) F^- and (5) none.

Table A.2. Groups in the bromide scheme of qualitative analysis

4	3														
Li	Be														
5	4	3													
Na	Mg	Al													
							2				1				
K	Ca	Sc	Ti	V	Cr	Mn	Fe	Co	Ni	Cu	Zn	Ga	Ge	As	Se
Rb	Sr	Y	Zr	Nb	Mo	Tc	Ru	Rh	Pd	Ag	Cd	In	Sn	Sb	Te
Cs	Ba	Ln	Hf	Ta	W	Re	Os	Ir	Pt	Au	Hg	Tl	Pb	Bi	Po
Fr	Ra	Ac	Th	Pa	U										

Next we consider a taxonomy of relationships, periodic in a broad sense and otherwise, with more explanation in the original [5]. Most useful chemical comparisons assume the constancy of at least three of the parameters: **F**, isofamilial (CO_3^{2-}, Pb^{2+}); **P**, isoperiodic (PH_3, ClO^-); **R**, isoradial (F^-, Au^+); **S**, isostructural (CF_4, $[FeCl_4]^-$); **V**, isovalent (NO_2, $ThCl_4$); **T**, isotypical (Hg, Sb^{3+}); and **C**, isocharacteristic, a catch-all, in Fig. A.10.

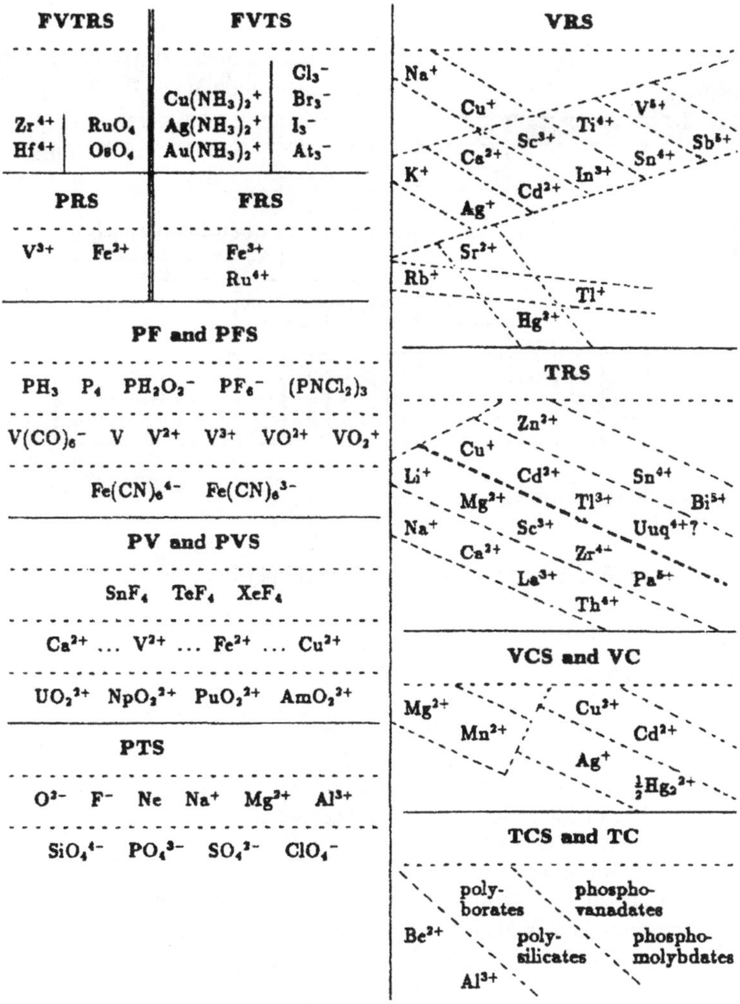

Fig. A.10. A taxonomy of relationships

One more periodic chart [6] shows in one place some additional similarities, mentioned separately in this text. It has the pseudohalogen CN with the halogens, NH_4 with the alkali metals, and some **p**-block elements and early actinoids with the **d**-block. It also suggests the similarity of Al^{III} to Fe (as Fe^{III}) (although it could well include Cr^{III} too), as well as some diagonal and "knight's move" (Cu through In, Bi etc.) relationships. We now know Uun as Ds and Uuu as Rg. See Fig. A.11.

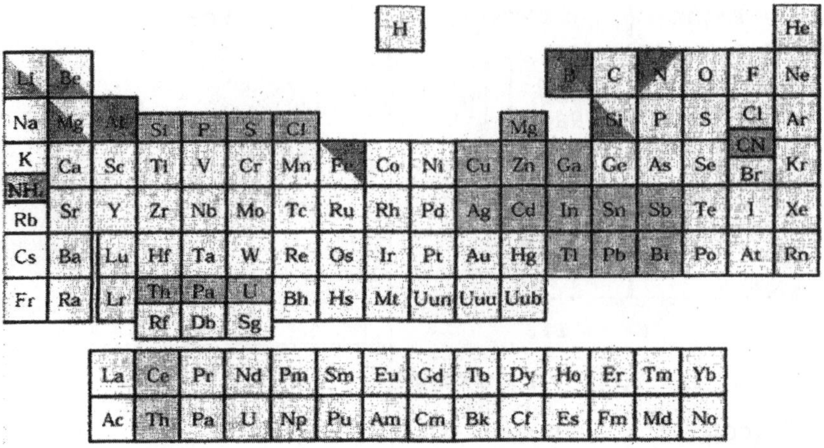

Fig. A.11. The Rayner-Canham chart

The Internet offers a highly flexible and useful periodic chart [7].

References

1. Rich RL (1985) J Chem Educ 62:44
2. Corbino OM (1928) Riv Nuovo Cimento 5:LXI
3. Martell AE, Hancock RD (1996) Metal complexes in aqueous solution. Plenum, New York, p 152,with kind permission of Springer Science and Business Media
4. Rich RL (1984) J Chem Educ 61:53
5. Rich RL (1991) J Chem Educ 68:828
6. Rayner-Canham G, Overton T (2003) Descriptive inorganic chemistry, 3rd edn. W.H. Freeman, New York, cover, with permission
7. www.synergycreations.com/periodic

Appendix B: Atomic and Ionic Energy Levels

The energy levels in the isolated atoms' various orbitals in the ground state are shown in Fig. B.1 [1a]. A special feature here is the plotting of $(-E/\text{eV})^{-1/2}$ to equalize, for legibility, the vertical spacings of the upper parts of the curves. This equalizes the spacings, of course, for the energies in hydrogen-like ions, which are proportional to n^{-2}.

The **d**- and **f**-block elements do not add **d** and **f** electrons smoothly as Z rises, and Figs. B.2 and B.3 give simplified explanations of this [2].

The much greater relativistic stabilization of **s** and destabilization of **d** electrons in the high-Z elements often yield $\mathbf{6s^2 5d}^{n-2}$ configurations even along with the $\mathbf{5s^1 4d}^{n-1}$ for the corresponding lower-Z ones.

Also available [1b] (but unfortunately disordered editorially after proofreading) are graphs of the energy levels of: (1) H^+, He^{2+}, Li^{3+} etc.; (2) $\mathbf{1s^2}$ in Li^+, Be^{2+}, B^{3+} etc.; (3) [He]$\mathbf{2s^2 2p^6}$ in Na^+, Mg^{2+}, Al^{3+} etc.; (4) [Ne]$\mathbf{3s^2 3p^6}$ in K^+, Ca^{2+}, Sc^{3+} etc.; (5) [Ar]$\mathbf{3d^{10} 4s^2 4p^6}$ in Rb^+, Sr^{2+}, Y^{3+} etc.; (6) [Kr]$\mathbf{4d^{10} 5s^2 5p^6}$ in Cs^+, Ba^{2+}, La^{3+} and Ce^{4+}; (7) [Xe]$\mathbf{4f^{14} 5d^{10} 6s^2 6p^6}$ in Ra^{2+}, Ac^{3+} and Th^{4+}; (8) [Ar]$\mathbf{3d^{10}}$ in Cu^+, Zn^{2+}, Ga^{3+} etc.; (9) [Kr]$\mathbf{4d^{10}}$ in Ag^+, Cd^{2+}, In^{3+} etc. and (10) [Xe]$\mathbf{4f^{14} 5d^{10}}$ in Au^+, Hg^{2+}, Tl^{3+} etc. These are omitted here, consonant with the overall emphasis on actual aqueous reactions.

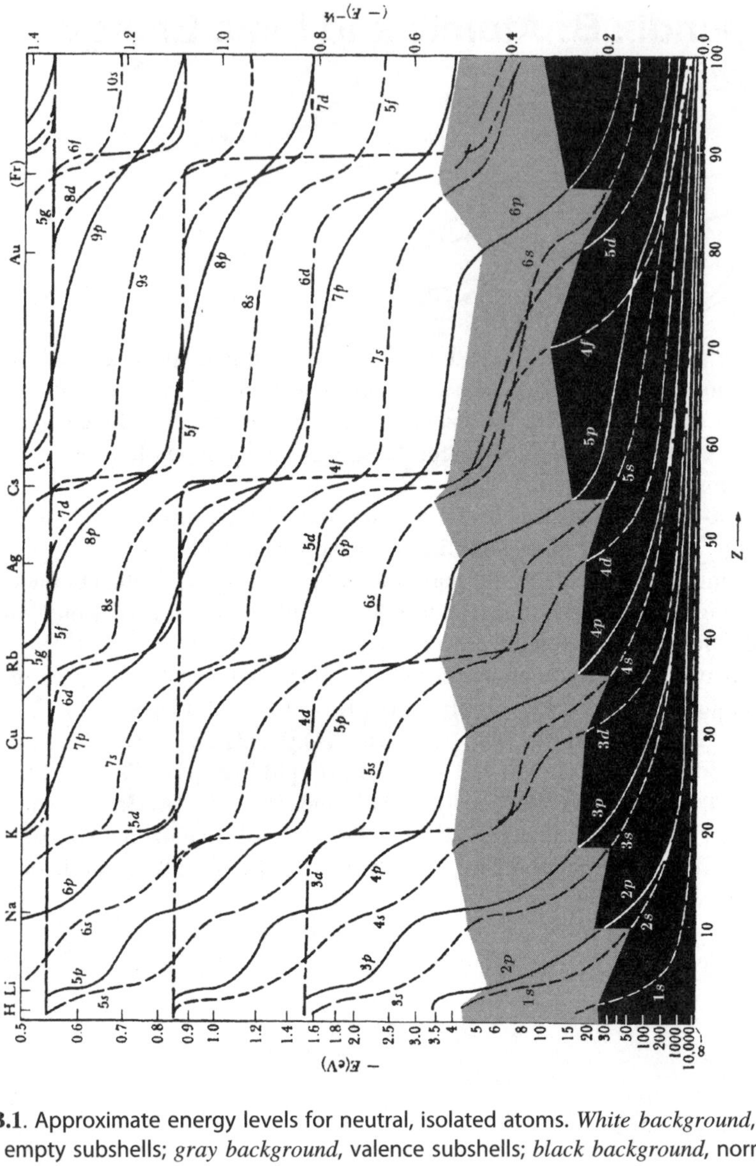

Fig. B.1. Approximate energy levels for neutral, isolated atoms. *White background*, normally empty subshells; *gray background*, valence subshells; *black background*, normally full subshells

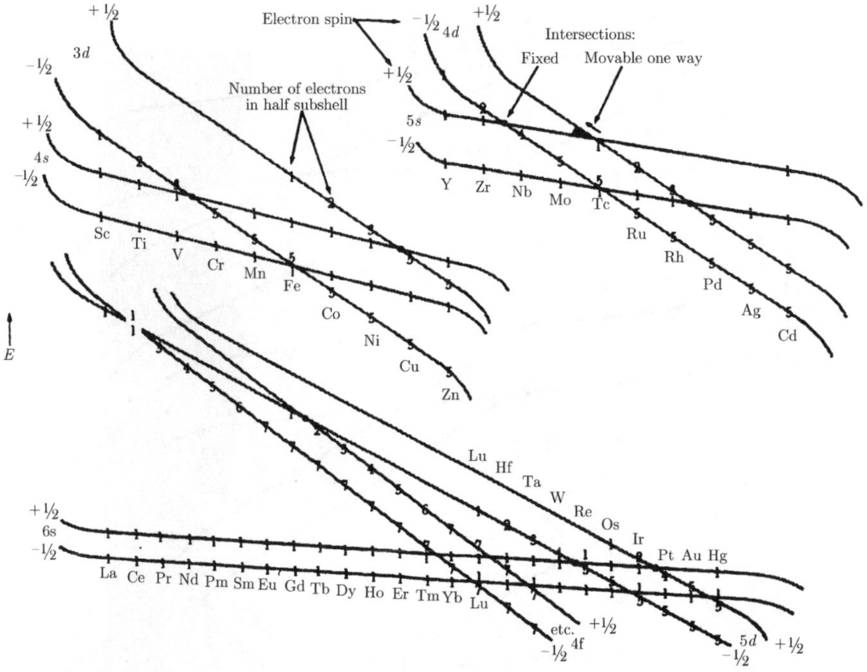

Fig. B.2. Schematic interpretation of electron configurations for **d**- and **f**-block atoms in the ground state, allowing for the intra-orbital repulsions and the trends in subshell energies

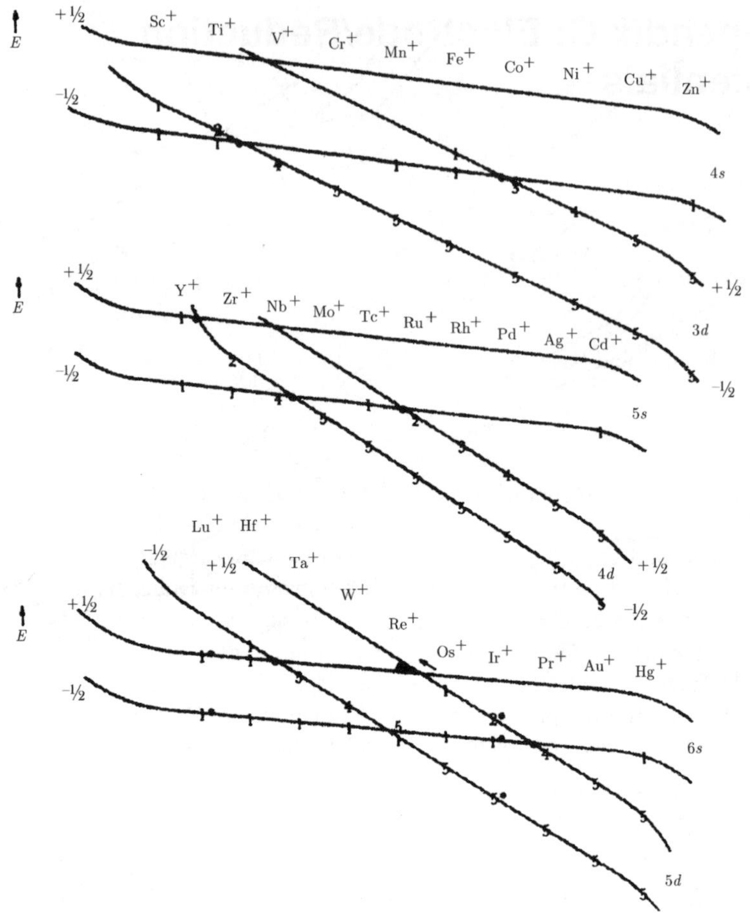

Fig. B.3. Schematic interpretation of electron configurations for **d**-block ions M+

References

1. (a) Rich RL (1965) Periodic correlations. Benjamin, New York, p 6; (b) ibid pp 12–18
2. (a) Rich RL, Suter RW (1988) J Chem Educ 65:702 (b) Rich RL (1965) Periodic correlations. Benjamin, New York, pp 9–10 (c) Schmid R (2003) J Chem Educ 80:931

Appendix C: Electrode/Reduction Potentials

Redox, but not acid-base, reactions, change the oxidation states in

$$^{1}/_{m}\, X_m + n\, H_3O^+ + n\, e^- \leftrightarrows H_nX + n\, H_2O$$

$$\text{but not } H_nX + H_2O \leftrightarrows H_{n-1}X^- + H_3O^+$$

Oxidants and acids, however, are both rather well described as taking electron density more or less than half way, respectively, from reductants and bases, in turn, although Table C.1 [1] shows that the simple trends for the two processes operate somewhat differently. We note that strong reductants have low (i.e. negative) $E°$, as in the tables below, and strong acids have low pK_a (i.e. high K_a), as in Table 1.1.

Table C.1. General qualitative trends in $E°$ and pK_a with some simple hydrides

Period Group		14	15	16	17	
	highest pK_a					highest $E°$
2		CH_4	NH_3	H_2O	HF	
3		SiH_4	PH_3	H_2S	HCl	
4		GeH_4	AsH_3	H_2Se	HBr	
5		SnH_4	SbH_3	H_2Te	HI	
	lowest $E°$					lowest pK_a

We also note briefly an instructive four-way classification [2] of various reagents as primarily electrophilic (acids and oxidants) and electrodotic (bases and reductants). Typical are: MnO_4^- (oxidant); BF_3 (acid); Ag^+ (both oxidant and acid); Cs (reductant); F^- (base); S^{2-} (both reductant and base); and H_2O (all four).

The following presentations are arranged to promote comparisons within various series, rather than to duplicate the numerous, published, practical compilations, and in order to promote the (re)discovery of the regularities and irregularities there, as well as possible explanations for some of them. Some data are relevant to several series and are therefore repeated. Others are not included in this limited treatment; some important oxidation states, of nitrogen, for example, are not matched by most of its congeners and hence do not allow comparisons.

Tables C.2 through C.14, with data mainly from [3] and some from [4] [5] and [6], provide greater precision than do graphs, albeit with less immediate visual impact. Our arrangement for the **s-** and **p-**blocks reflects the frequently greater interest in comparisons within the groups in the periodic chart, but for the **d-** and **f-**blocks, comparisons within the periods.

The superscripts to the left of each element in the **d-** or **f-**blocks tell how many **d-** or **f-**electrons are in the valence-shells of the respective *idealized* cations. This unconventional symbolism is intended to identify the isoelectronic series (from the upper left to the lower right in each case) for interesting and instructive comparisons and contrasts there, and to call attention especially to the empty, half-full and full subshells. In all cases we also easily see the isovalent series in the columns or rows and the isoelemental series in the rows or columns for those relevant comparisons and contrasts. Isoelectronic series are considered to be of somewhat less interest in the **s-** and **p-**blocks.

Where there are two rows of data for each element, the upper value for each set is with pH 0, the lower one with pH 14. Where there is only one row of data, the pH is 0, because the data at high pH are too sparse to elucidate the desired comparisons. The overall or cumulative potentials, labeled "X^j", are $E°/V$ for:

$$[X^{j+}L_m] + j\,e^- \leftrightarrows X + m\,L$$

where the ligands L may be H_2O, OH^- or O^{2-}, or a mixture. The negative ligands may of course react further on the right side of the equation to form H_2O or OH^-. Because good data for Cl^-, NH_3, CN^- etc. are few, perhaps surprisingly at this late date, we have no such tables for them.

Table C.2. Groups 1 and 2: overall standard electrode potentials; and Group 13: overall and stepwise, standard electrode potentials

${}_{1}H^{+}$					
0.000					
–0.828					
${}_{3}Li^{+}$	${}_{4}Be^{II}$	${}_{5}B$		B^{III}	
–3.04	–1.97			–0.89	
–3.04	–2.62			–1.24	
${}_{11}Na^{+}$	${}_{12}Mg^{II}$	${}_{13}Al$		Al^{III}	
–2.713	–2.356			–1.676	
–2.713	–2.687			–2.31	
${}_{19}K^{+}$	${}_{20}Ca^{II}$	${}_{31}Ga$		Ga^{III}	
–2.925	–2.84			–0.529	
–2.925	–3.02			–1.22	
${}_{37}Rb^{+}$	${}_{38}Sr^{II}$	${}_{49}In$	In^{I}	In^{III}	$In^{III\text{-}I}$
–2.924	–2.89		–0.126	–0.338	–0.444
–2.924	–2.99				
${}_{55}Cs^{+}$	${}_{56}Ba^{II}$	${}_{81}Tl$	Tl^{I}	Tl^{III}	$Tl^{III\text{-}I}$
–2.923	–2.92		–0.336	0.72	1.25
–2.923	–2.81!				
${}_{87}Fr^{+}$	${}_{88}Ra^{II}$				
–2.9	–2.92				
	–1.32!				

The stepwise potentials then are for "$X^{j\text{-}k}$" (with $j - k = 1$ or 2):

$$[X^{j+}L_m] + (j-k)\ e^- + n\ L \leftrightarrows [X^{k+}L_n] + m\ L$$

In Table C.2, as in others, we note that the value for M^{2+} in base is more negative than that in acid, because the reaction, e.g. for Mg:

$$Mg(OH)_2 + 2\ e^- \leftrightarrows Mg + 2\ OH^-$$

is displaced to the left by the insolubility of the hydroxide in base. We therefore question (with "!") the data in authoritative compilations [3, 6] for $Ba(OH)_2$ and $Ra(OH)_2$, or BaO and RaO, even with the somewhat greater solubilities for $Ba(OH)_2$ and $Ra(OH)_2$.

Other discrepancies such as that for Ge in Table C.3 may arise for example from the existence of several different forms of GeO and GeO_2, so that we find:

$$2(0.225) + 2(-0.370) = -0.290 \neq -0.144 = 4(-0.036)$$

We try, however, to choose the most stable forms, but note P_4 in Table C.4.

Table C.3. Group 14: overall and stepwise standard electrode potentials

$_6C$	$C^{\bar{\mathrm{IV}}}$	C^{II}: CO	C^{IV}	$C^{\mathrm{IV\text{-}II}}$
	0.132	0.517	0.206	–0.106
	–0.70			
$_{14}Si$	$Si^{\bar{\mathrm{IV}}}$	Si^{II}	Si^{IV}	$Si^{\mathrm{IV\text{-}II}}$
	0.102	–0.808	–0.909	–1.010
	–0.73		–1.69	
$_{32}Ge$	$Ge^{\bar{\mathrm{IV}}}$	Ge^{II}	Ge^{IV}	$Ge^{\mathrm{IV\text{-}II}}$
	< –0.3	0.225	–0.036	–0.370
	< –1.1		–0.89	
$_{50}Sn$	$Sn^{\bar{\mathrm{IV}}}$	Sn^{II}	Sn^{IV}	$Sn^{\mathrm{IV\text{-}II}}$
	–1.071	–0.137	–0.096	0.154
		–0.909	–0.92	–0.93
$_{82}Pb$		Pb^{II}	Pb^{IV}	$Pb^{\mathrm{IV\text{-}II}}$
		–0.125	0.787	1.698
		–0.50	–0.127	0.28

Table C.4. Group 15: some overall and stepwise standard electrode potentials

$_{7}N_2$	N^{-III}	N^{III}	N^{V}	$N^{V\text{-}III}$
	0.278	1.45	1.25	0.94
	−0.74	0.41	0.25	0.01
$_{15}P_4$	P^{-III}	P^{III}	P^{V}	$P^{V\text{-}III}$
	−0.063	−0.502	−0.412	−0.276
	−0.89	−1.73	−1.49	−1.12
$_{33}As$	As^{-III}	As^{III}	As^{V}	$As^{V\text{-}III}$
	−0.225	0.240	0.368	0.560
	−1.37	−0.68	−0.68	−0.67
$_{51}Sb$	Sb^{-III}	Sb^{III}	Sb^{V}	$Sb^{V\text{-}III}$
	−0.51	0.150	0.370	0.699
	−1.338	−0.639	−0.569	−0.465
$_{83}Bi$	Bi^{-III}	Bi^{III}	Bi^{V}	$Bi^{V\text{-}III}$
	−0.97	0.317	1.	2.
	<−1.6	−0.452		

Many more data are available for P_4 than for the more stable black P; therefore P_4 is taken here as the zero oxidation state.

All oxidation states in these tables are chosen for clarity and simplicity, rather than for consistency with better definitions. The electronegativities in SbH_3, for example, point to Sb^{III}, although Sb^{-III} leads to valid comparisons with, say, PH_3 (P^{-III}).

Table C.5. Group 16: overall and stepwise standard electrode potentials

$_{8}O_2$	$\bar{O}^{II}$			
	1.229			
	0.401			
$_{16}S_8$	$\bar{S}^{II}$	S^{IV}	S^{VI}	$S^{VI\text{-}IV}$
	0.144	0.500	0.386	0.158
	–0.476	–0.659	–0.751	–0.936
$_{34}Se$	$\bar{Se}^{II}$	Se^{IV}	Se^{VI}	$Se^{VI\text{-}IV}$
	–0.40	0.74	0.88	1.15
	–0.92	–0.366	–0.23	0.03
$_{52}Te$	$\bar{Te}^{II}$	Te^{IV}	Te^{VI}	$Te^{VI\text{-}IV}$
	–0.69	0.57	0.69	0.93
	–1.143	–0.42	–0.26	0.07
$_{84}Po$	$\bar{Po}^{II}$	Po^{IV}	Po^{VI}	$Po^{VI\text{-}IV}$
	~ –1.0	0.73	0.99	1.524
	~ –1.4	–0.5	0.16	1.474

Additional discrepancies, even with, say, Cl in Table C.6, are troubling, and we find:

$$5(1.458) + 2(1.201) = 9.692 \neq 9.905 = 7(1.415)$$

We cannot yet say with certainty which datum/data should at least be shown with fewer significant digits.

Table C.6. Groups 17 and 18: overall and stepwise standard electrode potentials

$_9F_2$	$\bar{F}$				
	3.053				
	2.866				
$_{17}Cl_2$	$\bar{Cl}$	Cl^{I}	Cl^{V}	Cl^{VII}	$Cl^{VII\text{-}V}$
	1.358	1.630	1.458	1.415	1.201
	1.358	0.421	0.474	0.457	0.374
$_{35}Br_2$	$\bar{Br}$	Br^{I}	Br^{V}	Br^{VII}	$Br^{VII\text{-}V}$
	1.065	1.604	1.478	1.585	1.853
	1.065	0.455	0.485	0.639	1.025
$_{53}I_2$	$\bar{I}$	I^{I}	I^{V}	I^{VII}	$I^{VII\text{-}V}$
	0.535	0.988	1.19	1.31	1.60
	0.535	0.48	0.20	0.33	0.65
$_{85}At_2$	$\bar{At}$	At^{I}	At^{V}		
	0.25	0.7	1.3		
	0.25	0.0	0.1		

$_{54}Xe$			Xe^{VI}	Xe^{VIII}	$Xe^{VIII\text{-}VI}$
			2.12	2.18	2.42
			1.24	1.18	0.99

The following **d**-block tables all include, of Group 3, only those members immediately to the left of Group 4 in the periodic chart, which are therefore also properly part of the **d**-block.

In Table C.10 the value for the important dimer, Hg_2^{2+}, 0.789 V, is omitted as not being comparable to the others and because Hg^{I} cannot have 11 outer **d**-electrons, as its position in the chart would suggest.

Some potentials in tables such as Table C.14, from thermodynamic or indirect calculations, are clearly far too large for achievable equilibria in water, and later data seem to change a few values moderately. Some small discrepancies among these tables remain unresolved.

Then the white areas in Fig. C.1 show the water-stable regions for **s**- and **p**-block elements, and in Fig. C.2 the **d**-block, as $E°$ vs pH, although regions below pH 0 and above pH 14 should also be white.

Table C.7. The **3d**-block: overall standard electrode potentials

$_{21}Sc$	$_{22}Ti$	$_{23}V$	$_{24}Cr$	$_{25}Mn$	$_{26}Fe$	$_{27}Co$	$_{28}Ni$	$_{29}{}^{10}Cu^{I}$	$_{30}Zn$
								0.521	
								-0.358	
	$^{2}Ti^{II}$	$^{3}V^{II}$	$^{4}Cr^{II}$	$^{5}Mn^{II}$	$^{6}Fe^{II}$	$^{7}Co^{II}$	$^{8}Ni^{II}$	$^{9}Cu^{II}$	$^{10}Zn^{II}$
	-1.638	-1.186	-0.913	-1.180	-0.440	-0.277	-0.257	0.340	-0.763
	-2.13	-0.820		-1.55	-0.877	-0.733	-0.72	-0.219	-1.285
$^{0}Sc^{III}$	$^{1}Ti^{III}$	$^{2}V^{III}$	$^{3}Cr^{III}$	$^{4}Mn^{III}$	$^{5}Fe^{III}$	$^{6}Co^{III}$	$^{7}Ni^{III}$		
-2.03	-1.208	-0.876	-0.744	-0.28	-0.036	0.414	0.52		
-2.6	-2.07	-0.709	-1.33	-1.12	-0.81	-0.432	0.32		
	$^{0}Ti^{IV}$	$^{1}V^{IV}$	$^{2}Cr^{IV}$	$^{3}Mn^{IV}$		$^{5}Co^{IV}$	$^{6}Ni^{IV}$		
	-0.882	-0.567	-0.03	0.025		>0.76	0.711		
	-1.90	-0.396		-0.80		-0.15	-0.12		
		$^{0}V^{V}$	$^{1}Cr^{V}$	$^{2}Mn^{V}$					
		-0.254	0.24	0.60					
		-0.119		-0.47					
			$^{0}Cr^{VI}$	$^{1}Mn^{VI}$	$^{2}Fe^{VI}$				
			0.293	0.71	1.08				
			-0.72	-0.33	-0.13				
				$^{0}Mn^{VII}$					
				0.74					
				-0.20					

Table C.8. The **3d**-block: stepwise standard electrode potentials

$_{21}$Sc	$_{22}$Ti	$_{23}$V	$_{24}$Cr	$_{25}$Mn	$_{26}$Fe	$_{27}$Co	$_{28}$Ni	$_{29}$Cu	$_{30}$Zn
								Cu^{II-I}	
								0.159	
								-0.080	
	Ti^{III-II}	V^{III-II}	Cr^{III-II}	Mn^{III-II}	Fe^{III-II}	Co^{III-II}			
	-0.369	-0.256	-0.408	1.51	0.738	1.808			
	-1.95	-0.486	-1.33	-0.25	-0.69	0.170			
	Ti^{IV-III}	V^{IV-III}	Cr^{IV-III}	Mn^{IV-III}		Co^{IV-III}			
	0.099	0.359	2.10	0.95		> 1.8			
	-1.38	0.542?		0.15		0.7			
		V^{V-IV}	Cr^{V-IV}	Mn^{V-IV}					
		1.000	1.34	2.90					
		0.991		0.85					
			Cr^{VI-V}	Mn^{VI-V}					
			0.55	1.28					
				0.35					
				Mn^{VII-VI}					
				0.90					
				0.564					

Table C.9. The **4d**-block: overall standard electrode potentials

$_{39}Y$	$_{40}Zr$	$_{41}Nb$	$_{42}Mo$	$_{43}Tc$	$_{44}Ru$	$_{45}Rh$	$_{46}Pd$	$_{47}Ag$	$_{48}Cd$
								$^{10}Ag^{I}$	
								0.799	
					$^{6}Ru^{II}$		$^{8}Pd^{II}$	$^{9}Ag^{II}$	$^{10}Cd^{II}$
					0.81		0.915	1.390	-0.403
$^{0}Y^{III}$		$^{2}Nb^{III}$	$^{3}Mo^{III}$		$^{5}Ru^{III}$	$^{6}Rh^{III}$		$^{8}Ag^{III}$	
-2.37		-1.099	-0.20		0.623	0.76		1.6	
	$^{0}Zr^{IV}$		$^{2}Mo^{IV}$	$^{3}Tc^{IV}$	$^{4}Ru^{IV}$		$^{6}Pd^{IV}$		
	-1.55		-0.152	0.28	0.68		1.05		
		$^{0}Nb^{V}$	$^{1}Mo^{V}$						
		-0.644	-0.1						
			$^{0}Mo^{VI}$	$^{1}Tc^{VI}$	$^{2}Ru^{VI}$				
			0.0	0.46	1.11				
				$^{0}Tc^{VII}$	$^{1}Ru^{VII}$				
				0.48	1.04				
					$^{0}Ru^{VIII}$				
					1.03				

Table C.10. The **5d**-block: overall standard electrode potentials

$_{71}Lu$	$_{72}Hf$	$_{73}Ta$	$_{74}W$	$_{75}Re$	$_{76}Os$	$_{77}Ir$	$_{78}Pt$	$_{79}Au$	$_{80}Hg$
								$^{10}Au^{I}$	
								1.691	
							$^{8}Pt^{II}$		$^{10}Hg^{II}$
							0.980		0.860
$^{0}Lu^{III}$				$^{4}Re^{III}$		$^{6}Ir^{III}$		$^{8}Au^{III}$	
-2.30				0.3		1.156		1.498	
	$^{0}Hf^{IV}$		$^{2}W^{IV}$	$^{3}Re^{IV}$	$^{4}Os^{IV}$	$^{5}Ir^{IV}$	$^{6}Pt^{IV}$		
	-1.70		-0.119	0.276	0.687	0.923	1.01		
		$^{0}Ta^{V}$	$^{1}W^{V}$						
		-0.812	-0.102						
			$^{0}W^{VI}$	$^{1}Re^{VI}$	$^{2}Os^{VI}$				
			-0.090	0.35	0.99				
				$^{0}Re^{VII}$	$^{1}Os^{VII}$				
				0.415	0.95				
					$^{0}Os^{VIII}$				
					0.846				

Table C.11. The **6d**-block: overall and stepwise standard electrode potentials

$_{103}Lr$	$_{104}Rf$	$_{105}Db$	$_{106}Sg$	$_{103}Lr$	$_{104}Rf$	$_{105}Db$	$_{106}Sg$
$^{1}Lr^{II}$	$^{2}Rf^{II}$	$^{3}Db^{II}$	$^{4}Sg^{II}$				
-1.6	-2.1	-0.24	0.46				
$^{0}Lr^{III}$	$^{1}Rf^{III}$	$^{2}Db^{III}$	$^{3}Sg^{III}$	Lr^{III-II}	Rf^{III-II}	Db^{III-II}	Sg^{III-II}
-1.96	-1.97	-0.56	0.27	-2.6	-1.7	-1.20	-0.11
	$^{0}Rf^{IV}$	$^{1}Db^{IV}$	$^{2}Sg^{IV}$		Rf^{IV-III}	Db^{IV-III}	Sg^{IV-III}
	-1.95	-0.87	-0.134		-1.5	-1.38	-1.34
		$^{0}Db^{V}$	$^{1}Sg^{V}$			Db^{V-IV}	Sg^{V-IV}
		-0.81	-0.13			-1.0	-0.11
			$^{0}Sg^{VI}$				Sg^{VI-V}
			-0.12				-0.046

Table C.12. The lanthanoids: overall and stepwise standard electrode potentials

$_{57}La$	$_{58}Ce$	$_{59}Pr$	$_{60}Nd$	$_{61}Pm$	$_{62}Sm$	$_{63}Eu$	$_{64}Gd$	$_{65}Tb$	$_{66}Dy$	$_{67}Ho$	$_{68}Er$	$_{69}Tm$	$_{70}Yb$	$_{71}Lu$
					${}^{6}Sm^{II}$	${}^{7}Eu^{II}$						${}^{13}Tm^{II}$	${}^{14}Yb^{II}$	
					-2.67	-2.80						-2.3	-2.8	
${}^{0}La^{III}$	${}^{1}Ce^{III}$	${}^{2}Pr^{III}$	${}^{3}Nd^{III}$	${}^{4}Pm^{III}$	${}^{5}Sm^{III}$	${}^{6}Eu^{III}$	${}^{7}Gd^{III}$	${}^{8}Tb^{III}$	${}^{9}Dy^{III}$	${}^{10}Ho^{III}$	${}^{11}Er^{III}$	${}^{12}Tm^{III}$	${}^{13}Yb^{III}$	${}^{14}Lu^{III}$
-2.38	-2.34	-2.35	-2.32	-2.29	-2.30	-1.99	-2.28	-2.31	-2.29	-2.33	-2.32	-2.32	-2.22	-2.30
-2.80	-2.78	-2.79	-2.78	-2.76	-2.80	-2.51	-2.82	-2.82	-2.80	-2.85	-2.84	-2.83	-2.74	-2.83
	${}^{0}Ce^{IV}$	${}^{1}Pr^{IV}$						${}^{7}Tb^{IV}$	${}^{8}Dy^{IV}$					
	-1.33	-0.96						-1.0	-0.29					
	-2.26	-1.89						-1.9						
					Sm^{III-II}	Eu^{III-II}						Tm^{III-II}	Yb^{III-II}	
					-1.55	-0.35						-2.3	-1.05	
	Ce^{IV-III}	Pr^{IV-III}						Tb^{IV-III}	Dy^{IV-III}					
	1.72	3.2						3.1	5.7					

Table C.13. The actinoids: overall standard electrode potentials

${}_{89}\mathrm{Ac}$	${}_{90}\mathrm{Th}$	${}_{91}\mathrm{Pa}$	${}_{92}\mathrm{U}$	${}_{93}\mathrm{Np}$	${}_{94}\mathrm{Pu}$	${}_{95}\mathrm{Am}$	${}_{96}\mathrm{Cm}$	${}_{97}\mathrm{Bk}$	${}_{98}\mathrm{Cf}$	${}_{99}\mathrm{Es}$	${}_{100}\mathrm{Fm}$	${}_{101}\mathrm{Md}$	${}_{102}\mathrm{No}$	${}_{103}\mathrm{Lr}$
${}^{1}\mathrm{Ac}^{\mathrm{II}}$	${}^{2}\mathrm{Th}^{\mathrm{II}}$	${}^{3}\mathrm{Pa}^{\mathrm{II}}$	${}^{4}\mathrm{U}^{\mathrm{II}}$	${}^{5}\mathrm{Np}^{\mathrm{II}}$	${}^{6}\mathrm{Pu}^{\mathrm{II}}$	${}^{7}\mathrm{Am}^{\mathrm{II}}$	${}^{8}\mathrm{Cm}^{\mathrm{II}}$	${}^{9}\mathrm{Bk}^{\mathrm{II}}$	${}^{10}\mathrm{Cf}^{\mathrm{II}}$	${}^{11}\mathrm{Es}^{\mathrm{II}}$	${}^{12}\mathrm{Fm}^{\mathrm{II}}$	${}^{13}\mathrm{Md}^{\mathrm{II}}$	${}^{14}\mathrm{No}^{\mathrm{II}}$	$\mathrm{Lr}^{\mathrm{II}}$
-0.9	0.7	0.4	-0.1	-0.3	-1.3	-1.95	-1.3	-1.6	-2.1	-2.3	-2.5	-2.5	-2.61	-1.6
${}^{0}\mathrm{Ac}^{\mathrm{III}}$	${}^{1}\mathrm{Th}^{\mathrm{III}}$	${}^{2}\mathrm{Pa}^{\mathrm{III}}$	${}^{3}\mathrm{U}^{\mathrm{III}}$	${}^{4}\mathrm{Np}^{\mathrm{III}}$	${}^{5}\mathrm{Pu}^{\mathrm{III}}$	${}^{6}\mathrm{Am}^{\mathrm{III}}$	${}^{7}\mathrm{Cm}^{\mathrm{III}}$	${}^{8}\mathrm{Bk}^{\mathrm{III}}$	${}^{9}\mathrm{Cf}^{\mathrm{III}}$	${}^{10}\mathrm{Es}^{\mathrm{III}}$	${}^{11}\mathrm{Fm}^{\mathrm{III}}$	${}^{12}\mathrm{Md}^{\mathrm{III}}$	${}^{13}\mathrm{No}^{\mathrm{III}}$	${}^{14}\mathrm{Lr}^{\mathrm{III}}$
-2.21	-1.2	-1.3	-1.65	-1.77	-2.00	-2.07	-2.06	-2.00	-1.91	-1.98	-2.07	-1.74	-1.26	-2.06
	${}^{0}\mathrm{Th}^{\mathrm{IV}}$	${}^{1}\mathrm{Pa}^{\mathrm{IV}}$	${}^{2}\mathrm{U}^{\mathrm{IV}}$	${}^{3}\mathrm{Np}^{\mathrm{IV}}$	${}^{4}\mathrm{Pu}^{\mathrm{IV}}$	${}^{5}\mathrm{Am}^{\mathrm{IV}}$	${}^{6}\mathrm{Cm}^{\mathrm{IV}}$	${}^{7}\mathrm{Bk}^{\mathrm{IV}}$	${}^{8}\mathrm{Cf}^{\mathrm{IV}}$	${}^{9}\mathrm{Es}^{\mathrm{IV}}$	${}^{10}\mathrm{Fm}^{\mathrm{IV}}$			
	-1.83	-1.5[a]	-1.37	-1.27	-1.24	-0.90	-0.8	-1.08	-0.6	-0.4	-0.3			
		${}^{0}\mathrm{Pa}^{\mathrm{V}}$	${}^{1}\mathrm{U}^{\mathrm{V}}$	${}^{2}\mathrm{Np}^{\mathrm{V}}$	${}^{3}\mathrm{Pu}^{\mathrm{V}}$	${}^{4}\mathrm{Am}^{\mathrm{V}}$								
		-1.2[a]	-1.01	-0.90	-0.79	-0.55								
			${}^{0}\mathrm{U}^{\mathrm{VI}}$	${}^{1}\mathrm{Np}^{\mathrm{VI}}$	${}^{2}\mathrm{Pu}^{\mathrm{VI}}$	${}^{3}\mathrm{Am}^{\mathrm{VI}}$								
			-0.83	-0.55	-0.50	-0.19								
				${}^{0}\mathrm{Np}^{\mathrm{VII}}$	${}^{1}\mathrm{Pu}^{\mathrm{VII}}$	${}^{2}\mathrm{Am}^{\mathrm{VII}}$								
				-0.18[a]	-0.1[a]	0.3[a]								
					${}^{0}\mathrm{Pu}^{\mathrm{VIII}}$									
					?									

[a] based partly or wholly on 1.0-M H_3O^+ (non-standard E)

Table C.14. The actinoids: stepwise standard electrode potentials

$_{89}Ac$	$_{90}Th$	$_{91}Pa$	$_{92}U$	$_{93}Np$	$_{94}Pu$	$_{95}Am$	$_{96}Cm$	$_{97}Bk$	$_{98}Cf$	$_{99}Es$	$_{100}Fm$	$_{101}Md$	$_{102}No$	$_{103}Lr$
Ac^{III-II}	Th^{III-II}	Pa^{III-II}	U^{III-II}	Np^{III-II}	Pu^{III-II}	Am^{III-II}	Cm^{III-II}	Bk^{III-II}	Cf^{III-II}	Es^{III-II}	Fm^{III-II}	Md^{III-II}	No^{III-II}	Lr^{III-II}
-4.9	-4.9	-4.7	-4.7	-4.7	-3.5	-2.3	-3.5	-2.8	-1.6	-1.3	-1.2	-0.15	1.45	-2.6
	Th^{IV-III}	Pa^{IV-III}	U^{IV-III}	Np^{IV-III}	Pu^{IV-III}	Am^{IV-III}	Cm^{IV-III}	Bk^{IV-III}	Cf^{IV-III}	Es^{IV-III}	Fm^{IV-III}	Md^{IV-III}	No^{IV-III}	
	-3.8	-2.0	-0.55	0.22	1.05	2.6	3.0	1.67	3.2	4.5	4.9	5.4	6.5	
		Pa^{V-IV}	U^{V-IV}	Np^{V-IV}	Pu^{V-IV}	Am^{V-IV}								
		-0.0_5 [a]	0.45	0.60	1.03	0.84								
			U^{VI-V}	Np^{VI-V}	Pu^{VI-V}	Am^{VI-V}								
			0.09	1.16	0.94	1.60								
				Np^{VII-VI}	Pu^{VII-VI}	Am^{VII-VI}								
				2.04[a]	2.3[a]	2.5[a]								
					$Pu^{VIII-VII}$									
					?									

[a] based partly or wholly on 1.0-M H_3O^+ (non-standard E)

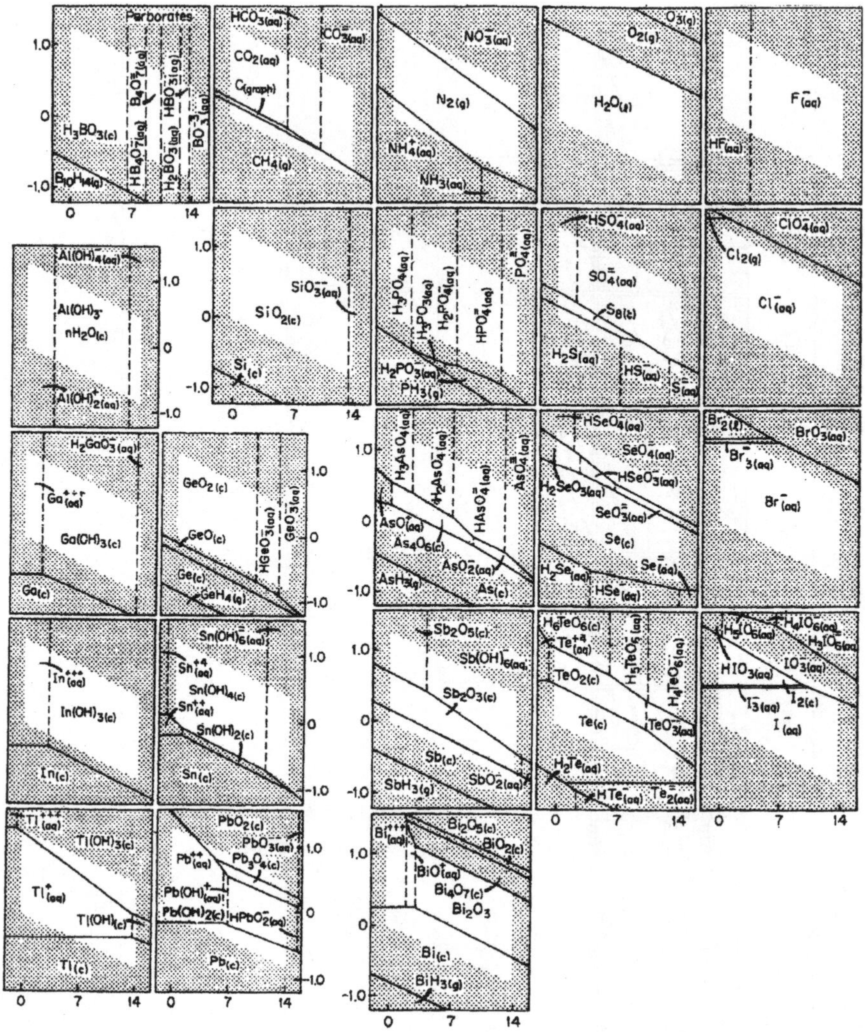

Fig. C.1. Standard electrode potentials $E°$ vs pH for **p**-block species [7]. The gap is unexplained. We omit Groups 1, 2 and 18 for paucity of data and to permit enlargement for the others

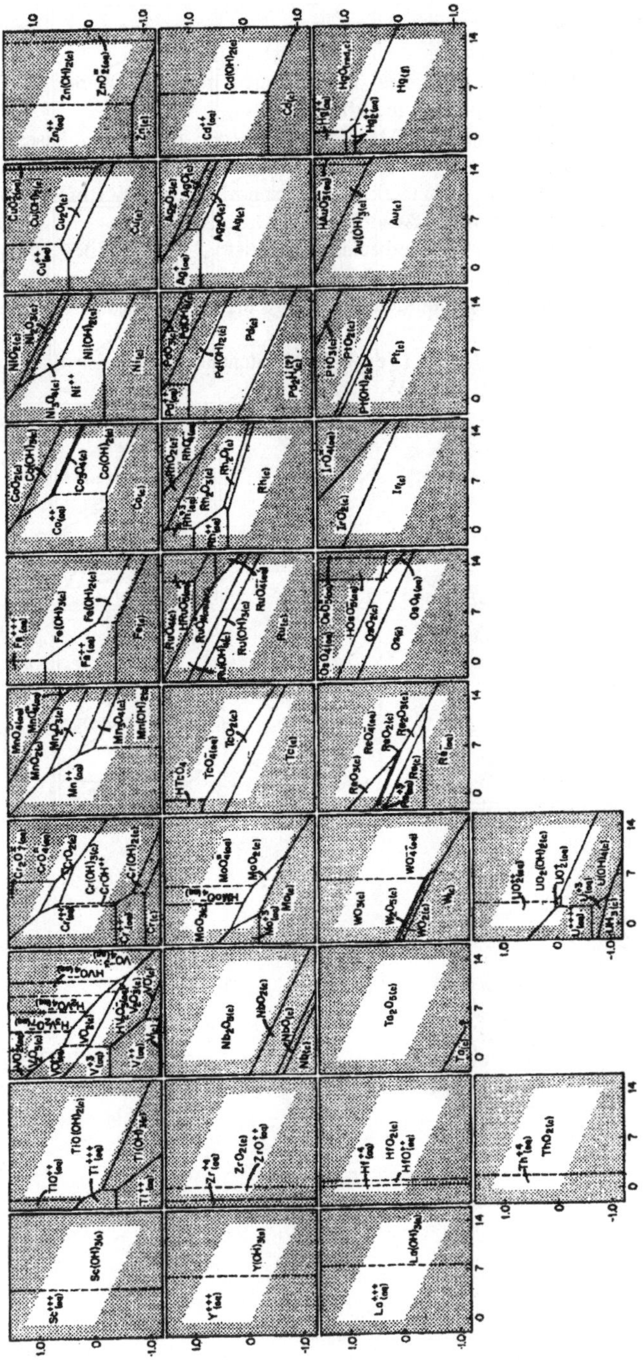

Fig. C.2. Standard electrode potentials E° vs pH for **d**-block species [7]

References

1. Wulfsberg G (1987) Principles of descriptive inorganic chemistry. Brooks/Cole, Monterey CA, p 163, Fig 5.8, modified
2. Luder WF, Zuffanti S (1961) The electronic theory of acids and bases, 2nd edn. Dover, New York, p 72
3. Holleman AF, Wiberg E, Wiberg N; Eagleson M, Brewer W (trans); Aylett BJ (rev) (2001) Inorganic chemistry. Academic, San Diego, p 1761
4. Cotton S (2006) Lanthanide and actinide chemistry, 2nd edn. Wiley, West Sussex
5. Morss LR, Edelstein NM, Fuger J, Katz JJ (honorary) (eds) (2006) The chemistry of the actinide and transactinide elements, 3rd edn, vol 3. Springer, Berlin Heidelberg New York, p 1779
6. Bard AJ, Parsons R, Jordan J (eds) (1985) Standard potentials in aqueous solution. Dekker, New York, pp 717, 725
7. Campbell JA, Whiteker RA (1969) J Chem Educ 46:91, 92, with permission

Appendix D: Abbreviations and Definitions

Table D.1 lists abbreviations and definitions, some new and some old.

Table D.1. Abbreviations and definitions

Ae	Mg, Ca, Sr, Ba and/or Ra (alkaline-earth metals)
Alk	Li, Na, K, Rb, Cs and/or Fr (alkali metals)
An, actinoid	$_{89}$Ac through $_{103}$Lr
aq	indefinite hydration, ignored in equations
Aq	(H_2O) in e.g. $[M\{(MAq_3)_3S_4\}_2]^{8+}$ for the formula $[M\{[M(H_2O)_3]_3S_4\}_2]^{8+}$
aqua regia	concentrated HNO_3 and HCl, mixed, e.g. 1:3 vols.
(A,B)X	either AX or BX, or both in some cases
Ced, ceroid	$_{57}$La through $\sim_{62}$Sm (including abundant $_{58}$Ce)
con-mutation	reproportionation = comproportionation
dismutate	disproportionate; cf. dismutase in biochemistry
fixed alkali	AlkOH [or $Ae(OH)_2$], not (volatile) NH_3
ligancy	coordination number, c. n.
Ln, lanthanoid	$_{57}$La through $_{71}$Lu
"$(NH_4)_2CO_3$"	mixture: NH_3, NH_4^+, HCO_3^-, $CO_2NH_2^-$ and less CO_3^{2-}
"$(NH_4)_2S$"	equilibrium mixture: NH_3, NH_4^+, HS^- and less S^{2-}
post-uranoid	$_{96}$Cm through $_{103}$Lr
Rth	rare-earth element(s): Sc, Y and/or Ln
uranoid	$_{92}$U through $_{95}$Am, all able to form MO_2^{2+}
Ytd, yttroid	$\sim_{63}$Eu through $_{71}$Lu, plus (abundant) $_{39}$Y
3d^{2+}, e.g.	some or all of the **3d**-block dications, e.g.

Formulas like $(Alk,NH_4)_5[NpO_2(CO_3)_3]$ are not to imply that every alkali-metal cation (as an alternative to NH_4^+) is known therein; there are simply too many data to include them all.

Index